Dynamics of Soils and Their Engineering Applications

Swami Saran
Formerly Professor and Emeritus Fellow
Indian Institute of Technology Roorkee, India

CRC Press is an imprint of the
Taylor & Francis Group, an **informa** business
A SCIENCE PUBLISHERS BOOK

CRC Press
Taylor & Francis Group
6000 Broken Sound Parkway NW, Suite 300
Boca Raton, FL 33487-2742

Printed on acid-free paper
Version Date: 20190704

International Standard Book Number-13: 978-0-367-52987-1 (Hardback)
International Standard Book Number-13: 978-0-367-52989-5 (Paperback)

Visit the Taylor & Francis Web site at
http://www.taylorandfrancis.com

and the CRC Press Web site at
http://www.routledge.com

Dedication

I dedicate this book to my respected teacher
Prof. Shamsher Prakash
He was my true GURU who taught me
"The only thing that overcomes hard luck is hard work."

Preface

In the world scenario, seeing the frequency of occurring small and big earthquakes throughout the globe, and also industrial development made it mandatory to have the knowledge of dynamics of soils and machine foundations. Keeping these facts in view, courses on this theme exist at undergraduate and graduate levels in almost all Engineering Institutions.

The author taught the courses related to this subject at graduate level for almost thirty five years. The text of this book has been developed mainly out of the lecture notes of the author prepared for teaching the students. Main consideration in developing the text is that a reader understands the subject clearly. The material has been arranged logically so that one can follow the development sequence with relative ease. A number of solved examples have been included in each chapter. All the formulae, charts and examples are given in SI units.

Some of the material included in the text book has been drawn from the works of the other authors. Inspite of sincere efforts, some contributions may not have been acknowledged. The author apologises such omissions.

The author wishes to express his appreciation to his many colleagues, friends and students for their constructive suggestions in writing this book. Encouragement and patience of my family members is worth mentioning.

Blessings of Almighty God is above all and with all humbleness, the book is dedicated on His feet. Hope all the readers will be blessed by His kindness.

Swami Saran

Contents

CHAPTER

1

Introduction

1.1 Objective and Overview

Geotechnical engineers frequently come across two types of problem in relation to the analysis and design of foundations namely (i) foundations subjected to static loads and (ii) foundations subjected to dynamic loads. The characteristic feature of a static load is that for a given structure the load carried by the foundation at any given time is constant in magnitude and direction e.g. dead weight of the structure. Live loads such as weight of train on a bridge and assembly of people in a building are also classified as static load. The characteristic feature of a dynamic load is that it varies with time. Dynamic loads on foundations and engineering structures may act due to earthquakes, bomb blasts, operation of machines, pile driving, quarrying, fast moving traffic, wind or sea waves action. The nature of each dynamic load is different from another. Figure 1.1 shows the variation of

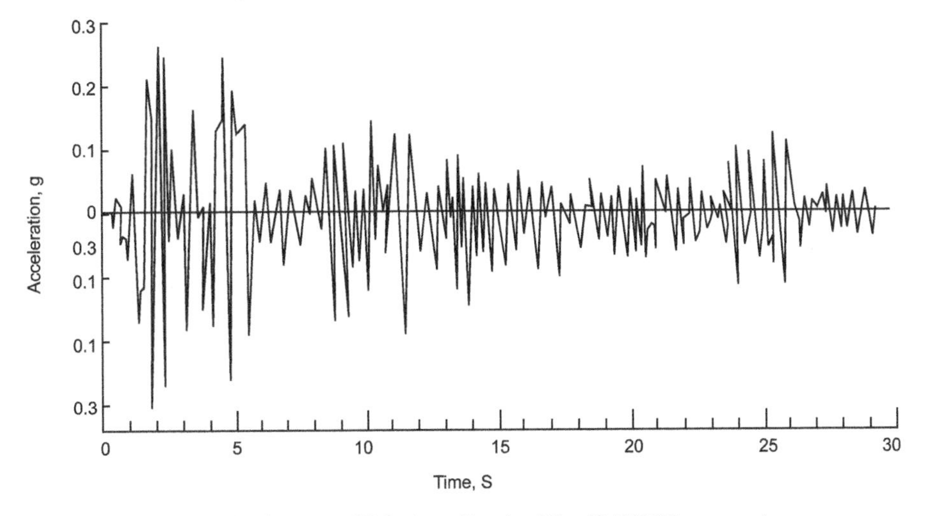

(a) Accelerogram of El Centro earthquake of May 18,1940 NS component

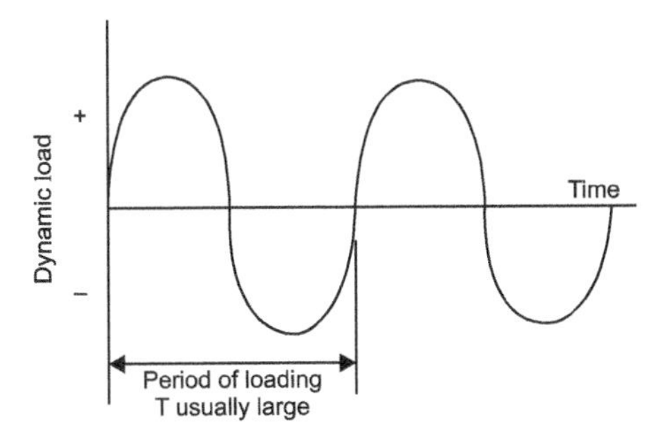

(b) Dynamic load due to steady state vibration

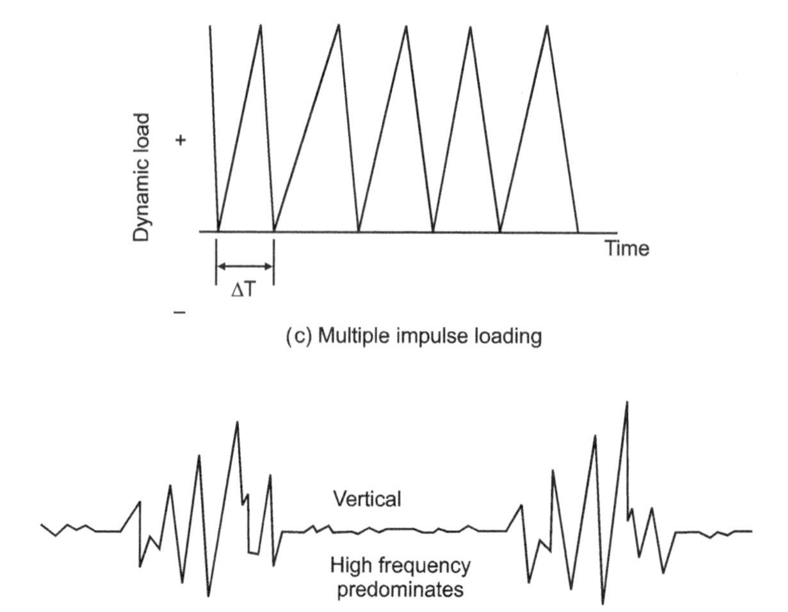

(d) Trace of vertical acceleration of ground due to pile driving

Fig. 1.1. Variation of dynamic load with time in some typical cases

dynamic load with time in some typical cases. Purely dynamic loads do not occur in nature. Loads are always combinations of static and dynamic loads. Static loads are caused by the dead weight of the structure, while dynamic loads may be caused through the sources mentioned above.

The basic objective of writing this book is to present the basic concepts of soil dynamics and their applications in designing of various engineering structures in lucid and simple way so that a reader understands with clarity. The author has taught this subject for more than thirty years to post graduate students. In general the contents of the book are the compilation of the lecture notes of the author prepared for delivering the material to the students.

The book has sixteen chapters. The first three chapters deal with general aspects, theory of vibrations and wave propagation .The next two chapters elaborate the concept of the dynamic soil properties and earth pressure. Their application in considering these aspects in designing shallow and pile foundations have been dealt in next two chapters. Very important topics on the slopes analysis and liquefaction of soils are covered in chapters 9 and 10. Next five chapters are devoted for clear understanding of analyzing and designing the various types of machine foundations. In last 2 chapters, seismic analysis and design of reinforced earth walls and wall with reinforced backfill have been discussed considering the huge importance and acceptance of reinforced earth.

1.2 Earthquake Loading

Vibrations of earth's surface caused by waves coming from a source of disturbance inside the earth are described as **Earthquakes** and are one of the most destructive forces that nature unleashes on earth. When, at any depth below the ground surface, the strain energy accumulated due to deformations in earth mass exceeds the resilience of the storing material, it gets release through rupture. The energy thus released is propagated in the form of waves which impart energy to the media through which they pass and vibrate the structures standing on the earth's surface in both the horizontal directions and also to some extent in vertical direction. The vertical motion is prominent in epicentral region, but it decreases in significance with distance from the epicenter.

The vibratory forces cause additional shear and moments in the structures and it may fail if they have not been designed adequately. The loss of property, life and utility facilities may occur which may disturb the whole system of living. It is, therefore necessary to consider this aspect in designing engineering structures.

The point inside the earth mass where slipping or fracture begins is termed as **focus** and the point just above the focus on the earth's surface is termed as **epicentre**. The position of the focus is determined with the help of seismograph records (Fig. 1.2) utilising the average velocities of different waves and time difference in reaching the waves at the ground surface. Figure 1.3 explains the various terms in simple manner.

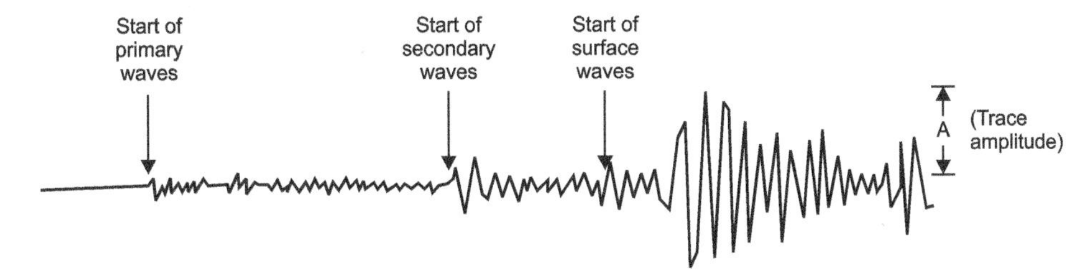

Fig. 1.2. A typical earthquake record

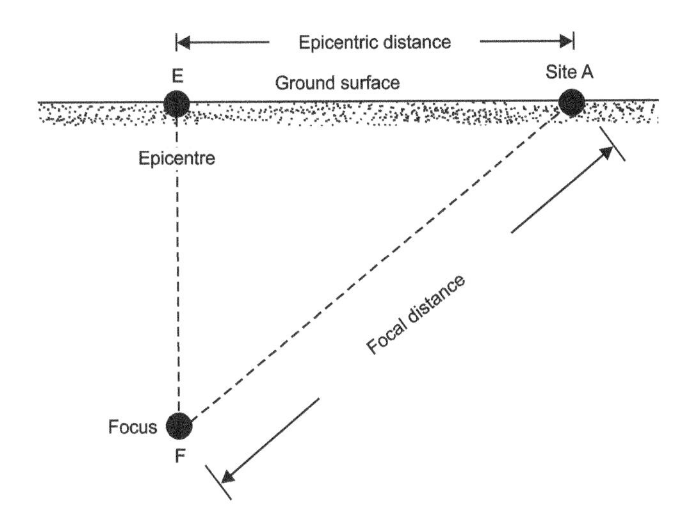

Fig. 1.3. Variation of dynamic load with time in some typical cases

1.2.1 Intensity

The severity of shaking of an earthquake as felt or observed through damage is described as intensity at a certain place on an arbitrary scale. For this purpose modified Mercalli scale is more common in use. It is divided into 12 degrees of intensity as presented in Table 1.1.

1.2.2 Magnitude

Magnitude of an earthquake is a measure of the size of an earthquake, based on the amplitude of elastic waves it generates. Richter (1958) suggested the following relation.

$$M = \log_{10} A - \log_{10} A_o \tag{1.1}$$

where M = Magnitude of earthquake

A = Trace amplitude in mm (Fig. 1.2)

A_o = Distance correction (Fig. 1.4)

Table 1.1. Modified Mercalli Intensity Scale (Abridged)

Class of earthquakes	*Description*
I	Not felt except by a very few under specially favourable circumstances.
II	Felt only by a few persons at rest, specially on upper floors of buildings; and delicately suspended objects may swing.
III	Felt quite noticeably indoors, specially on upper floors of buildings but many people do not recognize it as an earthquake; standing motor cars may rock slightly, and vibration may be felt like the passing of a truck.
IV	During the day felt indoors by many, outdoors by a few; at night some awakened, dishes, windows, doors disturbed, walls make cracking sound, sensation like heavy truck striking the building; and standing motor car rocked noticeably.
V	Felt by nearly everyone; many awakened; some dishes, windows, etc. broken; a few instances of cracked plasters; unstable objects overturned; disturbance of trees, poles and other tall objects noticed sometimes and pendulum clocks may stop.
VI	Felt by all; many frightened and run outdoors; some heavy furniture moved; a few instances of fallen plaster or damaged chimneys; damage slight.
VII	Everybody runs outdoors, damage negligible in buildings of good design and construction; slight to moderate in well built ordinary structures; considerable in poorly built or badly designed structues; some chimneys broken; noticed by persons driving motor cars.
VIII	Damage slight in specially designed structures; considerable in ordinary substantial buildings with partial collapse; very heavy in poorly built structures; panel walls thrown out of framed structure; heavy furniture overturned; sand and mud ejected in small amounts; changes in well water; and disturbs persons driving motor cars.
IX	Damage considerable in specially designed structures; well designed framed structures thrown out of plumb; very heavy in substantial buildings with partial collapse; buildings shifted off foundations; ground cracked conspicuously; and underground pipes broken.
X	Some well built wooden structures destroyed; most masonry and framed structures with foundations destroyed; ground badly cracked; rails bent; land-slides considerable from river banks and steep slopes; shifted sand and mud; and water splashed over banks.
XI	Few, if any, masonry structures remain standing; bridge destroyed; broad fissures in ground, underground pipe lines completely out of service; earth slumps and landslips in soft ground; and rails bent greatly.
XII	Total damage; waves seen on ground surface; lines of sight and level distorted; and objects thrown upward into the air

A relationship between strain energy released by an earthquake and its magnitude is given by Richter (1958) as follows:

$$\log_{10}E = 11.4 + 1.5M \tag{1.2}$$

where E = Energy released in earthquake in Ergs.

Energy released in earthquakes of different magnitude is given in Table 1.2, which would give an idea of their relative destructive power.

Table 1.2. Magnitudes of Earthquakes and Energy Released

M (Richter)	5.0	6.0	6.5	7.0	7.5	8.0	8.5	9.0
E (10^{20} Ergs)	0.08	2.5	14.1	80	446	2500	14125	80000

The maximum intensity of shaking attained during an earthquake of given magnitude depends upon the depth of focus as well as soil conditions below the foundation of structures. For shallow focus earthquakes of depth about 30 km or less, a comparison of the magnitude M of an earthquake with maximum intensity of the Modified Mercalli Scale is given in Table 1.3.

Table 1.3. Comparison of the Richter Scale Magnitude with the Modified Mercalli Scale

Richter scale magnitude (M)	*Maximum intensity, modified mercalli scale*
2	I, II
3	III
4	IV, V
5	VI, VII
6	VII, VIII
7	IX, X
8	XI

Earthquakes may have a magnitude from 2.0 to almost 8.0, but no shock smaller than 5.0 causes appreciable damage. The extent of damage depends upon the depth of focus, and a shock of magnitude 8.0 will envelop a vast area of earth. Very shallow shocks even of small size could cause damage locally.

Usually, earthquakes have their focus not shallower than about 5.0 km but go deeper than 300 km. An earthquake of magnitude of 5.0 causes damage within a radius of about 8.0 km but that of magnitude 7.0 may cause damage in a radius of 80 km and that of 8.0 over a radius of 250 km.

The fault length, affected area and duration of earthquake also depend on the magnitude of earthquake (Housner, 1965; Housner, 1970). Table 1.4 gives approximate idea about these.

Table 1.4. Fault Length. Affected Area and Duration of Earthquake

Magnitude of earthquake (richter scale)	*Fault length (km)*	*Affected area (km^2)*	*Duration of earthquake (S)*
5	1-2	20,000	5
6	2-5	60,000	15
7	25-50	1,20,000	25-30
8	>250	2,00,000	45-50

1.3 Equivalent Dynamic Load to an Actual Earthquake Load

Figure 1.1 (a) shows the variation of dynamic load with time observed during El Centro earthquake. The loading is not periodic and the peaks in any two cycles are different. For the analysis and design of foundations such a random variation is converted into equivalent number of cycles of uniformly varying load [Fig. 1.1 (b)]. It means that the structure-foundation-soil system subjected to N_S cycles of uniformly varying load will suffer same deformations and stresses as by the actual earthquakes. Most of the analyses and laboratory testing are carried out using this concept.

According to Seed and Idriss (1971), the average equivalent uniform acceleration is about 65 percent of the maximum acceleration. The number of significant cycles, N_S depends on the magnitude of earthquake. They recommended the values of N_S as 10, 20 and 30 for earthquakes of magnitudes 7, 7.5 and 8 respectively.

Lee and Chan (1972) suggested the following procedure for converting the irregular stress-time history to the equivalent number of cycles of cyclic shear stresses of maximum magnitude equal to $K\ \tau_{max}$, K being a constant less than unity:

(i) Let Fig. 1.5 shows a typical earthquake record. Divide the stress range (0 to τ_{max}) or acceleration range (0 to a_{max}) into convenient number of levels and note the mean stress or

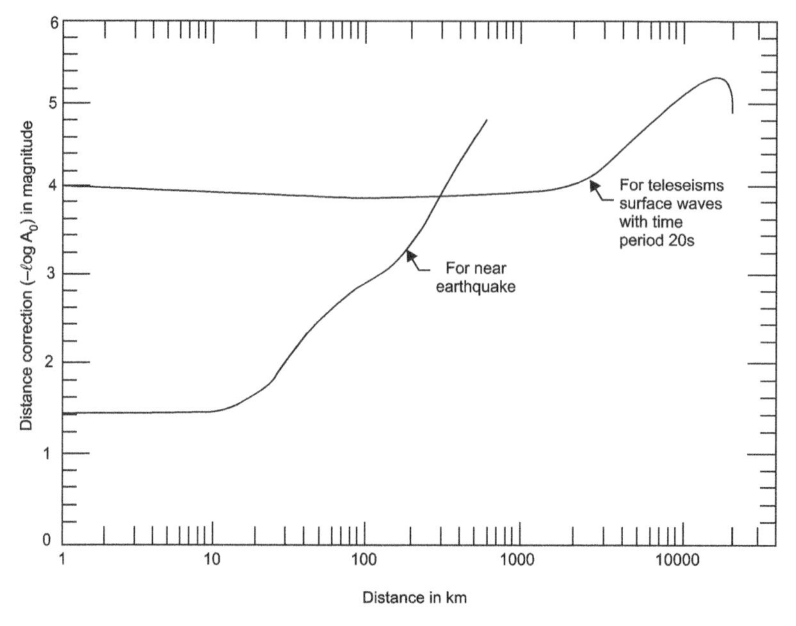

Fig. 1.4. Distance correction for magnitude determination

mean acceleration within each level as mentioned in column no. 2 of Table 1.5. Then the number of cycles with peaks which fall within each of these levels is counted and recorded. Note that because the actual time history is not symmetric about the zero stress axis, the number of peaks on both sides are counted and two peaks are equivalent to one cycle. For example, an earthquake record shown in Fig. 1.5 has number of cycles in various ranges of acceleration levels as listed in Col. 3 of Table 1.5.

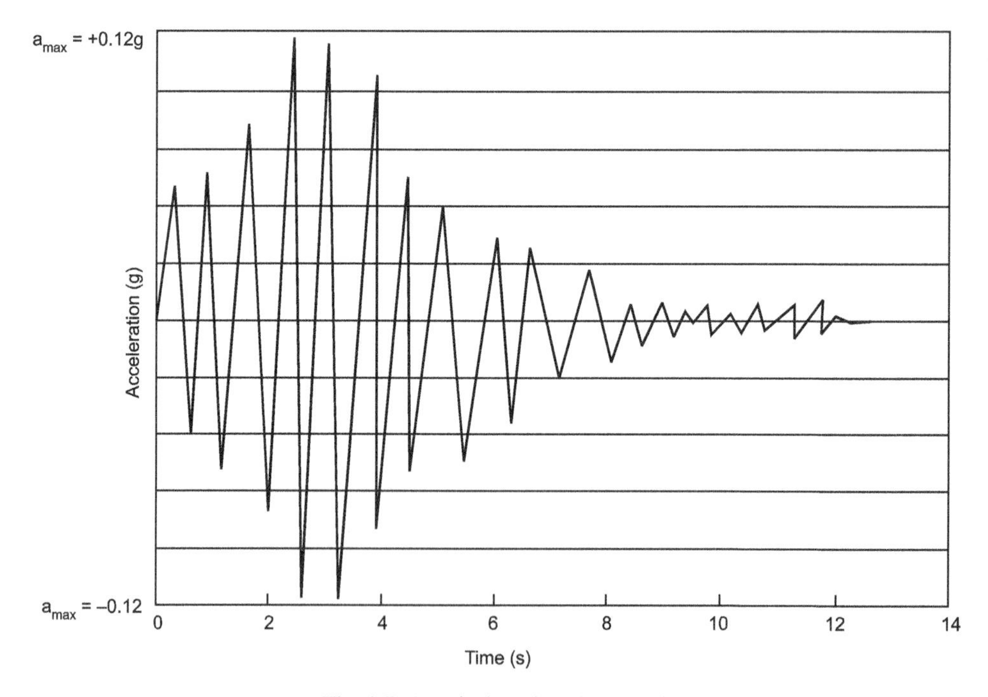

Fig. 1.5. A typical earthquake record

(ii) Seed et al. (1975) gave a plot between stress ratio and conversion factor as shown in Fig. 1.6. Conversion factor is defined as the ratio of equivalent number of cycles for 0.65 τ_{max} to equivalent number of cycles for $K \cdot \tau_{max}$. Referring to this curve (Fig. 1.6) determine the conversion factor to each average stress level (Col. 4 of Table 1.5).

(iii) Determine the equivalent number of uniform cycles at a maximum stress level of 0.65 τ_{max} by multiplying the values listed in Cols. 3 and 4. These are listed in Col. 5.

(iv) Determine the total number of equivalent stress cycles at 0.65 τ_{max} by adding the values listed in Col. 5.

Table 1.5. Equivalent Cycles for Anticipated Earthquake

Acceleration level in percent	*Average level in percent*	*Number of cycles*	*Conversion factor*	*Equivalent number of cycles at 0.65 τ_{max}*
(1)	**(2)**	**(3)**	**(4)**	**(5)**
100–80	90	5/2 = 2.5	2.6	6.5
80–60	70	3/2 = 1.5	1.2	1.8
60–40	50	7/2 = 3.5	0.20	0.70
40–20	30	5/2 = 2.5	Negligible	0.0
20–00	10	> 100	negligible	0.00
				Total number of cycles = 9.0

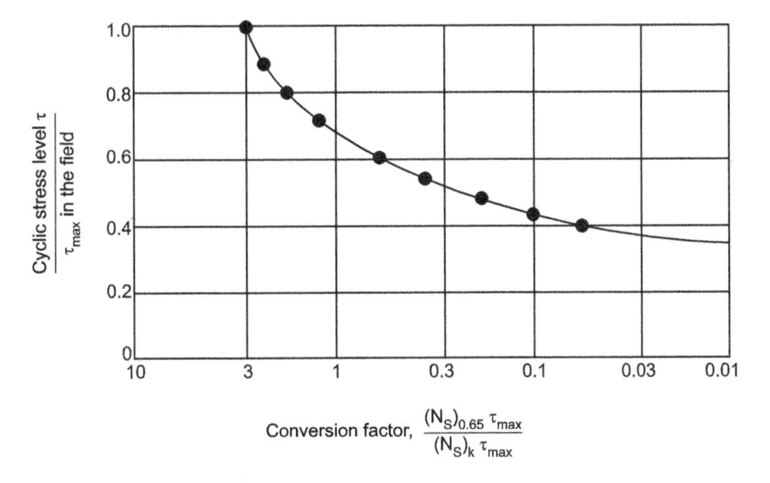

Fig. 1.6. Conversion factor versus shear stress ratio

For getting the equivalent number of cycles for 0.75 τ_{max}, read the value of conversion factor (Fig. 1.6) corresponding to an ordinate value of 0.75. It comes out as 1.5. The value of equivalent number of cycles obtained for 0.65 τ_{max} as illustrated in Table 1.4 is divided by this conversion factor to obtain equivalent number of cycles corresponding to 0.75 τ_{max} i.e. 9.0/1.5 = 6.0 cycles.

Seed and Idriss (1971) and Lee and Chan (1972) developed the above concepts specifically for liquefaction studies. More details of these procedures have been discussed in Chapter 7.

1.4 Seismic Force for Pseudo-Static Analysis

As per IS 1893 (Part-I)–2016, for the purpose of determining seismic force, the country is classified into five zones as shown in Fig. 1.7. Two methods namely (i) seismic coefficient method and (ii)

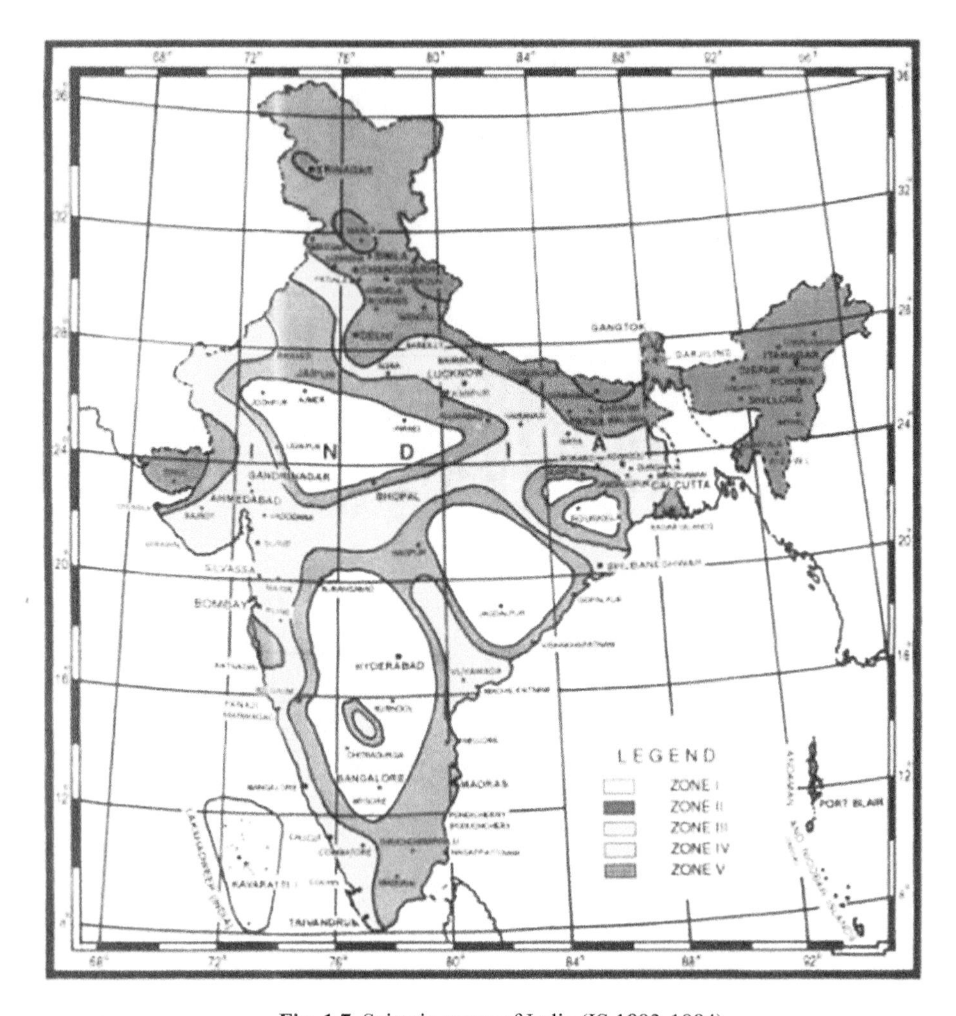

Fig. 1.7. Seismic zones of India (IS 1893-1984)

response spectrum method are used for computing the seismic force. For pseudo-static design of foundation of buildings, bridges and similar structures, seismic coefficient method is used. For the analysis of earth dams and dynamic designs, response spectrum method is used (IS 1893 (Part-I): 2016).

In seismic coefficient method, the design value of horizontal seismic coefficient α_h is obtained by the following expression:

$$\alpha_h = \beta I \alpha_o \tag{1.3}$$

where

α_o = Basic seismic coefficient, Table 1.6
I = Coefficient depending upon the importance of structure, Table 1.7
β = Coefficient depending upon the soil-foundation system, Table 1.8

The vertical seismic coefficient, α_v shall be considered in the case of structures in which stability is a criterion of design or for overall stability of structures. It may be taken as half of the horizontal seismic coefficient. Therefore,

$$\alpha_v = \frac{\alpha_h}{2} \tag{1.4}$$

Table 1.6. Values of Basic Seismic Coefficients α_0

Zone No.	α_0
V	0.08
IV	0.05
III	0.04
II	0.02

Table 1.7. Values of Importance Factor, I

S. No.	*Type of Structure*	*Value of I*
1.	Containment structure of seismic power reactor for preliminary design	3.0
2.	Dams (all types)	2.0
3.	Containers of inflammable or poisonous gases or liquids	2.0
4.	Important service and community structures, such as hospitals, water towers and tanks, schools, important bridges, emergency buildings like telephone exchange and fire brigades, large assembly structures like cinemas, assembly halls and subway stations	1.5
5.	All others	1.0

Table 1.8. Values of β for Different Soil-Foundation System

Soil	*Values of β for*				
	Isolated footings without tie beams	*Combined or isolated footings with tie beams*	*Raft foundation*	*Pile foundation*	*Well foundation*
Rock or Hard soil	1.0	1.0	1.0	1.0	1.0
Medium soils	1.2	1.0	1.0	1.0	1.2
Soft soils	1.5	1.2	1.0	1.2	1.5

In response spectrum method, the response acceleration coefficient is first obtained for the natural period and damping of the structure and the design value of horizontal seismic coefficient is computed using the following expression:

$$\alpha_h = \beta \cdot \mathrm{I} \cdot \mathrm{F}_0 \frac{S_a}{g} \tag{1.5}$$

where

F_0 = Seismic zoning factor for average acceleration spectra (Table 1.9)

$\frac{S_a}{g}$ = Average acceleration coefficient as read from Fig. 1.8 for appropriate natural period and damping of the structure.

Table 1.9. Values of Seismic Zoning Factor, F_0

Zone No.	F_0
V	0.40
IV	0.25
III	0.20
II	0.10

As per IS: 1893 (Part-I)-2016, for the purpose of determining seismic forces, the country is classified into four seismic zones as shown in Fig. 1.9. For pseudo-static design of foundations, horizontal and vertical seismic forces are obtained using the concept of seismic coefficients.

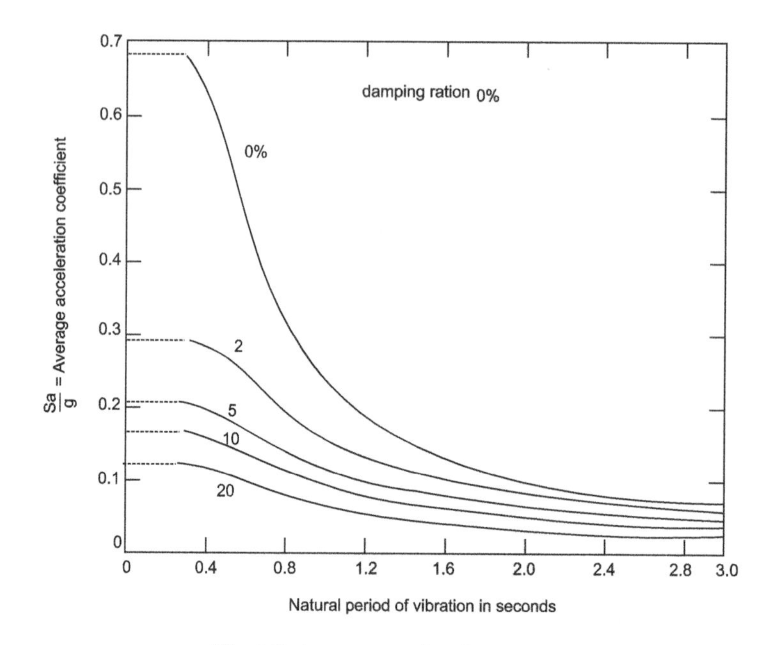

Fig. 1.8. Average acceleration spectra

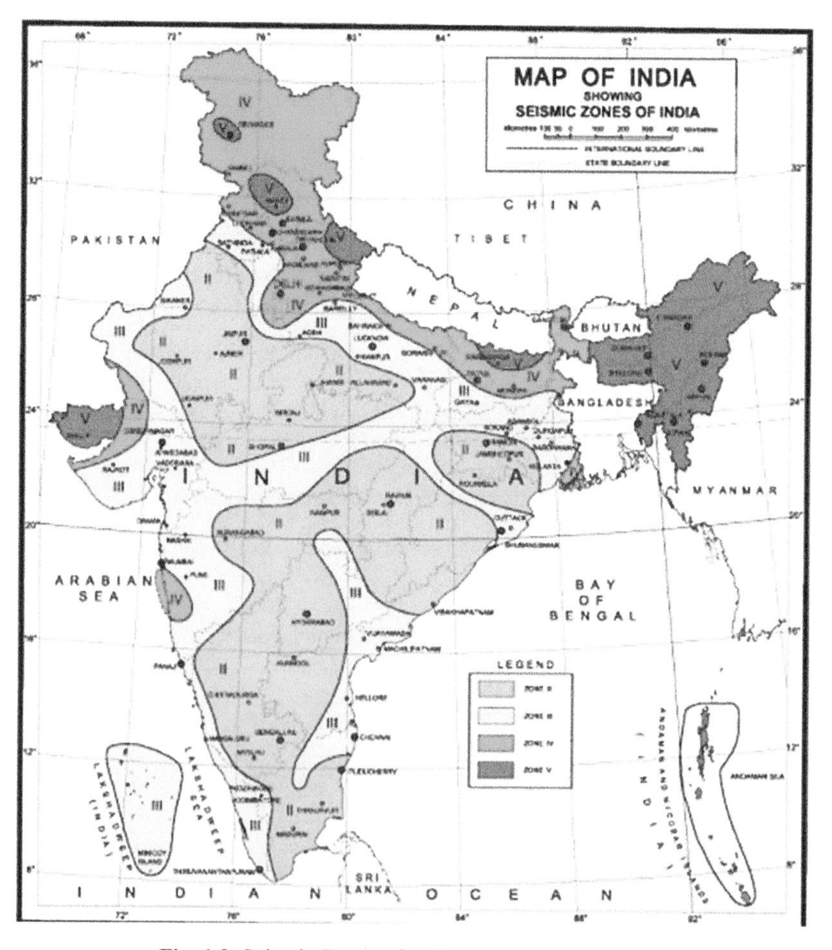

Fig. 1.9. Seismic Zones of India (IS: 1893 (part-I)2016)

The design horizontal seismic coefficient α_h for a structure is determined using the following expression:

$$\alpha_h = \frac{ZIS_a}{2Rg} \tag{1.6}$$

where

Z = Zone factor for the Maximum Considered Earthquake (MCE) and service life of the structure, is given in Table 1.10.

I = Importance factor, depending upon the functional use of the structures, characterized by hazardous consequences of its failure, post-earthquake functional needs, historical value, or economic importance, are given in Table 1.11.

R = Response reduction factor, depending on the perceived seismic damage performance of the structure, characterized by ductile or brittle deformation, is given in Table 1.12.

S_a/g = Average response acceleration coefficient for rock or soil given by Fig. 1.10 for 5% damping. To obtain the value, for other damping ratio, the value taken from this figure is to be multiplied with the factors given in Table 1.13.

The factor 2 in the denominator of expression (1.6) is used to reduce the MCE zone factor to the factor for Design Basis Earthquake (DBE).

The value of α_h will not be taken less than Z/2 provided that for any structure the natural time period (T) is less than 0.1s

Table 1.10. Zone Factor, Z

Seismic Zone	*II*	*III*	*IV*	*V*
Intensity Z	Low 0.1	Moderate 0.16	Severe 0.24	Very severe 0.36

Table 1.11. Values of Importance Factor

Sl. No.	*Structure*	*Importance factor, I*
1.	**Buildings**	
	(a) Important service and community buildings, such as hospitals; schools; monumental structures; emergency buildings like telephone exchange, television stations, radio stations, railway stations, fire station buildings; large community halls like cinemas, assembly halls and subway stations, power stations.	1.5
	(b) All other buildings.	1
2.	**Bridges**	
	(a) Important bridges	
	1. River bridge and flyovers inside cities	1
	2. Bridges of National Highway and State Highways	
	3. Bridges serving traffic near ports and the other centres of economical activities	
	4. Bridges crossing railway lines.	
	(b) Large critical bridges in all Seismic Zones	
	1. Long bridges more than 1 km length across perennial river and creeks	1.2
	2. Bridges for which alternative routes are not available	
	3. Railway bridges	
	(c) All other bridges	1.5

Note: The values of importance factor given above are for guidance. A designer may choose suitable value greater than those mentioned above. This does not apply to temporary structures like excavation, scaffolding etc. of short duration

Table 1.12. Response Reduction Factor R, for Building Systems

Sl. No.	*Lateral Load Resisting System*	*R*
1.	**Building Frame Systems**	
	(i) Ordinary RC moment-resisting frame (OMRF)	3
	(ii) Special RC moment-resisting frame (SMRF)	5
	(iii) Steel frame with	
	(a) Concentric braces	4
	(b) Eccentric braces	5
	(iv) Steel moment resisting frame designed as per SP 6 (6)	5
	(v) Load bearing masonry wall buildings	
	(a) Unreinforced	1.5
	(b) Reinforced with horizontal RC bands 2.5	
	(c) Reinforced with horizontal RC bands and vertical bars at corners of rooms and jambs of openings	3
	(vi) Ordinary reinforced concrete shear walls	3
	(vii) Ductile shear walls	4
	(viii) Ordinary shear wall with OMRF	
	(ix) Ordinary shear wall with SMRF	4
	(x) Ductile shear wall with OMRF	4.5
	(xi) Ductile shear wall with SMRF	5
2.	**Bridges**	
	(i) Superstructure, reinforced concrete	3
	(ii) Superstructure, steel, prestressed concrete	2.5
	(iii) Substructure	
	(a) Reinforced concrete piers with the ductile detailing cantilever type, wall type	3
	(b) Reinforced concrete piers without ductile detailing, cantilever type, wall type	2.5
	(c) Masonry piers (unreinforced) cantilever type, wall type	1.5
	(d) Reinforced concrete, framed construction in piers, with ductile detailing, columns of RCC bents, RC single column piers	4
	(e) Steel framed construction	2.5
	(f) Steel cantilever piers	1
	(g) Steel trussed arch	1.5
	(h) Reinforced concrete arch	3.5
	(i) Abutment of mass concrete and masonry	1
	(j) R.C.C. abutment	2.5
	(k) Integral frame with ductile detailing, and	4
	(l) Integral frame without ductile detailing	3.3

Table 1.13. Multiplying Factor for Obtaining Values for Other Damping

Damping, percent	0	2	5	7	10	15	20	25	30
Factors	3.2	1.4	1	0.9	0.8	0.7	0.6	0.55	0.5

The design acceleration spectrum for vertical motion shall be taken as two-third of the design horizontal acceleration spectrum. Therefore, the vertical seismic coefficient (α_v) will be

$$\alpha_v = \frac{2}{3}\alpha_h \tag{1.7}$$

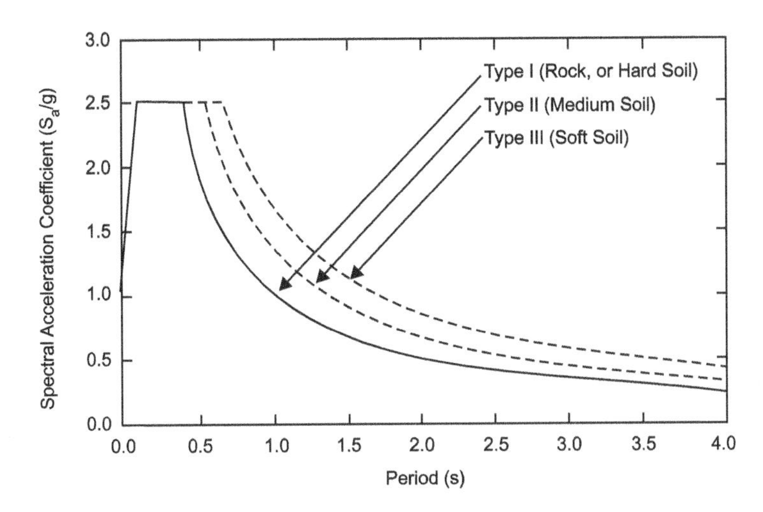

Fig. 1.10. Response spectra for rock and soil sites for 5 percent damping

Illustrative Examples

Example 1.1

The standard torsion seismograph recorded an average trace amplitude of 8.0 mm. The distance to the epicentre is estimated about 100 km. Determine the magnitude of earthquake.

Solution

From Fig. 1.4, the distance correction (i.e. – log A_0 for 100 km is 3.0.
Hence,

$$M = \log_{10} A - \log_{10} A_0 = \log_{10} 8.0 + 3 = 3.9$$

Example 1.2

Determine the value of horizontal seismic coefficient for a community centre situated in Delhi. The geotechnical exploration carried out at the site indicated soft clay upto 10.0 m depth, and therefore needing pile foundation for the structure. The natural time period is likely to be smaller than 0.4 s with damping ratio of 5%.

Solution

(i) Evaluation of α_h by Seismic coefficient method (IS 1893-1984)

$$\alpha_h = \beta.I.\alpha_0$$

Being Delhi in Zone IV, $\alpha_0 = 0.05$ (Table 1.6)

$$I = 1.5 \text{ (Table 1.7)}; \beta = 1.2 \text{ (Table 1.8)}$$

Therefore,

$$\alpha_h = 1.2 \times 1.5 \times 0.05 = 0.09$$

(ii) Evaluation of α_h by response spectrum method (IS 1893-1984)

$$\alpha_h = \beta\, I\, F_0 (S_a/g)$$

$$F_0 = 0.25 \text{ (Table 1.9)};$$

$$(S_a/g) = 0.2 \text{ (Fig. 1.8, for T = 0.4 s and } \xi = 5\%)$$

Therefore, $\alpha_h = 1.2 \times 1.5 \times 0.25 \times 0.2 = 0.09$

(iii) Evaluation of α_h using IS 1893 (Part 1) - 2002

$$\alpha_h = \frac{Z I S_a}{2 R g}$$

$R = 4.0$, $Z = 0.24$, $I = 1.5$, $S_a/g = 2.5$ [Fig. 1.10, for T = 0.95]

$$\alpha_h = \frac{0.24 \times 1.5 \times 2.5}{2 \times 4} = 0.11$$

References

Housner. G. W. (1965), "Intensity of earthquake ground shaking near the causative fault", Proceedings 3rd World Conference on Earthquake Engineering, New Zealand, Vol. I.

Housner, G. W. (1970), "Design spectrum", in Earthquake Engineering (R. W. Wiegel, Ed.), Prentice-Hall, Englewood Cliffs, New Jersey, pp. 97-106.

IS: 1893 Part 1-2016, "Criteria for earthquake resistant design of structures - General provisions and Buildings", I.S.I., New Delhi.

Lee, K. L. and Chan, K. (1972), "Number of equivalent significant cycles in strong motion earthquakes", Proceedings, International Conference on Microzonation, Seattle, Washington, vol. II, pp. 609-627.

Richter, C.F. (1958), "Elementary seismology", W.H. Freeman, San Francisco, California.

Seed, H. B., Idriss, J. M., Makdisi, F. and Banerjee, N. (1975), "Representation of irregular stress - time histories by equivalent uniform stress series in liquefaction analysis", Report No. EERC 75-29, Earthquake Engineering Research Center, University of California, Berkeley.

Seed, H. B. and Idriss, I. M. (1971), "Simplified procedure for evaluating soil liquefaction potential", J. Soil Mech. Found. Engg., ASCE, Vol. 97, No. SM9, pp. 1249-1273.

Practice Problems

1.1 Explain the terms 'Intensity' and 'Magnitude' in relation to earthquake. How are fault length and duration of earthquake depend on magnitude?

1.2 Describe a method of getting equivalent number of cycles of uniformly varying load for an actual earthquake record.

1.3 Determine the equivalent number of cycles for 0.75 t_{max} for El Centro earthquake.

CHAPTER

2

Theory of Vibrations

2.1 Objective and Overview

In order to understand the behaviour of a structure subjected to dynamic load lucidly, one must study the mechanics of vibrations caused by the dynamic load. The pattern of variation of a dynamic load with respect to time may be either periodic or transient. The periodical motions can be resolved into sinusoidally varying components e.g. vibrations in the case of reciprocating machine foundations. Transient vibrations may have very complicated non-periodic time history e.g. vibrations due to earthquakes and quarry blasts.

A structure subjected to a dynamic load (periodic or transient) may vibrate in one of the following four ways of deformation or a combination there-of:

(i) Extensional (Fig. 2.1a)
(ii) Shearing (Fig. 2.1b)
(iii) Bending (Fig. 2.1c)
(iv) Torsional (Fig. 2.1d)

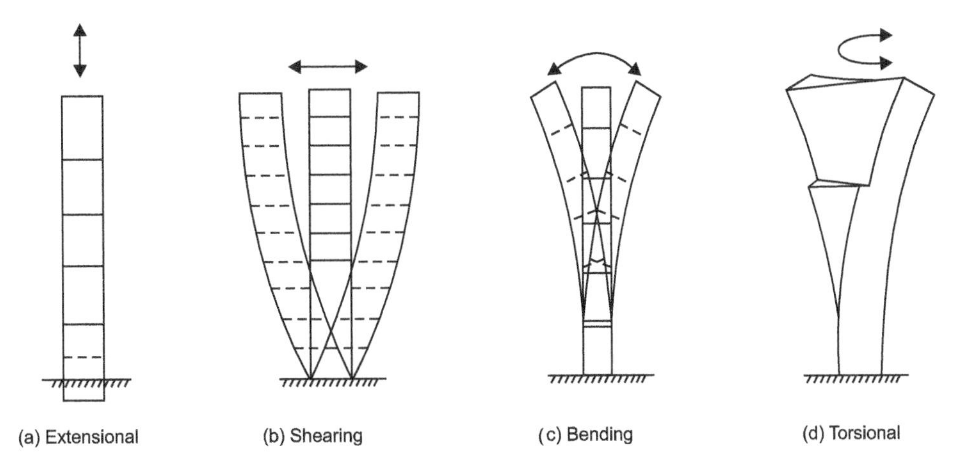

Fig. 2.1. Different types of vibrations

The forms of vibration mainly depend on the mass, stiffness distribution and end conditions of the system.

To study the response of a vibratory system, in many cases it is satisfactory to reduce it to an idealized system of lumped parameters. In this regard, the simplest model consists of mass, spring and dashpot. This chapter is framed to provide the basic concepts and dynamic analysis of such systems. Actual field problems which can be idealized to mass-spring-dashpot systems, have also been included.

2.2 Definitions

2.2.1 Vibrations

If the motion of the body is oscillatory in character, it is called vibration.

2.2.2 Degrees of Freedom

The number of independent co-ordinates which are required to define the position of a system during vibration, is called degrees of freedom (Fig. 2.2).

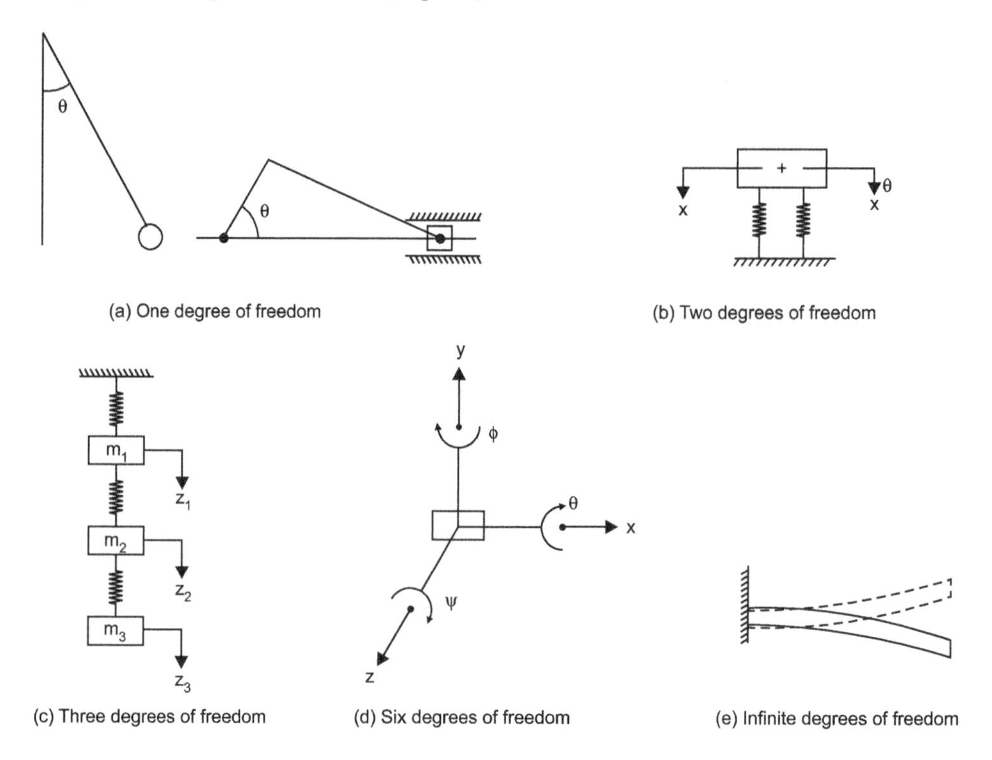

Fig. 2.2. Systems with different degrees of freedom

2.2.3 Periodic Motion

If motion repeats itself at regular intervals of time, it is called periodic motion.

2.2.4 Free Vibration

If a system vibrates without an external force, then it is said to undergo free vibrations. Such vibrations can be caused by setting the system in motion initially and allowing it to move freely afterwards.

2.2.5 Natural Frequency

This is the property of the system and corresponds to the number of free oscillations made by the system in unit time.

2.2.6 Forced Vibrations

Vibrations that are developed by externally applied exciting forces are called forced vibrations. These vibrations occur at the frequency of the externally applied exciting force.

2.2.7 Forcing Frequency

This refers to the periodicity of the external forces which act on the system during forced vibrations. This is also termed as operating frequency.

2.2.8 Frequency Ratio

The ratio of the forcing frequency and natural frequency of the system is referred as frequency ratio.

2.2.9 Amplitude of Motion

The maximum displacement of a vibrating body from the mean position is amplitude of motion.

2.2.10 Time Period

Time taken to complete one cycle of vibration is known as time period.

2.2.11 Resonance

A system having n degrees of freedom has n natural frequencies. If the frequency of excitation coincides with any one of the natural frequencies of the system, the condition of resonance occurs. The amplitudes of motion are very excessive at resonance.

2.2.12 Damping

All vibration systems offer resistance to motion due to their own inherent properties. This resistance is called damping force and it depends on the condition of vibration, material and type of the system. If the force of damping is constant, it is termed Coulomb damping. If the damping force is proportional to the velocity, it is termed viscous damping. If the damping in a system is free from its material property and is contributed by the geometry of the system, it is called geometrical or radiation damping.

2.2.13 Principal Modes of Vibration

In a principal mode, each point in the system vibrates with the same frequency. A system with n degrees of freedom possesses n principal modes with n natural frequencies.

2.2.14 Normal Mode of Vibration

If the amplitude of a point of the system vibrating in one of the principal modes is made equal to unity, the motion is then called the normal mode of vibration.

2.3 Harmonic Motion

Harmonic motion is the simplest form of vibratory motion. It may be described mathematically by the following equation:

$$Z = A \sin(\omega t - \theta) \tag{2.1}$$

Equation (2.1) is plotted as function of time in Fig. 2.3. The various terms of this equation are as follows:

Z = Displacement of the rotating mass at any time t

A = Displacement amplitude from the mean position, sometimes referred as single amplitude. The distance $2A$ represents the peak-to-peak displacement amplitude, sometimes referred to as double amplitude, and is the quantity most often measured from vibration records.

ω = Circular frequency in radians per unit time. Because the motion repeats itself after 2π radians, the frequency of oscillation in terms of cycles per unit time will be $\omega/2\pi$. It is denoted by f.

θ = Phase angle. It is required to specify the time relationship between two quantities having the same frequency when their peak values having like sign do not occur simultaneously. In Eq. (2.1) the phase angle is a reference to the time origin.

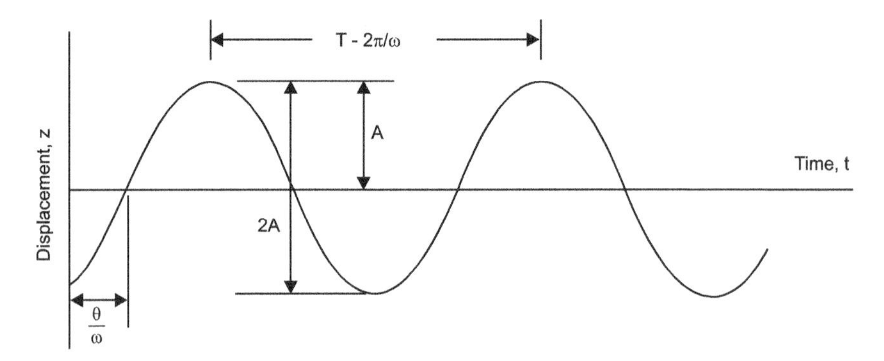

Fig. 2.3. Quantities describing harmonic motion

More commonly, the phase angle is used as a reference to another quantity having the same frequency. For example, at some reference point in a harmonically vibrating system, the motion may be expressed by

$$Z_1 = A_1 \sin \omega t \tag{2.2}$$

Motion at any other point in the system might be expressed as

$$Z_i = A_i \sin (\omega t - \theta_i) \tag{2.3}$$

with $\pi \geq \theta \geq -\pi$.

For positive values of θ the motion at point i reaches its peak within one half cycle after the peak motion occurs at point 1. The angle θ is then called **phase lag.** For negative values of θ the peak motion at i occurs within one half cycle ahead of motion at 1, and θ is called as **phase lead.**

The time period, T is given by

$$T = \frac{1}{f} = \frac{2\pi}{\omega} \tag{2.4}$$

The velocity and acceleration of motion are obtained from the derivatives of Eq. (2.1).

$$\text{Velocity} = \frac{dZ}{dt} = \dot{Z} = \omega A \cos(\omega t - \theta) \tag{2.5}$$

$$= \omega A \sin(\omega t - \theta + \frac{\pi}{2})$$

$$\text{Acceleration} = \frac{d^2 Z}{dt^2} = \ddot{Z} = -\omega^2 A \cos(\omega t - \theta) \tag{2.6}$$

$$= \omega^2 A \sin(\omega t - \theta + \pi)$$

Equations (2.5) and (2.6) show that both velocity and acceleration are also harmonic and can be represented by vectors ωA and $\omega^2 A$, which rotate at the same speed as A, i.e. ω rad/unit time. These, however, lead the displacement and acceleration vectors by $\pi/2$ and π respectively. In Fig. 2.4 vector representation of harmonic displacement, velocity and acceleration is presented considering the displacement as the reference quantity ($\theta = 0$).

When two harmonic motions having little different frequencies are superimposed, a non-harmonic motion as shown in Fig. 2.5 occurs. It appears to be harmonic except for a gradual increase and decrease in amplitude. The displacement of such a vibration is given by:

$$Z = A_1 \sin(\omega_1 t - \theta_1) + A_2 \sin(\omega_2 t - \theta_2) \tag{2.7}$$

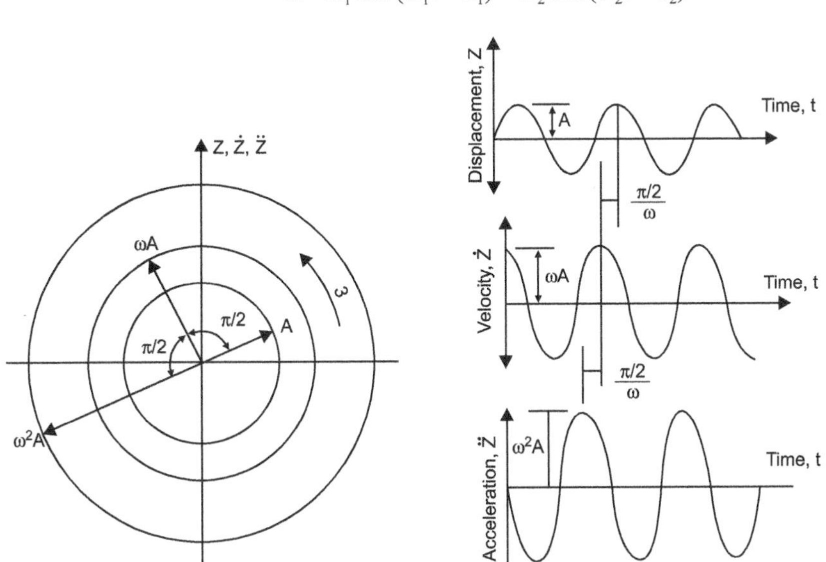

Fig. 2.4. Vector representation of harmonic displacement, velocity and acceleration

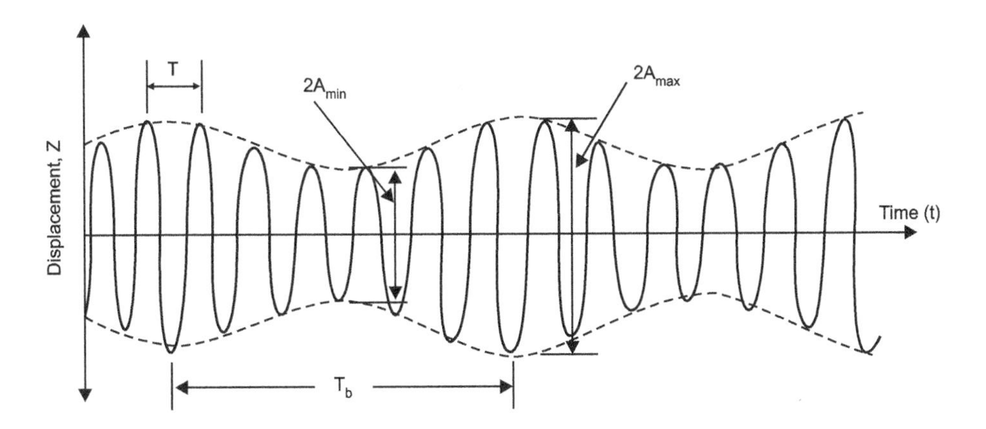

Fig. 2.5. Motion containing a beat

The dashed curve (Fig. 2.5) representing the envelop of the vibration amplitudes oscillates at a frequency, called the beat frequency, which corresponds to the difference in the two source frequencies:

$$f_b = \frac{1}{T_b} = \frac{|\omega_1 - \omega_2|}{2\pi} \tag{2.8}$$

The frequency of the combined oscillations is the average of the frequencies of the two components and is given by

$$f = \frac{1}{T} = \left(\frac{1}{2}\right)\left(\frac{\omega_1 + \omega_2}{2\pi}\right) \tag{2.9}$$

The maximum and minimum amplitudes of motion are the sum and difference of the amplitudes of the two sources respectively.

$$Z_{max} = A_1 + A_2 \tag{2.10a}$$

$$Z_{min} = |A_1 - A_2| \tag{2.10b}$$

If the drive systems of two machines designed to operate at the same speed are not synchronized, they may result vibrations having the beat frequency.

2.4 Vibrations of a Single Degree Freedom System

The simplest model to represent a single degree of freedom system consisting of a rigid mass *m* supported by a spring and dashpot is shown in Fig. 2.6a. The motion of the mass *m* is specified by one co-ordinate Z. Damping in this system is represented by the dashpot, and the resulting damping force is proportional to the velocity. The system is subject to an external time dependent force *F*(*t*).

Figure 2.6b shows the free body diagram of the mass *m* at any instant during the course of vibrations. The forces acting on the mass *m* are:

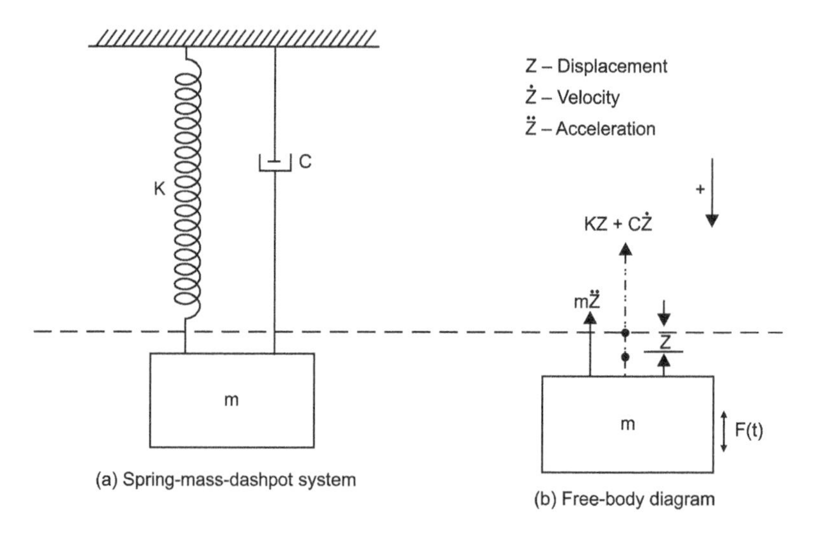

Fig. 2.6. Single degree freedom system

(i) Exciting force, $F(t)$: It is the externally applied force that causes the motion of the system.
(ii) Restoring force, F_r: It is the force exerted by the spring on the mass and tends to restore the mass to its original position. For a linear system, restoring force is equal to $K \cdot Z$, where K is the spring constant and indicates the stiffness. This force always acts towards the equilibrium position of the system.
(iii) Damping force, F_d: The damping force is considered directly proportional to the velocity and given by $C \cdot \dot{Z}$ where C is called the coefficient of viscous damping; this force always opposes

the motion. In some problems in which the damping is not viscous, the concept of viscous damping is still used by defining an equivalent viscous damping which is obtained so that the total energy dissipated per cycle is same as for the actual damping during a steady state of motion.

(iv) Inertia force, F_i: It is due to the acceleration of the mass and is given by $m\ddot{Z}$. According to De-Alembert's principle, a body which is not in static equilibrium by virtue of some acceleration which it possess, can be brought to static equilibrium by introducing on it an inertia force. This force acts through the centre of gravity of the body in the direction opposite to that of acceleration.

The equilibrium of mass m gives

$$m\ddot{Z} + C\dot{Z} + KZ = F(t) \tag{2.11}$$

which is the equation of motion of the system.

2.4.1 Undamped Free Vibrations

For undamped free vibrations, the damping force and the exciting force is equal to zero. Therefore the equation of motion of the system becomes

$$m\ddot{Z} + KZ = 0 \tag{2.12a}$$

or

$$\ddot{Z} + \left(\frac{K}{m}\right)Z = 0 \tag{2.12b}$$

The solution of this equation can be obtained by substituting

$$Z = A_1 \cos \omega_n t + A_2 \sin \omega_n t \tag{2.13}$$

where A_1 and A_2 are both constants and ω_n is undamped natural frequency.

Substituting Eq. (2.13) in Eq. (2.12b), we get,

$$-\omega_n^2(A_1 \cos \omega_n t + A_2 \sin \omega_n t) + \left(\frac{K}{m}\right)(A_1 \cos \omega_n t + A_2 \sin \omega_n t) = 0$$

or

$$\omega_n = \pm\sqrt{\frac{K}{m}} \tag{2.14}$$

The values of constants A_1 and A_2 are obtained by substituting proper boundary conditions. We may have the following two boundary conditions:

(i) At time $t = 0$, displacement $Z = Z_0$, and

(ii) At time $t = 0$, velocity $\dot{Z} = V_0$

Substituting the first boundary condition in Eq. (2.13)

$$A_1 = Z_0 \tag{2.15}$$

Now

$$\dot{Z} = -A_1 \omega_n \sin \omega_n t + A_2 \omega_n \cos \omega_n t \tag{2.16}$$

Substituting the second boundary condition in Eq. (2.16)

$$A_2 = \frac{V_0}{\omega_n} \tag{2.17}$$

Hence

$$Z = Z_0 \cos \omega_n t + \frac{V_0}{\omega_n} \sin \omega_n t \tag{2.18}$$

Now let

$$Z_0 = A_Z \cos\theta \tag{2.19}$$

and
$$\frac{V_0}{\omega_n} = A_Z \sin\theta \tag{2.20}$$

Substitution of Eqs. (2.19) and (2.20) into Eq. (2.18) yields

$$Z = A_Z \cos(\omega_n t - \theta) \tag{2.21}$$

where
$$\theta = \tan^{-1}\left(\frac{V_0}{\omega_n Z_0}\right) \tag{2.22}$$

$$A_z = \sqrt{Z_0^2 + \left(\frac{V_0}{\omega_n}\right)^2} \tag{2.23}$$

The displacement of mass given by Eq. (2.21) can be represented graphically as shown in Fig. 2.7. It may be noted that

At time t equal to	*Displacement Z is*
0	$A_Z \cos\theta$
$\frac{\theta}{\omega_n}$	A_Z
$\frac{\frac{\pi}{2}+\theta}{\omega_n}$	0
$\frac{\pi+\theta}{\omega}$	$-A_Z$
$\frac{\frac{3}{2}\pi+\theta}{\omega_n}$	0
$\frac{2\pi+\theta}{\omega_n}$	A_Z

It is evident from Fig. 2.7 that nature of foundation displacement is sinusoidal. The magnitude of maximum displacement is A_Z. The time required for the motion to repeat itself is the period of vibration, T and is therefore given by

$$T = \frac{2\pi}{\omega_n} \tag{2.24}$$

The natural frequency of oscillation. f_n is given by

$$f_n = \frac{1}{T} = \frac{\omega_n}{2\pi} = \frac{1}{2\pi}\sqrt{\frac{K}{m}} \tag{2.25}$$

Now
$$\frac{mg}{K} = \frac{W}{K} = \delta_{st} \tag{2.26}$$

where g = Acceleration due to gravity, 9.81 m/s^2

W = Weight of mass m

δ_{st} = Static deflection of the spring

Therefore
$$f_n = \frac{1}{2\pi}\sqrt{\frac{g}{\delta_{st}}} \tag{2.27}$$

Equation (2.27) shows that the natural frequency is a function of static deflection. The relation of f_n and δ_{st} given by Eq. (2.27) gives a curve as shown in Fig. 2.8.

The nature of variation of the velocity and acceleration of the mass is also shown in Fig. 2.7.

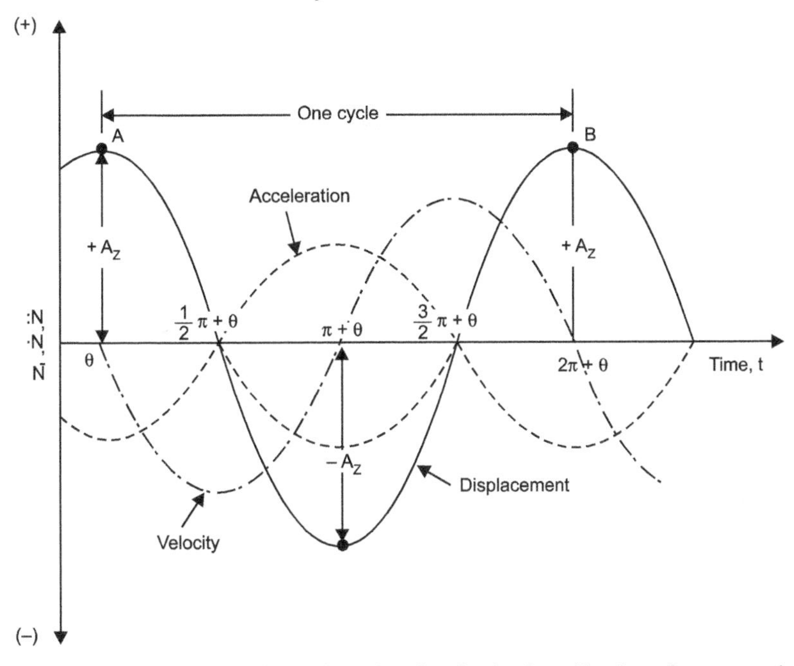

Fig. 2.7. Plot of displacement, velocity and acceleration for the free vibration of a mass-spring system

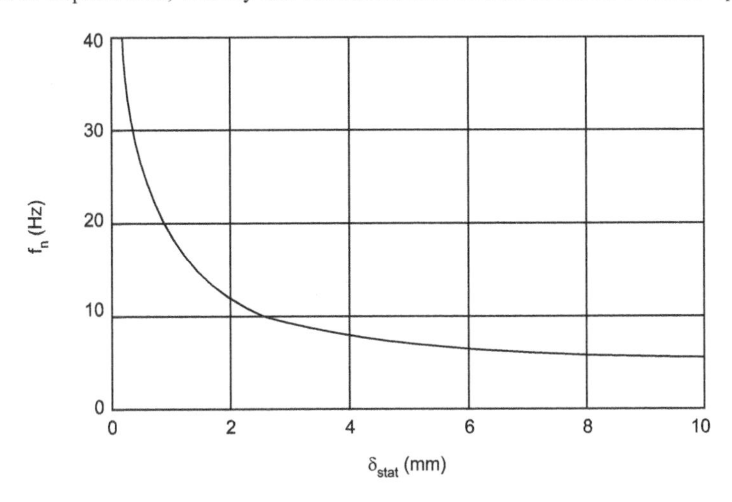

Fig. 2.8. Relationship between natural frequency and static deflection

2.4.2 Free Vibration with Viscous Damping

For damped free vibration system (i.e., the excitation force $F_0 \sin \omega t$ on the system is zero), the differential equation of motion can be written as

$$m\ddot{Z} + C\dot{Z} + KZ = 0 \tag{2.28}$$

where C is the damping constant or force per unit velocity. The solution of Eq. (2.28) may be written as

$$Z = Ae^{\lambda t} \tag{2.29}$$

where A and λ are arbitrary constants. By substituting the value of Z given by Eq. (2.29) in Eq. (2.28), we get

$$m\,A\,\lambda^2 e^{\lambda t} + C\,A\,\lambda e^{\lambda t} + K\,A\,e^{\lambda t} = 0$$

or

$$\lambda^2 + \left(\frac{C}{m}\right)\lambda + \frac{K}{m} = 0 \tag{2.30}$$

By solving Eq. (2.30)

$$\lambda_{1,2} = -\frac{C}{2m} \pm \sqrt{\left(\frac{C}{2m}\right)^2 - \frac{K}{m}} \tag{2.31}$$

The complete solution of Eq. (2.28) is given by

$$Z = A_1 e^{\lambda_1 t} + A_2 e^{\lambda_2 t} \tag{2.32}$$

The physical significance of this solution depends upon the relative magnitudes of $(C/2m)^2$ and (K/m), which determines whether the exponents are real or complex quantities.

To proceed further, it is convenient to define here a new term critical damping coefficient (C_c). It is that value of the damping coefficient (C) that makes the square root term of Eq. (2.31) equal to zero. Therefore, when

$$\left(\frac{C}{2m}\right)^2 = \frac{K}{m};\ C = C_c \tag{2.32a}$$

Then

$$C_c = 2\sqrt{Km} \tag{2.32b}$$

The ratio of the actual damping constant C to the critical damping constant C_c is defined as damping ratio.

$$\xi = \frac{C}{C_c} \tag{2.33a}$$

$$\frac{C}{2m} = \frac{C}{C_c}\cdot\frac{C_c}{2m} = \frac{C\cdot 2\sqrt{Km}}{C_c\cdot 2m} = \frac{C}{C_c}\cdot\sqrt{\frac{K}{m}} = \omega_n \xi \tag{2.33b}$$

$$\lambda_{1,2} = \left(-\xi \pm \sqrt{\xi^2 - 1}\right)\omega_n \tag{2.34}$$

Therefore, Eq. 2.32 may be written as:

$$Z = e^{-\xi\omega_n t}\left[A_1 e^{\sqrt{\xi^2-1}\omega_n t} + A_2\, e^{\sqrt{\xi^2-1}\,\omega_n t}\right] \tag{2.35}$$

The physical significance of this solution depends upon the relative magnitude of ξ with respect to 1, which determines whether the exponents are real or complex quantities.

Case 1: ξ > 1 (Over damped system)

In this system, the damping is comparatively large and the exponents of Eq. (2.35) are real. The values of arbitrary constants A_1 and A_2 may be obtained consideing that the body (Fig. 2.6) is displaced by distance Z_0 from equilibrium position and released without any initial velocity. Then, boundary conditions become:

(i) At $t = 0$; $Z = Z_0$ (2.36a)

(ii) At $t = 0$; $\dot{Z} = 0$ (2.36b)

$$\dot{Z} = \frac{dZ}{dt} = e^{-\xi\omega_n t} \times (-\xi\omega_n)\left[A_1 e^{\sqrt{\xi^2-1}\,\omega_n t} + A_2 e^{\sqrt{\xi^2-1}\,\omega_n t}\right]$$

$$+ e^{-\xi\omega_n t}\left[A_1\omega_n\sqrt{\xi^2-1}\, e^{\sqrt{\xi^2-1}\,\omega_n t} - A_2\omega_n\sqrt{\xi^2-1}\, e^{-\sqrt{\xi^2-1}\,\omega_n t}\right] \quad (2.37)$$

Apply boundary conditions in Eqs. (2.36) and (2.37),

$$A_1 + A_2 = Z_0 \quad (2.38a)$$

$$-(A_1 + A_2)\xi\,\omega_n + (A_1 - A_2)\sqrt{\xi^2-1}\;\omega_n = 0 \quad (2.38b)$$

Solving Eqs. (2.38a) and (2.38b)

$$A_1 = \frac{(\xi + \sqrt{\xi^2-1})}{2\sqrt{\xi^2-1}} Z_0 \quad (2.39a)$$

$$A_2 = \frac{(-\xi + \sqrt{\xi^2-1})}{2\sqrt{\xi^2-1}} Z_0 \quad (2.39b)$$

Equation (2.35) can be written as

$$Z = \frac{e^{-\xi\omega_{nt}}}{2\sqrt{\xi^2-1}}\left[\xi\left(e^{\sqrt{\xi^2-1}\,\omega_n t} - e^{-\sqrt{\xi^2-1}\,\omega_n t}\right) + \sqrt{\xi^2-1}\left(e^{\sqrt{\xi^2-1}\,\omega_n t} + e^{-\sqrt{\xi^2-1}\,\omega_n t}\right)\right] Z_0 \quad (2.40a)$$

or

$$\frac{Z}{Z_0} = \frac{1}{2\sqrt{\xi^2-1}}\left[\left(\xi + \sqrt{\xi^2-1}\right)e^{(-\xi+\sqrt{\xi^2-1})\omega_n t} + \left(-\xi + \sqrt{\xi^2-1}\right)e^{(-\xi-\sqrt{\xi^2-1})\omega_n t}\right] \quad (2.40b)$$

Since the power of e is negative in both the terms of Eq. (2.40b), they both decreases exponentially with t; therefore Z will decrease exponentially with the increase in t. Z will become zero when t will become infinity. Plots of $\frac{Z}{Z_0}$ versus $\omega_n t$ are shown by firm lines in Fig. 2.9 for different values of damping ratio ($\xi > 1$). In all these curves the system does not cross the equilibrium position. Higher the damping, more sluggish is the response of the system. Hence if $\xi > 1$, the motion is aperiodic and no oscillation occurs.

Case II: $\xi = 1$ (Critically damped system)
In this case, from Eq. (2.34)

$$\lambda_1 = \lambda_2 = -\omega_n$$

The solution of Eq. (2.30) is given by

$$Z = (A_1 + A_2 t)e^{-\omega_n t} \quad (2.41)$$

The above equation is the solution of a system having critical damping.
Using same boundary conditions as given in Eq. (2.36),

$$A_1 = Z_0, \text{ and } A_2 = \omega_n Z_o \quad (2.41)$$

Therefore,

$$Z = (1 + \omega_n t)e^{-\omega_n t}.Z_o \quad (2.42a)$$

or

$$\frac{Z}{Z_0} = (1 + \omega_n t)e^{-\omega_n t} \quad (2.42b)$$

The value of Z/Z_0 in the above equation also decreases with the increase in t as shown by dotted line curve in Fig. 2.9. This is also an aperiodic motion, and in this case the displacement-time curve lies below any of the curves for over-damped system.

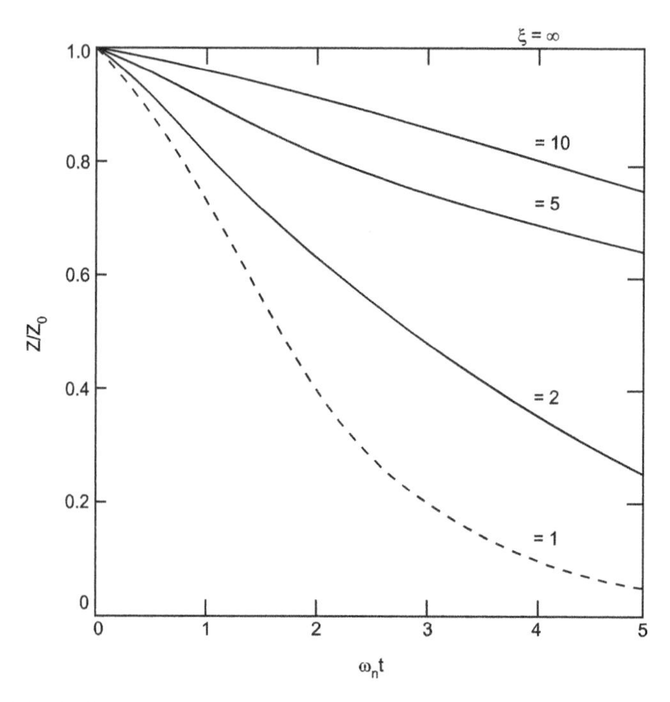

Fig. 2.9. Displacement time plots of over-damped and critically damped systems

Case III: $\xi < 1$ (Underdamped system)
The roots λ_1 and λ_2 are complex and are given by

$$\lambda_{1.2} = \left[-\xi \pm i\sqrt{1-\xi^2}\right] \tag{2.43}$$

The complete solution of Eq. (2.28) is given by

$$Z = A_1 e^{\left(-\xi \pm i\sqrt{1-\xi^2}\right)\omega_n t} + A_2 e^{\left(-\xi - i\sqrt{1-\xi^2}\right)\omega_n t} \tag{2.44a}$$

or

$$Z = e^{-\xi\omega_n t}\left[A_1 e^{i\sqrt{1-\xi^2}\,\omega_n t} + A_2 e^{-i\sqrt{1-\xi^2}\,\omega_n t}\right] \tag{2.44b}$$

The Eq. (2.44b) can be written as

$$Z = e^{-\xi\omega_n t}\left[C_1 \sin\left(\omega_n\sqrt{1-\xi^2}\,t\right) + C_2 \cos\left(\omega_n\sqrt{1-\xi^2}\,t\right)\right] \tag{2.44c}$$

or

$$Z = e^{-\xi\omega_n t}\,[C_1 \sin\omega_{nd}t + C_2 \cos\omega_{nd}t] \tag{2.44d}$$

where $\omega_{nd} = \omega_n\sqrt{1-\xi^2}$ = Damped natural frequency

$C_1 = A_1 + A_2$ and $C_2 = i(A_1 - A_2)$

In Eq. (2.44d), constants C_1 and C_2 are real which make A_1 and A_2 complex conjugate quantities.

The motion of the system is oscillatory (Fig. 2.10) and the amplitude of vibration goes on decreasing in an exponential fashion.

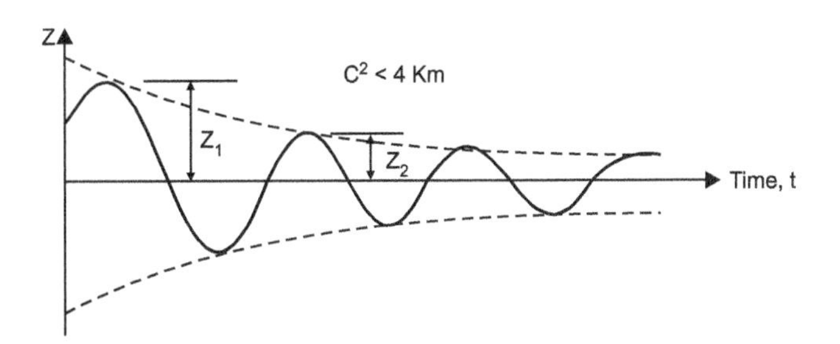

Fig. 2.10. Free vibrations of a viscously underdamped system

As a convenient measure of damping, we may compute the ratio of amplitudes of the successive cycles of vibration

$$\frac{Z_1}{Z_2} = \frac{e^{-\omega_n \xi t}}{e^{-\omega_n \xi (t + 2\pi/\omega_{nd})}} \tag{2.45a}$$

or
$$\frac{Z_1}{Z_2} = e^{\omega_n \xi . 2\pi/\omega_{nd}} \tag{2.45b}$$

or
$$\frac{Z_1}{Z_2} = e^{2\pi\xi/\sqrt{1-\xi^2}} \tag{2.45c}$$

or
$$\log_e \frac{Z_1}{Z_2} = \frac{2\pi\xi}{\sqrt{1-\xi^2}} \tag{2.45d}$$

Natural logarithm of ratio of two successive peak amplitudes $\left\{ i.e. \log_e \left(\frac{Z_1}{Z_2} \right) \right\}$ is called as **logarithmie decrement.**

or
$$\xi = \frac{1}{2\pi} \log_e \frac{Z_1}{Z_2}, \text{ as for small values of } \xi, \sqrt{1-\xi^2} = 1 \tag{2.45e}$$

Thus, damping of a system can be obtained from a free vibration record by knowing the successive amplitudes which are one cycle apart.

If the damping is very small, it may be convenient to measure the differences in peak amplitudes for a number of cycles, say *n*.

In such a case, if Z_n is the peak amplitudes of the nth cycle, then

$$\frac{Z_0}{Z_1} = \frac{Z_1}{Z_2} = \frac{Z_2}{Z_3} = \ldots = \frac{Z_{n-1}}{Z_n} = e^{\delta} \text{ where } \delta = 2\pi\xi$$

Therefore,
$$\frac{Z_0}{Z_n} = \left[\frac{Z_0}{Z_1}\right]\left[\frac{Z_1}{Z_2}\right]\left[\frac{Z_2}{Z_3}\right] \cdots \left[\frac{Z_{n-1}}{Z_n}\right] = e^{n\delta}$$

Hence
$$\delta = \frac{1}{n} \log_e \frac{Z_0}{Z_n} \tag{2.45f}$$

or
$$\xi = \frac{1}{2\pi n} \log_e \frac{Z_0}{Z_n} \tag{2.45g}$$

Variation of damped natural frequency with damping is shown in Fig. 2.11. The decrease in damped natural frequency with increase in damping ratio is small initially but is very steep as ξ increases further. Therefore the damped natural frequency may be taken approximately equal to undamped natural frequency for lower values of ξ (<0.4).

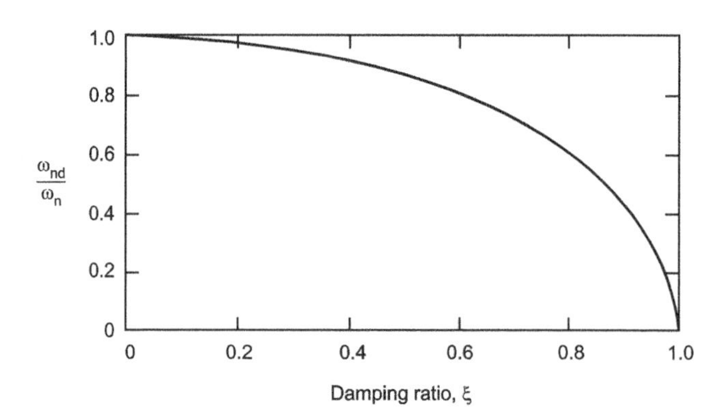

Fig. 2.11. Variation of damped natural frequency with damping

Therefore, a system is:

over damped if $\xi > 1$;
critically damped if $\xi = 1$ and
under damped if $\xi < 1$.

2.4.3 Forced Vibrations of Single Degree Freedom System

In many cases of vibrations caused by rotating parts of machines, the systems are subjected to periodic exciting forces. Let us consider the case of a single degree freedom system which is acted upon by a steady state sinusoidal exciting force having magnitude F and frequency ω (i.e. $F(t) = F_0 \sin \omega t$). For this case the equation of motion (Eq. 2.11) can be written as:

$$m\ddot{Z} + C\dot{Z} + K Z = F_0 \sin \omega t \quad (2.46)$$

Equation (2.46) is a linear, non-homogeneous, second order differential equation. The solution of this equation consists of two parts namely (i) complementary function, and (ii) particular integral. The complementary function is obtained by considering no forcing function. Therefore the equation of motion in this case will be:

$$m\ddot{Z}_1 + C\dot{Z}_1 + K Z_1 = 0 \quad (2.47)$$

The solution of Eq. (2.47) has already been obtained in the previous section and is given by,

$$Z_1 = e^{-\xi \omega_n t}(C_1 \sin \omega_{nd} t + C_2 \cos \omega_{nd} t) \quad (2.48)$$

Here Z_1 represents the displacement of mass m at any instant t when vibrating without any forcing function.

The particular integral is obtained by rewriting Eq. (2.45) as

$$m\ddot{Z}_2 + C\dot{Z}_2 + K Z_2 = F_0 \sin \omega t \quad (2.49)$$

where Z_2 is the displacement of mass m at any instant t when vibrating with forcing function.

The solution of Eq. (2.49) is given by

$$Z_2 = A_1 \sin \omega t + A_2 \cos \omega t \quad (2.50)$$

where A_1 and A_2 are two arbitrary constants.

Substituting Eq. (2.50) in Eq. (2.49)

$$m(-A_1\omega^2 \sin \omega t - A_2 \omega^2 \cos \omega t) + C(A_1 \omega \cos \omega t - A_2 \omega \sin \omega t) + K(A_1 \sin \omega t + A_2 \cos \omega t) = F_0 \sin \omega t \quad (2.51)$$

Considering Sine and Cosine functions in Eq. (2.50) separately,

$$(-mA_1\,\omega^2 + KA_1 - CA_2\,\omega)\sin\omega t = F_0 \sin\omega\, t \tag{2.52a}$$

$$(-mA_2\,\omega^2 + KA_2 - CA_1\,\omega)\cos\omega t = 0 \tag{2.52b}$$

From Eq. (2.52a),

$$A_1\left(\frac{K}{m} - \omega^2\right) - A_2\left(\frac{C}{m} - \omega\right) = \frac{F_0}{m} \tag{2.53a}$$

and from Eq. (2.52b)

$$A_1\left(\frac{K}{m} - \omega\right) + A_2\left(\frac{C}{m} - \omega^2\right) = 0 \tag{2.53b}$$

Solving Eqs. (2.53a) and (2.53b), we get

$$A_1 = \frac{(K - m\omega^2)F_0}{(K - m\omega^2) + C^2\omega^2} \tag{2.53c}$$

and

$$A_2 = \frac{-C\omega F_0}{(K - m\omega^2)^2 + C^2\omega^2} \tag{2.53d}$$

By substituting the values of A_1 and A_2 in Eq. (2.50),

$$Z_2 = \frac{F_0}{(K - m\omega^2)^2 + C^2\omega^2}\{(K - m\omega^2)\sin\omega t - C\omega\cos\omega t\} \tag{2.54}$$

let,

$$\tan\theta = \frac{C\omega}{K - m\omega^2} \tag{2.55}$$

By substituting Eq. (2.55) in Eq. (2.54), one can obtain

$$Z_2 = \frac{F_0}{\sqrt{(K - m\omega^2)^2 + C^2\,\omega^2}}.\sin(\omega t - \theta) \tag{2.56}$$

Eq. (2.56) may be written as

$$Z_2 = \frac{F_0 / K}{\sqrt{(1-\eta^2) + (2\eta\xi)^2}}.\sin(\omega t - \theta) \tag{2.57}$$

where η = frequency ratio $= \dfrac{\omega}{\omega_n}$

ξ = Damping ratio $= \dfrac{C}{C_c} = \dfrac{C}{2\sqrt{Km}}$

The complete solution is obtained by adding the complimentary function and the particular integral. Since the complimentary function is an exponentially decaying function, it will die out soon and the motion will be described by only the particular integral (Fig. 2.12). The system will vibrate harmonically with the same frequency as the forcing frequency, and the peak amplitude is given by

$$A_Z = \frac{F_0 / K}{\sqrt{(1-\eta^2)^2 + (2\eta\xi)^2}} \tag{2.58}$$

The quantity F_0/K is equal to the static deflection of the mass under force F_0. Dynamic magnification factor, μ is defined as the ratio of the dynamic amplitude A_Z to the static deflection, and is given by

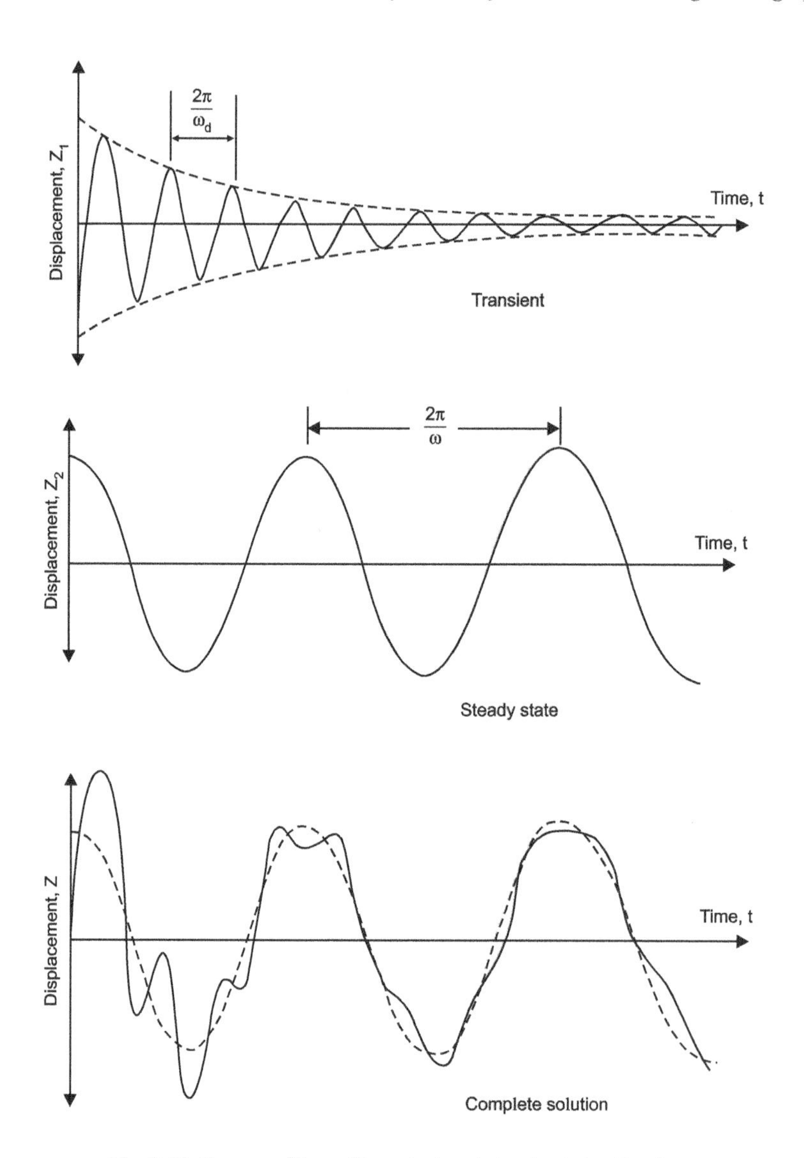

Fig. 2.12. Superposition of transient and steady state vibrations

$$\mu = \frac{1}{\sqrt{(1-\eta^2)^2 + (2\eta\xi)^2}} \tag{2.59}$$

The variation of μ versus η is shown in Fig. 2.13 for different values of damping ratio ξ. It would be seen that near $\eta = 1$, the value of μ is maximum. This is called resonance and the forcing frequency *f* at which it occurs is called the resonant frequency.

Differentiating Eq. (2.59) with respect to η and equating to zero, it can be shown that resonance will occur at a frequency ratio given by

$$\eta = \sqrt{1-2\xi^2} \tag{2.60a}$$

which is approximately equal to unity for small values of ξ.

or $$\omega_{nd} = \omega_n \sqrt{1-2\xi^2} \tag{2.60b}$$

where ω_{nd} = Damped resonant frequency

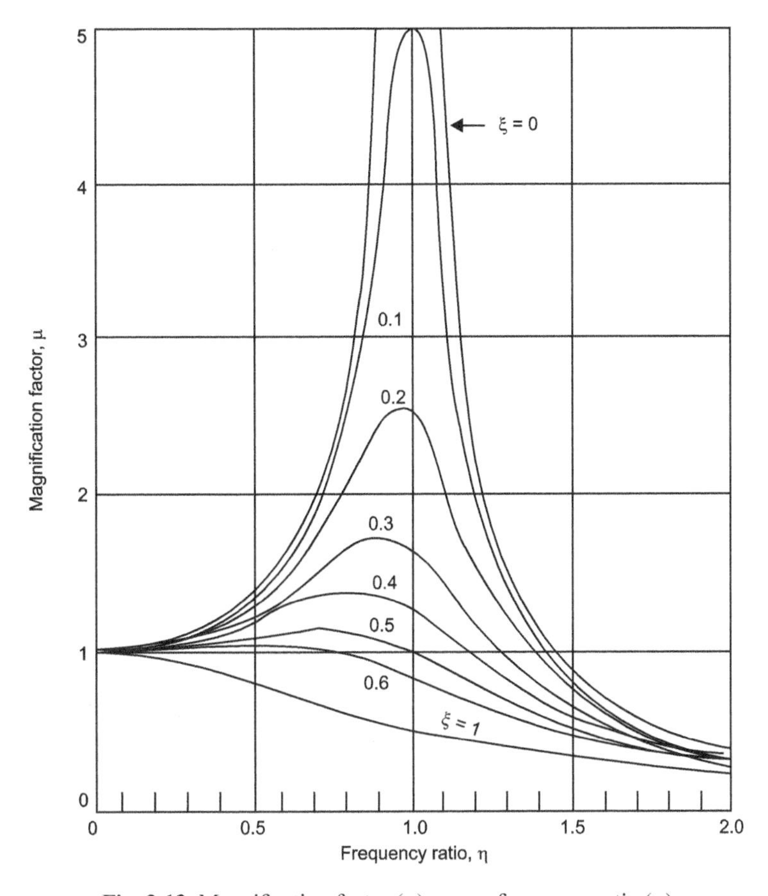

Fig. 2.13. Magnification factor (μ) versus frequency ratio (η)

By substituting Eq. (2.60) in Eq. (2.59), the maximum value of magnification factor is obtained. It is given by

$$\mu_{max} = \frac{1}{2\xi\sqrt{1-\xi^2}} \tag{2.61}$$

$$= \frac{1}{2\xi} \text{ (For small values of } \xi) \tag{2.62}$$

Assuming a damping of 5% in a structure, its amplitude at resonance will be 10 times the static deflection. This indicates that systems will be subjected to very large amplitudes at resonance which should be avoided.

The phase angle θ given by Eq. (2.55) indicates the phase difference between the motion and the exciting force. It can be written as

$$\theta = \tan^{-1}\left(\frac{2\eta\xi}{1-\eta^2}\right) \tag{2.63}$$

Variation of θ with respect to η is shown in Fig. 2.14.

It is important to remember that the damping ratio and any of these parameters are simply the parameters used to understand the effects of damping. They allow the effects of energy dissipation to be represented in a convenient manner mathematically. For most soils and structures, however, energy is dissipated hysteorically (i.e. by yielding or plastic straining of material).In such cases the behavior is more accurately characterized by evaluating the non-linear response of the system.

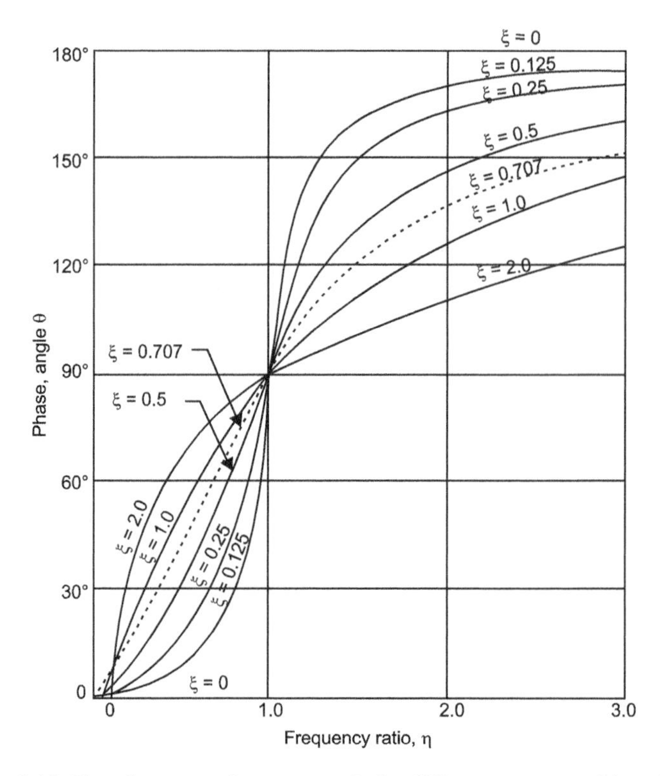

Fig. 2.14. Phase lag versus frequency ratio for different amounts of damping.

2.4.3.1 Rotating Mass Type Excitation

Machines with rotating masses develop alternating force as shown in Fig. 2.15a. Since horizontal forces on the foundation at any instant cancel, the net vibrating force on the foundation is vertical and equal to 2 $m_e e\omega^2 \sin \omega t$, where m_e is the mass of each rotating element, placed at eccentricity e from the centre of rotating shaft and ω is the angular frequency of masses. Figure 2.15b shows such a system mounted on elastic supports with dashpot representing viscous damping.

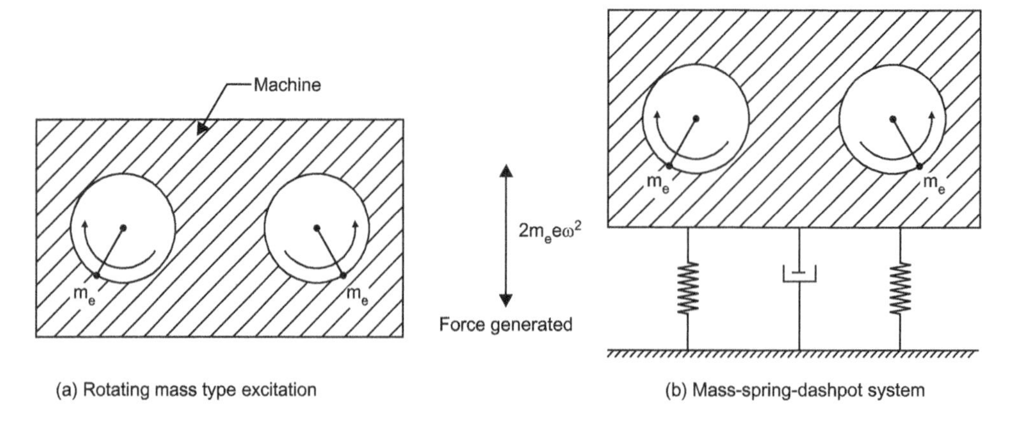

Fig. 2.15. Single degree freedom system with rotating mass type excitation

The equation of motion can be written as

$$m\ddot{Z} + C\dot{Z} + K\,Z = 2\,m_e\,e\omega^2 \sin \omega t \tag{2.64}$$

where m is the mass of foundation including 2 m_e. Equations (2.64) and (2.46) are similar, except that 2 $m_e\, e\, \omega^2$ appears in Eq. (2.64) in place of F_o. The solution of Eq. (2.64) may therefore be written as

$$Z = A_Z \sin(\omega t + \theta) \quad (2.65)$$

where

$$A_Z = \frac{(2m_e e/m)\cdot\eta^2}{\sqrt{(1-\eta^2)^2 + (2\xi\eta)^2}} \quad (2.66)$$

Since

$$F_o = 2\, m_e \cdot e\omega^2$$

or

$$\frac{F_o}{K} = 2m_e \cdot e\frac{\omega^2}{K} = 2m_e \cdot e\frac{\omega^2}{(m\omega_n^2)} = \left(2m_e \frac{e}{m}\right)\cdot\eta^2$$

$$\theta = \tan^{-1}\left(\frac{2\eta\xi}{1-n}\right) \quad (2.67)$$

Equation (2.66) can be expressed in non-dimensional form as given below:

$$\frac{A_z}{(2m_e e/m)} = \frac{\eta^2}{\sqrt{(1-\eta^2)^2 + (2\eta\xi)^2}} \quad (2.68)$$

The value of $A_z/(2m_e e/m)$ is plotted against frequency ratio η in Fig. 2.16a. The curves are similar in shape to those in Fig. 2.13 except that these start from origin. The variation of phase angle θ with η is shown in Fig. 2.16 *b*. Differentiating Eq. (2.68) with respect to η and equating to zero, it can be shown that resonance will occur at a frequency ratio given by

$$\eta = \frac{1}{\sqrt{1-2\xi^2}} \quad (2.69a)$$

$$\omega_{nd} = \frac{\omega_n}{\sqrt{1-2\xi^2}} \quad (2.69b)$$

By substituting Eq. (2.69a) in Eq. (2.68), we get

$$\left(\frac{A_z}{2m_e e/m}\right)_{max} = \frac{1}{2\xi\sqrt{1-\xi^2}} \quad (2.70)$$

$$\approx \frac{1}{2\xi} \text{ for small damping} \quad (2.71)$$

2.5 Vibration Isolation

In case a machine is rigidly fastened to the foundation, the force will be transmitted directly to the foundation and may cause objectionable vibrations. It is desirable to isolate the machine from the foundation through a suitable designed mounting system in such a way that the transmitted force is reduced. For example, the inertial force developed in a reciprocating engine or unbalanced forces produced in any other rotating machinery should be isolated from the foundation so that the adjoining structure is not set into heavy vibrations. Another example may be the isolation of delicate instruments from their supports which may be subjected to certain vibrations. In either case the effectiveness of isolation may be measured in terms of the force or motion transmitted to the foundation. The first type is known as **force isolation** and the second type as **motion isolation.**

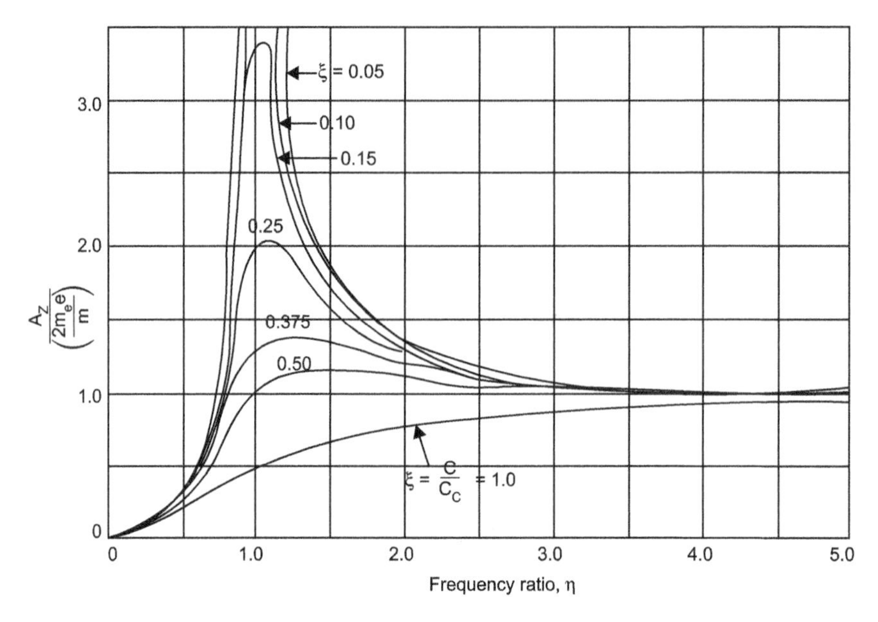

(a) $A_z/(2m_e e/m)$ versus frequency ratio η

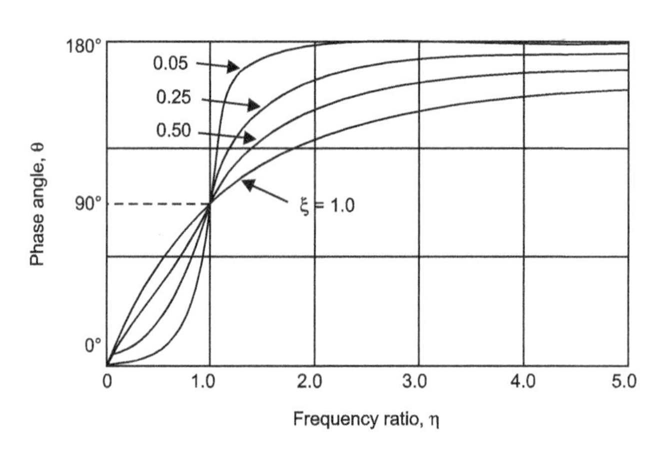

(b) Phase angle versus frequency ratio η

Fig. 2.16. Response of a system with rotating unbalance

2.5.1 Force Isolation

Figure 2.17 shows a machine of mass m supported on the foundation by means of an isolator having an equivalent stiffness K and damping coefficient C. The machine is excited with unbalanced vertical force of magnitude 2 $me\,e\omega^2 \sin \omega\, t$. The equation of motion of the machine can be written as:

$$m\ddot{Z} + C\dot{Z} + KZ = 2m_e\, e\omega^2 \sin \omega t \tag{2.72}$$

The steady state motion of the mass of machine can be worked out as

$$Z = \frac{2m_e e\omega^2 / K}{\sqrt{(1-\eta^2)^2 + (2\eta\xi)^2}} \cdot \sin(\omega t - \theta) \tag{2.73}$$

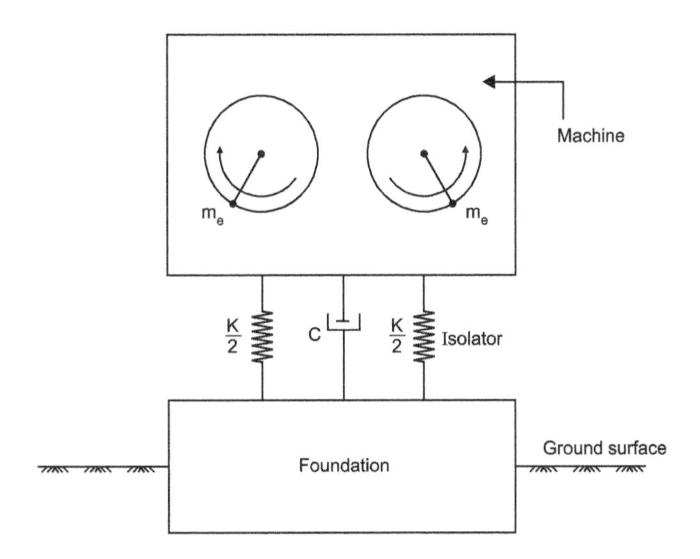

Fig. 2.17. Machine-isolator-foundation system

where $$\theta = \tan^{-1}\left[\frac{2\eta\xi}{1-\eta^2}\right] \tag{2.74}$$

The only force which can be applied to the foundation is the spring force KZ and the damping force $C\dot{Z}$; hence the total force transmitted to the foundation during steady state forced vibration is

$$F_t = KZ + C\dot{Z} \tag{2.75}$$

Substituting Eq. (2.73) in Eq. (2.75), we get

$$F_t = \frac{2m_e e\omega^2}{\sqrt{(1-\eta^2)^2+(2\eta\xi)^2}} \cdot \sin(\omega t - \theta) + \frac{C \cdot 2m_e e\omega^2 / K}{\sqrt{(1-\eta^2)^2+(2\eta\xi)^2}} \cdot \omega \cos(\omega t - \theta) \tag{2.76}$$

Equation (2.76) can be written as:

$$F_t = 2m_e e\omega^2 \frac{\sqrt{1+(2\eta\xi)^2}}{\sqrt{(1-\eta^2)^2+(2\eta\xi)^2}} \cdot \sin(\omega t - \beta) \tag{2.77}$$

where β is the phase difference between the exciting force and the force transmitted to the foundation and is given by

$$\beta = \theta - \tan^{-1}\left[\frac{C\omega}{K}\right] \tag{2.78}$$

Since the force $2\, m_e\, e\, \omega^2$ is the force which would be transmitted if springs were infinitely rigid, a measure of the effectiveness of the isolation mounting system is given by

$$\mu_T = \frac{F_t}{2m_e e\omega^2} = \frac{\sqrt{1+(2\eta\xi)^2}}{\sqrt{(1-\eta^2)^2+(2\eta\xi)^2}} \tag{2.79}$$

μ_T is called the transmissibility of the system. A plot of μ_T versus η for different values of ξ is shown in Fig. 2.18. It will be noted from the figure that for any frequency ratio greater than $\sqrt{2}$, the force transmitted to the foundation will be less than the exciting force. However in this case, the presence of damping reduces the effectiveness of the isolation system as the curves for damped case are above the undamped ones for $\eta > \sqrt{2}$. A certain amount of damping, however, is essential to maintain stability under transient conditions and to prevent excessive amplitudes when

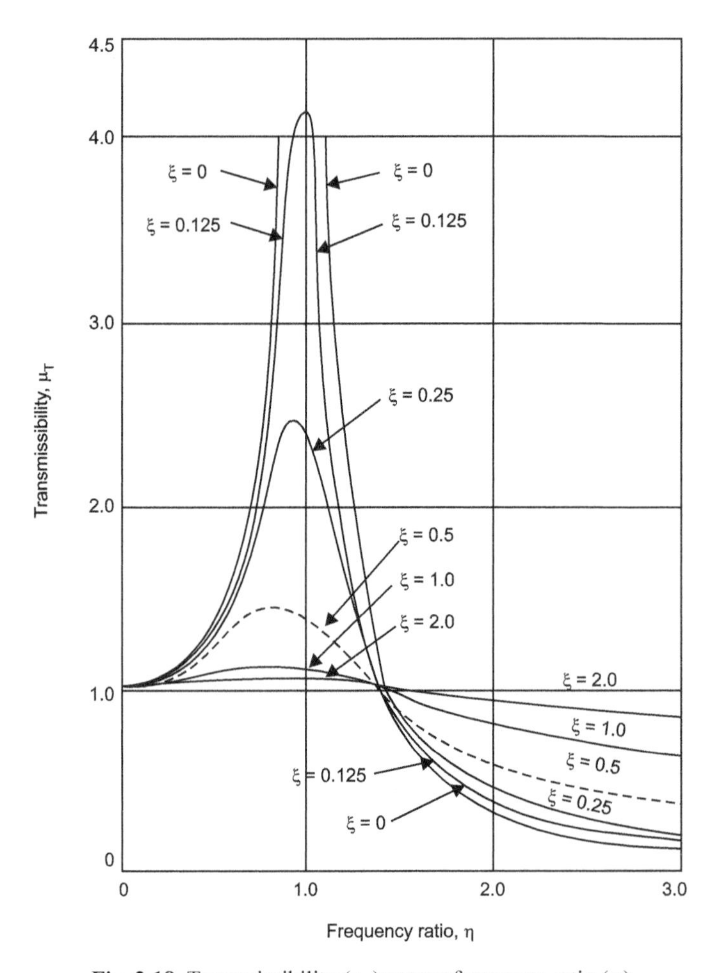

Fig. 2.18. Transmissibility (μ_T) versus frequency ratio (η)

the vibrations pass through resonance during the starting or stopping of the machine. Therefore, for the vibration isolation system to be effective η should be greater than $\sqrt{2}$.

2.5.2 Motion Isolation

In many situations, it would be necessary to isolate structure or mechanical systems from vibrations transmitted from the neighbouring machines. Again we require a suitable mounting system so that least vibrations are transmitted to the system due to the vibrating base. We consider a system mounted through a spring and dashpot and attached to the surface which vibrates harmonically with frequency ω and amplitude Y_0 as shown in Fig. 2.19.

Let Z be the absolute displacement of mass; the equation of motion of the system can be written as:

$$m\ddot{Z} + C(\dot{Z} - \dot{Y}) + K(Z - Y) = 0 \tag{2.80}$$

or
$$m\ddot{Z} + K\dot{Z} + KZ = C\dot{Y} + KY = C\omega Y_0 \cos\omega t + K Y_0 \sin\omega t$$

or
$$m\ddot{Z} + C\dot{Z} + KZ = Y_0\sqrt{K^2 + (C\omega)^2}\sin(\omega t + \alpha) \tag{2.81}$$

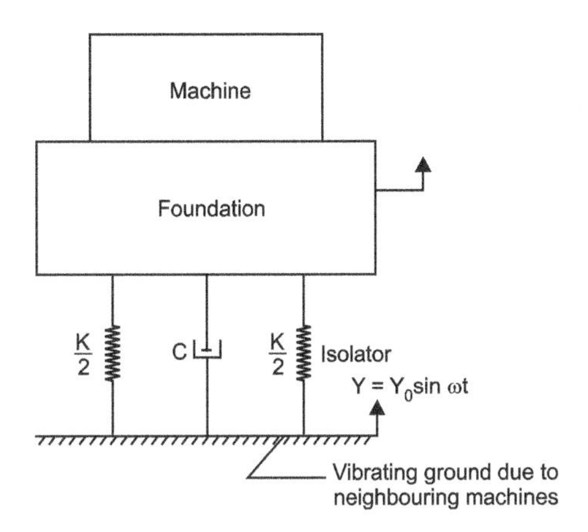

Fig. 2.19. Motion isolation system

where
$$\alpha = \tan^{-1}\frac{C\omega}{K} \tag{2.82}$$

The solution of Eq. (2.81) will give the maximum amplitude as:

$$Z_{\max} = Y_0 \cdot \frac{\sqrt{1+(2\eta\xi)^2}}{\sqrt{(1-\eta^2)^2+(2\eta\xi)^2}} \tag{2.83}$$

The effectiveness of the mounting system (transmissibility) is given by

$$\mu_T = \frac{Z_{\max}}{Y_0} = \frac{\sqrt{1+(2\eta\xi)^2}}{\sqrt{(1-\eta^2)^2+(2\eta\xi)^2}} \tag{2.84}$$

Equation (2.84) is the same expression as Eq. (2.79) obtained earlier. Transmissibility of such system can also be studied from the response curves shown in Fig. 2.18. It is again noted that for the vibration isolation to be effective, it must be designed in such a way that $\eta > \sqrt{2}$.

2.5.3 Materials Used in Vibration Isolation

Materials used for vibration isolation are rubber, felt, cork and metallic springs. The effectiveness of each depends on the operating conditions.

2.5.3.1 Rubber

Rubber is loaded in compression or in shear, the latter mode gives higher flexibility. With loading greater than about 0.6N per sq mm, it undergoes much faster deterioration. Its damping and stiffness properties vary widely with applied load, temperature, shape factor, excitation frequency and the amplitude of vibration. The maximum temperature upto which rubber can be used satisfactorily is about 65°C. It must not be used in presence of oil which attacks rubber. It is found very suitable for high frequency vibrations.

2.5.3.2 Felt

Felt is used in compression only and is capable of taking extremely high loads. It has very high damping and so is suitable in the range of low frequency ratio. It is mainly used in conjunction with metallic springs to reduce noise transmission.

2.5.3.3 Cork

Cork is very useful for acoustic isolation and is also used in small pads placed underneath a large concrete block. For satisfactory working it must be loaded from 10 to 25 N/sq mm. It is not affected by oil products or moderate temperature changes. However, its properties change with the frequency of excitation.

2.5.3.4 Metallic Springs

Metallic springs are not affected by the operating conditions or the environments. They are quite consistent in their behaviour and can be accurately designed for any desired conditions. They have high sound transmissibility which can be reduced by loading felt in conjunction with it. It has negligible damping and so is suitable for working in the range of high frequency ratio.

2.6 Theory of Vibration Measuring Instruments

The purpose of a vibration measuring instrument is to give an output signal which represents, as closely as possible, the vibration phenomenon. This phenomenon may be displacement, velocity or acceleration of the vibrating system and accordingly the instrument which reproduces signals proportional to these are called vibrometers, velometers or accelerometers.

There are essentially two basic systems of vibration measurement. One method is known as the directly connected system in which motions can be measured from a reference surface which is fixed. More often such a reference surface is not available. The second system, known as "Seismic system" does not require a fixed reference surface and therefore is commonly used for vibration measurement.

Figure 2.20 shows a **Vibration measuring instrument** which is used to measure any of the vibration phenomena. It consists of a frame in which the mass m is supported by means of a spring K and dashpot C. The frame is mounted on a vibrating body and vibrates along with it. The system reduces to a spring mass dashpot system having base on support excitation as discussed in Section 2.5.2 illustrating motion isolation.

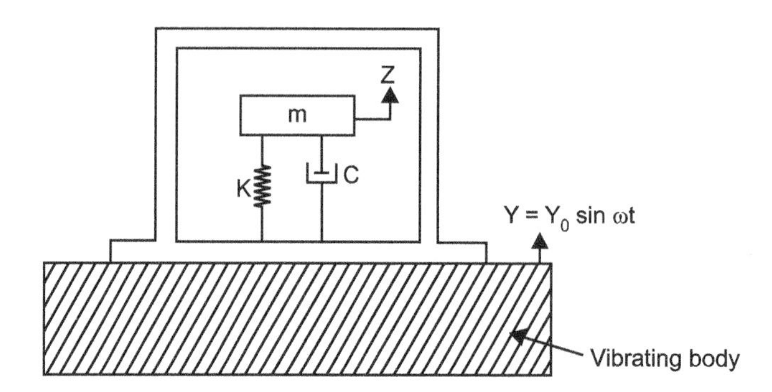

Fig. 2.20. Vibration measuring instrument

Let the surface S of the structure be vibrating harmonically with an unknown amplitude Y_0 and an unknown frequency ω. The output of the instrument will depend upon the relative motion between the mass and the structure, since it is this relative motion which is detected and amplified. Let Z be the absolute displacement of the mass, then the output of the instrument will be proportional to $X = Z - Y$. The equation of motion of the system can be written as:

$$m\ddot{Z} + C(\dot{Z} - \dot{Y}) + K(Z - Y) = 0 \tag{2.85}$$

Subtracting $m\ddot{Y}$ from both sides

$$m\ddot{X} + C\dot{X} + KX = -m\ddot{Y} = mY_0\,\omega^2 \sin \omega t \tag{2.86}$$

The solution can be written as:

$$X = \frac{\eta^2}{\sqrt{(1-\eta^2)^2 + (2\eta\xi)^2}} Y_0 \sin(\omega t - \theta) \tag{2.87}$$

where $\eta = \dfrac{\omega}{\omega_n}$ = frequency ratio

ξ = damping ratio

and $\theta = \tan^{-1}\left(\dfrac{2\eta\xi}{1-\eta^2}\right)$

Equation (2.87) can be rewritten as:

$$X = \eta^2 \mu\, Y_0 \sin(\omega t - \theta) \tag{2.88}$$

where $\mu = \dfrac{1}{\sqrt{(1-\eta^2)^2 + (2\eta\xi)^2}}$

2.6.1 Displacement Pickup

The instrument will read the displacement of the structure directly if $\eta^2 \mu = 1$ and $\theta = 0$. The variation of $\eta^2 \mu$ with η and ξ is shown in Fig. 2.21. The variation of θ with η is already given in Fig. 2.14. It is seen that when η is large, $\eta^2 \mu$ is approximately equal to 1 and θ is approximately equal to 180°. Therefore to design a displacement pickup, η should be large which means that the natural frequency of the instrument itself should be low compared to the frequency to be measured. Or in other words, the instrument should have a soft spring and heavy mass. The instrument is sensitive, flimsy and can be used in a weak vibration environment. The instrument cannot be used for measurement of strong vibrations.

2.6.2 Acceleration Pickup (Accelerometer)

Equation (2.88) can be rewritten as:

$$X = \frac{1}{\omega_n^2} \mu Y_0\, \omega^2 \sin(\omega t - \theta) \tag{2.89}$$

The output of the instrument will be proportional to the acceleration of the structure if μ is constant. Figure 2.13 shows the variation of μ with η and ξ. It is seen that μ is approximately equal to unity for small values of η. Therefore to design an acceleration pickup, η should be small which means that the natural frequency of the instrument itself should be high compared to the frequency to be measured. In other words, the instrument should have a stiff spring and small mass. The instrument is less sensitive and suitable for the measurement of strong motion. The instrument size is small.

2.6.3 Velocity Pickup

Equation (2.88) can be rewritten as:

$$X = \frac{1}{\omega_n} \eta \mu Y_0\, \omega \sin(\omega t - \theta) \tag{2.90}$$

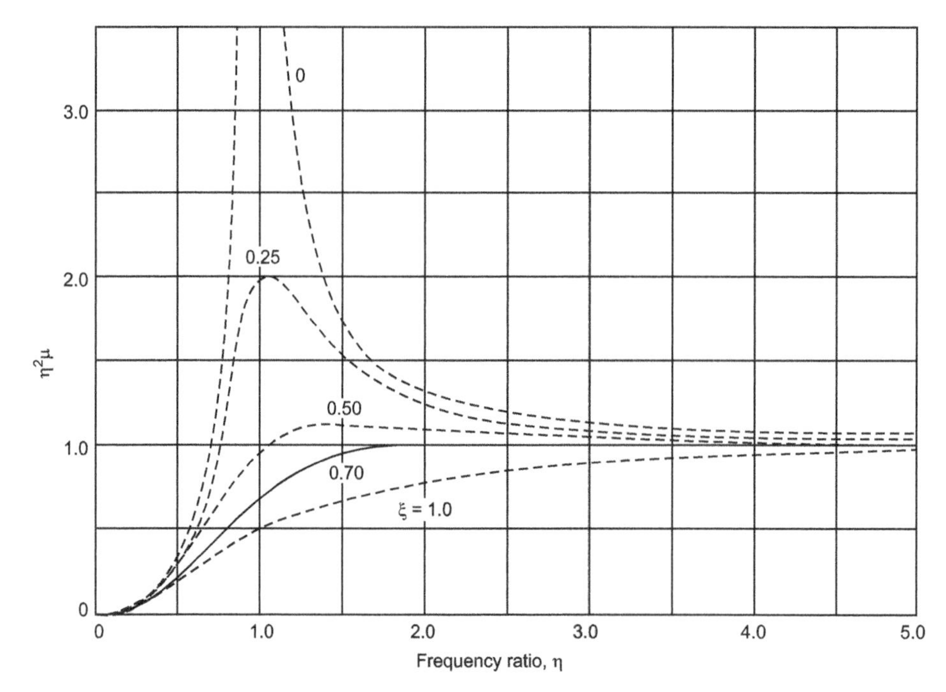

Fig. 2.21. Response of a vibration measuring instrument to a vibrating base

The output of the instrument will be proportional to velocity of the structure if $\frac{1}{\omega_n}\eta\ \mu$ is constant. At $\eta = 1$. Eq. (2.90) can be written as

$$X = \frac{1}{\omega_n}\frac{1}{2\xi}Y_0\,\omega\sin(\omega t - \theta) \qquad \because \text{ at } \eta = 1,\ \mu = \frac{1}{2\xi} \tag{2.91}$$

Since ω_n and ξ are constant, the instrument will measure the velocity at $\eta = 1$.

It may be noted that the same instrument can be used to measure displacement, acceleration and velocity in different frequency ranges.

$X \alpha Y$ if $\eta >> 1$ Displacement pickup (Vibrometers)

$X \alpha \ddot{Y}$ if $\eta << 1$ Acceleration pickup (Accelerometers)

$X \alpha \dot{Y}$ if $\eta = 1$ Velocity pickup (Velometers)

Displacement and velocity pickups have the disadvantage of having rather a large size if motions having small frequency of vibration are to be measured. Calibration of these pickups is not simple. Further, corrections have to be made in the observations as the response is not flat in the starting regions. From the point of view of small size, flat frequency response, sturdiness and ease of calibration acceleration pickups are to be favoured. They are relatively less sensitive and this disadvantage can easily be overcome by high gain electronic instrumentation.

2.6.4 Design of Acceleration Pickup

The relative displacement between the mass and the support would be a measure of the support acceleration if η is less than 0.75 and ξ is of the order of 0.6 to 0.7. Of the various methods of measurement of relative displacement, electrical gauging, in which the mechanical quantity is converted into an equivalent electrical quantity is best suited for acceleration pickups. Electrical gauging offers the possibility of high magnification of the signals which are usually weak because the spring is stiff and the displacements are small. The mechanical quantity alters either the resistance, or capacitance or the inductance of the circuit which consequently alters the current in the circuit.

2.7 Vibration of Multiple Degree Freedom Systems

In the preceding sections, vibrations of systems having single degree of freedom have been discussed. In many engineering problems, one may come across the systems which may have more than one degree of freedom. Two degrees freedom cases arise when the foundation of the system is yielding thus adding another degree of freedom or a spring mass system is attached to the main system to reduce its vibrations. In systems when there are a number of masses connected with each other, even if each mass is constrained to have one degree of freedom, the system as a whole has as many degrees of freedom as there are masses. Such an idealisation is done for carrying out dynamic analysis of multistoreyed buildings.

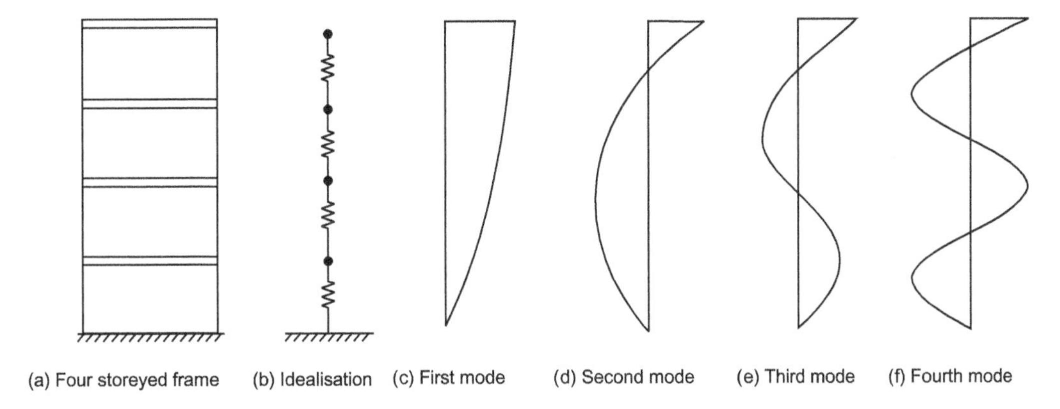

Fig. 2.22. A four storeyed frame with mode shapes

Figure 2.22a shows the frame work of a four storeyed building. It is usual to lump the masses at the floor levels and the lumped mass has a value corresponding to weight of the floor, part of the supporting system (columns) above and below the floor and effective live load. The restoring forces are provided by the supporting systems. Figure 2.22b shows such an idealisation and it gives a four degrees of freedom system. In free vibration a system having four degrees of freedom has four natural frequencies and the vibration of any point in the system, in general, is a combination of four harmonics of these four natural frequencies respectively. Under certain conditions, any point in the system may execute harmonic vibrations at any of the four natural frequencies, and these are known as the principal modes of vibration. Figures 2.22c to 2.22f show the four modes of vibration. If all the masses vibrate in phase (Fig. 2.22c), the mode is termed the first or lowest or fundamental mode of vibration and the frequency associated with this mode would be the lowest in magnitude compared to other modes. If all adjacent masses vibrate out of phase with each other (Fig. 2.22f), the mode is termed the highest mode of vibration and the frequency associated with this mode would be highest in magnitude compared to other modes.

In many aspects, the response of multi-degrees of freedom system, procedure of analysis is analogous to those described above for single degree freedom system. Although the additional degrees of freedom complicate the mathematics, the procedure is conceptually quite similar. In fact, a very useful approach to the response of linear multi-degrees freedom system allows their response to be computed as the sum of the responses as the series of single degree of freedom systems. One has to keep in mind that in evaluating the response of a multi- degree freedom system, dynamic equilibrium of all masses must be ensured simultaneously.

2.7.1 Two Degrees of Freedom Systems

2.7.1.1 Undamped Free Vibration

Figure 2.23 shows a mass-spring system with two degrees of freedom. Let Z_1 be the displacement of mass m_1 and Z_2 the displacement of mass m_2. The equations of motion of the system can be written:

$$m_1 \ddot{Z}_1 + K_1 Z_1 + K_2 (Z_1 - Z_2) = 0 \tag{2.92}$$

$$m_2 \ddot{Z}_2 + K_3 Z_2 + K_2 (Z_2 - Z_1) = 0 \tag{2.93}$$

The solution of Eqs. (2.92) and (2.93) will be of the following form:

$$Z_1 = A_1 \sin \omega_n t \tag{2.94}$$

$$Z_2 = A_2 \sin \omega_n t \tag{2.95}$$

Substitution of Eqs. (2.94) and (2.95), into Eqs. (2.92) and (2.93), yields:

$$(K_1 + K_2 - m_1 \omega_n^2) A_1 - K_2 A_2 = 0 \tag{2.96}$$

$$-K_2 A_1 + (K_2 + K_3 - m_2 \omega_n^2) A_2 = 0 \tag{2.97}$$

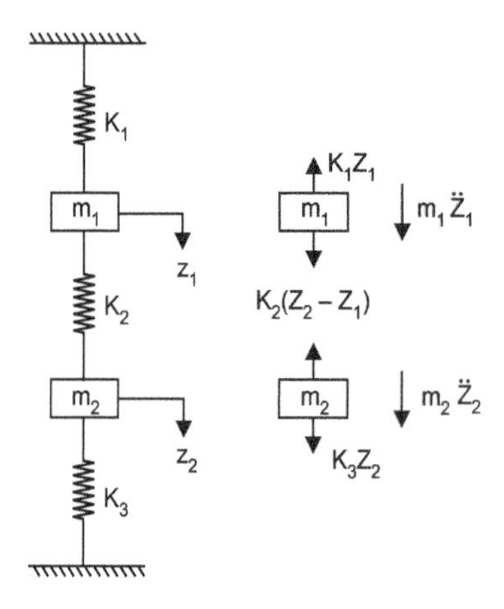

Fig. 2.23. Free vibration of a two degrees freedom system

For nontrivial solutions of ω_n in Eqs. (2.96) and (2.97),

$$\begin{vmatrix} K_1 + K_2 - m_1\omega_n^2 & -K_2 \\ -K_2 & K_2 + K_3 - m_2 \omega_n^2 \end{vmatrix} = 0 \tag{2.98}$$

or

$$\omega_n^4 - \left[\frac{K_1 + K_2}{m_1} + \frac{K_2 + K_3}{m_2}\right]\omega_n^2 + \frac{K_1K_2 + K_2K_3 + K_3K_1}{m_1m_2} = 0 \tag{2.99}$$

Equation (2.99) is quadratic in ω_2^n, and the roots of this equation are:

$$\omega_n^2 = \frac{1}{2}\left[\left(\frac{K_1 + K_2}{m_1} + \frac{K_2 + K_3}{m_2}\right) \mp \left\{\left(\frac{K_1 + K_2}{m_1} - \frac{K_2 + K_3}{m_2}\right)^2 + \frac{4K_2^2}{m_1 m_2}\right\}^{1/2}\right] \tag{2.100}$$

From Eq. (2.100), two values of natural frequencies ω_{n1} and ω_{n2} can be obtained. ω_{n1} is corresponding to the first mode and ω_{n2} is of the second mode.

The general equation of motion of the two masses can now be written as

$$Z_1 = A_1^{(1)} \sin \omega_{n1} t + A_1^{(2)} \sin \omega_{n2} t \tag{2.101}$$

and

$$Z_2 = A_2^{(1)} \sin \omega_{n1} t + A_2^{(2)} \sin \omega_{n2} t \tag{2.102}$$

The superscripts in A represent the mode.

The relative values of amplitudes A_1 and A_2 for the two modes can be obtained using Eqs. (2.96) and (2.97).

Thus,

$$\frac{A_1^{(1)}}{A_2^{(1)}} = \frac{K_2}{K_1 + K_2 - m_1\omega_{n1}^2} = \frac{K_2 + K_3 - m_2\omega_{n1}^2}{K_2} \tag{2.103}$$

$$\frac{A_1^{(2)}}{A_2^{(2)}} = \frac{K_2}{K_2 + K_2 - m_1\omega_{n2}^2} = \frac{K_2 + K_3 - m_2\omega_{n2}^2}{K_2} \tag{2.104}$$

2.7.1.2 Undamped Forced Vibrations

Consider the system shown in Fig. 2.24 with excitation force $F_o \sin \omega t$ acting on mass m_1. In this case, equations of motion will be:

$$m_1 \ddot{Z}_1 + K_1 Z_1 + K_2(Z_1 - Z_2) = F_0 \sin \omega t \tag{2.105}$$

$$m_2 \ddot{Z}_2 + K_3 Z_2 + K_2(Z_2 - Z_1) = 0 \tag{2.106}$$

For steady state, the solutions will be as

$$Z_1 = A_1 \sin \omega \text{ t} \tag{2.107}$$

$$Z_2 = A_2 \sin \omega \text{ t} \tag{2.108}$$

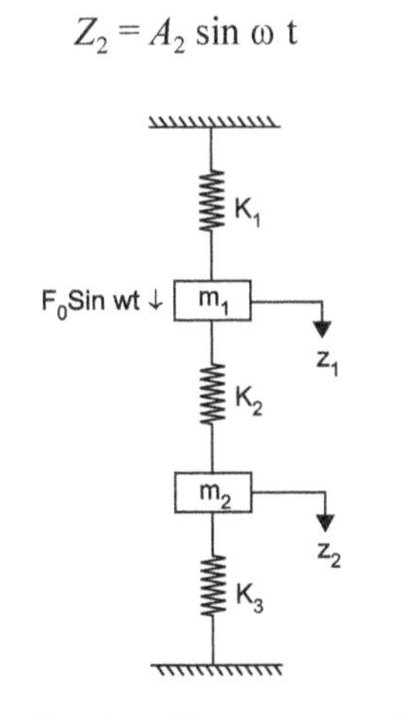

Fig. 2.24. Forced vibration of a two degrees freedom system

Substituting Eqs. (2.107) and (2.108) in Eqs. (2.105) and (2.106), we get

$$(K_1 + K_2 - m_1 \omega^2) A_1 - K_2 A_2 = F_0 \tag{2.109}$$

$$-K_2 A_1 + (K_2 + K_3 - m_2\omega^2) A_2 = 0 \tag{2.110}$$

Solving for A_1 and A_2 from the above two equations, we get

$$A_1 = \frac{(K_2 + K_3 - m_2 \omega^2) F_0}{m_1 m_2 \left[\omega^4 - \left(\frac{K_1 + K_2}{m_1} + \frac{K_2 + K_3}{m_2} \right) \omega^2 + \frac{K_1 K_2 + K_2 K_3 + K_3 K_1}{m_1 m_2} \right]} \tag{2.111a}$$

and $$A_2 = \frac{K_3 F_0}{m_1 m_2 \left[\omega^4 - \left(\frac{K_1 + K_2}{m_1} + \frac{K_2 + K_3}{m_2} \right) \omega^2 + \frac{K_1 K_2 + K_2 K_3 + K_3 K_1}{m_1 m_2} \right]} \tag{2.111b}$$

The above two equations give steady state amplitude of vibration of the two masses respectively, as a function of ω. The denominator of the two equations is same. It may be noted that:

(i) The expression inside the bracket of the denominator of Eqs. (2.111a) and (2.111b) is of the same type as the expression of natural frequency given by Eq. (2.99), Therefore at $\omega = \omega_{n1}$ and $\omega = \omega_{n2}$ values of A_1 and A_2 will be infinie as the denominator will become zero.

(ii) The numerator of the expression for A_1 becomes zero when

$$\omega = \sqrt{\frac{(K_2 + K_3)}{m_2}} \tag{2.112}$$

Thus it makes the mass m_1 motionless at this frequency. No such stationary condition exists for mass m_1. The fact that the mass which is being excited can have zero amplitude of vibration under certain conditions by coupling it to another spring-mass system forms the principle of dynamic vibration absorbers which will be discussed in Art. 2.8.

2.7.2 System with *n* Degrees of Freedom

2.7.2.1 Undamped Free Vibrations

Consider a system shown in Fig. 2.25 having n degrees of freedom. If Z_1, Z_2, Z_3, ... Z_n are the displacements of the respective masses at any instant, then equations of motion are:

$$m_1 \ddot{Z}_1 + K_1 Z_1 + K_2(Z_1 - Z_2) = 0 \tag{2.113}$$

$$m_2 \ddot{Z}_2 - K_2 (Z_1 - Z_2) + K_3(Z_2 - Z_3) = 0 \tag{2.114}$$

$$m_3 \ddot{Z}_3 - K_3 (Z_2 - Z_3) + K_4(Z_3 - Z_4) = 0 \tag{2.115}$$

....

....

$$m_n \ddot{Z}_n - K_n (Z_{n-1} - Z_n) = 0 \tag{2.116}$$

The solution of Eqs. (2.113) to (2.116) will be of the following form:

$$Z_1 = A_1 \sin \omega_n t \tag{2.117}$$

$$Z_2 = A_2 \sin \omega_n t \tag{2.118}$$

$$Z_3 = A_3 \sin \omega_n t \tag{2.119}$$

....

....

$$Z_n = A_n \sin \omega_n t \tag{2.120}$$

Substitution of Eqs. (2.117) to (2.120) into Eqs. (2.113) to (2.116), yields:

$$\left[(K_1 + K_2) - m_1 \omega_n^2\right] A_1 - K_2 A_2 = 0 \tag{2.121}$$

$$-K_2 A_1 + \left[(K_2 + K_3) - m_1 \omega_n^2\right] A_2 - K_3 A_3 = 0 \tag{2.122}$$

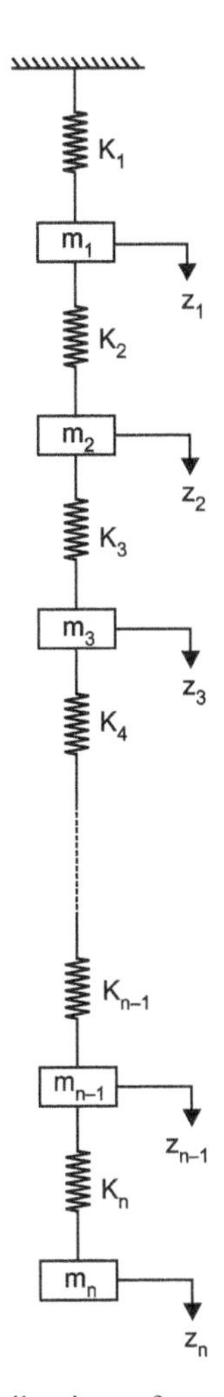

Fig. 2.25. Undamped free vibrations of a multi-degree freedom system

$$-K_3A_2 + \left[(K_3+K_4)-m_3\,\omega_n^2\right]A_3 - K_4A_4 = 0 \tag{2.123}$$

....

....

$$-K_nA_{n-1} + (K_n - m_n\,\omega_n^2)\,A_n = 0 \tag{2.124}$$

For nontrivial solutions of ω_n in Eqs. (2.121) to (2.124),

$$\begin{vmatrix} \left[(K_1+K_2)-m_1\omega_n^2\right] & -K_2 & \ldots & 0 & 0 \\ -K_2 & \left[(K_2+K_3)-m_2\,\omega_n^2\right] & \ldots & 0 & 0 \\ 0 & -K_3 & \ldots & 0 & 0 \\ \ldots & \ldots & \ldots & \ldots & \ldots \\ 0 & 0 & \ldots & -K_n & (K_n - m_n\,\omega_n^2 \end{vmatrix} = 0 \tag{2.125}$$

Equation (2.125) is of nth degree in ω_n^2 and therefore gives n values of ω_n corresponding to n natural frequencies. The mode shapes can be obtained from Eq. (2.121) to (2.124) by using, at one time, one of the various values of ω_n as obtained from Eq. (2.125).

When the number of degrees of freedom exceeds three, the problem of forming the frequency equation and solving it for determination of frequencies and mode shapes becomes tedius. Numerical techniques are invariably resorted to in such cases.

Holzer's numerical technique is a convenient method of solving the problem for the system idealized as shown in Fig. 2.26. By summing forces at free end,

Inertia force at a level below mass m_{i-1}

$$\sum_{j=1}^{i-1} m_j \ddot{Z}_j \tag{2.126}$$

Spring force at that level corresponding to the difference of adjoining masses

$$= K_{i-1}(Z_i - Z_{i-1}) \tag{2.127}$$

Equating Eqs. (2.126) and (2.127)

$$\sum_{j=1}^{i-1} m_j \ddot{Z}_i = K_{i-1}(Z_i - Z_{i-1}) \tag{2.128}$$

Putting $Z_i = A_i \sin \omega t$ in Eq. (2.128), we get

$$\sum_{j=1}^{i-1} m_j(-A_j \omega_n^2 \sin \omega_n t) = K_{i-1}(A_i \sin \omega_n t - A_{i-1} \sin \omega_n t)$$

or

$$A_i = A_{i-1} - \frac{\omega_n^2}{K_{i-r}} \sum_{j=1}^{i-1} m_j A_j \tag{2.129}$$

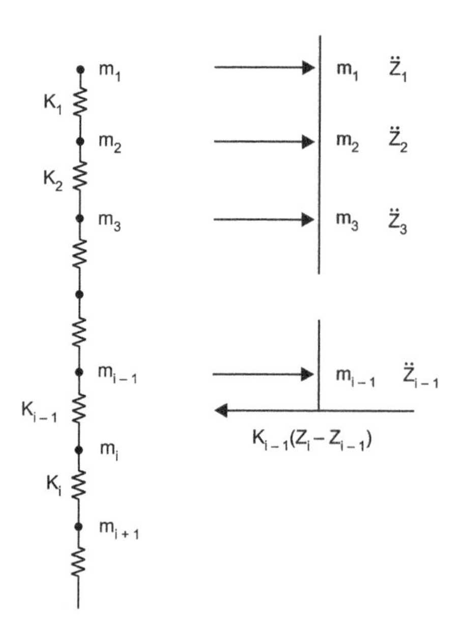

Fig. 2.26. An idealised multi-degree freedom system

Equation (2.129) gives a relationship between any two successive amplitudes. Starting with any arbitrary value of A_1, amplitude of all other masses can be determined. A plot of A_{n+1} versus ω_n^2 would have the shape as shown in Fig. 2.27. Finally A_{n+1} should worked out to zero due to fixity at the base. The intersection of the curve with ω_n^2 axis would give various values of ω_n^2. Mode shape can be obtained by substituting the correct value of ω_n^2 in Eq. (2.129).

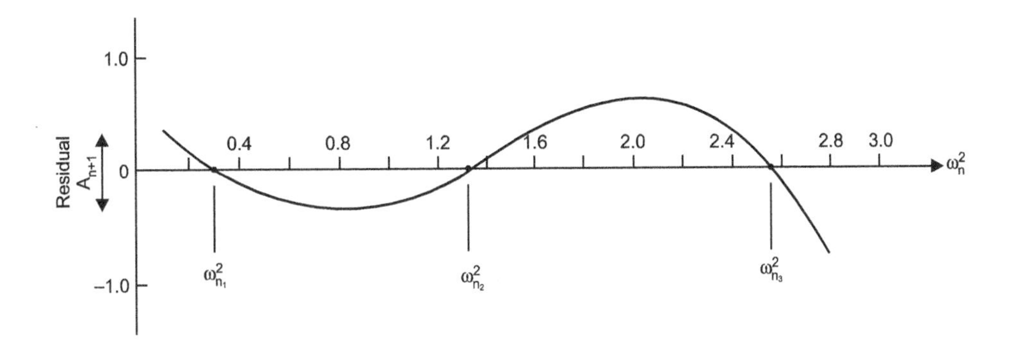

Fig. 2.27. Residual as a function of frequency in Hotzer method

2.7.2.2 Forced Vfibration

Let an undamped n degree of freedom system be subjected to forced vibration, and $F_i(t)$ represents the force on mas m_i. The equation of motion for the mass m_i will be

$$m_i \ddot{Z}_i + \sum_{i=1}^{n} K_{ij} Z_j = F_i(t) \tag{2.130}$$

where $i = 1, 2, 3, \ldots, n$

The amplitude of vibration of a mass is the algebraic sum of the amplitudes of vibration in various modes. The individual modal response would be some fraction of the total response with the sum of fractions being equal to unity. If the factors by which the modes of vibration are multiplied are represented by the coordinates d, then for mass m_i

$$Z_i = A_i^{(1)} d_i + A_i^{(2)} d_2 + \ldots + A_i^{(r)} d_r + \ldots + A_i^{(n)} d_n \tag{2.131}$$

Equation (2.131) can be written as

$$Z_i = \sum_{r=1}^{n} A_i^{(r)} d_r \tag{2.132}$$

Substituting Eq. (2.132) in Eq. (2.130)

$$\sum_{r=1}^{n} m_i A_i^{(r)} \ddot{d}_r + \sum_{r=1}^{n} \sum_{j=1}^{n} K_{ij} A_j^{(r)} d_r = F_i(t) \tag{2.133}$$

Under free vibrations, it can be shown

$$\sum_{i=1}^{n} K_{ij} A_j^{(r)} = \omega_{nr}^2 m_i A_i^{(r)} \tag{2.134}$$

Substituting Eq. (2.134) in Eq. (2.133), we get

$$\sum_{r=1}^{n} m_i A_i^{(r)} \ddot{d}_r + \sum_{r=1}^{n} \omega_{nr}^2 m_i A_i^{(r)} d_r = F_i(t) \tag{2.135}$$

or

$$\sum_{r=1}^{n} m_i A_i^{(r)} (\ddot{d}_r + \omega_{nr}^2 . d_r) = F_i(t) \tag{2.136}$$

Since the left hand side is a summation involving different modes of vibration, the right hand side should also be expressed as a summation of equivalent force contribution in corresponding modes.

Let $F_i(t)$ be expanded as:

$$F_i(t) = \sum_{r=1}^{n} m_i A_i^{(r)} f_r(t) \tag{2.137a}$$

where $f_r(t)$ is the modal force and given by

$$f_r(t) = \frac{\sum_{i=1}^{n} F_i(t) \cdot A_i^{(r)}}{\sum_{i=}^{n} m\left[(A_i^{(r)})\right]^2} \tag{2.137b}$$

Substituting Eq. (2.137a) in Eq. (2.136), we get

$$\ddot{d}_r + \omega_{nr}^2 d_r = f_r(t) \tag{2.138}$$

Equation (2.138) is a single degree freedom equation and its solution can be written as

$$d_r = \frac{1}{\omega_{nr}} \int_0^t f_r(\tau) \sin \omega_{nr}(t-\tau) d\tau \text{ where } 0 < \tau < 1 \tag{2.139}$$

It is observed that the co-ordinate d_r uncouples the n degree of freedom system into n systems of single degree of freedom. The d's are termed as **normal co-ordinates** and this approach is known as **normal mode theory.** Therefore the total solution is expressed as a sum of contribution of individual modes.

2.8 Undamped Dynamic Vibration Absorber

A system on which a steady oscillatory force is acting may vibrate excessively, specially when close to resonance. Such excessive vibrations can be eliminated by coupling a properly designed spring mass system to the main system. This forms the principle of undamped dynamic vibration absorber where the excitation is finally transmitted to the auxiliary system, bringing the main system to rest.

Let the combination of K and M be the schematic representation of the main system under consideration with the force F_0 sin ωt acting on it. A spring-mass (auxiliary) absorber system is attached to the main system as shown in Fig. 2.28. The equations of motion of the complete system can be written as:

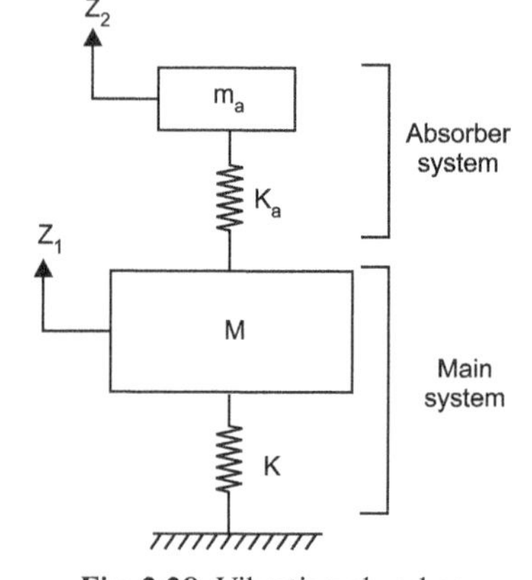

Fig. 2.28. Vibration absorber

$$M\ddot{Z}_1 + KZ_1 + K_a(Z_1 - Z_2) = F_0 \sin \omega t \tag{2.140}$$

$$m_a\ddot{Z}_2 + K_a(Z_1 - Z_2) = 0 \tag{2.141}$$

The forced vibration solution will be of the form

$$Z_1 = A_1 \sin \omega t \tag{2.142}$$

$$Z_2 = A_2 \sin \omega t \tag{2.143}$$

Substitution of Eqs. (2.142) and (2.143) in Eqs. (2.140) and (2.141) yields

$$A_1(-M\omega^2 + K + K_a) - K_a A_2 = F_0 \tag{2.144}$$

$$-K_a A_1 + A_2(-m_a\omega^2 + K_a) = 0 \tag{2.145}$$

Substituting:

$$Z_{st} = \frac{F_o}{K} = \text{Static deflection of main system}$$

$$\omega_{na}^2 = \frac{K_a}{m_a} = \text{Natural frequency of the absorber}$$

$$\omega_n^2 = \frac{K}{M} = \text{Natural frequency of main system}$$

$$\mu_m = \frac{m_a}{M} = \text{Mass ratio} = \text{Absorber mas/Main mass}$$

Equations (2.144) and (2.145) can be written as

$$A_1\left(1+\frac{K_a}{K}-\frac{\omega^2}{\omega_n^2}\right)-A_2\frac{K_a}{K} = Z_{st} \tag{2.146}$$

and

$$A_2 = \frac{A_1}{\left(1-\frac{\omega^2}{\omega_{na}^2}\right)} \tag{2.147}$$

Solving Eqs. (2.146) and (2.147) for A_1 and A_2, we get

$$\frac{A_1}{Z_{st}} = \frac{1-\frac{\omega^2}{\omega_{na}^2}}{\left(1-\frac{\omega^2}{\omega_{na}^2}\right)\left(1+\frac{K_a}{K}-\frac{\omega^2}{\omega_n^2}\right)-\frac{K_a}{K}} \tag{2.148}$$

$$\frac{A_2}{Z_{st}} = \frac{1}{\left(1-\frac{\omega^2}{\omega_{na}^2}\right)\left(1+\frac{K_a}{K}-\frac{\omega^2}{\omega_n^2}\right)-\frac{K_a}{K}} \tag{2.149}$$

If the natural frequency ω_{na} of the absorber is chosen equal to ω i.e.frequency of the excitation force, it is evident from Eq. (2.148) that $A_1 = 0$ indicating that the main mass does not vibrate at all. Further Eq. (2.149) gives

$$\frac{A_2}{Z_{st}} = \frac{-K}{K_a}$$

or

$$A_2K_a = -K\,Z_{st} \tag{2.150}$$

Thus the absorber system vibrate in such a way that its spring force at all instants is equal and opposite to $F_0 \sin \omega t$. Hence, there is no net force acting on main mass M and the same therefore does not vibrate.

The addition of a vibration absorber to a main system is not much meaningful unless the main system is operating at resonance or at least near it. Under these conditions, $\omega = \omega_n$. But for the absorber to be effective, ω should be equal to ω_{na}.

Therefore, for the effectiveness of the absorber at the operating frequency corresponding to the natural frequency of the main system alone, we have

or

$$\omega_{na} = \omega_n \tag{2.151a}$$

$$\frac{K_a}{m_a} = \frac{K}{M} \tag{2.151b}$$

or

$$\frac{K_a}{K} = \frac{m_a}{M} = \mu_m \tag{2.151c}$$

When the condition enumerated in Eqs. (2.151) is fulfilled, the absorber is known as a **tuned absorber.**

For a tuned absorber, Eqs. (2.148) and (2.149) become:

$$\frac{A_1}{Z_{st}} = \frac{1-\left(\dfrac{\omega^2}{\omega_{na}^2}\right)}{\left(1-\dfrac{\omega^2}{\omega_{na}^2}\right)\left(1+\mu_m-\dfrac{\omega^2}{\omega_{na}^2}\right)-\mu_m} \tag{2.152}$$

$$\frac{A_2}{Z_{st}} = \frac{1}{\left(1-\dfrac{\omega^2}{\omega_{na}^2}\right)\left(1+\mu_m-\dfrac{\omega^2}{\omega_{na}^2}\right)-\mu_m} \tag{2.153}$$

The denominators of Eqs. (2.152) and (2.153) are identical. At a value of ω when these denominators are zero the two masses have infinite amplitudes of vibration. Let when $\omega = \omega_{nl}$, the denominators become zero. For this condition the expression for the denominators can be written as

$$\left(\frac{\omega_{nl}}{\omega_{na}}\right)^4 - (2+\mu_m)\left(\frac{\omega_{nl}}{\omega_{na}}\right)^2 + 1 = 0 \tag{2.154}$$

Equation (2.154) is quadratic in ω_{nl}^2, and therefore there are two values of ω_{nl} for which the denominators of Eqs. (2.152) and (2.153) become zero. These two frequencies are the natural frequencies of the system. Solution of Eq. (2.154) gives:

$$\left(\frac{\omega_{nl}}{\omega_{na}}\right)^2 = \left(1+\frac{\mu_m}{2}\right) \pm \sqrt{\mu_m + \frac{\mu_m^2}{4}} \tag{2.155}$$

The relationship of Eq. (2.155) is plotted in Fig. 2.29. From this plot, it is evident that greater the mass ratio, greater is the spread between the two resonant frequencies. The frequency response

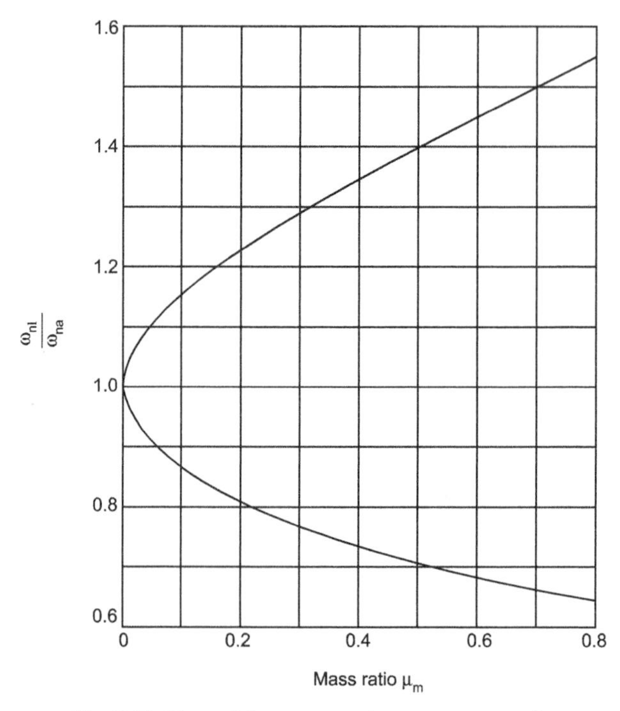

Fig. 2.29. Natural frequency ratio versus mass ratio

curve for the main system is shown in Fig. 2.30 for a value of $\mu_m = 0.2$. The dotted curves shown actually mean that the amplitude is negative or its phase difference with respect to the exciting force is 180°. It can be noticed from this figure that by attaching a vibration absorber ($\omega_{na} = \omega_n$) to the main system vibrating at resonance reduces its vibration to zero. Now if the exciting frequency is absolutely constant, the system will work efficiently. Any change in the exciting frequency will shift the operating point from the optimum point and the main system response will no longer be zero. It may be noted that by adding the vibration absorber, we have introduced two resonant points instead of one in the original system. Now if the variation of the exciting frequencyis such that the operating point shifts near one of the new resonant points, then amplitudes will be excessive. Thus depending upon the variation of the exciting frequencies the spread between the two resonant frequencies has to be decided to remain reasonably away from the resonant points. After deciding the spread between the resonant frequencies, a proper value of μ_m can be chosen from the curve of Fig. 2.29. Undamped dynamic vibration absorbers are not suitable for varying forcing frequency excitation. To make the vibration absorber effective over an extended range of frequencies of the disturbing force, it is advantageous to introduce a damping device in the absorber system. Such an absorber system is called a **damped dynamic vibration absorber.**

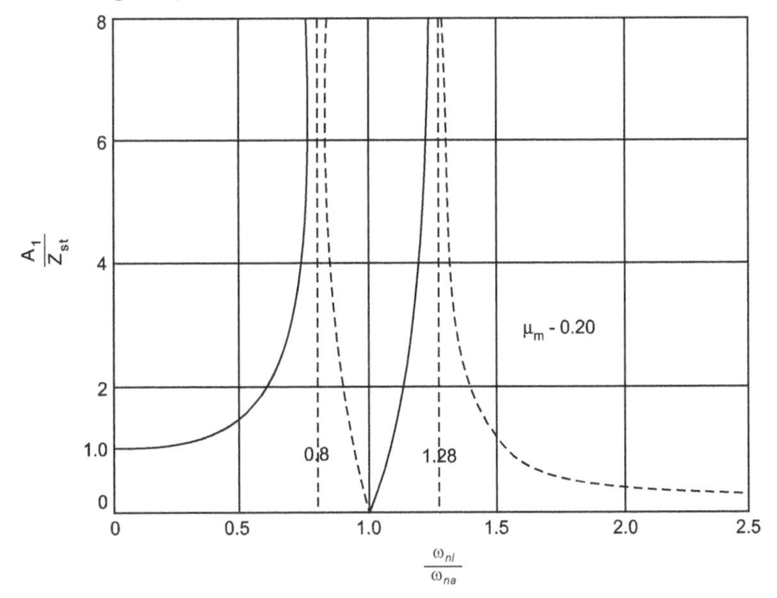

Fig. 2.30. Response versus frequency of a vibration absorber

Equation (2.154) can also be written as

$$\mu_m = \frac{\left\{\left(\frac{\omega_{nl}^2}{\omega_{na}}\right) - 1\right\}^2}{\left(\frac{\omega_{nl}}{\omega_{na}}\right)^2} \tag{2.156}$$

2.9 Dry Friction or Coulomb Damping

Systems discussed in Sections 2.4 to 2.6 have been analysed considering viscous damping which is the most important type of damping, and it occurs for small velocities in lubricated sliding surfaces, pistons with small clearances, etc. Eddy current damping is also of viscous nature. Dry friction or coulomb damping occurs when two machine parts rub against each other, dry or un-lubricated. The damping resistance in this case is practically constant, and is independent of the rubbing velocity.

Consider a single degree freedom system having a mass m and spring K, and let it move along a rough horizontal plane which subjects to a constant force F_D always opposing the motion (Fig. 2.31). The equation of motion becomes:

$$m\ddot{Z} + K\,Z \pm F_D = 0 \tag{2.157}$$

or
$$\ddot{Z} + (K/m)[Z \pm (F_D/K)] = 0 \tag{2.158}$$

Let,
$$Z \pm (F_D/K) = y$$

$$Z = Y$$

Equation (2.158) can be written as

$$Y + (K/m)y = 0 \tag{2.159}$$

This is similar to Eq. (2.12*b*), therefore

$$\omega_{nd} = \omega_n = \sqrt{K/m} \tag{2.160}$$

Hence the natural frequency of vibration of a system having Coulomb damping is the same as that of undamped system.

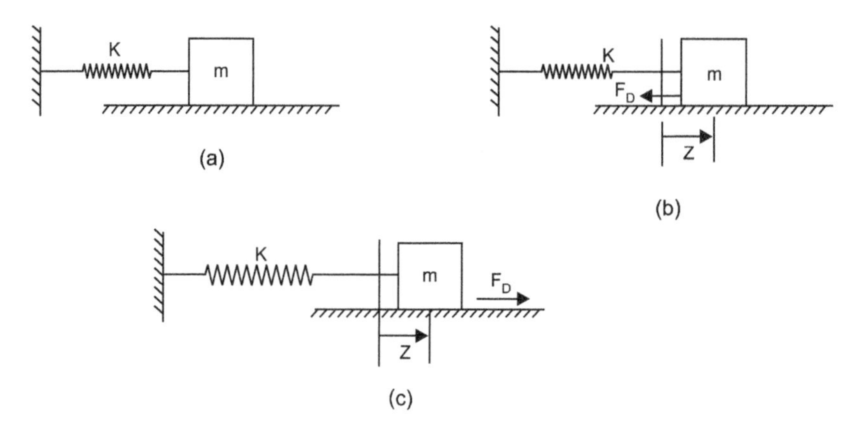

Fig. 2.31. Free vibration with Coulomb damping

Equation (2.158) is non-linear because of sign change of F_D. The period of free vibration and variation of amplitude Z with time can be obtained as below. Let the mass be given an initial displacement Z_o and then released. The time required for the mass to attain its maximum displacement on the other side of the equilibrium will be just one half of the period of vibration. The equation of motion will be as follows:

$$m\ddot{Z} + KZ = F_D \tag{2.161}$$

The solution of Eq. (2.161) is

$$Z = A\cos\omega_n t + B\sin\omega_n t + (F_D/K) \tag{2.162}$$

Using the boundary conditions (At $t = 0$, $Z = Z_o$ and $\dot{Z} = 0$) we get, $A = Z_o - (F_D/K)$, and $B = 0$.

Therefore
$$Z = [Z_o - (F_D/K]\cos\omega_n t + (F_D/K) \tag{2.163}$$

Now
$$\dot{Z} = -[Z_o - (F_D/K)]\omega_{n'}\sin\omega_n t \tag{2.164}$$

At the extreme position on the opposite side, the velocity will be again zero and time required for this half cycle will be:

$$\omega_n t = \pi \tag{2.165a}$$

or
$$t = \pi/\omega_n \tag{2.165b}$$

Hence the period of free vibration will be twice the above value i.e.

$$T = 2\pi/\omega_n \tag{2.166}$$

The coulomb damping therefore does not change the natural frequency and time period, and these remain the same as in undamped case. Equation (2.163) gives the variation of Z with time as below:

t	Z
0	Z_0
$\pi/(2\omega_n)$	F_D/K
π/ω_n	$-Z_o+2F_D/K$

Similar analysis can be done for the other half cycle considering that the spring is compressed by an amount $(Z_o - 2F_D/K)$ and then left. It can be seen that after this half cycle, the displacement becomes $(Z_o - 4F_D/K)$. Hence after full one cycle the displacement reduces from Z to $(Z_o - 4F_D/K)$.

Figure 2.32 shows a displacement-time plot of a system having free vibrations with coulomb damping. It indicates that amplitude loss per cycle is $4F_D/K$.

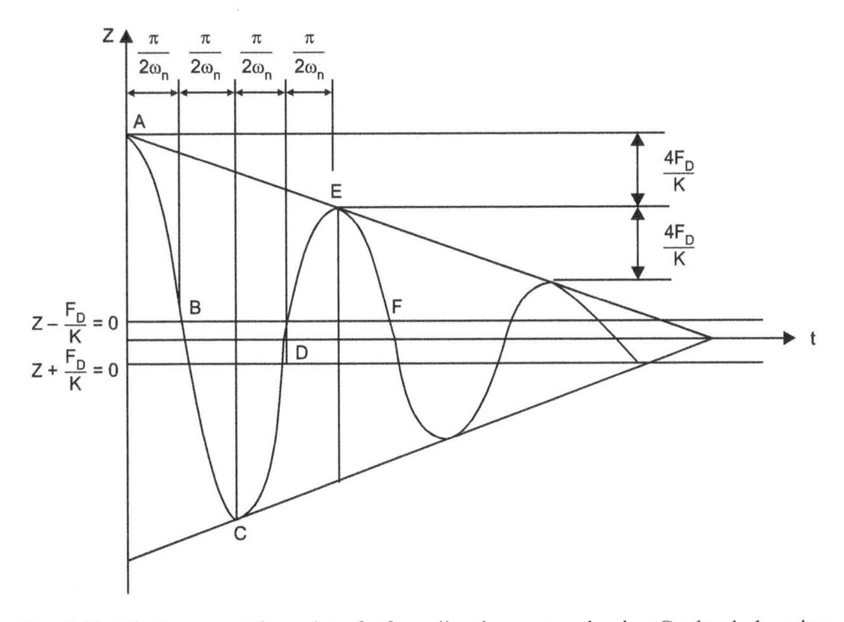

Fig. 2.32. Displacement-time plot of a free vibration system having Coulomb damping

2.10 Systems under Transient Vibrations

As mentioned earlier, a system subjected to periodic excitation has two components of motion, the transient and the steady state (Fig. 2.12). In most of such cases the transient part is not important as it dies out soon and the steady part is the one that persists. However, where the excitation is of non-periodic nature as caused by earthquakes, blasts, impacts, suddenly dropping of loads, etc. the response of the system is purely transient. The maximum motion occurs within a relatively short time after the application of force. For this reason, damping may be of little importance in transient loads. However in the cases where force consists of series of pulses like in earthquakes, responses are influenced by damping. In the subsequent paragraphs undamped transient vibrations have been dealt in few typical cases.

Case 1: Suddenly Applied Load

Consider a mass-spring system shown in Fig. 2.33 subjected to a forcing function $F(t) = F_D$. The equation of motion for the mass m is given by:

$$m\ddot{Z} + KZ = F_0 \tag{2.167}$$

As explained in the earlier section, the system will have natural frequency ω_n as

$$\omega_n = \sqrt{K/m} \tag{2.168}$$

The solution for displacement Z is given by

$$Z = A \cos \omega_n t + B \sin \omega_n t + F_0/K \tag{2.169}$$

The values of arbitrary constants A and B can be obtained considering two boundary conditions:

(i) At $t = 0, Z = 0$

(ii) At $t = 0, \dot{Z} = 0$

These give, $A = -F_0/K$ and $B = 0$
Therefore Eq. (2.169) becomes

$$Z = F_0/K\,(1 - \cos \omega_n t) \tag{2.170}$$

Magnification factor, μ is given by

$$\mu = \frac{Z}{F_0/K} = 1 - \cos \omega_n t \tag{2.171}$$

A plot of μ versus time is given in Fig. 2.33c. It has a maximum value of 2, which occurs when $\cos \omega_n t = -1$. The first peak is reached when $\omega_n t = \pi$ or $t = T_n/2$, where T_n is the natural period of vibration.

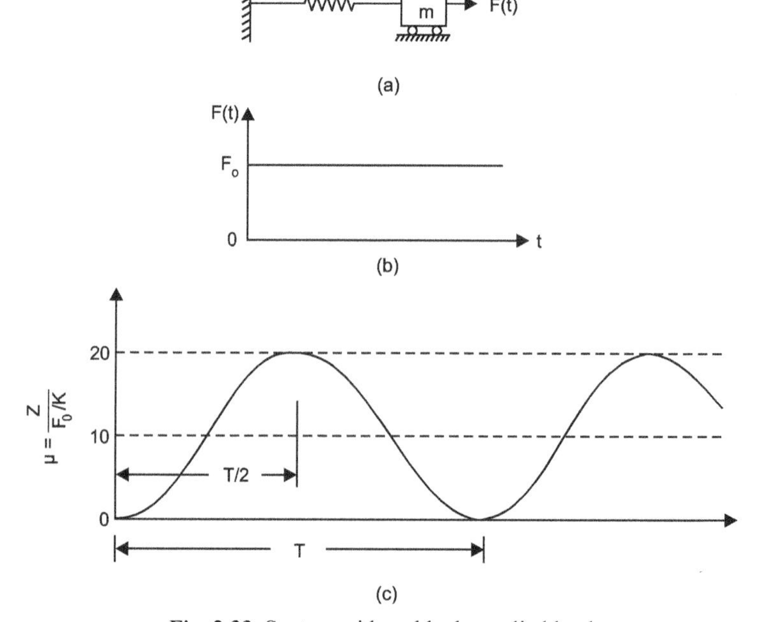

Fig. 2.33. System with suddenly applied load

Case 2: Rectangular Pulse

Consider a mass-spring system shown in Fig. 2.34a subjected to a rectangular pulse of uniform force F_0 for a given duration t_p (Fig. 2.34b).

When $t < t_p$, the equation of motion and μ will be same as given by Eqs. (2.167) and (2.171) respectively. At $t = t_p$, the values of displacement and velocity are given by

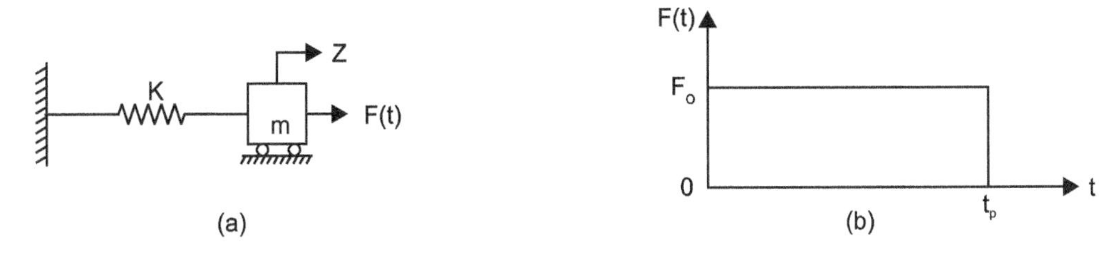

Fig. 2.34. System with load having rectangular pulse

$$Z_p = F_0/K\,(1 - \cos \omega_n.t_p) \tag{2.172a}$$

$$\dot{Z}_p = [(F_o\omega_n)/K]\sin \omega_n t_p \tag{2.172b}$$

For $t > t_p$, the equation of motion will be

$$m\ddot{Z} + K\,Z = 0 \tag{2.173}$$

The solution for displacement Z is given by

$$Z = A \cos \omega_n t + B \sin \omega_n t \tag{2.174}$$

The constants A and B can be obtained by using boundary conditions that at $t = t_p$, $Z = Z_p$, $\dot{Z} = \dot{Z}_p$. Equation (2.174) becomes:

$$Z = F_0/K\,[\cos \omega_n(t - t_p) - \cos \omega_n t] \tag{2.175a}$$

$$\mu = Z/(F_0/K) = [\cos \omega_n (t - t_p) - \cos \omega_n t] \tag{2.175b}$$

Plots of μ versus time for two typical cases namely

(i) $t_p = T_n/3 = 2\pi/(3\,\omega_n)$ and

(ii) $t_p = T_n/4 = \pi/(2\omega_n)$ are shown in Fig. 2.35.

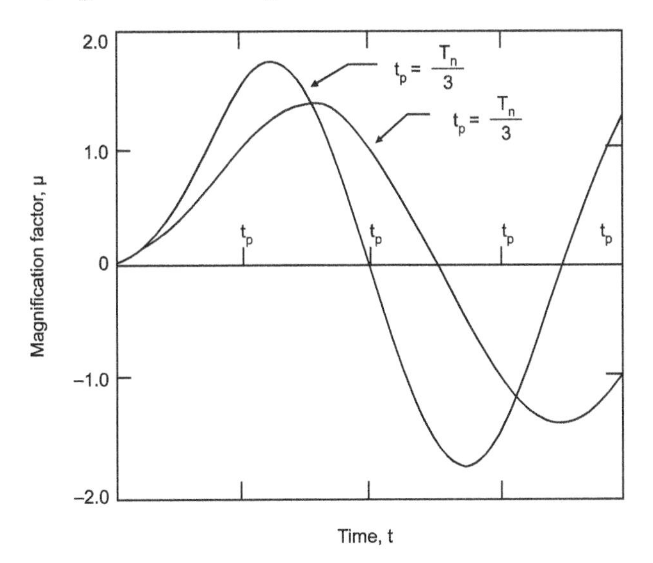

Fig. 2.35. Response of systems with natural periods of a rectangular pulse

Illustrative Examples

Example 2.1

The motion of a particle is represented by the equation $Z = 20 \sin \omega t$. Show the relative positions and magnitudes of the displacement, velocity and acceleration vectors at time $t = 0$, and $\omega = 2.0$ rad/s and 0.5 rad/s.

Solution

$$Z = 20 \sin \omega t$$

$$\dot{Z} = 20\, \omega \cos \omega\, t = 20\omega \sin\left(\omega t + \frac{\pi}{2}\right)$$

$$\ddot{Z} = -20\, \omega^2 \sin \omega\, t = 20\, \omega^2 \sin(\omega\, t + \pi)$$

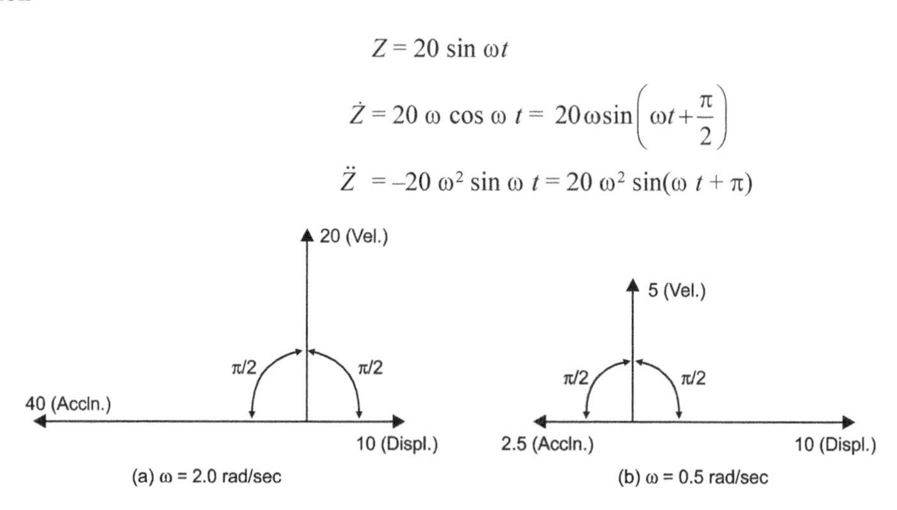

Fig. 2.36. Vector diagram (Example 2.1)

The magnitudes of displacement, velocity and acceleration vectors are 10, 10 ω and 10 ω^2 respectively. The phase difference is such that the velocity vector leads the displacement vector by $\pi/2$ and the acceleration vector leads the velocity vector by another $\pi/2$. Figures 2.36 *a* and 2.36 *b* show the three vectors for $\omega = 2.0$ and 0.50 rad/s respectively.

$$\text{Time period} = \frac{2\pi}{\omega} = \frac{2\pi}{2} = \pi\, s \qquad \text{for } \omega = 2.0 \text{ rad/s}$$

$$\text{Time period} = \frac{2\pi}{\omega} = \frac{2\pi}{(0.5)} = 4\pi\, s \qquad \text{for } \omega = 0.5 \text{ rad/s}$$

Example 2.2

A body performs, simultaneously the motions

$$Z_1 \text{ (mm)} = 20 \sin 8.0\, t$$

$$Z_2 \text{ (mm)} = 21 \sin 8.5\, t$$

Determine the maximum and minimum amplitude of the combined motion, and the time period of the periodic motion.

Solution

$$Z_{max} = 21 + 20 = 41 \text{ mm}$$

$$Z_{min} = 21 - 20 = 1 \text{ mm}$$

The beat frequency is given by

$$f = \frac{8.5 - 8.0}{2\pi} = \frac{0.5}{2\pi} = 0.0795 \text{ Hz, and}$$

$$T = \frac{1}{f} = \frac{2\pi}{0.5} = 4\pi = 12.57\,\text{s}$$

Example 2.3

A mass of 20 kg when suspended from a spring, causes a static deflection of 20 mm. Find the natural frequency of the system.

Solution Stiffness of the spring, $K = \dfrac{W}{\delta_{st}}$

$$K = \frac{20 \times 9.81}{20 \times 10^{-3}} \approx 10^4 \text{ N/m}$$

$$\text{Natural frequency, } f_n = \frac{1}{2\pi}\sqrt{\frac{K}{m}}$$

$$= \frac{1}{2\pi}\sqrt{\frac{10^4}{20}} = 3.6\,\text{Hz}$$

Example 2.4

For the system shown in Fig. 2.37, determine the natural frequency of the system if

$K_1 = 1000$ N/m $K_2 = 500$ N/m

$K_3 = 2000$ N/m $K_4 = K_5 = 750$ N/m`

Mass of the body = 5 kg

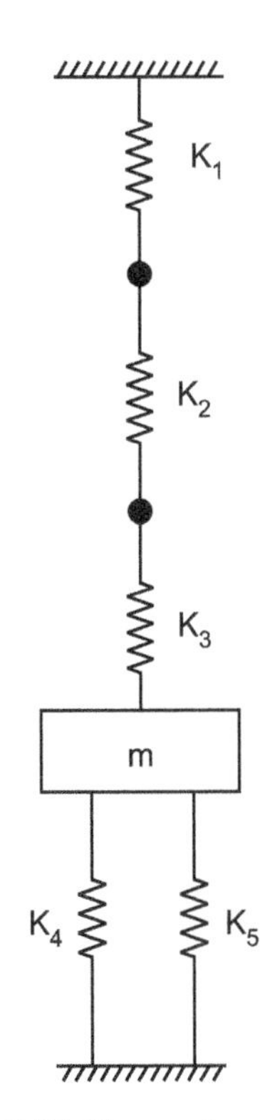

Fig. 2.37. Mass-springs system

Solution

Let K_{e1} and K_{e2} represent respectively the effective stiffnesses of the top three springs and the lower two springs, then

$$\frac{1}{K_{el}} = \frac{1}{K_1} + \frac{1}{K_2} + \frac{1}{K_3} = \frac{1}{1000} + \frac{1}{500} + \frac{1}{2000} = 0.0035$$

$$K_{e1} = 285.7 \text{ N}/m$$

$$K_{e2} = K_4 + K_5 = 750 + 750 = 1500 \text{ N}/m$$

Now K_{e1} and K_{e2} are two springs in parallel, therefore effective stiffness,

$$K_e = K_{e1} + K_{e2} = 285.7 + 1500 = 1785.7 \text{ N}/m$$

$$f_n = \frac{1}{2\pi}\sqrt{\frac{K}{m}} = \frac{1}{2\pi}\sqrt{\frac{1785.7}{5.0}} = 3.0\,\text{Hz}$$

Example 2.5

A vibrating system consists of a mass of 5 kg, a spring stiffness of 5 N/mm and a dashpot with a damping coefficient of 0.1 N-s/m. Determine (*i*) damping ratio and (ii) logarithmic decrement.

Solution

(i)

$$C_c = 2\sqrt{km} = 2\sqrt{5\times10^{-3}\times5} = 0.319 N - s/m$$

$$\xi = \frac{C}{C_c} = \frac{0.1}{0.319} = 0.313$$

(ii) Logarithmic decrement $= \dfrac{2\pi\xi}{\sqrt{1-\xi^2}} = \dfrac{2\pi \times 0.313}{\sqrt{1-0.313^2}} = 2.07$

$$\log_e \frac{Z_1}{Z_2} = 2.07$$

i.e. $\dfrac{Z_1}{Z_2} = 7.92$

Therefore the free amplitude in the next cycle decreases by 7.92 times.

Example 2.6

A mass attached to a spring of stiffness of 5 N/mm has a viscous damping device. When the mass was displaced and released, the period of vibration was found to be 2.0 s, and the ratio of the consecutive amplitudes was 10/3. Determine the amplitude and phase angle when a force F = 3 sin 4 t acts on the system. The unit of the force is Newton.

Solution

(i)
$$\frac{2\pi\xi}{1-\xi} = \log_e \frac{Z_1}{Z_2} = \log_e \frac{10}{3} = 1.2$$

or
$$\xi = 0.195$$

(ii)
$$T_n = 2.0 \text{ s}$$

$$\omega_n = \frac{2\pi}{T} = \frac{2\pi}{2} = 3.14 \text{ rad/s}$$

$$\omega = 4.0 \text{ rad/s}$$

$$\eta = \frac{\omega}{\omega_n} = \frac{4.0}{3.14} = 1.273$$

$$F_0 = 3.0\,N;\ A_{st} = \frac{F_0}{K} = \frac{3.0}{5.0} = 0.6 \text{ mm}$$

From Eq. (2.58),

$$A_z = \frac{A_{st}}{\sqrt{(1-\eta^2)^2 + (2\xi\eta)^2}};\ A_{st} = \text{Static Deflection}$$

$$= \frac{0.6}{\sqrt{(1-1.273^2)^2 + (2 \times 0.195 \times 1.273)^2}} = 0.755 \text{ mm}$$

$$q = \tan^{-1}\left(\frac{2\eta\xi}{1-\eta^2}\right) = \tan^{-1}\left(\frac{2 \times 1.273 \times 0.195}{1-1.273^2}\right) = 141.4^\circ$$

Example 2.7

Show that in frequency-dependent excitation the damping factor ξ is given by the following expression:

$$\xi = \frac{1}{2}\left(\frac{f_2 - f_1}{2f_n}\right)$$

where f_1 and f_2 are frequencies at which the amplitude is $1/\sqrt{2}$ times the peak amplitude.

Solution

In a forced vibration test, the system is excited with constant force of excitation and varying frequencies. A response curve as shown in Fig. 2.38 is obtained.

At resonance, $\eta = 1$ and $A_z/Z_{st} = 1/2\xi$ (for small values of ξ). If the frequency ratio is η when amplitude of motion is $1/\sqrt{2}$ times the peak amplitude, then from Eq. 2.59, we get

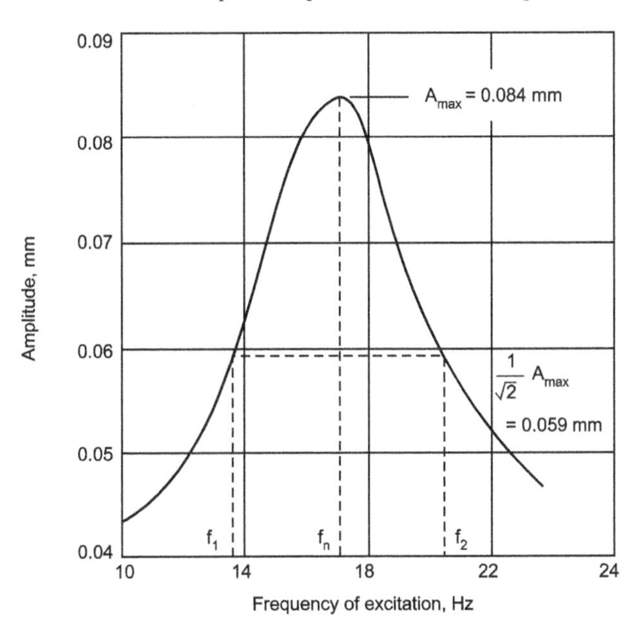

Fig. 2.38. Determination of viscous damping in forced vibrations by Bandwidth method

$$\frac{1}{\sqrt{2}} \cdot \frac{1}{2\xi} = \frac{1}{\sqrt{(1-\eta^2)^2 + 4\eta^2\xi^2}}$$

or $$\eta^4 - 2\eta^2(1 - 2\xi^2) + (1 - 8\xi^2) = 0$$

or $$\eta_{1,2}^2 = \frac{1}{2}\left[2(1-2\xi^2) \pm \sqrt{4(1-2\xi^2)^2 - 4(1-8\xi^2)}\right]$$

$$= (1-2\xi^2) \pm 2\xi\sqrt{1+\xi^2}$$

Now $$\eta_2^2 - \eta_1^2 = 4\xi\sqrt{1-\xi^2} \approx 4\xi \text{ [for small values of } \xi]$$

Also, $$\eta_2^2 - \eta_1^2 = \frac{f_2^2 - f_1^2}{f_n^2} = \left(\frac{f_2 - f_1}{f_n}\right)\left(\frac{f_2 + f_1}{f_n}\right)$$

$$= 2\left(\frac{f_2 - f_1}{f_n}\right) \text{ since } \frac{f_2 + f_1}{f_n} \approx 2$$

Therefore $$\xi = \frac{1}{2}\left(\frac{f_2 - f_1}{f_n}\right)$$

This method for determining viscous damping is known as the **band width method.**

Example 2.8

A machine of mass 100 kg is supported on springs of total stiffness of 784 N/mm. The machine produces an unbalanced disturbing force of 392 N at a speed 50 c/s. Assuming a damping factor of 0.20, determine

(i) the amplitude of motion due to unbalance,
(ii) the transmissibility, and
(iii) the transmitted force.

Solution

(i)
$$\omega_n = \sqrt{K/m} = \sqrt{\frac{784 \times 10^3}{100}} = 87.7 \text{ rad/s}$$

$$\omega_n = 2\pi \times 50 = 314 \text{ rad/s}$$

$$\eta = \frac{\omega}{\omega_n} = \frac{314}{87.7} = 3.58$$

$$Z_{st} = \frac{F_0}{K} = \frac{392}{784} = 0.5 \text{ mm}$$

Now
$$A_Z = \frac{Z_{st}}{\sqrt{(1-\eta^2)^2 + (2\eta\xi)^2}} = \frac{0.5}{\sqrt{(1-3.58^2)^2 + (2\times 3.58\times 0.2)^2}}$$
$$= 0.042 \text{ mm}$$

(ii) Transmissibility
$$\mu_T = \frac{\sqrt{(1+(2\eta\xi)^2}}{\sqrt{(1-\eta^2)^2 + (2\eta\xi)^2}} = \frac{\sqrt{1+(2\times 3.58\times 0.2)^2}}{\sqrt{(1-3.58^2)^2 + (2\times 3.58\times 0.2)^2}}$$
$$= 0.1467$$

(iii) Force transmitted = 392 × 0.1467 = 57.5 N

Example 2.9

The rotor of a motor having mass 2 kg was running at a constant speed of 30 c/s with an eccentricity of 160 mm. The motor was mounted on an isolator with damping factor of 0.25. Determine the stiffness of the isolator spring such that 15% of the unbalanced force is transmitted to the foundation. Also determine the magnitude of the transmitted force.

Solution

(i) Maximum force generated by the motor
$$= 2\, m_e e\omega^2 = 2 \times 2.0 \times 0.16 \times (2\pi \times 30)^2 = 22716 \text{ N} = 22.72 \text{ kN}$$

(ii)
$$\mu_T = \frac{\text{Force transmitted}}{\text{unbalanced force}} = 0.15$$

i.e.
$$\frac{\sqrt{1+4\eta^2\xi^2}}{\sqrt{(1-\eta^2)^2 + (2\eta\xi)^2}} = 0.15$$

or
$$1 + 4\eta^2 \times (0.25)^2 = (0.15)^2 [(1-\eta^2)^2 + (2\eta \times 0.25)^2]$$

or
$$\eta^4 - 12.84\,\eta^2 - 43.44 = 0$$

It gives
$$\eta = 3.95 \quad \textit{i.e.} \frac{\omega}{\omega_n} = 3.95$$

Therefore
$$\omega_n = \sqrt{K/m} = \frac{\omega}{3.95} = \frac{60\pi}{3.95} = 47.7 \text{ rad/s}$$

$$K = m(47.7)^2 = 2.0 \times (47.7)^2 = 4639 \text{ N/m}$$

(iii) Force transmitted to the foundation = 0.15 × 22.72 = 3.4 kN.

Example 2.10

A seismic instrument with a natural frequency of 6 Hz is used to measure the vibration of a machine running at 120 rpm. The instrument gives the reading for the relative displacement of the seismic mass as 0.05 mm. Determine the amplitudes of displacement, velocity and acceleration of the vibrating machine. Neglect damping.

Solution

(i)

$$\omega_n = 6 \text{ Hz} = 37.7 \text{ rad/s}$$

$$\omega = 120 \text{ rpm} = \frac{120 \times 2\pi}{60} = 12.57 \text{ rad/s}$$

$$\eta = \frac{12.57}{37.7} = 0.333$$

$$\mu = \left[\frac{1}{1-\eta^2} \text{ for } \xi = 0\right] = \frac{1}{1-(0.333)^2} = 1.125$$

(ii) For displacement pickup, Eq. (2.88) gives

$$X = \eta^2 \mu Y_0$$

$$0.05 = (0.333)^2 \times 1.125 \times Y_0$$

or

$$Y_0 = 0.40 \text{ mm}$$

(iii) For velocity pickup, Eq. (2.91) gives

$$X = \frac{1}{\omega_n} \eta \mu (Y_0 \omega)$$

$$0.05 = \frac{0.333}{(37.7)} \times 1.125 \times Y_0$$

or

$$(Y_0 \omega) = \text{velocity} = 5.03 \text{ mm/s}$$

(iv) For acceleration pickup, Eq. (2.89) gives

$$X = \frac{\mu}{\omega_n^2}(Y_0 \omega^2)$$

or

$$0.05 = \frac{1.125}{(37.7)^2}(Y_0 \omega^2)$$

i.e.

$$(Y_0\omega^2) = \text{Acceleration} = \frac{(37.7)^2 \times 0.05}{1.125} = 63.17 \text{ mm/s}^2$$

Example 2.11

Determine the natural frequencies and mode shapes of the system represented by a mathematical model shown in Fig. 2.39a.

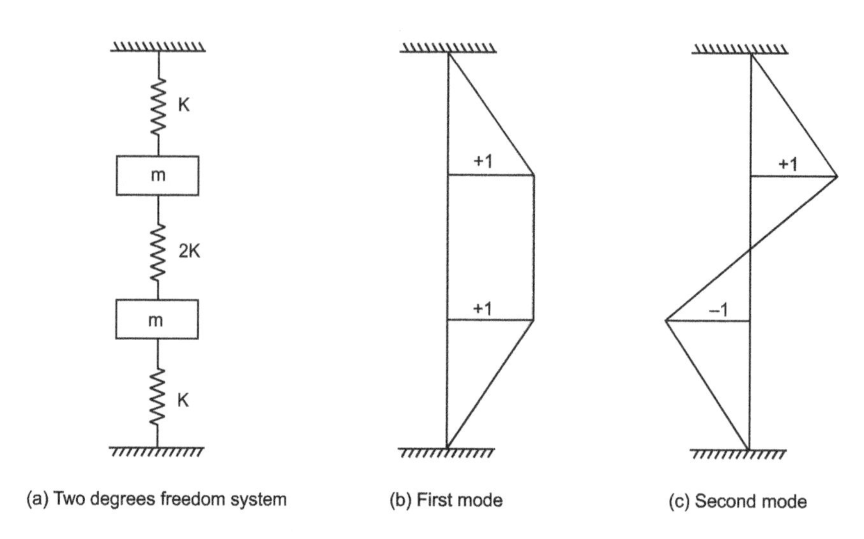

Fig. 2.39. Two degrees freedom system with mode shapes

Solution

(i) The system shown in Fig. 2.39a is a two degrees freedom system. The solution of such a system has already been described in Section 2.7

(ii) The two natural frequencies of the system can be obtained using Eq. (2.100) by putting $K_1 = K$, $K_2 = 2K$ and $K_3 = K$, and $m_1 = m_2 = m$. By doing this, we get

$$\omega_{n1}^2 = \frac{1}{2}\left[\left(\frac{3K}{m}+\frac{3K}{m}\right)-\left\{\frac{4\times(2K)^2}{m^2}\right\}^{1/2}\right]=\frac{K}{m}$$

$$\omega_{n2}^2 = \frac{1}{2}\left[\frac{6K}{m}+\frac{4K}{m}\right]=\frac{5K}{m}$$

Hence, $\omega_{n1} = \sqrt{K/m}$ and $\omega_{n2} = \sqrt{5K/m}$

(iii) The relative values of amplitudes A_1 and A_2 for the two modes can be obtained using Eqs. (2.103) and (2.104).

$$\frac{A_1^{(1)}}{A_2^{(1)}} = \frac{K_2}{K_1+K_2-m_1\omega_{n1}^2} = \frac{2K}{K+2K-m\times K/m} = +1$$

$$\frac{A_1^{(2)}}{A_2^{(2)}} = \frac{2K}{K+2K-m\times 5K/m} = -1$$

The mode shapes are shown in Figs. 2.39b and 2.39c.

Example 2.12

Determine the natural frequencies and mode shapes of the system represented by the mathematical model shown in Fig. 2.40a.

Solution

(i) Equations of motion for the three masses can be written as

$$m\ddot{Z}_1 + KZ_1 + 2K(Z_1 - Z_2) = 0 \tag{2.157a}$$

$$m\ddot{Z}_2 + 2K(Z_2 - Z_1) + K(Z_2 - Z_3) = 0 \tag{2.157b}$$

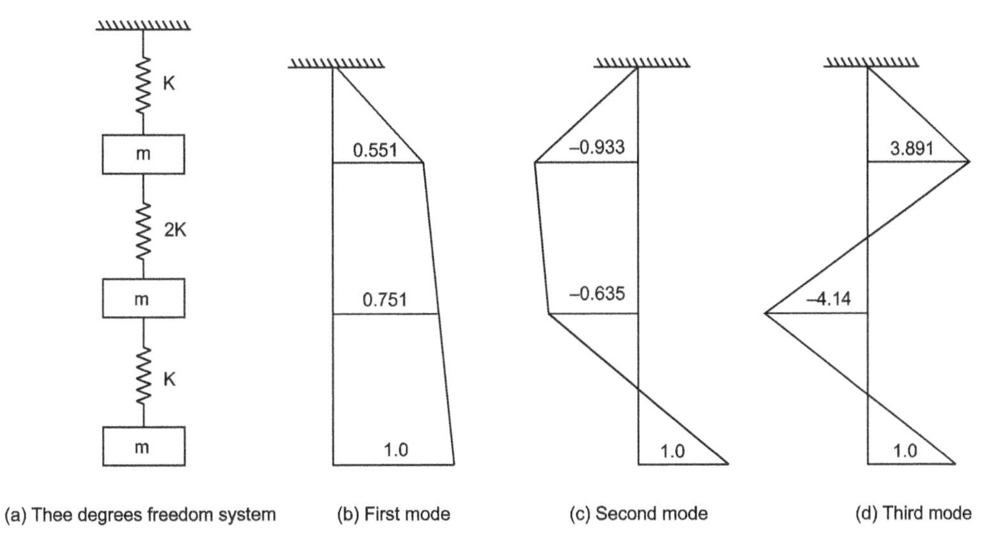

Fig. 2.40. Three degrees freedom system with mode shapes

$$m\ddot{Z}_3 + K(Z_3 - Z_2) = 0 \tag{2.157c}$$

For steady state, the solutions will be as

$$Z_1 = A_1 \sin \omega_n t \tag{2.158a}$$

$$Z_2 = A_2 \sin \omega_n t \tag{2.158b}$$

$$Z_3 = A_3 \sin \omega_n t \tag{2.158c}$$

Substituting Eqs. (2.158) in Eqs. (2.157), we get

$$(3K - m\omega_n^2)A_1 - 2K\,A_2 = 0 \tag{2.159a}$$

$$-2K\,A_1 + (3K - m\omega_n^2)A_2 - K\,A_3 = 0 \tag{2.159b}$$

$$-K\,A_2 + (K - m\omega_n^2)A_3 = 0 \tag{2.159c}$$

For nontrivial solutions of ω_n in Eqs. (2.159)

$$\begin{vmatrix} 3K - m\omega_n^2 & -2K & 0 \\ -2K & 3K - m\omega_n^2 & -K \\ 0 & -K & K - m\omega_n^2 \end{vmatrix} = 0 \tag{2.160}$$

Putting $\lambda = \dfrac{m\omega_n^2}{K}$, Eq. (2.160) becomes

$$\begin{vmatrix} 3-\lambda & -2 & 0 \\ -2 & 3-\lambda & -1 \\ 0 & -1 & 1-\lambda \end{vmatrix} = 0 \tag{2.161a}$$

or

$$\lambda^3 - 7\lambda^2 + 10\lambda - 2 = 0 \tag{2.161b}$$

Equation (2.161b) is cubic in λ. The values of λ are worked out as

$$\lambda_1 = 0.238,\ \lambda_2 = 1.637;\ \text{and}\ \lambda_3 = 5.129$$

Therefore, $\omega_{n1} = \sqrt{0.238K/m}$; $\omega_{n2} = \sqrt{1.637K/m}$; and $\omega_{n3} = \sqrt{5.129K/m}$

(ii) Eqs. (2.159) in terms of λ can be written as

$$(3 - \lambda) A_1 - 2A_2 = 0 \quad (2.162a)$$

$$-2A_1 + (3 - \lambda) A_2 - A_3 = 0 \quad (2.162b)$$

$$- A_2 + (1 - \lambda) A_3 = 0 \quad (2.162c)$$

For I mode: $\lambda = 0.238$

Equation (2.162a) gives

$$(3 - 0.238)A_1 - 2A_2 = 0 \text{ or } \frac{A_1}{A_2} = 0.724$$

Equation (2.162b) gives $-2 \times 0.724\, A_2 + (3 - 0.238)A_2 - A_3 = 0$

or

$$\frac{A_2}{A_3} = 0.761$$

Assuming $A_1 = a$, $A_2 = 1.381\, a$ and $A_3 = 1.815\, a$

$$A_1 : A_2 : A_3 = a : 1.381\, a : 1.815\, a$$
$$= 1 : 1.381 : 1.815$$
$$= 0.551 : 0.761 : 1$$

Similarly,

For II mode; $\lambda = 1.637$

$$A_1 : A_2 : A_3 = -0.933 : -0.635 : 1, \text{ and}$$

For III mode: $\lambda = 5.129$

$$A_1 : A_2 : A_3 = 3.891 : -4.14 : 1$$

The mode shapes are plotted in Figs. 2.40b, 2.40c and 2.40d.

Example 2.13

A small reciprocating machine weighs 50 kg and runs at a constant speed of 6000 rpm. After it was installed, it was found that the forcing frequency is very close to the natural frequency of the system. What dynamic absorber should be added if the nearest natural frequency of the system should be at least 20 percent from the forcing frequency.

Solution

(i)
$$\omega = \frac{2\pi N}{60} = \frac{2\pi \times 6000}{60} = 628 \text{ rad/s}$$

At the time of installation of machine,

Forcing frequency $\approx$ Natural frequency of system

Therefore,
$$\sqrt{\frac{K}{m}} = 628$$

or
$$K = m \times 628^2$$
$$= 50 \times 628^2 = 201 \times 10^5 \text{ N/m}$$

(ii) After adding the vibration absorber to the system, the natural frequency becomes $(1 \pm 0.2)\, 628$ i.e. 753.6 rad/s or 502.4 rad/s

For tuned absorber:

$$\frac{m_a}{M} = \frac{K_a}{K} = \mu_M$$

Now from Eq. (2.156)
$$\mu_m = \frac{\left\{\left(\frac{\omega_{nl}}{\omega_{na}}\right)^2 - 1\right\}^2}{\left(\frac{\omega_n}{\omega_{na}}\right)}$$

for $\frac{\omega_{nl}}{\omega_{na}} = 0.8$

$$\mu_m = \frac{\{(0.8)^2 - 1\}^2}{0.8^2} = 0.2025$$

and when $\frac{\omega_{nl}}{\omega_{na}} = 1.2$

$$\mu_m = \frac{\{(1.2)^2 - 1\}^2}{1.2^2} = 0.134$$

Adopting the higher value of μ_m

$$K_a = 0.2025 \times 201 \times 10^5 = 40.7 \times 10^5 \text{ N/m}$$

$$m_a = 0.2025 \times 50 = 10.12 \text{ kg}$$

Example 2.14

A horizontal mass-spring system with coulomb damping has a mass of 8.0 kg attached to a spring of stiffness 1200 N/m. The coefficient of friction is 0.04. Determine (i) frequency of oscillation, (ii) number of cycles corresponding to 60% reduction in amplitude if the initial amplitude is 60 mm, and (iii) time taken to achieve this reduction.

Solution

(i) $F_D = \mu \text{ mg} = 0.04 \times 8.0 \times 9.81 = 3.14\ N$

$$\omega_n = \sqrt{K/m} = \sqrt{1200/8} = 12.25 \text{ rad/s}$$

$$f_n = \omega_n/2\pi = 12.25/2\pi = 1.95 \text{ Hz}$$

(ii) Initial amplitude = 60 mm = 0.06 m
Amplitude after 60% reduction = 0.036 m
Total reduction in amplitude = 0.06 – 0.036
= 0.024 m

Reduction in amplitude per cycle = $4F_D/K = (4 \times 3.14)/1200$
= 0.0105 m

Number of cycles for 60% reduction = 0.024/0.0105 = 2.3 cycles

(iii) Time taken to achieve 60% reduction = Time taken to perform 2.3 cycles
$= 2.3 \times (2\pi/\omega_n)$
$= 2.3 \times (2\pi/12.25) = 1.18$s

Practice Problems

2.1 A single degree (mass-spring-dashpot) system is subjected to a frequency dependent oscillatory force ($m\ e_0\ \omega^2 \sin \omega\ t$). Proceeding from fundamentals, derive the expression of the amplitude of the system.

2.2 'Presence of damping reduces the effectiveness of the isolation system'. Is this statement true? If yes, explain with neat sketches.

2.3 Give two methods of determining 'damping factor' of a single degree freedom system.

2.4 Starting from fundamentals, explain the principles involved in the design of (i) Displacement pickup, (ii) Velocity pickup, and (iii) Acceleration pickup. Illustrate your answer with neat sketches.

2.5 Describe the principles involved in a 'tuned dynamic vibration absorber'. Illustrate your answer with neat sketches. Discuss clearly its limitations.

2.6 A mass of 25 kg when suspended from a spring, causes a static deflection of 25 mm. Find the natural frequency of the system.

Ans. (20 rad/s)

2.7 A spring mass system (K_1, m) has a natural frequency of f_1. If a second spring of stiffness K_2 is attached in series with the first spring, the natural frequency becomes $f_1/2$. Determine K_2 in terms of K_1.

Ans. ($K_1/3$)

2.8 A mass of 5 kg is attached to the lower end of a spring whose upper end is fixed. The natural period of this system is 0.40 s. Determine the natural period when a mass of 2.5 kg is attached to the mid point of this spring with the upper and lower ends fixed.

Ans. (0.14 sec)

2.9 Determine the differential equation of motion of the system shown in Fig. 2.41. The moment of inertia of weight W about the point O is J_0. Show that the system becomes unstable when:

$$b \geq \frac{K \cdot a}{W}$$

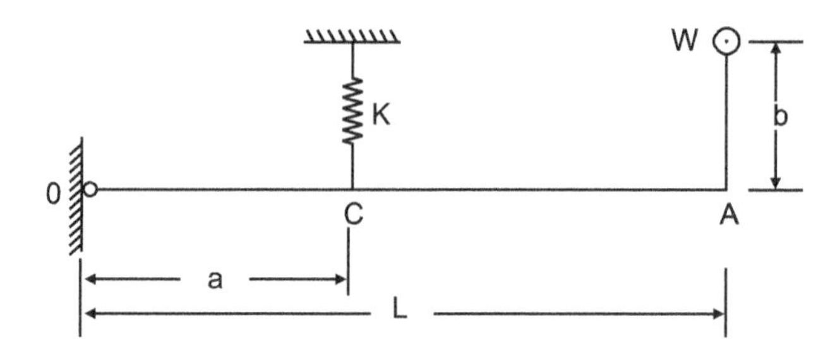

Fig. 2.41. Mass-spring system

2.10 A body vibrating in a viscous medium has a period of 0.30 s and an inertial amplitude of 30 mm. Determine the logarithmic decrement if the amplitude after 10 cycles is 0.3 mm.

Ans. (0.46)

2.11 A vibration system consists of mass of 6 kg, a spring stiffness of 0.7 N/m and a dashpot with a damping coefficient of 2 N-s/m. Determine

(a) Damping ratio

(b) Logarithmic decrement

Ans. (0.488, 3.55)

2.12 Write a differential equation of motion for the system shown in Fig. 2.42 and determine the natural frequency of damped oscillations and the critical damping coefficient.

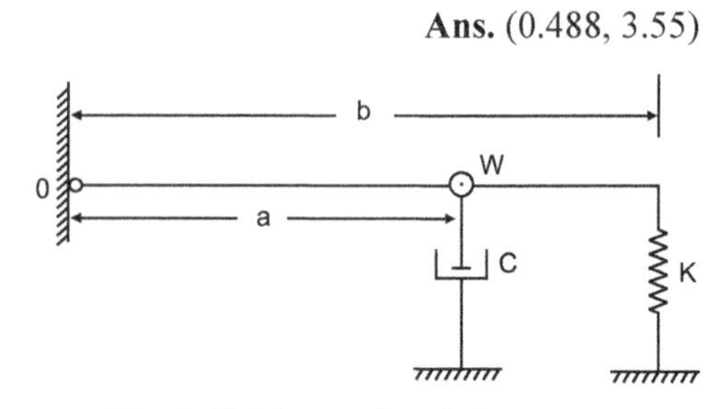

Fig. 2.42. Mass-spring dashpot system

2.13 A mass is attached to a spring of stiffness 6 N/mm with a viscous damping device. When the mass was displaced and released, the period of vibration was found to be 1.8 s and the ratio of consecutive amplitude was 4.2 to 1. Determine the amplitude and phase angle when a force $F = 2 \sin 3t\ N$ acts on the system.

Ans. (0.708 mm, 56.4°)

2.14 A spring mass system is excited by a force $F_0 \sin \omega\, t$. At resonance the amplitude was measured to be 100 mm. At 80% resonant frequency the amplitude was measured 80 mm. Determine the damping factor of the system.

Ans. (0.1874)

2.15 Assuming small amplitudes, set up differential equation of motion for double pendulum using the coordinates shown in Fig. 2.43. Show that the natural frequencies of the system as given by the equation

$$\omega_{n1,2} = \sqrt{\frac{g}{l}(2 \pm \sqrt{2})}$$

Determine the ratio of the amplitudes x_1/x_2.

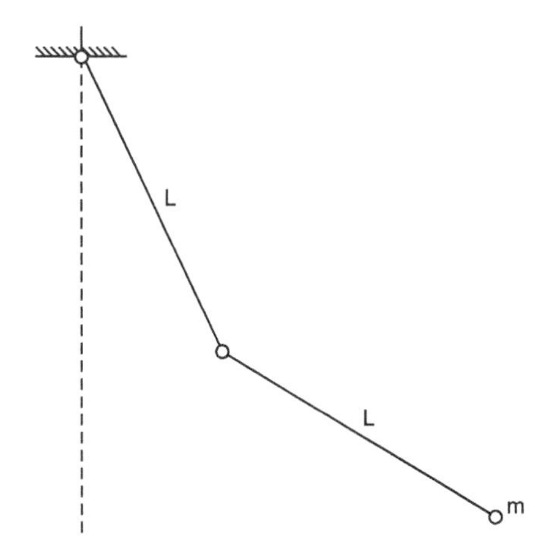

Fig. 2.43. Double pendulum system

2.16 A motor weighs 220 kg and has rotating unbalance of 3000 N-mm. The motor is running at constant speed of 2000 rpm. For vibration isolation, springs with damping factor of 0.25 is used. Specify the springs for mounting such that only 20 percent of the unbalanced force is transmitted to the foundation. Also determine the magnitude of the transmitted force.

Ans. (K_a = 931.22 kN/m, 26.3 kN)

2.17 A small reciprocating machine weighs 60 kg and runs at a constant speed of 5000 rpm. After it was installed, it was found that the forcing frequency is very close to the natural frequency of the system. What dynamic vibration absorber should be added if the nearest natural frequency of the system should be at least 25 percent from the forcing frequency?

Ans. (15.3 kg, 4.2×10^6 N/m)

2.18 A mass of 1 kg is to be supported on a spring having a stiffness of 980 N/m. The damping coefficient is 6.26 N-s/m. Determine the natural frequency of the system. Find also the logarithmic decrement and the amplitude after three cycles if the initial displacement is 0.3 mm.

Ans. (31.14 rad/s, 0.628, 0.0456 mm)

2.19 A machine having a mass of 100 kg and supported on springs of total stiffness 7.84×10^5 N/m has an unbalanced rotating element which results in a disturbing force of 392 N at a speed of 3000 rpm. Assuming a damping factor of 0.20, determine

(a) the amplitude of motion due to the unbalance,
(b) the transmissibility, and
(c) the transmitted force.

Ans. (0.043 mm, 0.148, 58.2 N)

2.20 The static deflection of the vibrometer mass is 20 mm. The instrument when attached to a machine vibrating with a frequency of 125 cpm records a relative amplitude of 0.03 mm. Find out for the machine,

(a) the amplitude of vibration.
(b) the maximum velocity of vibration, and
(c) the maximum acceleration of vibration.

Ans. (0.0576 mm, 0.754 mm/sec, 9.86 mm/sec^2)

CHAPTER

3

Wave Propagation in an Elastic Homogeneous and Isotropic Medium

3.1 General

A sudden load applied to a body does not disturb the entire body at the instant of loading. The parts closest to the source of disturbance are affected first, and the deformations produced by the disturbance subsequently spread throughout the body in the form of stress waves. The propagation speed of seismic waves through the earth depends on the elastic properties and density of materials. The phenomenon of wave propagation in an elastic medium is of great importance in the study of foundations subjected to dynamic loads.

In this chapter wave propagation in (i) an elastic bar, (ii) an elastic infinite medium and (iii) an elastic half space have been discussed.

3.2 Stress, Strain and Elastic Constants

3.2.1 Stress

The external forces acting on a body constitute what is called the "load". No material is perfectly rigid, therefore the application of a load on a body causes deformation. In all cases internal forces are called into play in the material to resist the load and are referred to as "stresses". The intensity of the stress is estimated as the force acting on unit area of cross-section, and is expressed in such units as N/mm^2, KN/m^2 etc. At right angles to the direction of the stress, the body dilates or contracts, depending on whether the stress is compressional or tensile. The stresses preserve the shape of the body but change the volume. Shear stress is said to exist on a section of body if on opposite faces of the section equal and opposite forces exist.

3.2.2 Strain

Strain is a measure of the deformation produced by the application of the external forces. In Fig. 3.la, the deformation is an elongation of a bar by the amount ΔL, and if L is the initial length of the bar, then

$$\text{Tensile strain} = \varepsilon_t = \frac{\Delta L}{L} \tag{3.1a}$$

Similarly in Fig. 3.1b the deformation is a shortening of the bar by the amount ΔL, therefore

$$\text{Compressive strain} = \varepsilon_c = \frac{\Delta L}{L} \tag{3.1b}$$

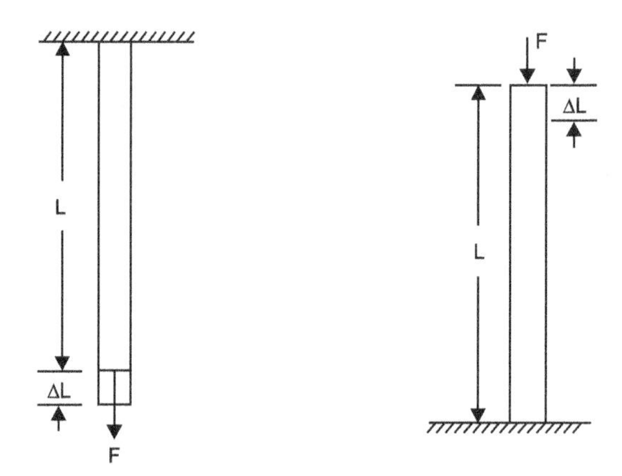

Fig. 3.1. Axial strain

3.2.2.1 A Transverse Strain

A transverse strain ε_w is defined as the ratio of the expansion or contraction Δw perpendicular to the direction of the stress to the original width W of the body (Fig. 3.2). Thus

$$\varepsilon_w = \frac{\Delta W}{W} \tag{3.2}$$

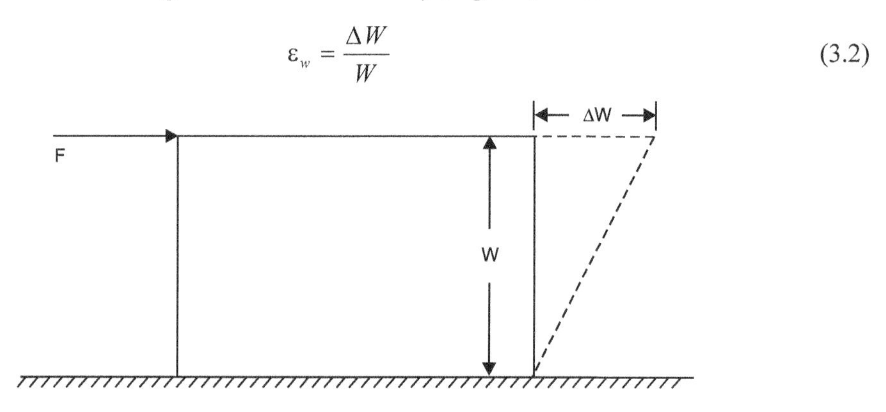

Fig. 3.2. Transverse strain

3.2.3 Elastic Constants

An elastic material is one which obeys Hook's Law of proportionally between stress and strain. For an isotropic elastic material subjected to normal stress σ_x in the x-direction, the strains in x, y, z directions are:

$$\varepsilon_x = \frac{\sigma_x}{E} \tag{3.3a}$$

$$\varepsilon_y = \varepsilon_z = -\frac{\mu\sigma_x}{E} \tag{3.3b}$$

If the element of material is subjected to normal stress σ_x, σ_y, σ_z, then by superposition we obtain

$$\varepsilon_x = \frac{1}{E}[\sigma_x - \mu(\sigma_y + \sigma_z)] \tag{3.4a}$$

$$\varepsilon_y = \frac{1}{E}[\sigma_y - \mu(\sigma_x + \sigma_z)] \tag{3.4b}$$

$$\varepsilon_z = \frac{1}{E}[\sigma_z - \mu(\sigma_x + \sigma_y)] \tag{3.4c}$$

In the above expressions, E is the modulus of elasticity and μ is Poisson's ratio. It may be noted that here E is dynamic modulus of elasticity and not the static modulus.

Equations (3.4a), (3.4b), and (3.4c) can be rearranged so that the stresses are expressed in terms of the strains as follows (Timoshenko and Goodier, 1951; Kolsly, 1963):

$$\sigma_x = \frac{\mu E}{(1+\mu)(1-2\mu)}(\varepsilon_x + \varepsilon_y + \varepsilon_z) + \frac{E}{1+\mu}\varepsilon_x \tag{3.5a}$$

$$\sigma_y = \frac{\mu E}{(1+\mu)(1-2\mu)}(\varepsilon_x + \varepsilon_y + \varepsilon_z) + \frac{E}{1+\mu}\varepsilon_y \tag{3.5 b}$$

$$\sigma_z = \frac{\mu E}{(1+\mu)(1-2\mu)}(\varepsilon_x + \varepsilon_y + \varepsilon_z) + \frac{E}{1+\mu}\varepsilon_z \tag{3.5c}$$

For simplicity the equations may be written

$$\sigma_x = \lambda\bar{\varepsilon} + 2G\varepsilon_x \tag{3.6a}$$

$$\sigma_y = \lambda\bar{\varepsilon} + 2G\varepsilon_y \tag{3.6b}$$

$$\sigma_z = \lambda\bar{\varepsilon} + 2G\varepsilon_z \tag{3.6c}$$

in which

$$\bar{\varepsilon} = \varepsilon_x + \varepsilon_y + \varepsilon_z \tag{3.7}$$

$$\lambda = \frac{\mu E}{(1+\mu)(1-2\mu)} \tag{3.8}$$

$$G = \frac{E}{2(1+\mu)} \tag{3.9}$$

Similarly in an isotropic elastic material, there exists linear relation between shear stress and shear strain. Thus

$$\gamma_{xy} = \frac{\tau_{xy}}{G} \tag{3.10a}$$

$$\gamma_{yz} = \frac{\tau_{yz}}{G} \tag{3.10b}$$

$$\gamma_{zx} = \frac{\tau_{zx}}{G} \tag{3.10c}$$

G is the shear modulus or rigidity modulus and is the same as given by Eq. (3.9).

Equations (3.6) and (3.10) comprise six equations that define the stress-strain relationship.

3.3 Longitudinal Elastic Waves in a Rod of Infinite Length

Consider the free vibration of a rod with cross-sectional area A, Young's modulus E and unit weight γ (Fig. 3.3). Now let the stress along section *aa* is σ_x and the stress on section *bb* is $\left(\sigma_x + \frac{\partial \sigma_x}{\partial x}\cdot\Delta x\right)$. Assuming that the stress is uniform over the entire cross-sectional area and the cross-section remains plane during the vibration, the summation of forces in x-direction is given by:

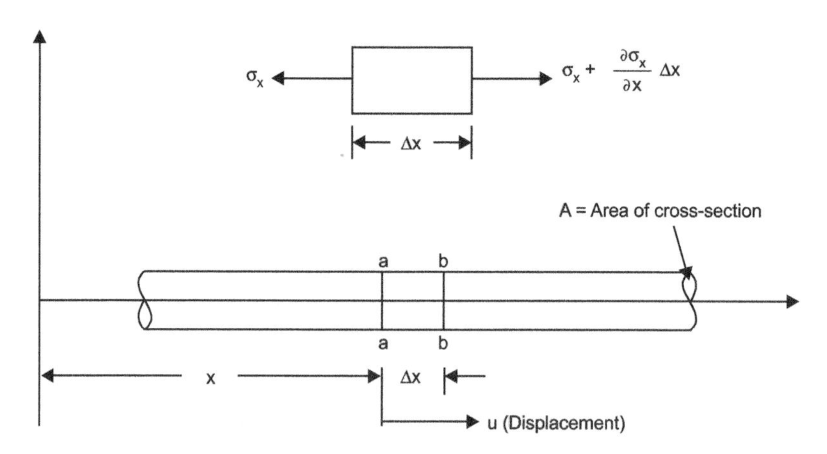

Fig. 3.3. Longitudinal vibration of a rod

$$\Sigma F_x = -\sigma_x A + \left(\sigma_x + \frac{\partial \sigma_x}{\partial x} \cdot \Delta x\right) A = \frac{\partial \sigma_x}{\partial x} \cdot \Delta x \cdot A \tag{3.11}$$

If the displacement of the element in x-direction is u, the equation of motion for element can be written by applying Newton's second law of motion as given below:

$$\frac{\partial \sigma_x}{\partial x} \cdot \Delta x \cdot A = \left(\Delta x \cdot A \cdot \frac{\gamma}{g}\right) \cdot \frac{\partial^2 u}{\delta t^2}$$

or

$$\frac{\partial \sigma_x}{\partial x} = \frac{\gamma}{g} \cdot \frac{\partial^2 u}{\partial t^2} \tag{3.12}$$

From stress-strain relationship $E = \dfrac{\text{stress in } x\text{-direction}}{\text{strain in } x\text{-direction}} = \dfrac{\sigma_x}{\partial u / \partial x}$

$$\sigma_x = E \cdot \frac{\partial u}{\partial x} \tag{3.13}$$

Differentiating Eq. (3.13) with respect to x

$$\frac{\partial \sigma_x}{\partial x} = E \frac{\partial^2 u}{\partial x^2} \tag{3.14}$$

Therefore,

$$\frac{\gamma \, \partial^2 u}{g \, \partial t^2} = E \frac{\partial^2 u}{\partial x^2} \tag{3.15}$$

Using the mass density $\rho = \dfrac{\gamma}{g}$ Eq. (3.15) can be written

$$\frac{\partial^2 u}{\partial t^2} = \frac{E}{\rho} \frac{\partial^2 u}{\partial x^2} \tag{3.16}$$

or

$$\frac{\partial^2 u}{\partial t^2} = v_c^2 \frac{\partial^2 u}{\partial x^2} \tag{3.17}$$

where

$$v_c^2 = \frac{E}{\rho} \tag{3.18}$$

v_c is defined as the longitudinal-wave-propagation-velocity in the rod. Equation (3.17) has the exact form of the wave equation, and it indicates that during longitudinal vibrations, displacement patterns are propagated in the axial direction at the velocity v_c.

The solution of Eq. (3.17) may be written in the form

$$u = f_1(v_c t + x) + f_2(v_c t - x) \tag{3.19}$$

where f_1 and f_2 are arbitrary functions. In this equation, the first term represents the wave travelling in the positive x direction and the second term represents the wave travelling in the negative x direction.

If the wave propagation in a rod is considered at some intermediate point in the bar, it can be noted that at the instant a wave is generated, there is compressive stress of the face in the positive direction of x and tensile stress in the negative direction of x. Hence, when the compressive wave travels in one direction, the tensile waves travel in the opposite direction.

To understand the difference between the wave propagation velocity v_c and the velocity of particles in the stressed zone $\dot{u}$, consider the stressed zone at the end of the bar as shown in Fig. 3.4a. When a uniformly distributed compressive stress pulse of intensity σ_x and duration t_n (Fig. 3.4b) is applied to the end of the bar, initially only a small zone of the rod will experience the compression. This compression will be transmitted to the successive zones of the bar as time increases. The transmission of the compressive stress from one zone to another occurs at the velocity of the wave propagated in the medium, i.e. v_c. During a time interval Δt, the compressive stress will travel along the bar a distance ($\Delta x = v_c \cdot \Delta t$). At any time after t_n, a segment of the bar of length, $x_n = v_c t_n$, constitutes the compressed zone. The amount of the elastic shortening of this zone is given by

$$u = \frac{\sigma_x}{E} x_n = \frac{\sigma_x}{E} v_c \cdot t_n \tag{3.20}$$

or

$$\frac{u}{t_n} = \frac{\sigma_x}{E} v_c \tag{3.21}$$

The displacement u divided by t_n represents the velocity of the end of the bar or particle velocity. Hence

$$\dot{u} = \frac{\sigma_x}{E} v_c \tag{3.22}$$

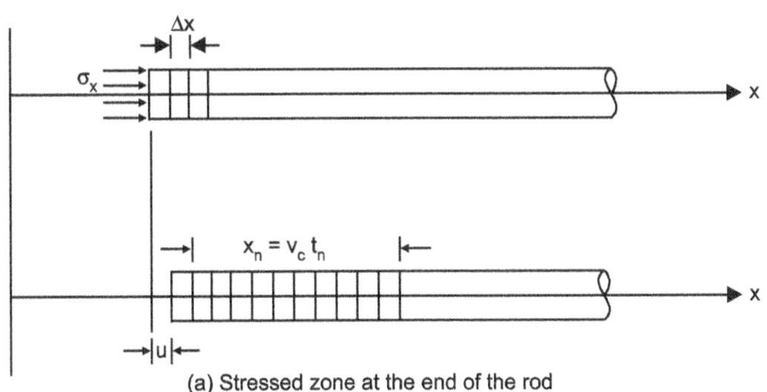

(a) Stressed zone at the end of the rod

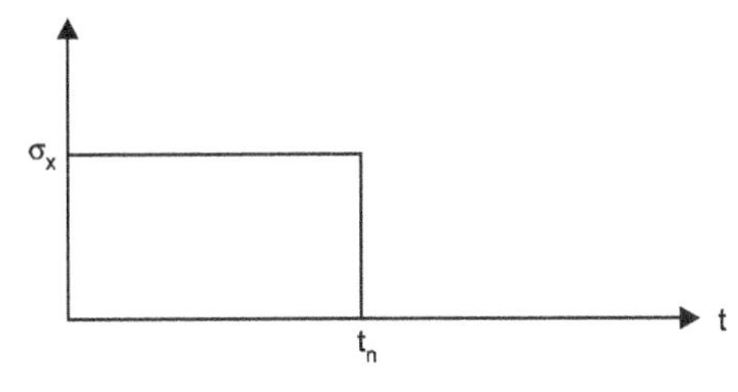

(b) Uniformly distributed compressive stress

Fig. 3.4. Wave propagation velocity and particle velocity in a rod

It is evident from Eq. (3.22), that the particle wave velocity $\dot{u}$ depends on the intensity of stress but the wave propagation velocity v_c (Eq. 3.18), is only a function of material properties. Further both wave propagation velocity and particle velocity are in the same direction when a compressive stress is applied but the wave propagation velocity is opposite to the particle velocity when a tensile stress is applied.

3.4 Torsional Vibration of a Rod of Infinite Length

In Fig. 3.5a, a rod subjected to a torque T which produces angular rotation θ is shown. The expression for the torque can be written as

$$T = GI_p \frac{\partial \theta}{\partial x} \tag{3.23}$$

where G = Shear modulus of the material of rod

I_p = Polar moment of inertia of the cross-section of rod

$\frac{\partial \theta}{\partial x}$ = Angle of twist per unit length of rod.

The torque due to rotational inertia of an element of rod of length Δx can be written as

$$T = \rho I_p \Delta x \frac{\partial^2 \theta}{\partial t^2} \tag{3.24}$$

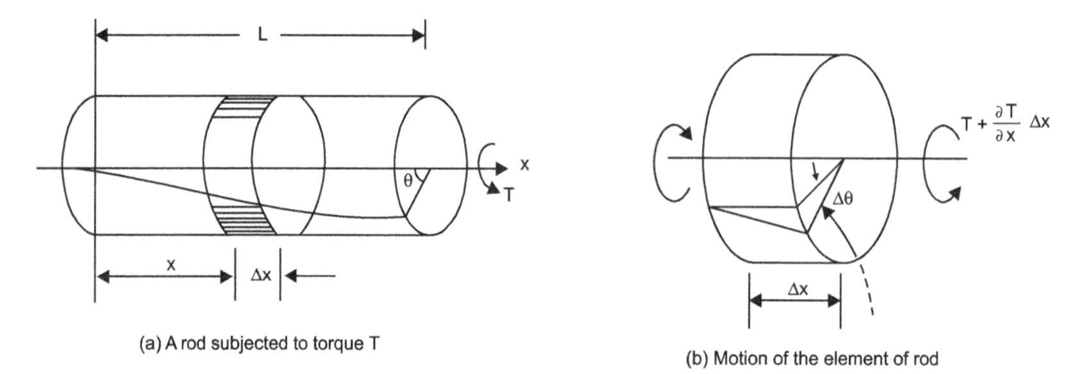

Fig. 3.5. Torsional vibration of a rod

By applying Newton's second law of motion to an element of length Δx shown in Fig. 3.5b,

$$-T + \left(T + \frac{\partial T}{\partial x}\Delta x\right) = \rho I_p \Delta x \frac{\partial^2 \theta}{\partial t^2}$$

or

$$\frac{\partial T}{\partial x} = \rho I_p \frac{\partial^2 \theta}{\partial t^2} \tag{3.25}$$

From Eq. (3.23),

$$\frac{\partial T}{\partial x} = GI_p \frac{\partial^2 \theta}{\partial x^2} \tag{3.26}$$

Therefore,

$$GI_p \frac{\partial^2 \theta}{\partial x^2} = \rho I_p \frac{\partial^2 \theta}{\partial t^2}$$

or

$$\frac{\partial^2 \theta}{\partial t^2} = \frac{G}{\rho} \frac{\partial^2 \theta}{\partial x^2} \tag{3.27}$$

$$\frac{\partial^2 \theta}{\partial t^2} = v_s^2 \frac{\partial^2 \theta}{\partial x^2} \tag{3.28}$$

v_s is the shear wave velocity of the material of the rod.

3.5 End Conditions

Free End Conditions. Consider an elastic rod in which a compression wave is travelling in the positive x-direction and an identical tension wave is travelling in the negative x-direction (Fig. 3.6a). When the two waves pass by each other in the crossover zone, the portion of the rod in which the two waves are superposed has zero stress with twice the particle velocity of either wave (Fig. 3.6b). After the two waves have passed the crossover zone the stress and velocity return to zero at the crossover point and both the compressive and tensile waves return to their initial shape and magnitude (Fig. 3.6c). It will thus be seen that on the centreline cross-section, the stress is zero at all time. This stress condition is the same as that which exists at the free end of the rod. By removing one-half of the rod, the centreline cross-section can be considered a free end (Fig. 3.6d). Hence it can be seen that a compression wave is reflected from a free end as a tension wave of the same magnitude and shape. Similarly, it can be observed that a tension wave is reflected from a free end as a compression wave of the same magnitude and shape.

Fixed End Conditions. Now consider an elastic rod in which a compression wave is travelling in positive x-direction and an identical compression wave is travelling in the negative x-direction (Fig. 3.7a). When the two waves pass by each other in the crossover zone, the centreline cross-section has stress equal to twice the stress in each wave and zero particle velocity (Fig. 3.7b). After the waves pass each other, they return to their original shape and magnitude. The centreline cross-section remains stationary during the entire process and hence, behaves like a fixed end of the rod. Considering left half of the rod (Fig. 3.7d), it can be observed that a compression wave is reflected from a fixed end of a rod as a compression wave of the same magnitude and shape, and that at the fixed end the stress is doubled.

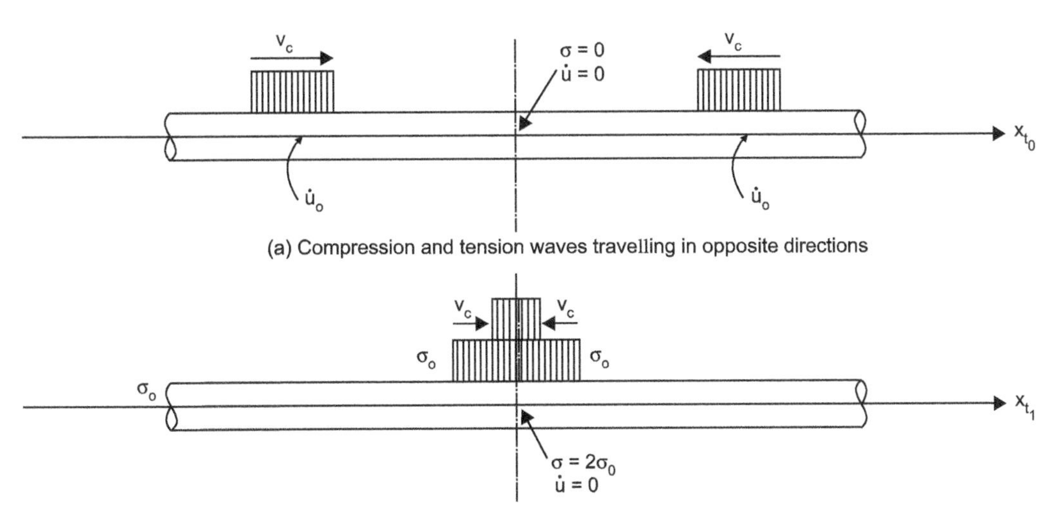

(a) Compression and tension waves travelling in opposite directions

(b) Waves at the crossover zone

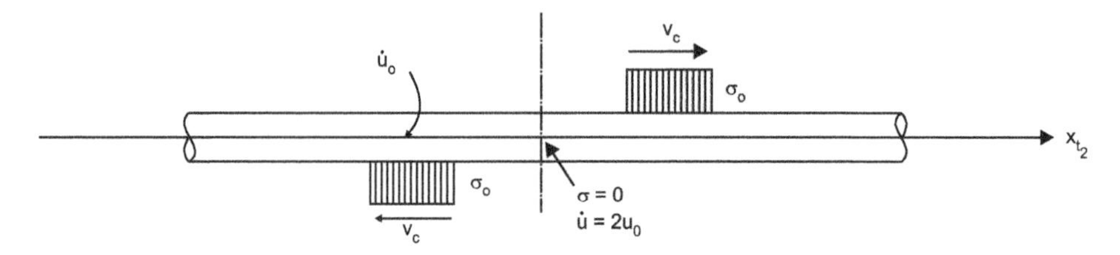

(c) Waves after passing the crossover zone

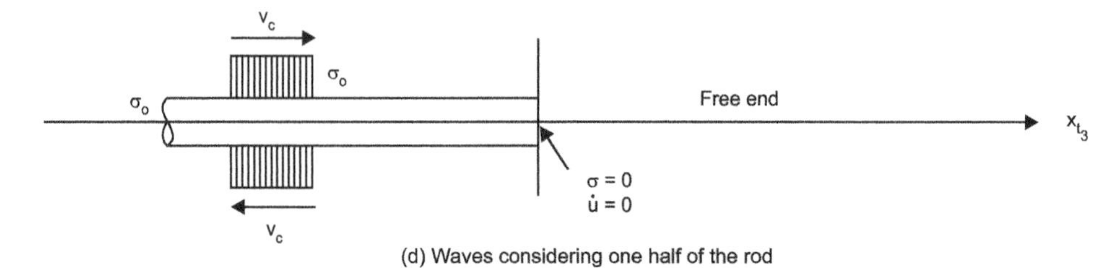

(d) Waves considering one half of the rod

Fig. 3.6. Elastic waves in a rod with free end conditions

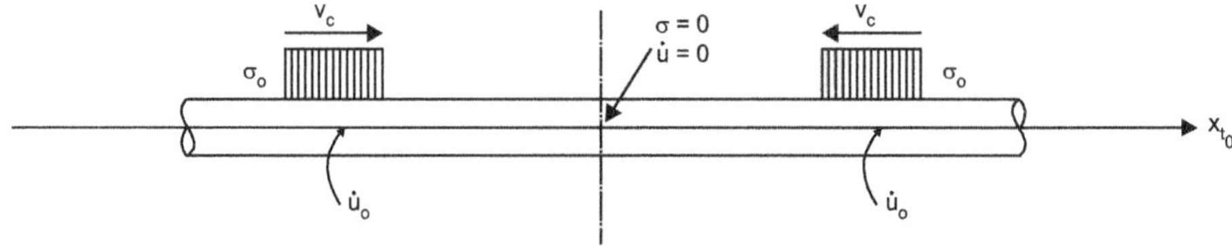

(a) Two identical waves travelling in opposite directions

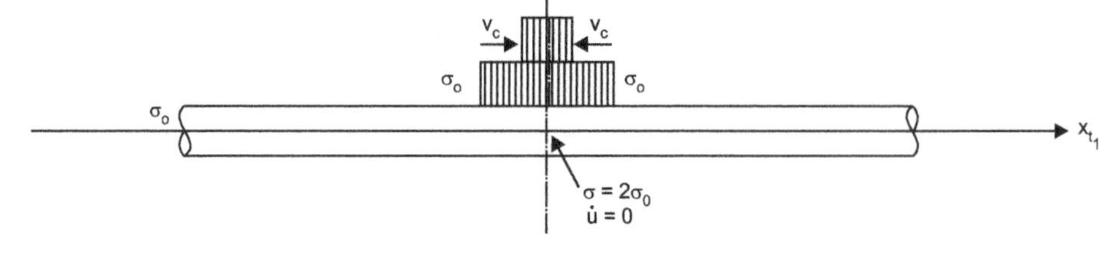

(b) Waves at the crossover zone

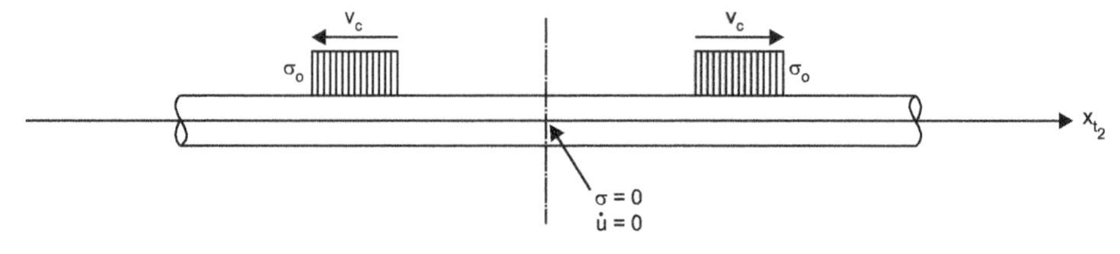

(c) Waves after passing the crossover zone

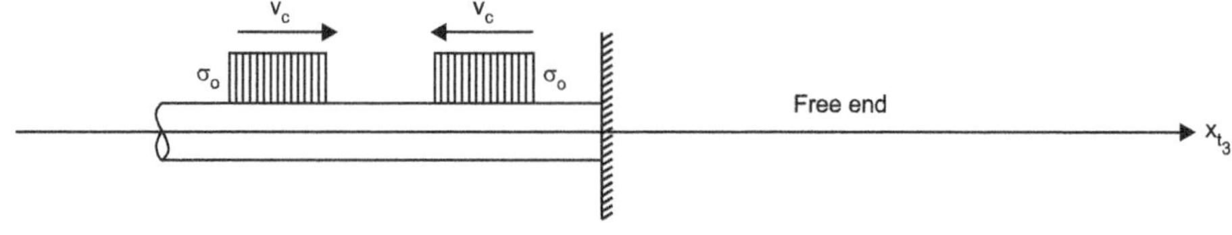

(d) Waves considering one half of the rod

Fig. 3.7. Elastic waves in a rod with fixed end conditions

3.6 Longitudinal Vibrations of Rods of Finite Length

Consider a rod of length L vibrating in one of its normal mode. The solution of the wave equation (Eq. 3.17) can be written as

$$u = U(A_1 \cos \omega_n t + A_2 \sin \omega_n t) \tag{3.29}$$

where U = Displacement amplitude along the length of rod
A_1, A_2 = Arbitrary constants
ω_n = Natural frequency of the rod

Substituting Eq. (3.29) in Eq. (3.17), we get

$$\frac{d^2U}{dx^2} + \frac{\omega_n^2}{v_c^2} U = 0 \tag{3.30}$$

The solution of Eq. (3.30) is $$U = A_3 \cos \frac{\omega_n x}{v_c} + A_4 \sin \frac{\omega_n x}{v_c} \tag{3.31}$$

A_3 and A_4 are arbitrary constants which are determined by satisfying the boundary conditions at the ends of the rod. Three possible end conditions are:

1. Both ends free (free-free)
2. One end fixed and one end free (fixed-free)
3. Both ends fixed (fixed-fixed)

Free-Free Condition

The stress and strain at both ends of a rod of finite length in free-free condition (Fig. 3.8a) will be zero. This means that $dU/dx = 0$ at $x = 0$ and $x = L$.

Differentiating Eq. (3.31) w.r.t. x, we get

$$\frac{dU}{dx} = \frac{\omega_n}{v_c}\left(-A_3 \sin \frac{\omega_n x}{v_c} + A_4 \cos \frac{\omega_n x}{v_c} \right) \tag{3.32}$$

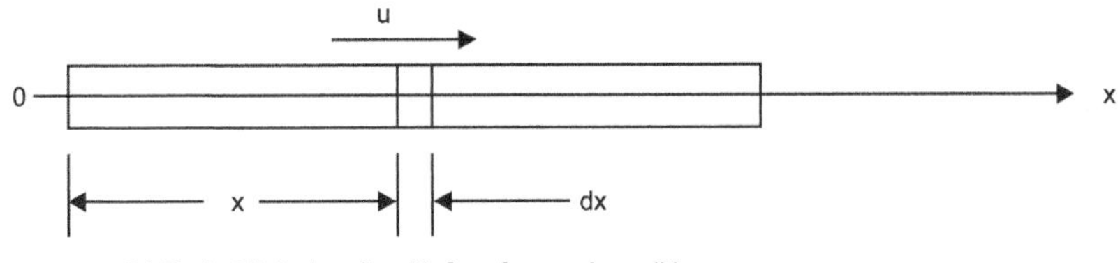

(a) Rod of finite length with free-free end conditions

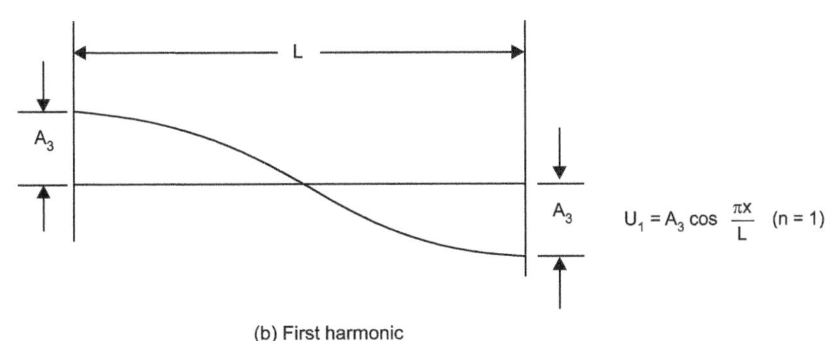

(b) First harmonic

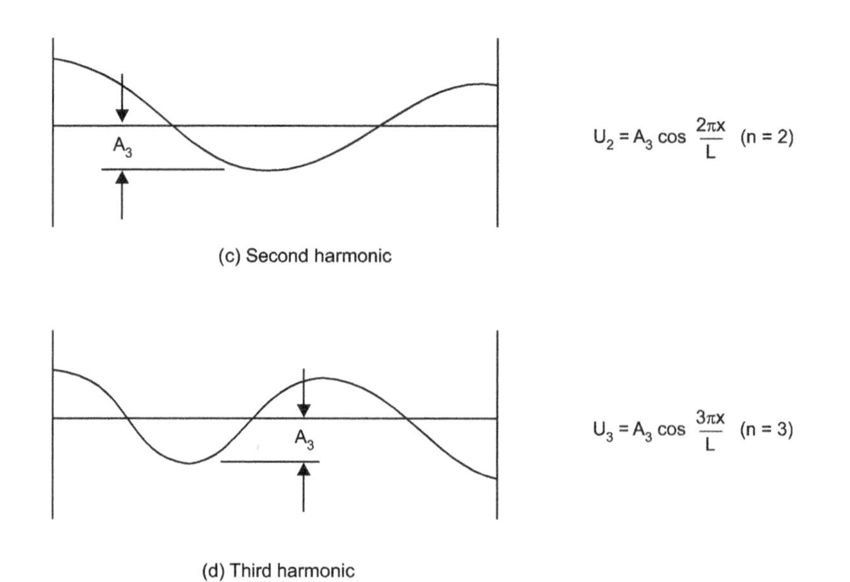

(c) Second harmonic

(d) Third harmonic

Fig. 3.8. Normal modes of vibration of a rod of finite length with free-free end conditions

The condition $\frac{dU}{dx} = 0$ at $x = 0$ gives $A_4 = 0$

Putting $\frac{dU}{dx} = 0$ at $x = 1$, we get

$$A_3 \sin\frac{\omega_n L}{v_c} = 0 \tag{3.33}$$

For a nontrivial solution, $\frac{\omega_n L}{v_c} = n\,\pi$

or
$$\omega_n = \frac{n\pi v_c}{L},\ n = 1, 2, 3... \tag{3.34}$$

Equation (3.34) is the frequency equation for the rod in free-free case. By substituting Eq. (3.34) in Eq. (3.31), we get

$$U_n = A_3 \cos\frac{n\pi x}{L} \tag{3.35}$$

Using Eq. (3.35), the distribution of displacement along the rod can be found for any harmonic. The first three harmonics are shown in Figs. 3.8 b, c and d.

Fixed-Free Condition

In Fixed-free case (Fig. 3.9a), the end conditions of the rod are:

(i) At $x = 0$, Displacement i.e. $U = 0$, and

(ii) At $x = L$, Strain i.e. $dU/dx = 0$

Putting the first end condition in Eq. 3.31, yields $A_3 = 0$. By substituting the second end condition in Eq. 3.32,

$$A_4 \cos\frac{\omega_n L}{v_c} = 0 \tag{3.36}$$

or
$$\frac{\omega_n L}{v_c} = (2n-1)\frac{\pi}{2}$$

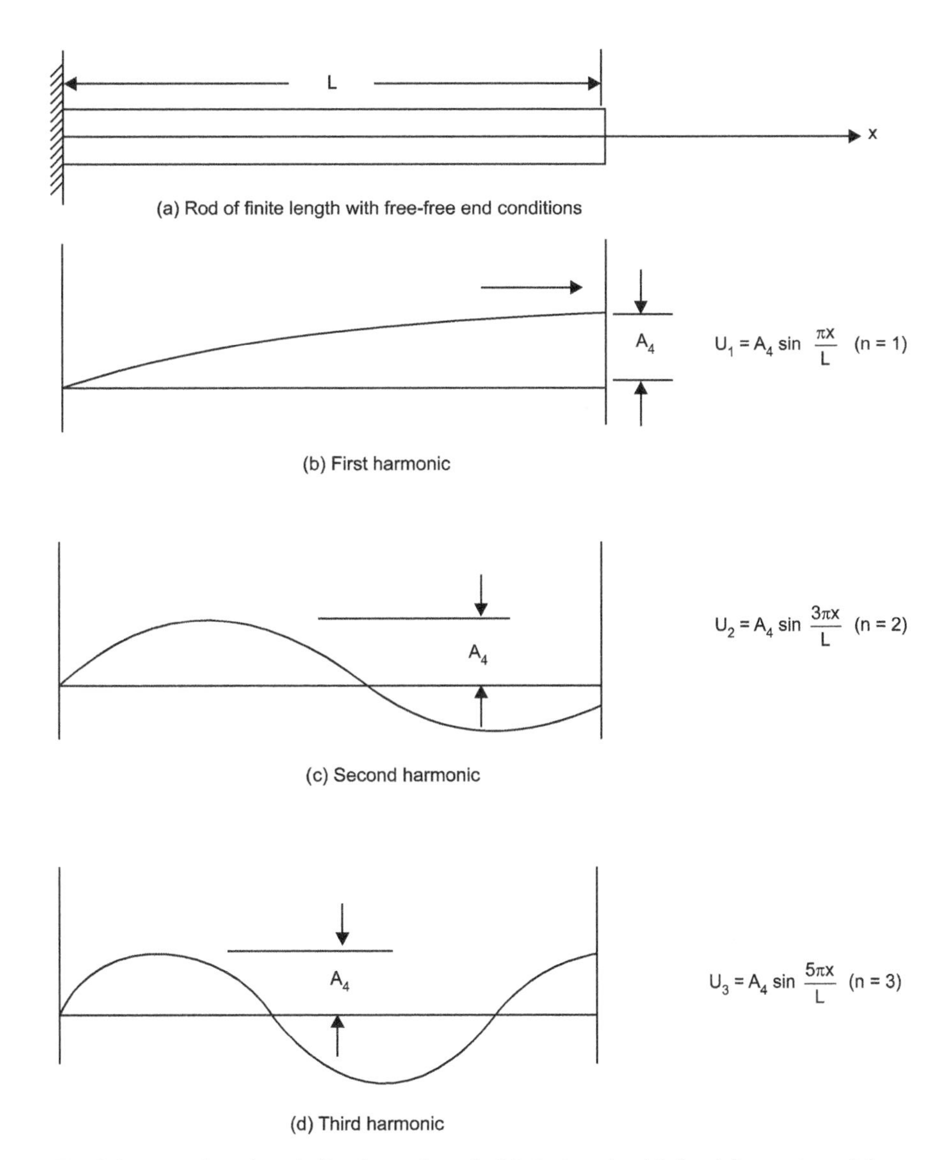

Fig. 3.9. Normal modes of vibrations of a rod of finite length with fixed-free end conditions

or
$$\omega_n = (2n-1)\frac{\pi v_c}{2L} \tag{3.37}$$

The displacement amplitude can be written as

$$U_n = A_4 \sin\frac{(2n-1)\pi x}{2L} \tag{3.38}$$

The first three harmonics described by Eq. (3.38) are shown in Figs. 3.9b, c and d.

Fixed-Fixed Condition

In this case (Fig. 3.10a) the end conditions are:

(i) At $x = 0$, $U = 0$ and

(ii) At $x = L$, $U = 0$

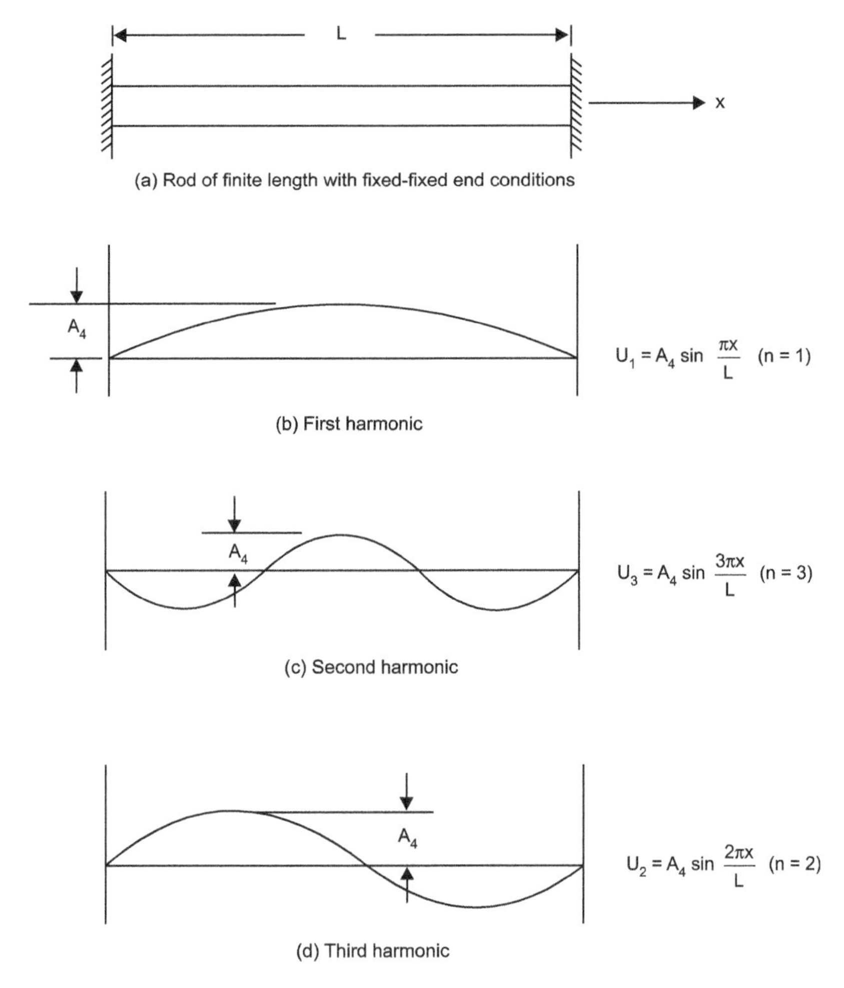

Fig. 3.10. Normal modes of vibration of a rod of finite length with fixed-fixed end conditions

Putting these end conditions in Eq. (3.31), we get,

$$A_3 = 0 \text{ and}$$

$$A_4 \sin \frac{\omega_n L}{v_c} = 0 \tag{3.39}$$

or

$$\omega_n = \frac{n\pi x}{L}, \; n = 1, 2, 3, \ldots \tag{3.40}$$

The displacement is given by $\quad U_n = A_4 \sin \dfrac{n\pi x}{L} \quad (3.41)$

The first three harmonics described by Eq. (3.41) are shown in Figs. 3.10b, c and d.

3.7 Torsional Vibrations of Rods of Finite Length

As the wave Eq. (3.28) is identical to wave Eq. (3.17), the problem of torsional vibrations of rods of finite length can be solved in the same manner as for the case of longitudinal vibrations discussed in the previous section. The solution of Eq. (3.28) can be written as:

$$\theta = \theta_A(A_1 \cos \omega_n t + A_2 \sin \omega_n t) \tag{3.42}$$

where θ_A = Rotational amplitude of angular vibration

A_1, A_2 = Arbitrary constants

ω_n = Natural frequency of the rod

Substituting Eq. (3.42) in Eq. (3.28), we get

$$\frac{d^2\theta_A}{dx^2} + \frac{\omega_n^2}{v_s^2}\theta_A = 0 \tag{3.43}$$

The solution of Eq. (3.43) is

$$\theta_A = A_3 \cos\frac{\omega_n x}{v_s} + A_4 \sin\frac{\omega_n x}{v_s} \tag{3.44}$$

The values of arbitrary constants A_3 and A_4 can be obtained by putting appropriate end conditions. The solution for the three types of end conditions are given below.

Free-Free Condition

$$\omega_n = \frac{n\pi v_s}{L} \tag{3.45}$$

$$\theta_{An} = A_3 \cos\frac{n\pi x}{L} \tag{3.46}$$

Fixed-Free Condition

$$\omega_n = \frac{(2n-1)\pi v_s}{2L} \tag{3.47}$$

$$\theta_{An} = A_4 \sin\frac{(2n-1)\pi x}{2L} \tag{3.48}$$

Fixed-Fixed Condition

$$\omega_n = \frac{n\pi v_s}{L} \tag{3.49}$$

$$\theta_{An} = A_4 \frac{\sin n\pi x}{L} \tag{3.50}$$

3.8 Wave Propagation in an Infinite, Homogeneous, Isotropic, and Elastic Medium

In this section, the propagation of stress waves in an infinite, homogeneous, isotropic, elastic medium is presented. Figure 3.11 shows the stresses acting on a soil element with sides measuring *dx*, *dy* and *dz*. The solid vectors are acting on the visible faces of the element and the dotted vectors are acting on the hidden faces. For obtaining the differential equations of motion, the sum of the forces acting parallel to each axis is considered. In the *x*-direction the equilibrium equation is

$$\left[\sigma_x - \left(\sigma_x + \frac{\partial \sigma_x}{\partial x}dx\right)\right](dy \cdot dz) + \left[\tau_{zx} - \left(\tau_{zx} + \frac{\partial \tau_{zx}}{\partial z}\right)dz\right](dx \cdot dy)$$

$$+\left[\tau_{yx}-\left(\tau_{yx}+\frac{\partial \tau_{xy}}{\partial y}dy\right)\right](dx\cdot dz)+\rho(dx\cdot dy\cdot dz)\frac{\partial^2 u}{\partial t^2}=0 \tag{3.51}$$

or

$$\rho\frac{\partial^2 u}{\partial \tau^2}=\frac{\partial \sigma_x}{\partial x}+\frac{\partial \tau_{xy}}{\partial y}+\frac{\partial \tau_{xz}}{\partial z} \tag{3.52a}$$

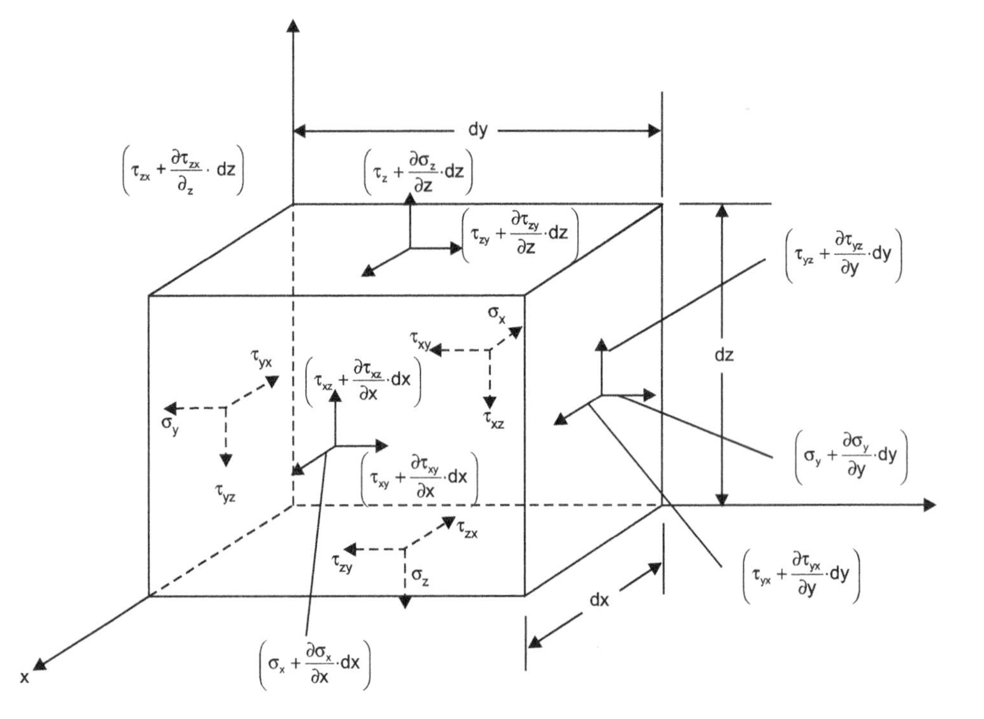

Fig. 3.11. Stress on an element of an infinite elastic medium

Equations similar to Eq. (3.51) can be written for the *y*- and *z*-directions. These will give

$$\rho\frac{\partial^2 v}{\partial t^2}=\frac{\partial \tau_{yx}}{\partial x}+\frac{\partial \sigma_y}{\partial y}+\frac{\partial \tau_{yz}}{\partial z} \tag{3.52b}$$

$$\rho\frac{\partial^2 w}{\partial t^2}=\frac{\partial \tau_{zx}}{\partial x}+\frac{\partial \tau_{zy}}{\partial y}+\frac{\partial \sigma_z}{\partial z} \tag{3.52c}$$

In the above expressions, ρ is the mass density of the soil, u, v and w are displacements in the x, y and z directions respectively. To express the right hand sides of Eqs. (3.52), the relationship for an elastic medium given by Eqs. (3.6) to Eqs. (3.10) are used. The equations for strains and rotations of elastic and isotropic materials in terms of displacements are as follows:

Axial strains

$$\varepsilon_x=\frac{\partial u}{\partial x} \tag{3.53a}$$

$$\varepsilon_y=\frac{\partial v}{\partial y} \tag{3.53b}$$

$$\varepsilon_z=\frac{\partial w}{\partial z} \tag{3.53c}$$

Shear Strains

$$\gamma_{xy} = \frac{\partial v}{\partial x} + \frac{\partial u}{\partial y} \tag{3.54a}$$

$$\gamma_{yz} = \frac{\partial w}{\partial y} + \frac{\partial v}{\partial z} \tag{3.54b}$$

$$\gamma_{zx} = \frac{\partial u}{\partial z} + \frac{\partial w}{\partial x} \tag{3.54c}$$

Rotations

$$2\bar{w}_x = \frac{\partial w}{\partial y} - \frac{\partial v}{\partial z} \tag{3.55a}$$

$$2\bar{w}_y = \frac{\partial u}{\partial z} - \frac{\partial w}{\partial x} \tag{3.55b}$$

$$2\bar{w}_z = \frac{\partial v}{\partial x} - \frac{\partial u}{\partial y} \tag{3.55c}$$

In Eqs. (3.55), $\bar{w}_x, \bar{w}_y$ and $\bar{w}_z$ represent the rotations about x, y and z axes respectively.

3.8.1 Compression Waves

Substitution of Eqs. (3.6a), (3.10a), and (3.10b) into Eq. (3.52a) gives

$$\rho\frac{\partial^2 u}{\partial t^2} = \frac{\partial}{\partial x}(\lambda\bar{\varepsilon} + 2G\varepsilon_x) + \frac{\partial}{\partial y}(G\gamma_{xy}) + \frac{\partial}{\partial z}(G\gamma_{zx}) \tag{3.56}$$

Now on substitution of Eqs. (3.54a) and (3.54b) in Eq. (3.56), we get

$$\rho\frac{\partial^2 u}{\partial t^2} = \frac{\partial}{\partial x}(\lambda\bar{\varepsilon} + 2G\varepsilon_x) + G\frac{\partial}{\partial y}\left(\frac{\partial v}{\partial x} + \frac{\partial u}{\partial y}\right) + G\frac{\partial}{\partial z}\left(\frac{\partial u}{\partial z} + \frac{\partial w}{\partial x}\right)$$

or

$$\rho\frac{\partial^2 u}{\partial t^2} = \lambda\frac{\partial\bar{\varepsilon}}{\partial x} + G\left[\frac{\partial^2 u}{\partial x^2} + \frac{\partial^2 v}{\partial x\cdot\partial y} + \frac{\partial^2 w}{\partial x\cdot\partial z} + \frac{\partial^2 u}{\partial x^2} + \frac{\partial^2 u}{\partial y^2} + \frac{\partial^2 u}{\partial z^2}\right] \tag{3.57}$$

As

$$\frac{\partial^2 u}{\partial x^2} + \frac{\partial^2 v}{\partial x\cdot\partial y} + \frac{\partial^2 w}{\partial x\cdot\partial z} = \frac{\partial\bar{\varepsilon}}{\partial x} \tag{3.58}$$

Equation (3.57) can be written as

$$\rho\frac{\partial^2 u}{\partial t^2} = (\lambda + G)\frac{\partial\bar{\varepsilon}}{\partial x} + G\nabla^2 u \tag{3.59}$$

where

$$\nabla^2 u = \frac{\partial^2 u}{\partial x^2} + \frac{\partial^2 u}{\partial y^2} + \frac{\partial^2 u}{\partial z^2} \tag{3.60}$$

Similarly Eqs. (3.52b) and (3.52c) can be expressed as

$$\rho\frac{\partial^2 v}{\partial t^2} = (\lambda + G)\frac{\partial\bar{\varepsilon}}{\partial y} + G\nabla^3 v \tag{3.61}$$

and

$$\rho\frac{\partial^2 w}{\partial t^2} = (\lambda + G)\frac{\partial\bar{\varepsilon}}{\partial z} + G\nabla^2 w \tag{3.62}$$

Equations (3.59), (3.61) and (3.62) are the equations of motion of an infinite homogeneous, isotropic, and elastic medium. On differentiating these equations with respect to x, y and z respectively, and adding

$$\rho\frac{\partial^2}{\partial t^2}\left(\frac{\partial u}{\partial x}+\frac{\partial v}{\partial y}+\frac{\partial w}{\partial z}\right) = (\lambda+G)\left(\frac{\partial^2\bar{\varepsilon}}{\partial x^2}+\frac{\partial^2\bar{\varepsilon}}{\partial y^2}+\frac{\partial^2\bar{\varepsilon}}{\partial z^2}\right)+G\nabla^2\left(\frac{\partial u}{\partial x}+\frac{\partial v}{\partial y}+\frac{\partial w}{\partial z}\right)$$

or

$$\rho\frac{\partial^2\bar{\varepsilon}}{\partial t^2} = (\lambda+G)(\nabla^2\bar{\varepsilon})+(G\nabla^2\bar{\varepsilon})$$

Hence

$$\rho\frac{\partial^2\bar{\varepsilon}}{\partial t^2} = (\lambda+2G)\nabla^2\bar{\varepsilon} \tag{3.63a}$$

or

$$\frac{\partial^2\bar{\varepsilon}}{\partial t^2} = \frac{(\lambda+2G)}{\rho}(\nabla^2\bar{\varepsilon}) = v_p^2\nabla^2\bar{\varepsilon} \tag{3.63b}$$

where

$$v_p^2 = \frac{\lambda+2G}{\rho} \tag{3.64}$$

v_p is the velocity of compression waves which are also referred as primary wave or P-wave.

It is important to note the difference in the wave velocities for an infinite elastic medium with those obtained for an elastic rod. In the rod $v_c=\sqrt{E/\rho}$ but in the infinite medium $v_p=\sqrt{(\lambda+2G)/\rho}$. This means that $v_p > v_c$ that is compression wave travels faster in infinite medium. It is due to the fact that in infinite medium, there are no lateral displacements, while in the rod lateral displacements are possible.

3.8.2 Shear Waves

Differentiating Eq. (3.61) with respect to z and Eq. (3.62) with respect to y, we get

$$\rho\frac{\partial^2}{\partial t^2}\left(\frac{\partial v}{\partial z}\right) = (\lambda+G)\frac{\partial\bar{\varepsilon}}{(\partial y)(\partial z)}+G\nabla^2\frac{\partial v}{\partial z} \tag{3.65}$$

and

$$\rho\frac{\partial^2}{\partial t^2}\left(\frac{\partial w}{\partial y}\right) = (\lambda+G)\frac{\partial\bar{\varepsilon}}{\partial y\partial z}+G\nabla^2\frac{\partial w}{\partial y} \tag{3.66}$$

Subtracting Eq. (3.65) from Eq. (3.66), we get

$$\rho\frac{\partial^2}{\partial t^2}\left(\frac{\partial w}{\partial y}-\frac{\partial v}{\partial z}\right) = G\nabla^2\left(\frac{\partial w}{\partial y}-\frac{\partial v}{\partial z}\right) \tag{3.67}$$

From Eq. (3.55a), $\frac{\partial w}{\partial y}-\frac{\partial v}{\partial z} = 2\bar{w}_x$. Therefore

$$\rho\frac{\partial^2\bar{w}}{\partial t^2} = G\nabla^2\bar{w}_x$$

or

$$\frac{\partial^2\bar{w}_x}{\partial t^2} = \frac{G}{\rho}\nabla^2\bar{w}_x = v_s^2\nabla^2\bar{w}_x \tag{3.68}$$

Similar expressions can be obtained for $\bar{w}_y$ $\bar{w}$and $\bar{w}_z$ as below:

$$\frac{\partial^2\bar{w}_y}{\partial t^2} = \frac{G}{\rho}\nabla^2\bar{w}_y = v_s^2\nabla^2\bar{w}_y \tag{3.69}$$

$$\frac{\partial^2 \bar{w}_z}{\partial t^2} = \frac{G}{\rho}\nabla^2 \bar{w}_z = v_s^2 \nabla^2 \bar{w}_z \tag{3.70}$$

The above expressions indicate that the rotation is propagated with velocity v_s which is equal to $\sqrt{G/\rho}$. Shear wave is also referred as distortion wave or s-wave. It may be noted that shear wave propagates at the same velocity in both the rod and the infinite medium. Figure 3.12 shows plots of shear wave velocity and void ratio at several confining pressures for sands (Hardin and Richart, 1963).

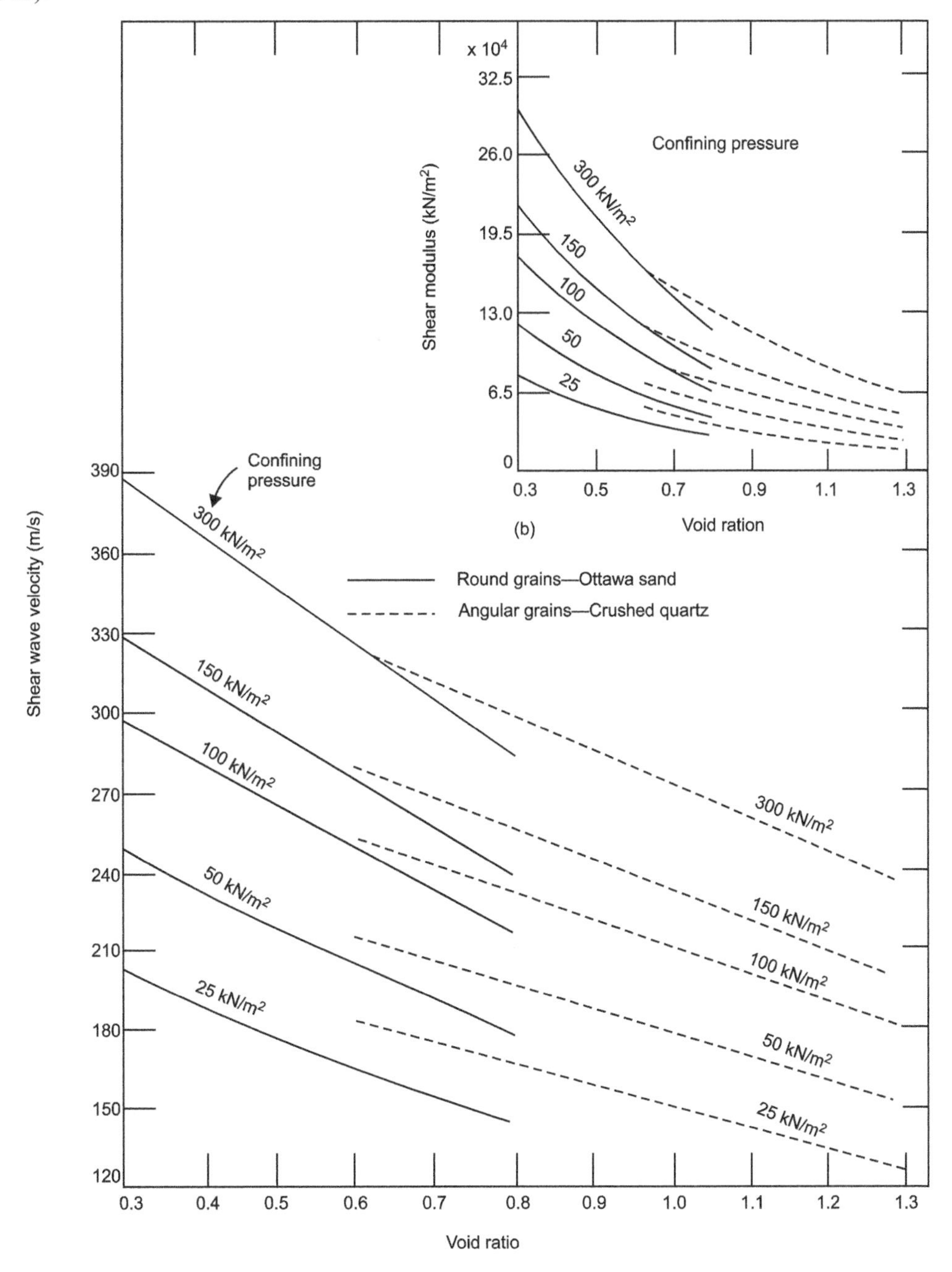

Fig. 3.12. Variation of shear wave velocity and shear modulus with void ratio and confining pressure for dry sands (Hardin and Richart, 1963)

3.9 Wave Propagation in Elastic Half Space

In an elastically homogeneous ground, stressed suddenly at a point 'S' near its surface (Fig. 3.13), three elastic waves travel outwards at different speeds. Two are body waves; that is, they are propagated as spherical fronts affected only a minor extent by the free surface of the ground, and the third is a surface wave which is confined to the region near this free surface.

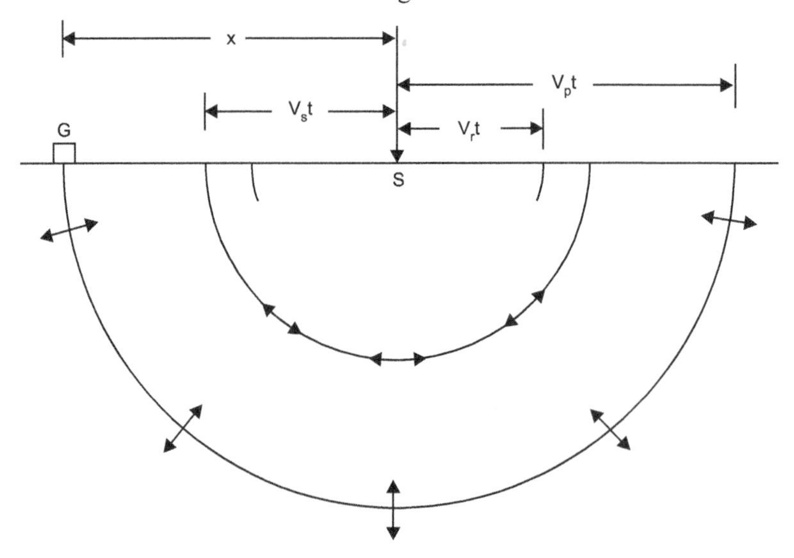

Fig. 3.13. Pulse fronts of the P, S and R waves

The two body waves as already described in the previous section differ in that the ground motion within the pulse is in the direction of propagation (i.e. radial) is the faster 'P' (primary) wave, but normal to it (i.e. tangential to the pulse front) is the slower 'S' (secondary) wave (Fig. 3.13). The stresses in the P wave, which is a longitudinal wave like a sound wave in air, are thus due to uniaxial compression, while during the passage of an S wave the medium is subjected to shear stress. The surface wave travels more slowly than either body wave, and is generally complex. This wave was first studied by Rayleigh (1885) and later was described in detail by Lamb (1904). It is referred as Rayleigh wave or R-wave. The influence of Raleigh wave decreases rapidly with depth.

Equations (3.59), (3.61) and (3.62) may be used to study the characteristics of Rayleigh wave. The half space is defined as the *x*-*y* plane with *z* assumed to be positive toward the interior of the half-space (Fig. 3.14). Let *u* and *w* represent the displacements in the directions *x* and *z*, respectively and are independent of *y*, then

$$u = \frac{\partial \phi}{\partial x} + \frac{\partial \psi}{\partial z} \tag{3.71}$$

$$w = \frac{\partial \phi}{\partial z} - \frac{\partial \psi}{\partial x} \tag{3.72}$$

where ϕ and ψ are two potential functions. As $\frac{\partial v}{\partial y} = 0$, the dilation $\bar{\varepsilon}$ of the wave can be written as

$$\bar{\varepsilon} = \frac{\partial u}{\partial x} + \frac{\partial w}{\partial z} = \left(\frac{\partial^2 \phi}{\partial x^2} + \frac{\partial^2 \psi}{\partial x \partial z}\right) + \left(\frac{\partial^2 \phi}{\partial z^2} - \frac{\partial^2 \psi}{\partial x \partial z}\right)$$

or

$$\bar{\varepsilon} = \frac{\partial \phi}{\partial x^2} + \frac{\partial^2 \phi}{\partial z^2} = \nabla^2 \phi \tag{3.73}$$

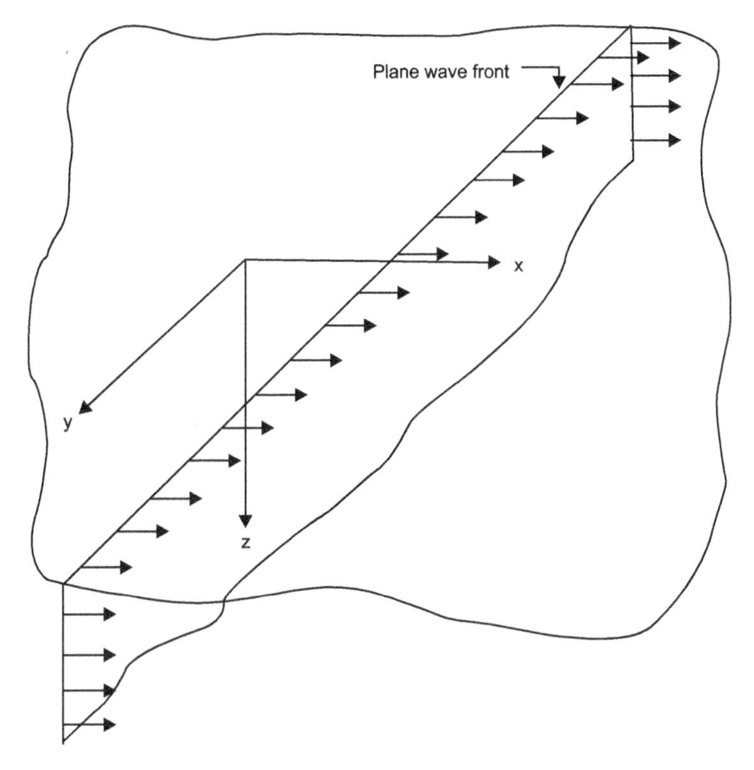

Fig. 3.14. Wave propagation in elastic half space

Similarly the rotation in *x-z* plane is given by

$$2\bar{w}_y = \frac{\partial u}{\partial z} - \frac{\partial w}{\partial x} = \frac{\partial^2 \psi}{\partial x^2} + \frac{\partial^2 \psi}{\partial z^2} = \nabla^2 \psi \tag{3.74}$$

Substituting u and w from Eqs. (3.71) and (3.72) in Eqs. (3.59) and (3.62), we get

$$\rho \frac{\partial}{\partial x}\left(\frac{\partial^2 \phi}{\partial t^2}\right) + \rho \frac{\partial}{\partial z}\left(\frac{\partial^2 \psi}{\partial t^2}\right) = (\lambda + 2G)\frac{\partial}{\partial x}(\nabla^2 \phi) + G\frac{\partial}{\partial z}(\nabla^2 \phi) \tag{3.75}$$

and

$$\rho \frac{\partial}{\partial z}\left(\frac{\partial^2 \phi}{\partial t^2}\right) - \rho \frac{\partial}{\partial x}\left(\frac{\partial^2 \psi}{\partial t^2}\right) = (\lambda + 2G)\frac{\partial}{\partial z}(\nabla^2 \phi) - G\frac{\partial}{\partial x}(\nabla^2 \psi) \tag{3.76}$$

Equations (3.75) and (3.76) are satisfied if

$$\frac{\partial^2 \phi}{\partial t^2} = \frac{\lambda + 2G}{\rho}\nabla^2 \phi = v_p^2 \nabla^2 \phi \tag{3.77}$$

and

$$\frac{\partial^2 \psi}{\partial t^2} = \frac{G}{\rho}\nabla^2 \psi = v_s^2 \nabla^2 \psi \tag{3.78}$$

Now, consider a sinusoidal wave travelling in positive x direction. Let the solutions of ϕ and ψ be expressed as

$$\phi = F(z)\, e^{i(\omega t - nx)} \tag{3.79}$$

$$\psi = G(z) e^{i(\omega t - nx)} \tag{3.80}$$

$F(z)$ and $G(z)$ are the functions which describe the variation in amplitude of the wave with depth, and n is the wave number given by

$$n = \frac{2\pi}{L} \tag{3.81}$$

where L is the wave length.

Substituting Eq. (3.79) in Eq. (3.77), and Eq. (3.80) in Eq. (3.78), we get

$$-\omega^2 F(z) = v_p^2[F''(Z) - n^2 F(z)] \tag{3.82}$$

$$-\omega^2 G(z) = v_p^2[G''(Z) - n^2 G(z)] \tag{3.83}$$

Equations (3.82) and (3.83) can be rearranged as

$$F''(z) - q^2 F(z) = 0 \tag{3.84}$$

$$G''(z) - s^2 F(z) = 0 \tag{3.85}$$

where

$$q^2 = n^2 - \frac{\omega^2}{v_p^2} \tag{3.86}$$

$$s^2 = n^2 - \frac{\omega^2}{v_s^2} \tag{3.87}$$

The solution of Eqs. (3.84) and (3.85) can be expressed in the form

$$F(z) = A_1 e^{-qz} + A_2 e^{qz} \tag{3.88}$$

$$G(z) = B_1 e^{-sz} + B_2 e^{sz} \tag{3.89}$$

where A_1, A_2, B_1 and B_2 are constants.

A solution that allows the amplitude of the wave to become infinity is not possible; therefore $A_2 = B_2 = 0$. Hence Eqs. (3.79) and (3.80) can be written as

$$\phi = A_1 e^{[-qz + i(\omega t - nx)]} \tag{3.90}$$

$$\psi = B_1 e^{[-sz + i(\omega t - nx)]} \tag{3.91}$$

Now at the surface of the half space i.e. at $z = 0$, σ_z, τ_{zx} and τ_{zy} are equal to zero

Therefore,

$$\sigma_z = \lambda\bar{\varepsilon} + 2G\varepsilon_z = \lambda\bar{\varepsilon} + 2G\frac{\partial w}{\partial z} = 0 \tag{3.92}$$

and

$$\tau_{zx} = G\gamma_{zx} = G\left(\frac{\partial w}{\partial x} + \frac{\partial u}{\partial z}\right) = 0 \tag{3.93}$$

Combining Eqs. (3.71) and (3.72) and the solutions of ϕ and ψ from Eqs. (3.90) and (3.91), Eqs. (3.92) and (3.93) can be written as

$$\frac{A_1}{B_1} = \frac{2iGns}{(\lambda + 2G)q^2 - \lambda n^2} \tag{3.94}$$

and

$$\frac{A_1}{B_1} = \frac{-(n^2 + s^2)}{2inq} \tag{3.95}$$

Equating the right hand sides of Eqs. (3.94) and (3.95)

$$16 G^2 n^4 s^2 q^2 = (s^2 + n^2)^2 [(\lambda + 2G) q^2 - \lambda n^2]^2 \tag{3.96}$$

Substituting q and s from Eqs. (3.86) and (3.87) in Eq. (3.96), and dividing both sides by $G^2 n^8$, we get

$$16\left(1-\frac{\omega^2}{v_p^2 n^2}\right)\left(1-\frac{\omega^2}{v_s^2 n^2}\right) = \left(2-\frac{(\lambda+2G)}{G}\cdot\frac{\omega^2}{v_p^2 n^2}\right)^2\left(2-\frac{\omega^2}{v_s^2 n^2}\right) \tag{3.97}$$

From Eq. (3.81)

$$\text{Wave length} = \frac{2\pi}{n} = \frac{\text{velocity of wave}}{(\omega/2\pi)} = \frac{v_r}{(\omega/2\pi)} \tag{3.98}$$

or
$$n = \frac{\omega}{v_r} \tag{3.99}$$

where v_r is the Rayleigh wave velocity.

Using the following relationships:

$$\frac{\omega^2}{v_p^2 n^2} = \frac{\omega^2}{v_s^2(\omega^2/v_r^2)} = \frac{v_r^2}{v_p^2} = \alpha^2\beta^2 \tag{3.100}$$

and
$$\frac{\omega^2}{v_s^2 n^2} = \frac{\omega^2}{v_s^2(\omega^2/v_r^2)} = \frac{v_r^2}{v_s^2} = \beta^2 \tag{3.101}$$

where
$$\alpha^2 = \frac{v_s^2}{v_p^2} \tag{3.102a}$$

$$= \frac{G/\rho}{(\lambda+2G)/\rho}$$

or
$$\alpha^2 = \frac{G}{(\lambda+2G)} \tag{3.102b}$$

Using the relations (3.8) and (3.9), Eq. (3.102) can be written as

$$\alpha^2 = \frac{1-2\mu}{2-2\mu} \tag{3.103}$$

Sustituting Eqs. (3.100), (3.101) and (3.103) in Eq. (3.97),

$$16(1-\alpha^2\beta^2)\ (1-\beta^2) = (2-\beta^2)^2\,(2-\beta^2)^2 \tag{3.104}$$

or
$$\beta^6 - 8\beta^4 - (16\alpha^2 - 24)\,\beta^2 - 16\,(1-\alpha^2) = 0 \tag{3.105}$$

Equation (3.105) is cubic in β^2. For a given value of Poisson's ratio μ, the value of β^2 can be determined. Using Eqs. (3.100) and (3.101), we may then obtain the values of v_r/v_p and v_r/v_s. It may be noted that the value of β^2 is independent to the frequency of the wave. Therefore the Rayleigh wave velocity is also independent to the frequency and dependent only on the elastic properties of the medium.

Figure 3.15 shows the variation of v_r/v_s and v_p/v_s with Poisson ratio μ.

The three types of wave appear in order on idealized seismogram (Fig. 3.16), which is a graph of ground motion against time at a particular geophone G at a distance x from the source 'o'. The time zero is, of course, the time of the shot, and it is clear that the three velocities v_p, v_s and v_r could be found from this record. In practice. this determination is made by combining the information from several geophones at various distances from the source on time-distance graph as shown in Fig. 3.17.

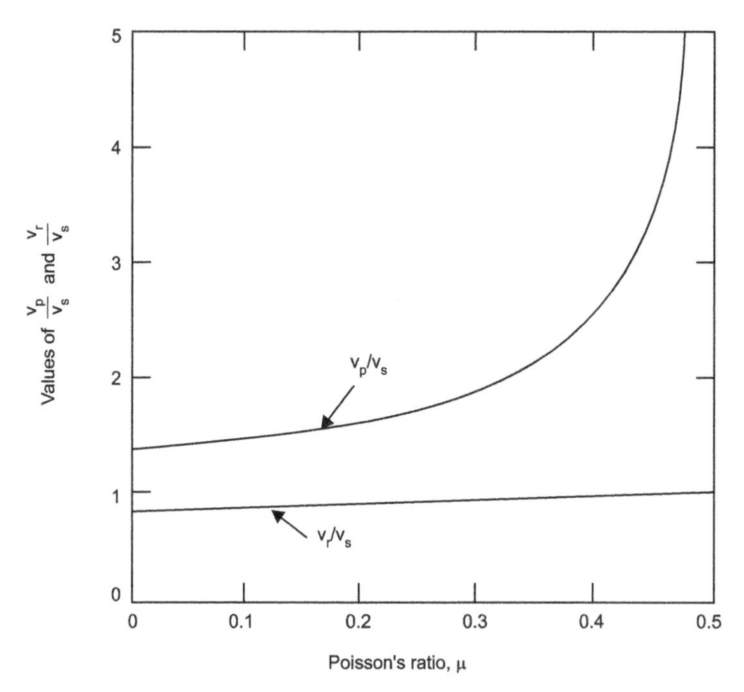

Fig. 3.15. Variations of v_r/v_s and v_p/v_s with Poisson's ratio, μ

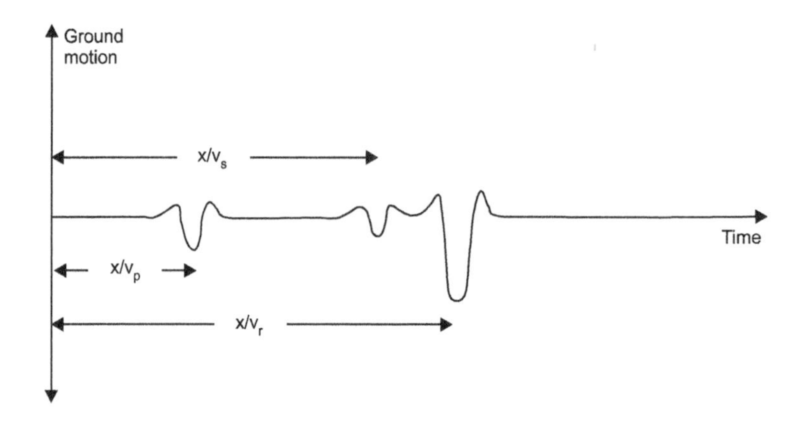

Fig. 3.16. Idealised seismogram of the ground motion at a distance x from the source

3.9.1 Displacement of Rayleigh Waves

Substituting the relations developed for ϕ and ψ {Eqs. (3.90) and (3.91)} in Eqs. (3.72), we get

$$u = -(in A_1 e^{-qz} + B_1 s e^{-sz})\, e^{i(\omega t - nx)} \tag{3.106}$$

$$w = -(A_1 q e^{-qz} - B_1\, ine^{-sz})\, e^{i(\omega t - nx)} \tag{3.107}$$

Now, substituting the value of B_1 in terms of A_1 from Eq. (3.95) the above expressions can be written as

$$u = A_1\, ni\left(-e^{-qz} + \frac{2qs}{s^2+n^2} e^{-sz}\right) e^{i(\omega t - nx)} \tag{3.108}$$

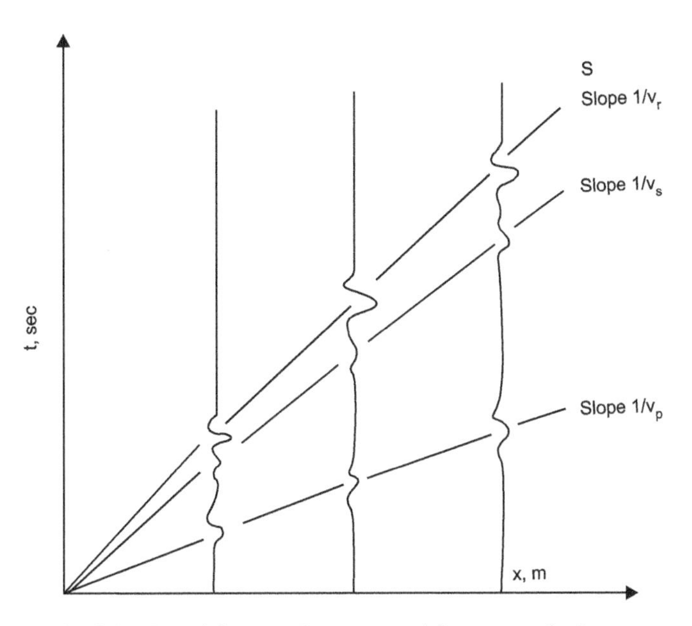

Fig. 3.17. Travel time graph constructed from a set of seismograms

$$w = A_1 q\left(-e^{-qz} + \frac{2n^2}{s^2+n^2}e^{-sz}\right)e^{i(\omega t - nx)} \tag{3.109}$$

From Eqs. (3.108) and (3.109), the variatin of u and w with depth can be expressed as

$$U(z) = e^{(-q/n)(nz)}\left(\frac{2(q/n)(s/n)}{s^2/n^2+1}\right)e^{-(s/n)(nz)} \tag{3.110}$$

$$W(z) = -e^{(-q/n)(nz)} + \frac{2}{s^2/n^2+1}e^{-(s/n)(nz)} \tag{3.111a}$$

Using Eqs. (3.99), (3.100) and (3.101), Eqs. (3.86) and (3.87) can be written as

$$\frac{q^2}{n^2} = 1-\frac{\omega^2}{n^2 v_p^2} = 1-\frac{v_r^2}{v_p^2} = 1-\alpha^2\beta^2 \tag{3.111b}$$

and

$$\frac{s^2}{n^2} = 1-\frac{\omega^2}{n^2 v_s^2} = 1-\frac{v_r^2}{v_s^2} = 1-\beta^2 \tag{3.112}$$

For a given value of Poisson's ratio, the values of v_p/v_r and v_r/v_s can be obtained from Fig. 3.15 (Richart, 1962). Hence values of q/n and s/n are determinable for a given values of Poisson's ratio. As $n = (2\pi/\text{wave length})$, the variation of $U(z)$ and $W(z)$ can then be studied with respect to a non-dimensional term (z/wave length). Such variation is shown in Fig. 3.18 for Poisson's ratio of 0.25, 0.33, 0.40 and 0.50.

The amplitude of body waves, which spread out along a hemispherical wave front, are proportional to $1/x$, x being the distance from the source. The amplitude of the Rayleigh waves, which spread out in a cylindrical wave front, are proportional to $1/\sqrt{x}$. Thus the attenuation of the amplitude of the Rayleigh waves is slower than that of the body waves.

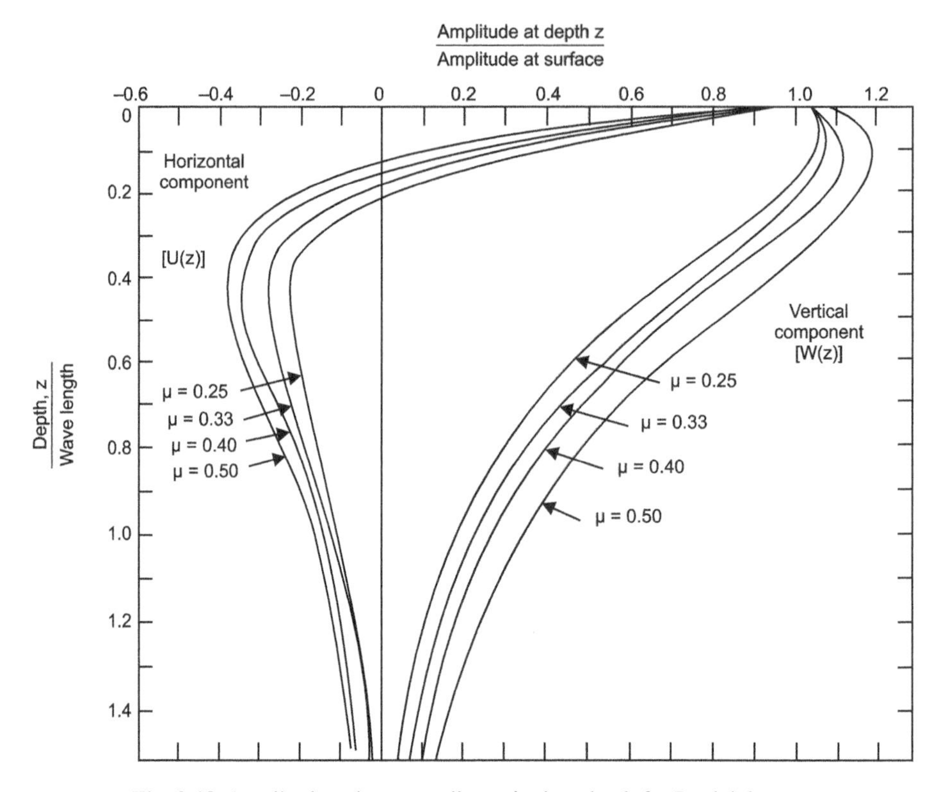

Fig. 3.18. Amplitude ratio versus dimensionless depth for Rayleigh wave

3.10 Geophysical Prospecting

3.10.1 General

Geophysical exploration is relatively new area of technology. It involves measurement of some physical field such as electric, magnetic etc. on the earth's surface and interpreting the data so obtained in terms of properties of subsurface layers of soil. The geophysical techniques most widely employed for exploration work are the seismic, gravity, magnetic and electrical methods. Less common methods involve the measurement of radioactivity and temperature at or near the earth's surface and in air.

In this section, the seismic method of geophysical exploration had been discussed. This method utilises the propagation of electric waves through the earth and is based on three fundamental principles namely (a) the waves are propagated with different velocities in different geological strata, (b) the contrast between the velocities is large, and (c) the strata velocities increase with depth. It consists of generating an elastic pulse or a more extended elastic vibration at shallow depth, and the resulting motion of the ground at nearby points on the surface is detected by seismic instruments known as geophones. Measurements of the travel-time of the pulse to geophones at various distances give the velocity of propagation of the pulse in the ground. The ground is generally not homogeneous in its elastic properties and this velocity therefore vary both with depth and laterally.

The real stratum, which in fact often consists of stratified material, is usually best approximated by a layered medium, each layer having a constant velocity or one changing in a simple and regular way with depth. The interfaces between layers may be inclined at an angle to the horizontal and to each other. In this section few simple cases have been discussed.

Let us consider the case of one horizontal interface at a depth h_1 between media in which the compression wave (P-wave) velocities are v_{p1} and v_{p2}, v_{p2} being greater. Figure 3.19 shows the possible paths of the body waves generated from the source S.

The first path as indicated by ray 1 is the same as the path of surface wave (i.e. Rayleigh wave). A compression or shear wave (ray 2) striking an interface will generate two reflected (P and S) and two refracted (P and S) waves (Fig. 3.19). According to the laws of reflection and refraction:

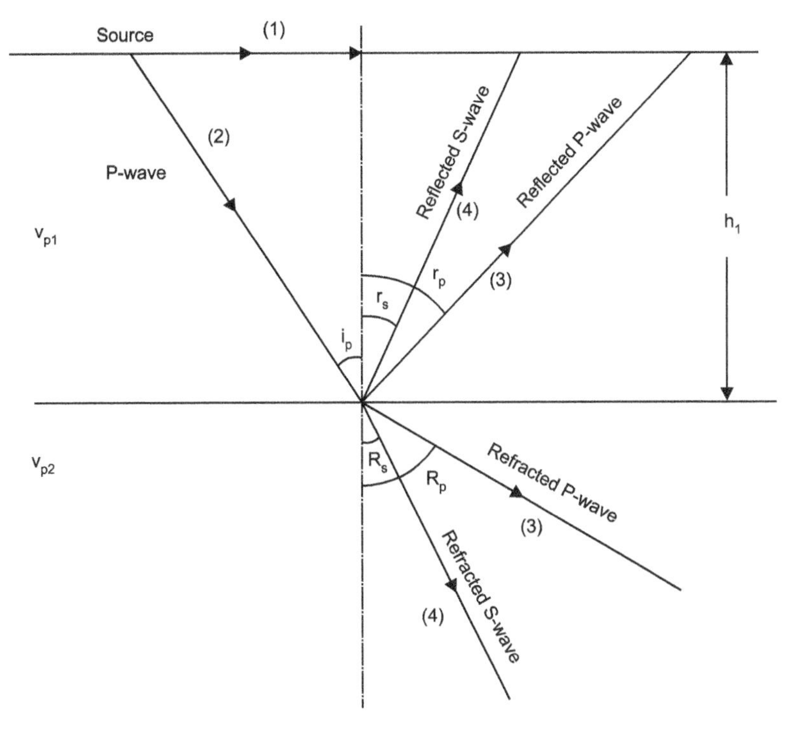

Fig. 3.19. Possible path of body waves

$$\frac{\sin i_p}{v_{p1}} = \frac{\sin r_p}{v_{p1}} = \frac{\sin r_s}{v_{s1}} = \frac{\sin R_p}{v_{p2}} = \frac{\sin R_s}{v_{s2}} \tag{3.113}$$

The equality of the angles of incidence and reflection (e.g. $i_p = r_p$) hold only if incident and reflected waves are of the same type.

In **seismic refraction** only compressional waves (P-waves) are considered and the interpretation is based mainly on the first arrival times derived from the seismograms. This is due to the fact that P-wave travels much faster than any other wave. Therefore for this case, Eq. (3.113) can be written as

$$\frac{\sin i_p}{\sin R_p} = \frac{v_{p1}}{v_{p2}} \tag{3.114}$$

The above equation is Snell's law or the law of refraction. Since $v_{p2} > v_{p1}$, angle of refractin R_p is greater than angle of incidence i_p. When i_p increases, there is a unique case where $R_p = 90°$ and $\sin R_p = 1$. Then

$$\sin i_p = \frac{v_{p1}}{v_{p2}} = \sin i_{pc} \tag{3.115}$$

Angle i_{pc} is called critical angle of incidence. For $i_p > i_{pc}$, the energy is totally reflected in the upper layer. If v_{p2} is less than v_{p1} so that the ray path is refracted away from the normal this critical refraction cannot occur.

It can be shown that the trajectory based on critical angles give the shortest time. Let a geophone is placed at a distance x from the source (Fig. 3.20).

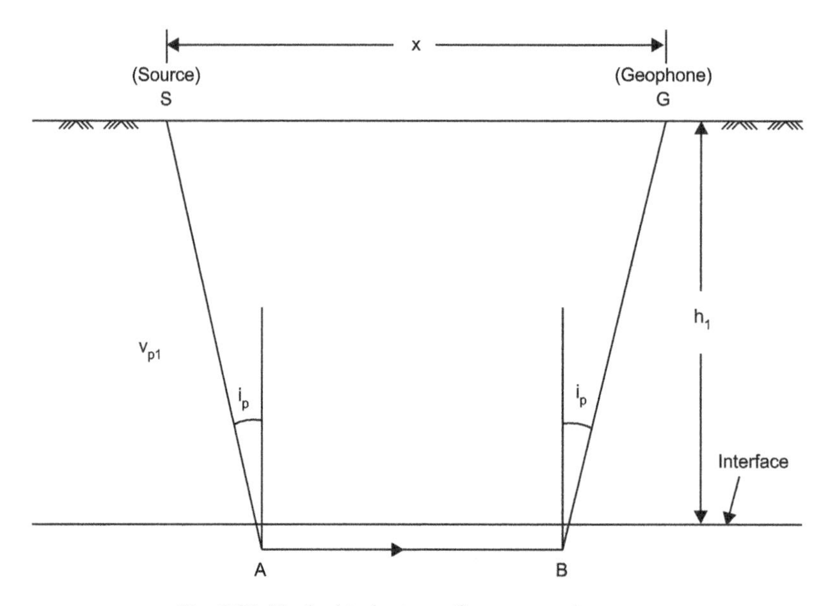

Fig. 3.20. Typical trajectory of a compression wave

$$SA = BG = \frac{h_1}{\cos i_p} \tag{3.116}$$

$$AB = x - 2\,h_1 \tan i_p \tag{3.117}$$

Total time taken by the wave reaching from S to G will be:

$$T = \frac{2h_1}{v_{p1}\cos i_p} + \frac{x - 2h_1 \tan i_p}{v_{p2}} \tag{3.118}$$

The time T will be minimum when

$$\frac{dT}{di_p} = \frac{2h_1 \sin i_p}{v_{p1}\cos^2 i_p} - \frac{2h_1}{v_{p2}\cos^2 i_p} = 0 \tag{3.119}$$

or

$$\sin i_p = \frac{v_{p1}}{v_{p2}} = \sin i_{pc} \tag{3.120}$$

Therefore the travel time from S to G via the second layer is minimum when slant ray paths through v_{p1} layer make angle i_{pc} with the normal to the surface. Hence a geophone on the surface at any distance greater than the critical range $2\,h_1 \tan i_{pc}$ from S will lie on one of these rays and will record the arrival of the wave at the appropriate time (Fig. 3.21). The refracted waves shown by dotted lines are known as **head waves.**

3.10.2 Depth Formulae

3.10.2.1 Two-layer Soil Medium

If the first arrivals of the elastic waves are recorded by detectors planted in the ground, the times from the impact instant to the detectors can be plotted on a time distance graph as shown in

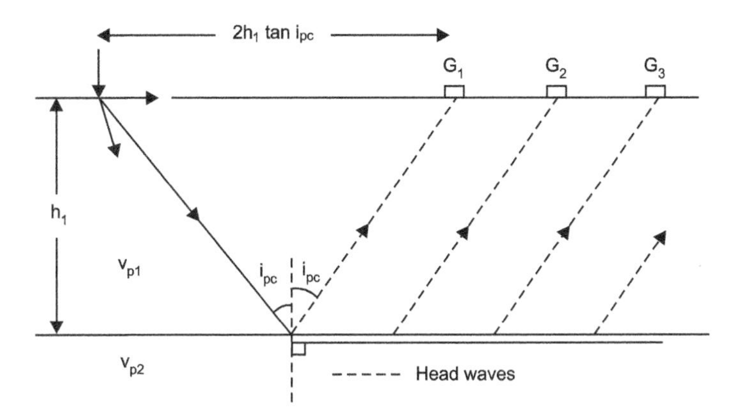

Fig. 3.21. Critically refracted ray path and head waves

Fig. 3.22b. The slope of the lines yield the reciprocal of velocities, namely $1/v_p$. Therefore, the lower the velocity, the steeper the slope of time-distance line.

The intersection, **break point** between the two velocity lines is obviously the point where the times are equal. The distance between the impact point S and the break point is called the **critical distance**. The break point corresponds to the emergence of the wave front contact at the ground surface. **Intercept time** is the total arrival time of the refracted waves minus the time x/v_{p2}, x being the distance between the impact point and the receiving station. It is the intersection between the prolongation of the time-distance segment corresponding to the second medium and the time axis through the impact point.

For calculating the depth at an impact point two different approaches are available either using the intercept time or critical distance (Milton, 1960; Parasnis, 1962).

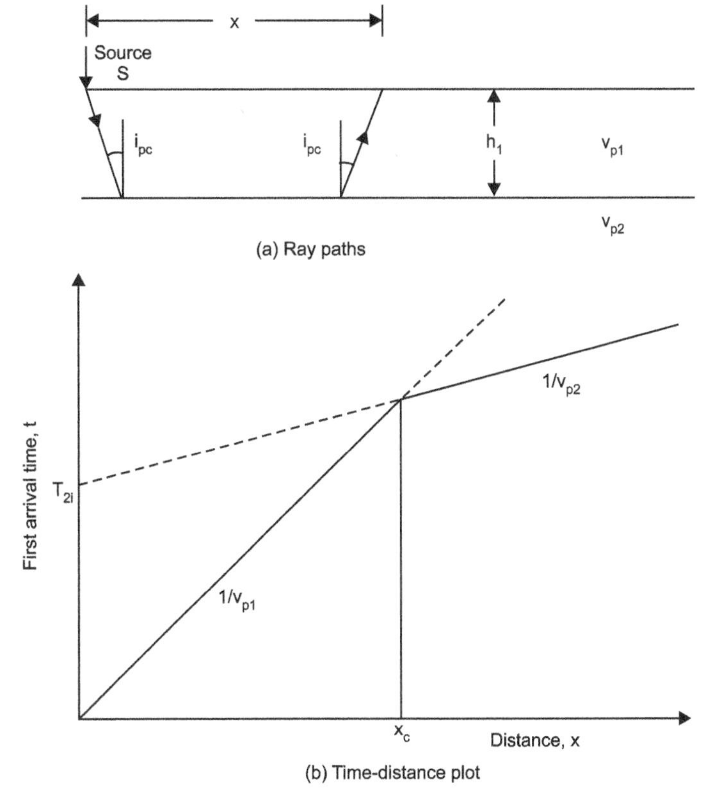

Fig. 3.22. Fundamental principle of refraction shooting

Intercept Time (Refer Fig. 3.22a)

$$\text{Arrival time } T_1 \text{ of direct surface waves} = \frac{x}{v_{p1}} \tag{3.121}$$

$$\text{Arrival time } T_2 \text{ of refracted waves} = \frac{2h_1}{v_{p1}\cos i_{pc}} + \frac{x - 2h_1 \tan i_{pc}}{v_{p2}} \tag{3.122}$$

$$\sin i_{pc} = \frac{v_{p1}}{v_{p2}} \tag{3.123}$$

$$T_2 = \frac{2h_1}{v_{p1}\cos i_{pc}} - \frac{2h_1 \sin i_{pc} \sin i_{pc}}{v_{p1}\cos i_{pc}} + \frac{x}{v_{p2}}$$

$$= \frac{2h_1}{v_{p1}\cos i_{pc}} - \frac{2h_1(1-\cos^2 i_{pc})}{v_{p1}\cos i_{pc}} + \frac{x}{v_{p2}}$$

or

$$T_2 = \frac{x}{v_{p2}} + \frac{2h_1 \cos i_{pc}}{v_{p1}} \tag{3.124a}$$

or

$$T_2 = \frac{x}{v_{p2}} + \frac{2h_1}{v_{p1}v_{p2}}\sqrt{v_{p2}^2 - v_{p1}^2} \tag{3.124b}$$

$$\cos i_{pc} = \sqrt{1 - \frac{v_{p1}^2}{v_{p2}^2}}$$

This is the equation of a straight line with slope $1/v_{p2}$ and an intercept on the time axis through impact point is given by putting $x = 0$.

$$T_{2i} = \frac{2h_1 \cos i_{pc}}{v_{p1}} \tag{3.125}$$

or

$$h_1 = \frac{T_{2i}\, v_{p1}}{2\cos i_{pc}} \tag{3.126}$$

or

$$h_1 = \frac{T_{2i}\, v_{p1}\, v_{p2}}{2\cdot\sqrt{v_{p2}^2 - v_{p1}^2}} \tag{3.127}$$

Critical Distance: Let the critical distance is x_c at which $T_1 = T_2$ (Fig. 3.22b)

$$\frac{2h_1}{v_{p1}\cos i_{pc}} + \frac{x_c - 2h_1 \tan i_{pc}}{v_{p2}} = \frac{x_c}{v_{p1}} \tag{3.128}$$

or

$$\frac{2h_1}{v_{p1}\cos i_{pc}} - \frac{2h_1(1-\cos^2 i_{pc})}{v_{p1}\cos i_{pc}} = x_c\left(\frac{1}{v_{p1}} - \frac{1}{v_{p2}}\right)$$

or

$$2h_1 \cos i_{pc} = x_c\left(1 - \frac{v_{p1}}{v_{p2}}\right)$$

or

$$h_1 = \frac{x_c(1 - \sin i_{pc})}{2\cos i_{pc}} \tag{3.129}$$

or

$$h_1 = \frac{x_c}{2}\sqrt{\frac{v_{p2} - v_{p1}}{v_{p2} + v_{p1}}} \tag{3.130}$$

3.10.2.2 Three-layer Soil Medium

Consider a three-layered soil medium with compression wave velocities of layers 1, 2 and 3 as v_{p1}, v_{p2} and v_{p3} with the condition that $v_{p1} < v_{p2} < v_{p3}$ (Fig. 3.23a). If *S* is a source of disturbance, the direct wave travelling through layer 1 will arrive first at *A*, which is located a small distance away from *S*. The travel time for this can be given by Eq. (3.121) as $T_1 = x/v_{p1}$. At a greater distance *x*, the first arrival will correspond to the wave taking the path *SDEB*. The travel time for this can be obtained from Eq. (3.123b) as

$$T_2 = \frac{x}{v_{p2}} + \frac{2h_1\sqrt{v_{p2}^2 - v_{p1}^2}}{v_{p1}\, v_{p2}}$$

At a still larger distance, the first arrival corresponds to the path *SFGHIC*. According to Snell's law

$$\frac{\sin i}{\sin i_{c2}} = \frac{v_{p1}}{v_{p2}} \tag{3.131}$$

$$\sin i_{c2} = \frac{v_{p2}}{v_{p3}} \tag{3.132}$$

Therefore,

$$\sin i = \frac{v_{p1}}{v_{p2}} \cdot \frac{v_{p2}}{v_{p3}} = \frac{v_{p1}}{v_{p3}} \tag{3.133}$$

The equation of the time of arrival, T_3 for the wave from *S* to *C* through third layer is

$$T_3 = \frac{2h_1}{v_{p1}\cos i} + \frac{2h_2}{v_p \cos i_{c2}} + \frac{x - 2h_1 \tan i - 2h_2 \tan i_{c2}}{v_{p3}} \tag{3.134}$$

Substituting the values of cos *i* and cos i_{c2} from Eqs. (3.132) and (3.133) in Eq. (3.134). We get

$$T_3 = \frac{x}{v_{p3}} + \frac{2h_1\sqrt{v_{p3}^2 - v_{p1}^2}}{v_{p3}\, v_{p1}} = \frac{2h_2\sqrt{v_{p3}^2 - v_{p2}^2}}{v_{p3}\, v_{p2}} \tag{3.135}$$

The records of geophones placed at various distances from the source may be plotted on time versus distance graph as shown in Fig. 3.23b. The line *OA* corresponds to Eq. (3.121), *OB* to Eq. (3.123b), and *BC* to Eq. (3.135). The thickness of the first layer can be obtained either by Eq. (3.127) using intercept time T_{2i} or by Eq. (3.130) using critical distance x_{c1}. The thickness of the second layer h_2 can be obtained by the two approaches as given below.

Intercept time (Fig. 3.23)

$$T_{3i} = \frac{2h_1 \cos i}{v_{p1}} + \frac{2h_2 \cos i_{c2}}{v_{p2}} \tag{3.136}$$

or

$$h_2 = \frac{1}{2}\left(T_{3i} - \frac{2h_1\sqrt{v_{p3}^2 - v_{p1}^2}}{v_{p3}\, v_{p1}}\right)\frac{v_{p3}\, v_{p2}}{\sqrt{v_{p3}^2 - v_{p2}^2}} \tag{3.137}$$

Critical distance (Fig. 3.23)

At distance x_{c2}, $T_2 = T_3$. Therefore

$$\frac{x_{c2}}{v_{p2}} + \frac{2h_1\sqrt{v_{p2}^2 - v_{p1}^2}}{v_{p1}\, v_{p2}} = \frac{x_{c2}}{v_{p3}} + \frac{2h_1\sqrt{v_{p3}^2 - v_{p1}^2}}{v_{p3}\, v_{p1}} + \frac{2h_2\sqrt{v_{p3}^2 - v_{p2}^2}}{v_{p3}\, v_{p2}} \tag{3.138}$$

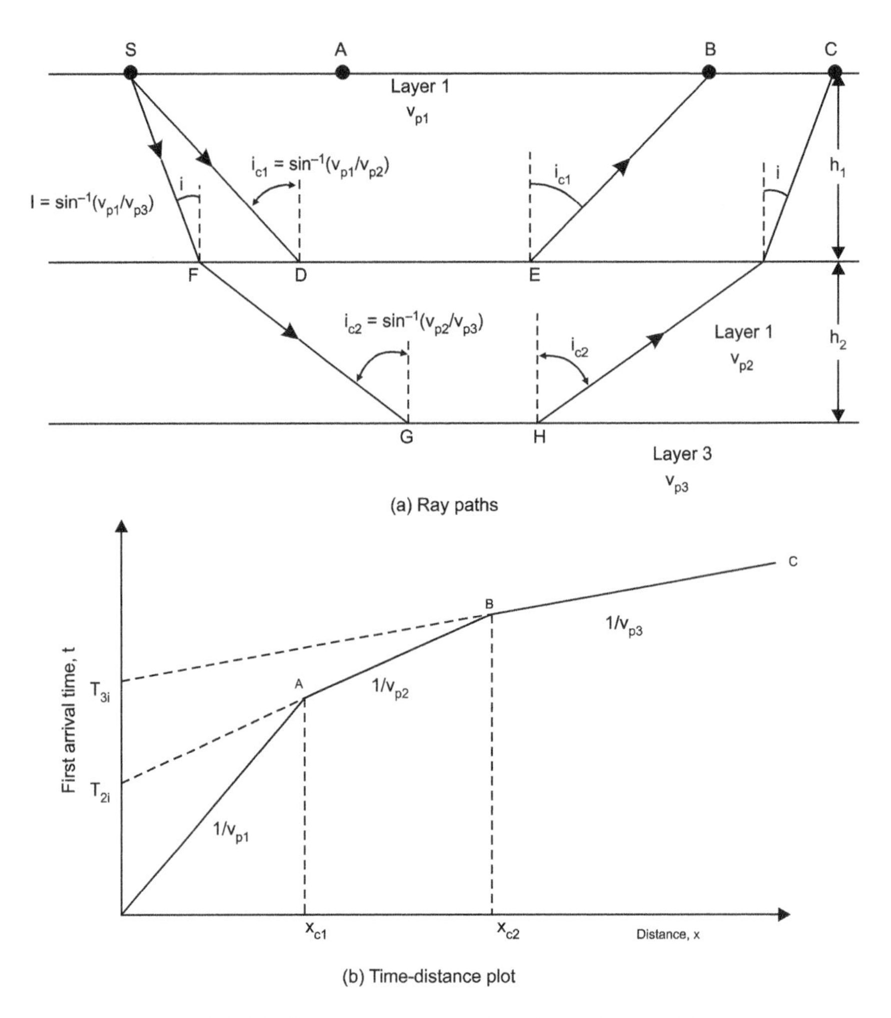

Fig. 3.23. Refraction shooting in three layered soil medium

or

$$h_2 = \frac{x_{c2}}{2}\sqrt{\frac{v_{p3}-v_{p2}}{v_{p3}+v_{p2}}} - \frac{h_1\left(v_{p2}\sqrt{v_{p3}^2-v_{p1}^2} - v_{p3}\sqrt{v_{p2}^2-v_{p1}^2}\right)}{v_{p1}\sqrt{v_{p3}^2-v_{p2}^2}} \tag{3.139}$$

3.10.2.3 Multilayer Soil System

If there are n number of layers, the first arrival time at various distances from the sources of disturbance will plot as shown in Fig. 3.24. There will be n segments on the t versus x plot. Using either intercept time or critical distance approach, the thickness of various layers can be obtained by determining $h_1, h_2, ..., h_{n-1}$ in sequence. The general equation can be written as:

$$h_{n-1} = \frac{T_{ni}\, v_{pn}\, v_{p(n-1)}}{2\sqrt{v_{pn}^2 - v_{p(n-1)}^2}} - \frac{v_{pn}\, v_{p(n-1)}}{\sqrt{v_{pn}^2 - v_{p(n-1)}^2}} \sum_{j=1}^{j=n-2} h_j \sqrt{\frac{1}{v_{pj}^2} - \frac{1}{v_{pn}^2}} \tag{3.140}$$

Equation (3.140) is based on intercept time approach.

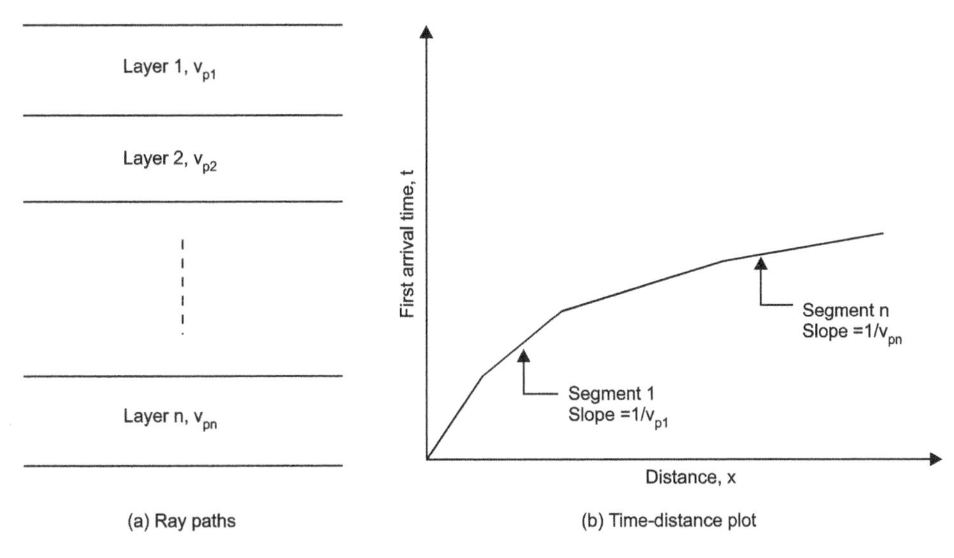

Fig. 3.24. Refraction shooting in multilayer soil

3.10.2.4 Sloping Layer System

Figure (3.25a) shows two soil layers. The interface of soil layers 1 and 2 is inclined at an angle α with respect to horizontal. The rays such as *ABCD* making the critical angle i_{pc} ($\sin i_{pc} = v_{p1}/v_{p2}$) with the normal to the refractor, take the shortest time from *A* to *D* and are therefore **first arrivals.**

Referring Fig. 3.25a,

$$AB = CI = \frac{Z_d}{\cos i_{pc}} \tag{3.141}$$

$$DK = x \sin \alpha \tag{3.142}$$

$$ID = \frac{DK}{\cos i_{pc}} = \frac{x \sin \alpha}{\cos i_{pc}} \tag{3.143}$$

$$CD = \frac{(Z_d + x \sin \alpha)}{\cos i_{pc}} \tag{3.144}$$

$$AK = EG = x \cos \alpha \tag{3.145}$$

$$EB = Z_d \tan i_{pc} \tag{3.146}$$

$$CG = (Z_d + x \sin \alpha) \tan i_{pc} \tag{3.147}$$

If we assume that point *A* to be the energy source and *D* the detector station, the time from *A* to *D* for the ray *ABCD*, i.e. the downdip time T_{2d} is

$$T_{2d} = \frac{AB}{v_{p1}} + \frac{BC}{v_{p2}} + \frac{CD}{v_{p1}} \tag{3.148}$$

$$= \frac{Z_d}{v_{p1} \cos i_{pc}} + \frac{x \cos \alpha - Z_d \tan i_{pc} - (Z_d + x \sin \alpha) \tan i_{pc}}{v_{p2}} + \frac{Z_d + x \sin \alpha}{v_{p1} \cos i_{pc}} \tag{3.149}$$

or

$$T_{2d} = \frac{2 Z_d \cos i_{pc}}{v_{p1}} + \frac{x}{v_{p1}} \sin (i_{pc} + \alpha) \tag{3.150}$$

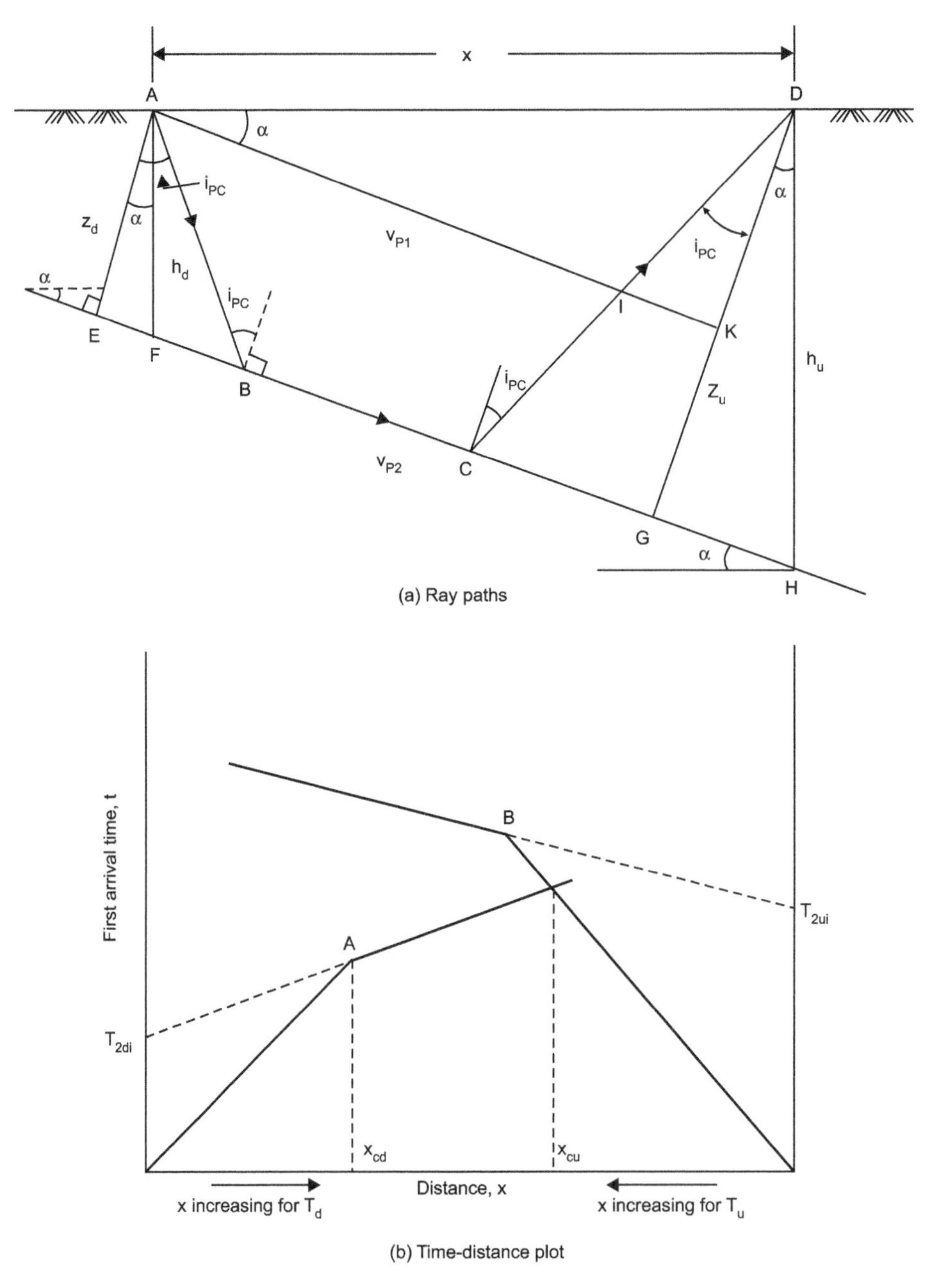

Fig. 3.25. Refraction survey in soils with inclined layering

Equation (3.150) represents a straight line with slope $\sin(i_{pc} + \alpha)/v_{p1}$, and when $x = 0$, intercept is T_{2di},

where $$T_{2di} = \frac{2 z_d \cos i_{pc}}{v_{p1}} \tag{3.151}$$

The apparent velocity $v_{p1}/\sin(i_{pc} + \alpha)$ is equal to $v_{p2} \sin i_{pc}/\sin(i_{pc} + \alpha)$ and is smaller than the true velocity v_{p2} since $\sin i_{pc}/\sin(i_{pc} + \alpha) < 1.0$.

For up-dip recording the equation of time can be obtained by replacing Z_d by Z_u, and α by $-\alpha$, then we get,

$$T_{2u} = \frac{2z_u \cos i_{pc}}{v_{p1}} + \frac{x}{c_{p1}} \sin(i_{pc} - \alpha) \tag{3.152a}$$

In this case, apparent velocity is $v_1/\sin(i_{pc} - \alpha)$ which is equal to $v_2 \sin i_{pc}/\sin(i_{pc} - \alpha)$. This velocity will be greater than the true velocity v_{p2}

For $x = 0$, intercept is T_{2ui}, where

$$T_{2ui} = \frac{2Z_u \cos i_{pc}}{v_{p1}} \tag{3.152b}$$

Therefore

$$Z_d = \frac{T_{2di}\, v_{p1}}{2\cos i_{pc}} \tag{3.153a}$$

$$h_d = \frac{T_{2di}\, v_{pi}}{2\cos i_{pc} \cos\alpha} \tag{3.153b}$$

$$Z_u = \frac{T_{ui}\, v_{p1}}{2\cos i_{pc}} \tag{3.154a}$$

$$h_u = \frac{T_{ui}\, v_{p1}}{2\cos i_{pc} \cos\alpha} \tag{3.154b}$$

The vertical depths h_d and h_u can be obtained by dividing Z_d and Z_u by cos α.

For critical distances: (Fig. 3.25b)

$$\frac{x_{cd}}{v_{p1}} = \frac{2Z_d \cos i_{pc}}{v_{p1}} + \frac{x_{cd} \sin(i_{pc} + \alpha)}{v_{p1}} \tag{3.155}$$

$$Z_d = \frac{x_{cd}\{1 - \sin(i_{pc} + \alpha)\}}{2\cos i_{pc}} \tag{3.156}$$

or

$$h_d = \frac{x_{cd}\{1 - \sin(i_{pc} + \alpha)\}}{2\cos i_{pc} \cos\alpha} \tag{3.157}$$

Similarly

$$h_u = \frac{x_{cu}\{1 - \sin(i_{pc} - \alpha)\}}{2\cos i_{pc} \cos\alpha} \tag{3.158}$$

It is evident from Eq. (3.150), that apparent down dip velocity in the second layer is given by

$$v_{2d} = \frac{v_{p1}}{\sin(i_{pc} - \alpha)} \tag{3.159}$$

Similarly the apparent up dip velocity will be

$$v_{2u} = \frac{v_{p1}}{\sin(i_{pc} - \alpha)} \tag{3.160}$$

The true velocity v_{p2} in the refractor can be derived as follows:

$$\sin(i_{pc} + \alpha) = \frac{v_{p1}}{v_{2d}} = \sin i_{pc} \cos\alpha + \cos i_{pc} \sin\alpha \tag{3.161}$$

$$\sin(i_{pc} - \alpha) = \frac{v_{p1}}{v_{2u}} = \sin i_{pc} \cos\alpha - \cos i_{pc} \sin\alpha \tag{3.162}$$

or

$$2 \sin i_{pc} \cos \alpha = \frac{v_{p1}}{v_{2d}} + \frac{v_{p1}}{v_{2u}} \tag{3.163}$$

or

$$2\frac{v_{p1}}{v_{p2}} \cos\alpha = \frac{v_{p1}(v_{2d} + v_{2u})}{v_{2u}\, v_{2d}} \quad \because \left(\sin i_{pc} = \frac{v_{p1}}{v_{p2}} \right) \tag{3.164}$$

or

$$v_{p2} = 2\cos\alpha \frac{v_{2u}\, v_{2d}}{v_{2u} + v_{2d}} \tag{3.165}$$

3.10.2.5 Refraction in a Medium having Continuous Change of Speed with Depth

The problem may be solved by dividing the strata in infinitesimally thin layers and each of higher speed than the previous one (Fig. 3.26).

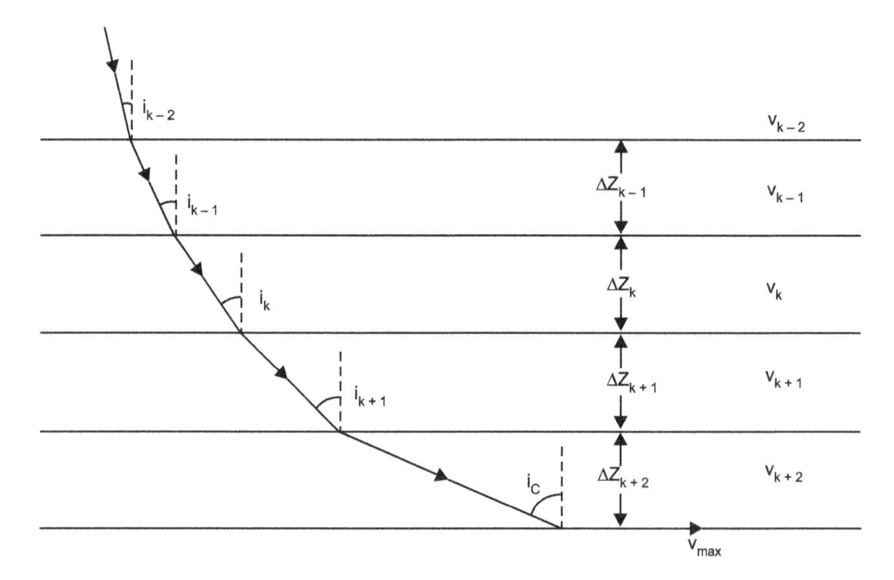

Fig. 3.26. Refraction survey in soil having continuous change of velocity with depth

The solution may be obtained for a given velocity-depth functions such as

$$v = v_o + k'z \tag{3.166}$$

where v = speed at depth z, v_o = speed at zero depth, and k' = constant.

For any particular ray there will be a layer having a speed v_{max} in which the path becomes horizontal. Therefore v_{max} is actually a parameter that itself identifies the particular ray under consideration.

Consider k^{th} layer.

Travel time for passing the ray through k^{th} layer

$$\Delta T_k = \frac{\Delta Z_k}{v_k \cos i_k} = \frac{\Delta Z_k}{v_k \sqrt{1 - \sin^2 i_k}} \tag{3.167}$$

As

$$i_k = \sin^{-1}\left(\frac{v_k}{v_{\max}} \right) \tag{3.168}$$

$$\Delta T_k = \frac{\Delta Z_k}{v_k\sqrt{1-\left(\frac{v_k}{v_{\max}}\right)^2}} \tag{3.169}$$

Horizontal distance traversed by this ray in k^{th} layer

$$\Delta x_k = \Delta Z_k \tan i_k \tag{3.170 a}$$

or

$$\Delta x_k = \Delta Z_k \frac{\frac{v_k}{v_{\max}}}{\sqrt{1-\frac{v_k^2}{v_{\max}^2}}} \tag{3.170 b}$$

Total time for a wave to travel through N such layers:

$$t = \sum_{k=1}^{N} \frac{\Delta Z_k}{v_k\sqrt{1-\left(\frac{v_k}{v_{\max}}\right)^2}} \tag{3.171}$$

and net horizontal distance

$$x = \sum_{k=1}^{N} \frac{\Delta Z_k \frac{v_k}{v_{\max}}}{\sqrt{1-\left(\frac{v_k}{v_{\max}}\right)^2}} \tag{3.172}$$

According to Snell's law:

$$\frac{\sin i_k}{\sin 90} = \frac{v_k}{v_{\max}} \tag{3.173}$$

or

$$\frac{1}{v_{\max}} = \frac{\sin i_k}{v_k} = \frac{\sin i_k}{v_0} = p\,(\text{constant})\ \text{ray parameter}$$

Considering velocity-depth function i.e. $v = v_0 + k'z$, and replacing v_k by v and $1/v_{max}$ by p.

$$t = \int_0^{Z_{\max}} \frac{dZ}{v\sqrt{1-p^2v^2}} \tag{3.174}$$

and

$$x = \int_0^{Z_{\max}} \frac{pv\,dZ}{v\sqrt{1-p^2v^2}} \tag{3.175}$$

Then

$$t = \int_0^{Z_{\max}} \frac{pv\,dZ}{(v_0+k'Z)\sqrt{1-p^2(v_0+k'Z)^2}} \tag{3.176}$$

and

$$x = \int_0^{Z_{\max}} \frac{p(v_0+k'Z)dZ}{\sqrt{1-p^2(v_0+k'Z)^2}} \tag{3.177}$$

Integrations of the above two expressions are as under:

$$x = \frac{1}{k'p}[(1-p^2 v_0^2)^{1/2} - \{1-p^2(v_0+k'Z)^2\}^{1/2}] \tag{3.178}$$

and

$$t = \frac{1}{k'} \ln \frac{(v_0+k'Z)\left(1+\sqrt{1-p^2 v_0^2}\right)}{v_0\left(1+\sqrt{1-p^2(v_0+k'Z)^2}\right)} \tag{3.179}$$

Equation of x can be written as:

$$\left(x - \frac{1-p^2 v_0^2}{k'p}\right)^2 + \left(Z + \frac{v_0}{k'}\right)^2 = \frac{1}{k'^2 p^2} \tag{3.180}$$

This is a equation of a circle of radius $1/k'p$ and centre at $-v_0/k'$ and $\sqrt{1-p^2 v_0^2}/k'p$ from x and z-axis respectively. Figure 3.27a shows a family of such circles for a number of rays having different angles of emergence into the earth.

$$\text{Radius of circle} = \frac{1}{k'p} = \frac{v_0}{k'} + Z_{max} \tag{3.181}$$

or

$$Z_{max} = \frac{1}{k'p} - \frac{v_0}{k'} \tag{3.182a}$$

$$= \frac{v_0}{k' \sin i_0} - \frac{v_0}{k'} \qquad \left\{p = \frac{\sin i_0}{v_0}\right\}$$

or

$$Z_{max} = \frac{v_0}{k'}(\operatorname{cosec} i_0 - 1) \tag{3.182b}$$

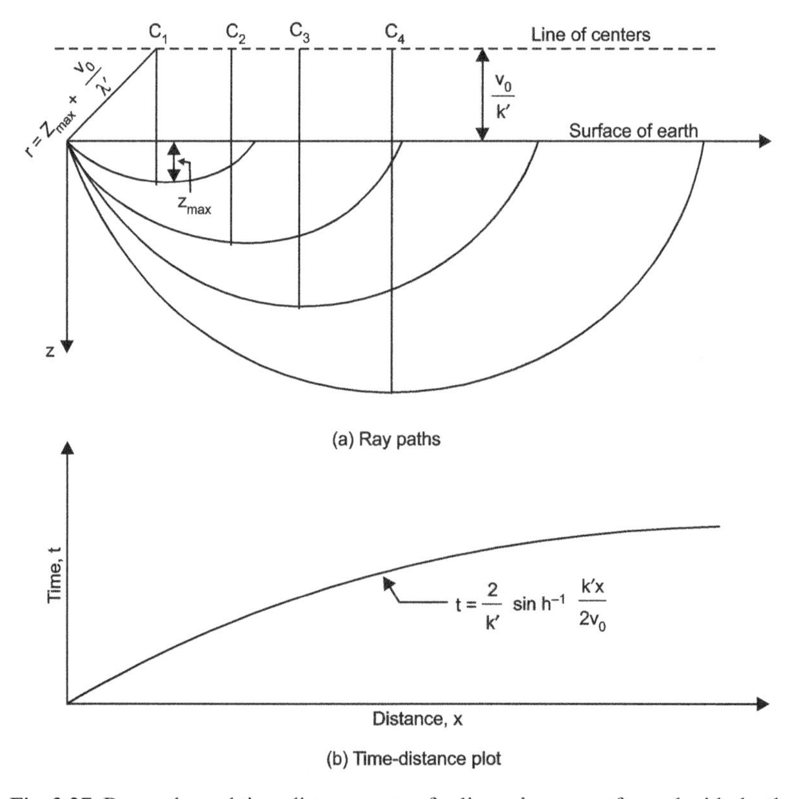

Fig. 3.27. Ray paths and time distance curve for linear increase of speed with depth

Putting the value of z and z_{max} in Eqs. (3.178) and (3.179) and eliminating p using Eq. (3.181), we get

$$t = \frac{2}{k}\ln\frac{v_{max}\left(1+\sqrt{p^2 v_0^2}\right)}{v_0}$$

$$= \frac{2}{k}\ln\frac{v_{max}\left(1+\sqrt{1-\left(\frac{v_0}{v_{max}}\right)^2}\right)}{v_0} \quad (3.183a)$$

$$x = \frac{2v_{max}}{k}\sqrt{1-\left(\frac{v_0}{v_{max}}\right)^2} = \frac{2v_0}{k'}\cot i_0 \quad (3.183b)$$

where $$v_{max} = v_0 + k'Z_{max} \quad (3.184)$$

The time-distance relation for such a circular ray between entry into and emergence from the earth can be obtained by eliminating p and z from Eqs. (3.178) and (3.179). This can be expressed as

$$t = \frac{2}{k'}\sinh^{-1}\frac{k'x}{2v_0} \quad (3.185)$$

The time-distance curve applying for a linear increase of speed with depth is shown in Fig. 3.27b. The inverse slope of the time-distance curve at any point is equal to the velocity at the depth of maximum penetration for the ray reaching the surface at that point.

3.11 Typical Values of Compression Wave and Shear Wave Velocities

Some typical values of compression wave and shear wave velocities (v_p and v_s) are given in Table 3.1.

Table 3.1. Compression and Shear Wave Velocities

Material	v_p, *m/s*	v_s, *m/s*
1. Moist clay and wet soils	600-1750	80-160
2. Sand	300-800	100-500
3. Sand stone	1500-4500	750-2200
4. Lime stone	3500-6500	1800-3800
5. Granite	4600-7000	2500-4000

Illustrative Examples

Example 3.1

Wave propagation tests were conducted near Mathura Refinery, Mathura, for determining the in situ velocities of wave propagation and dynamic elastic moduli. Seismic waves were generated by the impact of a hammer falling through a height of 2.0 m. Two geophones were placed in the ground respectively at 1.0 m and 5.0 m from the source. The analysis of data gave the velocity of compression wave, v_p as 300 m/s. The soil at the site was cohesionless and the position of water table was at 1.0 m depth below the ground surface. Determine E, G, v_s and v_r.

Solution

1. Assume suitable values of Poisson ratio, μ and submerged density of soil, ρ_b say

$$\mu = 0.25 \text{ and } \gamma_b = 10 \text{ kN/m}^3$$

2. From Eq. (3.64)

$$v_p^2 = \frac{\lambda + 2G}{\rho}$$

Putting the values of λ and G from Eqs. (3.8) and (3.9) in the above equation, we get

$$E = \frac{\rho(1+\mu)(1-2\mu)}{(1-\mu)} v_p^2$$

$$= \frac{10(1+0.25)(1-2\times 0.25)}{9.81\times(1-0.25)}\times(300)^2 = 76453000 \text{ N/m}^2$$

$$= 76453 \text{ kN/m}^2$$

From Eq. (3.9)

$$G = \frac{E}{2(1+\mu)} = \frac{76453}{5(1+0.25)} = 30581 \text{ kN/m}^2$$

3. From Eq. (3.103),

$$\alpha^2 = \frac{1-2\mu}{2-2\mu} = \frac{1-2\times 0.25}{2-2\times 0.25} = \frac{1}{3}$$

From Eq. (3.105),

$$\beta^6 - 8\beta^4 - (16\alpha^2 - 24)\beta^2 - 16(1-\alpha^2) = 0$$

For $\mu = 0.25$

$$3\beta^6 - 24\beta^4 + 56\beta^2 - 32 = 0$$

or

$$(\beta^2 - 4)(3\beta^4 - 12\beta^2 + 8) = 0$$

Therefore,

$$\beta^2 = 2,\ 2+\frac{2}{\sqrt{3}},\ 2-\frac{2}{\sqrt{3}}$$

From Eq. (3.112),

$$\frac{s^2}{n^2} = 1 - \beta^2$$

For $\beta^2 = 2$ and $(2+2/\sqrt{3})$, value of s/n will work out imaginary. Therefore

$$\beta^2 = (2-2/\sqrt{3}) = 2-1.155 = 0.845$$

$$\alpha^2\beta^2 = 0.845\times\frac{1}{3} = 0.282$$

From Eq. (3.100),

$$\frac{v_r}{v_p} = \sqrt{\alpha^2\beta^2} = 0.531$$

$$v_r = 300 \times 0.531 = 160 \text{ m/s}$$

From Eq. (3.101),

$$\frac{v_r}{v_s} = \sqrt{\beta^2} = \sqrt{0.845} = 0.9192$$

$$v_s = \frac{r}{0.9192} = \frac{160}{0.9192} = 174 \text{ m/s}$$

Example 3.2

The data of a refractor test is given below:

Distance of geophones from source (m)	*Travel time (milli s)*
0	0
4	2
8	4
12	6
20	8
25	9

Determine the depth of the refractor using time intercept approach and critical distance approach.

Solution

1. The data given in the example has been plotted on time-distance graph as shown in Fig. 3.28. From this figure,

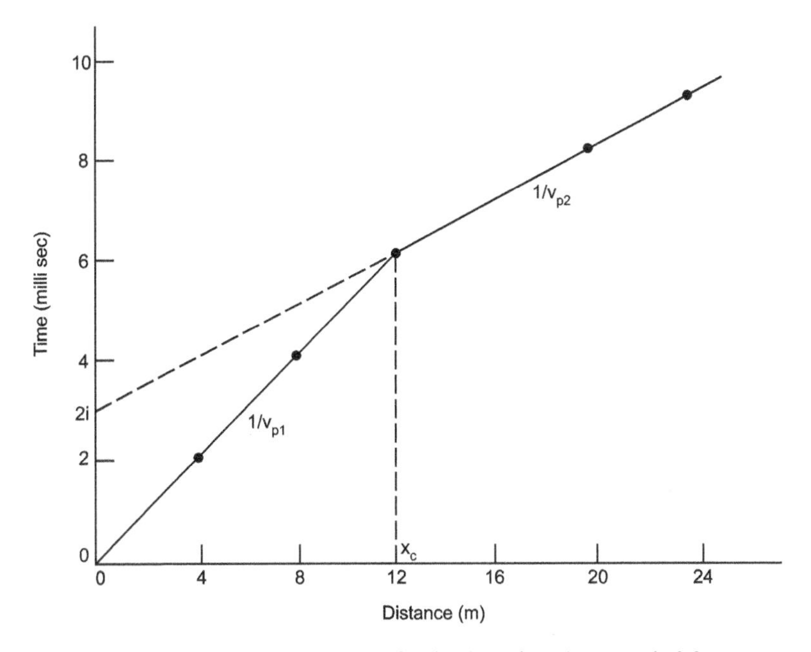

Fig. 3.28. Time-distance plot for the data given in example 3.2

$$T_{2i} = 3 \text{ milli sec} = 0.003\text{s and } x_c = 12 \text{ m}$$

Also
$$v_{p1} = \frac{12}{0.006} = 2000 \text{ m/s}$$

$$v_{p2} = \frac{8}{0.002} = 4000 \text{ m/s}$$

2. From time-intercept approach (Eq. 3.127)

$$h_1 = \frac{T_{2i}\, v_{p1}\, v_{p2}}{2\sqrt{v_{p2}^2 - v_{p1}^2}} = \frac{0.003 \times 2000 \times 4000}{2\sqrt{4000^2 - 2000^2}} = 3.46 \text{ m}$$

3. From critical distance approach (Eq. 3.130)

$$h_1 = \frac{x_c}{2}\sqrt{\frac{v_{p2} - v_{p1}}{v_{p2} + v_{p1}}} = \frac{12}{2}\sqrt{\frac{4000 - 2000}{4000 + 2000}} = 3.46 \text{ m}$$

Example 3.3

In Fig. 3.29, determine the time of the first arrival wave from source S and geophone G.

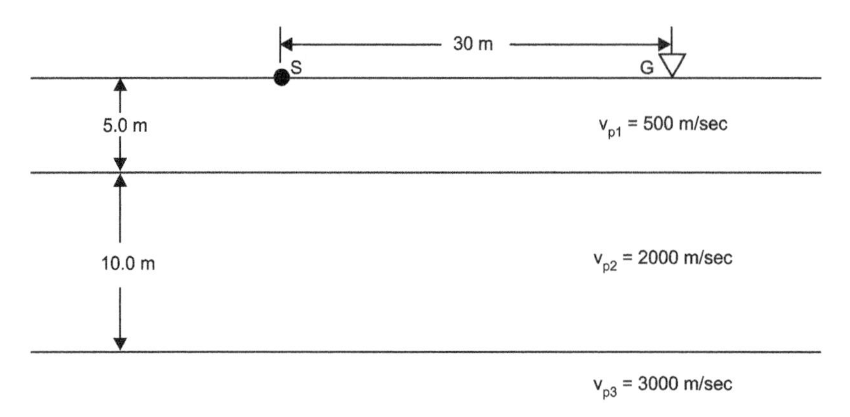

Fig. 3.29. Data for example 3.3

Solution

(i) Arrival time T_1 of the direct surface wave $= \frac{x}{v_{p1}} = \frac{30}{500} = 0.06$ s

(ii) Arrival time T_2 of refracted wave from first layer (Eq. 3.124 *b*)

$$= \frac{x}{v_{p2}} + \frac{2h_1}{v_{p1}\, v_{p2}}\sqrt{v_{p2}^2 - v_{p1}^2}$$

$$= \frac{30}{2000} + \frac{2 \times 5}{500 \times 2000}\sqrt{2000^2 - 500^2} = 0.015 + 0.01936$$

$$= 0.0344 \text{ s}$$

(iii) Arrival time T_3 of refracted wave from second layer (Eq. 3.135)

$$= \frac{x}{v_{p3}} + \frac{2h_1\sqrt{v_{p3}^2 - v_{p1}^2}}{v_{p3}\, v_{p1}} + \frac{2h_2\sqrt{v_{p3}^2 - v_{p2}^2}}{v_{p3}\, v_{p2}}$$

$$= \frac{30}{3000} + \frac{2 \times 5\sqrt{3000^2 - 500^2}}{3000 \times 500} + \frac{2 \times 10\sqrt{3000^2 - 2000^2}}{3000 \times 2000}$$

$$= 0.01 + 0.01972 + 0.00745$$

$$= 0.03715 \text{ s}$$

$\therefore$ Minimum travel time = 0.0344 s

Example 3.4

Between two points A and B (Fig. 3.30) two seismic refraction tests were performed and the data obtained is given below.

Source at A		*Source at B*	
Distance of Geophones from A (m)	*Travel time (milli sec)*	*Distance of Geophones from B (m)*	*Travel time (milli sec)*
0	0	0	0
5	3.1	5	3.1
10	6.3	10	6.1
15	9.4	15	9.3
20	12.6	20	11.1
25	15.0	25	12.9
30	15.9	30	14.8
35	16.5	35	16.5

Determine the slope of the refractor, depth of refractor at points A and B, and the two velocities using critical distance approach and time intercept approach.

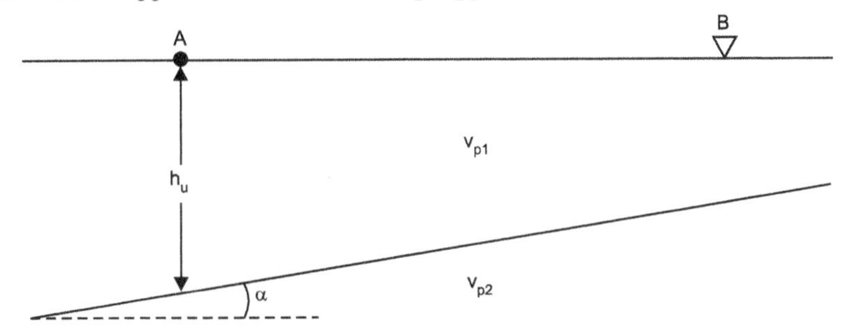

Fig. 3.30. Sloping layer (example 3.4)

Solution

1. Plot the data on time-distance graph as shown in Fig. 3.31. In this figure

$$v_{p1} = \text{Inverse slope of AC} = \frac{15}{9.4 \times 10^{-3}} = 1600 \text{ m/s, or}$$

$$= \text{Inverse slope of BE} = \frac{10}{6.1 \times 10^{-3}} = 1640 \text{ m/s}$$

Adopt $$v_{p1} = \frac{1600+1640}{2} = 1620 \text{ m/s}$$

$$v_{2u} = \text{Inverse slope of } CD = \frac{35}{(16.5 - 11.2) \times 10^{-3}} = 6603 \text{ m/s}$$

$$v_{2d} = \text{Inverse slope of } EF = \frac{35}{(16.5-3.8) \times 10^{-3}} = 2756 \text{ m/s}$$

$$T_{2ui} = 11.2 \times 10^{-3} \text{ s}, x_{cu} = 23.5 \text{ m}$$

$$T_{2di} = 3.8 \times 10^{-3} \text{ s}, x_{cd} = 14.75 \text{ m}$$

2. From Eq. (3.160), $$\sin(i_{pc} - \alpha) = \frac{v_{p1}}{v_{2u}} = \frac{1600}{6603} = 0.2423$$

or $$i_{pc} - \alpha = 14°$$

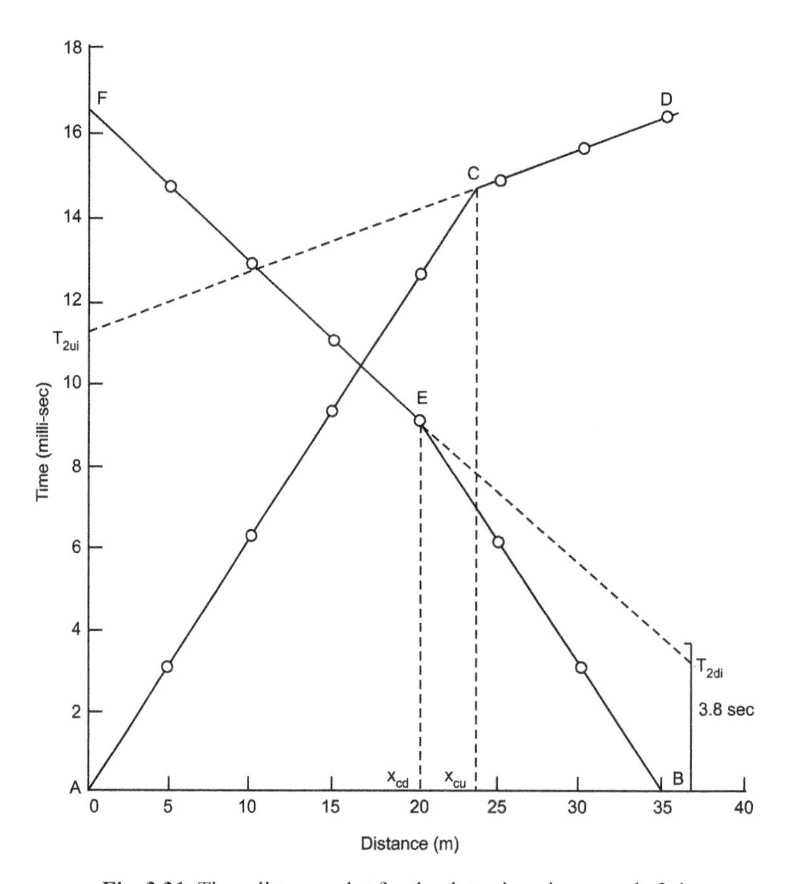

Fig. 3.31. Time-distance plot for the data given in example 3.4

From Eq. (3.159) $\sin(i_{pc} + \alpha) = \frac{v_{p1}}{v_{2d}} = \frac{1600}{2756} = 0.5805$

or $i_{pc} + \alpha = 35.5°$

Therefore, $i_{pc} = 24.7°$ and $\alpha = 10.75°$

3. Time-intercept approach:

From Eqs. (3.153b) and (3.154b)

$$h_d = \frac{T_{2di}\, v_{p1}}{2\cos i_{pc} \cos\alpha}$$

$$= \frac{3.8\times10^{-3} \times 1620}{2\cos 24.7° \cos 10.75°} = 3.4 \text{ m}$$

$$h_u = \frac{T_{2ui}\, v_{p1}}{2\cos i_{pc} \cos\alpha} = \frac{11.2\times10^{-3} \times 1620}{2\cos 24.7° \cos 10.75°} = 10.03 \text{ m}$$

4. Critical distance approach

From Eqs. (3.157) and (3.158)

$$h_d = \frac{x_{cd}\{1-\sin(i_{pc}+\alpha)\}}{2\cos i_{pc} \cos\alpha}$$

$$= \frac{14.75(1-0.5805)}{2\cos 24.7^\circ \cos 10.75^\circ} = 3.4 \text{ m}$$

$$h_u = \frac{x_{cu}\{1-\sin(i_{pc}-\alpha)\}}{2\cos i_{pc} \cos\alpha} = \frac{23.5(1-0.2423)}{2\cos 24.7^\circ \cos 10.75^\circ} = 9.97 \text{ m}$$

5. True velocity (Eq. 3.165), $v_{p2} = \frac{v_{2u}\, v_{2d}}{v_{2u} + v_{2d}} 2\cos\alpha = \frac{6603 \times 2756}{6603 + 2756} \times 2 \times \cos 10.75^\circ$

$$= 3820 \text{ m/sec.}$$

Example 3.5

Following is the data obtained from a seismic refraction test.

Geophone No.	*Distance from shot point (m)*	*Travel time (milli sec)*
G0	0	0
G1	100	200
G2	200	399
G3	300	592
G4	400	780
G5	500	963
G6	600	1138
G7	700	1306
G8	800	1465
G9	900	1618
G10	1000	1763

Assuming that the velocity is varying linearly with depth, determine the values of depth of deepest point of the wave paths from ground surface corresponding to the waves reaching to geophones $G7$ and $G9$. Also determine the coefficient 'k' in both cases.

Solution

(i) Initial velocity $v_0 = \frac{100}{200 \times 10^{-3}} = 500 \text{ m/s}$

Velocity of the wave reaching geophone $G7$ i.e.

$$v_7 = \frac{200}{(1465-1138)\times 10^{-3}} = 611 \text{ m/s}$$

$$\therefore \quad i_0 = \sin^{-1}\frac{500}{611} = 54.9^\circ$$

From Eq. (3.183), $k' = \frac{2v_0}{x}\cot i_0 = \frac{2\times 500}{700}\cot 54.9^\circ = 1.0$

From Eq. (3.182b) $Z_{max} = \frac{v_0}{k'}(\text{cosec } i_0 - 1) = \frac{500}{1.0}(\text{cosec } 54.9^\circ - 1) = 111 \text{ m}$

(ii) Similarly $v_9 = \frac{200}{(1763 - 1465) \times 10^{-3}} = 671 \text{ m/s}$

$$i_o = \sin^{-1}\frac{500}{671} = 48.16^\circ$$

$$k' = \frac{2v_0 \cot i_0}{x} = \frac{2 \times 500 \cot 48.16^\circ}{900} = 0.995$$

$$Z_{max} = \frac{v_o}{k'}(\text{cosec } i_o - 1) = \frac{500}{0.995}(\text{cosec}\, 48.16^\circ - 1) = 172 \text{ m}$$

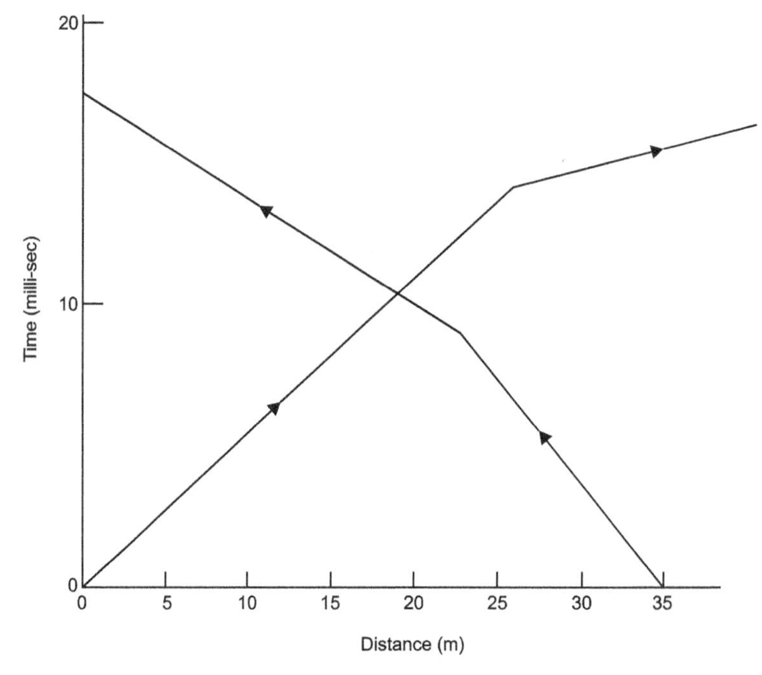

Fig. 3.32. Time distance plots

References

Hardin, B.O. and Richart, F.E. Jr. (1963), "Elastic wave velocities in granular soils." J. Soil, Mech. Found. Engg. Div., Am. Soc. Civ. Engg., vol. 89, SM-I, pp. 33-65.

Koisky, H. (1963). "Stress waves in solids." Dover, New York.

Lamb. H. (1904), "On the propagation of tremors over the surface of an elastic solid," Philos. Trans. Royal Soc. London, Ser. A. 203, pp. 1-42.

Milton, D.B. (1960). "Introduction to geophysical prospecting", McGraw-Hill, New York.

Parasnis, D.S. (1962). "Principles of applied geophysics", Methuen.

Rayleigh, L. (1885), "On waves propagated along the plane surface of an elastic solid," Proc. London Math. Soc., vol. 17, pp. 4-11.

Richart, F.E., Jr. (1962) "Foundation vibrations", Trans. Am. Soc. Civ Engg., vol. 127, Part 1 pp. 863-898.

Timoshenko, S. and Goodier, J.N. (1951), "Theory of elasticity", McGraw Hill Book Co. Inc., Tokyo.

Practice Problems

3.1 Explain the generation of (a) compression wave, (b) shear wave and (c) Rayleigh wave. Describe their relative magnitudes.

3.2 Describe the wave propagation in an infinite, homogeneous, isotropic and elastic medium.

3.3 Assuming Poisson's ratio (μ) as 0.35 and density of soil as 1800 kg/m^3, determine E, G, v_s and v_r if compression wave velocity is 450 m/s.

3.4 What are the informations obtained by a seismic survey? Give its economic aspects.

3.5 Give precisely the three principles which make the basis of seismic wave propagation theory.

3.6 A shot is fired at the ground surface on a particular location and observations from geophones gave the following data.

Distance of gephone from shot point (m)	*Travel time (milli sec)*
5	42.50
10	85.00
15	127.50
20	170.00
25	187.50
30	197.50
40	215.00

Plot the time-distance graph and determine:

(i) Velocities of two layers

(ii) Values of E and G of the two layers assuming $\mu = 0.25$

(iii) Time intercept and critical distance

(iv) Depth of layer from time intercept and critical distance

(v) Verification of travel time for 30 m geophone.

3.7 Prove that in a dipping layer interface trajectory based on critical angles give shortest time.

3.8 In a dipping interface travel time-distance plot is obtained as shown in Fig. 3.32. Determine the following:

(a) Apparent velocities

(b) Slope of interface

(c) Depth of layer at points A and B using critical distance concept

(d) Depth of layer at points A and B using time intercept concept

(e) True velocity V_2

CHAPTER

4

Dynamic Soil Properties

4.1 General

The problems involving the dynamic loading of soils can be divided into two categories:

(i) Having large strain amplitude response—Strong motion earthquakes, blasts and nuclear explosions can develop large strain amplitudes of the order of 0.01% to 0.1%; and
(ii) Having small strain amplitude response—Foundations of the machines have usually low strain amplitude of the order of 0.0001% to 0.001%.

The soil properties which are needed in analysis and design of a structure subjected to dynamic loading are:

(a) Dynamic moduli, such as Young's modulus E, shear modulus G, and bulk modulus K
(b) Poisson's ratio μ
(c) Dynamic elastic constants, such as coefficient of elastic uniform compression C_u, coefficient of elastic uniform shear C_τ. coefficient of elastic non-uniform compression C_ϕ and coefficient of elastic non-uniform shear C_ψ
(d) Damping ratio ξ
(e) Liquefaction parameters, such as cyclic stress ratio, cyclic deformation and pore pressure response
(f) Strength-deformation characteristics in terms of strain rate effects.

Since the dynamic properties of soils are strain dependent, various laboratory and field techniques have been developed to measure these properties over a wide range of strain amplitudes.

4.2 Laboratory Techniques

The laboratory methods used for determining the dynamic properties of soils are:
(i) Resonant column test,
(ii) Ultrasonic pulse test,
(iii) Cyclic simple shear test,
(iv) Cyclic torsional simple shear test, and
(v) Cyclic triaxial compression test.

4.2.1 Resonant Column Test

The resonant column test is used to obtain the elastic modulus E, shear modulus G and damping characteristics of soils at low strain amplitudes. This test is based on the theory of wave propagation in prismatic rods (Richart, Hall and Woods, 1970). Either a cyclically varying axial load or torsional load is applied to one end of the prismatic or cylindrical specimen of soil. This in turn will propagate either a compression wave or a shear wave in the specimen. In this technique the excitation frequency generating the wave is adjusted until the specimen experiences resonance. The value of the resonant frequency is used in getting the value of E and G depending on the type of the excitation (axial or torsional).

The resonant column technique was used for testing of soils by many investigators (Ishimoto and Iida, 1937; Iida, 1938, 1940; Wilson and Dietrich, 1960; Hardin and Richart, 1963; Hall and Richart, 1963; Hardin and Music, 1965; Drnevich, 1967; Anderson, 1974; Lord et al, 1976; Woods, 1978). Several versions of torsional resonant column device using different end conditions to constraint the test specimen are available. Some common end conditions used in developing the equipment are discussed below.

(i) Fixed-free: Hall and Richart (1963) described the apparatus with fixed-free end condition. In this arrangement one end of the specimen is fixed against rotation and the other end is free to rotate under the applied torsion (Fig. 4.1a). A node occurs at the fixed end and the distribution of angular rotation θ along the specimen is a 1/4 sine wave.

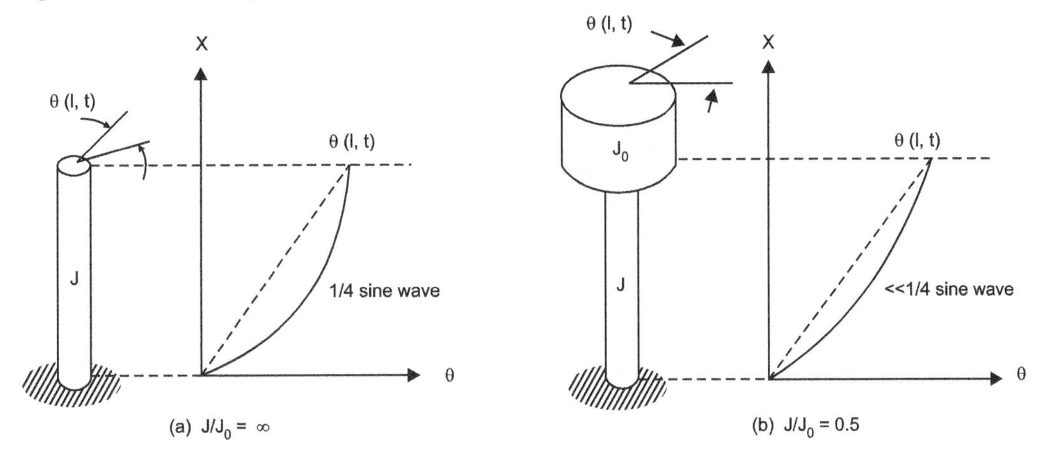

Fig. 4.1. Schematic of resonant column with fixed-free end conditions (After Hardin, 1970 and Drnevich, 1967)

As shown in Fig. 4.1b, by adding a mass at the free end, the variation of θ along the specimen becomes nearly linear. J and J_0 are respectively the polar moment of inertias of the specimen and the added mass respectively. Drnevich (1967) used the concept of added mass to obtain a uniform strain distribution throughout the length of the specimen.

(ii) Spring-base model: Figure 4.2 shows the configuration of an apparatus which can be described as the spring-base model. Depending on the stiffness of the spring compared to specimen's stiffness, it can represent either fixed-free arrangement or fixed-free configuration. In the case when the spring is stiff in comparison to specimen, the configuration may be considered as fixed-free. In such a case, a node will occur at mid height of the specimen, and the distribution of angular rotation would then be a 1/2 cosine wave.

(iii) Fixed-partially restrained: Figure 4.3 shows the configuration of an apparatus in which the top cap is partially restrained by springs acting against an inertial mass. The apparatus used by Hardin and Music (1965) is of this type.

Drnevich (1967, 1972) developed a hollow cylinder apparatus as shown in Fig. 4.4 to study the effect of shear strain amplitude on shear modulus and damping. It may be noted that in the usual torsional resonant column test, the shear strain is not constant but varies from zero at the centre of the sample to a maximum at the outer surface. In hollow specimen, the variation in shearing strain across the thickness of the cylinder wall becomes relatively very small. The configuration of this apparatus is similar to Fig. 4.1b and therefore the shearing strain is almost uniform along the length of specimen. Using Drnevich apparatus, Anderson (1974) tested clays upto 1% shearing strain and Woods (1978) tested dense sands upto shearing strain of 5%.

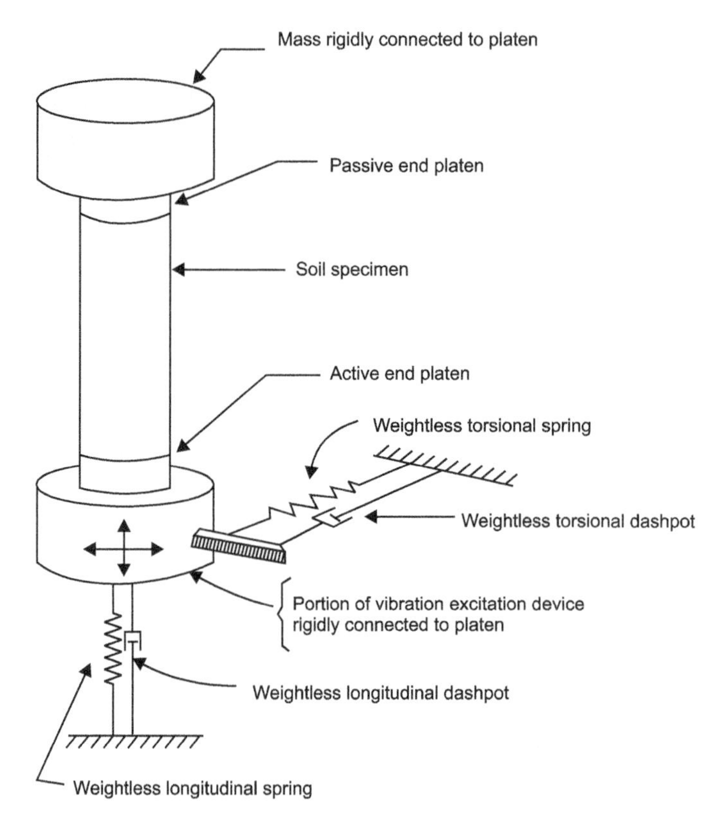

Fig. 4.2. Schematic of resonant column with spring-base model (Woods, 1978)

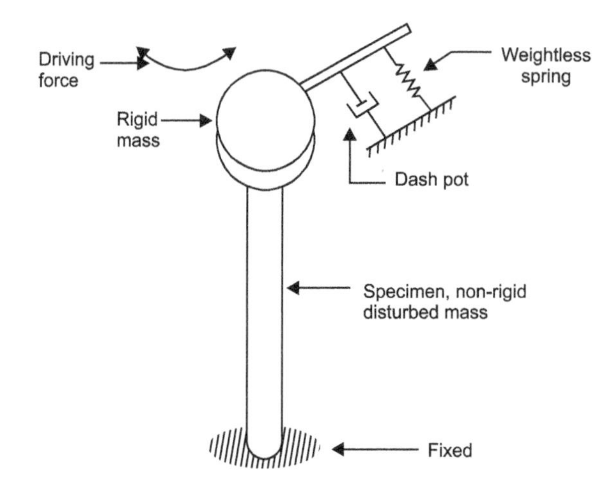

Fig. 4.3. Schematic of resonant column with fixed-partially restrained end conditions (Woods, 1978)

4.2.1.1 Calibration and Determination of G and ξ

Hardin (1970) suggested the following procedure of calibration of the apparatus described in Fig. 4.1b.

(i) For this model the vibration excitation device itself, without a specimen attached, is a single degree of freedom system. Firstly remove the specimen cap and the additional rigid mass, connect the sine wave generator to the vibration excitation device and vary the excitation frequency to determine the resonant frequency f_{n1} of the device.

(ii) Attach the additional rigid mass of polar moment of inertia J_0, and determine the resonant frequency f_{nA} of the new system.

The rotational spring constant (torque per unit rotation), K_0, of the spring about the axis of specimen can be obtained using Eq. 4.1.

$$K_0 = \frac{4\pi^2 - J_A f_{nA}^2}{\left[1 - \left(\frac{f_{nA}}{f_{n1}}\right)^2\right]} \tag{4.1}$$

(iii) With the added mass removed and with the specimen cap, specimen and all apparatus, determine the resonant frequency f_{n0}. The value of mass polar moment of inertia of the rigid mass, J_0 can be computed using Eq. 4.2.

$$J_0 = \frac{K_0}{4\pi f_{n0}^2} \tag{4.2}$$

Now at resonance cut off the power and record the decay curve for the vibration. From the decay curve compute the logarithmic decrement for the apparatus, δ_A, as follows:

$$\delta_A = \frac{1}{n} \log_e \frac{A_1}{A_n} \tag{4.3}$$

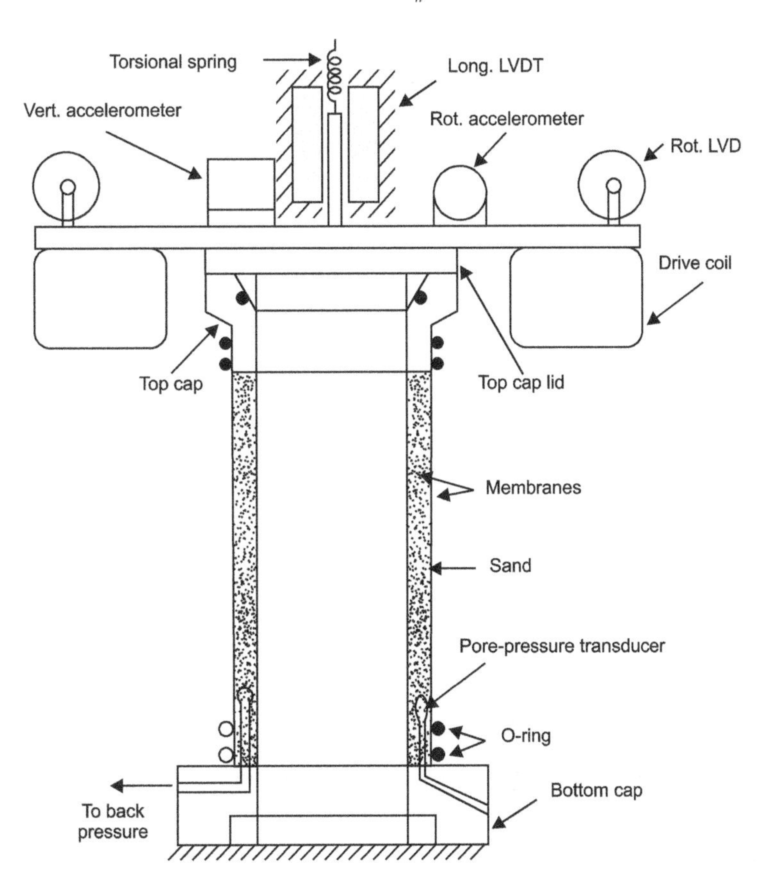

Fig. 4.4. Hollow specimen resonant column apparatus (Drnevich, 1972)

where A_1 = Amplitude of vibration for the first cycle
A_n = Amplitude of vibration for the nth cycle.
Under steady state vibrations, the apparatus damping constant, D_A is given by

$$D_A = \frac{\delta_A}{\pi}\sqrt{K_0 J_0} \tag{4.4}$$

(iv) To measure the torque current constant, K_t excite the apparatus successively at frequencies $(\sqrt{2}/2)f_{n0}$, $\sqrt{2}f_{n0}$ and $2f_{n0}$. During the steady state vibration at each of these frequencies measure the current flowing through the coils, C in amperes, and the displacement amplitude of vibration, θ in radians. For each frequency compute the torque-current constant K_τ, as follows:

$$K_t = \frac{\theta K_0}{CM_f} \tag{4.5}$$

where M_f is given in Table 4.1.

Table 4.1. Value of M_f

Frequency	M_f
$\left(\frac{\sqrt{2}}{2}\right)f_{n0}$	2
$(\sqrt{2})\,f_{n0}$	1
$2f_{n0}$	$\frac{1}{3}$

The value of K_1 shall be taken as the average of three measured values. The three measured values should not differ by more than 10%.

After calibration, the specimen shall be placed in the apparatus with minimum disturbance. A known value of ambient pressure is applied as done in triaxial compression test. With the power as low as is practical, the resonant frequency of the system, f_{nR}, is obtained by varying the frequency of excitation. At resonant frequency, the amplitude of vibration, θ_R, in radians and the current flowing through the coils of the vibration excitation device, C_R are measured. Now the power is cut off and the record of decay curve for the free vibration of the system with specimen is obtained. Using this decay curve, the value of the logarithmic decrement, δ_s can be obtained employing Eq. (4.3).

The procedure of obtaining G and ξ has been explained in the following steps:

(i) Calculate the mass density of the specimen, ρ, from Eq. (4.6).

$$\rho = \frac{4W}{\pi d^2 lg} \tag{4.6}$$

where W = Total weight of specimen
l = Length of specimen
d = Diameter of specimen
g = Acceleration due to gravity

(ii) Calculate the inertia of the specimen about its axis, J, as follows:

$$J = \frac{\pi \rho d^4 l}{32} \tag{4.7}$$

(iii) Calculate the system factor, T as follows:

$$T = \frac{J_0}{J} - \frac{K_0}{4\pi^2 J f_{nR}^2} \tag{4.8}$$

where
J_0 = Mass polar moment of inertia of the apparatus, defined by Eq. (4.2).
K_0 = Rotational spring constant, defined by Eq. (4.1)
J = Inertia of the specimen, defined by Eq. (4.7)
F_{nR} = Resonant frequency of the complete system.

(iv) Using Fig. 4.5, determine the dimensionless frequency F for the value of T computed in step (iii).

$$G = 4\pi^2 \rho \left(\frac{f_{nR} \cdot l}{F} \right)^2 \tag{4.9}$$

(v) For steady state vibrations, the damping factor of the system, D_s is given by

$$D_s = \frac{1}{4\pi^2 J^2 f_{nR}^2} \left(\frac{K_t C_R}{\theta_R} - 2\pi D_A F_{nR} \right) \tag{4.10}$$

(vi) Compute D_s/T. Using Fig. 4.6, determine the value of R corresponding to the value of T computed in step (iii). Then damping ratio is given by

$$\xi = 0.5 \frac{D_s}{TR} \tag{4.11}$$

(vii) For the free vibration method, using Fig. 4.7, determine the value of mode shape factor C_m for the value of T computed in step (iii).

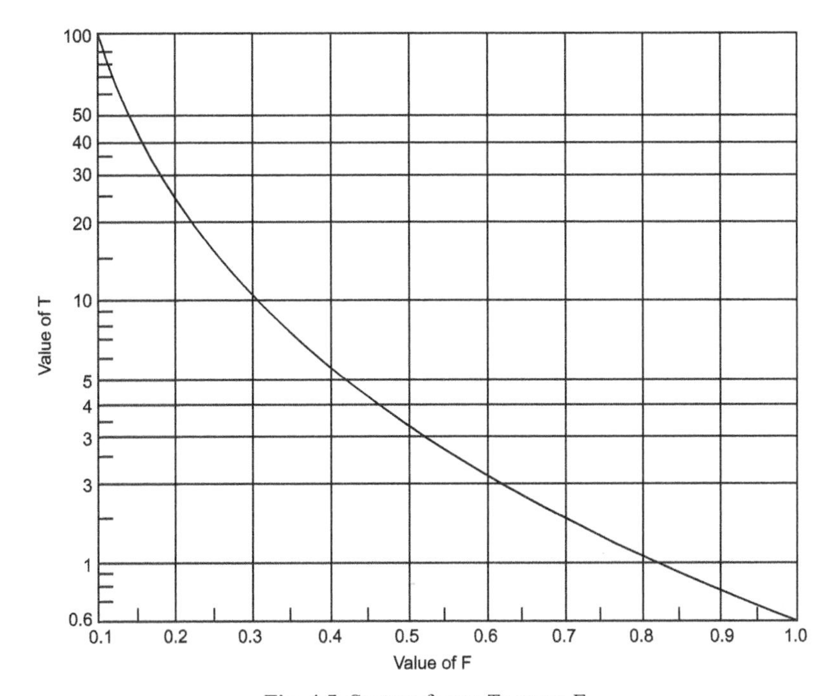

Fig. 4.5. System factor T versus F

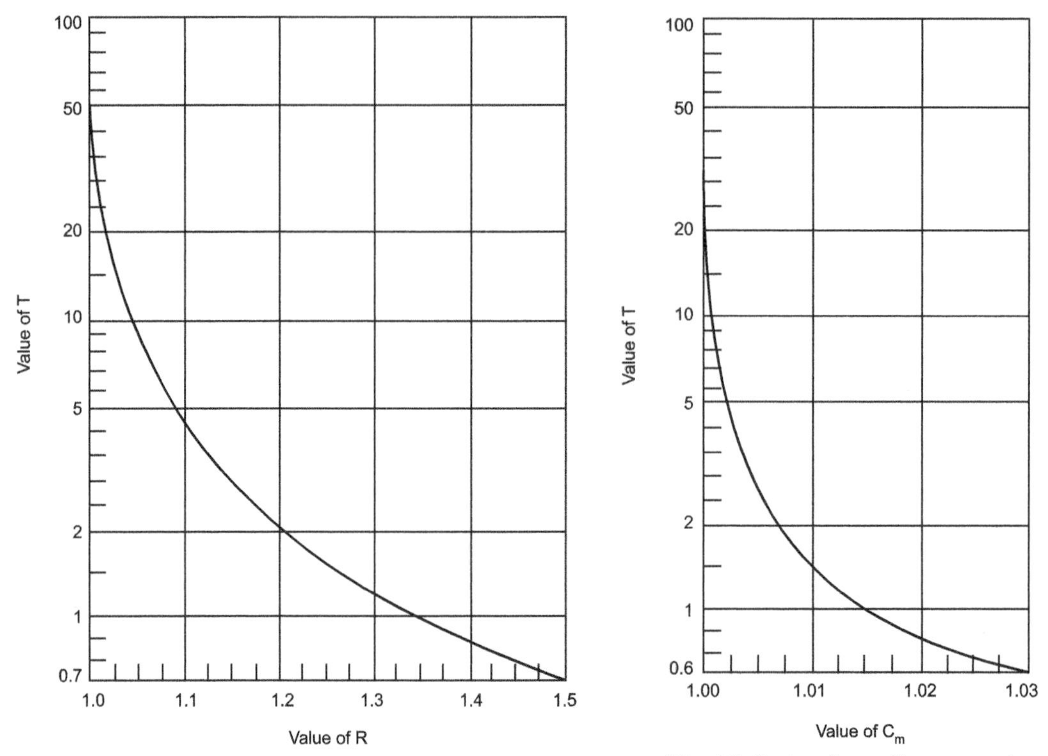

Fig. 4.6. System factor T versus R

Fig. 4.7. System factor T versus mode shape factor C_m

(iii) Compute the system energy ratio, S as follows.

$$S = \frac{32 K_0 l}{\pi C_m G d^4} \tag{4.12}$$

The value of damping ratio ξ is then given by

$$\xi = \frac{l}{2\pi}[\delta_s(1+S) - \delta_A S] \tag{4.13}$$

4.2.2 Ultra Sonic Pulse Test

The theory of ultrasound is similar to that of audible sound. Sound is the result of mechanical disturbance of a material, that is, a vibration. Ultrasonic pulses of either compression or shear waves can be generated and received by suitable piezoelectric crystals. Using elastic theory, a relationship between the speed of propogation and wave amplitude of these waves and certain properties of the media through which they are travelling can be determined as follows:

$$E = \rho v_c^2 \frac{(1+\mu)(1-2\mu)}{(1-\mu)} \tag{4.14}$$

$$G = \rho v_s^2 \tag{4.15}$$

$$\mu = \frac{1 - \frac{1}{2}\left(\frac{v_c}{v_s}\right)^2}{1 - \left(\frac{v_c}{v_s}\right)^2} \tag{4.16}$$

$$\delta = \frac{2.302}{n} \log_{10} \frac{A_0}{A_n} \tag{4.17}$$

where
v_c = Velocity of compression wave
v_s = Velocity of shear wave
μ = Poisson's ratio
E = Young's modulus
G = Shear modulus
ρ = Mass density = γ/g
δ = Logarithmic decrement
A_0 = Initial value of amplitude
A_n = Amplitude after n oscillations.

Lawrence (1963) described the basic apparatus required to measure the propagation velocities (i.e. v_c and v_s) through sand. Stephenson (1978) described an equipment for conducting the ultrasonic tests. His equipment includes a pulse generator, an oscilloscope, and two ultrasonic probes (transmitter and receiver). The pulse generator delivers a variable-voltage direct current pulse to the transmitting probe simultaneously with a 7 volt trigger pulse to the time base of the oscilloscope. The generator was designed such that the pulse interval and pulse width can be varied. Stephenson carried out the tests on silty clay samples.

In this test ultrasonic transmitters and receivers are attached to platens that can be placed at each end of a specimen with the distance separating them carefully measured. The transmitters and receivers are made of piezoelectric materials which exhibit changes in dimensions when subjected to a voltage across their faces, and which produce a voltage across their faces when distorted. A high-frequency electrical pulse applied to the transmitter causes it to deform rapidly and produce a stress wave that travels through the specimen towards the receiver. When the stress wave reaches the receiver, it generates a voltage pulse that is measured. The distance between the transmitter and receiver is divided by the time difference between the voltage pulses to obtain the wave propagation velocity.

One of the main advantages of ultrasonic test is that it can be performed on very soft sea floor sediments while still retained in the core liner.

The drawback of this approach is that it is very difficult to identify the exact wave arrival times. Secondly the strain amplitudes which can be achieved by this technique, are only in the very low region.

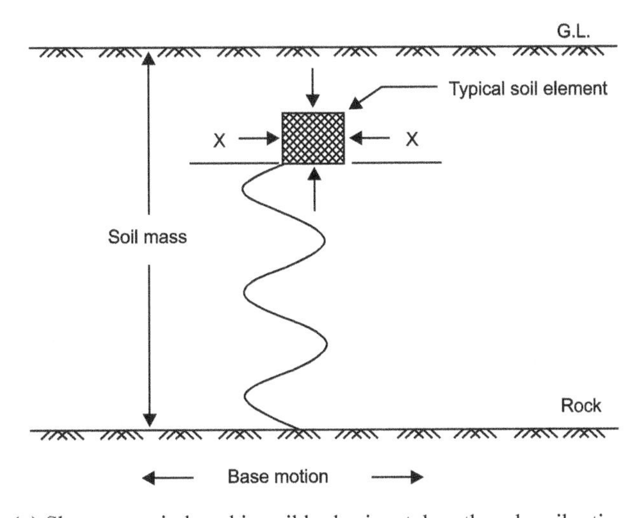

(a) Shear wave induced in soil by horizontal earthquake vibrations

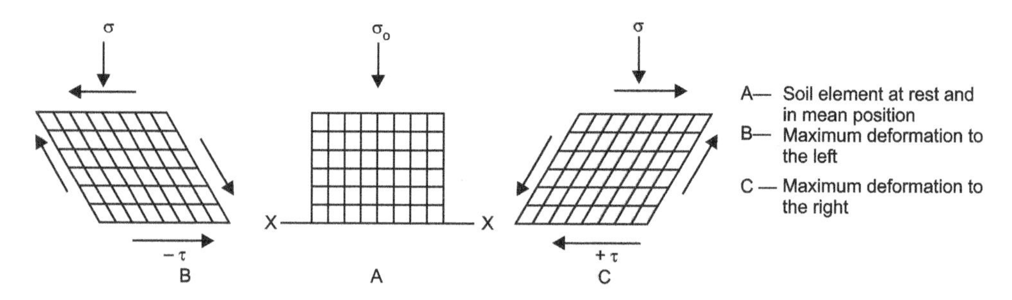

(b) Shear deformation resulting from propagated wave (for a single-cycle)

Fig. 4.8. Stress condition on an element of soil below ground surface

4.2.3 Cyclic Simple Shear Test

During an earthquake or other source of ground vibrations, a soil element below a foundation or in an embankment is subjected to an initial sustained stress together with a superimposed series of repeated and reversals of shear stresses (Fig. 4.8). The magnitude of induced shear stresses depend on the magnitude of acceleration of the dynamic force. In a direct shear box test, uniform state of shear strain occurs only on either side of failure plane. The simple shear device was designed to overcome this limitation of direct shear box by enabling a uniform state of shear strain throughout the specimen. This simulates the field conditions in a much better way.

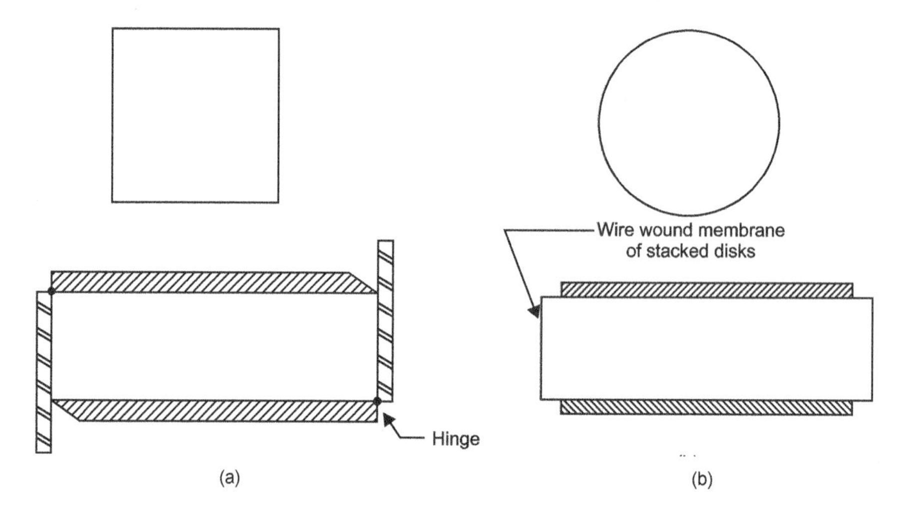

Fig. 4.9. Schematic arrangement of Roscoe simple shear apparatus

Kjellman (1951), Hvorslev and Kaufman (1952), Roscoe (1953), and Bjerrum and Landra (1966) have described simple shear apparatus. The Roscoe simple shear device has a box for a square shaped sample with side length of 60 mm and a thickness of 20 mm. The box is provided with two side walls and two hinged walls (Fig. 4.9). Peacock and Seed (1968) have modified Roscoe's simple shear apparatus for dynamic testing. The dynamic shear forces were applied by a double acting piston with controlled compressed air pressure using solenoid valves. Figure 4.10 illustrates how the end walls rotate simultaneously at the ends of the shearing chamber to deform the soil uniformly. Prakash, Nandkumaran and Joshi (1973) have also developed similar type of simple shear apparatus which has the facility of applying sustained normal stress, sustained shear stress and oscillatory shear stress.

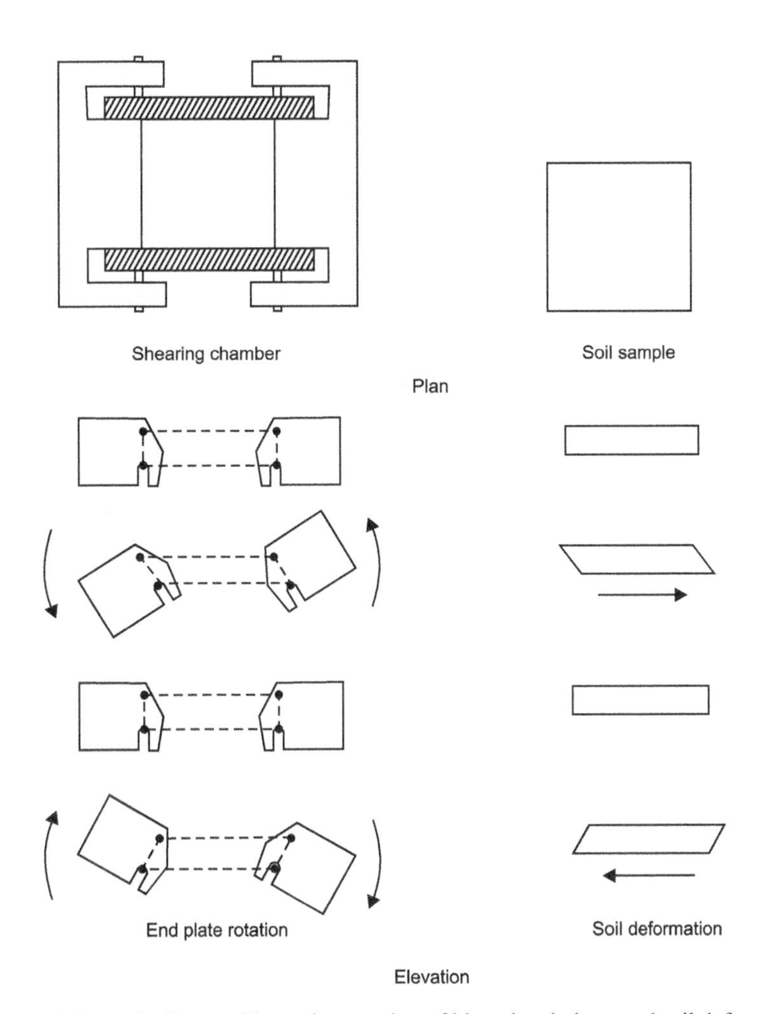

Fig. 4.10. Schematic diagram illustrating rotation of hinged end plates and soil deformation in oscillatory simple shear (After Peacock and Seed, 1968)

Prakash, Nandkumaran and Bansal (1974) have conducted tests on three artificial soils (SM, CL and CH) using oscillatory shear apparatus developed by Prakash et al (1973). A more systematic study has been done by Dass (1977) on clay of high compressibility (CH, LL = 65.5%. PL = 28.0%, q_u = 78 kN/m^2) using the same dynamic simple shear apparatus. The salient features of her work are reported here to understand the behaviour of soil under dynamic load. The static strength of soil was 36 kN/m^2. She performed the tests keeping the variation of various parameters as given below:

Sustained normal stress, ρ_n (kN/m^2) = 15, 21, 26, 29

Sustained shear stress, τ_{st} (% of static strength) = 10, 25, 50, 60

Oscillatory shear stress, τ_{dyn} (kN/m^2) = 13, 19, 22, 25

Number of cycles of oscillatory shear stress = 1 to 1100

The oscillatory shear stress was applied at the rate of one cycle per second. The failure criteria was chosen corresponding to 12 mm displacement.

A typical plot of test data in terms of number of cycles versus shear displacement is shown in Fig. 4.11 for τ_{dyn} equal to 13 kN/m^2. Similar plots were obtained for other values of τ_{dyn}. From these plots, number of cycles and oscillatory shear stress corresponding to 12 mm displacement have been obtained and plotted as shown in Figs. 4.12 and 4.13. It can be seen from Figs. 4.12 and 4.13

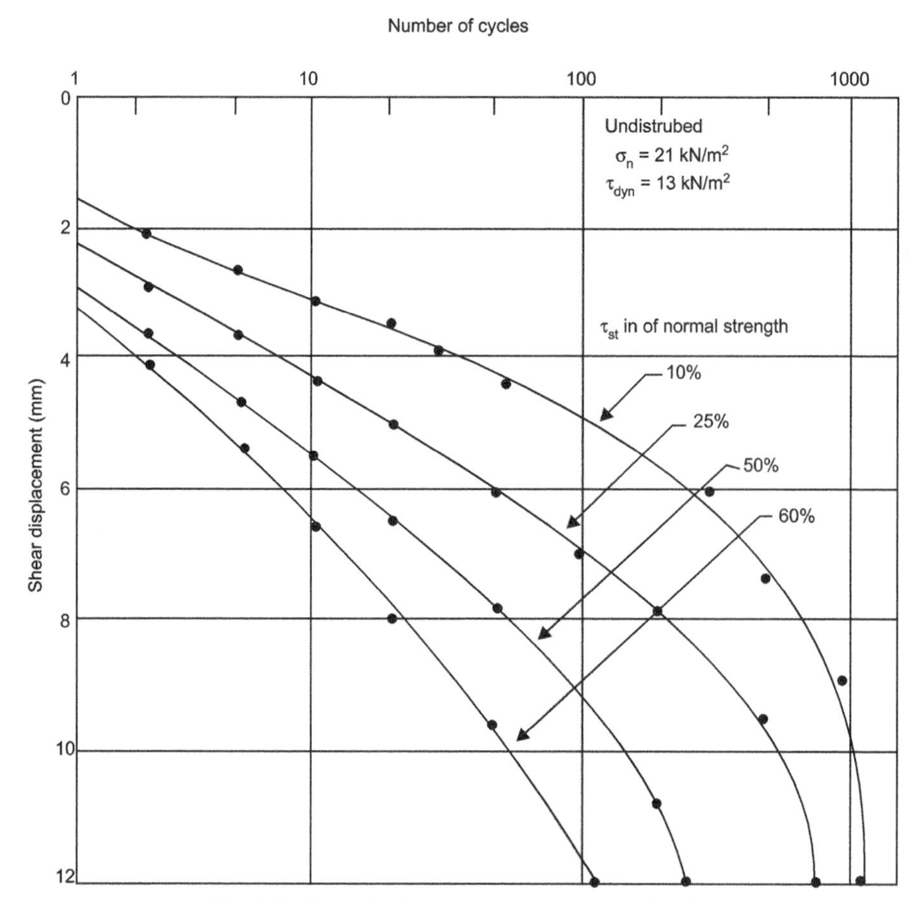

Fig. 4.11. Shear displacement versus number of cycles

that for a fixed sustained shear stress, the amount of dynamic shear stress decreases as number of cycles increases for causing 12 mm displacement. As sustained shear stress increases, less number of cycles and dynamic shear stress is needed to cause failure.

4.2.4 Cyclic Torsional Simple Shear Test

In cyclic simple shear apparatus, it is not possible to measure the confining pressure and the test is performed under K_0 consolidation conditions. Torsional simple shear devices have been developed to ovecome these difficulties. Ishihara and Li (1972) modified the triaxial apparatus to provide torsional strain capabilities. This has the disadvantage that the shear strain in the sample varies with maximum at the outer circumference and zero at the centre. This problem has been minimised by using hollow cylinder torsional shear apparatus (Hardin, 1971; Drnevich, 1972; Yoshimi and Oh-Oka, 1973; Ishibashi and Sherif, 1974; Ishihara and Yasuda, 1975; Cho et al, 1976; Iwasaki, Tatsuoka and Takagi, 1978). The apparatus used by Drnevich (1972) has already been described earlier and shown in Fig. 4.4. This has the advantage that both resonant column and cyclic torsional shear tests can be performed in the same device and on the same sample.

Some of the above mentioned investigators have used long hollow cylinder to obtain uniform conditions at the test section (Hardin, 1971; Drnevich, 1972; Ishihara and Yasuda, 1975; Iwasaki et al, 1977). It is difficult to prepare samples for long hollow cylinder devices. Keeping this fact in view, Yoshini and Oh-Oka (1973), Ishibashi and Sherif (1974) and Cho et al (1976) used a short cylinder in which the taper was proportional to the inside and outside radii (Fig. 4.14). The internal

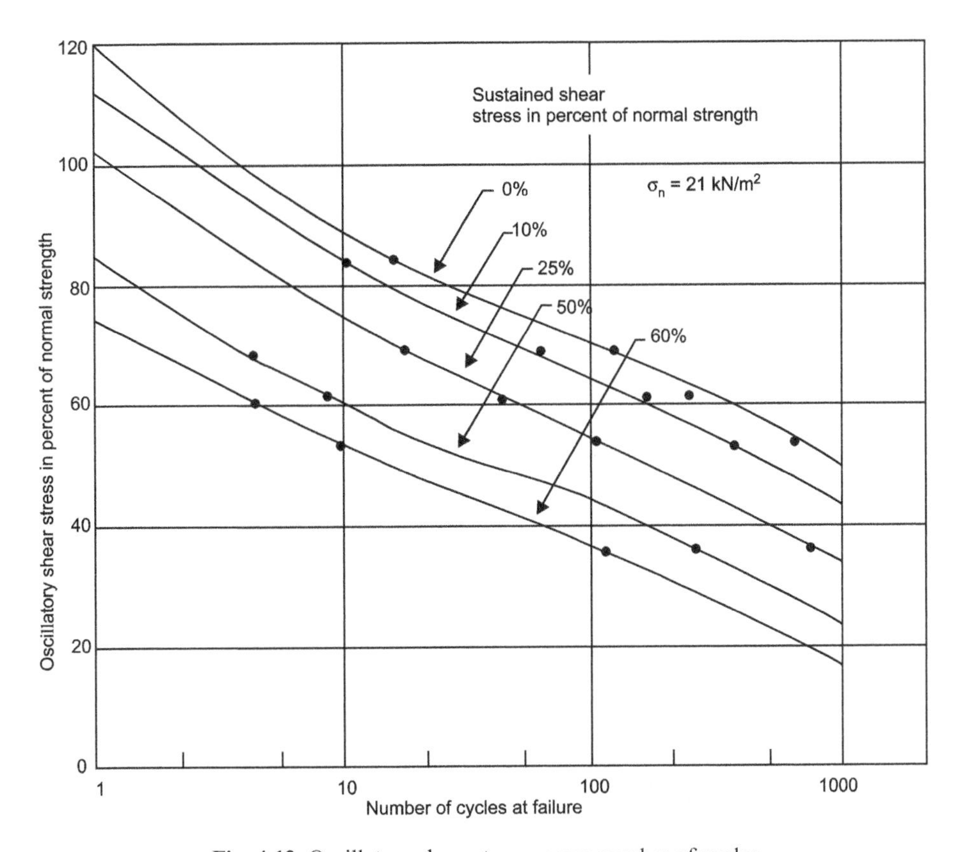

Fig. 4.12. Oscillatory shear stress versus number of cycles

and external radii r_1 and r_2 are selected such that the difference in average shear stresses computed considering two extreme conditions is minimum. The two extreme conditions are:

(i) Shear stress varies linearly with radius as for an elastic material (Eq. 4.18)
(ii) Shear stress is constant as for pure plastic deformation (Eq. 4.19)

$$\tau_{ave} = \frac{4T}{3\pi}\left[\frac{r_2^3 - r_1^3}{(r_2^2 - r_1^2)(r_2^4 - r_1^4)}\right] \tag{4.18}$$

$$\tau_{avp} = \frac{3T}{2\pi}\left(\frac{1}{r_2^3 - r_1^3}\right) \tag{4.19}$$

where T is the applied torque.

4.2.5 Cyclic Triaxial Compression Test

In general, the stress-deformation and strength characteristics of a soil depend on the following factors:

1. Type of soil
2. Relative density in case of cohesionless soil; consistency limits, water content and state of disturbance in cohesive soils
3. Initial static stress level i.e. sustained stress
4. Magnitude of dynamic stress
5. Number of pulses of dynamic load
6. Frequency of loading

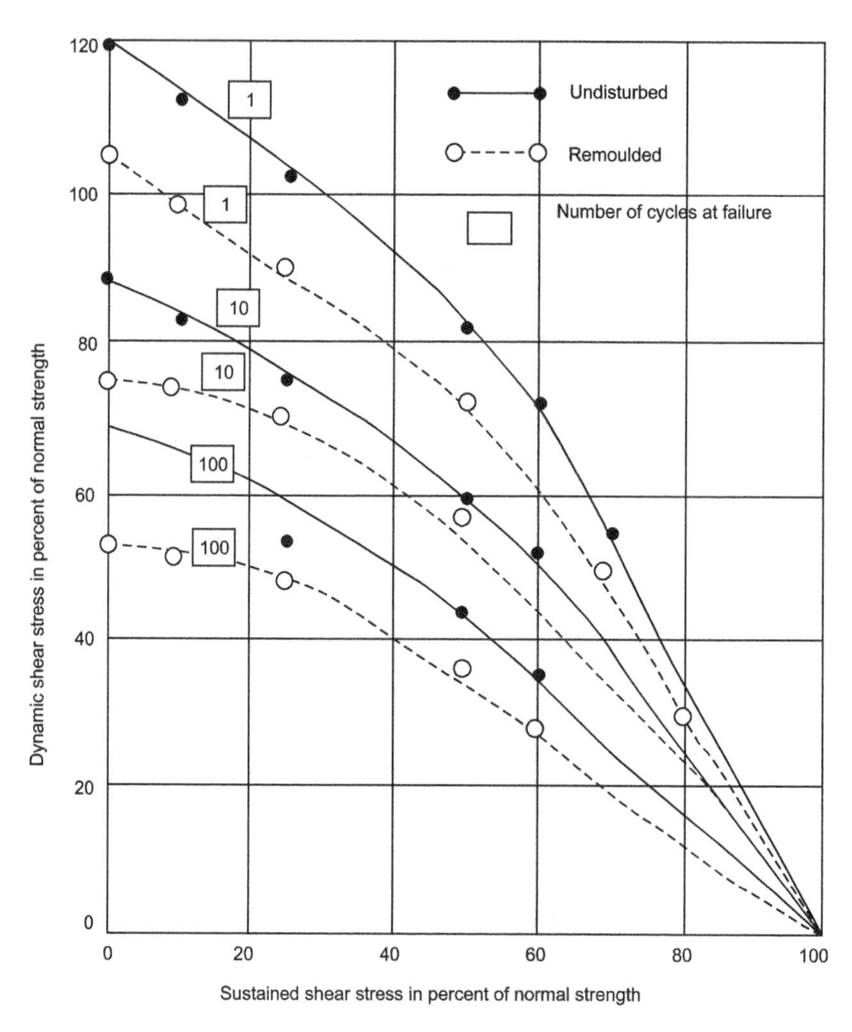

Fig. 4.13. Dynamic shear stress versus sustained shear stress

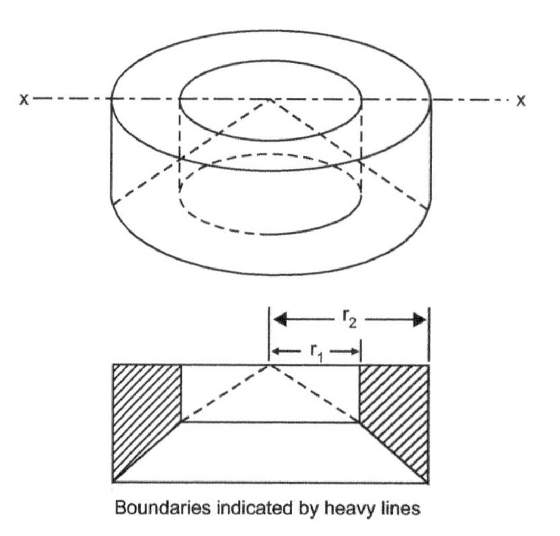

Fig. 4.14. Cross-section for short specimen

7. Shape of wave form of loading
8. One directional or two directional loading

In one directional loading only compression of the sample is done while in two directional loading both compression and extension is done. All the factors listed above can be studied lucidly on a triaxial set up.

Prof. Arthur Casagrande of Harvard University was refered a problem of studying the effect of vibrations created by bomb explosion on the stability of the Panama canal projects. For this Casagrande and Shannon (1948, 1949) developed the following three types of apparatus for studying the strength of soils under transient loading (Table 4.2).

Table 4.2. Type of Apparatus

Type of apparatus	*Time of loading (seconds)*	*Remarks*
(i) Pendulum loading	0.05 to 0.01	Suitable for performing fast transient tests
(ii) Falling beam	0.5 to 300	
(iii) Hydraulic loading	0.05 to any desired larger value	

Time of loading was defined as the time between the beginning of test and the point at which the maximum compressive stress is reached (Fig. 4.15). The pendulum loading apparatus (Fig. 4.16) utilizes the energy of a pendulum which, when released from a selected height, strikes a spring connected to the piston rod of a hydraulic lower cylinder. This lower cylinder is connected hydraulically to an upper cylinder, which is mounted on a loading frame.

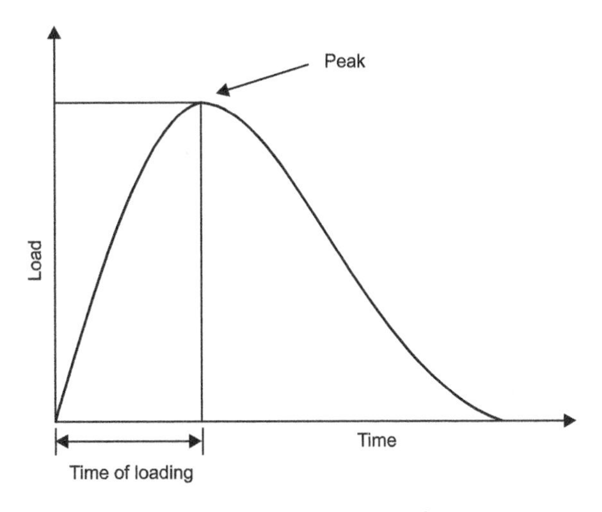

Fig. 4.15. Time of loading in transient tests

The falling beam apparatus consists essentially of a beam with a weight and rider, a dashpot to control the velocity of the fall of the beam, and a yoke for transmitting the load from the beam to the specimen (Fig. 4.17). A small beam mounted above the yoke counter-balances the weight of the beam.

The hydraulic loading apparatus (Fig. 4.18) consists of a constant volume vane-type hydraulic pump connected to a hydraulic cylinder through valves by which either the pressure in the cylinder or the volume of the liquid delivered to the cylinder can be controlled. The peak load that can be produced by this apparatus is much greater than can be obtained by either the pendulum type or falling beam apparatus.

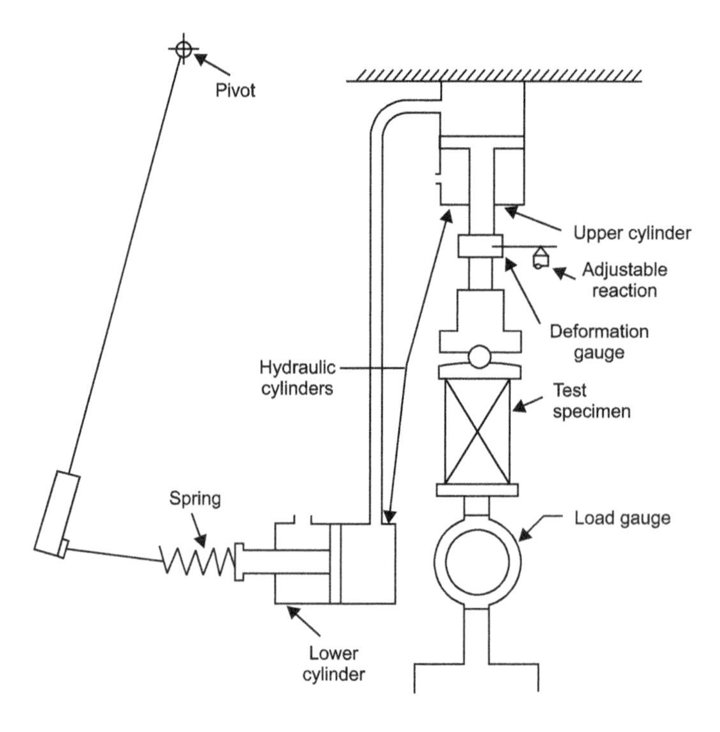

Fig. 4.16. Pendulum loading apparatus (Casagrande and Shannon, 1948)

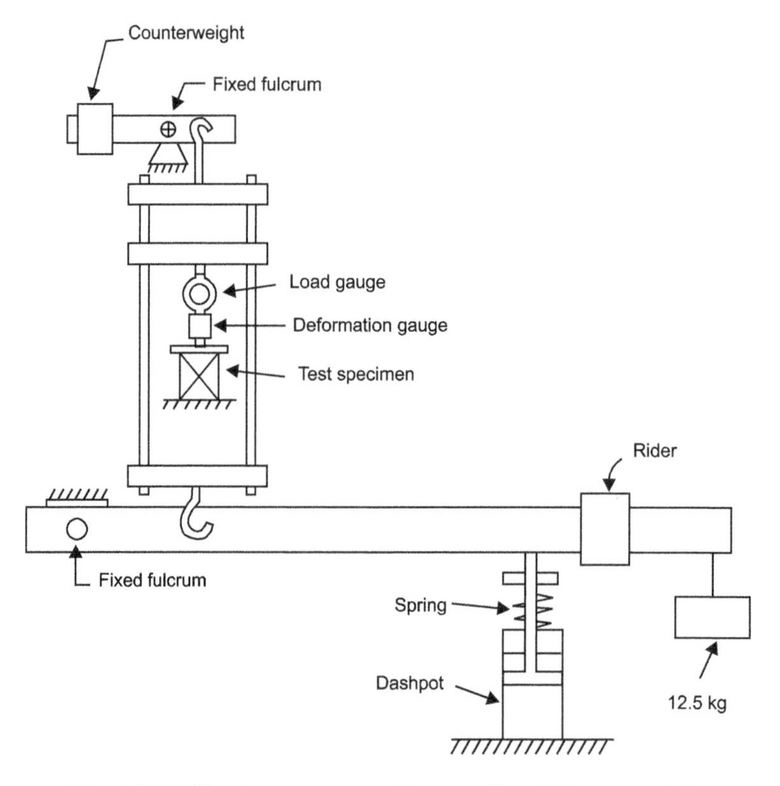

Fig. 4.17. Falling beam apparatus (Casagrande and Shannon, 1948)

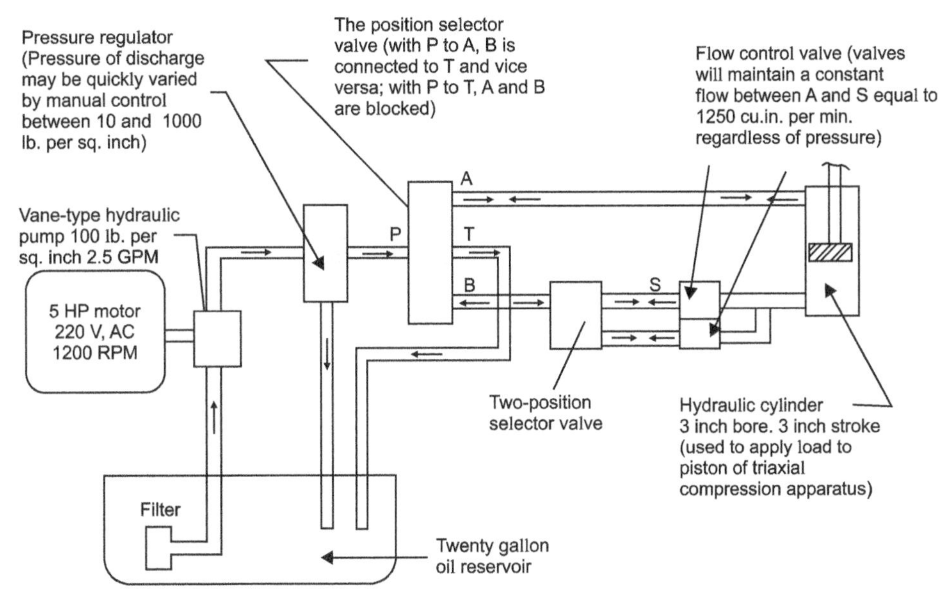

Fig. 4.18. Hydraulic loading apparatus (Casagrande and Shannon, 1948)

For measuring load, a load gauge of rectangular or cylindrical shape is used, with four strain gauges mounted on the inside face. For measuring deflection, a thin flexible steel spring cantilever is used with strain gauges mounted on the cantilever, the base of which is clamped to the loading piston.

Casagrande and Shannon (1948) performed both static and transient compression tests on six different materials. Out of these, two typical materials named as Cambridge clay and Manchester sand having properties as given in Table 4.3 are selected for illustration. The transient compression tests were performed with different time of loading both in confined and unconfined states.

Table 4.3. Properties of Soils used in Tests

Cambridge clay		*Manchester sand*	
Natural water Content	30-50%	Grain Size	0.21 mm to 0.42 mm
Liquid limit	37-59%	e_{max}	0.88
Plastic limit	20-27%	e_{min}	0.61

In Fig. 4.19, a simultaneous plot of stress and strain versus time from an unconfined compression test with a time of loading of 0.02 s on Cambridge clay is shown. Similar plots were prepared for other times of loading and on Manchester sand. Using this data, stress-strain plots were obtained as shown in Figs. 4.20a and 4.20b. In these figures, stress-strain curves for corresponding static tests are also shown. Typical plots of maximum compressive stress versus time of loading for unconfined and confined transient tests on Cambridge clay are shown in Fig. 4.21a and b respectively. A typical plot in terms of principal stress ratio at failure and time of loading for Manchester sand is shown in Fig. 4.22.

From the typical test data presented above, it may be concluded that the strength of clay decreases with the increases in time of loading and for time of loading equal to 0.02 s, the strength of clay is approximately 1.75 to 2.0 times greater than the static strength. The strength of sand is almost independent to the time of loading. Transient strength of sand increased only about 10 percent.

Modulus of deformation is defined as the slope of a line drawn from the origin through the point on the stress-deformation curve and corresponding to stress of one-half the strength. It is found that

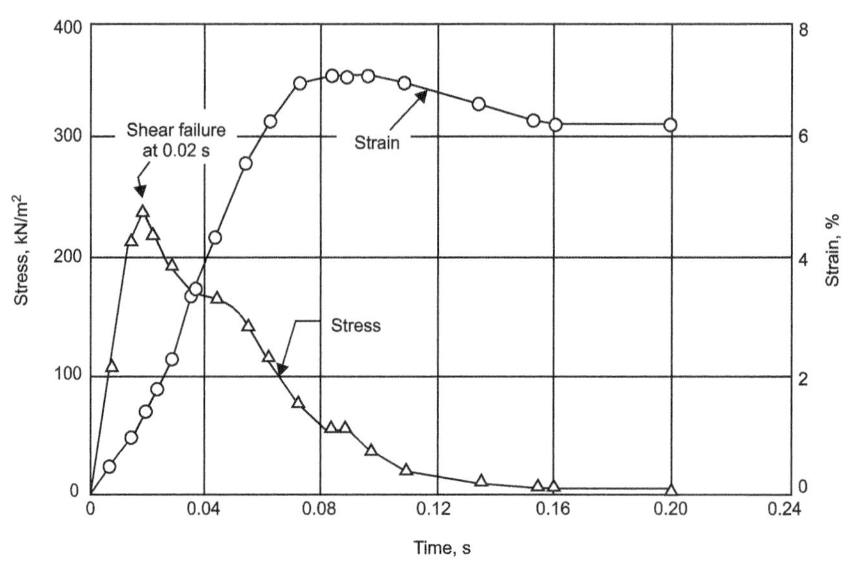

Fig. 4.19. Time vs stress and strain in an unconfined transient test on Cambridge clay (Cassagrande and Shannon, 1948)

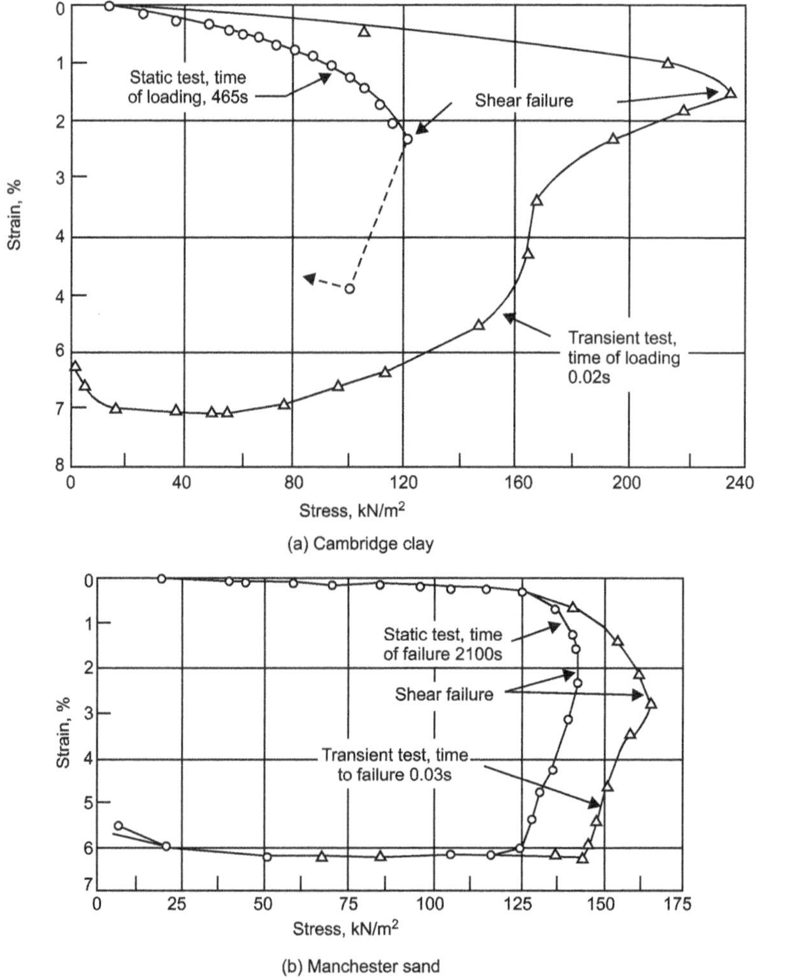

Fig. 4.20. Stress vs strain curves

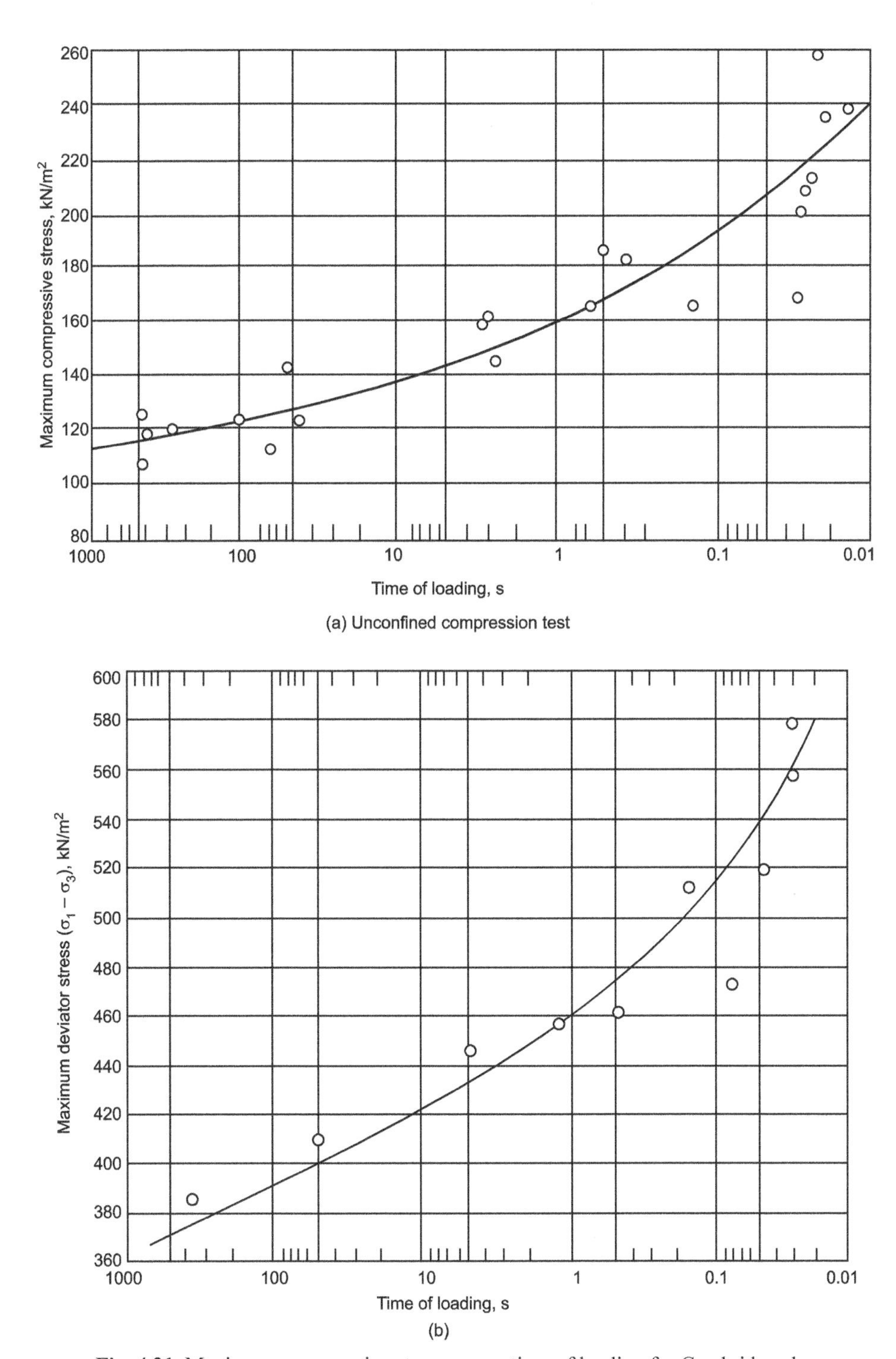

Fig. 4.21. Maximum compressive stress versus time of loading for Cambridge clay

in case of clays, modulus of deformation in fast transient tests was about two times that obtained in static tests. In case of sands, modulus of deformation was found independent to the time of loading.

As evident from above discussions, in transient loading test, an initially unstressed sample of soil is loaded to failure in a short period of time. Under earthquake loading conditions, an initially stressed soil element is subjected to a series of stress pulses, none of which would necessarily cause

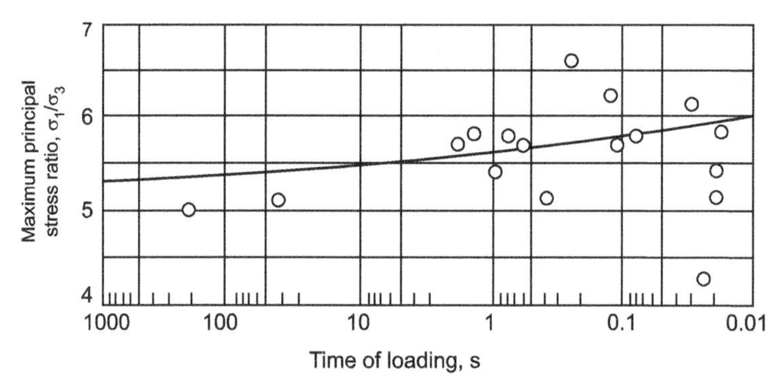

Fig. 4.22. Maximum principal stress ratio versus time of loading for Manchester sand

failure by itself, but the cumulative effect of which is to induce failure or significant deformations. Seed and Fead (1959) developed the oscillatory triaxial apparatus as shown in Fig. 4.23 to study the effect of number of stress reversals and other factors on the deformation in soils.

Seed (1960) performed tests on Vicksburg silty clay (w_n = 22%, S = 93%) for studying the effect of various factors on its dynamic strength characteristics. A typical stress strain-relationship is shown in Fig. 4.24. It pertains to the sustained static stress of 200 kN/m². The magnitude of dynamic

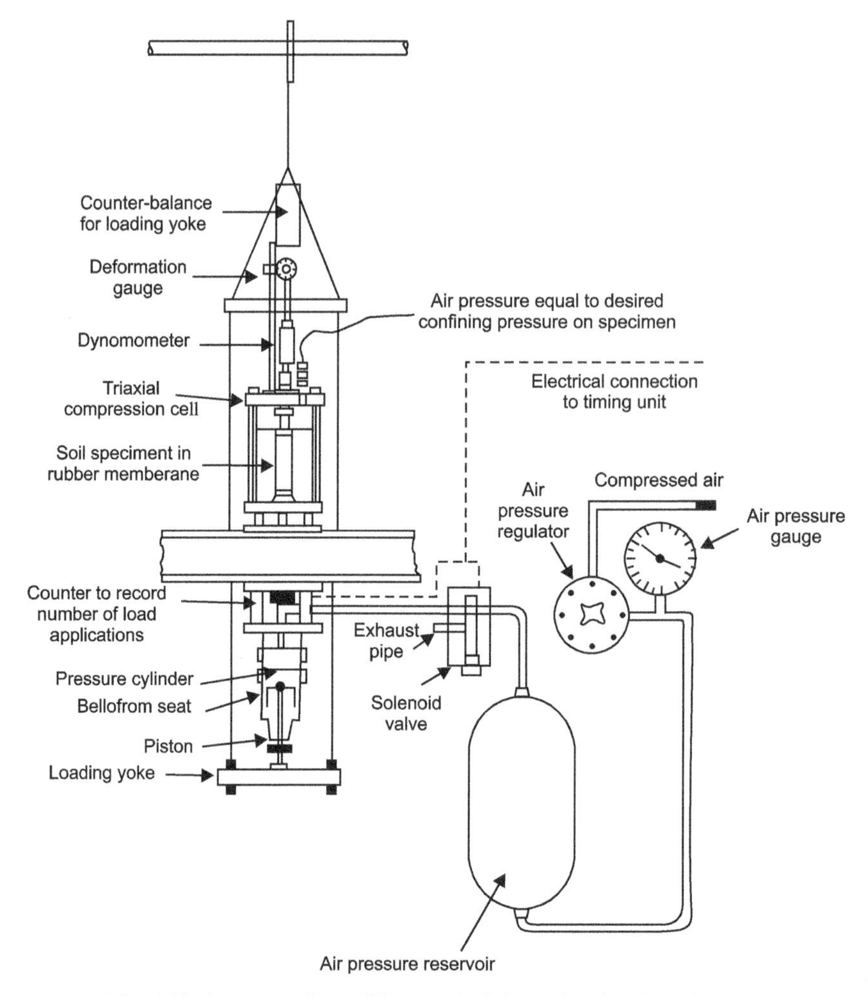

Fig. 4.23. Apparatus for oscillatory triaxial test (Seed and Fead, 1959)

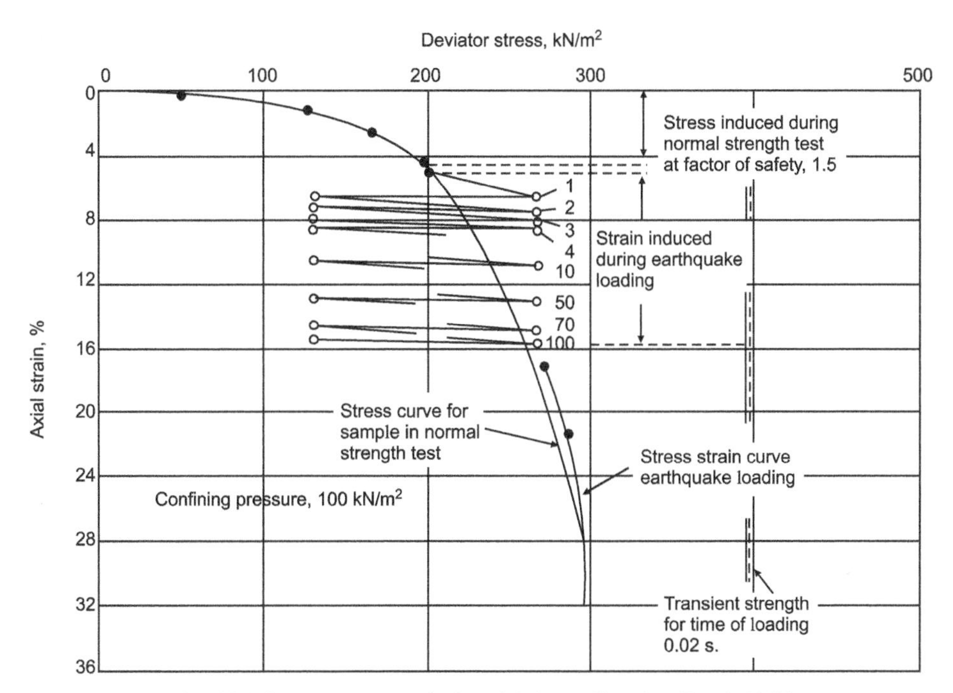

Fig. 4.24. Stress versus strain for Vicksburg silty clay (Seed, 1960)

stress was 35% of the sustained static stress. It gives the magnitude of dynamic stress as ±70 kN/m^2. The static stress-strain curve indicating the static strength as 300 kN/m^2 is also shown in the figure. Therefore the initial factor of safety for this typical case will be 300/200 = 1.5. Transient strength of the soil corresponding to 0.02 s time of loading was found as 400 kN/m^2. In this figure, *B* represents the point which is obtained after 100 cycles of loading. It may be noted that at point '*B*', the strains are increased but factor of safety still remains more than unity (3.0/2.7). It means that failure will not occur but the strains may become excessive. Hence in design, one should examine whether these strains are within permissible limits.

The deviator stress versus strain curves similar to as shown in Fig. 4.24 were drawn for different values of sustained static stress and dynamic oscillatory stress. Figure 4.25 shows the effects of single transient stress applications of the various intensities for initial factors of safety ranging between 1.0 and 2.0. The shaded portion of the figure shows the deformation of the specimens induced at stress levels corresponding to the different factors of safety in normal strength test. The upper curves show the increase in deformation caused by single transient pulses corresponding to 20, 40 and 60 percent of the initial sustained stress. Figures 4.26 and 4.27 show similar data for a series of 30 and 100 pulses. It is interesting to note that for this soil a single transient pulse equal to 20 percent of initial sustained stress causes no appreciable deformation even though the factor of safety may be as low as 1.1. However it may be seen that a series of 100 such pulses for the same initial factor of safety will cause an increase in axial strain of 10 percent. The significance of increased numbers of stress pulses in producing increased deformation of soil is readily apparent from these figures.

It may be seen from Fig. 4.24 that in a normal strength test the maximum resistance of the soil is reached at an axial strain of about 25 percent. If this strain is adopted as a criterion of failure, then a variety of combinations of initially sustained stress and earthquake stress intensities which will cause failure of the soil may be obtained from Figs. 4.25, 4.26 and 4.27. Figure 4.28 shows the combination of initial stress and earthquake stress intensity causing 25% axial strain (i.e. failure). For single transient stress pulse the combinations of sustained and transient stresses would have to approach about 140 percent of the normal strength to induce failure. For earthquake inducing

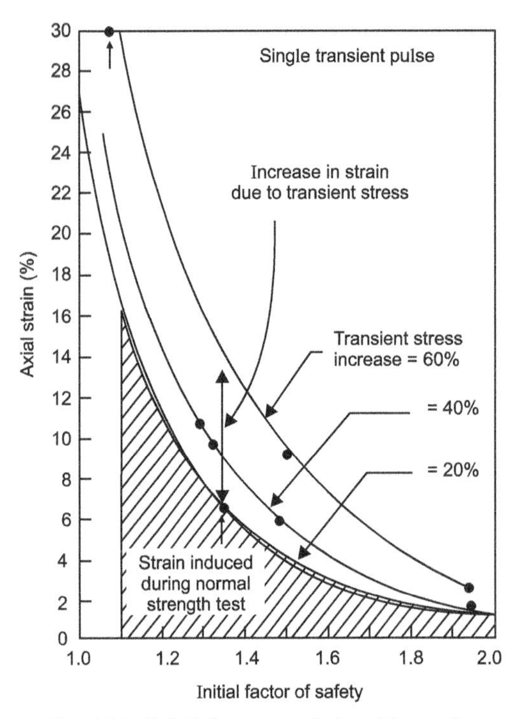

Fig. 4.25. Soil deformations induced by various combinations of sustained and transient stresses (Seed, 1960)

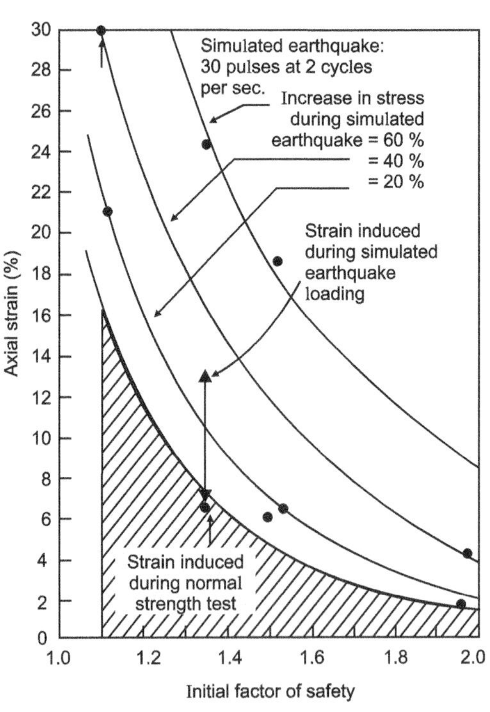

Fig. 4.26. Soil deformation induced by various combination of sustained and vibratory stresses for 30 pulses (Seed, 1960)

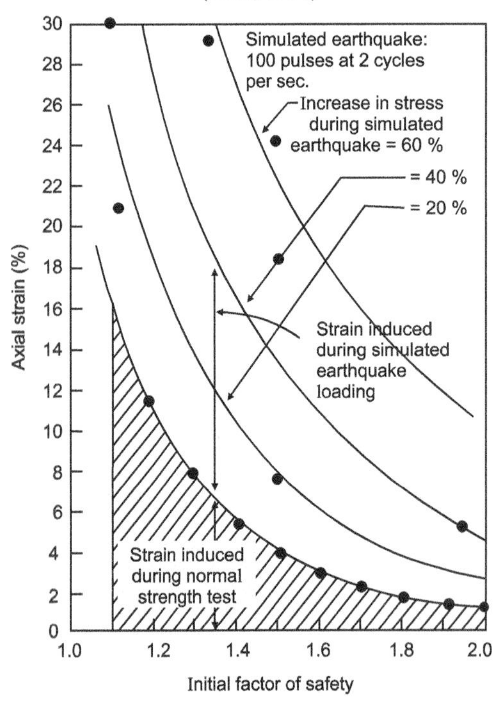

Fig. 4.27. Soil deformation induced by various combination of sustained and vibratory stresses for 100 pulses (Seed, 1960)

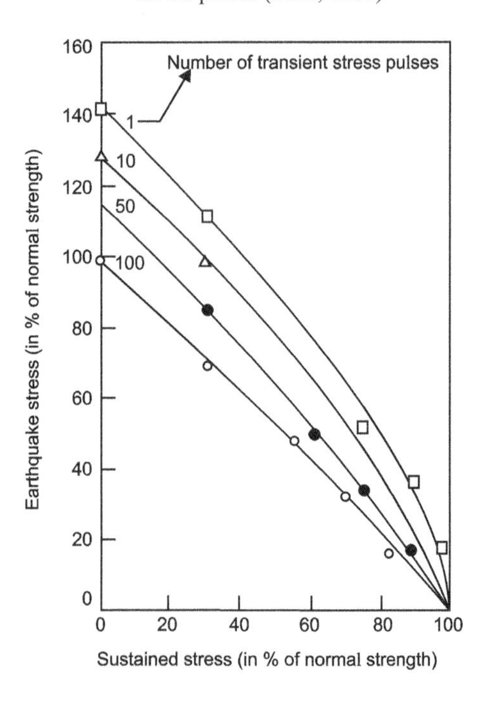

Fig. 4.28. Combination of sustained and vibratory stress intensities causing failure in compacted silty clay (Seed, 1960)

100 major pulses failure would occur when the combined stresses totalled only 100 percent of the normal strength.

Seed (1960) reported that the conditions producing failure shown in Fig. 4.28 for compacted silty clay do not in any way represent the characteristics of conditions producing failure in other types of soils. For example in sensitive undisturbed clays, repeated deformations will lead to increase in pore pressure and a resulting reduction in strength. Consequently a series of transient pulses is likely to induce failure at lower stress levels than for the silty clay.

On the same oscillatory triaxial test set up, Seed and Chan (1966) carried out more elaborate study on different types of soils. Here, few typical results are presented. Figure 4.29a compares combinations of sustained and pulsating stresses that induce failure in soft and compacted clays in one transient pulse. The strength exhibited by undisturbed silty clay was greater than that displayed by compacted soils. As the number of pulses increases to 30 (Fig. 4.29b), failure occurs in sensitive soils at considerably lower stress level than that in compacted soils.

Symmetrical stress pulses of two directional loadings resulted in a reduction in strength of all the soils tested. Typical results with San Francisco Bay mud are shown in Fig. 4.30. Below the dotted line drawn at 45° from origin the stress conditions in one- or two-directional loadings are same since the pulsating stress is either smaller than or equal to sustained stress.

Figure 4.31 shows the results of pulsating load tests with one-directional loading on duplicate specimens of San Francisco Bay mud using the two forms of stress pulse. The longer dwell period under maximum stress for the flat peaked pulse causes larger deformations and induces failure in a smaller number of stress applications than in comparable tests using the triangular pulse form.

A typical total stress versus total strain curve under pulsating loads is shown in Fig. 4.32. For comparison, the stress strain relationship obtained from normal strength test is also shown. It may be noted that in situations involving 10 stress pulses, total stress versus total strain is somewhat higher than the stress strain relationship of a normal test; if there are 100 stress pulses, this is slightly below

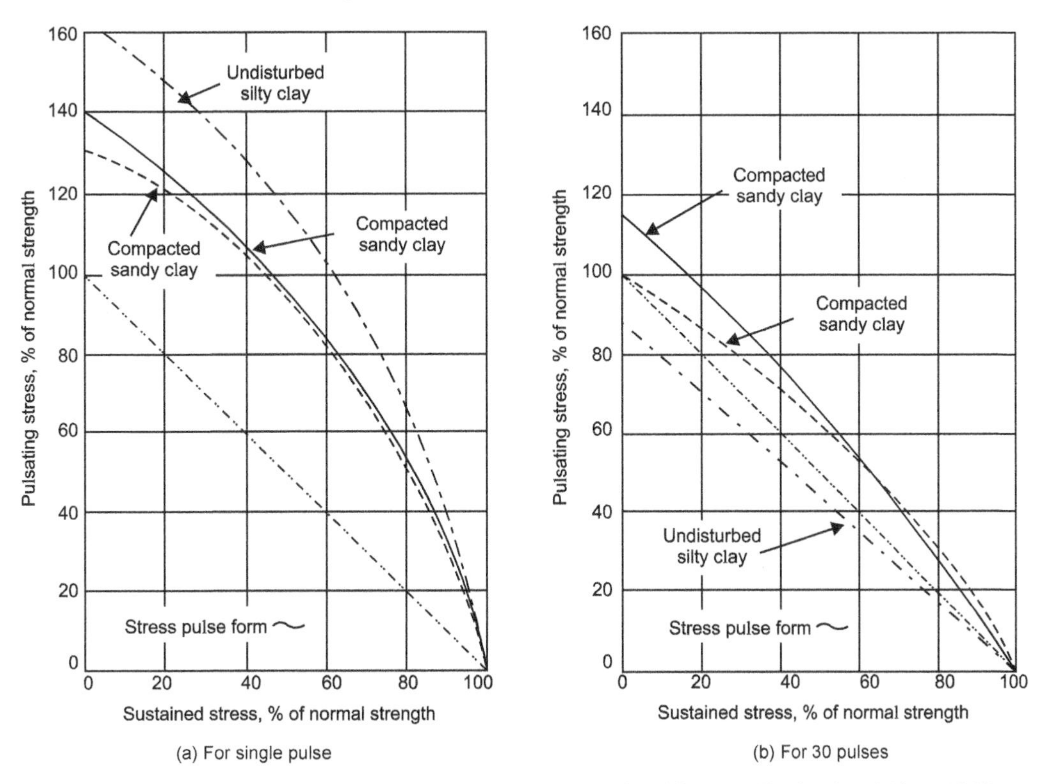

Fig. 4.29. Comparison of stress conditions causing failure for different soils (Seed and Chan, 1960)

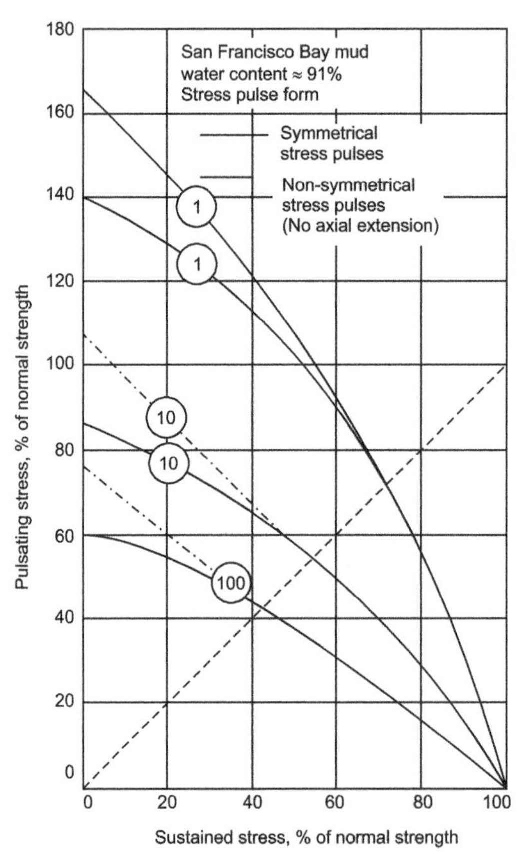

Fig. 4.30. Combination of sustained and pulsating stresses causing failure in one- and two-directional loadings (Seed and Chan, 1966)

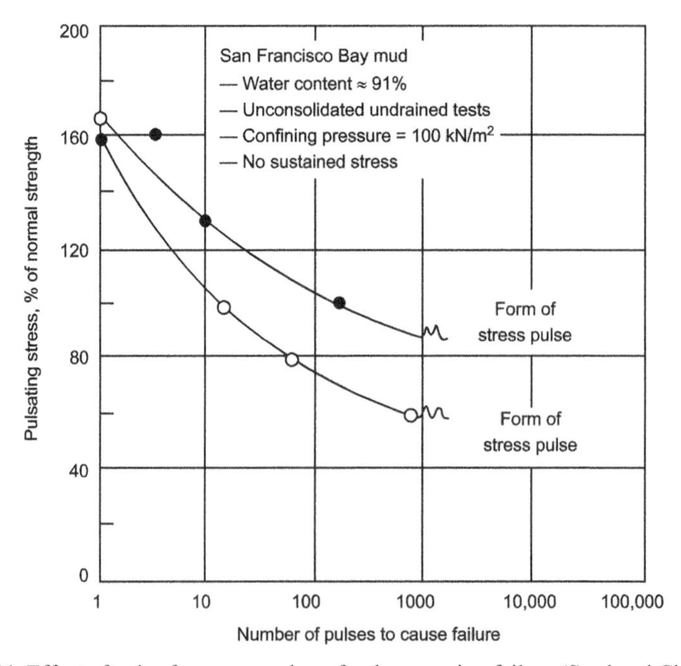

Fig. 4.31. Effect of pulse form on number of pulses causing failure (Seed and Chan, 1966)

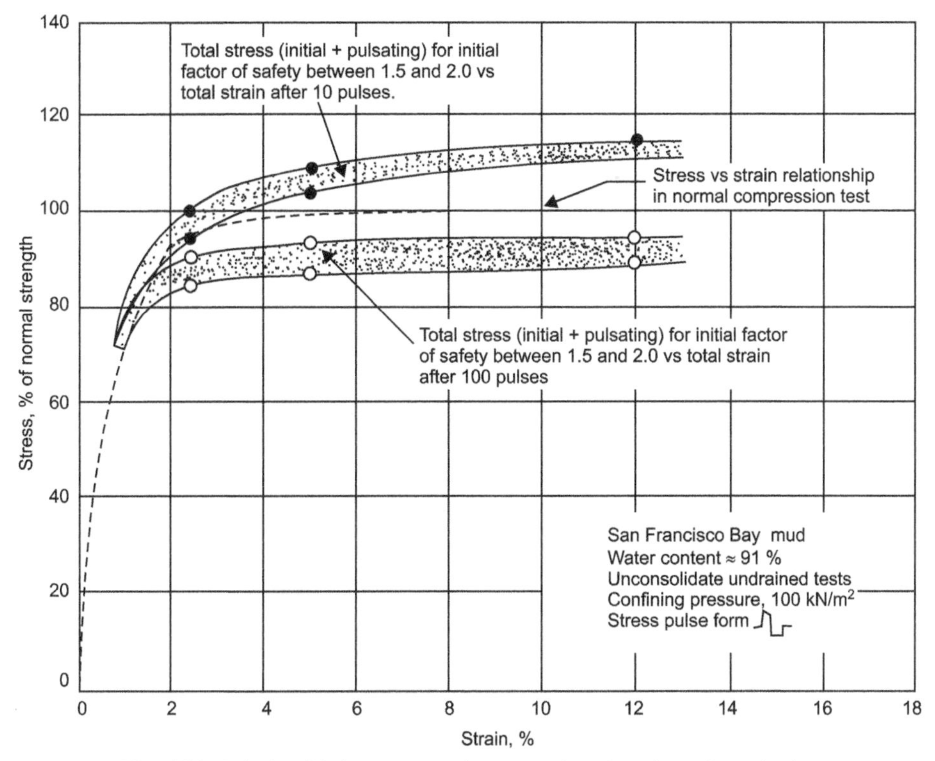

Fig. 4.32. Relationship between total stress and total strain under pulsating load conditions (Seed and Chan, 1966)

the normal plot. Further it was noted that with an initial factor of safety between 1.5 and 2, and 30 pulses of dynamic stress, the relationship between total stress and total strain will coincide almost exactly with the normal stress versus strain relationship.

4.2.6 Final Comments on Laboratory Testing

The laboratory techniques discussed above and more common in use are listed in Table 4.4. The specific soil properties measured by each test are also indicated. The range of strain amplitude over which each technique is applicable is shown in Fig. 4.33.

Table 4.4. Laboratory Techniques for Measuring Dynamic Soil Properties (Woods, 1978)

Techniques	*Shear modulus*	*Young's modulus*	*Material damping*	*Cyclic stress behaviour*	*Attenuation*
Resonant column with adaption	×	×	×	—	—
Ultrasonic pulse	×	×	—	—	×
Cyclic triaxial	—	×	×	×	—
Cyclic simple shear	×	—	×	×	—
Cyclic torsional shear	×	—	×	×	—
Dynamic I D compression	×	—	—	—	—

Each laboratory test has some merit when compared with the other. The resonant column device is better suited for determining shear modulus at low strains and the hollow cylinder device at higher strain levels. The cyclic simple shear device is suitable for determining shear modulus and damping

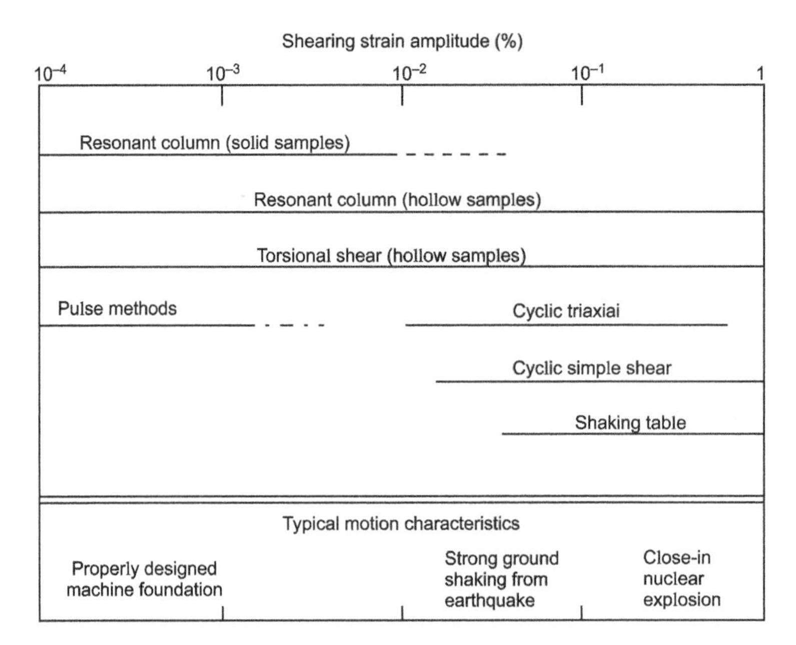

Fig. 4.33. Shearing strain amplitude capabilities of laboratory apparatus (Woods, 1978)

characterstics of soils. The cyclic triaxial test is more suited to obtain the Young's modulus of the material.

4.3 Field Tests

Field methods generally depend on the measurement of velocity of waves propagating through the soil or on the response of soil structure systems to dynamic excitation. The following methods are in use for determining dynamic properties of soils:

1. Seismic cross-borehole survey
2. Seismic up-hole survey
3. Seismic down-hole survey
4. Seismic refraction survey
5. Vertical block resonance test
6. Horizontal block resonance test
7. Cyclic plate load test
8. Standard penetration test

In this section, the above listed field methods have been described briefly along with the typical test setups and methods of interpretation.

4.3.1 Seismic Cross-borehole Survey

This method is based on the measurement of velocity of wave propagation from one borehole to another. Figure 4.34 shows the essentials of seismic cross-hole survey as outlined by Stoke and Woods (1972). A source of seismic energy is generated at the bottom of one borehole, and the time of travel of the shear wave from this borehole to another at known distance is measured. Shear wave velocity is then computed by dividing the distance between the boreholes by the travel time.

As discussed above, seismic cross-borehole survey can be done using two boreholes: one having the source for causing wave generation and another having geophone for recording travel

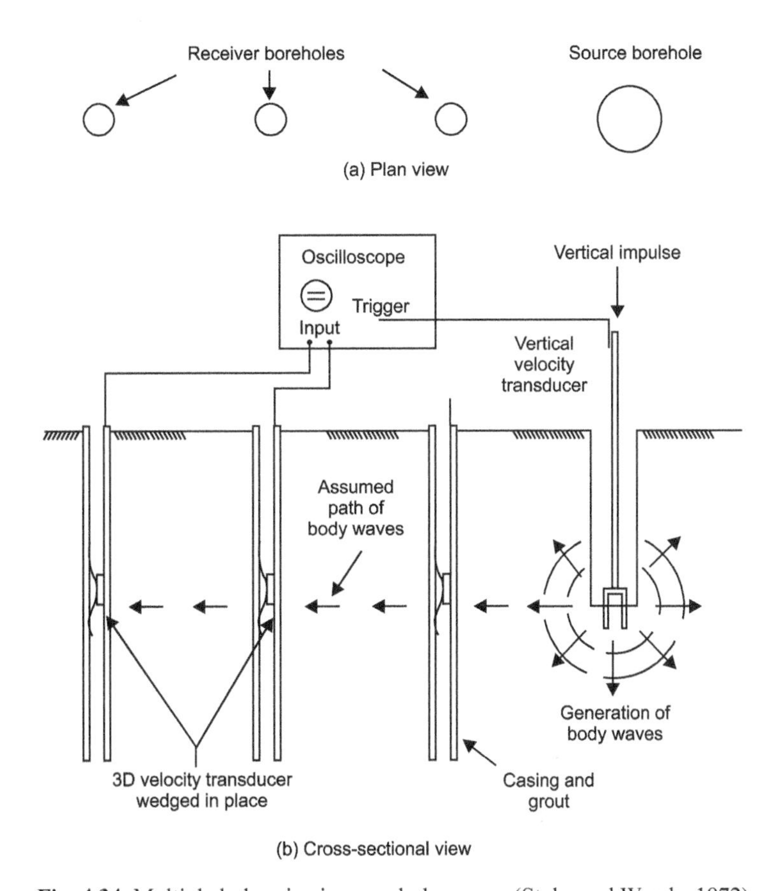

Fig. 4.34. Multiple hole seismic cross-hole survey (Stoke and Woods, 1972)

time. However, for extensive investigations and better accuracy, three or more boreholes arranged in a straight line should be used.

In this case the wave velocities can be calculated from the time intervals between succeeding pairs of holes, eliminating most of the concern over triggering the timing instruments and the effects of borehole casing and backfilling (Stokoe and Hoar, 1978). Also this arrangement of boreholes in a straight line overcomes problems of site anisotropy by examining one direction only at a time.

For better results, the following points should be kept in view:

(i) The diameter of the boreholes should be small to cause least disturbance in the soil. Casing in the boreholes will provide good coupling with the soil and transmission of waves. Void spaces around the casing should be filled with weak cement slurry grout or dry sand.

(ii) Boreholes should be vertical for the travel distance to be measured properly. In general any borehole 10 m or more in depth should be surveyed using an inclinometer or other logging device for determining verticality (Woods, 1978).

(iii) Boreholes should be spaced as close as possible within the time resolution characteristics of the recording equipment. Large spacings can lead to difficulties with refracted waves arriving before the direct transmission through the intervening soil. Spacings as close as 2-3 m can be used satisfactorily (Woods, 1978).

(iv) The seismic source must be capable of generating predominantly one kind of wave. Further it must also be capable of repeating desired characteristics at a predetermined energy level. Miller, Troncosco and Brown (1975) have described a source which is capable of developing high amplitude shear waves. It consists of a falling weight which impacts on an hydraulically expanded borehole anchor.

(v) The receivers must be oriented in the shearing mode and should be securely coupled to the sides of the borehole.

4.3.2 Seismic Up-hole Survey

Seismic up-hole survey is done by using only one borehole. In this method, the receiver is placed at the surface, and shear waves are generated at different depths within the borehole. Figure 4.35 shows the schematic presentation of the arrangement used in seismic up-hole survey (Goto et al., 1977). This method gives the average value of wave velocity for the soil between the excitation and the receiver if one receiver is used, or between the receivers.

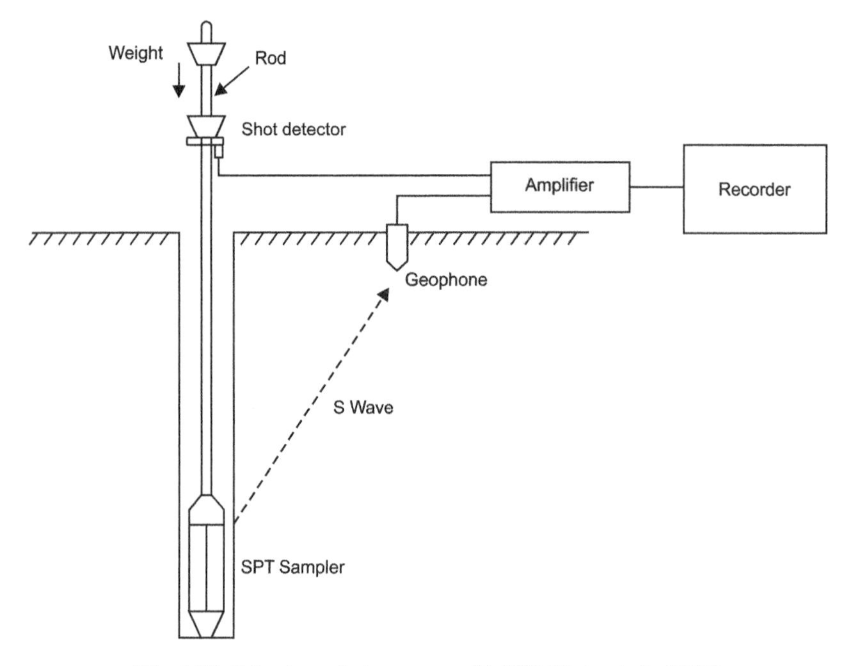

Fig. 4.35. Seismic up-hole survey with SPT (Goto et al., 1973)

The major disadvantage in seismic up-hole survey is that it is more difficult to generate waves of the desired type.

4.3.3 Seismic Down-hole Survey

In this method, seismic waves are generated at the surface of the ground near the top of the borehole and travel times of the body waves between the source and the receivers which have been clamped to the borehole wall at predetermined depths are obtained. The arrangement used in seismic down-hole survey is shown schematically in Fig. 4.36. This also requires only one borehole.

The main advantage of this method is that low velocity layers can be detected even if trapped between layers of greater velocity provided the geophone spacings are close enough.

4.3.4 Seismic Refraction Survey

The seismic refraction survey is frequently used for site investigations. It enables the determination of elastic wave velocity in each layer, the thickness of each layer, and the dip angle of each layer as long as the wave velocities increase in each suceedingly deep layer.

The details of this method has already been described in section 3.10 of Chapter 3.

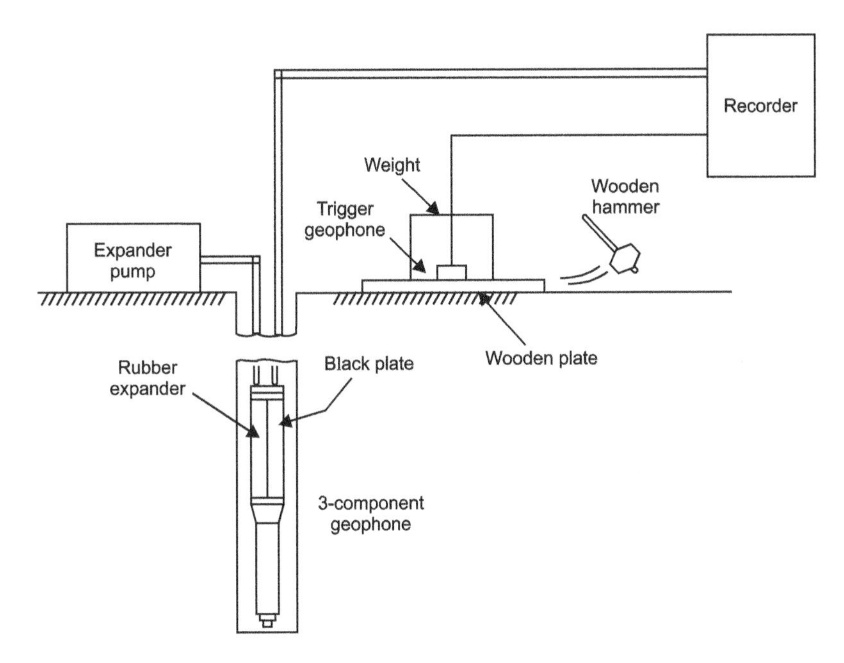

Fig. 4.36. Seismic down-hole survey (Woods, 1978)

4.3.5 Vertical Block Resonance Test

The vertical block resonance test is used for determining the values of coefficient of elastic uniform compression (C_u), Young's modulus (E) and damping ratio (ξ) of the soil. According to IS 5249: 1992, a test block of size 1.5 m × 0.75 m × 0.70 m high is casted in M20 concrete in a pit of plan dimensions 4.5 m × 2.75 m and depth equal to the proposed depth of foundation. Foundation bolts should be embedded into the concrete block at the time of casting for fixing the oscillator assembly. The oscillator assembly is mounted on the block so that it generates purely vertical sinusoidal vibrations. The line of action of vibrating force should pass through the centre of gravity of the block. Two acceleration or displacement pickups are mounted on the top of the block as shown in Fig. 4.37 such that they sense the vertical motion of the block. A schematic diagram of the set up is shown in Fig. 4.37.

The mechanical oscillator works on the principle of eccentric masses mounted on two shafts rotating in opposite directions (Fig. 2.15). The force generated by the oscillator is given by

$$F_d = 2\, m_e\, e\, \omega^2 \tag{4.20}$$

The oscillator is first set at a particular eccentricity (e). As evident from Eq. (4.20) higher the eccentricity more will be the force level. It is then operated at constant frequency, and the acceleration of the oscillatory motion of the block is monitored. The oscillator frequency is increased in steps, and the signals of monitoring pickups are recorded. At any eccentricity and frequency the dynamic force should not exceed 20 percent of the total mass of the block and oscillator assembly. The amplitude of vibration (A_z) at a given frequency f is given by

$$A_z(\text{mm}) = \frac{a_z}{4\pi^2 f^2} \tag{4.21}$$

where a_z = vertical acceleration of the block, mm/s²
f = frequency, Hz.

Amplitude versus frequency curves are plotted for each eccentricity to determine the natural frequency of the foundation-soil system (Fig. 4.38). The natural frequency, f_{nz}, at different eccentricity

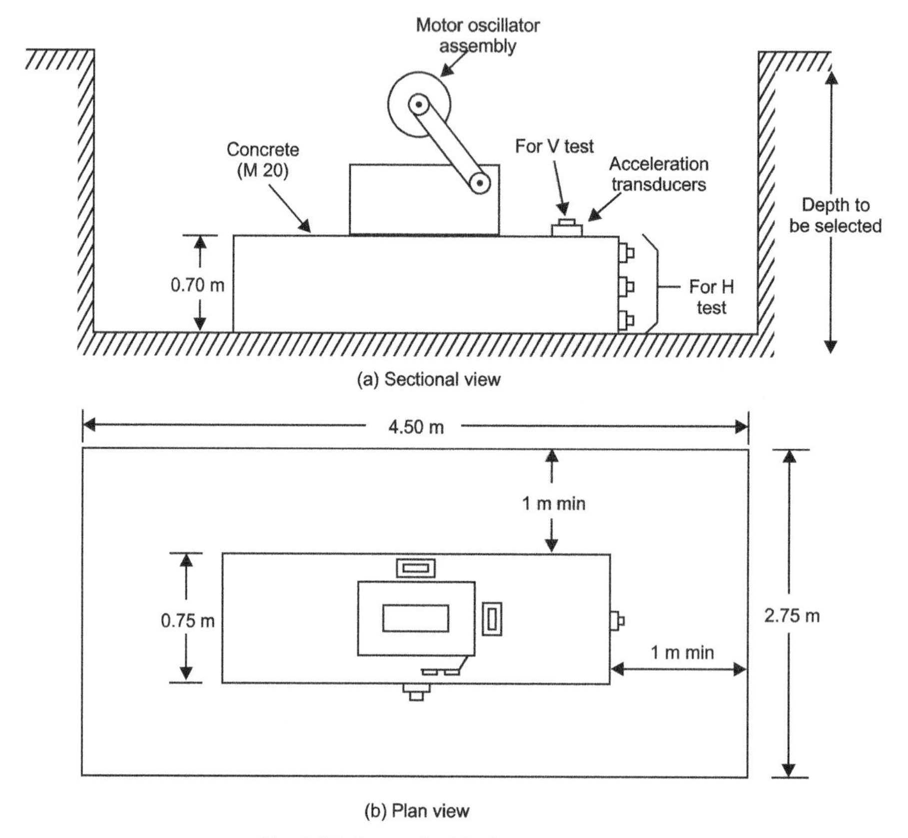

Fig. 4.37. Set up for block resonance test

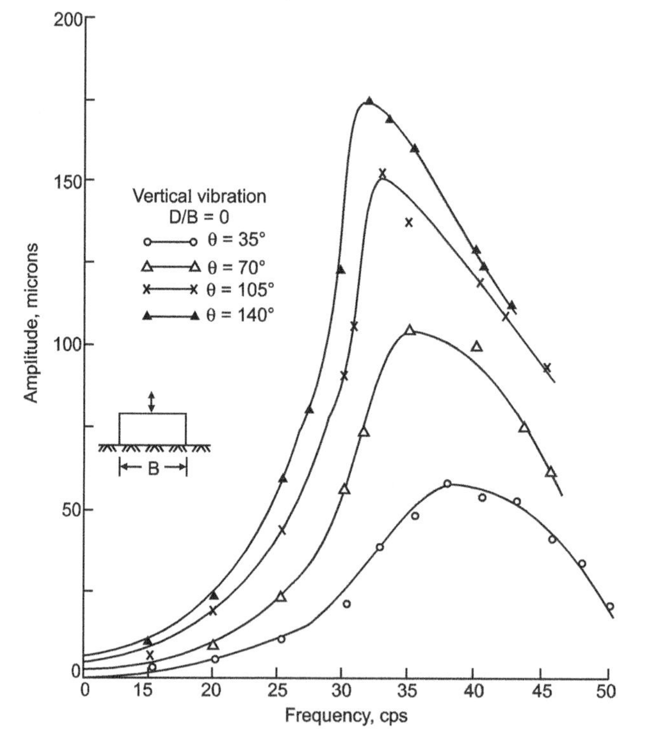

Fig. 4.38. Amplitude versus frequency plots from vertical resonance test

(i.e. force level) is different because different forces cause different strain levels of the block which may be accounted for when appropriate design parameters are being chosen.

The coefficient of elastic uniform compression (C_u) in N/m^3 of the soil is then determined using Eq. (4.22)

$$C_u = \frac{4\pi^2 f_{nz}^2 m}{A} \tag{4.22}$$

where f_{nz} = Natural frequency of foundation-soil system, Hz
m = Mass of the block oscillator and motor, kg
A = Base contact area of the block, m^2

From the value of C_u obtained from Eq. (4.22) for the test block of contact area A the value of C_{u1} for the actual foundation having contact area A_1 may be obtained from Eq. (4.23)

$$C_{u1} = C_u\sqrt{\frac{A}{A_1}} \tag{4.23}$$

Equation (4.23) is valid for base areas of foundations upto 10 m^2. For areas larger than 10 m^2, the value of C_u obtained for 10 m^2 is used.

The coefficient of elastic uniform compression (C_u) is related to the elastic Young's modulus (E) by Eq. (4.24) which is in the form of Boussinesq relationship for the elastic settlement of a surface footing.

$$C_u = \frac{E}{(1-\mu^2)} \cdot \frac{C_s}{\sqrt{BL}} \tag{4.24}$$

where μ = Poisson's ratio
B = Width of base of the block
L = Length of base of the block
C_s = Coefficient depending on L/B ratio

Barkan (1962) recommended the values of C_s for various L/B ratios as listed in Table 4.5.

Table 4.5. Values of C_s (After Barkan, 1962)

L/B	C_s
1.0	1.06
1.5	1.07
2.0	1.09
3.0	1.13
5.0	1.22
10.0	1.41

The value of damping ratio ξ is determined using Eq. (4.25)

$$\xi = \frac{f_2 - f_1}{2 f_{nz}} \tag{4.25}$$

where f_2, f_1 = Two frequencies at which amplitudes is equal to $\frac{A_{max}}{\sqrt{2}}$
A_{max} = Maximum amplitude
F_{nz} = Resonant frequency

This is also illustrated in example 2.6.

4.3.6 Horizontal Block Resonance Test

Horizontal block resonance test is also performed on the block set up as shown in Fig. 4.37. In this test, the mechanical oscillator is mounted on the block so that horizontal sinusoidal vibrations are generated in the direction of the longitudinal axis of the block. Three acceleration or displacement pickups are mounted along the vertical centreline of the transverse face of the block to sense horizontal vibrations (Fig. 4.37a). The oscillator is excited in steps starting from rest condition. The signal of each acceleration pickup is amplified and recorded. Rest of the procedure is same as described for vertical block resonance test. Similar tests can be performed by exciting the block in the direction of transverse axis.

The amplitude of horizontal vibrations (A_x) is obtained using Eq. (4.26).

$$A_x = \frac{a_x}{4\pi^2 f^2} \tag{4.26}$$

where A_x(mm) = Horizontal resonant frequency of block soil system, Hz
f = Frequency in Hz

Amplitude versus frequency curves are plotted for each force level to obtain the natural frequency, f_{nx}, of the block soil system as done in vertical resonance test. A typical frequency versus amplitude plot is shown in Fig. 4.39. It may be noted that the case of horizontal vibrations is problem of two degrees of freedom. The mode of vibration is obtained by plotting amplitude versus height of the block at natural frequency of the system from the analysis of data from the pickups mounted at different locations at the face of the block. Typical plots are shown in Fig. 4.40. If the plot corresponds to Fig. 4.40a, then the natural frequency corresponds to first mode or lower natural frequency. On the other hand, if it corresponds to that in Fig. 4.40b, then natural frequency corresponds to second mode or higher natural frequency.

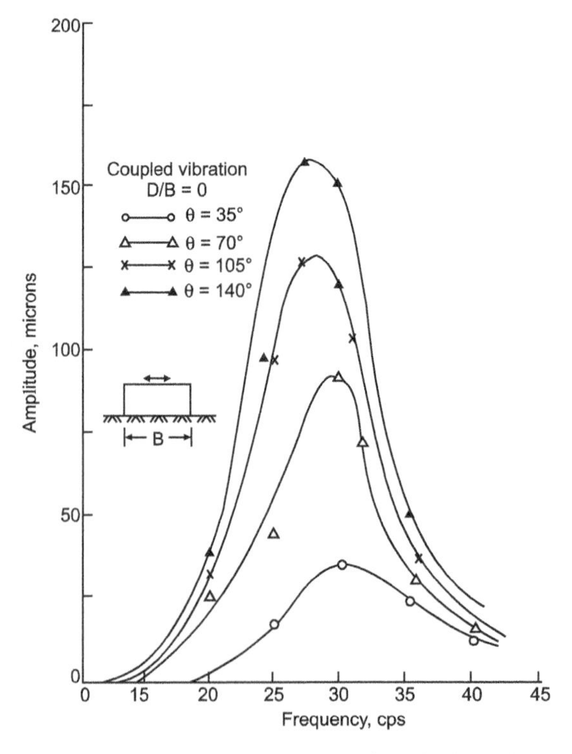

Fig. 4.39. Amplitude versus frequency plots from horizontal resonance test

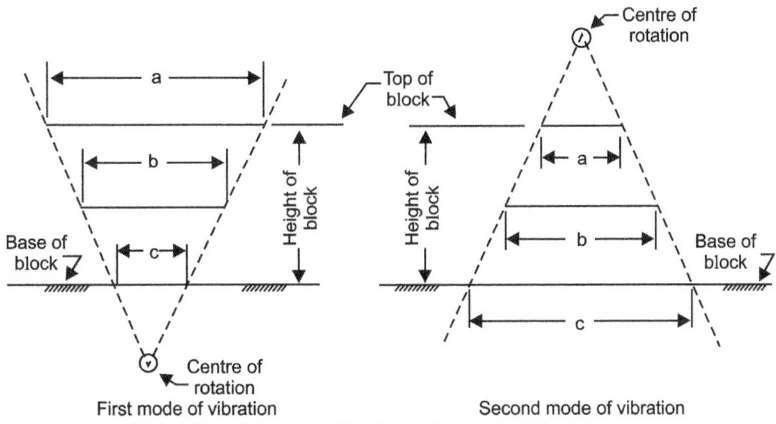

Fig. 4.40. Determination of mode of vibration

The coefficient of elastic uniform shear (C_τ) in N/m^3 of soil is given by the following equation:

$$C_\tau = \frac{8\pi^2 r f_{nx}^2}{(A_0 + I_0) \pm \sqrt{(A_0 + I_0)^2 - 4rA_0 I_0}} \tag{4.27}$$

where $r = \dfrac{M_m}{M_{mo}}$

f_{nx} = Horizontal resonant frequency of block soil system, H_z
A_0 = A/M, m^2/kg
A = Contact area of block with soil, m^2
M = Mass of block, oscillator and soil, kg
I_0 = 3.46 I/M_{mo}, m^2/kg
M_m = Mass moment of inertia of block, oscillator, motor, etc. about the horizontal axis passing through the centre of gravity of block and perpendicular to the direction of vibration, kg-m^2
M_{mo} = Mass moment of inertia of the block, oscillator, motor etc. about the horizontal axis passing through the centre of contact area of block and soil and perpendicular to the direction of vibration, kg-m^2.
I = Moment of inertia of the foundation contact area about the horizontal axis passing through the centre of gravity of area and perpendicular to the direction of vibration, m^4.

In Eq. (4.27), negative sign is taken when the system vibrates in first mode and positive sign when the system vibrates in second mode.

For the size of the block recommended in IS 5249-1992 and for first natural frequency f_{nx1}, the Eq. (4.27) reduces to

$$C_\tau = 92.3 f_{nx1} \tag{4.28}$$

In Eq. (4.28), C_t is in kN/m^3.

The coefficient of elastic uniform shear ($C_{\tau 1}$) for actual area of foundation (A_1) is given by

$$C_{\tau 1} = C_\tau \sqrt{\frac{A}{A_1}} \tag{4.29}$$

IS 5249: 1977 recommends the following relations between C_u and C_τ, C_ϕ and C_ψ:

$$C_u = 1.5 \text{ to } 2.0\, C_\tau \tag{4.30}$$

$$C_\phi = 3.46\, C_\tau \tag{4.31}$$

$$C_\psi = 0.75\, C_u \tag{4.32}$$

C_ϕ and C_ψ are respectively the coefficient of elastic non-uniform compression and the coefficient of elastic non-uniform shear. The coefficients C_u, C_τ, C_ϕ and C_ψ are known as dynamic elastic constants; their use is explained in detail in Chapter 9 on machine foundation.

4.3.7 Cyclic Plate Load Test

The cyclic plate load test is performed in a test pit dug upto the proposed base level of foundation. The equipment is same as used in static plate load test. Circular or square bearing plates of mild steel not less than 25 mm thickness and varying in size from 300 to 750 mm with chequered or grooved bottom are used. The test pit should be at least five times the width of the plate. The equipment is assembled according to details given in IS 1988-1982. A typical set up is shown in Fig. 4.41.

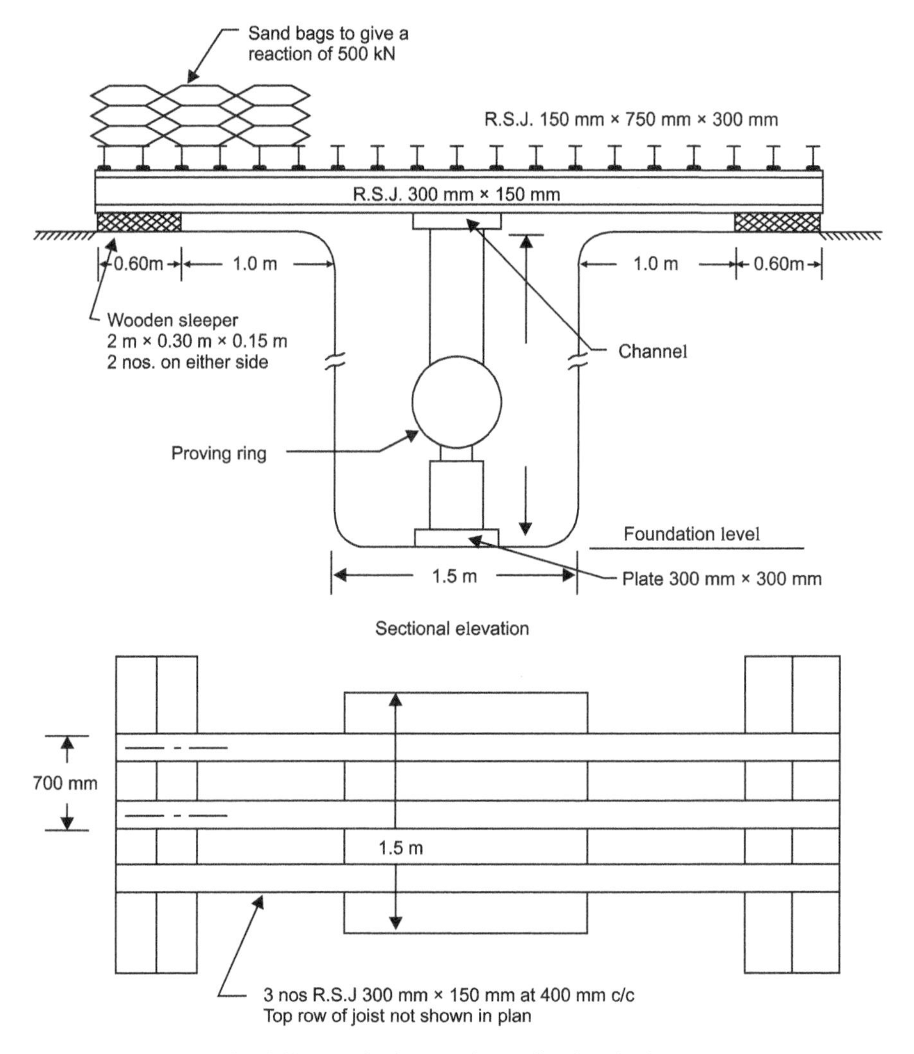

Fig. 4.41. A typical set up for cyclic plate load test

To commence the test, a seating pressure of about 7 kPa is first applied to the plate. It is then removed and dial gauges are set to read zero. Load is then applied in equal cumulative increments of not more than 100 kPa or of not more than one fifth of the estimated allowable bearing pressure. In cyclic plate load test, each incremental load is maintained constant till the settlement of the plate is complete. The load is then released to zero and the plate is allowed to rebound. The reading of final settlement is taken. The load is then increased to next higher magnitude of loading and maintained constant till the settlement is complete, which again is recorded. The load is then reduced to zero and the settlement reading taken. The next increment of load is then applied. The cycles of unloading and reloading are continued till the required final load is reached.

The data obtained from a cyclic plate load test is shown in Fig. 4.42. From this data, the load intensity versus elastic rebound is plotted as shown in Fig. 4.43, and the slope of the line is coefficient of elastic uniform compression.

$$C_u = \frac{p}{S_e} \text{ (kN/m}^3\text{)} \tag{4.33}$$

where p = Load intensity in kN/m^2

S_e = Elastic rebound corresponding to p in m.

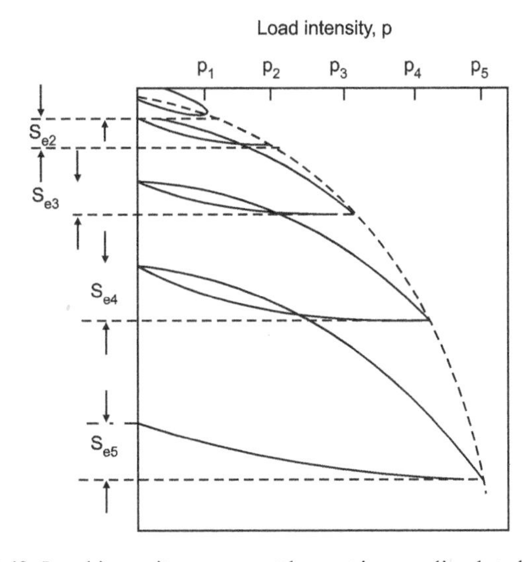

Fig. 4.42. Load intensity versus settlement in a cyclic plate load test

4.3.8 Standard Penetration Test (SPT)

The standard penetration test (SPT) is the most extensively used in situ test in India and many other countries. This test is carried in a bore hole using a split spoon sampler. As per IS: 2131-1981, steps involved in carrying out this test are as follows:

(i) The borehole is advanced to the depth at which the SPT has to be performed. The bottom of the borehole is cleaned.

(ii) The split-spoon, attached to standard drill rods of required length, is lowered into the borehole and rested at the bottom.

(iii) The split-spoon sampler is seated 150 mm by blows of a drop hammer of 65 kg falling vertically and freely from a height of 750 mm. Thereafter, the split spoon sampler shall be further driven 300 mm in two steps each of 150 m. The number of blows required to effect each 150 mm of penetration shall be recorded. The first 150 mm of drive may be considered to be seating drive. The total blows required for the second and third 150 mm of penetration is termed the penetration resistance N.

If the split spoon sampler is driven less than 450 mm (total), then N-value shall be for the last 300 mm penetration. In case, the total penetration is less than 300 mm for 50 blows, it is entered as refusal in the borelog.

(iv) The split spoon sampler is then withdrawn and is detached from the drill rods. The split barrel is disconnected from the cutting shoe and the coupling. The soil sample collected inside the barrel is collected carefully and preserved for transporting the same to the laboratory for further tests.

(v) Standard penetration tests shall be conducted at every change in stratum or intervals of not more than 1.5 m whichever is less. Tests may be done at lesser intervals (usually 0.75 m) if specified or considered necessary.

The penetration test in gravelly soils requires careful interpretation since pushing a piece of gravel can greatly change the blowcount.

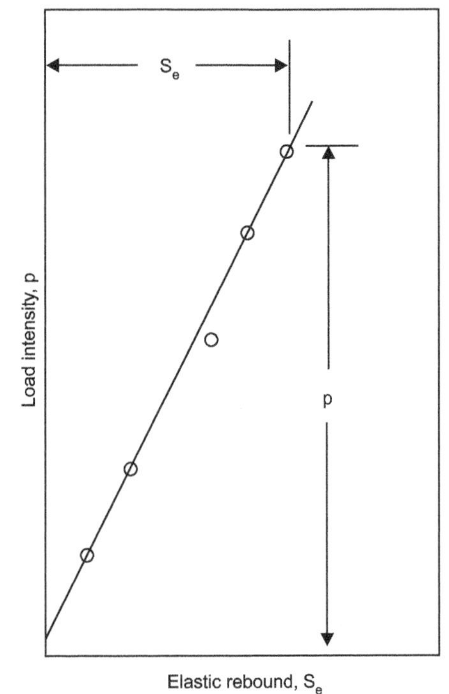

Fig. 4.43. Load intensity versus elastic rebound from cyclic plate load test

4.3.8.1 Corrections to Observed SPT Values (N) in Cohesionless Soils

Following two types of corrections are normally applied to the observed SPT values (N) in cohesionless soils:

Corrections due to dilatancy

In very fine, or silty, saturated sand, Terzaghi and Peck (1967) recommend that the observed N-values be corrected to N' if N was greater than 15 as

$$N' = 15 + \frac{1}{2}(N - 15) \tag{4.34}$$

Bazaraa (1967) recommended the correction as

$$N' = 0.6\,N \quad (\text{for } N > 15) \tag{4.35}$$

This correction is introduced with the view that in saturated dense sand ($N > 15$), the fast rate of application of shear through the blows of drop hammer, is likely to induce negative pore pressures and thus temporary increase in shear strength will occur. This will lead to a N-value higher than the actual one. Since sufficient experimental evidence is not available to confirm this correction, many engineers are not applying this correction. However this correction has also been recommended in IS: 2131 - 1981.

Correction due to overburden pressure

On the basis of field tests, corrections to the N-value for overburden effects were proposed by many investigators (Gibbs and Holtz 1957; Teng 1965; Bazaraa 1967; Peck, Hanson and Thornburn 1974). The methods normally used are:

Bazaraa (1967)

For $\bar{\sigma}_0 < 75$ kPa

$$N' = \frac{4N}{1 + 0.04\bar{\sigma}_0} \tag{4.36}$$

For $\bar{\sigma}_0 > 75$ kPa

$$N' = \frac{4N}{3.25 + 0.01\bar{\sigma}_0} \tag{4.37}$$

where $\bar{\sigma}_0$ = Effective over burden pressure, kPa

Peck, Hanson and Thornburn (1974)

$$N' = 0.77\,N \log_{10} \frac{2000}{\bar{\sigma}_0} \tag{4.38}$$

Figure 4.44 gives the correction factor based on Eq. (4.38). Use of this figure has been recommended in IS 2131-1981. In this figure.

$$C_N = \text{Correction factor} = 0.77 \log_{10} \frac{2000}{\bar{\sigma}_0} \tag{4.39}$$

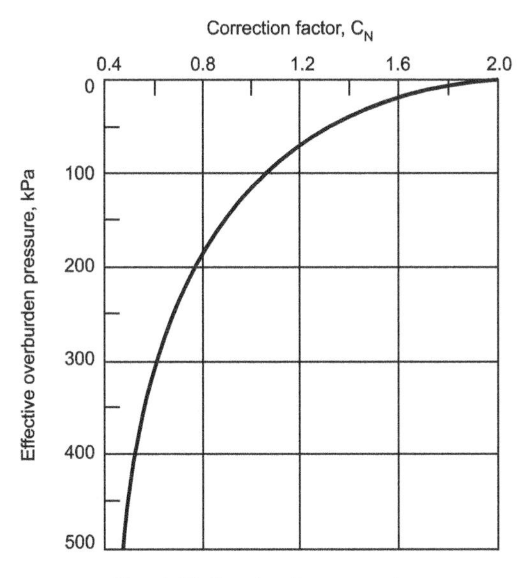

Fig. 4.44. Overburden correction

There is a controversy whether the correction due to dilatancy should be applied first and then the correction due to overburden pressure or vice versa. However in IS 2131 - 1981, it is recommended that the correction due to overburden should be applied first.

A typical set of observed *N*-values are shown in Fig. 4.45. Corrected *N*-values as per IS code recommendations are also shown in the figure.

The SPT is essentially undrained test for the duration of each blow, and the energy generated by the SPT hammer is principally shearing energy. Therefore the test may be useful to predict the dynamic behaviour of soils. Seed et al. (1983) presented correlations between SPT and observed liquefaction. Ohasaki (1970) describes a useful Japanese rule of thumb that says liquefaction is not a problem if the blow count from a SPT exceeds twice the depth of sample in metres.

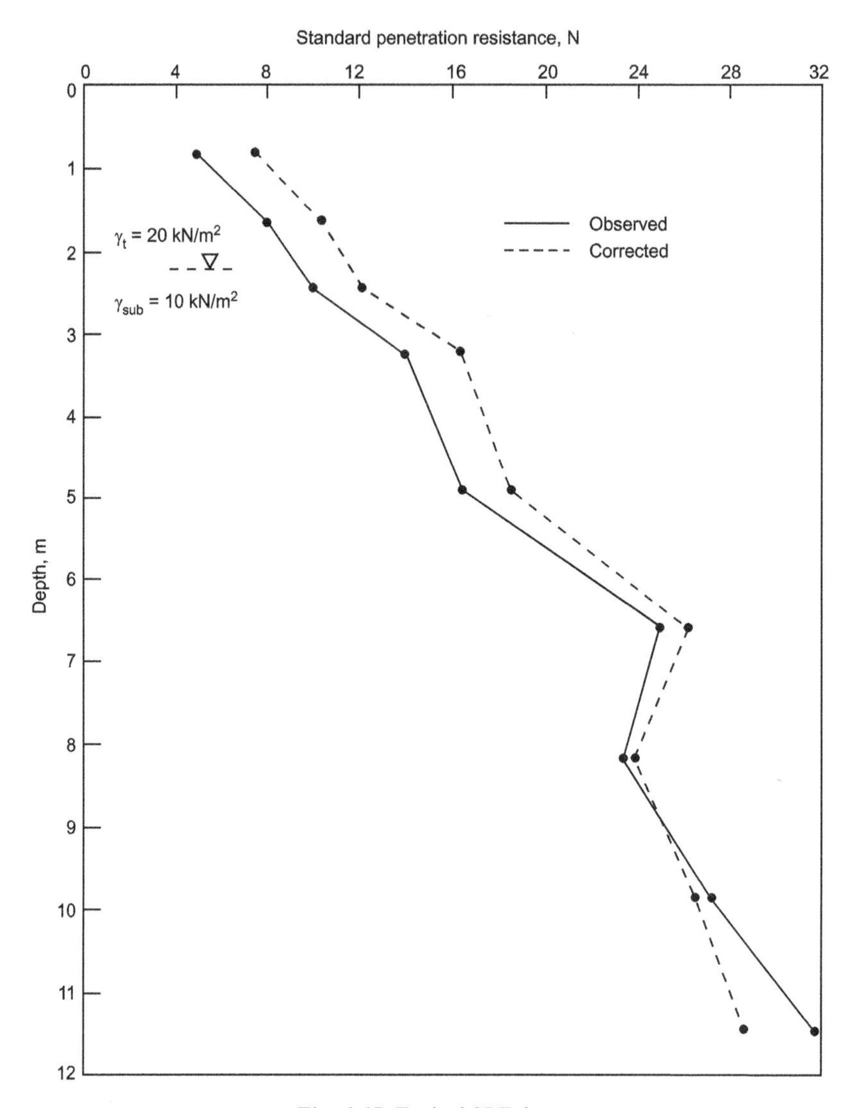

Fig. 4.45. Typical SPT data

Imai (1977) reported the following correlation between N (observed) and shear wave velocity V_s (m/s):

$$V_s = 91\ N^{0.337} \tag{4.40}$$

Bowles (1982) has given a number of equations to obtain stress-strain modulus E_s on the basis of SPT and cone-penetration test (CPT). These equations are given in Table 4.6.

Table 4.6. Equations for Stress-Strain Modulus E_s by SPT and CPT (After Bowles, 1982)

	SPT	*CPT*
Sand	$E_s = 500\ (N + 15)$ $E_s = 18000 + 750\ N$ $E_s = (15000 \text{ to } 22000) \ln N$	$E_s = 2 \text{ to } 4q_c$ $E_s = 2(1 + D_r^2)q_c$
Clayey sand	$E_s = 320\ (N + 15)$	$E_s = 3 \text{ to } 6q_c$

(*Contd.*)

Silty sand	$E_s = 300(N + 6)$	$E_s = 1$ to $2q_c$
Gravelly sand	$E_s = 1200(N + 6)$	
Soft clay		$E_s = 6$ to $8q_c$
Using the undrained shear strength, c_u		
	$I_p > 30$, or organic	$E_s = 100$ to $500\ c_u$
	$I_p < 30$, or stiff	$E_s = 500$ to $1500\ c_u$
	$1 < OCR < 2$	$E_s = 800$ to $1200\ c_u$
	$OCR > 2$	$E_s = 1500$ to $2000\ c_u$

Notes: 1. Unit of E_S is kPa in correlations with N.
2. E_S will have the same unit as of q_c
3. E_S will have the same unit as of c_u

4.4 Factors Affecting Shear Modulus, Elastic Modulus and Elastic Constants

Hardin and Black (1968) have given the following factors which influence the shear modulus, elastic modulus and elastic constants:

(i) Type of soil including grain characteristics, grain shape, grain size, grading and mineralogy;
(ii) Void ratio;
(iii) Initial average effective confining pressure;
(iv) Degree of saturation;
(v) Frequency of vibration and number of cycles of load;
(vi) Ambient stress history and vibration history;
(vii) Magnitude of dynamic stress; and
(viii) Time effects.

The shear modulus G, elastic modulus E, and dynamic elastic constants (C_u, C_ϕ, C_τ and C_ψ) are related with each other directly as evident from Eqs. 3.9, 4.24 and 4.32. Therefore the factors listed from (i) to (viii) will effect G, E, C_u, C_ϕ, C_τ and C_ψ in similar way. Keeping this in view, the effect of the factors have been discussed only on dynamic shear modulus G and damping ratio.

Soil behaviour over a wide range of strain amplitudes is nonlinear and on unloading follows a different stress-strain path forming a hysteresis loop as shown in Fig. 4.46. The area inside this loop represents the energy absorbed by the soil during its deformation and is a measure of the internal damping within the soil. At very low strain amplitudes (<0.0001%) the soil acts essentially as a linear elastic material with little or no loss of energy. The shear modulus under these conditions is maximum but as the strain amplitudes is increased, the shear modulus decreases and the damping within the soil increases.

Ishihara (1971) presented Fig. 4.47, which shows strain levels associated with different phenomenon in the field and in corresponding field and laboratory tests. Prakash and Puri (1980) presented the data of G from different in situ tests as shown in Fig. 4.48. It is evident from this figure that G decreases significantly when strain amplitude is higher than 10^{-5}. For lower strain amplitude ($< 10^{-5}$), G may be considered constant.

The machine foundations are usually designed for very low strain amplitudes so that the behaviour of soil is elastic under vibrations. It is to avoid the building up of any residual strains in the soil due to the operation of the machine. Large strain amplitudes may be developed by commercial blasting, earthquakes, nuclear blasts, pile driving operations, compaction devices or excessive vibrations of the machinery.

The subsequent discussions have been made under two heads: (i) Shear modulus for low strain amplitudes, and (ii) Shear modulus for large strain amplitudes.

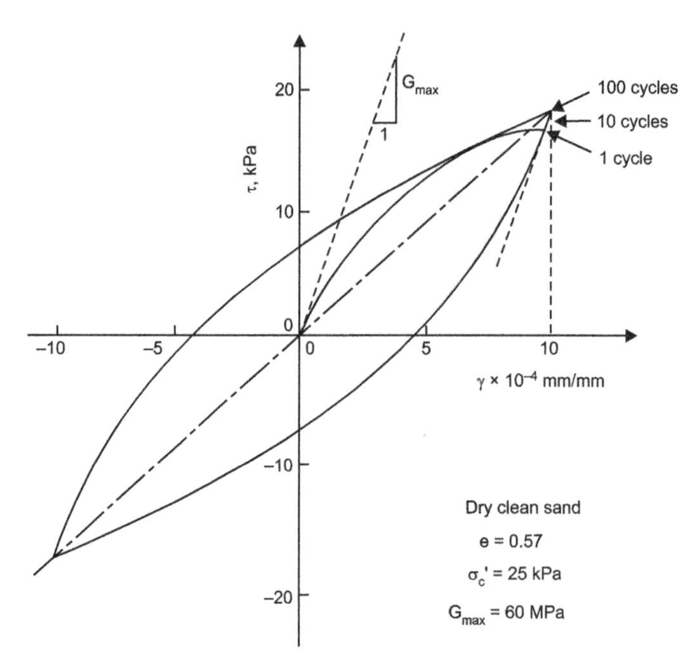

Fig. 4.46. Stress-strain loop and points at Ist, 10th and 100th cycles of loading (After Hoadley, 1985)

Magnitude of strain		10^{-6} — 10^{-5} — 10^{-4}	10^{-4} — 10^{-3} — 10^{-2}	10^{-2} — 10^{-1}
Phenomena		Wave propagation, vibration	Cracks, differential settlement	Slide, compaction, liquefaction
Mechanical characteristics		Elastic	Elastic-plastic	Failure
Constants		Shear modulus, Poisson's ratio, damping ratio	Shear modulus, Poisson's ratio, damping ratio	Angle of internal friction cohesion
In situ measurement	Seismic wave method			
	In situ vibration test			
	Repeated loading test			
Laboratory measurement	Wave propagation test			
	Resonant column test			
	Repeated loading test			

Fig. 4.47. Strain levels associated with different in situ laboratory tests (After Ishihara, 1971)

4.4.1 Shear Modulus for Low Strain Amplitudes in Cohesionless Soil

In the case of cohesionless soils, shear modulus G is dependent on effective confining pressure $\bar{\sigma}_0$ and void ratio e. The effect of other factors on G is negligible (Hardin and Richart, 1963). They have reported the results of several resonant column tests in dry Ottawa sands as shown in Fig. 4.49. Straight lines have been fitted through the test points corresponding to different confining pressures.

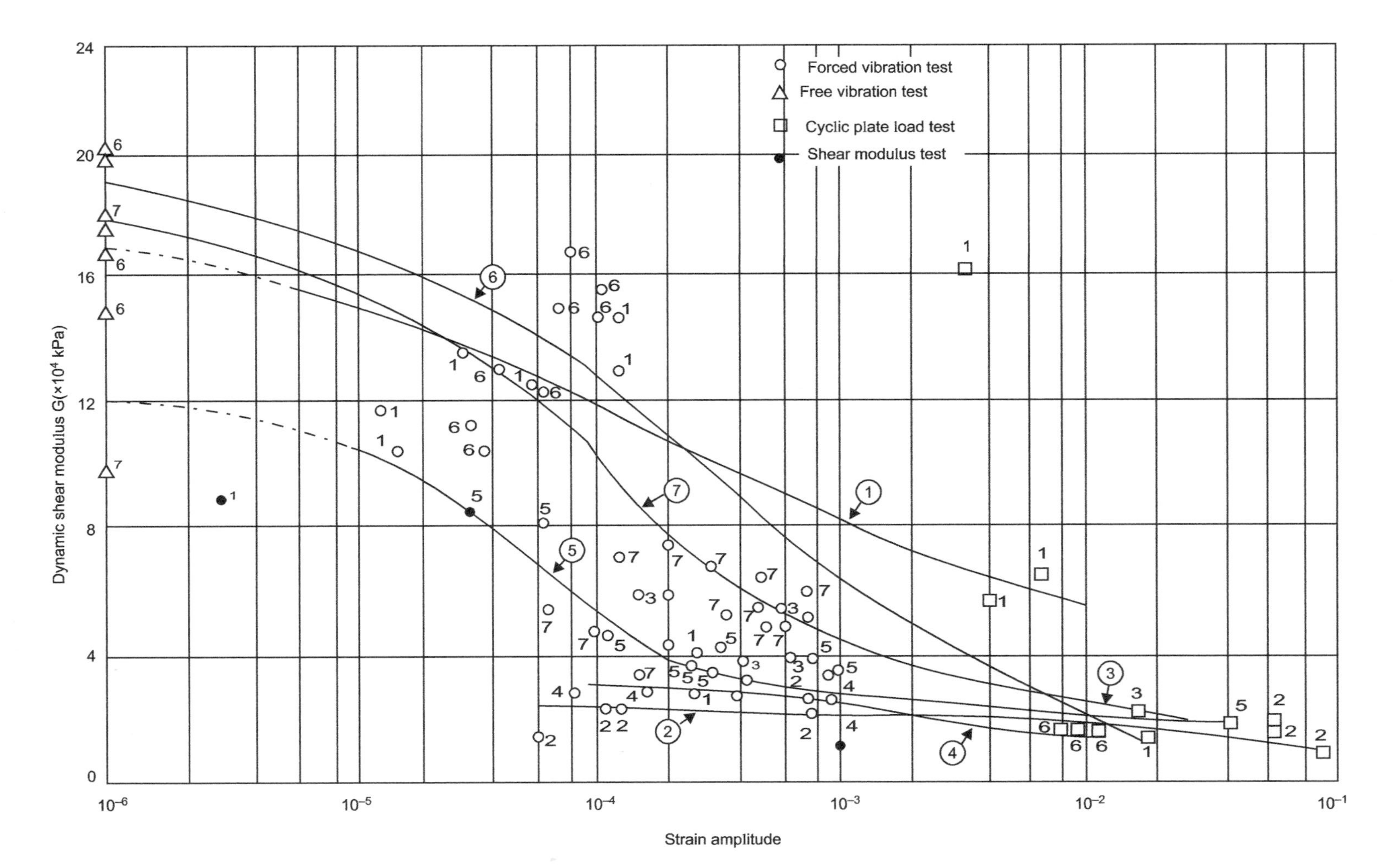

Fig. 4.48. Dynamic shear modulus versus strain (Prakash and Puri; 1981)

Similar results are also shown in Fig. 3.12 (of chapter 3) as solid lines. In order to extend the lines for wider range of void ratios, dotted lines have been drawn in Fig. 3.12 to represent the results from tests using clean angular grained materials. The peak to peak shear strain amplitude for these tests was 10^{-3} rad. It can be seen from these figures that v_s and G are independent of the gradation and grain size distribution. The effect of confining pressure and void ratio on v_s and G is significant.

Following empirical expressions have been developed for v_s and G for round-grained sands and angular-grained crushed quartz (Hardin and Black, 1968).

For round grained sands ($e < 0.8$)

$$v_s = (11.36-5.35e)(\bar{\sigma}_o)^{0.3} \tag{4.41}$$

$$G = \frac{6908(2.17-e)^2}{1+e}(\bar{\sigma}_o)^{0.5} \tag{4.42}$$

For angular grained sands ($e \geq 0.8$)

$$v_s = (18.43-6.2e)\,(\bar{\sigma}_o)^{0.25} \tag{4.43}$$

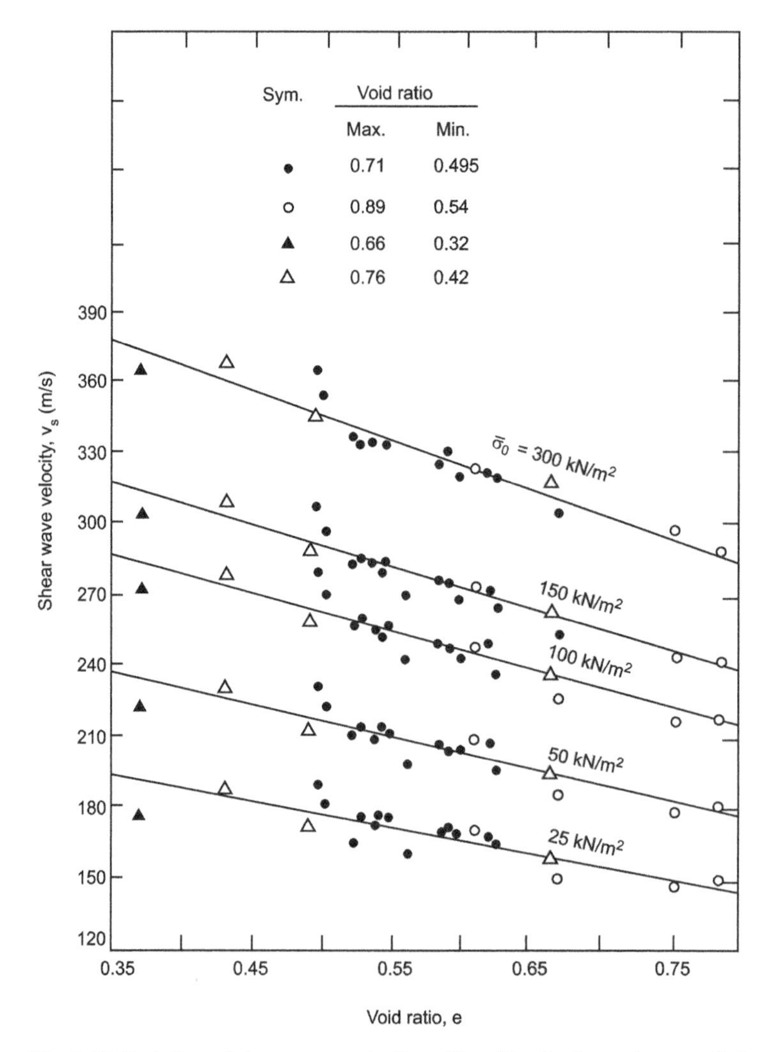

Fig. 4.49. Variation of shear wave velocity with void ratio for various confining pressures, grain size and gradation in dry Ottawa sand (Hardin and Richart, 1963)

$$G = \frac{3230(2.97-e)^2}{1+e}(\bar{\sigma}_o)^{0.5} \tag{4.44}$$

where v_s = Shear wave velocity in m/s

G = Shear modulus in kN/m²

$\bar{\sigma}_o$ = Mean effective confining pressure in N/m² for Eqs. (4.41) and (4.43); and in kN/m² for Eqs. (4.42) and (4.44)

e = Void ratio

Hardin and Richart (1963) have shown that the effect of degree of saturation on v_s is insignificant (Fig. 4.50).

Seed and Idriss (1970) have suggested the following equation

$$G = 1000\,K\,(\bar{\sigma}_o)^{0.5} \tag{4.45}$$

where G is in kN/m² units, K is an empirical factor which varies according to relative density of sand, and $\bar{\sigma}_o$ is the mean effective confining stress in kN/m² units. Table 4.7 gives some values of K obtained from field measured values of shear modulus.

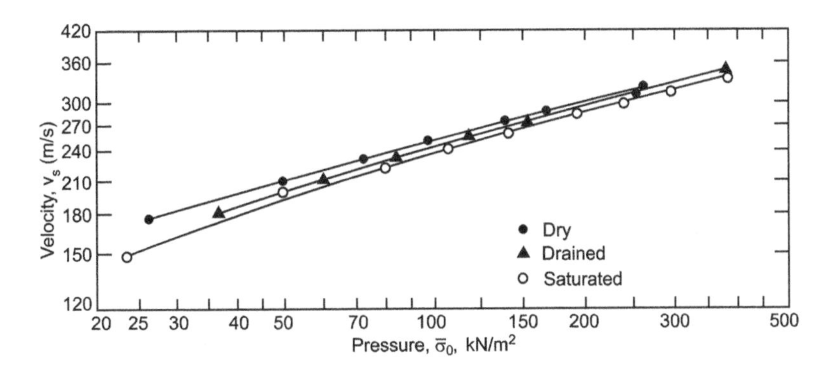

Fig. 4.50. Variation of shear wave velocity with confining pressure for a specimen of Ottawa sand in the dry, saturated and drained conditions (Hardin and Richart, 1963)

Table 4.7. Field Measured Values of K (Seed and Idriss, 1970)

Soil	K
Loose moist sand	7.5
Dense dry sand	10.0
Dense saturated sand	13.0
Dense saturated silty sand	14.0
Dense saturated silty sand	16.0
Extremely dense silty sand	19.0
Dense dry sand (slightly cemented)	36.0
Moist clayey sand	26.0

4.4.2 Cohesive Soils

Few investigators (Lawrence, 1965; Hardin and Black, 1968; Humphries and Wahls, 1963) have performed tests on cohesive soils using resonant column devices to get the shear modulus. On the basis of analysis of experimental data, Hardin and Black (1968) have obtained the variation of G with e and $\bar{\sigma}_o$ as shown in Fig. 4.51 for normally loaded clays. They have also reported the shear modulus for some undisturbed clay specimens collected from the field. The following relation has been suggested:

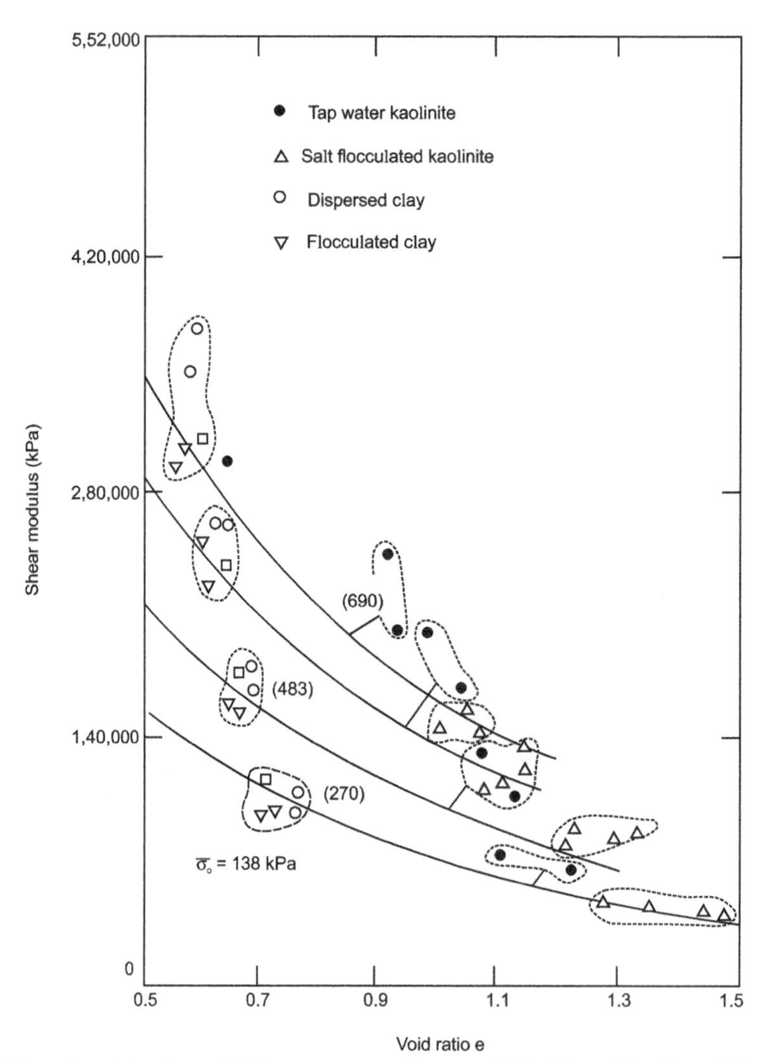

Fig. 4.51. Experimental values of G for some normally consolidated clays (Hardin and Black, 1968)

$$G = C_1 \frac{(2.97-e)^2}{1+e} (\bar{\sigma}_o)^{0.5} \tag{4.46}$$

where C_1 is constant; and G and $\bar{\sigma}_o$ are both in kN/m² units.

Equation 4.47 has been suggested by Hardin and Drnevich (1972 *b*) for both clays and sand.

$$G = \frac{476.4(2.97-e)^2}{(1+e)} (OCR)^k (\bar{\sigma}_o)^{0.5} \tag{4.47}$$

where k is function of plasticity index (Table 4.8), and OCR is over consolidation ratio $\bar{\sigma}_o$ and G are in kN/m².

Hardin (1978) has suggested the following expression:

$$G = \frac{625\,(OCR)^k}{0.3+0.7e^2} \left(\frac{\bar{\sigma}_o}{p_a}\right)_{0.5} \tag{4.48}$$

where p_a is the atmospheric pressure in the same unit as of $\bar{\sigma}_o$. G is in kN/m² units.

Table 4.8. Values of k (Hardin and Drnevich, 1972 *b*)

Plasticity index, PI	k
00	0.00
20	0.18
40	0.30
60	0.41
80	0.48
> 100	0.50

For clays, Seed and Idriss (1970) suggested an equation of the form:

$$G/c_u = K \tag{4.49}$$

where c_u is the undrained shearing strength of soil. K is a constant whose value lies between 1000 and 3000.

Ohasaki and Iwasaki (1973) developed a relationship by correlating the shear modulus obtained in a crosshole survey (SCS) with SPT 'N' values.

$$G = 12000\ N^{0.8} \tag{4.50}$$

where G is in kN/m^2 units, and N-value recorded in standard penetration test. This equation applies to both sands and clays.

4.4.3 Shear Modulus for Large Strain Amplitudes

Figure 4.52 shows a plot between shear stress (τ) and shear strain (γ). The stress-strain curve is approximated by a hyperbolic function defined in terms of initial shear modulus $G_{\max}$ and a reference shear strain γ_r which is defined by Eq. (4.55)

$$\gamma_r = \frac{\tau_{\max}}{G_{\max}} \tag{4.51}$$

The reference strain is equal to the strain at which a line drawn through the origin with a slope equal to G_{max} intersects the horizontal line at $\tau = \tau_{\max}$. The $\tau_{\max}$ is the shear stress at failure in the soil. It can be obtained in the following manner. Figure 4.53 a shows a soil element at a given depth being subjected to vertical and horizontal effective stresses of $\bar{\sigma}_v$ and $K_o\,\bar{\sigma}_v$ respectively. K_o is the coefficient of earth pressure at rest. The Mohr circle corresponding to stresses $\bar{\sigma}_v$ and $K_o\,\bar{\sigma}_v$ is shown as circle 1 in Fig. 4.53c.

From the geometry of the circles 1 and 2, we get

$$\tau_{max} = \left[\left\{\frac{1}{2}(1+K_o)\bar{\sigma}_v \sin\phi + c\cos\phi\right\}^2 - \left\{\frac{1}{2}(1-K_o)\bar{\sigma}_v\right\}^2\right] \tag{4.52}$$

G_{max} is the value of G applicable for very low strain amplitude, and therefore can be obtained using the appropriate equation from Eqs. (4.42), (4.44) to (4.50). Thus the value of reference shear strain γ_r can be evaluated.

The hyperbolic stress-strain curve which defines the initial loading curve and also the end point of the complete stress reversal loop is given by Eq. (4.53).

$$\tau = \frac{\gamma}{\dfrac{1}{G_{\max}} + \dfrac{\gamma}{\tau_{\max}}} \tag{4.53}$$

As

$$G = \frac{\tau}{\gamma} \tag{4.54}$$

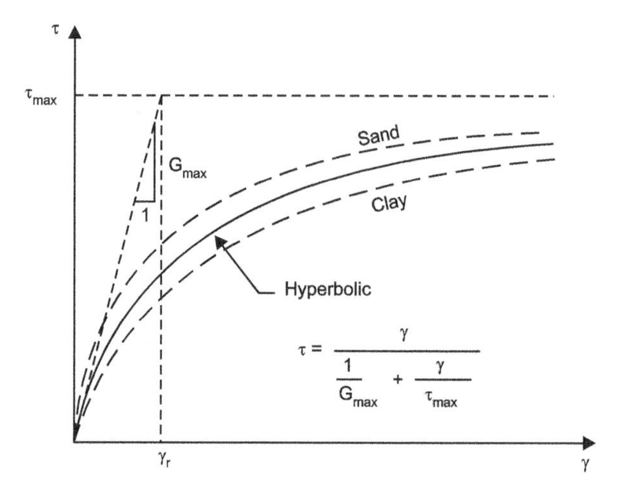

Fig. 4.52. Hyperbolic stress-strain relationship (Hardin and Drnevich, 1972b)

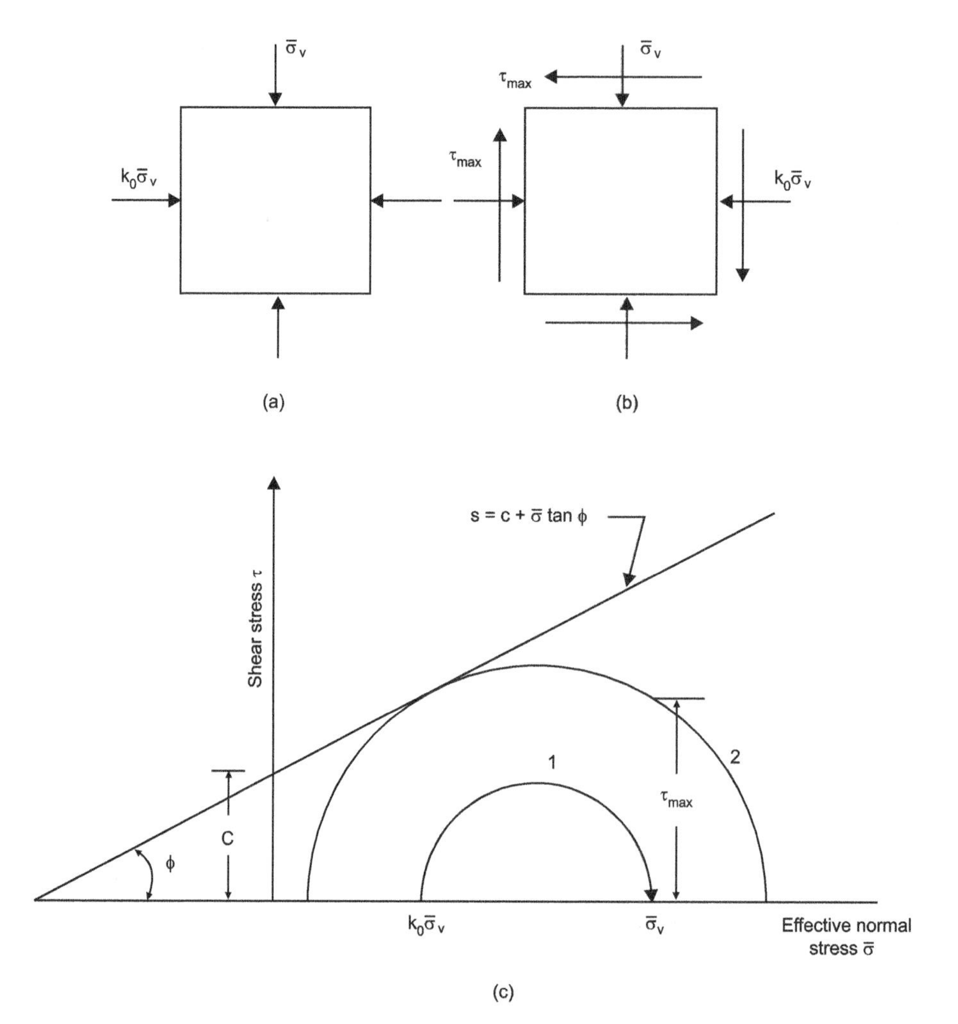

Fig. 4.53. Determination of τ_{max}

Combining Eqs. (4.51), (4.53) and (4.54), we get

$$G = \frac{G_{\max}}{1+\dfrac{\gamma}{\gamma_r}} \tag{4.55}$$

Using Eq. (4.55), one can obtain the value of G at any strain amplitude, γ. For very small strain amplitudes, $\gamma/\gamma_r = 0$; and Eq. (4.55) reduces to $G = G_{max}$.

4.4.4 Estimation of Damping Ratio

Hardin and Drnevich (1972) presented a relation between damping ratio ξ and the maximum value of damping ratio ξ_{max} as below:

$$\xi = \xi_{\max}\left(1-\frac{G}{G_{\max}}\right) \tag{4.56}$$

Combining Eqs. (4.55) and (4.56), we get,

$$\xi = \left(\frac{\dfrac{\gamma}{\gamma_r}}{1+\dfrac{\gamma}{\gamma_r}}\right) \tag{4.57}$$

Typical values of ξ_{max} are given in Table 4.9.

Table 4.9. Typical values of ξ_{max} (Hardin & Drnevich, 1972)

Soil type	*Value of ξ_{max} (%)*
Clean dry sands	33 – 1.5 (log n)
Clean saturated sands	28 – 1.5 (log n)
Saturated silt	$26-4\bar{\sigma}_o^{1/2}+0.7f^{1/2}-1.5(\log n)$
Saturated cohesive soil	$31-(3+0.03f)\bar{\sigma}_o^{1/2}+1.5f^{1/2}-1.5(\log n)$

n - Number of cycles
$\bar{\sigma}_o$ - Mean effective principal stress, kg/cm^2
f - Frequency of loading, Hz

For cohesionless soils, the value of ξ_{max} is dependent only on the number of cycles of loading n while for silts and clays, the frequency of loading f (Hertz) and the mean effective principal stress ($\bar{\sigma}_o$ in kg/cm^2) also influence the maximum damping ratio.

Illustrative Examples

Example 4.1

A soil specimen was tested in a resonant column device (torsional vibration, Fixed-free condition) for determination of shear modulus. Given a specimen length of 90 mm, diameter 35 mm, mass of 160 g, and a frequency at a normal mode of vibration (n = 1) of 800 cps, determine the shear modulus of the specimen.

Solution

1. From Eq. (3.47)

$$\omega_n = \frac{1}{2}\frac{(2n-1)\pi v_s}{L}$$

Putting $n = 1$

$$v_s = \frac{2L\omega_n}{\pi} = \frac{2\cdot(0.090)\cdot(2\pi\cdot 800)}{\pi} = 288 \text{ m/s}$$

2. Mass density of soil in the specimen

$$\rho = \frac{0.160}{\pi\left(\frac{0.035}{2}\right)^2\cdot 0.090} = 1848.7 \text{ kg/m}^3$$

3.
$$G = \rho v_s^2 = 1848.7(288^2)$$
$$= 1.533 \times 10^8 \text{ N/m}^2$$

Example 4.2

A vertical vibration test was conducted on a 1.5 m × 0.75 m × 0.70 m high concrete block in an open pit having depth 2.0 m which is equal to the anticipated depth of actual foundation. The test was repeated at different settings (θ) eccentric masses.

The data obtained from the tests are given below:

S. No.	θ *(Deg)*	f_{nz}	*Amplitude at resonance (Microns)*
1.	36	41.0	13.0
2.	72	40.0	24.0
3.	108	34.0	32.0
4.	144	31.0	40.0

The soil is sandy in nature having angle of internal friction $\phi = 35°$ and saturated density $\gamma_{sat} = 20$ kN/m^3. The water table lies at a depth of 3.0 m below the ground surface. Probable size of the actual foundation 4.0 × 3.0 × 3.5 m high. Determine the values of C_u, E and G to be adopted for the design of actual foundation. Limiting vertical amplitude of the machine is 150 microns.

Solution

1. Area of block = 1.5 × 0.75 = 1.125 m^2
   Mass of block = (1.125 × 0.70) × 2400 = 1890 kg
   Mass of oscillator and motor = 100 kg (assumed)
   Mass of block, oscillator and motor = 1890 + 100
   = 1990 kg

2. Computations of C_u, E and G for test block

$$C_u = \frac{4\pi^2 f_{nz}^2 m}{A}$$

$$= \frac{4\pi^2 f_{nz}^2\cdot 1990}{1.125\cdot 1000} = 71.1 f_{nz}^2 \text{ kN/m}^3$$

The calculated values of C_u for different observed resonant frequencies are listed in column No. 5 of Table 4.10.

$$C_u = \frac{1.13E}{(1-\mu^2)}\cdot\frac{1}{\sqrt{A}}$$

Assuming $\mu = 0.35$ $$E = \frac{\sqrt{1.125}\,(1-0.35^2)}{1.13} C_u = 0.8236\,C_u \text{ kN/m}^2$$

$$G = \frac{E}{2(1+\mu)} = \frac{0.8236\,C_u}{2(1+0.35)} = 0.3050\,C_u \text{ kN/m}^2$$

For different values of C_u (column No. 5), calculated values of E and G are listed in Cols. 6 and 7 of Table 4.10 respectively.

3. Corrections for confining pressure and area

The mean effective confining pressure $\bar{\sigma}_{01}$ at depth of one-half the width below the centre of block is given by

$$\bar{\sigma}_{01} = \bar{\sigma}_v \frac{(1+2K_0)}{3}$$

where $\bar{\sigma}_v = \bar{\sigma}_{v1} + \bar{\sigma}_{v2}$

$\bar{\sigma}_{v1}$ = Effective overburden pressure at the depth under consideration

$\bar{\sigma}_{v2}$ = Increase in vertical pressure due to the weight of block

Assuming that the top 2.0 m soil has a moist unit weight of 18 kN/m², and the next 1.0 m soil i.e. upto water table is saturated having saturated unit weight equal to 20 kN/m³, then

$$\bar{\sigma}_{v1} = 18 \times 2.0 + 20 \times \frac{0.70}{2} = 43.5 \text{ kN/m}^2$$

From Taylor (1948):

$$\bar{\sigma}_{v2} = \frac{4q}{4\pi}\left[\frac{2mn\sqrt{m^2+n^2+1}}{m^2+n^2+1+m^2n^2}\cdot\frac{m^2+n^2+2}{m^2+n^2+1} + \sin^{-1}\frac{2mn\sqrt{m^2+n^2+1}}{m^2+n^2+1+m^2n^2}\right]$$

$$m = \frac{L/2}{Z} = \frac{1.5/2}{0.70/2} = 2.14$$

$$n = \frac{B/2}{Z} = \frac{0.75/2}{0.70/2} = 1.07$$

$$q = 24 \times 0.70 = 16.8 \text{ kN/m}^2$$

[Assuming unit weight of concrete = 24 kN/m³]

Substituting the above values of m, n and q in the expression of $\bar{\sigma}_{v2}$, we get

$$\bar{\sigma}_{v2} = 16.8 \times 0.82 = 13.73 \text{ kN/m}^2$$

$$\bar{\sigma}_v = 43.5 + 13.73 = 57.23 \text{ kN/m}^2$$

$$K_o = 1 - \sin\phi = 1 - \sin 35° = 0.426$$

$$\bar{\sigma}_{01} = 57.23\left(\frac{1 + 2 \times 0.426}{3}\right) = 35.30 \text{ kN/m}^2$$

For the actual foundation

$$\bar{\sigma}_{v1} = 18 \times 2.0 + 20 \times 1.0 + (20 - 10) \times 0.5 = 61 \text{ kN/m}^2$$

$$m = \frac{4.0/2}{3.0/2} = 1.334$$

$$n = \frac{3.0/2}{3.0/2} = 1.0$$

$$q = 24 \times 3.5 = 84 \text{ kN/m}^2$$

Substituting the above values of m, n and q in the expression of $\bar{\sigma}_{v2}$, we get

$$\bar{\sigma}_{v2} = 63.76 \text{ kN/}m^2$$

$$\bar{\sigma}_v = 61 + 63.76 = 124.76 \text{ kN/m}^2$$

$$\bar{\sigma}_{02} = 124.76\left(\frac{1 + 2 \times 0.426}{3}\right) = 77.01 \text{ kN/m}^2$$

Area of actual foundation = 4.0 × 3.0 = 12.0 m² (> 10 m²)

$$\frac{C_{u2}}{C_{u1}} = \left(\frac{\bar{\sigma}_{02}}{\bar{\sigma}_{01}}\right)^{0.5} \cdot \left(\frac{A_1}{A_2}\right)^{0.5} = \left(\frac{77.01}{35.30}\right)^{0.5} \left(\frac{1.125}{10}\right)^{0.5} = 0.50$$

The values of C_u of the actual foundation at different strain levels (= amplitude at resonance / width of test block) are given in column 8 of Table 4.10 respectively. The corresponding values of strain levels are listed in column 11.

Values of E and G are obtained using following relations:

$$E = \frac{\sqrt{10}\ (1 - 0.35^2)}{1.13} C_u = 2.455 C_u \text{ kN/m}^2$$

$$G = \frac{2.455\ C_u}{2(1+0.35)} = 0.9095 C_u \text{ kN/m}^2$$

Values of E and G so obtained are listed in cols. 9 and 10 of Table 4.10.

4. Strain level correction

$$\text{Strain in actual foundation} = \frac{150 \times 10^{-6}}{3.0} = 0.5 \times 10^{-4}$$

The values of C_u, E and G corresponding to strain level of 0.5×10^{-4} can be obtained by interpolation.

$$C_u = \left(4.10 - (4.10 - 3.40)\frac{0.50 - 0.427}{0.533 - 0.427}\right) \times 10^4$$

$$= 4.10 - 0.7 \times 0.6886 = 3.62 \times 10^4 \text{ kN/m}^3$$

$$E = 2.455 \times 3.62 \times 10^4 = 8.89 \times 10^4 \text{ kN/m}^2$$

$$G = 0.9095 \times 3.62 \times 10^4 = 3.29 \times 10^4 \text{ kN/m}^2$$

Hence the response of the proposed foundation block should be checked using

$$C_u = 3.62 \times 10^4 \text{ kN/m}^3$$

$$E = 8.89 \times 10^4 \text{ kN/m}^2$$

$$G = 3.29 \times 10^4 \text{ kN/m}^2$$

Table 4.10. Analysis of Data for C_u, E and G

S. No.	θ *(Deg.)*	f_{nz}	*Amplitude at resonance (microns)*	*For test block*			*For actual foundation*			
				$C_u \times 10^4$ kN/m^3	$E \times 10^4$ kN/m^2	$G \times 10^4$ kN/m^2	$C_u \times 10^4$ kN/m^3	$E \times 10^4$ kN/m^2	$G \times 10^4$ kN/m^2	*Strain level* $\times 10^{-4}$
1.	36	41	13	11.95	9.84	3.64	5.96	14.65	5.43	0.173
2.	72	40	24	11.38	9.84	3.47	5.67	13.97	5.17	0.320
3.	108	34	32	8.22	6.77	2.51	4.10	10.09	3.74	0.427
4.	144	31	40	6.83	5.62	2.08	3.40	8.37	3.10	0.533

Example 4.3

(a) Determine the expressions of coefficient of elastic uniform shear, C_τ in terms of resonant frequency for the block of size 1.5 m × 0.75 m × 0.70 m high, tested under horizontal vibrations.

(b) Determine the value of C_τ for the foundation mentioned in example 4.2, if the block tested in horizontal vibrations give the following results:

S. No.	θ *(Deg)*	f_{nz} *(Hz)*	*Amplitude at resonance (Microns)*
1.	36	35	10
2.	72	34	16
3.	108	30	21
4.	144	27	28

Premissible value of horizontal amplitude is 100 microns.

Solution

(a) Value of C_τ is given by Eq. 4.3

$$C_\tau = \frac{8\pi^2 r\, f_{nx}^2}{(A_o + I_o) \pm \sqrt{(A_o + I_o)^2 - 4r\,A_o\,I_o}}$$

Moment of inertia of the base contact area of block about axis of rotation

$$I = \frac{0.75 \times 1.5^3}{12} = 0.2109 \text{ m}^4$$

Height of combined centre of gravity of block, oscillator and motor from base

$$L = \frac{1890 \times 0.35 + 100 + 0.85}{19.9} = 0.375 \text{ m}$$

The centre of gravity of oscillator and motor is assumed at a height of 0.15 m from the top of the block. Mass moment of inertia M_m about an axis passing through the combined centre of gravity and perpendicular to the plane of vibrations is given by

$$M_m = 1890\left(\frac{1.5^2 + 0.70^2}{12}\right) + 1890(0.375 - 0.35)^2 + 100(0.85 - 0.375)^2$$

$$= 431.55 + 1.18 + 22.56 = 455.29 \text{ Kg-m}^2$$

$$M_{mo} = M_m + mL^2 = 455.29 + 1990 \times (0.375)^2 = 735.13 \text{ Kg-m}^2$$

$$r = \frac{M_m}{M_{mo}} = \frac{455.29}{735.13} = 0.6194$$

$$A_0 = \frac{A}{m} = \frac{1.125}{1990} = 0.5546 \times 10^{-3} \text{ m}^3\text{/kg}$$

$$I_0 = 3.46\left(\frac{I}{M_{mo}}\right) = 3.46\left(\frac{0.2109}{735.13}\right) = 0.9926 \times 10^{-3} \text{ m}^2\text{/kg}$$

$$C_\tau = \frac{8\pi^2 \times 0.6194\ f_{nx}^2}{(0.5546+0.9926)\times 10^{-3} \pm \sqrt{(0.5546+0.9926)^2 \times 10^{-6} - 4\times 0.6194\times 0.5546\times 0.9926\times 10^{-6}}}$$

$$= \frac{48.856\ f_{nx}^2}{(1.5283) \pm (0.9989)} \times 10^3\ \text{N/m}^3$$

Therefore

$$C_\tau = 92.3\, f_{nx1}^2\ \text{kN/m}^3\ \text{(First mode)}$$

$$C_\tau = 19.3\, f_{nx2}^2\ \text{kN/m}^3\ \text{(Second mode)}$$

(b) $C_\tau = 92.3\, f_{nx1}^2$

The calculated values of C_τ for different observed resonant frequencies are listed in column 5 of Table 4.11. As in example 4.2

$$\frac{C_{\tau 2}}{C_{\tau 1}} = 0.50$$

The values of C_τ for the actual foundation are given in column 6. The corresponding strain levels are listed in column 7.

Table 4.11. Analysis of Data for C_τ

S. No.	θ *(Deg.)*	f_{nx1} (Hz)	*Amplitude (microns)*	$C_{\tau 1} \times 10^4$ kN/m²	$C_{\tau 2} \times 10^4$ kN/m²	*Strain level* $\times 10^{-4}$
(1)	(2)	(3)	(4)	(5)	(6)	(7)
1	36	25	10	5.77	2.87	0.133
2	72	23	16	4.88	2.43	0.213
3	108	21	21	4.07	2.03	0.280
4	144	19	28	3.33	1.66	0.373

$$\text{Strain in actual foundation} = \frac{100 \times 10^{-6}}{3.0} = 0.333 \times 10^{-4}$$

Therefore, the value of C_τ for actual foundation

$$= \left[2.03 - (2.03-1.66) \times \frac{0.373-0.333}{0.373-0.280}\right] \times 10^4 = 1.87 \times 10^4\ \text{kN/m}^3$$

Example 4.4

The soil profile at a site is shown in Fig. 4.54. Two cross borehole tests were conducted at the site to determine the values of shear wave velocities in the small areas around points *A* and *B*. The average values of shear wave velocities were observed as 110 m/s and 130 m/s respectively. Determine the values of dynamic shear modulus *G* for points *A*, *B* , *C* and *D*.

Solution

1. Fine grained soil stratum (0.0 to – 4.0 m)
   Observed shear wave velocity at point *A*

$$v_s = 110\ \text{m/s}$$

$$G = \rho v_s^2$$

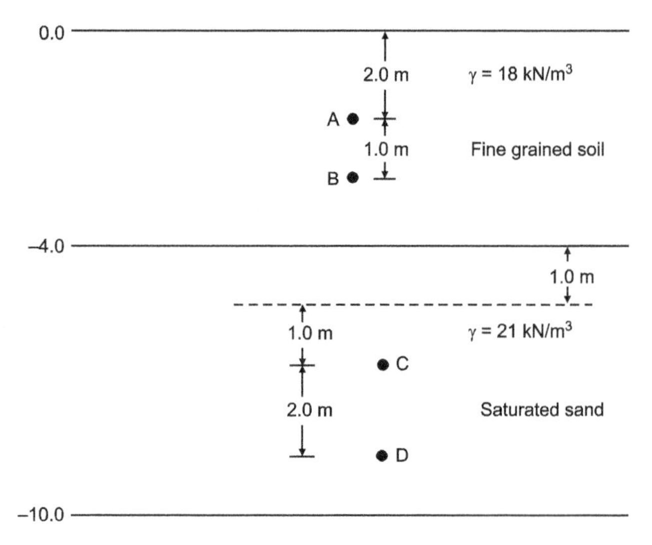

Fig. 4.54. Soil profiles (example 4.4)

$$G = \frac{18}{9.81} \times (110)^2 = 2.2 \times 10^4 \text{ kN/m}^2$$

$$\bar{\sigma}_{vA} = 18 \times 2 = 36 \text{ kN/m}^2$$

$$\bar{\sigma}_{vC} = 18 \times 3 = 54 \text{ kN/m}^2$$

$$\frac{(G)_C}{(G)_A} = \left(\frac{\bar{\sigma}_{vC}}{\bar{\sigma}_{vA}}\right)^{0.5} = \left(\frac{54}{36}\right)^{0.5} = 1.2247$$

$$(G)_C = 2.2 \times 10^4 \times 1.2247 = 2.7 \times 10^4 \text{ kN/m}^4$$

2. Saturated sand (–4.0 m to –10.00 m)

$$(v_s)_B = 130 \text{ m/s}$$

$$(G)_B = \frac{21}{9.81} \times (130)^2 = 3.6 \times 10^4 \text{ kN/m}^2$$

$$\bar{\sigma}_{vB} = 18 \times 4.0 + 21 \times 1.0 + (21 - 10) \times 1.0 = 104 \text{ kN/m}^2$$

$$\bar{\sigma}_{vD} = 104 + (21 - 10) \times 2.0 = 126 \text{ kN/m}^2$$

$$(G)_D = 3.6 \times 104 \times \left(\frac{26}{104}\right)^{0.5} = 2.96 \times 10^4 \text{ kN/m}^2$$

Example 4.5

At a particular site, the top 10.0 m soil is medium grained sand having dry unit weight as 17 kN/m³. The water table is 6.0 m below the ground surface. The value of specific gravity of soil grains is 2.67. The direct shear test gave the value of ϕ as 36°. Determine the value of shear modulus of the soil at depth of 7.0 m below ground surface.

Solution

1.
$$\gamma_d = 17 = \frac{2.67 \times 10}{1 + e}$$

$$e = 0.57$$

$$\gamma_{sat} = \frac{(2.67 + 0.57)}{1 + 0.57} \times 10 = 20.6 \text{ kN/m}^3$$

$$(\bar{\sigma}_v)_{7.0m} = 17 \times 6.0 + (20.6 - 10) \times 1.0$$

$$= 112.6 \text{ kN/m}^2$$

$$K_o = 1 - \sin\phi = 1 - \sin 36^\circ = 0.412$$

$$\bar{\sigma}_o = \left(\frac{1+2K_o}{3}\right)\bar{\sigma}_o = \left(\frac{1+2\times 0.412}{3}\right) \times 112.6$$

$$= 68.46 \text{ kN/m}^2$$

2. From Eq. (4.46)
$$G = \frac{6908(2.17 - e)^2}{1 + e} \cdot (\bar{\sigma}_o)^{0.5}$$

$$G = \frac{6908(2.17 - 0.57)^2}{1 + 0.57} \cdot (68.46)^{0.5}$$

$$= 9.3 \times 10^4 \text{ kN/m}^2$$

References

Anderson, D. G. (1974). "Dynamic modulus of cohesive soils", Ph.D. Dissertation, University of Michigan, Ann Arbor.

Bakran, D. D. (1962). "Dynamics of bases and foundations." McGraw-Hill, New York.

Bjerrum, L., and Landra, A. (1966). "Direct simple shear tests on a Norwegian quick clay", Geotechnique 26(1), pp. 1-20.

Casagrande, A. and Shanan, W.L. (1948a), "Stress deformation and strength characteristics of soils under dynamic loads". Proc. Second Int. Conf. Soil Mech. Foundation Engg., Vol. 5, pp. 29-34.

Casagrande, A. and Shanan, W.L. (1948b), "Research on stress deformation and strength characteristics of soils and soft rocks under transient loading," Harvard Soil Mechanics Series No. 31.

Cho, Y;, Rizzo, P. C., and Humphries, W.K. (1976), "Saturated sand and cyclic dynamic tests", Am. Soc. Civ. Eng., Ann. Conv. Expo., Philadelphia, PA, Meet. Prepr. 2752, pp. 285-312.

Dass, B (1977), "Behavior of clay under oscillatory loading", M.E. Thesis, U.O.R. Roorkee.

Drnevich. V.P. (1967), "Effect of strain history on the dynamic properties of sand", Ph.D. Dissertation, University of Michigan, Ann. Arbor.

Drnevich, V.P. (1972), "Undrained cyclic shear of saturated sand", J. Soil. Mech. Found. Div., Am., Soc. Civ. Eng. 98 (SM-8), pp. 807-825.

Goto, N., Kagami, H., Shiono, K. and Ohta, Y. (1977), "An easy-capable and high precise shear wave measurement by means of the standard penetration test". Proceedings sixth world conference on earthquake engineering, pp. 171-176.

Hall, J.R., Jr., and Richart, F.E., Jr. (1963), "Dissipation of elastic wave energy in granular soils", J. Soil Mech. Found. Div. Am. Soc. Civ. Engg., 89 (SM-6), pp. 27-56.

Hardin, B.C. (1971), "Progrm of simple shear testing of soils", Univ. Ky., Soil Mech., Ser. No. 8, pp. 1-14.
Hardin, B.O. and Black, W.L., (1968), "Vibration modulus of normally consolidated clay", Journal of the Soil Mechanics and Foundations Division, ASCE, 94 (SM2).
Hardin, B. O. and Richart, F. E. Jr., (1963), " Elastic wave velocities in granular soils", Journal of the Soil Mechanics and Foundations Division, ASCE 89 (SMI), pp. 33-65.
Hardin, B.O. and Music, J. (1965), "Apparatus for vibration of soil specimens during the triaxial test." Instruments and apparatus for soil and rock mechanics, ASTM, STP 392, pp. 55-74.
Hoadley, P.J. (1985), "Measurement of dynamic soil properties", Chapter of a book on Analysis and Design of foundations for vibration, pp. 349-420.
Hvorslev, M. J., and Kaufman, R.I. (1952), "Torsion shear apparatus and testing procedure", USAE Waterways exp. Stn., Bull 38, pp. 1-76.
Iida, K. (1938), "The velocity of elastic waves in sand," Bull. Earthquake Res. Inst., Tokyo Imp. Univ., 16, pp. 131-144.
Iida, K. (1940), "On the elastic properties of soil particularly in relation to its water content", Bull. Earthquake Res. Inst., Tokyo Imp. Univ., p. 18, 675-690.
Imai. T. (1977), "Velocity of P- and S-waves in subsurface layers of ground in Japan", Proc. 9th Int. Conf. Sol Mech. Found., Tokyo, Vol. 2, pp. 257-260.
IS: 5249 (1992), "Determination of dynamic properties of soil - method of test", ISI, New Delhi.
IS: 1888 (1982), "Method of load test on soils", ISI, New Delhi.
IS: 2131 (1981), "Method for standard penetration test for soils", ISI New Delhi.
Ishibashi, I. and Sherif, M.A. (1974), "Soil liquefaction by torsional simple shear device", J. Geotech. Eng. Div., Am. Soc. Civ. Eng., 100(GT-8), pp. 871-888.
Ishihara, K. (1971), "Factors affecting dynamic properties of soil", Proc. Asian Reg. Conf. Soil Mech. Found. Eng., 4th, Bangkok, Vol. 2.
Ishihara, K., and Li, S. (1972), "Liquefaction of saturated sand in triaxial torsion shear test", Soils Found (Jpn.) 12(2), pp. 1939.
Ishihara, K., and Yasuda, S. (1975), "Sand liquefaction in hollow cylinder torsion under irregular excitation", Soils Found. (Jpn.), 15(1), pp. 45-59.
Ishimoto. M., and Iida, K. (1937), "Determination of elastic constants of soils by means of vibration methods", Bull. Earthquake Res. Inst., Tokyo Imp. Univ., 15, p. 67.
Iwasaki, T., Tatsuoka, F., and Takagi, Y. (1977), "Shear moduli of sands under cyclic torsional shear loading," Tech. Memo. No. 1264, Public Works Res. Inst., Ministry of Construction, Chiba-Shi, Japan.
Kjellman. W. (1951), Testing of shear strength in Sweden, Geotechnique, 2, pp. 225-232.
Lawrence, F. G., Jr., (1963), "Propagation velocity of ultrasonic waves through sand", MIT Research Report, R63-8.
Lord, A.F., Jr., Curran, J. W., and Koerner, R.M. (1976), "New transducer system for determining dynamic mechanical properties and attenuation in soil", J. Acoust. Soc. Am. 60(2), pp. 517-520.
Miller, R.P., Troncosco J.H. and Brown, F.R. (1975), "In situ impulse test for dynamic shear modulus of soils", Proceedings of the conference on in situ measurement of soil properties, Geotechnical Engineering Division, ASCE. Specialty conference, Raleigh. North Carolina, Vol. I, pp. 319-335.
Peacock, W.H., and Seed, H.B. (1968), "Sand liquefaction under cyclic loading simple shear conditions", J. Soil Mech. Found. Div., Am. Soc. Civ. Eng. 94 (SM-3), pp. 689-708.
Prakash, S., and Puri, V.K. (1981), "Dynamic properties of soils from in situ tests", J. Geotech. Eng. Div., Am. Soc. Civ. Eng. 107 (GT-7), pp. 943-963.
Prakash, S., Nandkumaran, P., and Joshi, V.H. (1973), "Design and performance of an oscillatory shear box", J. Indian Geotech. Soc., 3(2), pp. 101-112.

Prakash, S., Nandkumaran, P., and Joshi, V.H. (1974). "Behaviour of soils under oscillatory shear stress", Proc. Vth Symposium on Earthquake Eng., Roorkee, Vol. 1, pp. 101-102.
Richart, F.E., Jr., Hall, J.R., and Woods, R.D. (1970), "Vibrations of soils and foundations." Prentice-Hall, Englewood Cliffs, New Jersey.
Roscoe, K.H. (1953), "An apparatus for the application of simple shear to soil samples," Proc. Int. Conf. Soil Mech. Found. Eng. 3rd, Zurich, vol. 1. pp. 186-191.
Seed, H.B., Idriss, I.M., and Arango, L. (1983), "Evaluation of liquefaction potential using field performance data", J. Geotech. Eng. Div., Am. Soc. Civ. Eng., 109 (GT-3), pp. 458-482.
Seed, H.B. (1960), "Soil strength during earthquakes", Proc. Second World Conf. Earthquake Engg., vol. 1, pp. 183-194.
Seed, H. B. and Chan, C. K. (1966), "Clay strength under earthquake loading conditions", J. Soil Mech. Found. Div., ASCE, vol. 92, SM 2, p. 53-78.
Seed, H. B. and Fead, J. W. N. (1959), "Apparatus for repeated load tests on soils," Special Technical Publication No. 204, ASTM, Philadelphia.
Silver, M.L., Chan, C.K. et al. (1976), "Cyclic triaxial strength of standard test sand", J. Geot. Engg. Div. ASCE, vol. 102, no. GT 5, pp. 511-523.
Stephenson, R.W., (1978), "Ultrasonic testing for determining dynamic soil moduli", Denver: Dynamic Geotechnical Testing, ASTM, STP 654., pp. 179-195.
Stokoe, K.H., II, and Hoar, R.J. (1978), "Variables affecting in situ seismic measurements", Proc. Am. Soc. Civ. Eng. Spec. Conf. Earthquake Eng. Soil Dyn., Pasadena, CA, vol. 2.
Stokoe, K.H., and Woods, R.D. (1972), "In situ shear wave velocity by cross-hole method", J. Soil Mech. Found. Div., Am. Soc. Civ. Eng. 98 (SM-5), pp. 443-460.
Taylor, D.W. (1948): "Fundamentals of soil mechanics," John Wiley Sons, Inc., New York.
Terzaghi, K. and Peck, R.B. (1967), "Soil mechanics in engineering practice", John Willey and Sons, New York.
Wilson, S.D., and Dietrich, R.J. (1960), "Effect of consolidation pressure on elastic and strength properties of clay", Proc. Am. Soc. Civ. Eng. Res. Conf. Shear Strength Cohesive Soils, Boulder, Co, pp. 419-435.
Woods, R.D. (1978), "Measurement of dynamic soil properties: State-cf-the-Art", Proc. Am. Soc. Civ. Eng. Spec. Conf. Earthquake Eng. Soil Dyn., Pasadena, CA, Vol. 1, p. 91-180.
Yoshimi, Y., and Oh-Oka, H. (1973), "A ring torsion apparatus for simple shear tests", Proc. Int. Conf. Soil Mech. Found. Eng., Moscow, Vol. 1, Pt. 2, pp. 501-506.

Practice Problems

4.1 Describe the salient features of a resonant column aparatus. How is calibration done and the value of shear modulus determined?

4.2 A clayey soil specimen was tested in a resonant column device (torsional vibration, free-free end condition) for determination of shear modulus. Given a specimen length of 100 mm, diameter 36 mm, mass of 180 g, and a frequency at normal mode of vibration of 900 cps, determine the shear modulus.

4.3 Explain the difference between simple shear and direct shear tests. What is the principle involved in oscillatory shear box test? Give the salient features of a study made on clay under dynamic loads using oscillatory shear box.

4.4 List the factors affecting shear strength of cohesive soils under static and dynamic load. Explain with neat sketches the effect of dynamic stress level, initial factor of safety and number of pulses.

4.5 Draw typical transient strength (single impulse) characteristics of sand and clay tested for the following time of loading:

(i) 0.02 s

(ii) 0.08 s and
(iii) Static

4.6 Describe briefly the following:
(a) Seismic cross-borehole survey
(b) Seismic up-hole survey
(c) Seismic down-hole survey

4.7 A vertical vibration test was conducted on a M15 concrete block 1.5 × 0.75 × 0.75 m high using different eccentricities (θ) of the rotating mass of the oscillator. The data obtained are given as follows:

S. No.	*θ (Deg.)*	f_{nz}	*Amplitude at resonance (Microns)*
1.	36	41	13
2.	72	40	24
3.	108	34	32
4.	144	31	40

Determine the value of coefficient of elastic uniform compression. C_u for confining pressure of 100 kN/m^2 and base contact area of 10 m^2. Also give the values of strain levels at different eccentricities.

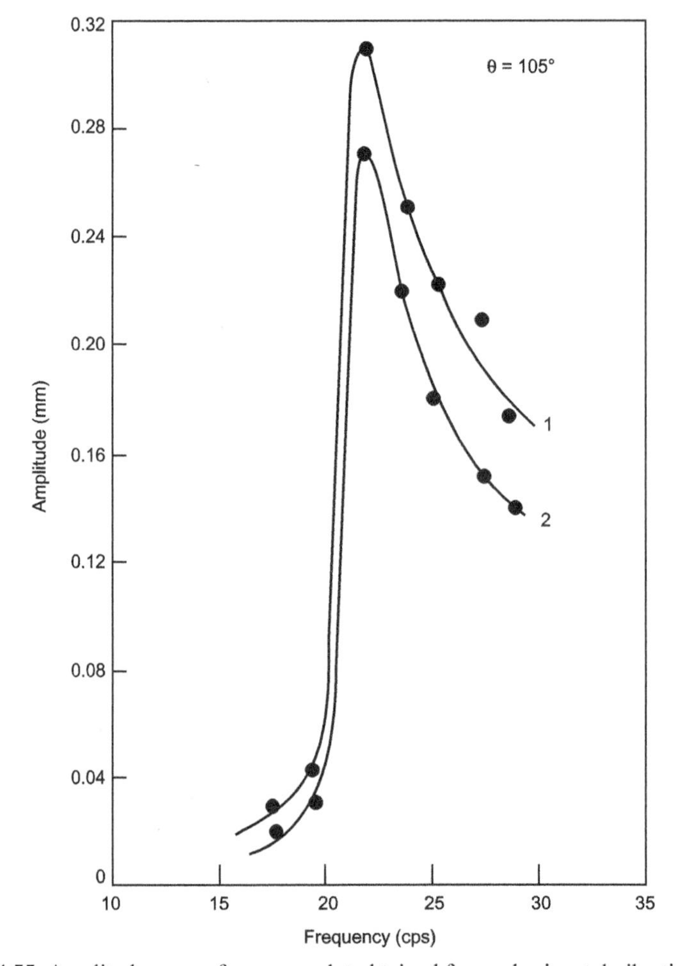

Fig. 4.55. Amplitude versus frequency plot obtained from a horizontal vibration test

4.8 A horizontal vibration test was conducted on a M15 concrete block 1.5 × 0.75 × 0.75 m high. The data obtained is shown in Fig. 4.55. Determine the value of coefficient of elastic uniform shear, C_τ and the corresponding strain level.

4.9 A cyclic plate load test was performed on a plate of 600 mm × 600 mm size. The elastic settlement corresponding to a loading intensity of 80 kN/m^2 was 2 mm. Using this data, determine the coefficient of elastic uniform compression C_u for a foundation block of base area 15 m^2.

4.10 Discuss the factors affecting shear modulus and damping. Illustrate the procedure of obtaining the shear modulus and damping at given strain amplitude from G_{max} and ξ_{max}.

CHAPTER

5

Dynamic Earth Pressure

5.1 General

In the seismic zones, the retaining walls are subjected to dynamic earth pressure, the magnitude of which is more than the static earth pressure due to ground motion. Since a dynamic load is repetitive in nature, there is a need to determine the displacement of the wall due to earthquakes and their damage potential. This becomes more essential if the frequency of the dynamic load is likely to be close to the natural frequency of the wall-backfill-foundation-base soil system. This essentially consists in writing down the equation of motion of the system under free and forced vibrations. This requires the information on the distribution of backfill soil mass and base soil mass participating in vibrations. It is often difficult to assess these. Therefore, more often, pseudo-static analysis is carried out for getting dynamic earth pressure. In this method, the dynamic force is replaced by an equivalent static force.

In this chapter, firstly various methods of computing the magnitude and point of application of dynamic earth pressure based on pseudo-static analysis have been discussed. It is followed with the methods of predicting the displacement of the retaining wall.

5.2 Pseudo-Static Methods

5.2.1 Mononobe-Okabe Theory

Mononobe-Okabe (1992) modified classical Coulomb's theory for evaluating dynamic earth pressure by incorporating the effect of inertia force.

Figure 5.1 shows a wall of height H and inclined vertically at an angle α retaining cohesionless soil with unit weight γ and angle of shearing resistance ϕ. BC_1 is the trial failure plane which is inclined to vertical by an angle θ. The backfill is inclined and making an angle i with horizontal.

During an earthquake the inertia force may act on the assumed failure wedge ABC_1 both horizontally and vertically. If a_h and a_v are the horizontal and vertical accelerations caused by the earthquake on the wedge ABC_1, the corresponding inertial forces are $W_1 \cdot a_h/g$ horizontally and $W_1 \cdot a_v/g$ vertically, W_1 being the weight of the wedge ABC_1. During the worst condition, $W_1 \cdot a_h/g$ acts towards the fill and $W_1 \cdot a_v/g$ may act vertically either in the downward or upward direction. Therefore the direction that gives the maximum increase in earth pressure is adopted in practice.

If α_h and α_v are respectively the horizontal and vertical seismic coefficients, then

$$\alpha_h = \frac{a_h}{g} \tag{5.1}$$

$$\alpha_h = \frac{a_v}{g} \tag{5.2}$$

For the failure condition the soil wedge ABC_1 is in equilibrium under the following forces:

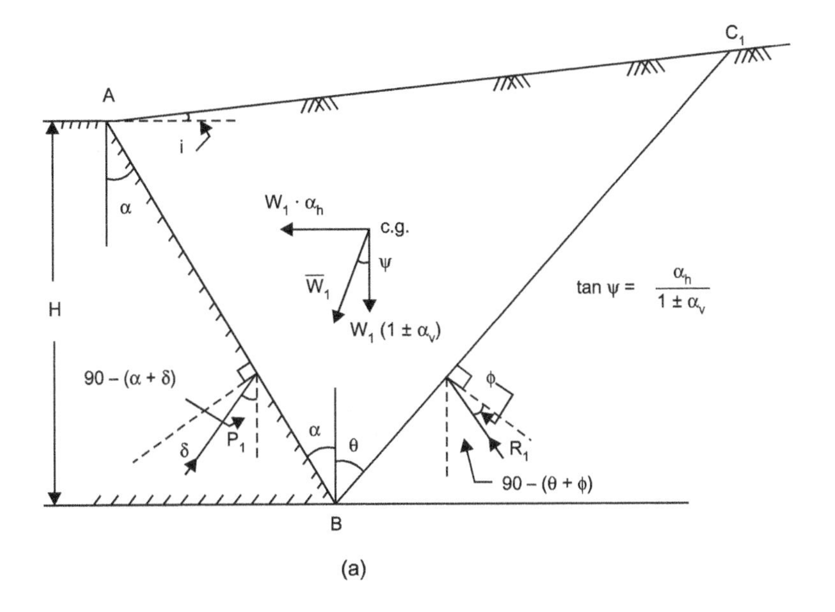

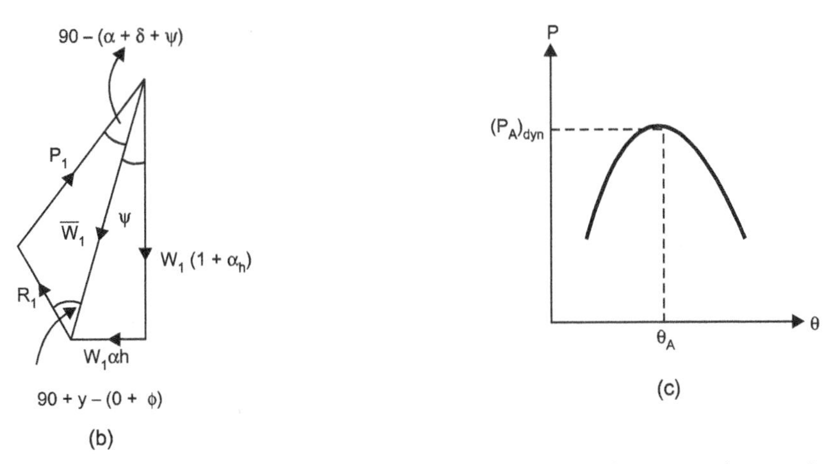

Fig. 5.1. (a) Forces acting on failure wedge in active state; (b) Force polygon and (c) Dynamic earth pressure versus wedge angle plot

(i) W_1, weight of the wedge acting at centre of gravity of ABC_1.
(ii) Earth pressure P_1 inclined at an angle δ to normal to the wall in the anticlockwise direction.
(iii) Soil reaction R_1 inclined at an angle ϕ to the normal on the face BC_1.
(iv) Horizontal inertia force ($W_1\ \alpha_h$) acting at the centre of gravity of the wedge ABC_1.
(v) Vertical inertia force $\pm\ W_1\ \alpha_v$ acting at centre of gravity of the wedge ABC_1.

Weight W_1 and the inertia forces $W_1\ \alpha_h$ and $\pm\ W_1\ \alpha_v$ can be combined to give a resultant $\overline{W}_1$ where

$$\overline{W}_1 = W_1\left[(1 \pm \alpha_v)^2 + \alpha_h^2\right]^{1/2} \tag{5.3}$$

The resultant $\overline{W}_1$ is inclined with vertical at angle ψ, such that

$$\psi = \tan^{-1}\left(\frac{\alpha w}{1 \pm \alpha v}\right) \tag{5.4}$$

The directions of all the three forces $\overline{W}_1$, P_1 and R_1 are known but the magnitude of only one force $\overline{W}_1$ is known. The magnitude of the other forces can be obtained by considering the force polygon as shown in Fig. 5.1b. P_1 is the value of dynamic earth pressure corresponding to the trial wedge ABC_1. More trials are made and the values of P_2, P_3 etc. are obtained. Variation of P and θ is shown in Fig. 5.1c. The maximum value of P is the dynamic active earth pressure $(P_A)_{dyn}$.

Mononobe and Okabe (1929) gave the following relation for the computation of dynamic active earth pressure $[(P_A)_{\text{dyn}}]$:

$$(P_A)_{\text{dyn}} = \frac{1}{2}\gamma H^2 (K_A)_{dyn} \tag{5.5}$$

where $(K_A)_{dyn}$ is coefficient of dynamic active earth pressure and given by:

$$(K_A)_{dyn} = \frac{(1 \pm \alpha_v)\cos^2(\phi - \psi - \alpha)}{\cos\psi\cos^2\alpha\cos(\delta + \alpha + \psi)} \times \left[\frac{1}{1 + \left\{\frac{\sin(\phi + \delta)\sin(\phi - i - \psi)}{\cos(\alpha - i)\cos(\delta + \alpha + \psi)}\right\}^{1/2}}\right] \tag{5.6}$$

The expression of $(K_A)_{dyn}$ gives two values depending on the sign of α_v. For design purposes the higher of the two values shall be taken.

Mononobe and Okabe also gave the expression for the computation of dynamic passive earth pressure $(P_P)_{dyn}$ which is

$$(P_P)_{dyn} = \frac{1}{2}\gamma H^2 (K_p)_{dyn} \tag{5.7}$$

where $(K_P)_{dyn}$ is coefficent of dynamic passive earth pressure and given by:

$$(K_P)_{dyn} = \frac{(1 \pm \alpha_v)\cos^2(\phi + \alpha - \psi)}{\cos\psi\cos^2\alpha\cos(\delta - \alpha + \psi)} \times \left[\frac{1}{1 + \left\{\frac{\sin(\phi - \delta)\sin(\phi + i - \psi)}{\cos(\alpha - i)\cos(\delta - \alpha + \psi)}\right\}^{1/2}}\right]^2 \tag{5.8}$$

For design purposes, the lesser value of $(K_P)_{dyn}$ will be taken out of its two values corresponding to $\pm \alpha_v$.

5.2.2 Effect of Uniform Surcharge

The additional active and passive dynamic earth pressure $[(P_{Aq})_{dyn}$ and $(P_{Pq})_{dyn}]$ against the wall due to uniform surcharge of intensity q per unit area on the inclined earth fill surface shall be:

$$(P_{Aq})_{dyn} = \frac{qH\cos\alpha}{\cos(\alpha - i)}(K_A)_{dyn} \tag{5.9}$$

$$(P_{Pq})_{dyn} = \frac{qH\cos\alpha}{\cos(\alpha - i)}(K_P)_{dyn} \tag{5.10}$$

5.2.3 Effect of Saturation on Lateral Dynamic Earth Pressure

For saturated earth fill, the saturated unit weight of soil shall be adopted.

For submerged earth fill, the dynamic active and passive earth pressure during earthquakes shall be found with the following modifications (IS:1893-2016):

(i) The value of δ shall be taken as 1/2 the value of the δ for dry backfill.
(ii) The value of ψ shall be taken as

$$\psi = \tan^{-1}\left(\frac{\gamma_s}{\gamma_s - 1}\right)\frac{\alpha_h}{1 \pm \alpha_v} \tag{5.11}$$

where γ_s = Saturated unit weight of the soil.

(iii) Submerged unit weight shall be adopted.

Hydrodynamic pressure on account of water contained in earth fill shall not be considered separately as the effect of acceleration on water has been taken indirectly.

5.2.4 Partially Submerged Backfill in Active Case

The ratio of the lateral dynamic increment in active case due to backfill to the vertical effective pressure at various depths along the height of wall may be taken as shown in Fig. 5.2. The pressure distribution of dynamic increments in active case due to backfill may be obtained by multiplying the vertical effective pressures by the coefficients in Fig 5.2 at corresponding depths.

The value of lateral dynamic increment due to backfill in active case is obtained by integrating it in the portion above water level and below water level separately. By doing this we get (Ref. Fig. 5.2):

$$(P_{A\gamma})_{di} = \int_0^{H-H_w} 3[(K_A)_{dyn} - K_A]\cdot\frac{H-z}{H}\gamma z dz$$

$$+\int_0^{H_w}\frac{3[(K'_A)_{dyn} - K'_A]\cdot H_w}{H}\cdot\frac{H_w - z'}{H_w}[\gamma(H-H_w)+\gamma_b z']dz'$$

or

$$(P_{A\gamma})_{di} = \frac{\gamma}{2H}\left[[(K_A)_{dyn} - K_A](H-H_w)^2(H+2H_w)\right]$$

$$+\frac{H_w^2}{2H}[(K'_A)_{dyn} - K'_A][3\gamma(H-H_w)+\gamma_b H_w] \tag{5.12}$$

For getting the point of application of dynamic increment, firstly the moments of portions above water table and below water are taken about AB (Ref. Fig. 5.2):

$$(M_{A\gamma})_{di} = \int_0^{H-H_w} 3[(K_A)_{dyn} - K_A]\cdot\frac{H-z}{H}\gamma z.zdz$$

$$+\int_{H-H_w}^{H}\frac{3[(K'_A)_{dyn} - K'_A]\cdot H_w}{H}\cdot\frac{H-z''}{H_w}[\gamma(H-H_w)+\gamma_b(z''-H+H_w)]z''\, dz''$$

or

$$(M_{A\gamma})_{di} = \frac{\gamma}{4H}\left[\{(K_A)_{dyn} - K_A\}(H-H_w)^3(H+3H_w)\right]$$

$$+\frac{\gamma[(K'_A)_{dyn} - K'_A]}{2H}(H-H_w)[H^3-(H-H_w)^2(H+2H_w)]$$

$$+\frac{\gamma_b[(K'_A)_{dyn} - K'_A]}{4H}[2H^3(2H_w - H)-H_w^3(2H-H_w)] \tag{5.13}$$

Height of point of application of the dynamic increment $(P_{A\gamma})_{di}$ is then given by Eq. (5.14),

$$H_{\gamma di} = H - \frac{(M_{A\gamma})_{di}}{(P_{A\gamma})_{di}} \tag{5.14}$$

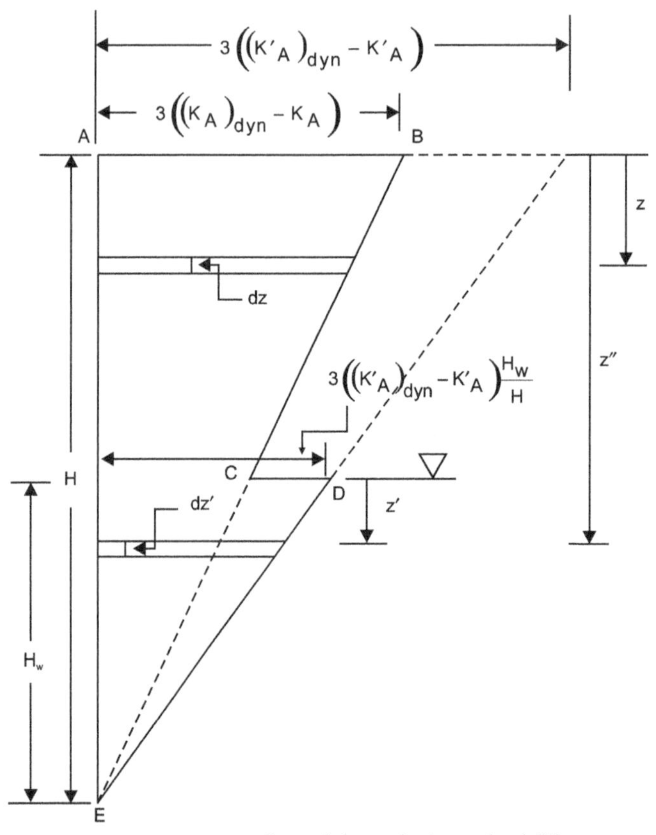

Fig. 5.2. Distribution of the ratio $\frac{\text{Lateral dynamic due to backfill}}{\text{Vertical effective pressure}}$ with height of wall

The ratio of lateral dynamic increments in active earth pressure due to surcharge on the backfill at various depths along the height of the wall may be taken as shown in Fig 5.3. The pressure distribution of dynamic increment in active case due to surcharge on the backfill may be obtained by multiplying the surcharge intensity 'q' by the coefficients in Fig. 5.3 at corresponding depths. The value of lateral dynamic increment due to surcharge on the backfill in active case by integrating pressure in the portion above water level and below water level. By doing this we get (Fig. 5.3):

$$(P_{Aq})_{di} = \int_0^{H-H_w} 2[(K_A)_{dyn} - K_A]\frac{\cos\alpha}{\cos(\alpha-i)}\cdot\frac{H-z}{H}qdz$$

$$+\int_0^{H_w} \frac{2[(K'_A)_{dyn} - K'_A]\cdot H_w}{H}\frac{\cos\alpha}{\cos(\alpha-i)}\cdot\frac{H_w-z'}{H_w}qdz'$$

$$\text{or}\quad (P_{Aq})_{di} = [(K_A)_{dyn} - K_A]\frac{\cos\alpha}{\cos(\alpha-i)}q.\frac{H^2-H_w^2}{2H} + [(K'_A)_{dyn} - K'_A]\frac{\cos\alpha}{\cos(\alpha-i)}q.\frac{H_w^2}{2H} \tag{5.15}$$

For getting the point of application of dynamic increment, firstly the moments of portions above water table and below water table are taken about AB (Fig. 5.3).

$$(M_{Aq})_{di} = \int_0^{H-H_w} 2[(K_A)_{dyn} - K_A]\frac{\cos\alpha}{\cos(\alpha-i)}\cdot\frac{H-z}{H}qdz$$

$$+\int_{H-H_w}^{H} \frac{2[(K'_A)_{dyn} - K'_A]\cdot H_w}{H}\frac{H-z''}{H_w}\frac{\cos\alpha}{\cos(\alpha-i)}qz''dz''$$

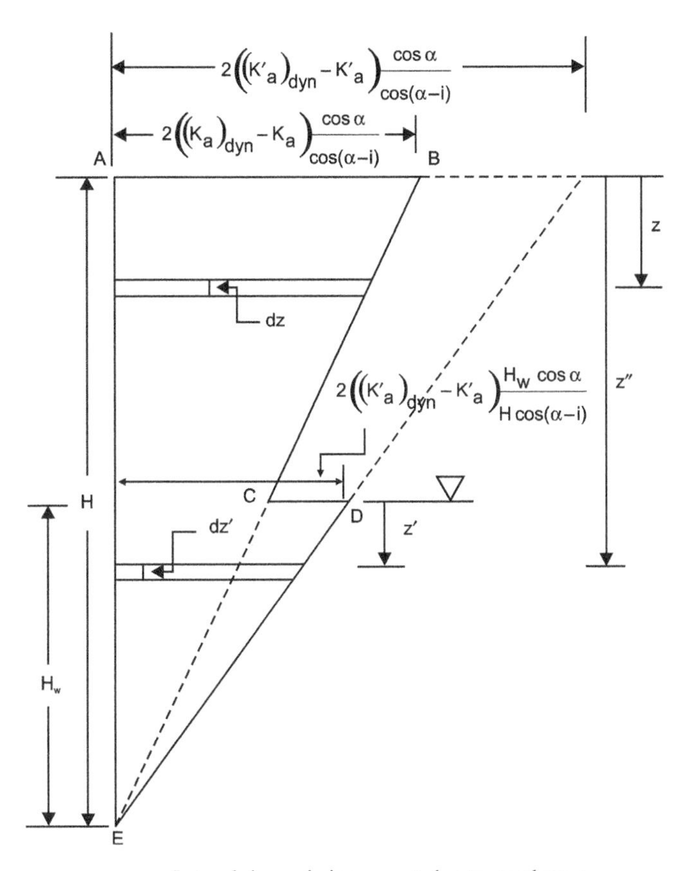

Fig. 5.3. Distribution of the $\frac{\text{Lateral dynamic increment due to surcharge}}{\text{Vertical effective pressure}}$ ratio with height of wall

or

$$(M_{Aq})_{di} = \frac{q}{3H}[(K_A)_{dyn} - K_A]\frac{\cos\alpha}{\cos(\alpha-i)}(H-H_w)^2(H+2H_w)$$

$$+\frac{q[(K'_A)_{dyn} - K'_a]}{3H}\frac{\cos\alpha}{\cos(\alpha-i)}[H^3-(H-H_2)^2(H+2H_w)] \tag{5.16}$$

Height of point of application of the dynamic increment $(P_{Aq})_{di}$ from the base of wall taken is given by

$$H_{qdi} = H - \frac{(M_{Aq})_{di}}{(P_{Aq})_{di}} \tag{5.17}$$

It may be noted that by putting in expressions (5.12), (5.13) and (5.14), H_w= 0 (i.e. fully dry or moist backfill) or H_w=H (i.e. fully submerged backfill), $H_{d\gamma i}$ from Eq, (5.14) works out as $H/2$. It means that dynamic increment in these cases will act at the mid height of wall. Similarly, solving Eqs. (5.15), (5.16) and (5.17) for H_w= 0 and H_w = H cases, value of H_{dqi} comes 2/3 H.

5.2.5 Partially Submerged Backfill in Passive Case

The ratio of the lateral dynamic decrement in passive case due to the vertical pressures at various depths along the height of wall may be taken as shown in Fig. 5.4. The pressure distribution of dynamic decrements in passive case due to backfill may be obtained by multiplying the vertical effective pressures by the coefficients on Fig 5.4 at corresponding depths. The value of lateral dynamic decrement due to backfill in passive case is obtained by integrating it in the portion above water level and below water level separately. By doing this we get (Ref. Fig. 5.4):

$$(P_{P\gamma})_{dd} = \int_0^{H-H_w} 3[K_P - (K_P)_{dyn}]\cdot\frac{H-z}{H}\gamma z dz$$

$$+\int_0^{H_w} \frac{3[K_P' - (K_P')_{dyn}]\cdot H_w}{H}\cdot\frac{H_w - z'}{H_w}[\gamma(H-H_w)+\gamma_b z']dz'$$

or

$$(P_{P\gamma})_{dd} = \frac{\gamma}{2H}[\{K_P - (K_P)_{dyn}\}(H-H_w)^2(H+2H_w)]$$

$$+\frac{H_w^2}{2H}\{(K_P' - (K_P')_{dyn})\}[3\gamma(H-H_w)+\gamma_b H_w] \tag{5.18}$$

For getting the point of application of dynamic increment, firstly the moments of portions above water table and below water table are taken about AB.

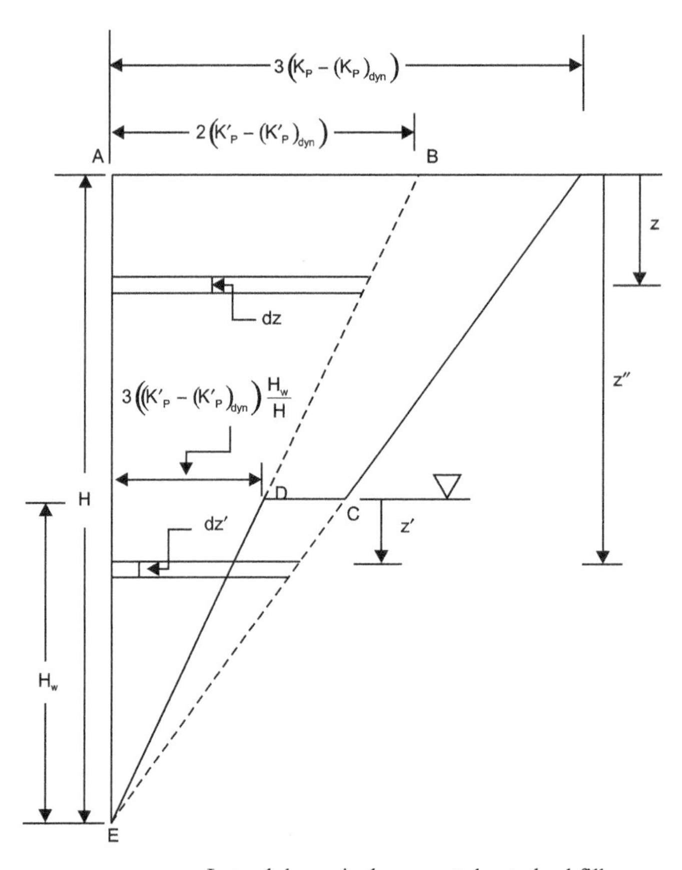

Fig. 5.4. Distribution of the ratio $\frac{\text{Lateral dynamic decrement due to backfill}}{\text{Vertical effective pressure}}$ with height of wall

$$(M_{P\gamma})_{dd} = \int_0^{H-H_w} 3[K_P - (K_P)_{dyn}]\cdot\frac{H-z}{H}\gamma z.zdz$$

$$+\int_{H-H_w}^{H} \frac{3[K'_P-(K'_P)_{dyn}]\cdot H_w}{H}\cdot\frac{H-z''}{H_w}[\gamma(H-H_w)+\gamma_b(Z''-H+H_w)]z''dz''$$

$$(M_{P\gamma})_{dd} = \frac{\gamma[K_p-(K_p)_{dyn}]}{4H}[(H-H_w)^3(H+3H_w)]$$

$$+\frac{\gamma[K'_p-(K'_p)_{dyn}]}{2H}(H-H_w)[H_w^2(3H-2H_w)]$$

$$+\frac{\gamma[K'_p-(K'_p)_{dyn}]}{4H}[H_w^3(2H-H_w)] \tag{5.19}$$

Height of point of application of the dynamic decrement $(P_{P\gamma})_{dd}$ from the base of wall is then given by Eq. (5.20),

$$H_{\gamma dd} = H - \frac{(M_{P\gamma})_{dd}}{(P_{P\gamma})_{dd}} \tag{5.20}$$

The ratio of lateral dynamic decrement in passive earth pressure due to surcharge on the backfill at various depths along the height of wall may be taken as shown in Fig. 5.5. The pressure distribution of dynamic decrement in passive pressure due to surcharge on the backfill may be obtained by multiplying the surcharge intensity 'q' by the coefficients in Fig. 5.5 at corresponding depths. The value of lateral dynamic decrement due to surcharge on the backfill in active case by integrating pressures in the portions above water level and below water level. By doing this we get (Fig. 5.5):

$$(P_{Pq})_{dd} = \int_0^{H-H_w} 2[K_P-(K_P)_{dyn}]\frac{\cos\alpha}{\cos(\alpha-i)}\cdot\frac{H-z}{H}qdz$$

$$+\int_0^{H_w}\frac{2[K'_P-(K'_P)_{dyn}]\cdot H_w}{H}\frac{\cos\alpha}{\cos(\alpha-i)}\cdot\frac{H_w-z'}{H_w}qdz'$$

or $$(P_{Pq})_{dd} = [K_P-(K_P)_{dyn}]\frac{\cos\alpha}{\cos(\alpha-i)}q\cdot\frac{H^2-H_w^2}{H} + [K'_p-(K'_p)_{dyn}]\frac{\cos\alpha}{\cos(\alpha-i)}q\cdot\frac{H_w^2}{H} \tag{5.21}$$

For getting the point of application of dynamic increment, firstly the moments of portions above water table and below water are taken about AB.

$$(M_{Pq})_{dd} = \int_0^{H-H_w} 2[K_P-(K_P)_{dyn}]\frac{\cos\alpha}{\cos(\alpha-i)}\cdot\frac{H-z}{H}q.zdz$$

$$+\int_{H-H_w}^{H}\frac{2[K'_P-(K'_P)_{dyn}]\cdot H_w}{H}\cdot\frac{H-z''}{H_w}\frac{\cos\alpha}{\cos(\alpha-i)}qz''dz''$$

or $$(M_{Aq})_{dd} = \frac{q}{3H}[K_P-(K_P)_{dyn}]\frac{\cos\alpha}{\cos(\alpha-i)}(H-H_w)^2(H+2H_w)$$

$$+\frac{q[K'_P-(K'_P)_{dyn}]}{3H}\frac{\cos\alpha}{\cos(\alpha-i)}[H_w^3-(3H-2H_w)] \tag{5.22}$$

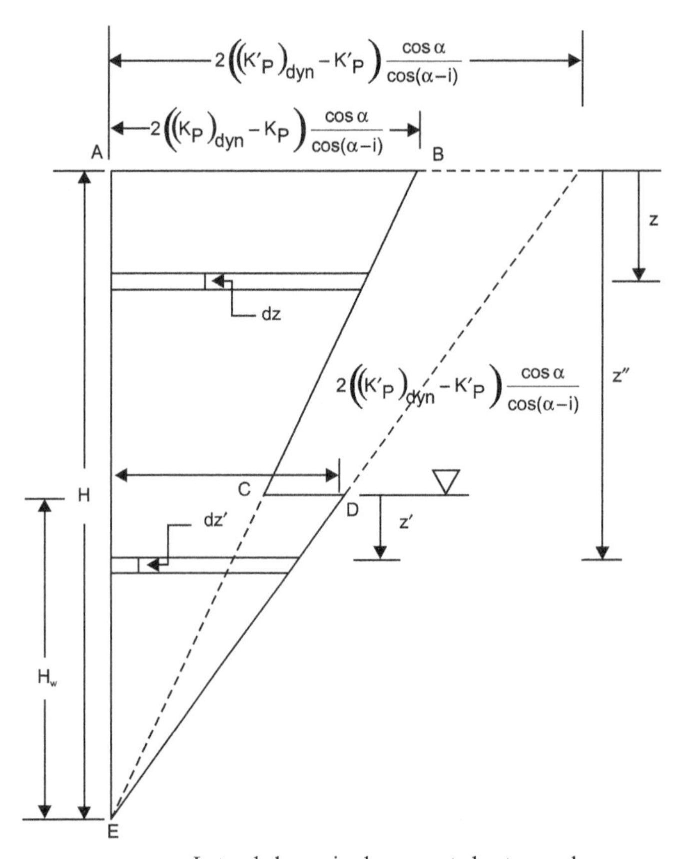

Fig. 5.5. Distribution of the ratio $\frac{\text{Lateral dynamic decrement due to surcharge}}{\text{Vertical effective pressure}}$ with height of wall

Height of point of application of the dynamic increment $(P_{Pq})_{dd}$ from the base of wall taken is given by,

$$H_{qdd} = H - \frac{(M_{Pq})_{dd}}{(P_{Pq})_{dd}} . \tag{5.23}$$

5.2.6 Modified Culmann Construction

Kapila (1962) modified the Culmann's graphical for obtaining dynamic active and passive earth pressures.

Different steps in modified construction for determining dynamic active earth pressure are as follows (Fig. 5.6):

(i) Draw the wall section along with backfill surface on a suitable scale.
(ii) Draw *BS* at an angle $(\phi - \psi)$ with the horizontal.
(iii) Draw *BL* at an angle of $(90 - \alpha - \delta - \psi)$ below *BS*.
(iv) Intercept BD_1 equal to the resultant of the weight W_1 of first trial wedge ABC_1 and inertial forces ($\pm W_1 \alpha_v$ and $W_1 \alpha_h$). The magnitude of this resultant is $\overline{W}_1$.

$$\overline{W}_1 = W_1\sqrt{(1 \mp \alpha_v)^2 + \alpha_h^2}$$

(v) Through D_1 draw D_1E_1 parallel to *BL* intersecting BC_1 at E_1.
(vi) Measure $D_1 E_1$ to the same force scale as BD_1. The $D_1 E_1$ is the dynamic earth pressure for trial wedge.

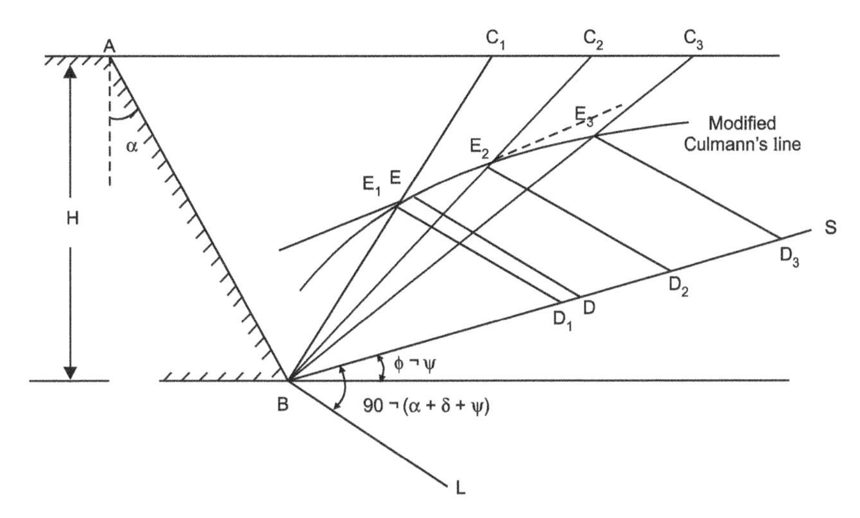

Fig. 5.6. Modified Culmann's construction for dynamic active earth pressure

(vii) Repeat steps (iv) to (vi) with BC_2, BC_3 etc. as trial wedges.

(viii) Draw a smooth curve through BE_1 E_2 E_3. This is the modified Culmann's line.

(ix) Draw a line parallel to BS and tangential to this curve. The maximum coordinate in the direction of BL is obtained from the point of tangency and is the dynamic active earth pressure, $(P_A)_{dyn}$.

For determining the passive earth pressure draw BS at $(\phi - \psi)$ below horizontal. Next draw BL at $(90 - \alpha - \delta - \psi)$ below BS. The other steps for construction remain unaltered (Fig. 5.7).

Effect of uniformly distributed load and line load on the backfill surface may be handled in the similar way as for the static case.

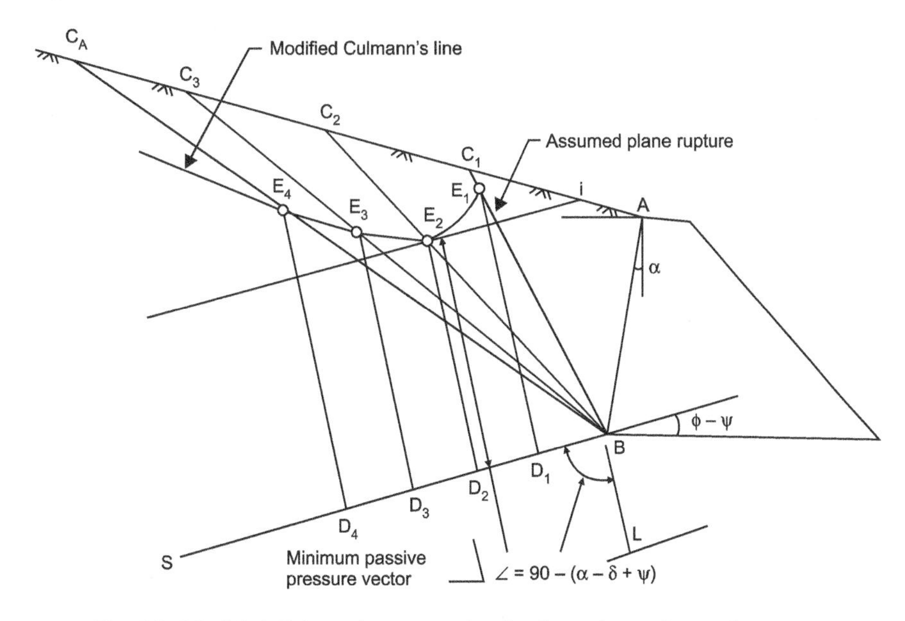

Fig. 5.7. Modified Culmann's construction for dynamic passive earth pressure

5.2.7 Dynamic Active Earth Pressure for *c*-ϕ Soils

The solutions so far discussed consider the soil to be cohesionless. A general solution for the determination of total (static plus dynamic) earth pressures for a *c*-ϕ soil has been developed by Prakash and Saran (1966) and Saran and Prakash (1968).

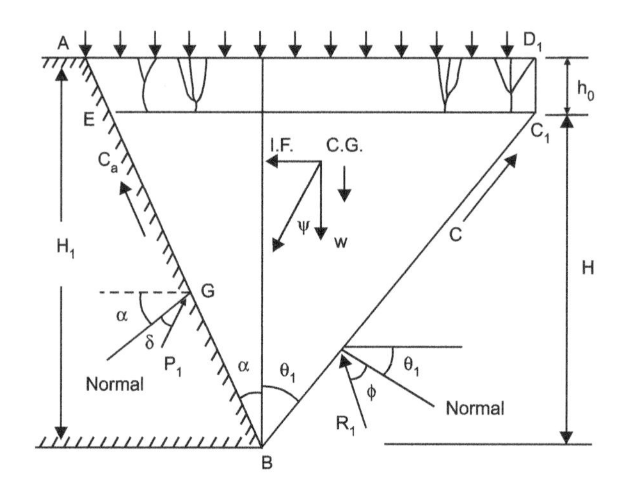

Fig. 5.8. Forces acting on failure wedge in active state for seismic condition in *c*-ϕ soil

Figure 5.8 shows a section of wall whose face *AB* is in contact with soil. The soil retained is horizontal and carries a uniform surcharge. The inclination of the wall *AB* with vertical is α and inclination of the trial failure surface is θ_1. AEC_1D_1 is the cracked zone in clayey soils, EC_1 being at depth h_o below AD_1. h_o is given by expression

$$h_o = n(H_1 - h_o) = nH \tag{5.24}$$

where H_1 = Total height of retaining wall

H = Height of retaining wall in which backfill is free from cracks

In this analysis only horizontal inertia force is considered. All the forces acting on the assumed failure wedge $AEBC_1D_1$ are listed in Table 5.1 along with their horizontal and vertical components.

Table 5.1. Computation of Forces Action on Wedge $AEBC_1D_1$ (Fig. 5.8)

Designation	*Vertical component*	*Horizontal component*
1. Weight of wedge, W	$\frac{1}{2}\gamma H^2 (\tan\alpha + \tan\theta_1)$	–
ABC_1D_1A	$+\gamma n H^2 (\tan\alpha + \tan\theta_1) + \frac{1}{2}\gamma n^2 H^2 (\tan\alpha)$	
2. Cohesion, $C = cH \sec\theta_1$	cH	$c H \tan\theta_1$
3. Adhesion, $C_a = c'H \sec\alpha$	$c'H$	$c'H \tan\alpha$
4. Surcharge, Q	$q[H(\tan\alpha + \tan\theta_1) + nH\tan\alpha]$	–
5. Soil reaction R_1	$R_1 \sin(\theta_1 + \phi)$	$R_1 \cos(\theta_1 + \phi)$
6. Inertia force	–	$(W + Q)\,\alpha_h$
7. Earth pressure P_1	$P_1 \sin(\alpha + \delta)$	$P_i \cos(\alpha + \delta)$

A summation of all the vertical components gives

$$\frac{1}{2}\gamma H^2(\tan\alpha + \tan\theta_1) + \gamma n H^2(\tan\alpha + \tan\theta_1) + \frac{1}{2}\gamma n^2 H^2 \tan\alpha - cH - c'H$$

$$+ qH(\tan\alpha + \tan\theta_1 + n\tan\alpha) = P_1\sin(\alpha + \delta) + R_1\sin(\theta_1 + \phi) \tag{5.25}$$

A summation of all the horizontal components gives

$$-cH\tan\theta_1 + c'H\tan\alpha + (W + Q)\alpha_h = P_1\cos(\alpha + \delta) - R_1\cos(\theta_1 + \phi) \tag{5.26}$$

Multiply Eq. (5.25) by cos $(\theta_1 + \phi)$, Eq. (5.26) by sin $(\theta_1 + \phi)$, substitute for W and Q from Table 5.1, assuming $c = c'$, and adding, we get

$$P_1 \sin(\beta + \delta) = \gamma H^2 [(n+1/2)(\tan\alpha + \tan\theta_1) + n^2 \tan\alpha][\cos(\theta_1 + \phi) + \alpha_h \sin(\theta_1 + \phi)] + qH[(n+1)\tan\alpha + \tan\theta_1][\cos(\theta_1 + \phi) + \alpha_h \sin(\theta_1 + \phi)] - cH[\cos\beta \sec\alpha + \cos\phi \sec\theta_1] \tag{5.27}$$

where $\beta = \theta_1 + \phi + \alpha$

Introducing the following dimensionless parameters:

$$(N_{ac})_{dyn} = \frac{\cos\beta \sec\alpha + \cos\phi \sec\theta_1}{\sin(\beta + \delta)} \tag{5.28}$$

$$(N_{aq})_{dyn} = \frac{[(n+1)\tan\alpha + \tan\theta_1][\cos(\theta_1 + \phi) + \alpha_h \sin(\theta_1 + \phi)]}{\sin(\beta + \delta)} \tag{5.29}$$

$$(N_{a\gamma})_{dyn} = \frac{[(n+1/2)(\tan\alpha + \tan\theta_1) + n^2 \tan\alpha][\cos(\theta_1 + \phi) + \alpha_h \sin(\theta_1 + \phi)]}{\sin(\beta + \delta)} \tag{5.30}$$

We get

$$(P_1)_{dyn} = \gamma H^2 (N_{a\gamma})_{dyn} + qH(N_{aq})_{dyn} - cH(N_{ac})_{dyn} \tag{5.31}$$

where $(N_{ac})_{dyn}$, $(N_{aq})_{dyn}$ and $(N_{a\gamma})_{dyn}$ are earth pressure coefficients which depend on α, n, ϕ, δ and θ_1. For given parameters of wall and soil, the values of $(N_{ac})_{dyn}$, $(N_{aq})_{dyn}$ and $(N_{a\gamma})_{dyn}$ are computed for different wedge angles θ_2, θ_3 etc. The variation of these coefficients with respect to wedge angle θ are shown in Fig. 5.6, and the maximum values of $(N_{aq})_{dyn}$ and $(N_{a\gamma})_{dyn}$ and minimum value of $(N_{ac})_{dyn}$ are obtained.

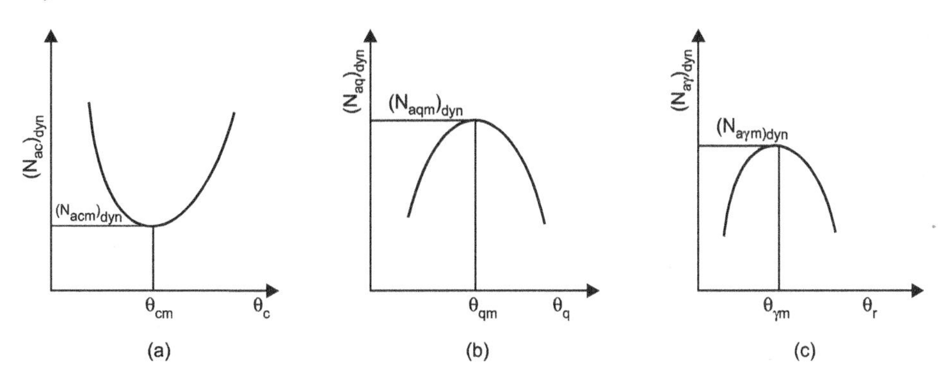

Fig. 5.9. (a) $(N_{ac})_{dyn}$ versus θ plot; (b) $(N_{aq})_{dyn}$ versus θ plot; (c) $(N_{a\gamma})_{dyn}$ versus θ plot

For such condition, the earth pressure corresponds to dynamic active earth pressure. Equation (5.31) can be written as

$$(P_A)_{dyn} = \gamma H^2 (N_{a\gamma m})_{dyn} + qH(N_{aqm})_{dyn} - cH(N_{acm})_{dyn} \tag{5.32}$$

For static case, $\alpha_h = 0$; earth pressure coefficients then become

$$(N_{ac})_{stat} = \frac{\cos\beta \sec\alpha + \cos\phi \sec\theta_1}{\sin(\beta + \delta)} \tag{5.33}$$

$$(N_{aq})_{stat} = \frac{[(n+1)(\tan\alpha + \tan\theta_1)]\cos(\theta_1 + \phi)}{\sin(\beta + \delta)} \tag{5.34}$$

$$(N_{a\gamma})_{stat} = \frac{[(n+1/2)(\tan\alpha + \tan\theta_1) + n^2 \tan\alpha]\cos(\theta_1 + \phi)}{\sin(\beta + \delta)} \tag{5.35}$$

Minimum value of $(N_{ac})_{stat}$, and maximum values of $(N_{aq})_{stat}$ and $(N_{a\gamma})_{stat}$ can be obtained in similar manner, as illustrated above, for getting the static earth pressure coefficients. It is found convenient to obtain the dynamic earth pressure coefficients from the following constants:

$$\lambda_1 = \frac{(N_{aqm})_{dyn}}{(N_{aqm})_{stat}} \tag{5.36}$$

$$\lambda_2 = \frac{(N_{a\gamma m})_{dyn}}{(N_{a\gamma m})_{stat}} \tag{5.37}$$

N_{acm} both for the static and dynamic case is same and has been plotted in Fig. 5.10 for different inclination of the wall varying +20° to –20° with the vertical. As evident from Eq. (5.33), N_{acm} factor is also independent of n.

Figures 5.11, 5.12 and 5.13 show plots of $(N_{aqm})_{stat}$ versus ϕ for n of 0, 0.2 and 0.4 respectively. These plots consider the inclination of the wall from +20° to –20°. Plots of $(N_{a\gamma m})_{stat}$ for the same range of n, ϕ and α have been drawn in Figs. 5.14, 5.15 and 5.16.

It is found that the value of $\lambda_1$and λ_2 alter slightly with increase in n. It is therefore recommended that the effect of n on λ_1 and λ_2 may not be considered. Secondly, it is observed that λ_1 and λ_2 are almost same (Prakash and Saran, 1966; Saran and Prakash, 1968). Hence only one value of $\lambda = (\lambda_1 = \lambda_2)$ is recommended (Fig. 5.17). Since λ is the ratio of earth pressure coefficients in (i) dynamic and (ii) static case, and both the coefficients decrease with ϕ, the shape of the curves for different α_h values indicate the rate of decrease of one in relation to the other.

5.2.8 Point of Application

According to Indian standard (IS: 1893-2016) specifications, the pressures are located as follows.

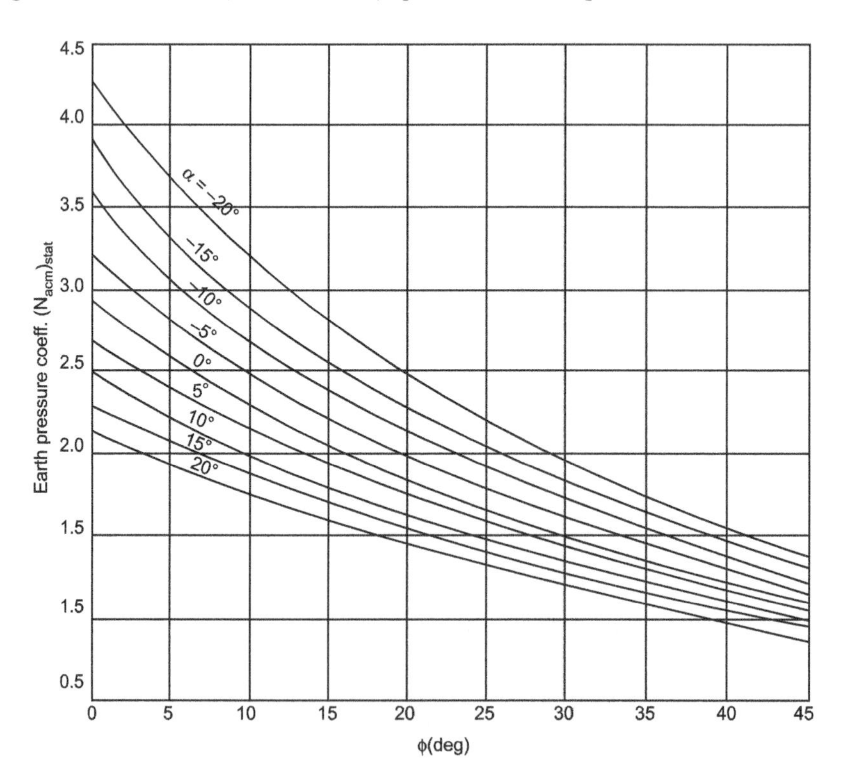

Fig. 5.10. $(N_{acm})_{stat}$ versus ϕ for all n (Prakash and Saran, 1966 and Saran and Prakash, 1968)

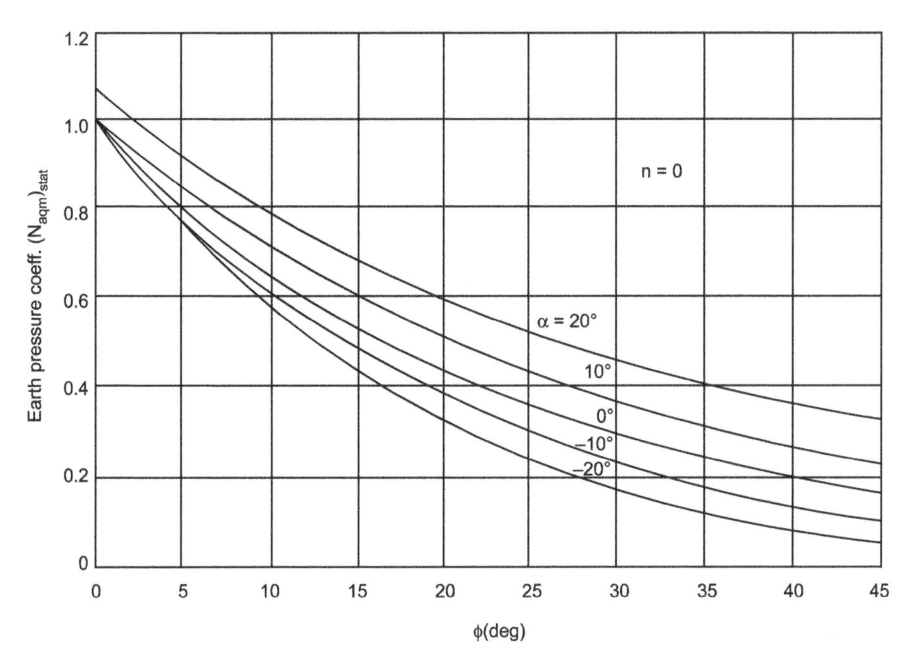

Fig. 5.11. $(N_{aqm})_{stat}$ versus ϕ for $n = 0$ (Prakash and Saran, 1966; Saran and Prakash, 1968)

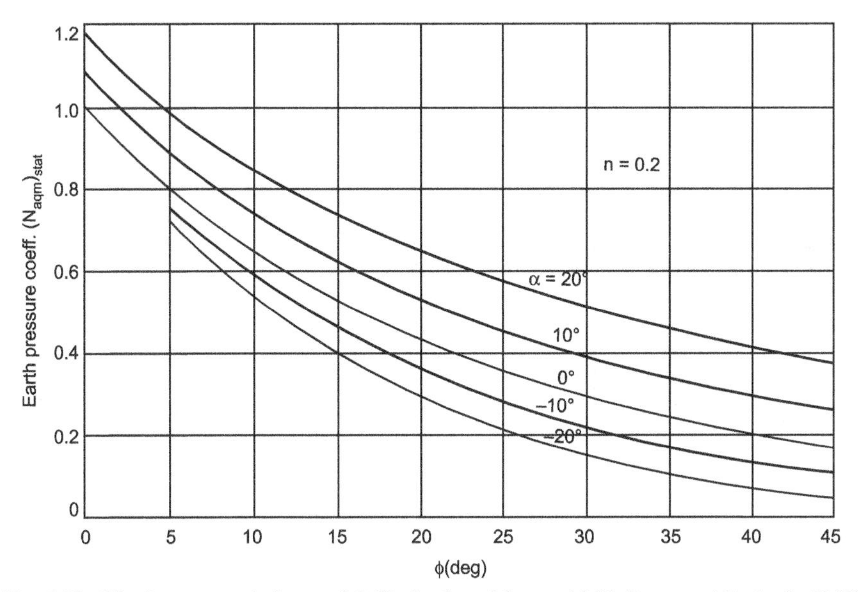

Fig. 5.12. $(N_{aqm})_{stat}$ versus ϕ for $n = 0.2$ (Prakash and Saran, 1966; Saran and Prakash, 1968)

From the total pressures computed from Eqs. (5.5) and (5.7) or from graphical construction, subtract the static pressure obtained by putting $\alpha_h = \alpha_v = 0$. The remainder is the dynamic increment in active case and dynamic decrement in passive case. The static component of the total pressure shall be applied at an elevation $H/3$ above the base of wall. The point of application of the dynamic increment and dynamic decrement shall be assumed to be at an elevation $H/2$ above the base of the wall.

The static component of total active and passive earth pressure due to uniformly distributed surcharge on the backfill surface obtained by putting $\alpha_h = \alpha_v = 0$ in Eqs. (5.9) and (5.10) shall be applied at $H/2$ above the base of the wall. The point of application of both the dynamic increment

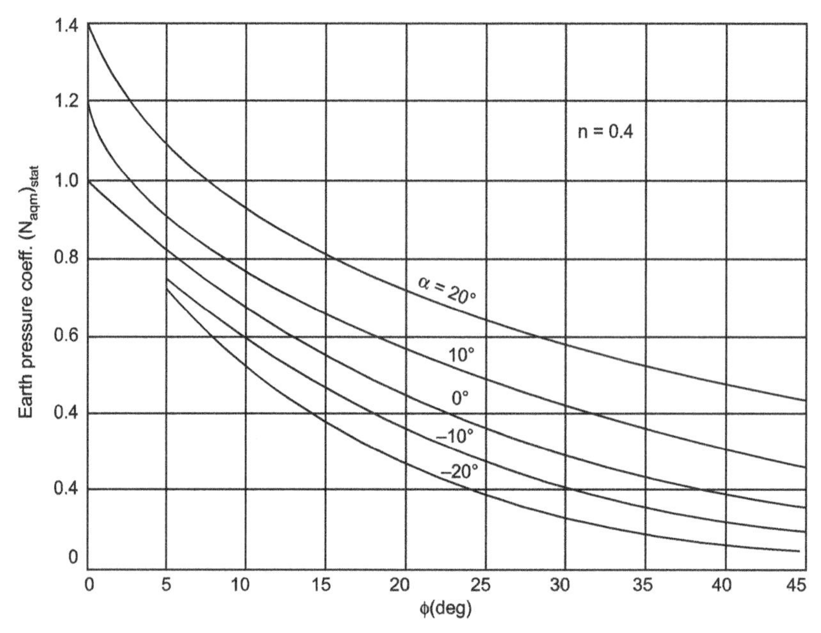

Fig. 5.13. $(N_{aqm})_{stat}$ versus ϕ for $n = 0.4$ (Prakash and Saran, 1966; Saran and Prakash, 1968)

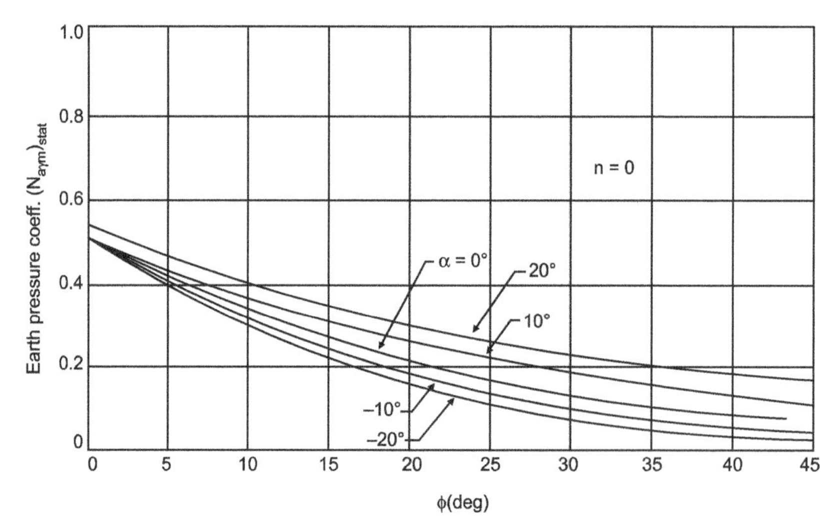

Fig. 5.14. $(N_{a\gamma m})_{stat}$ versus ϕ for $n = 0$ (Prakash and Saran, 1966; Saran and Prakash, 1968)

and dynamic decrement in this case shall be assumed to be at an elevation $2H/3$ above the base of the wall.

The static and dynamic active earth pressures due to cohesion only ($q = \gamma = 0$) are same. The point of application of this pressure shall be assumed to be at an elevation of $H/2$ above the base of the wall.

5.3 Displacement Analysis

There are very few methods available to compute displacements of rigid retaining walls during earthquakes. They are:

(i) Richard-Elms model based on Newmark's Approach

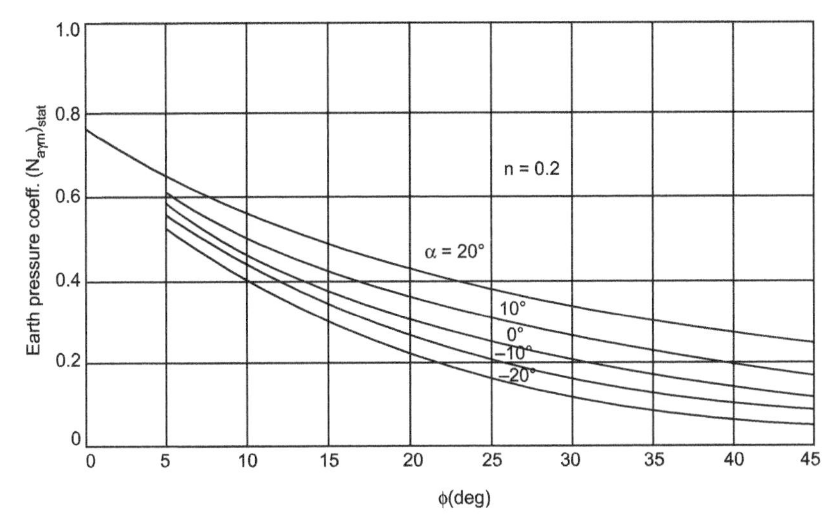

Fig. 5.15. $(N_{a\gamma m})_{stat}$ versus ϕ for $n = 0.2$ (Prakash and Saran, 1966; Saran and Prakash, 1968)

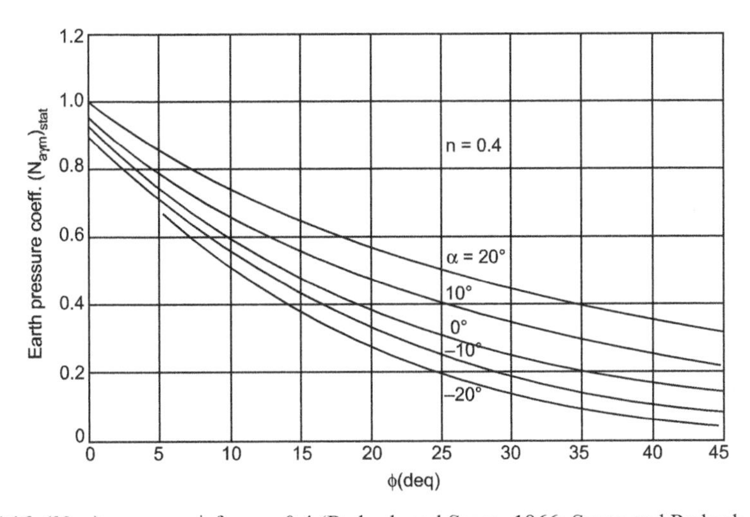

Fig. 5.16. $(N_{a\gamma m})_{stat}$ versus ϕ for $n = 0.4$ (Prakash and Saran, 1966; Saran and Prakash, 1968)

(ii) Solution in pure translation
(iii) Solution in pure rotation
(iv) Nadim-Whitman model
(v) Saran-Reddy-Viladkar model

5.3.1 Richard-Elms Model

Newmark (1965) proposed a basic procedure for evaluating the potential deformation that would be experienced by an embankment dam shaken by an earthquake by considering the sliding block-on-a-plane mode as shown in Fig. 5.18a. In this important development, it was envisaged that slope failure would be initiated and movements would begin to occur if the inertial forces on the potential sliding mass were reversed. Thus by computing an acceleration at which the inertial forces become sufficiently high to cause yielding to begin, and integrating the effective acceleration on the sliding mass in excess of this yield acceleration as a function of time (Fig. 5.18c), velocities and ultimately the displacements of the sliding mass could be evaluated.

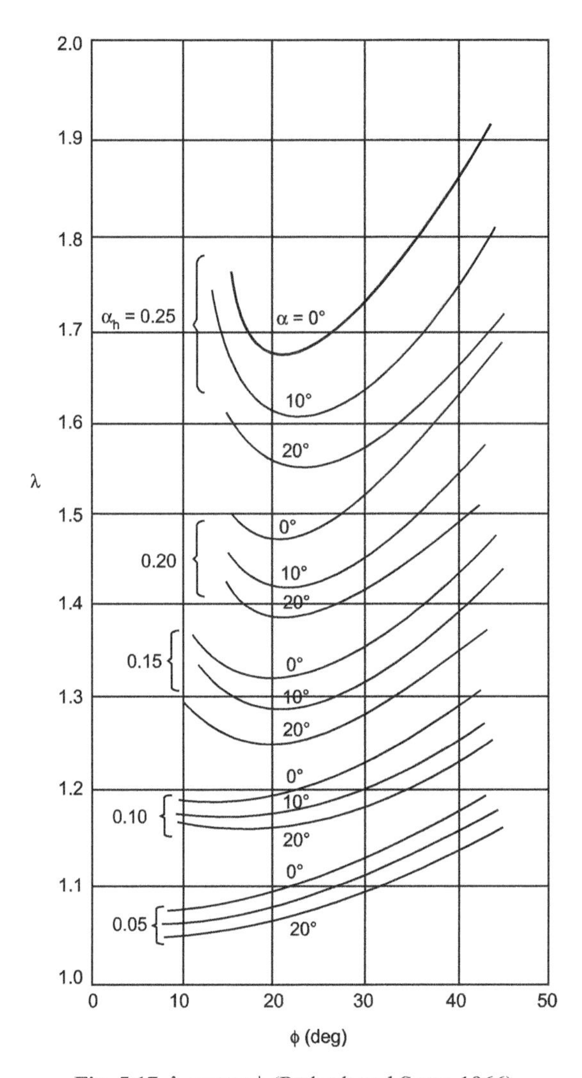

Fig. 5.17. λ versus φ (Prakash and Saran, 1966)

This analysis is based essentially upon the rigid plastic behaviour of materials (Fig. 5.18b). Though this method was developed for a sliding analysis of an earth dam, it has been used by Richard and Elms (1979) to compute the displacements of retaining walls. They have proposed a method for design of gravity retaining walls based on limiting displacement considering the wall inertia effect. The procedure developed them is described below.

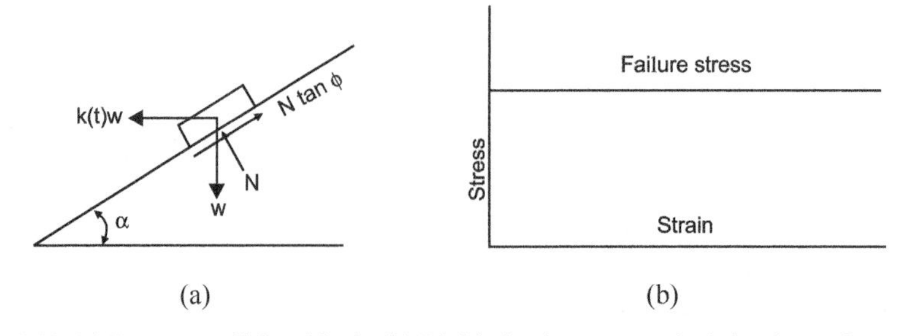

Fig. 5.18. (a) Forces on sliding block, (b) Rigid plastic stress strain behaviour of a material

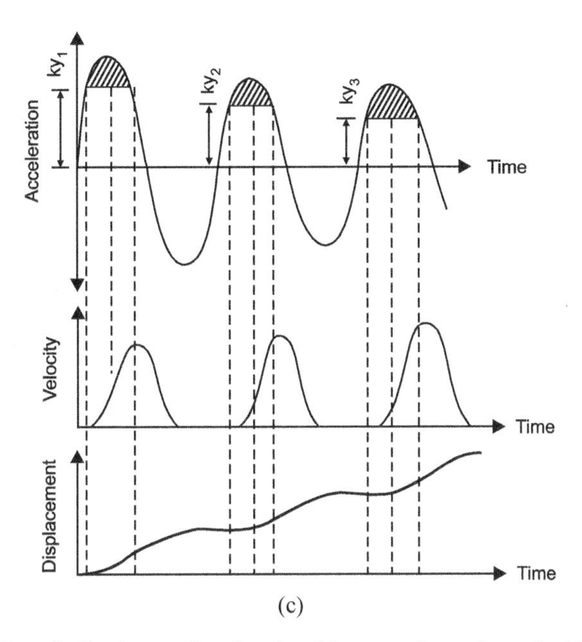

Fig. 5.18. (c) Integration of effective acceleration time history to determine velocities and displacements

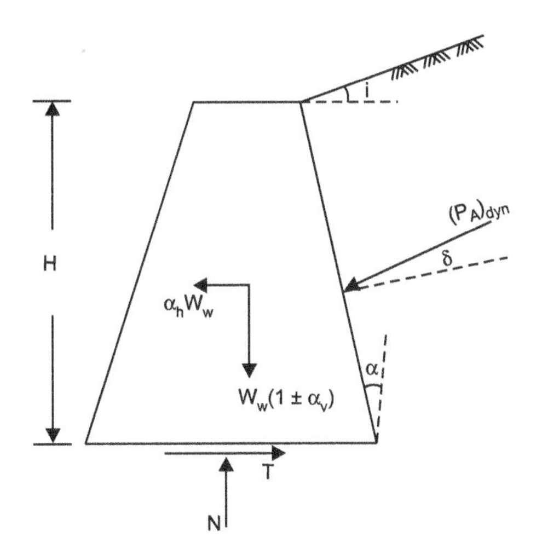

Fig. 5.19. Forces on a gravity wall

A gravity retaining wall is shown in Fig. 5.19, along with the forces acting on it during an earthquake. In this figure various terms used are:

W_w = Weight of the retaining wall

α_h, α_v = Horizontal and vertical seismic coefficients

$(P_A)_{dyn}$ = Dynamic active earth pressure, Eq. (5.5)

α = Inclination of wall face with vertical

δ = Angle of wall friction

ϕ_b = Soil-wall friction angle at the base of the wall

N = Vertical component of the reaction at the base of the wall

T = Horizontal component of the reaction at the base of the wall

Summing the forces in the vertical and horizontal directions, we get

$$N = W_w \pm \alpha_v W_w + (P_A)_{dyn} \sin(\alpha + \delta) \quad (5.38)$$

$$T = \alpha_h W_w + (P_A)_{dyn} \cos(\alpha + \delta) \quad (5.39)$$

At sliding $$T = N \tan \phi_b \quad (5.40)$$

Solving Eqs. (5.38), (5.39) and (5.40), we get

$$W_w = \frac{(P_A)_{dyn}[\cos(\alpha+\delta) - \sin(\alpha+\delta)\cdot\tan\phi_b]}{(1 \pm \alpha_v)\tan\phi_b - \alpha_h} \quad (5.41)$$

Putting $(P_A)_{dyn} = \frac{1}{2}\gamma H^2 (K_A)_{dyn}$ and $\alpha_h = (1 \pm \alpha_v)\tan\psi$, Eq. (5.41) can be written as

$$W_w = \frac{1}{2}\gamma H^2 (K_A)_{dyn} \cdot C_{IE} \quad (5.42)$$

where $$C_{IE} = \frac{\cos(\alpha+\delta) - \sin(\alpha+\delta)\cdot\tan\phi_b}{(1 \pm \alpha_v)(\tan\phi_b - \tan\psi)} \quad (5.43)$$

For static condition, the weight of wall W is given by:

$$W = \frac{1}{2}\gamma H^2 K_A \cdot C_1 \quad (5.44)$$

where $$C_I = \frac{\cos(\alpha+\delta) - \sin(\alpha+\delta)\cdot\tan\phi_b}{\tan\phi_b} \quad (5.45)$$

Therefore, $$\frac{W_w}{W} = \frac{(K_A)_{dyn}}{K_A}\cdot\frac{\tan\phi_b}{(1 \pm \alpha_v)(\tan\phi_b - \tan\psi)} \quad (5.46)$$

Substituting

F_T = Ratio of earth pressure coefficients in dynamic and static cases

i.e. $$F_T = \frac{(K_A)_{dyn}}{K_A} \quad (5.47)$$

and $$F_I = \text{Wall inertia factor} = \frac{\tan\phi_b}{(1 \pm \alpha_v)(\tan\phi_b - \tan\psi)} \quad (5.48)$$

in Eq. (5.46), $$\frac{W_w}{W} = F_T \cdot F_I = F_w \quad (5.49)$$

F_w is factor of safety applied to the weight of the wall to take into account the effect of soil present and wall inertia. Figure 5.20 shows a plot of F_T, F_I and F_w for various values of α_h. From this figure for $F_I = 1.0$ and $F_w = 1.5$, the value α_h works out to be 0.18. However, if the wall inertial factor is considered, the critical horizontal acceleration corresponding to $F_w = 1.5$ is equal to 0.105. Therefore, if a wall is designed such that $W_w = 1.5\ W$, the wall will start to move laterally at a value of $\alpha_h = 0.105$. Hence for no lateral movement, the weight of the wall has to be increased by a considerable amount over the static condition, which may prove to be uneconomical. Keeping this in view, the actual design is carried for some lateral displacement of wall.

Richards and Elms (1969) have given a design procedure based on a limited allowable wall movement, rather than on the assumption that the wall will not move at all. Such procedure is as follows:

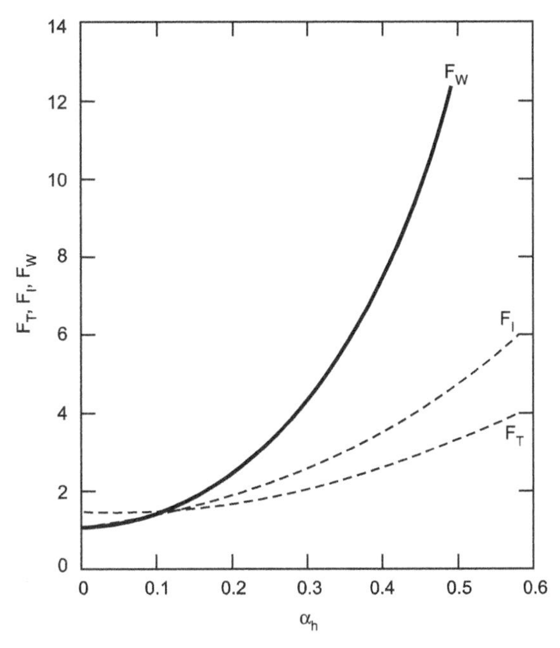

Fig. 5.20. Variation of F_T, F_I and F_w with α_h (Richards and Elms, 1979)

(i) Decide upon an acceptable maximum displacement, *d*.

(ii) Determine the design value of α_{hd} from Eq. (5.50) (Franklin and Chang, 1977)

$$\alpha_{hd} = \alpha_h \left(\frac{5\alpha_h}{d} \right)^{\frac{1}{4}} \tag{5.50}$$

where α_h = Acceleration coefficient from earthquake record

d = Maximum displacement in mm

(iii) Using α_{hd}, determine the required wall weight, W_w by substituting it in Eq. (5.41). The value of α_{vd} may be taken as $\dfrac{\alpha_{hd}}{2}$.

(iv) Apply a suitable safety factor, say 1.5, to W_w.

There are three limitation to Richard-Elms analysis (Prakash, 1981). These are:

1. The soil is assumed to be a rigid plastic material. The walls do undergo reasonable displacements before the limiting equilibrium condition (active) develop and experience very large displacements before the passive condition develop.
2. The physical properties of the system and its geometry (particularly its natural period) are not considered.
3. Walls may undergo displacements by either sliding or tilting or both. This method does not apparently consider this difference in their physical behaviour, although it is logical to conclude that displacements computed by this method are in sliding only.

5.3.2 Solution in Pure Translation

A method for computation of displacement in translation only, of rigid retaining wall under dynamic loads had been developed by Nandakumaran (1973).

The force-displacement relationships considered in this analysis are shown in Fig. 5.21. Figure 5.21(a) shows the variation of earth pressure with displacement. In Fig 5.21(b), variation of base resistance with displacement is given. The net force away from the fill is the difference of active earth pressure P_A and the base resistance, R_{BA} (Fig. 5.21(c)). The net force towards the wall is the sum of the passive earth pressure, P_P and the base resistance, R_{BP} (Fig. 5.21(c)). The resulting bilinear force-displacement relationship is shown in Fig. 5.21(d) and is characterized by the following parameters:

(i) Slope of force displacement relationship on the active and passive sides as K_1 and K_2 respectively, where $K_2 = n{\cdot}K_1$.
(ii) Yield displacement, Z_y

For the resistance of the base, it is assumed that a column of soil of height $(B/2)\tan\phi$ provides all of the resistance in a passive case (Fig. 5.21(e)), B being the width of the wall at its base.

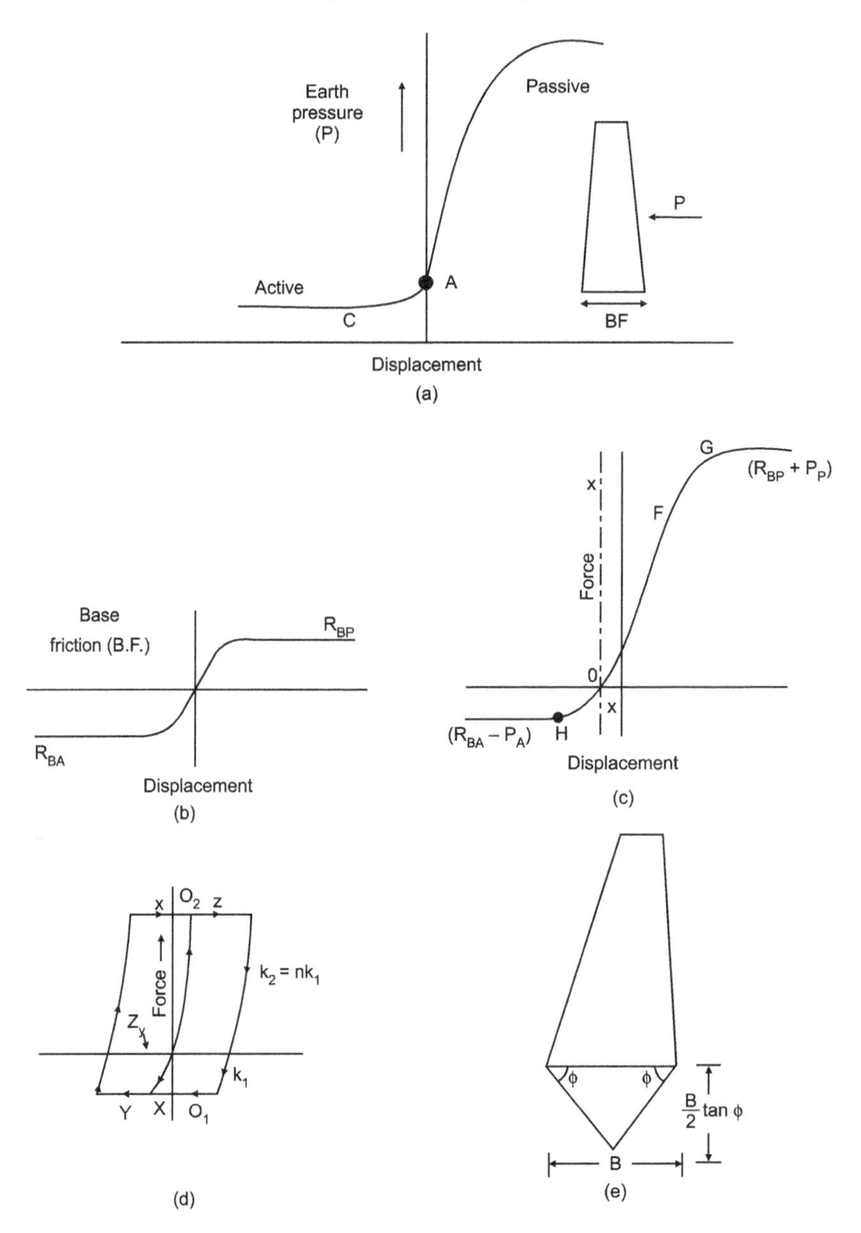

Fig. 5.21. (a) Eatrth pressure (P) versus displacement of wall (b) Base-Friction (B.F.) versus displacement, (c) Resultant of 'P' and B.F. versus displacement, (d) Simplified bilinear forces-displacement diagram, (e) Computation of base resistance

The mathematical model is shown in Fig. 5.22. The parameters that are needed to define the system for displacement analysis are: (1) the mass of system, m, (2) period of the wall-soils system, (3) yield displacement, (4) damping in the system, and (5) parameters of ground motion.

The vibrating mass of the system consists of the mass of the wall and that of the soil vibrating with the wall. Nandakumaran (1973) conducted vibratory tests on translating walls and found that for the purposes of matching the computed frequency of the wall with the measured natural frequency, the soil mass participating in the vibrations is 0.8 times the mass of soil on the Rankine failure wedge.

Yield displacement for a given wall can be determined by considering the force-displacement relationships.

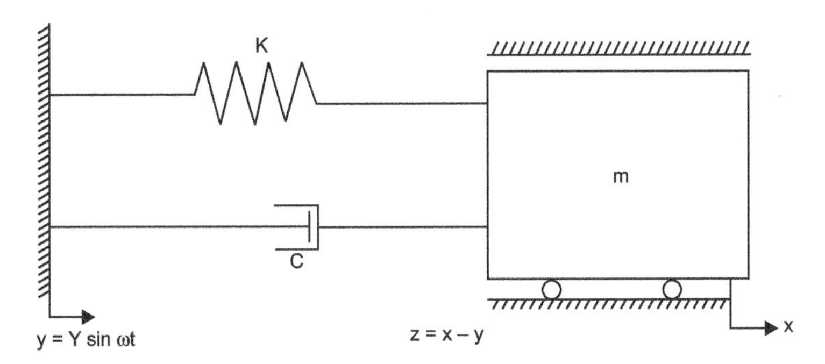

Fig. 5.22. Mathematical model considered for the analysis (Nandkumaran, 1973)

The ground motion is considered to be a sinusoidal motion of definite magnitude and period. The equation of motion can be written in the following form (Fig. 5.22):

$$m\ddot{x} + C(\dot{x} - \dot{y}) + K(x - y) = 0 \tag{5.51a}$$

$$m\ddot{z} + C\dot{z} + Kz = m\ddot{y} \tag{5.51b}$$

$$\ddot{z} + 2\eta\xi\dot{z} + \eta^2 z = -\ddot{y} \tag{5.51c}$$

where $z = (x - y)$,

$\eta^2 = K/m$ where K has been defined as the stiffness on the tension side and

ξ = Damping ratio = $\dfrac{C}{2\sqrt{Km}}$

For ease in computations, all the three equations obtained by linear acceleration method (Biggs, 1963) to be satisfied at each instant of time or at the end of each time interval selected, can be divided by Z_y, the relative displacement on the tension side at which the resistance becomes constant (yield displacement) to obtain the following relations:

$$\psi_{n+1} = \psi_n + \dot{\psi}_n \cdot t + \frac{t^2}{6}(\ddot{\psi}_{n-1} + 2\ddot{\psi}_n) \tag{5.52}$$

$$\dot{\psi}_{n+1} = \dot{\psi}_n + \frac{t}{2}(\ddot{\psi}_{n+1} + 2\ddot{\psi}_n) \tag{5.53}$$

$$\ddot{\psi}_{n+1} = -y_n - 2\eta\xi\ddot{\psi}_{n+1} - \eta^2 \cdot K \cdot Z_y(\psi_n + 1) \tag{5.54}$$

where

$$\psi = \frac{Z}{Z_y} \tag{5.55}$$

$$\dot{\psi} = \frac{\dot{Z}}{\dot{Z}_y} \tag{5.56}$$

$$\ddot{\psi} = \frac{\ddot{Z}}{\ddot{\bar{Z}}} \tag{5.57}$$

With these relationship, the analysis is performed for the range of variables listed in Table 5.2.

Table 5.2. Range of Variables Considered in Displacement Analysis in Translation

Variable	*Range of values*
Ground acceleration amplitude (a), gals	100, 200 and 300
Period of ground motion (T), s	0.5, 0.3, 0.2 and 0.1
Damping (ξ), %	5, 10, 15
Natural period (T_n), s	1.0, 0.5, 0.3 and 0.2
Yield displacement (Z_y), mm	1.0, 2.0, 3.0, 5.0 and 10.0

To study the response characteristics of the system, two cases were considered; one in which plastic deformation does not take place and the other in which it does. Figure 5.23 shows the response of the elastic system. It is evident from this figure that steady state conditions are attained in about six cycles and also that displacements on the tension side are larger than those on compression side. The response of the system wherein slips take place has been plotted in Fig. 5.24. This shows that even when plastic deformations occur, a sort of steady state is achieved in the sense that slip per cycle becomes a constant after about six cycles.

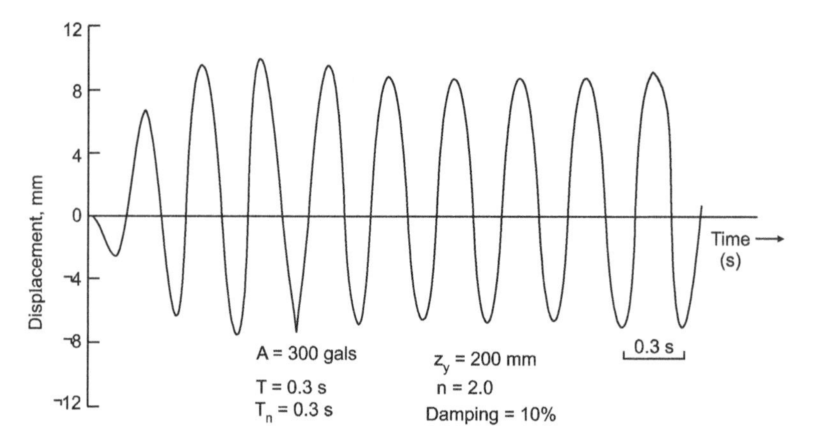

Fig. 5.23. Responses of an elastic system with different stiffness of tension and compression sides (Nandkumaran, 1973)

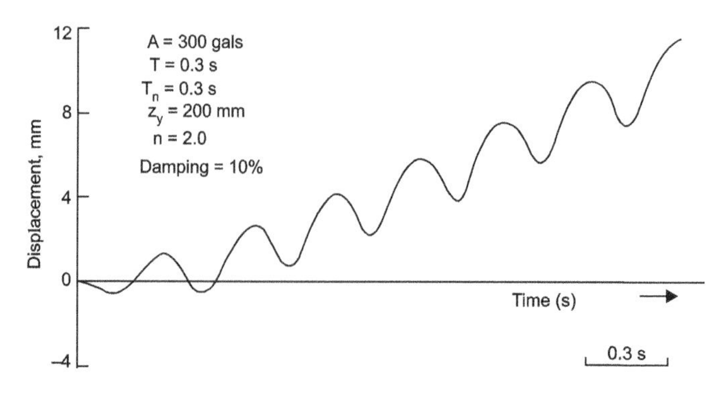

Fig. 5.24. Displacement versus time (Nandkumaran, 1973)

Figure 5.25 shows typical set of results in the form of slip per cycle versus the natural period of the wall in seconds for the yield displacement $Z_y = 5.0$ mm, and 10.0 mm, $\xi = 10\%$ and $n = 2$ for different ground motions. The ground motion is considered to be an equivalent motion of uniform peak acceleration of well defined cycles.

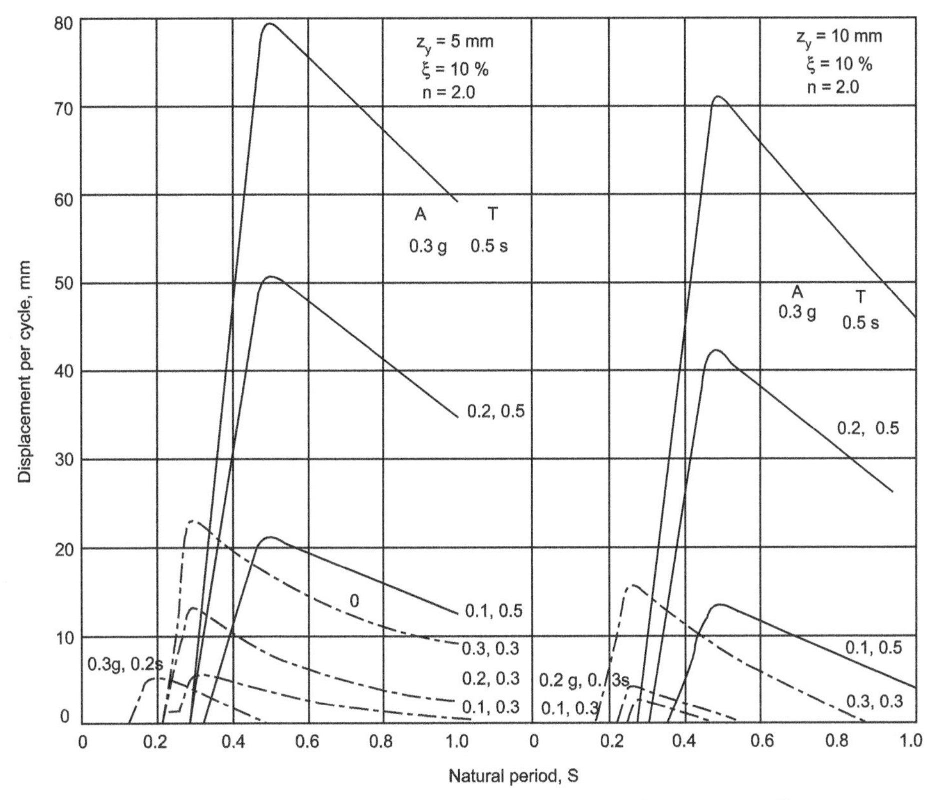

Fig. 5.25. Natural period versus slip per cycle (Nandkumaran, 1973)

Any problem can be solved with the following steps:

1. Determine the natural period of the wall using the following equation:

$$T = 2\pi\sqrt{\frac{m}{K}} \tag{5.58}$$

where K = stiffness on the tension side, and m = mass of soil and the wall.

2. Determine the yield displacement.
3. Determine the slip per cycle from Fig. 5.25 or similar other plots corresponding to the yield displacement, the natural period of the wall and the ground motion considered.
4. Compute the total slip during the ground motion.

This method of analysis is better than the one proposed by Richards and Elms (1979) in that (i) definite procedure for determining the natural period of the soil-wall system in translation has been formulated, and (ii) physical behaviour of the retaining wall is considered in developing the force-displacement relationships. The method, however, suffers from the fact that the tilting of the wall has not been considered.

5.3.3 Solution in Pure Rotation

A method of analysis for computing the rotational displacement of rigid retaining walls under dynamic loads has been presented by Prakash et al. (1981) and it is based on the following assumptions:

(i) Rocking vibrations are independent of sliding vibrations and the rocking stiffness is not affected by sliding of the wall.
(ii) The earthquake motion may be considered as an equivalent sinusoidal motion having constant peak acceleration.
(iii) Wall may be assumed to rotate about the heel.
(iv) Soil stiffness for rotational displacement of wall away from the backfill may be computed corresponding to average displacement for development of fully active conditions.
(v) Soil stiffness for rotational displacement of the wall towards the backfill may be computed corresponding to average displacement for development of fully passive conditions.
(vi) The stiffness values computed in (iv) and (v) remain unchanged during phases of wall rotation towards and away from backfill respectively.
(vii) Soil participating in vibrations may be neglected.

The mathematical model based upon these simplifying assumptions is shown in Fig. 5.26a. Figure 5.26b shows the scheme for calculation of side resistance corresponding to active and passive conditions. If fully active conditions are assumed to develop at a displacement of 0.25% of height of wall, then soil stiffness K_1 in active state is given by

$$K_1 = \frac{P_O - P_A}{\text{Average displacement}} = \frac{\dfrac{K_0 \gamma H^2}{2} - \dfrac{K_A \gamma H^2}{2}}{\left(\dfrac{0.25\,H}{100}\right)} \tag{5.59}$$

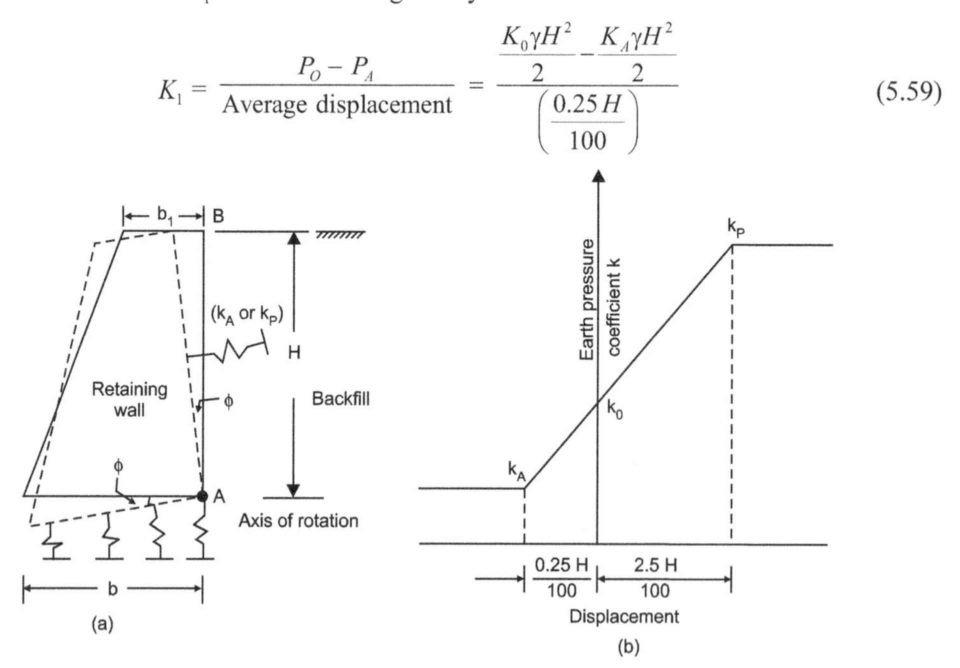

Fig. 5.26. (a) Mathematical model for rotation of rigid walls, (b) Scheme for computation of spring stiffness (After Prakash et al., 1981)

Similarly if fully passive conditions are assumed to develop at 2.5% of wall height, soil stiffness K_2 in passive state may be computed as:

$$K_2 = \frac{P_P - P_O}{\text{Average displacement}} = \frac{\dfrac{K_p \gamma H^2}{2} - \dfrac{K_0 \gamma H^2}{2}}{\left(\dfrac{2.5\,H}{100}\right)} \tag{5.60}$$

where P_A = Active earth pressure
P_P = Passive earth pressure
P_O = Earth pressure at rest

K_A = Coefficient of active earth pressure
K_P = Coefficient of passive earth pressure
K_O = Coefficient of earth pressure at rest

The rotational resistances of the base, in active and passive states (M_{RA} and M_{RP}) may be given by

$$M_{RA} = C_\phi \cdot I \cdot \phi_A \tag{5.61a}$$

$$M_{RP} = C_\phi \cdot I \cdot \phi_B \tag{5.61b}$$

in which C_ϕ is coefficient of elastic non-uniform compression; I is moment of inertia of the base about an axis through the heel of the wall and perpendicular to the plane of vibrations; and ϕ_A and ϕ_B are angles of rotation away and towards the backfill.

The equations of motion for rotation of wall away and towards the backfill are respectively:

$$M_{mo}\ddot{\phi}_A + \left(C_\phi I - \frac{K_1 H^2}{3}\right)\phi_A = M(t) \tag{5.62a}$$

and

$$M_{mo}\ddot{\phi}_P + \left(C_\phi I - \frac{K_2 H^2}{3}\right)\phi_P = M(t) \tag{5.62b}$$

Since the stiffness K_1 and K_2 are different, the period of the wall for the two conditions i.e. towards the backfill and away from backfill would be different. This would result in different values of f_A and ϕ_P for each half cycle of motion and net rotational displacement of $(\phi_A - \phi_P)$ for one cycle of ground motion. The maximum displacement of wall for any number of cycles may be computed as :

$$\phi_T = N(\phi_A - \phi_P) \cdot H \tag{5.63}$$

where N = Number of equivalent uniform cycles of ground motion
H = Height of wall

Based on the above, a parametric study was made considering the range of variables listed in Table 5.3.

Table 5.3. Range of Variables considered in Displacement Analysis in Rotation

Variable	*Range of values*
Height of wall (m)	3.0, 5.0, 7.5 and 10.0
Angle of internal friction for backfill (degrees)	30, 33, 36
Period of ground motion (s)	0.3
Damping (ξ)	0, 5, 10, 15
C_ϕ base (kN/m^3)	3, 4, 5, 6 and 8 ($\times 10^4$)
Base width/Height of wall	1/3

It was observed that the contribution of rotational displacement may be significant. The contribution of rotational displacement using the above approach was compared with the sliding displacement for a 3 m high wall with backfill having angle of internal friction, ϕ, equal to 36°, period of ground motion of 0.3 s, horizontal seismic coefficient α_h equal to 0.25 and C_ϕ equal to 3×10^4 kN/m^3. The total slip in 15 cycles due to sliding was 213 mm. Displacement of top of wall due to rotation found by this analysis was 147 mm.

This illustrates that the rotational displacement may not be negligible and an attempt should be made to account for it. The displacement analysis for rotational displacement is highly simplified. Nevertheless it shows explicitly that in some cases neglecting rotational displacement may seriously underestimate the total displacement. In actual practice it may be essential to account for combined effects of rocking and sliding that will affect the overall response of the system.

5.3.4 Nadim-Whitman Analysis

The Richard-Elms model assumes a constant value of wall acceleration ($\alpha_h \cdot g$) when slippage is taking place. But once the backfill begins to slip, compatability of movement requires the backfill to have a vertical acceleration, thus causing change in wall acceleration.

Zarrabi (1979) considered the equilibrium of the wall and the backfill wedge separately and satisfied the continuity requirements at failure surfaces as shown in Fig. 5.27a. An iterative procedure was developed for computing the instantaneous values of the inclination of failure plane, the dynamic active earth pressure and the acceleration of the wall, given the input of horizontal and vertical ground acceleration. The horizontal acceleration of the wall and the inclination of failure plane in the backfill are not constant in Zarrabi's model.

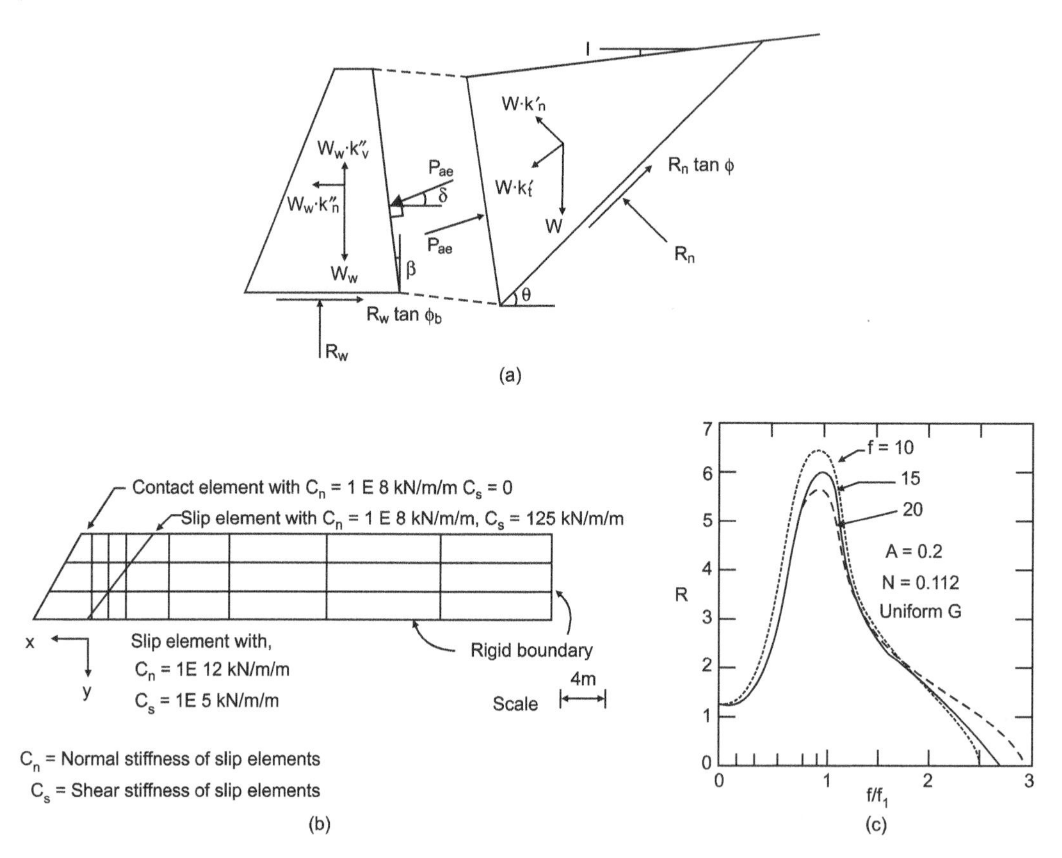

Fig. 5.27. (a) Force resolution of wall and soil wedge in Zarrabi's model, (b) Retaining wall and its finite element idealization, (c) Effect of ground motion amplification on permanent wall displacement (Zarrabi, 1979)

Generally, displacements computed with Zarrabi's model are slightly lower than those computed with the Richard-Elms model. Dynamic tests on model retaining walls performed by Lai (1979) show that Zarrabi's model predicts the movement of the wall more accurately than Richard-Elms model. Lai, also, observed a single rupture plane in the backfill in contrast to Zarrabi prediction. Later on, Zarrabi model has been modified to have a constant inclination of failure plane in the backfill.

The Richard-Elms model and Zarrabi's model assume a rigid-plastic behaviour of the backfill material. Hence the input ground acceleration is constant throughout the backfill. But due to more-or-less elastic behaviour of soil at stress level below failure, the input acceleration is not constant. Hence amplification of motion cannot be taken into account in these models.

Nadim and Whitman (1983) used a two-dimensional plane-strain finite element model for computing permanent displacements taking into account the ground motion amplification. The slip element at the base of the wall has been assigned a very large value of normal stiffness, thus restraining the wall from vertical and rotational movements relative to its base. Thus, the wall undergoes only translational movements. In their paper, the results of finite element mesh used by them is shown in Fig. 5.27b. To understand the effect of ground motion amplification, typical results are shown in Fig. 5.27c. In this figure, R is the ratio of permanent displacement from the *FE* model to the permanent displacement from rigid-plastic (Richard-Elms or Zarrabi's) model. f_l is the fundamental frequency of wall and f is the frequency of ground motion. It can be seen that effect of amplification of motion on displacement is greater when f/f_l is greater than 0.3. The *FE* model predicts zero permanent displacement at high frequency, because in the analysis only three cycles of base motion are considered during which steady-state conditions cannot be achieved. However, it can be said that large values of f/f_l are not of great practical interest because displacements are very small.

Nadim and Whitman (1983) suggested the following simple procedure for taking into account the effects of ground motion amplification in the seismic design:

(i) Evaluate the fundamental frequency f_1 of the backfill for the design earthquake using one-dimensional amplification theory by using the following equation and estimate the ground motion frequency, f.

$$f_1 = V_1/4H \tag{5.64}$$

where H = Height of retaining wall in m

V_1 = Peak velocity of earthquake in m/s

(ii) If f/f_1 is less than 0.25, neglect the amplification of ground motion. If f/f_1 is in the vicinity of 0.5, increase the peak acceleration, A and the peak velocity, V of the design earthquake by 25-30%. If f/f_1 is between 0.7 and 1, increase A and V by 50%. Obtain α_h as A/g.

(iii) Use the value of α_h from the previous step in Eq. (5.50) given by Richard-Elms model for getting α_{hd} for known value of the displacement.

(iv) The value of α_{hd} estimated in step (iii) is used as the value of horizontal seismic coefficient in the Mononobe-Okabe analysis to calculate the lateral thrust for which the wall is designed. The value of vertical seismic coefficient may be taken as $\frac{\alpha_{hd}}{2}$.

5.3.5 Saran, Reddy and Viladkar Model

Saran et al. (1985) have chosen the mathematical model in such way that it results translation and rotation simultaneously and therefore it has two degrees of freedom.

In practice, cross-section of rigid retaining wall varies to a great extent. A reasonable approximation is, therefore, made by lumping the mass of the rigid retaining wall at its centre of gravity. The backfill soil is replaced by closely spaced independent elastic springs shown in Fig. 5.28.

To determine the spring constants soil modulus values have been used. The soil modulus depends on the type of soil. It varies linearly with depth in sands and normally consolidated clays, but remains constant with depth in case of over consolidated clays. For linear form of variation $k = \eta_h \cdot h$, where η_h is the constant of horizontal subgrade reaction and h is the depth below ground surface. Value of η_h also depends on the type of movement namely (i) wall moving away from backfill (active) and (ii) wall moving towards backfill (passive). Probable range of η_h in cohesionless soils is given in Table 5.4.

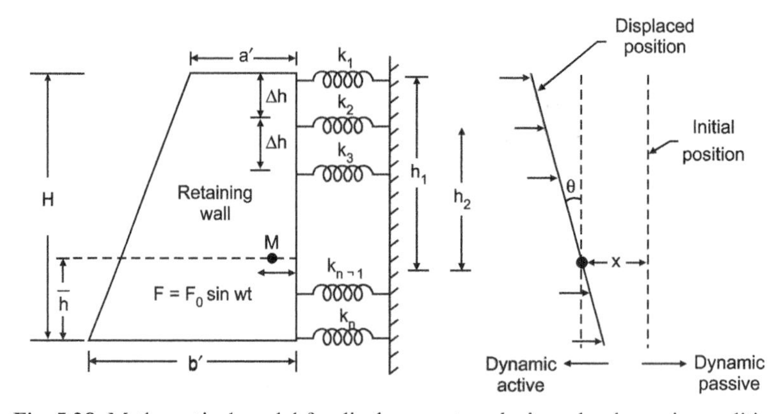

Fig. 5.28. Mathematical model for displacement analysis under dynamic condition

Table 5.4. Range of values of Modulus of Subgrade Reactions

Soil	η_h kN/m^3	
	Active	*Passive*
Loose sand	200-300	400-600
Medium dense sand	400-600	800-1200
Dense sand	800-1200	1600-2400

In case of soil modulus linearly varying with depth, the soil reaction is assumed to act as a loading intensity. Treating this load to be acting on a beam of length, equal to the height of retaining wall, the reactions at different points are evaluated treating this beam to be simply supported at the spring points. For the retaining wall of height H, divided into a convenient number of equal segments of height Δh, the reactions hence the spring constants values at various division points would be as under,

$$k_1 = \frac{1}{6}\eta_h(\Delta h)^2 \tag{5.65a}$$

$$k_2 = \eta_h(\Delta h)^2 \tag{5.65b}$$

$$k_3 = 2\eta_h(\Delta h)^2 \tag{5.65c}$$

$$\vdots$$

$$k_i = (i-1)\,\eta_h\,(\Delta h)^2 \tag{5.65d}$$

$$\vdots$$

$$k_n = \frac{1}{6}(3n-4)\eta_h\,(\Delta_h)^2 \tag{5.65e}$$

where k_1 and k_n are the spring constants at the top most and bottom most points, k_i, the spring constant at any division point 'i'.

In case of soil modulus constant with depth, the soil reaction is assumed to act as uniformly distributed loading intensity. Treating this uniformly distributed load to be acting on a beam of length, equal to the height of retaining wall, the reactions at different points are evaluated treating this to be simply supported at these points. The spring constants would be as under:

For the top most spring, $$k_1 = \frac{1}{2}\,k(\Delta h) \tag{5.66a}$$

For any intermediate spring, $k_i \; = k\,(\Delta\, h)$ (5.66b)

For the bottom most spring, $k_n \; = \dfrac{1}{2}k\,(\Delta\, h)$ (5.66c)

The method is based on the following assumptions:

1. The earthquake motion may be considered as an equivalent sinusoidal motion with uniform peak acceleration and the total displacement is equal to residual displacement per cycle multiplied by number of cycles.
2. Soil stiffnesses (or spring constants) for displacement of wall towards the backfill and away from the backfill are different.
3. Soil participating in vibration, damping of soil and base friction are neglected.

Assumptions 1 and 2 are usually made in such as analysis while assumption 3 needs justification. It is difficult to determine analytically the soil mass that would participate in vibrations along with wall when it undergoes translational and rotational motions simultaneously. Neglecting this mass, the method gives higher displacements and the solution is conservative. However, the mass of vibrating soil can be found out by carefully conducted experiments. For the case of pure translation, Nandakumaran (1973) has conducted experiments to determine the vibrating soil mass and concluded that it can be taken equal to 0.8 times the mass of Rankine's wedge. By adopting similar technique, the soil mass vibrating along with rigid retaining wall under combined rotational and translational motions can be found out. Then it is added to the mass of the wall to lump at centre of gravity and the analysis can be carried out without any changes.

In soils, it is customary to consider values of damping such as 15% or 20% of critical in view of larger energy absorption compared to other engineering structural materials. In the present analysis, however, energy absorption in the form of plastic displacement of the wall has been considered. Therefore smaller damping values would be appropriate. Neglecting even this smaller damping, the displacement of the wall by this method will be more than the actual displacement.

The displacement of retaining wall is greatly influenced by base friction. In case of walls in alluvial deposits and at the waterfront, translational motion is predominant. In some other cases, the walls may have predominant rotational motion. But in general for any type of foundation soil, retaining wall possesses translational and rotational motions simultaneously. For rigid retaining walls, the stability is mainly due to its gravity, hence base friction, the analysis will lead to an overestimation of the displacement.

However, refinment of the model by including vibrating soil mass, damping of soil and base friction is needed so that the analysis can predict displacement close to the actual displacements.

To study the response characteristics of the system, two cases are considered, one in which plastic deformations do not occur (elastic system) and the other in which plastic deformations do occur (plastic system).

5.3.5.1 Analysis of an Elastic System Active Condition

The equations of motion of the retaining wall using D'Alemberts principle can be written in general terms as follows:

$$M\ddot{x}+\sum_{n}^{i=1}k_i[x+\{(H-\bar{h})-(i-1)\Delta h\}\theta] \;= F_o \sin \omega\, t \tag{5.67}$$

$$J\ddot{\theta}+\sum_{n}^{i=1}k_i[x+\{(H-\bar{h})-(i-1)\Delta h\}\,\theta\,\{(H-\bar{h})-(i-1)\Delta h\}] \;= 0 \tag{5.68}$$

where M = Mass of reitaining wall

J = Polar mass moment of inertia of the wall about the axis of rotation

ω = Frequency of the excitation force

H = Height of retaining wall
$\bar{h}$ = Height of centre of gravity of wall from its base
x = Translatory displacement
θ = Rotational displacement

Letting:

$$\frac{\sum_{n}^{i=1} k_i}{M} = a \quad (5.69)$$

$$\frac{F_o}{M} = a_o \quad (5.70)$$

$$\frac{\sum_{n}^{i=1} k_i\{(H-\bar{h})-(i-1)\Delta h\}}{M} = b \quad (5.71)$$

and

$$\frac{\sum_{n}^{i=1} k_i\{(H-\bar{h})-(i-1)\Delta h\}^2}{J} = c \quad (5.72)$$

The equations of motion of the rigid retaining wall can thus be written as,

$$\ddot{x}+ax = b\,\theta + a_o \sin \omega t \quad (5.73)$$

$$\ddot{\theta}+c\theta = \left(\frac{b}{r^2}\right)x \quad (5.74)$$

where $J = Mr^2$, r being the radius gyration and 'b' can be called as coupling coefficient because if $b = 0$, the two equations become independent of each other.

The solutions of Eqs. (5.73) and (5.74) can be written as

$$x = X \sin \omega t \quad (5.75)$$

$$\theta = \beta \sin \omega t \quad (5.76)$$

where X and β are arbitrary constants.

Substituting Eqs. (5.75) and (5.76) in Eqs. (5.73) and (5.74), we get

$$(-\omega^2 + a)X = b\beta + a_o \quad (5.77)$$

$$(\omega^2 + c)\beta = \left(\frac{b}{r^2}\right)X \quad (5.78)$$

Solving Eqs. (5.77) and (5.78), we get

$$X = \frac{a_o}{(a-\omega^2)-\dfrac{b^2}{r^2(c-w^2)}} \quad (5.79)$$

$$\beta = \frac{a_o}{(a-\omega^2)(c-\omega^2)\cdot\dfrac{r^2}{b}-b} \quad (5.80)$$

Hence the solution becomes:

$$x = \frac{a_o}{(a-\omega^2) - \dfrac{b^2}{r^2(c-\omega^2)}} \sin \omega t \tag{5.81}$$

$$\theta = \frac{a_o}{(a-\omega^2)(c-\omega^2)\cdot\dfrac{r^2}{b} - b} \sin \omega t \tag{5.82}$$

Therefore, the displacement of the top of rigid retaining wall is given by:

$$x_{top} = x + (H - \bar{h})\theta \tag{5.83a}$$

or

$$x_{top} = \left\{ \frac{r^2(c-\omega)^2 + b(H-\bar{h})}{(a-\omega^2)(c-\omega^2)r^2 - b^2} \right\} a_0 \sin \omega t \tag{5.83b}$$

5.3.5.2 Natural Frequencies

Under free vibration condition, the equations of motion are:

$$\ddot{x} + ax = b\theta \tag{5.84}$$

$$\ddot{\theta} + c\theta = \left(\frac{b}{r^2}\right)x \tag{5.85}$$

Substituting the Solution

$$x = A \sin \omega_n t \tag{5.86}$$

$$\theta = B \sin \omega_n t \tag{5.87}$$

where A and B are arbitrary constants.

Equations (5.84) and (5.85) become:

$$(-\omega_n^2 + a)A = b \,.\, B \tag{5.88}$$

$$(-\omega_n^2 + c)B = \left(\frac{b}{r^2}\right)A \tag{5.89}$$

From these we get,

$$\frac{A}{B} = \frac{b}{a-\omega_n^2} \tag{5.90}$$

and

$$\frac{A}{B} = \frac{c-\omega_n^2}{\left(\dfrac{b}{r^2}\right)} \tag{5.91}$$

Equating,

$$\frac{b}{a-\omega_n^2} = \frac{c-\omega_n^2}{\left(\dfrac{b}{r^2}\right)}$$

$$\omega_n^4 - (a+c)\omega_n^2 + \left(ac - \frac{b^2}{r^2}\right) = 0 \tag{5.92}$$

and solving we get,
$$\omega_{n1}^2 = \frac{1}{2}(a+c)+\sqrt{\left(\frac{c-a}{2}\right)^2+\left(\frac{b}{r}\right)^2} \quad (5.93)$$

$$\omega_{n2}^2 = \frac{1}{2}(a+c)-\sqrt{\left(\frac{c-a}{2}\right)^2+\left(\frac{b}{r}\right)^2} \quad (5.94)$$

5.3.5.3 Passive Condition

The ratio of stiffnesses on the compression and tension sides is denoted by n. Hence in the passive condition, the values of a, b and c change and these can be given by:

$$a = n(a)_a \quad (5.95)$$

$$b = n(b)_a \quad (5.96)$$

$$c = n(c)_a \quad (5.97)$$

The solution for this condition is similar to active condition described above.

5.3.5.4 Analysis of a Plastic System-active Condition

Assume that Z_y and θ_y are the yield displacements occurring simultaneously in all springs, the equations of motion can be written as:

$$\ddot{x}+aZ = b\,\theta_y + a_o \sin \omega t \quad (5.98)$$

$$\ddot{\theta}+c\theta_y = \left(\frac{b}{r^2}\right)Z_y \quad (5.99)$$

Integrating the above equations twice, we get

$$x = (b\theta_y - aZ_y)\frac{t^2}{2} - \frac{a_0 \sin \omega t}{\omega^2} + C_1 t + C_2 \quad (5.100)$$

$$\theta = \left(\frac{b}{r^2}Z_y - c\theta_y\right)\frac{t^2}{2} + C_3 t + C_4 \quad (5.101)$$

Let 't_e' be the time after which displacement of top of wall (y_{top}) becomes greater than yield displacement (y_d) and plastic system starts. Let x_e, $\dot{x}_e$, θ_e, $\dot{\theta}_e$ be the values corresponding to time t_e and can be calculated by using the equations developed for elastic system. The following boundary conditions can be applied to evaluate the constants of integration:

(i) $t = t_e,\ \dot{x} = \dot{x}_e$ (5.102a)

(ii) $t = t_e,\ x = x_e$ (5.102b)

(iii) $t = t_e,\ \dot{\theta} = \dot{\theta}_e$ (5.102c)

(iv) $t = t_e,\ \theta = \theta_e$ (5.102d)

Therefore, we have

$$Z_y = x_e \quad (5.103)$$

$$\theta_y = \theta_e \quad (5.104)$$

$$C_1 = \dot{x}_e - (b\theta_y - aZ_y)t_e + \frac{a_o \cos \omega t_e}{\omega} \quad (5.105)$$

$$C_2 = x_e - \dot{x}_e t_e + (b\theta_y - aZ_y)\frac{t_e^2}{2} + \frac{a_o \sin \omega t_e}{\omega} - \frac{t_e a_o \cos \omega t_e}{\omega} \tag{5.106}$$

$$C_3 = \theta_e - \left(\frac{b}{r^2} Z_y - c\theta_y\right) t_e \tag{5.107}$$

$$C_4 = \theta_e + \left(\frac{b}{r^2} Z_y - c\theta_y\right)\frac{t_e^2}{2} - \theta_e t_e \tag{5.108}$$

Displacement of the top of rigid retaining wall is given by

$$x_{\text{top}} = x - (H - \bar{h})\theta \tag{5.109}$$

5.3.5.5 Passive Condition

The ratio of stiffnesses on the compression and tension sides is denoted by n. Hence in the passive condition, the values of a, b and c change and these can be given by :

$$a = n(a)_a \tag{5.110a}$$

$$b = n(b)_a \tag{5.110b}$$

$$c = n(c)_a \tag{5.110c}$$

The solution is similar to the above procedure for active condition except the values of Z_y and θ_y. In compression side (passive condition), the displacements for achieving yield condition are very large, hence in most of the cases plastic system for passive case is not considered.

Illustrative Examples

Example 5.1

A 6.0 m high retaining wall with back face inclined 20° with vertical retains cohesionless backfill (ϕ = 33°, γ_t = 18 kN/m³ and δ = 20°). The backfill surface is sloping at an angle 10° to the horizontal.

(a) Determine the total active earth pressure using Coulomb's theory and Culmann's graphical construction.

(b) If the retaining wall is located in a seismic region (α_h = 0.1), determine total active earth pressure using Mononobe's equation and modified Culmann's graphical construction.

Solution

(a) Static active earth pressure

$$P_A = \frac{1}{2}\gamma H^2 \frac{\cos^2(\phi - \alpha)}{\cos^2 \alpha \cos(\delta + \alpha)} \frac{1}{\left\{1 + \left[\frac{\sin(\phi+\delta)\sin(\phi - i)}{\cos(\alpha - i)\cos(\delta + \alpha)}\right]^{1/2}\right\}^2}$$

$$= \frac{1}{2} \times 18 \times 6.0^2 \times \frac{\cos^2(33-20)}{\cos^2 20 \cos(20+20)} \times \frac{1}{\left\{1 + \left[\frac{\sin(33+20)\sin(33-10)}{\cos(20-10)\cos(20+30)}\right]^{1/2}\right\}^2}$$

= 168.42 kN/m

Refer Fig. 5.29 for Culmann's graphical construction for getting static active pressure. D_5E_5 gives the total active earth pressure.

$$P_A = 17 \times 10 = 170 \text{ kN/m}$$

(b) Dynamic active earth pressure

$$(P_A)_{dyn} = \frac{1}{2}\gamma H^2 \frac{\cos^2(\phi-\psi-\alpha)(1\pm\alpha_v)}{\cos\psi\cos^2\alpha\cos(\delta+\alpha+\psi)} \times \frac{1}{\left\{1+\left[\frac{\sin(\phi+\delta)\sin(\phi-i-\psi)}{\cos(\alpha-i)\cos(\delta+\alpha+\psi)}\right]^{1/2}\right\}^2}$$

Assuming $\alpha_v = \frac{\alpha_h}{2} = \frac{0.1}{2} = 0.05$

$$\psi = \tan^{-1}\frac{\alpha_h}{1\pm\alpha_v} = \tan^{-1}\frac{0.1}{1\pm 0.05}$$

$= 5.44°$ with $+\alpha_v$ and

$= 6.0°$ with $-\alpha_v$.

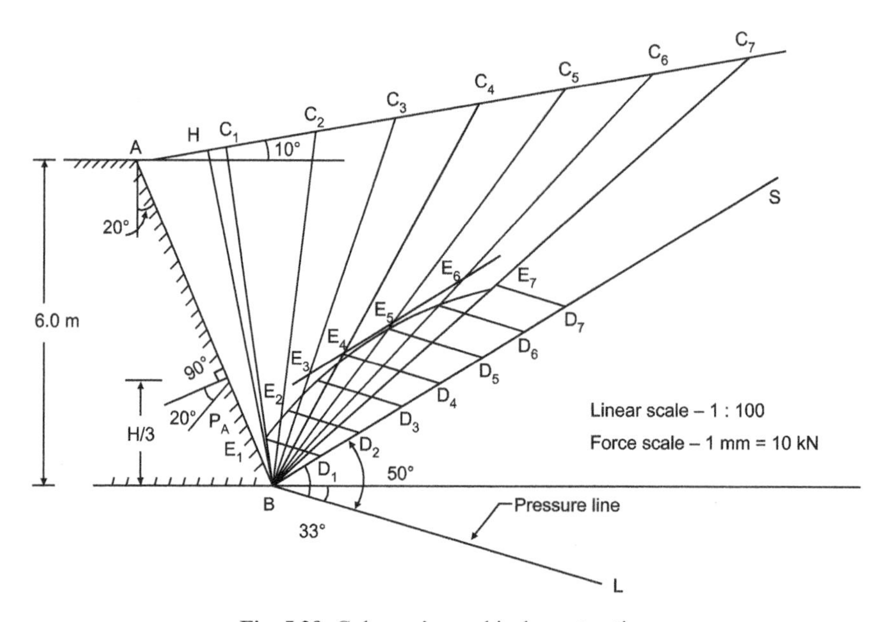

Fig. 5.29. Culmann's graphical construction

Value of $(P_A)_{dyn}$ with $(+)\alpha_v$

$$= \frac{1}{2}\times 18\times 6.0^2 \frac{\cos^2(33-5.44-20)(1+0.05)}{\cos 5.44\cos^2 20\cos(20+20+5.44)} \times \frac{1}{\left\{1+\left[\frac{\sin(33+20)\sin(33-10-5.44)}{\cos(20-10)\cos(20+20+5.44)}\right]^{1/2}\right\}^2}$$

$= 214.26$ kN/m

Value of $(P_A)_{dyn}$ with $(-)\,\alpha_v$

$$= \frac{1}{2} \times 18 \times 6.0^2 \frac{\cos^2(33-6-20)}{\cos 6 \cos^2 20 \cos(20+20+6)} \times \frac{1}{\left\{1+\left[\frac{\sin(33+20)\sin(33-10-6)}{\cos(20-10)\cos(20+20+6)}\right]^{1/2}\right\}^2}$$

= 198.05 kN/m

Therefore (+) α_v case governs the value of dynamic active earth pressure.

Hence, $(P_A)_{dyn} = 214.26$ kN/m

Refer Fig. 5.30 for modified Culmann's graphical construction for getting dynamic active earth pressure.

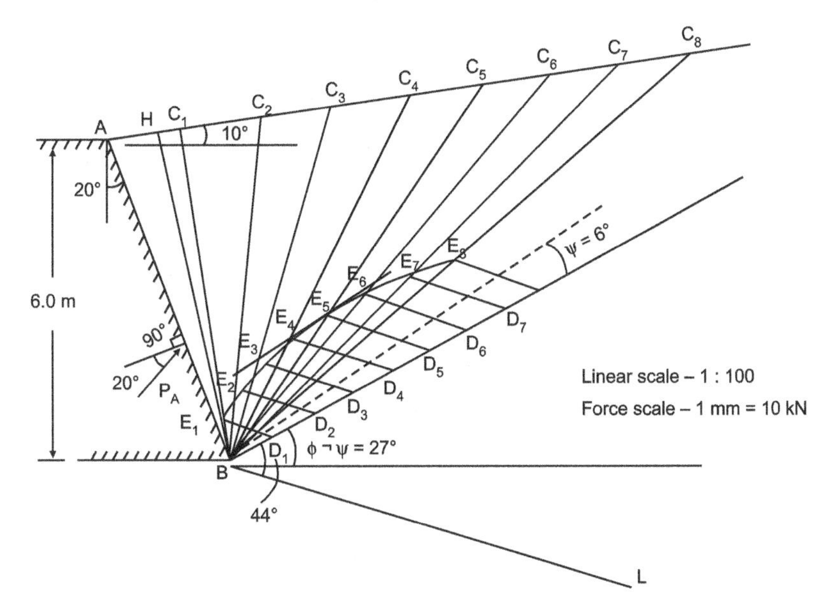

Fig. 5.30. Modified Culmann's graphical construction

D_5E_5 gives the total dynamic active earth pressure.

$$(P_A)_{dyn} = 21 \times 10 = 210 \text{ kN/m}$$

Example 5.2

A retaining wall 8.0 m high is inclined 20° to the vertical and retains horizontal backfill with following properties:

$$\gamma_t = 18 \text{ kN/m}^3, \phi = 30° \text{ and } c = 6.0 \text{ kN/m}^2$$

There is a superimposed load of intensity 15 kN/m² on the backfill. The wall is located in seismic region having horizontal seismic coefficient of 0.1. Compute the dynamic active earth pressure and determine the percentage increase in pressure over the static earth pressure.

Solution

(i)

$$h_0 = \frac{2c}{\gamma} \frac{1}{\sqrt{K_A}} \qquad \text{where } K_A = \frac{1-\sin 30}{1+\sin 30} = \frac{1}{3}$$

$$= \frac{2 \times 6.0}{18} \times \sqrt{3} = 1.15 \text{ m}$$

$$H = 8.0 - 1.15 = 6.85 \text{ m}$$

$$n = \frac{h_0}{H} = \frac{1.16}{6.85} = 0.167$$

(ii) For $\phi = 30°$, $\alpha = +20°$ and $n = 0.167$

Figures 5.10, 5.11, 5.12, 5.14 and 5.15 give:

$$(N_{a\gamma m})_{stat} = 0.33,\ (N_{aqm})_{stat} = 0.512 \text{ and } (N_{acm})_{stat} = 1.2$$

$$(P_A)_{stat} = 18 \times 6.85^2 \times 0.33 + 15 \times 6.85 \times 0.512 - 6 \times 6.85 \times 1.2$$

$$= 282.0 \text{ kN/m}$$

(iii) For $\phi = 30°$, $\alpha = +20°$, $\alpha_h = 0.1$, Fig. 5.17 gives

$$\lambda = 1.18$$

Therefore,

$$(N_{aqm})_{dyn} = 1.18 \times 0.512 = 0.609 \text{ and}$$

$$(N_{a\gamma m})_{dyn} = 1.18 \times 0.33 = 0.393$$

$$(P_A)_{dyn} = 18 \times 6.85^2 \times 0.393 + 15 \times 6.85 \times 0.609 - 6 \times 6.85 \times 1.2$$

$$= 345.18 \text{ kN/m}$$

(iv) Percentage increase over static pressure

$$= \frac{345.10 - 282.00}{282.01} \times 100 = 22.4\%$$

Example 5.3

A 5.0 m high retaining wall with backface inclined 20° with vertical retains cohesionless backfill ($\phi = 30°$, $\gamma_t = 18$ kN/m^3 and $\delta = 20°$). The backfill surface is sloping at an angle 15° to the horizontal. Determine the weight of the retaining wall.

(a) For static condition,
(b) For zero displacement condition under earthquake loading.
(c) For a displacement of 50 mm under earthquake loading.

Solution

(a) For static condition

$$K_A = \frac{\cos^2(\phi-\alpha)}{\cos^2\alpha(\cos\delta+\alpha)} \cdot \frac{1}{\left\{1+\left[\frac{\sin(\phi+\delta)\sin(\phi-i)}{\cos(\alpha-i)\cos(\delta+\alpha)}\right]^{1/2}\right\}^2}$$

$$= \frac{\cos(30-20)}{\cos^2 20\cos(20+20)} \cdot \frac{1}{\left\{1+\left[\frac{\sin(30+20)\sin(30-15)}{\cos(30-20)\cos(20+20)}\right]^{1/2}\right\}^2}$$

$$= 0.6225$$

From Eq. (5.45)

$$C_1 = \frac{\cos(\alpha+\delta) - \sin(\alpha+\delta)\tan\phi_b}{\tan\phi_b}$$

$$= \frac{\cos(20+20) - \sin(20+20)\tan 30}{\tan 30} = 0.6840$$

Thus from Eq. (5.44)

$$W = \frac{1}{2}\gamma H^2 K_A C_1$$

$$= \frac{1}{2} \times 1800 \times 5^2 \times 0.6225 \times 0.6840 = 9765 \text{ kg/m}$$

(b) For zero displacement condition

$$(K_A)_{dyn} = \frac{\cos^2(\phi - \psi - \alpha)}{\cos\psi \cos^2\alpha \cos(\delta + \alpha + \psi)} \times \frac{1}{\left\{1 + \left[\frac{\sin(\phi+\delta)\sin(\phi - i - \psi)}{\cos(\alpha - i)\cos(\delta + \alpha + \psi)}\right]^{1/2}\right\}^2}$$

Assuming $\alpha_v = \frac{\alpha_h}{2} = \frac{0.1}{2} = 0.05$

$$\psi = \tan^{-1}\frac{\alpha_h}{1 \pm \alpha_v} = \frac{0.1}{1 \pm 0.05}$$

$= 5.44°$ with $+\alpha_v$ and

$= 6.0°$ with $-\alpha_v$

With $+\alpha_v$:

$$(K_A)_{dyn} = \frac{\cos^2(30 - 5.44 - 20)(1 + 0.05)}{\cos 5.44 \cos^2 20 \cos(20 + 20 + 5.44)} \times \frac{1}{\left\{1 + \left[\frac{\sin(30+20)\sin(30 - 15 - 5.44)}{\cos(20 - 15)\cos(20 + 20 + 5.44)}\right]^{1/2}\right\}^2}$$

$= 0.8311$

With $-\alpha_v$:

$$(K_A)_{dyn} = \frac{\cos^2(30 - 6 - 20)(1 - 0.05)}{\cos 6 \cos^2 20 \cos(20 + 20 + 6)} \times \frac{1}{\left\{1 + \left[\frac{\sin(30+20)\sin(30 - 15 - 6)}{\cos(20 - 15)\cos(20 + 20 + 6)}\right]^{1/2}\right\}^2}$$

$= 0.7727$

From Eq. (5.46)
With $+\alpha_v$:

$$\frac{W_w}{W} = \frac{0.8311}{0.6225} \times \frac{\tan 30}{(1 + 0.05)(\tan 30 - \tan 5.44)}$$

$= 1.524$

With $-\alpha_v$:

$$\frac{W_w}{W} = \frac{0.7727}{0.6225} \times \frac{\tan 30}{(1 - 0.05)(\tan 30 - \tan 6)}$$

$= 1.596$

Therefore

$$W_W = 1.596 \times 9765 = 15585 \text{ kg/m}$$

(c) For displacement condition, $d = 50$ mm
From Eq. (5.50)

$$\alpha_{hd} = \left[\frac{5\alpha_h}{d}\right]^{1/2} = 0.1\left[\frac{5 \times 0.1}{50}\right]^{1/2}$$

$$= 0.03162$$

Assuming

$$\alpha_{vd} = \frac{\alpha_{hd}}{2} = \frac{0.03162}{2} = 0.01581$$

With (+) α_{vd}

$$\psi = \tan^{-1}\left[\frac{\alpha_{hd}}{1+\alpha_{vd}}\right] = \tan^{-1}\left[\frac{0.3162}{1+0.01581}\right]$$

$$= 1.78^\circ$$

$$(K_A)_{dyn} = \frac{\cos^2(30-1.78-20)(1+0.01581)}{\cos 1.78 \cos^2 20 \cos(20+20+1.78)} \times \frac{1}{\left\{1+\left[\frac{\sin(30+20)\sin(30-15-1.78)}{\cos(20-15)\cos(20+20+1.78)}\right]^{1/2}\right\}^2}$$

$$= 0.684$$

With (–) α_{vd}

$$\psi = \tan^{-1}\left[\frac{0.03162}{1-0.01581}\right] = 1.84^\circ$$

It gives

$$(K_A)_{dyn} = \frac{\cos^2(30-1.84-20)(1-0.01581)}{\cos 1.84 \cos^2 20 \cos(20+20+1.84)} \times \frac{1}{\left\{1+\left[\frac{\sin(30+20)\sin(30-15-1.84)}{\cos(20-15)\cos(20+20+1.84}\right]^{1/2}\right\}^2}$$

$$= 0.672$$

Therefore, for (+) ve α_{vd}

$$\frac{W_w}{W} = \frac{0.684}{0.6225} \cdot \frac{\tan 30}{(1+0.01581)(\tan 30 - \tan 1.78)} = 1.143$$

For (–) ve α_{vd}

$$\frac{W_w}{W} = \frac{0.672}{0.6225} \cdot \frac{\tan 30}{(1-0.01581)(\tan 30 - \tan 1.84)} = 1.162$$

Hence

$$W_W = 1.162 \times 9765 = 11347 \text{ kg/m}$$

If a factor of safety of 1.5 is used, then

$$W = 9765 \times 1.5 = 14647 \text{ kg (Static condition)}$$

W_W = 15585 × 1.5 = 23377 kg (Earthquake condition - zero displacement)

W_W = 11347 × 1.5 = 17020 kg/m (Earthquake condition - 50 mm displacement)

It may be noted that the weight of wall gets reduced significantly if the wall is designed for some displacement.

Example 5.4

Compute the displacement of a vertical retaining wall having section and backfill properties as shown in Fig . 5.31. The characteristics of the ground motion are:

Period = 0.50 s

Average ground acceleration = 0.2 *g*

Number of significant cycles = 10

Yield displacement = 5 mm; $n = 2$ and $\xi = 10\%$

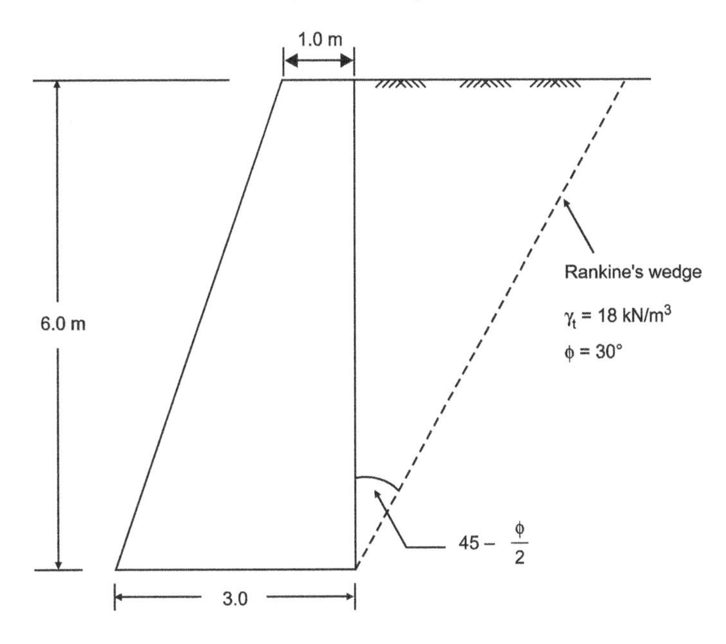

Fig. 5.31. Section of retaining wall

Solution

(i) Refer Fig. 5.31

$$\text{Weight of wall} = \frac{1}{2}(1.0+3.0)\times 6.0\times 24 = 288 \text{ kN}$$

Weight of soil vibrating along with wall

$$= 0.8\times\frac{1}{2}\times 6.0 \tan\left(45-\frac{30}{2}\right)\times 6.0\times 18 = 149.6 \text{kN}$$

$$\text{Total weight} = 288 + 149.6 = 437.6 \text{ kN}$$

$$m = \frac{437.6\times 10^3}{9.81} = 44.61\times 10^3 \text{kg}$$

Let the coefficient of base friction = 0.6

$$K_1 = \frac{288 \times 0.6 - \frac{1}{2} \times 18 \times \frac{1}{3} \times 6^2}{0.005} \quad \text{(Fig. 5.21d)}$$

$$= 12960 \text{ kN/m} = 12960 \times 10^3 \text{ N/m}$$

$$T_n = 2\pi \sqrt{\frac{m}{K_1}}$$

$$T_n = 2\pi \sqrt{\frac{44.61 \times 10^3}{12960 \times 10^3}} = 0.368 \text{ s}$$

(ii) From Fig. 5.25, for Z_y = 5 mm, T_n = 0.368 s, n = 2, T = 0.5 s, A = 0.2 g

Slip per cycle = 24 mm

Total slip = 24 × 10 = 240 mm

Example 5.5

Determine the displacement of a model wall shown in Fig. 5.32 retaining medium dense sand (ϕ = 36°, γ = 18 kN/m² and η_h = 520 kN/m³). The wall is subjected to following dynamic conditions:

Yield displacement = 6 mm
Ground acceleration = 0.25 g
Time period = 0.3 s

Solution

1. The wall is divided into four equal number of segments with h equal to 0.75 m and the backfill soil is idealized by using springs as shown in Fig. 5.33. The mass of retaining wall is assumed to be lumped at its c.g. which is at a distance of 1.23 m above the base (Fig. 5.33).

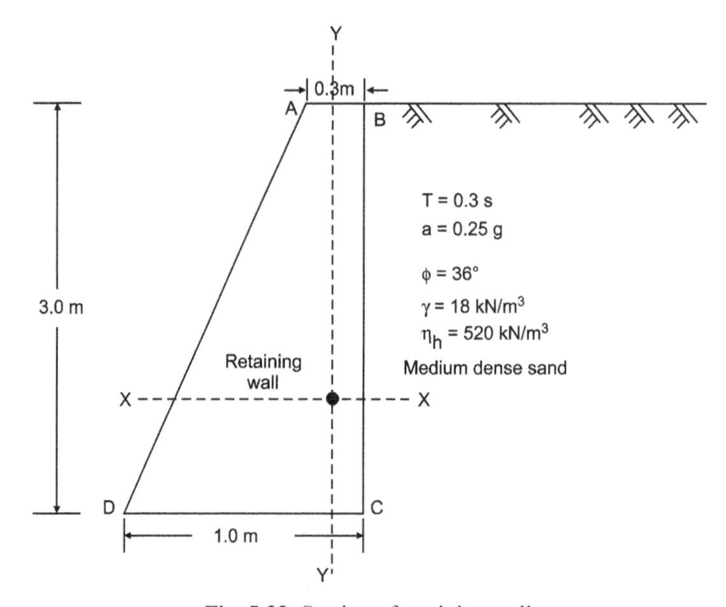

Fig. 5.32. Section of retaining wall

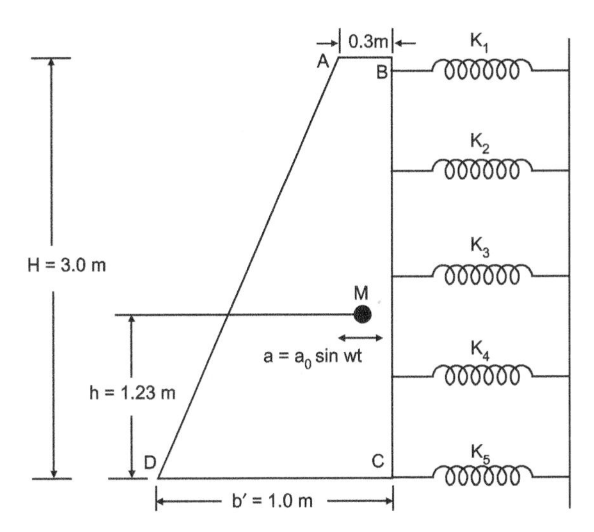

Fig. 5.33. Mathematical model adopted for solution

2. Consider the backfill characteristics, the spring constants in active and passive cases determined using Eqs. (5.65) and are given below in Table 5.5. The ratio of stiffnesses in passive and active states is taken as 2.0.

Table 5.5. Values of Spring Constants

Spring	*Spring location w.r. to c.g., h_i (m)*	*Spring constant active case, k_i* (kN/m)	*Spring constant passive case, k_i* (kN/m)
K_1	1.77	48.8	77.6
K_2	1.02	292.5	585.0
K_3	0.27	585.0	1170.0
K_4	– 0.48	877.5	1755.0
K_5	– 1.23	536.3	1072.6
		Σk_i= 2340 .1	Σk_i= 4680.2

3. Equation of Motion

The quantities required in the analysis are calculated as shown below (Fig. 5.32):

$$\text{Distance of c.g. from } CD = \frac{2\times0.3+1.0}{0.3+1.0}\times\frac{3}{3} = \frac{1.6}{1.3} = 1.23 \text{ m}$$

$$\text{Distance of c.g. from } BC = \frac{3\times0.3\times0.15+0.5\times3\times0.7\left(0.3+\frac{0.7}{3}\right)}{\frac{1+0.3}{2}\times3.0}$$

$$= \frac{0.135+0.56}{1.95} = 0.356 \text{ m}$$

$$I_{xx} = \frac{0.3\times3^3}{12}+0.3\times3.0(1.5-1.23)^2+\frac{0.7\times3^3}{36}+\frac{1}{2}\times0.7\times3\times(1.23-1)^2$$

$$= 0.675 + 0.06561 + 0.525 + 0.0555 = 1.321155 \text{ m}^4$$

$$I_{yy} = \frac{0.3^3 \times 3}{12} + 0.3 \times 3.0(0.356 - 0.15)^2 + \frac{0.7^3 \times 3}{36} + \frac{1}{2} \times 0.7 \times 3 \times \left(\frac{0.7}{3} - 0.056\right)^2$$

$$= 0.00675 + 0.0381924 + 0.0285833 + 0.0330194 = 0.1065451 \text{ m}^4$$

$$I_{xy} = 1.321155 + 0.1065451 = 1.4277002 \text{ m}^4$$

$$A = \frac{1+0.3}{2} \times 3.0 = 1.95 \text{ m}^2$$

$$r = \sqrt{\frac{1.4277}{1.95}} = 0.8556 \text{ m}$$

$$M = \left(\frac{0.3+1.0}{2}\right) \times 3.0 \times \frac{23.0 \times 10^3}{9.8} = 4.58 \times 10^3 \text{ kg}$$

$$J = Mr^2 = 3.35 \times 10^3 \text{ kg-m}^3$$

The values of a, b and c in the active and passive cases can be determined as below:

Active case

$$a = \frac{\Sigma k_1}{M} = 510.94$$

$$b = -\frac{\Sigma k_1 h_1}{M} = 117.51$$

$$c = \frac{\Sigma k_1 h_i^2}{J} = \frac{1513.393}{3.35} = 451.76$$

Passive case

$$a = -2.0 \times 510.94 = -1021.88$$

$$b = -2.0 \times 117.51 = -235.02$$

$$c = -2.0 \times 451.76 = -903.52$$

The natural frequencies of the wall by considering the tension side (i.e. active case) are given by:

$$\omega_{n1,2}^2 = \frac{1}{2}(a+c) \pm \sqrt{\left(\frac{c-a}{2}\right)^2 + \left(\frac{b}{2}\right)^2}$$

Putting the values of $a = 510.94$, $b = 117.51$ and $c = 451 .76$, we get

$$\omega_{n1} = 24.94 \text{ rad/s}$$

$$\omega_{n2} = 18.46 \text{ rad/s}$$

The natural time periods are, therefore,

$$T_{n1} = 0.25 \text{ s}$$

$$T_{n2} = 0.34 \text{ s}$$

Earthquake motions are erratic and no two accelerograms are similar. The two main parameters of any ground motion are the amplitude of acceleration and the number of zero crossing in unit time. A very simple and convenient form of ground motion including the two above parameters, is a sinusoidal motion. Moreover, while proposing a method for analysing the liquefaction potential of sand deposits, Seed and Idriss (1967) contended that any given accelerogram can be considered equivalent to some definite number of cycles of loading of equal magnitude. Such idealization have the advantage that after studying the effect of two parameters, the effect of a probable earthquake motion at any site, can be analysed. Because of the above advantages, sinusoidal ground motions are utilised in the present study.

Given $a_o = 0.25\text{ g} = 2.45\text{ m/s}^2$

$$T_P = 0.3\text{ s};\ \omega = \frac{2\pi}{0.3} = 20.94\text{ rad/s}$$

$$a = 2.45\sin(20.94\,t)$$

Prediction of Displacements in Elastic System

The displacements in passive case ($t = 0$ to $t_{p/2}$) can be calculated as shown below:

$$x = X\sin\omega t$$

$$\theta = \beta\sin\omega t$$

where

$$X = \frac{a_o}{(a-\omega^2) - \dfrac{b^2}{r^2(c-\omega^2)}}$$

$$= \frac{2.45}{(-1021.88 - 20.94^2) - \dfrac{(-235.02)^2}{0.8556^2(-903.52 - 20.94^2)}}$$

$$= -1.7448\times10^{-3}\text{ m}$$

$$\beta = \frac{a_o}{(a-\omega^2)\times(c-\omega^2)\dfrac{r^2}{b} - b}$$

$$= \frac{2.45}{(-1021.88 - 20.94^2)\times(-903.52 - 20.94^2)\dfrac{0.8556^2}{-235.02} - (-235.02)}$$

$$= -4.1741\times10^{-4}\text{ rad}$$

Hence, we have

$$x = -1.745\times10^{-3}\sin(20.94\,t)$$

$$\theta = -4.174\times10^{-4}\sin(20.94\,t)$$

The computed values of displacements from time 0 to 0.15 s are given in Table 5.6.

Table 5.6. Values of Displacements in Passive State (Elastic Condition)

Time t(s)	*Translational displacement x* (mm)	*Rotation* θ (rad)	*Disp. of wall at top due to rotation* x_θ (mm)	*Total disp. at top* x_{top} (mm)
0	0	0	0	0
0.0375	– 1.23	-2.9510×10^{-4}	– 0.52	– 1.75
0.0750	– 1.74	-4.1740×10^{-4}	– 0.74	– 2.48
0.1125	– 1.23	-2.9510×10^{-4}	– 0.52	– 1.75
0.15	0	0	0	0

The displacement in active case $\left(t = \frac{t_P}{2} \text{ to } t_P\right)$ can be calculated as shown below:

$$x = X \sin \omega t$$

$$\theta = \beta \sin \omega t$$

where
$$X = \frac{a_o}{(a-\omega^2) - \left(\frac{\frac{b^2}{r^2}}{c-\omega^2}\right)}$$

$$= \frac{2.45}{(510.94 - 20.94^2) - \frac{117.51^2}{0.8556^2(451.76 - 20.94^2)}}$$

$$= -1.817 \times 10^{-3} \text{ m}$$

$$\beta = \frac{a_o}{(a-\omega^2)\cdot(c-\omega^2)\frac{r^2}{b} - b}$$

$$= \frac{2.45}{(510.94 - 20.94^2)\cdot(451.76 - 20.94^2)\frac{0.8556^2}{117.51} - 117.51}$$

$$= -2.197 \times 10^{-2} \text{ rad}$$

Hence, we have
$$x = -1.817 \times 10^{-3} \times \sin(20.94\, t)$$

$$\theta = -2.197 \times 10^{-2} \times \sin(20.94\, t)$$

The values of displacements in active state considering elastic condition are given in Table 5.7.

Table 5.7. Values of Displacement in Active Condition (Elastic State)

Time t (s)	*Translational displacement* x (mm)	*Rotation* (rad)	*Disp. of wall at top due to rotation* x (mm)	*Total disp. at top* x_{top} (mm)
0.15	0	0	0	0
0.1875	1.28	0.02197	27.48 mm	28.76
0.2250	1.82	0.01552	38.89 mm	40.71
0.2625	1.29	0.02197	27.48 mm	28.76
0.3	0	0	0	0

The dynamic response of the retaining wall under elastic system is shown in Fig. 5.34 which indicates that the slip (permanent displacement) after one cycle of ground motion is zero. It can be concluded that in elastic system, after any number of cycles, the residual displacement would be zero.

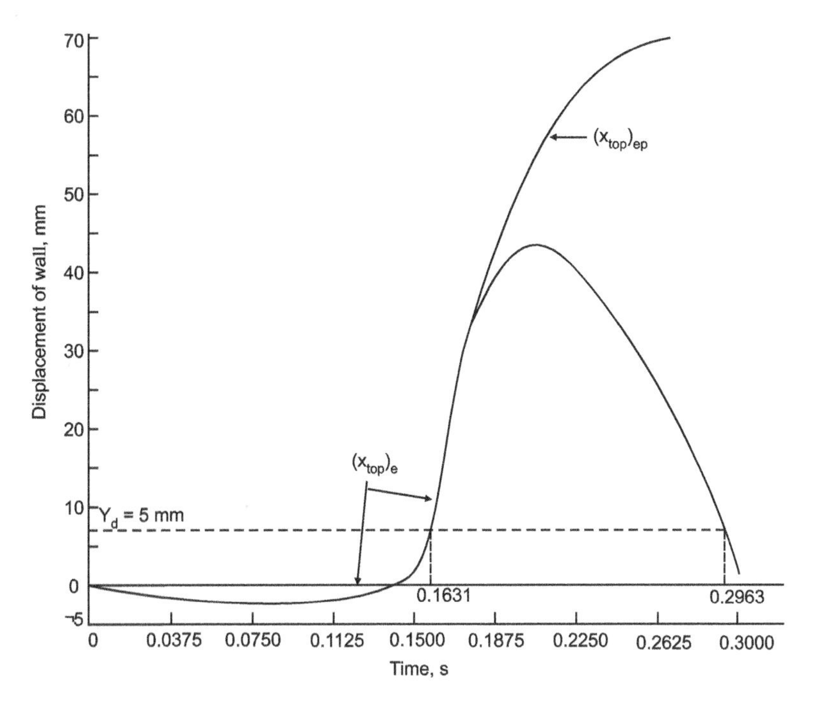

Fig. 5.34. Dynamic response of retaining wall

Check for Plastic Conditions

When the displacement of wall (x_{top}) is greater than yield displacement (y_d), the system would be plastic. Therefore the equations of plastic system should be used. In passive case, yield displacement (y_d) is so large that plastic condition do not arise and elastic system is considered. In active case, to identify the time 't_e' after which plastic conditions exist, a line has been drawn corresponding to Y_d = 5 mm (Fig. 5.34).

It gives,

$$t_e = 0.1631 \text{ sec}$$

$$x_e = -1.817 \times 10^{-3} \sin(20.94\, t_e) = 4.9116 \times 10^{-4} \text{ m}$$

$$\theta_e = -2.197 \times 10^{-2} \sin(20.94\, t_e) = 5.9388 \times 10^{-3} \text{ rad}$$

Based on the analysis,

$$Z_y = x_e = 4.9116 \times 10^{-4} \text{ m}$$

$$\theta_y = \theta_e = 5.9388 \times 10^{-3} \text{ rad}$$

$$x_e = 0.3663 \text{ m/s}$$

$$\theta_e = 0.4429 \text{ rad/sec}$$

$$C_1 = 0.3663 - (117.51 \times 5.9388 \times 10^{-3} - 510.94$$

$$\times 4.9116 \times 10^{-4})\, 0.1631 + \frac{2.45\cos(20.94 \times 0.1631)}{20.94}$$

$$= -0.1489$$

$$C_2 = 4.9116 \times 10^{-4} - 0.3663 \times 0.1631 + (117.51$$

$$\times\ 5.9388 \times 10^{-3} - 510.94 \times 4.9116 \times 10^{-4}) \times \frac{0.1631^2}{2}$$

$$+2.45 \frac{\sin(20.94 \times 0.1631)}{20.94^2} - 0.1631$$

$$\times 2.45 \frac{\cos(20.94 \times 0.1631)}{20.94} = -0.0173$$

$$C_3 = 0.4429 - \left[\frac{117.51}{0.8556^2} \times 4.9116 \times 10^{-4}\right.$$

$$\left. -451.76 \times 5.9388 \times 10^{-3}\right] \times 0.1631 = 0.8676$$

$$C_4 = 5.9388 \times 10^{-3} + (0.4247) \times \frac{0.1631}{2} - 0.4429 \times 0.1631$$

$$= -0.1009$$

Hence the governing equations for displacement become,

$$x = 0.2234\, t^2 - 5.587 \times 10^{-3} \sin(20.94\, t) - 0.1489\, t - 0.0173$$

$$\theta = -1.302\, t^2 + 0.8676\, t - 0.1009$$

From time 0.1631 s to 0.2963 s, the computations of displacements are given in Table 5.8.

Table 5.8. Values of Displacements in Active Condition (Plastic State)

Time t (s)	*Translational displacement x (mm)*	*Rotation (rad)*	*Disp. of wall at top due to rotation x (mm)*	*Total disp. at top x_{top} (mm)*
0.1531	0.47	5.97026×10^{-3}	10.57	11.04
0.1875	1.18	16.0015×10^{-3}	28.32	29.50
0.2250	0.69	8.396×10^{-3}	50.26	51.01
0.2625	–2.44	37.129×10^{-3}	65.72	63.28
0.2963	–6.77	41.8624×10^{-3}	74.10	67.33

From $t = 0.2963$ s to $t = 0.3$ s, the displacements are computed using the expressions obtained by solving the equations of motion under elastic condition taking boundary conditions satisfying the previously computed values $t = 0.2963$ s.

The complete solution can be expressed as:

$$x_e = A_1^1 \sin \omega t + A_2^1 \cos \omega t + A_3^1 \sin \omega t + A_4^1 \cos \omega t$$

$$\theta_e = B_1^1 \sin \omega t + B_2^1 \cos \omega t + B_3^1 \sin \omega t + B_4^1 \cos \omega t$$

Superscripts of A and B indicate the mode of vibration. Therefore constants A_1, A_2, B_1 and B_2 correspond to the mode when system is vibrating with w_{n1}, and A_3, A_4, B_3 and B_4 for the second mode.

$$x_e = A_1 \sin \omega t + A_2 \cos \omega t + A_3 \sin \omega t + A_4 \cos \omega t$$

$$\theta_e = \frac{A_1}{m} \sin \omega t + \frac{A_2}{m} \cos \omega t + \frac{A_3}{m} \sin \omega t + \frac{A_4}{m} \cos \omega t$$

Putting boundary conditions, we get

$$-0.0067669 = -0.0786\,A_1 - 0.9969\,A_2 + 0.0786\,A_3 + 0.9969\,A_4$$

$$0.0418624 = 0.0763\,A_1 - 0.9679\,A_2 - 0.1139\,A_3 + 1.445\,A_4$$

$$-0.133143 = 20.875\,A_1 + 1.6459\,A_2 + 20.075\,A_3 + 1.6459\,A_4$$

$$0.09603 = -20.267\,A_1 + 1.5979\,A_2 + 29.094\,A_3 + 2.385\,A_4$$

$$A_1 = -0.00376$$

$$A_2 = -0.02171$$

$$A_3 = -0.00187$$

$$A_4 = 0.01448$$

Therefore for range of t_p from 0.2963 to 0.3 s the equations of displacements will be:

$$x_e = -0.00376 \sin \omega t - 0.02171 \cos \omega t - 0.00187 \sin \omega t + 0.01448 \cos \omega t$$

$$\theta_e = 0.00365 \sin \omega t + 0.02107 \cos \omega t - 0.002710 \sin \omega t + 0.02099 \cos \omega t$$

Values of displacements in elastic condition from time 0.2963 s to 0.3 s are given in Table 5.9.

Table 5.9. Values of Displacement in Elastic State

Time t (s)	*Translational displacement x (mm)*	*Rotation* θ*(rad)*	*Disp. of wall at top due to rotation* x_θ *(mm)*	*Total disp. at top* x_{top} *(mm)*
0.2963	– 6.77	41.8624×10^{-3}	74.10	67.33
0.2975	– 6.85	41.916×10^{-3}	74.191	67.34
0.298	– 6.98	41.98359×10^{-3}	74.3109	67.33
0.299	– 7.10	42.0320×10^{-3}	74.39670	67.29
0.3	– 7.23	42.062×10^{-3}	74.4498	67.21

Hence it is found that after one cycle, the displacement is equal to 67.21 mm and it can be called as slip. The total displacement after N cycles is equal to N times the slip. The final translational displacement and rotation of the retaining wall are therefore known. The displacement curve in plastic state is also shown in Fig. 5.34.

References

Biggs, J.M. (1963), "Introduction to structural dynamics", McGraw Hill Book Co., New York.

Coulomb, C.A. (176), "Essai sur une application des regles des maximis et minimis a quelque problems de statique relalifs a l'architecture", Mem. acad. roy. press. disversavants, vol 7, Paris.

Culmann, K. (1866), "Die graphische statik", Mayer and Zeller, Zurich.

IS 1893-2016, "Criteria of earthquake resistant design of structures", I.S.I., New Delhi.

Kapila, I.P. (1962), "Earthquake resistant design of retaining wall", Proc. Symposium in Earthquake Engineering, University of Roorkee, Roorkee.

Franklin, A.G. and Chang, F.K. (1977), Permanent displacements of earth embankments by New mark sliding block analysis, Report 5, U.S. Army Corps of Engineers Waterways Experiment station, Vicksburg, Mississippi.

Lai, C.S. (1979), "Behaviour of retaining walls under seismic loading", M.E. Report, University of Canterbury, New Zealand.

Mononobe, N. (1929), "Earthquake-proof construction of masonry dam", Proceedings. World Engineering Congress, vol. 9, p. 275.

Nadim, F. and Whitman, R.V. (1983), "Seismically induced movement of retaining walls", Jour. of Geot. Engg. Divn. , ASCE, Vol. 109, No. 7, pp. 915-931.

Nandakumaran, P. (1973), "Behaviour of retaining walls under dynamic loads", Ph.D. Thesis, University of Roorkee, Roorke.

Newmark, N.M. (1965), "Effect of earthquakes on dams and embankments", Geotechnique, Vol. 15, No. 2, pp. 129-160.

Ohde, S. (1926), "General theory of earth pressures", Journal, Japanese Society of Civil Engineers, Tokyo, Japan, Vol. 12, No. 1.

Okabe, S. (1924), "General theory on earth pressure and seismic stability of retaining walls and dams", J. Japanese Society of Civil Engrs., Vol. 6.

Prakash, S., and Saran, S. (1966), "Static and dynamic earth pressures behind retaining walls'" Proc. 3rd Symposium an Earthquake Engineering, University of Roorkee, Vol. 1, pp. 277-288.

Prakash, S., Puri, V.K. and Khandoker J. U. (1981), "Rocking displacements of retaining walls during earthquakes", Int. Conf. on Recent Advances in Geotechanical Earthquake Engineering and Soil Dynamics, Vol. 3, St. Louis, U.S.A.

Prakash, S. (1981), "Analysis of rigid retaining walls during earthquakes", Int. Conf. on Recent Advances in Geotech. Earthquake Engg. and Soil Dynamics, Vol. 3, St. Louis U.S.A.

Reddy, R.K., Saran, S. and Viladkar, M.N. (1985), "Prediction of displacements of retaining walls under dynamic conditions", Bull. of Indian Soc. Earth. Tech., Paper No. 239, Vol. 22, No. 3.

Richard, R. Jr. and Elms, D.G. (1979), "Seismic behaviour of gravity retaining walls", Journ. Geotech. Engg. Divn., ASCE, Vol. 105, No. GT4, pp. 449-464.

Saran, S. and Prakash, S. (1968), "Dimensionless parameters for static and dynamic earth pressures behind retaining walls", Jour. Indian National Society of Soil Mech. and Found. Engg., July pp. 295-310.

Seed, H.B. and Whitman, R.V. (1970), "Design of earth retaining structures for dynamic loads", ASCE Speciality conference on Lateral Stresses in Ground and Design of Earth Retaining Structures, pp. 103-147, Ithaca, New York.

Zarrabi, K. (1979), "Sliding of gravity retaining wall during earthquakes considering vertical acceleration and changing inclination of failure surface", M.S. Thesis, MIT, USA.

Practice Problems

5.1 Explain with neat sketches the following:

(a) Mononobe-Okabe's approach, and

(b) Modified Culmann's graphical construction for getting dynamic active earth pressure.

5.2 How is the effect of partly submerged backfill considered in computing dynamic earth pressure?

5.3 Explain the salient features of the following:

(a) Richard-Elms model

(b) Nadim-Whitman model

(c) Reddy-Saran-Viladkar model

for getting displacement of rigid retaining wall.

5.4 A vertical retaining wall is 8 m high and retains non-cohesive backfill with γ = 18 kN/m^3, $\phi = 30°$, $\delta = 20°$. The backfill is inclined to the horizontal by 15°. The wall is located in a seismic area where the design seismic coefficients are

$$\alpha_h = 0.10; \ \alpha_v = 0.05$$

Compute the static and dynamic earth pressure on the wall using both modified Coulomb's approach, and Culmann's graphical construction.

5.5 If the retaining wall (Problem 5.4) is to incline at 10° with the vertical, would you recommend its inclination towards or away from the fill. Justify your answer fully.

5.6 The backfill of retaining wall (Problem 5.4) is carrying a surcharge of 50 kN/m^2. Estimate the increase in static and dynamic earth pressures.

5.7 The backfill of retaining wall (Problem 5.4) is submerged upto 4.0 m from the base of wall . Estimate the total pressure on wall both in static and dynamic cases.

5.8 Compute the displacement of the wall (Problem 5.4) for the following condition:

Period of wall = 0.25 s

Z_y = 5.0 mm

Period of ground motion = 0.40 s

Equivalent number of cycles in an earthquake of magnitude 7.0 will not exceed 15.

CHAPTER

6

Dynamic Bearing Capacity of Shallow Foundations

6.1 General

Foundations may be subjected to dynamic loads due to earthquakes, bomb blasts and operations of machines. The dynamic loads due to nuclear blasts are mainly vertical. Horizontal dynamic loads on foundations are mostly due to earthquakes. Basically there are two types of approaches namely (i) pseudo-static analysis and (ii) dynamic analysis for getting the solution. In this chapter, pseudo-static analysis is first presented and it is followed by dynamic analysis. Design of foundations of different types of machines have been given in detail in chapters 8 to 10.

6.2 Pseudo-Static Analysis

Pseudo-static analysis is more commonly used for designing foundations subjected to earthquake forces. Adopting appropriate values of horizontal and vertical seismic coefficients, equivalent seismic forces can be conveniently evaluated. These forces in combination of static forces make the foundation subjected to eccentric inclined load. In Secs 6.3 and 6.4, the procedure of determining bearing capacity, settlement, tilt and horizontal displacement of shallow foundations subjected to eccentric-inclined loads have been presented. It is preceded by brief description of fundamental concepts involved in bearing capacity analysis.

6.3 Bearing Capacity of Footings

6.3.1 Modes of Shear Failure

The maximum load per unit area that can be imposed on a footing without causing rupture of soil is its bearing capacity (some times termed critical or ultimate bearing capacity). It is usually denoted by q_u. This load may be obtained by carrying out a load test on the footing which will give a curve between average load per unit area and settlement of the footing. Based on pressure-settlement characteristics of a footing and pattern of shearing zones, three modes of shear failure have been identified as (i) general shear failure, (ii) punching shear failure and (iii) local shear failure (Caquot, 1934; Terzaghi, 1943; DeBeer and Vesic, 1958; Vesic, 1973).

In general shear failure, well defined slip lines extend from the edge of the footing to the adjacent ground. Abrupt failure is indicated by the pressure-settlement curve (Fig. 6.1a). Usually in this type, failure is sudden and catastrophic and bulging of adjacent ground occurs. This type of failure occurs in soils having brittle type stress-strain behaviour (e.g. dense sand and stiff clays).

In punching shear failure, there is vertical shear around the footing perimeter and compression of soil immediately under the footing, with soil on the sides of the footing remaining practically uninvolved. The pressure-settlement curve indicates a continuous increase in settlement with increasing load (Fig. 6.1b)

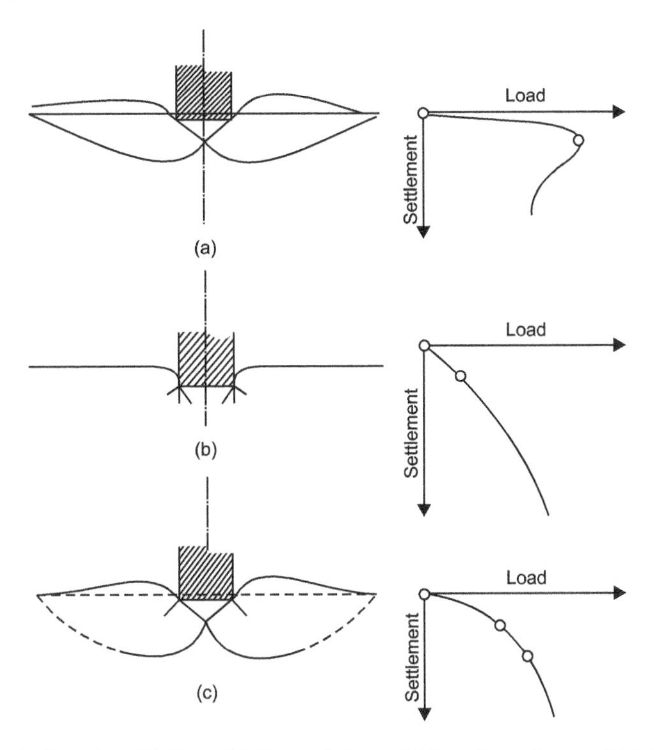

Fig. 6.1. Typical modes of failure: (a) General shear, (b) Punching shear and (c) Local shear

The local shear failure is an intermediate failure mode and has some of the characteristics of both the general shear and punching shear failure modes. Well defined slip lines immediately below the footing extend only a short distance into the soil mass. The pressure-settlement curve does not indicate the bearing capacity clearly (Fig. 6.1 *c*). This type of failure occurs in soils having plastic stress-strain characteristics (e.g. loose sand and soft clay).

In Fig. 6.2, types of failure modes that can be expected for a footing in a particular type of sand is illustrated (Vesic, 1973). This figure indicates that the type of failure depends on the relative density and depth-width ratio (D_f/B) of the footing. There is a critical value of (D_f/B) ratio below which only punching shear failure occurs.

The criteria given in Table 6.1 may also be followed for identification of type of failure.

Table 6.1. Identification of Type of Failure

Type of failure	*Relative density D_r (%)*	*ϕ (Deg.)*	*Void ratio e*
1. General shear failure	≥ 70	$\geq 36°$	≤ 0.55
2. Local shear failure or punching shear failure	≤ 20	$\leq 29°$	≥ 0.75

6.3.2 Generalized Bearing Capacity Equation

In the design of foundation usually net bearing capacity is computed and used. It is defined as the maximum net intensity of loading at the base of the foundation that the soil can support before failing in shear. It is denoted by q_{nu}. Therefore

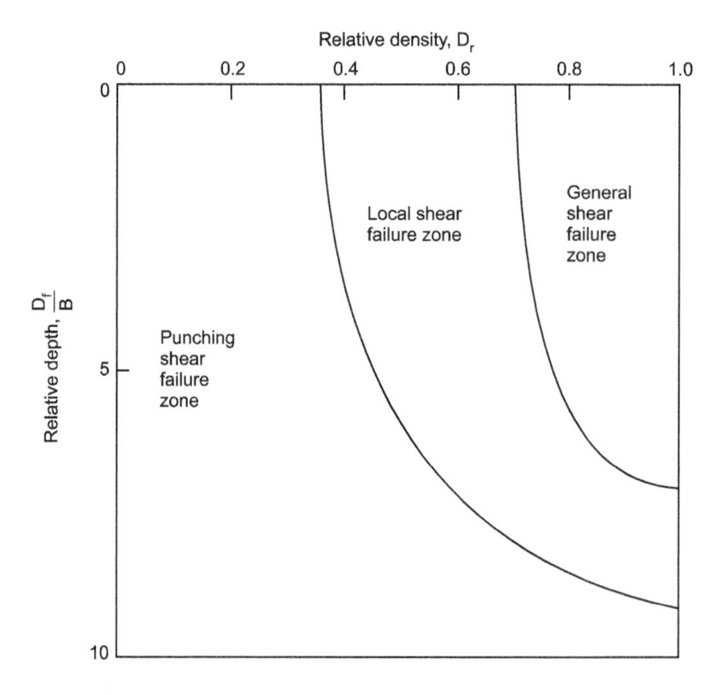

Fig. 6.2. Region for three different modes of failure

$$q_{nu} = q_u - \gamma_1 D_f \tag{6.1}$$

where q_u = Ultimate bearing capacity

γ_1 = Unit weight of soil above the base of the footing

The equation of net bearing capacity, q_{nu}, developed for strip footing considering general shear failure (Terzaghi, 1943; Meyerhof, 1951) is extended to consider variations from the basic assumptions by applying modification factors that account for the effect of each variation (Hansen, 1970). It may be written as:

$$q_{nu} = c N_c \cdot S_c \cdot d_c \cdot i_c \cdot b_c + \gamma_1 \cdot D_f \cdot (N_q - 1) \cdot S_q \cdot d_q \cdot i_q \cdot b_q \cdot r_w + \frac{1}{2} \cdot \gamma_2 \cdot B \cdot N_\gamma \cdot S_\gamma \cdot d_\gamma \cdot i_\gamma \cdot b_\gamma \cdot r'_\omega \tag{6.2}$$

where γ_2 = Unit weight of soil below the base of the footing

c = Undrained cohesion of soil

B = Width of footing

D_f = Depth of foundation below ground surface

N_c, N_q, N_γ = Bearing capacity factors

S_c, S_q, S_γ = Shape factors for square, rectangular and circular foundations

d_c, d_q, d_γ = Depth factors

i_c, i_q, i_γ = Inclination factors

b_c, b_q, b_γ = Ground inclination factors

r_w, r'_w = Ground water table factors

6.3.2.1 Bearing Capacity Factors

N_c, N_q and N_γ are non-dimensional factors which depend on angle of shearing resistance of soil (Terzaghi, 1943; Terzaghi and Peck, 1967). Their values may be obtained from Table 6.2.

Table 6.2. Bearing Capacity Factors

ϕ	N_c	N_q	N_γ
0	5.14	1.00	0.00
5	6.49	1.57	0.45
10	8.35	2.47	1.22
15	10.98	3.94	2.65
20	14.83	6.40	5.39
25	20.72	10.66	10.88
30	30.14	18.40	22.40
35	46.12	33.30	48.03
40	75.31	64.20	109.40
45	138.88	134.88	271.76
50	266.89	319.07	762.89

6.3.2.2 Shape Factors

Approximate values of shape factors which are sufficiently accurate for most practical purposes are given in Table 6.3.

Table 6.3. Shape Factors

S. No.	*Shape of footing*	S_c	S_q	S_γ
(i)	Continuous strip	1.0	1.0	1.0
(ii)	Rectangle	$1 + 0.2\ B/L$	$1 + 0.2\ B/L$	$1 - 0.4\ B/L$
(iii)	Square	1.3	1.2	0.8
(iv)	Circle (B = diameter)	1.3	1.2	0.6

6.3.2.3 Depth Factors

The bearing capacity factors given in Table 6.2 do not consider the shearing resistance of the failure plane passing through the soil zone above the level of the foundation base. If this upper soil zone possesses significant shearing strength, the ultimte value of bearing capacity would be increased (Meyerhof, 1951). For this case, depth factors are applied, whereby

$$d_c = 1 + 0.4\frac{D_f}{B} \tag{6.3}$$

$$d_q = 1 + 2\tan\phi(1 - \sin\phi)^2 \cdot \frac{D_f}{B} \tag{6.4}$$

$$d_\gamma = 1 \tag{6.5}$$

The use of depth factors is conditional upon the soil above foundation level being not significantly inferior in shear strength characteristics to that below this level.

6.3.2.4 Factors for Eccentric-inclined Loads

The effect of eccentricity can be conveniently and conservatively considered as follows:

One way eccentricity (Fig. 6.3a): If the load has an eccentricity e, with respect to the centroid of the foundation in only one direction, then the dimension of the footing in the direction of eccentricity shall be reduced by a length equal to $2e$. The modified dimension shall be used in the bearing capacity equation and in determining the effective area of the footing in resisting the load.

Two-way eccentricity (Fig. 66.3 *b*): If the load has double eccentricity (e_L and e_B) with respect to the centroid of the footing then the effective dimensions of the footing to be used in determining the bearing capacity as well as in computing the effective area of the footing in resisting the load shall be determined as given below:

$$L' = L - 2\,e_L \tag{6.6}$$

$$B' = B - 2\,e_B \tag{6.7}$$

$$A' = L' \times B' \tag{6.8}$$

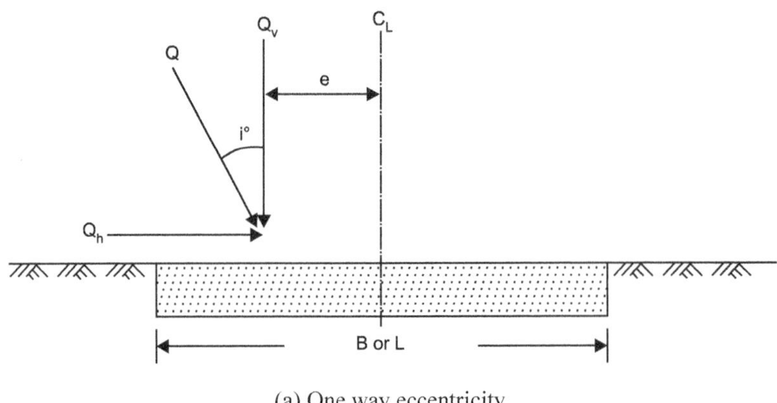

(a) One way eccentricity

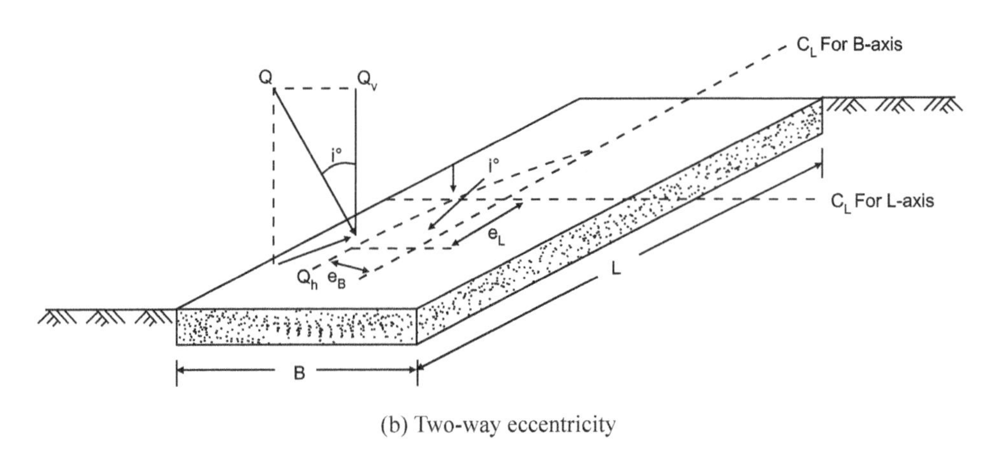

(b) Two-way eccentricity

Fig. 6.3. Eccentrically-obliquely loaded footings

In computing the shape and depth factors for eccentrically-obliquely loaded footings, effective width (B') and effective length (L') will be used in place of total width (B) and total length (L).

For a design, eccentricity should be limited to one-sixth of the foundation dimension to prevent the condition of uplift occurring under part of the foundation.

Inclination factors: Following inclination factors may be adopted in design:

$$i_\gamma = \left[1 - \frac{Q_h}{Q + BLc \cot \phi}\right]^{m+1} \tag{6.9}$$

$$i_q = \left[1 - \frac{Q_h}{Q + BLc \cot \phi}\right]^{m} \tag{6.10}$$

$$i_c = i_q - \left[\frac{1 - i_q}{N_c \tan\phi}\right] \quad \text{(when } \phi > 0^\circ\text{)} \tag{6.11}$$

$$= 1 - \frac{mQ_h}{cN_c BL} \quad \text{(when } \phi = 0^\circ\text{)} \tag{6.12}$$

where Q_h is the horizontal component of the load Q acting on the foundation at inclination i with vertical. Values of m are taken as given below:

(i) If the angle of inclination i is in the plane of L-axis

$$m = \left(2 + \frac{L}{B}\right)\left(1 + \frac{L}{B}\right) \tag{6.13}$$

(ii) If the angle of inclination i is in the plane of the B-axis

$$m = \left(2 + \frac{B}{L}\right)\left(1 + \frac{B}{L}\right) \tag{6.14}$$

As per IS 6403 : 1981, the inclination factors are given by

$$i_c = i_q = \left(1 - \frac{i}{90}\right)^2 \tag{6.15}$$

$$i_\gamma = \left(1 - \frac{i}{\phi}\right)^2 \tag{6.16}$$

i and ϕ are in degrees in Eqs. (6.15) and (6.16).

For more accurate estimation of bearing capacity of a eccentrically-obliquely loaded footing, bearing capacity factors as shown in Figs. 6.4 to 6.6 developed by Saran and Agarwal (1991) may be used. These factors have been obtained by carrying out a theoretical analysis based on limit equilibrium and limit analysis approaches.

As evident from these figures, bearing capacity factors ($N_{\gamma ei}$, N_{qei} and N_{cei}) depend on ϕ, i and e/B. Values of these bearing capacity factors are substituted in Eq. (6.2) in place of N_c, N_q and N_γ for getting the bearing capacity of eccentrically-obliquely loaded footing. If use of these bearing capacity factors is made, then inclination factors, and reduced dimensions of B' and L' for accounting the effect of eccentricity and inclination are not used. Therefore the modified bearing capacity equation will be as given below:

$$q_{nu} = cN_{cei} \cdot S_c \cdot d_c \cdot \gamma_1 D_f (N_{qei} - 1) S_q \cdot d_q \cdot r_w + \frac{1}{2} \cdot \gamma_2 \cdot B \cdot N_{\gamma ei} \cdot S_\gamma \cdot d_\gamma \cdot r'_w \tag{6.17}$$

6.3.2.5 Base Inclination Factors

If the base of foundation is inclined from the horizontal and an applied load acts normal to the base (Fig. 6.7), the pattern of rupture surface beneath the foundation will be different from the pattern that develop beneath the level footing carrying a vertical load (Meyerhof, 1953). For this condition base inclination factors as given below may be used:

$$b_c = b_q - \frac{(1 - b_q)}{N_c \tan\theta} \quad \text{for } \phi > 0^\circ \tag{6.18}$$

$$= 1 - 0.0067\,\alpha \quad \text{for } \phi = 0^\circ \tag{6.19}$$

$$b_q = b_\gamma = \left(1 - \frac{\alpha}{57.3} \tan\phi\right)^2 \tag{6.20}$$

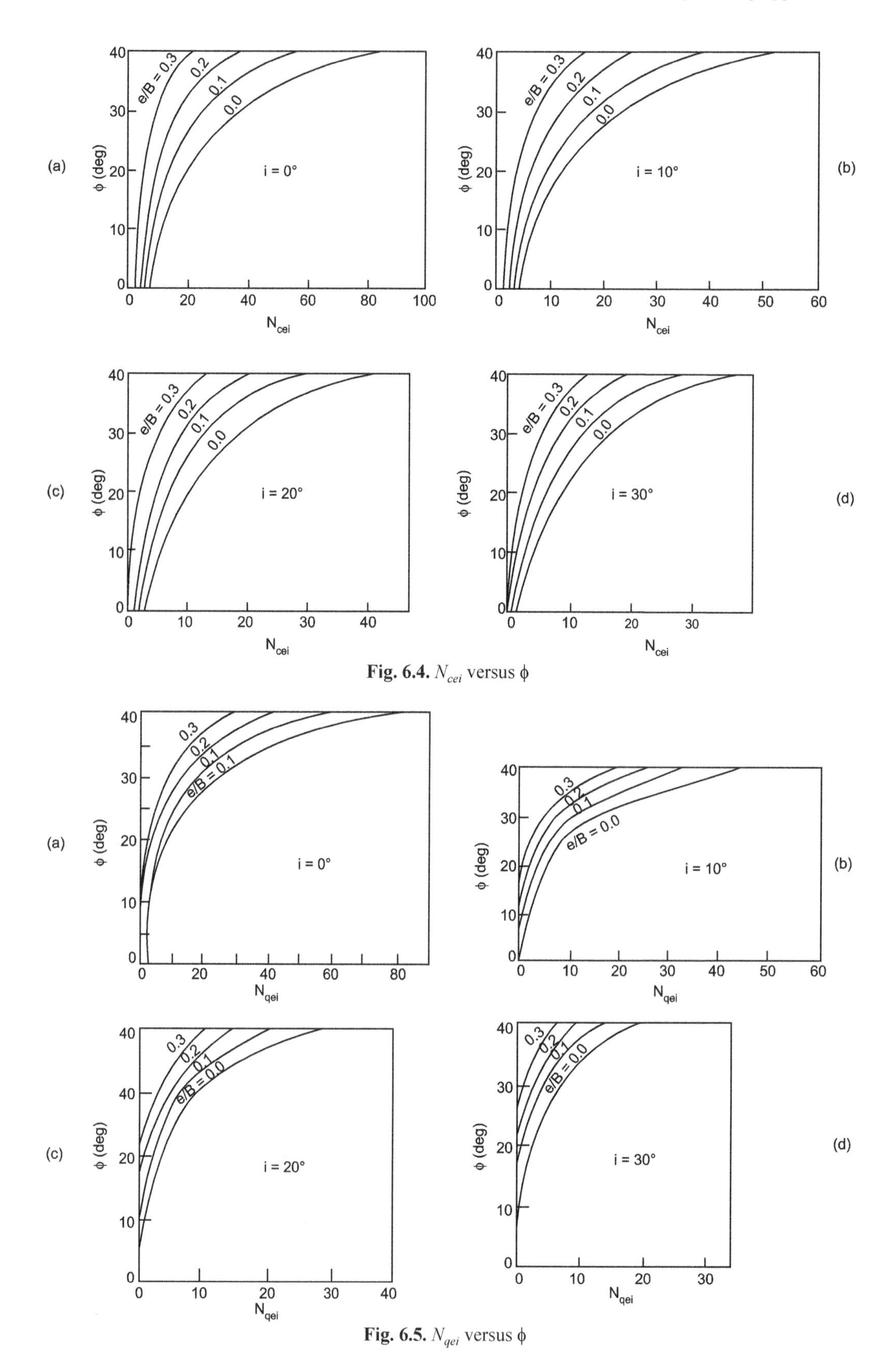

Fig. 6.4. N_{cei} versus ϕ

Fig. 6.5. N_{qei} versus ϕ

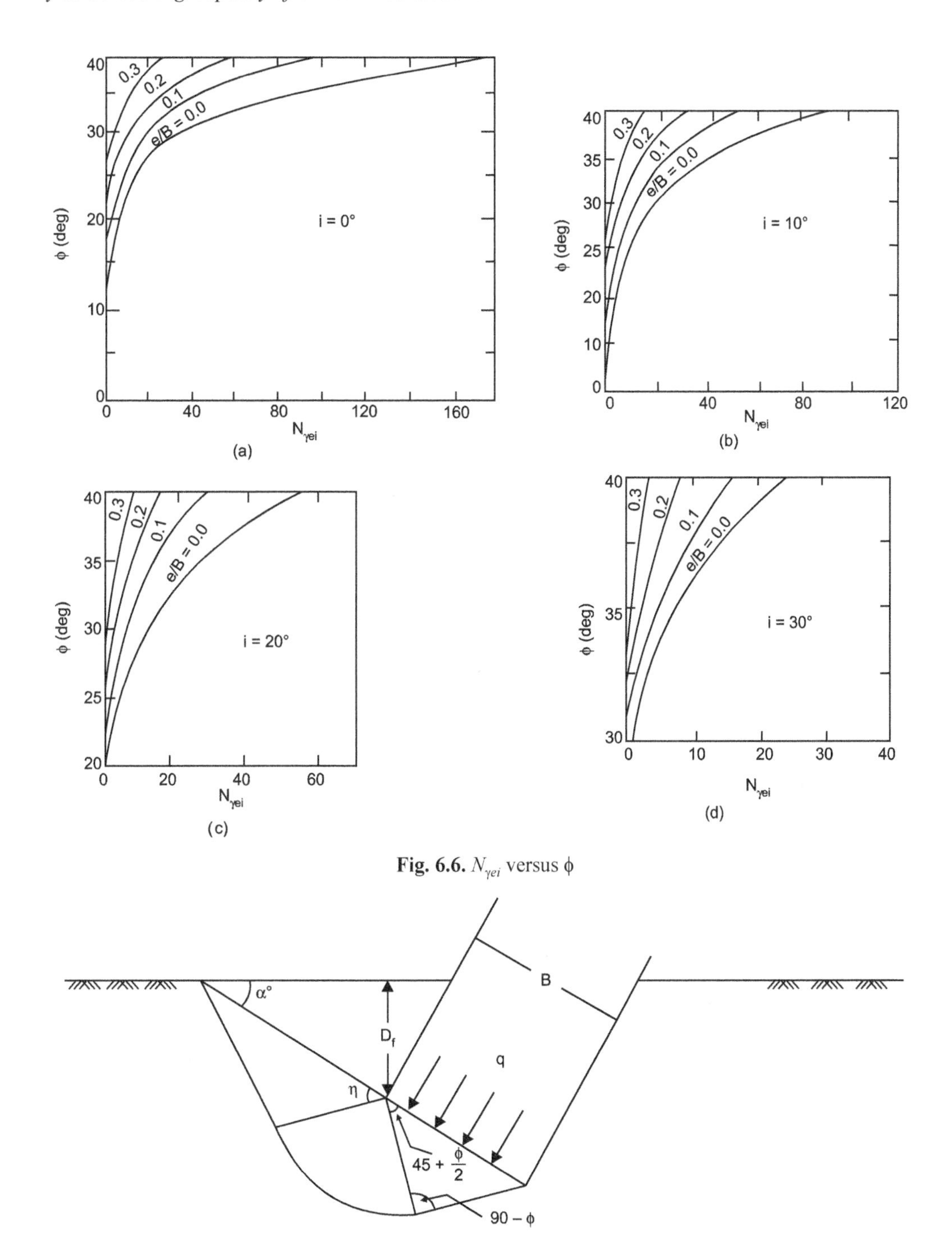

Fig. 6.6. $N_{\gamma ei}$ versus ϕ

Fig. 6.7. Inclined footing

where α represents the angle of the base inclination in degree with horizontal.

6.3.2.6 Water Table Factors

Correction factors r_w and r'_w may be computed using following Eqs. (6.21) and 6.22) or (Fig. 6.8).

$$r_w = 1.0 - 0.5\frac{d_a}{D_f} \qquad (\text{For } d_a \le D_f) \tag{6.21}$$

$$r'_w = 0.5 + 0.5\frac{d_b}{B} \qquad (\text{For } d_b \leq B) \tag{6.22}$$

where d_a and d_b represent the position of water table with respect to the base of the footing as shown in Fig. 6.8. For the position of water level below the base of footing, $d_a = 0$ *i.e.* $r_w = 1$; and for the position of water level at depth more than B, $d_b = B$ *i.e.* $r'_w = 1$.

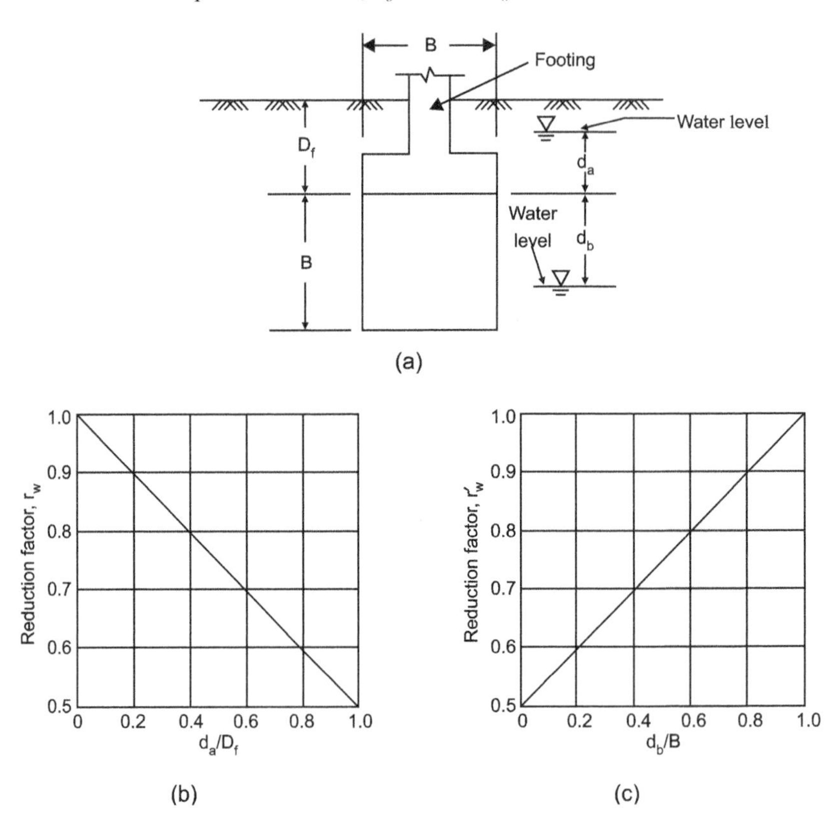

Fig. 6.8. Correction factors for position of water table

6.3.3 Local and Punching Shear Failure

The assumption that the soil behaves as a rigid material is satisfied for the case of general shear but is not appropriate for punching and local shear. Comparison of the relative pressure-settlement curves (Fig. 6.1) indicates that, for punching and local shear failure cases, the ultimate pressure is less and the settlement is greater than for the condition of general shear failure. For design purposes, the general shear, local shear and punching shear failures can be identified as per the criterion given in Table 6.1.

Terzaghi (1943) proposed empirical adjustments to shear strength parameters c and ϕ to cover the case of local and punching shear failure. Shear strength parameters c_m and ϕ_m should be used in the bearing capacity equation and the bearing capacity factors are obtained on the basis of ϕ_m instead of ϕ where

$$c_m = \frac{2}{3}c \tag{6.23a}$$

$$\phi_m = \tan^{-1}\left(\frac{2}{3}\tan\phi\right) \tag{6.23b}$$

If the failure lies between general shear and local shear failure, then linear interpolation is done to evaluate the value of bearing capacity factors. For example, value of N_γ for $\phi = 34°$ will be

$$(N_\gamma)_{\phi=34°} = (N_{\gamma m})_{\phi=29°} + \frac{(N_\gamma)_{\phi=36°} - (N_{\gamma m})_{\phi=29°}}{36° - 29°} \times (36° - 34°) \tag{6.24}$$

where $(N_{\gamma m})_{\phi=29°}$ is value of N_γ factor for $\phi = 29°$ considering local shear failure condition. Therefore its value will be obtained using Table 6.2 for $\phi_m = \tan^{-1}[(2/3)\tan 29°] = 20.29°$.

6.3.4 Factor of Safety

The net bearing capacity of the soil is divided by a safety factor to obtain the net safe bearing capacity. It is denoted by q_{nF}.

A factor of safety is used as a safeguard against

(i) natural variations in shear strength against,
(ii) assumptions made in theoretical methods,
(iii) inaccuracies of empirical methods and
(iv) excessive settlement of footings near shear failure.

A factor of safety of 2.5 to 3.0 is generally used to cover the variation or uncertainties listed above. Therefore

$$q_{nF} = \frac{q_{nu}}{F} \tag{6.25}$$

6.4 Settlement, Tilt and Horizontal Displacement

An eccentrically-obliquely loaded rigid footing settles as shown in Fig. 6.9 in which S_e and S_m represent respectively the settlement of the point under load and edge of the footing. If 't' is the footing, then S_m is given by:

$$S_m = S_e + (B/2 - e)\sin t \tag{6.26}$$

In Fig. 6.9, H_D represents the horizontal displacement of the footing.

Agarwal (1986) carried out model tests on eccentrically-obliquely loaded footings resting on sand. Footings of different widths and shapes were used. In each test, for a pressure increment observations were taken to record S_e, t and H_D. Effect of relative density of sand was also studied. In addition to these tests pressure-settlement and pressure-tilt characteristics of eccentrically-obliquely loaded footings resting on clay and sand beds were also obtained using non-linear constitutive laws of soils (Saran and Agarwal, 1989). From the model test data and results of analysis based on

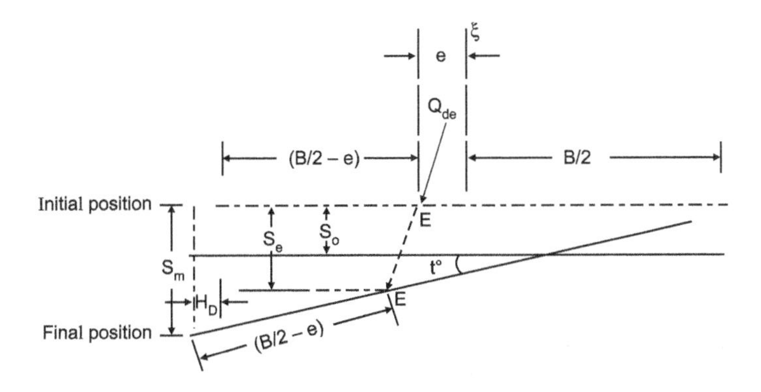

Fig. 6.9. Settlement, tilt and horizontal displacement of eccentrically-obliquely loaded footing

constitutive laws, plots of S_e/S_0 versus e/B and S_m/S_0 versus e/B were prepared for different load inclinations (Figs. 6.10 and 6.11). S_0 represents the settlement of the footing subjected to central vertical load (*i.e.* $e/B = 0 = i$) and obtained corresponding to the pressure intensity giving the same factor of safety at which S_e and S_m values are taken.

These plots were found independent to the type of soil, factor of safety, size and shape of footing. The average relationship can be represented by the following expressions:

$$\frac{S_e}{S_0} = A_0 + A_1\left(\frac{e}{B}\right) + A_2\left(\frac{e}{B}\right)^2 \tag{6.27}$$

$$\frac{S_m}{S_0} = B_0 + B_1\left(\frac{e}{B}\right) \tag{6.28}$$

where

$$A_0 = 1 - 0.56\left(\frac{i}{\phi}\right) - 0.82\left(\frac{i}{\phi}\right)^2 \tag{6.29}$$

$$A_1 = -3.51 + 1.47\left(\frac{i}{\phi}\right) + 5.67\left(\frac{i}{\phi}\right)^2 \tag{6.30}$$

$$A_2 = 4.74 - 1.38\left(\frac{i}{\phi}\right) - 12.45\left(\frac{i}{\phi}\right)^2 \tag{6.31}$$

$$B_0 = 1 - 0.48\left(\frac{i}{\phi}\right) - 0.82\left(\frac{i}{\phi}\right)^2 \tag{6.32}$$

$$B_1 = -1.80 + 0.94\left(\frac{1}{\phi}\right) + 1.63\left(\frac{i}{\phi}\right)^2 \tag{6.33}$$

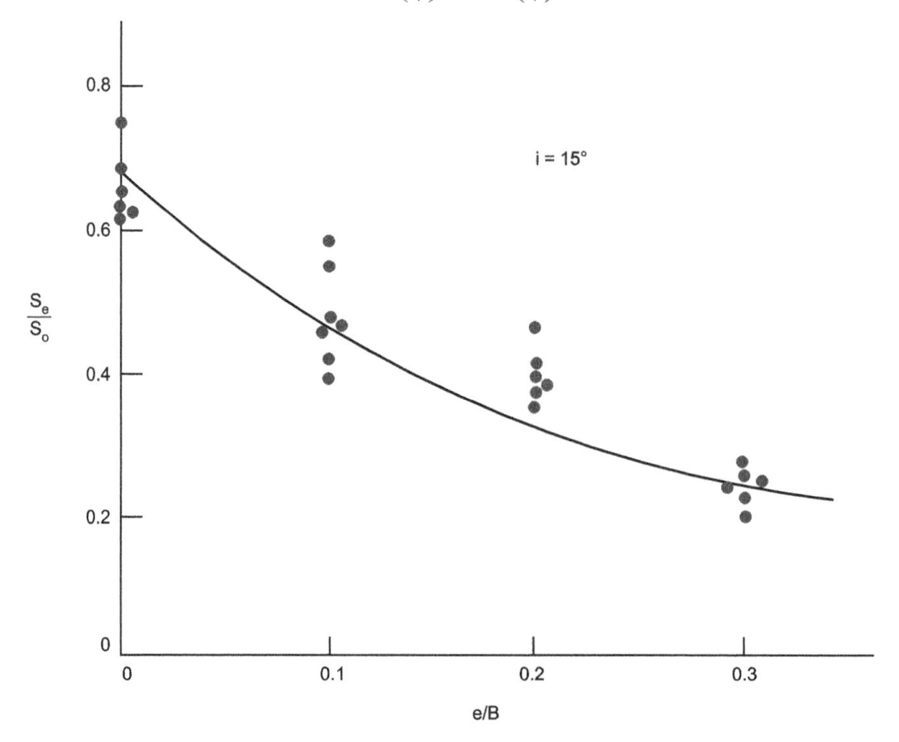

Fig. 6.10. S_e/S_0 versus e/B for $i = 15°$

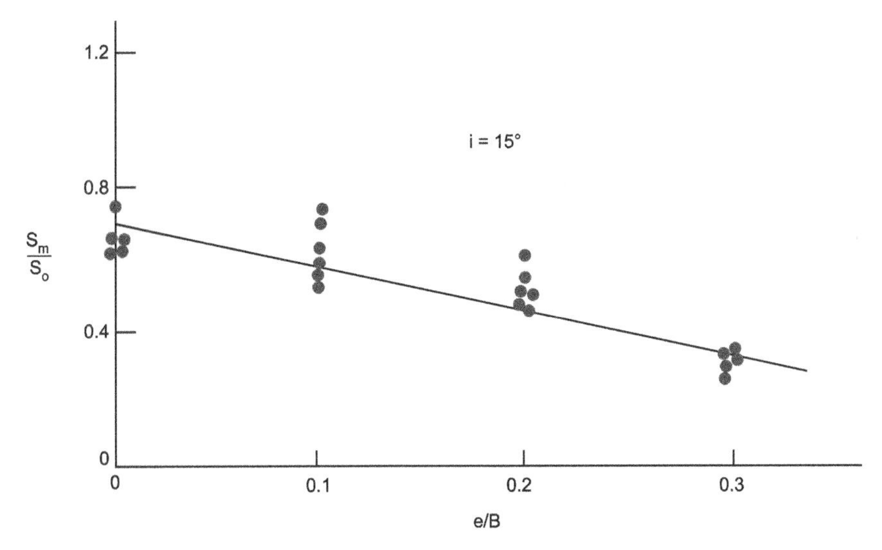

Fig. 6.11. S_m/S_0 versus e/B for $i = 15°$

Values of S_0 can be obtained using the data of plate load test or standard penetration test in cohesionless soils, and consolidation test data in clays in conventional manner.

Similarly a unique correlation was obtained between H_D/B and i/ϕ.

$$\frac{H_D}{B} = 0.121\left(\frac{i}{\phi}\right) - 0.682\left(\frac{i}{\phi}\right)^2 + 1.99\left(\frac{i}{\phi}\right)^3 - 2.01\left(\frac{i}{\phi}\right)^4 \tag{6.34}$$

The above correlation is also found independent to the type of soil, factor of safety, size and shape of the footing. The effect of e/B was found small and the displacement value decreased little with the increase in eccentricity. This effect is neglected, considering the results slightly on the safe side.

6.5 Dynamic Analysis

The dynamic bearing capacity problem attracted attention of the investigators in 1960 when the performance of foundations under transient loads became of concern to the engineering profession (Wallace, 1961; Cunny and Sloan, 1961; Fisher, 1962; Johnson and Ireland, 1963; Mckee and Shenkman, 1962; White, 1964; Chummar, 1965; Triandafilidis, 1965). All analytical approaches are based on the assumption that soil rupture under transient loads occurs along a static rupture surface. In this section the salient features of the analysis developed by Triandafilidis (1965) and Wallace (1961) for transient vertical load; and by Chummar (1965) for transient horizontal load have been presented.

6.5.1 Triandafilidis's Solution

Triandafilidis (1965) has presented a solution for dynamic response of continuous surface footing supporting by saturated cohesive soil ($\phi = 0$ condition) and subjected to vertical transient load. The analysis is based on the following assumptions:

(i) The failure surface of soil is cylinderical for evaluation of bearing capacity under static condition (Fig. 6.12).

(ii) The saturated cohesive soil ($\phi = 0$) behaves as a rigid plastic material (Fig. 6.13).

(iii) The forcing function is assumed to be an exponentially decaying pulse (Fig. 6.14).
(iv) The influence of strain rate on the shear strength is neglected.
(v) The dead weight of the foundation is neglected.

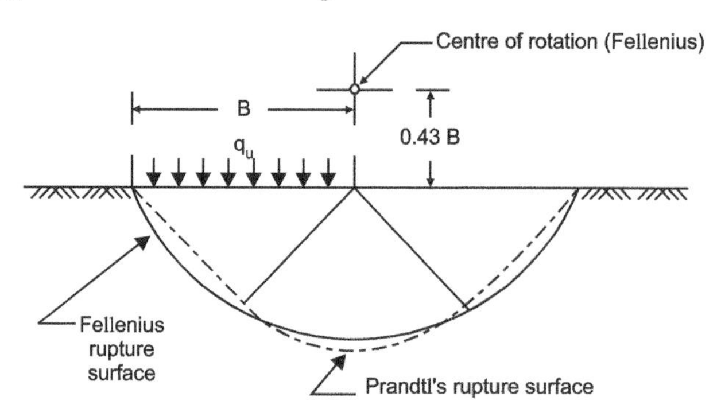

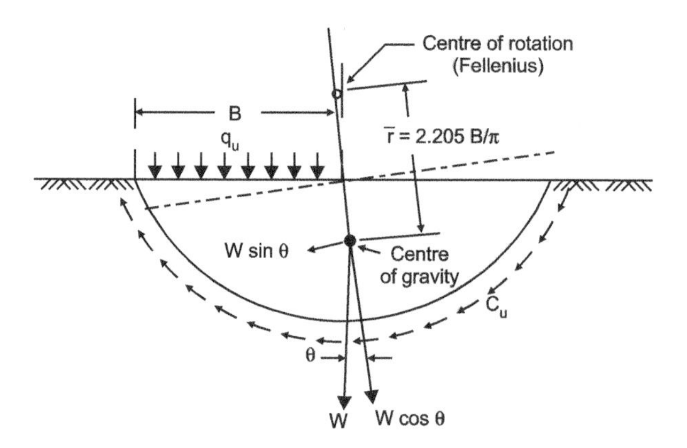

Fig. 6.12. Illustrations of mode of failure and dynamic equilibrium of moving soil mass

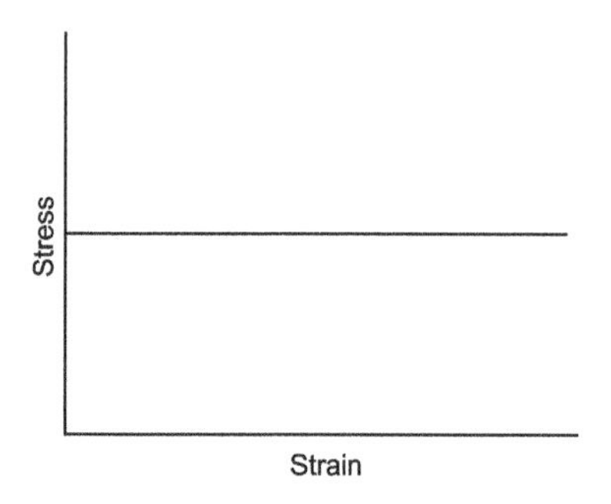

Fig. 6.13. Assumed stress-strain relationship

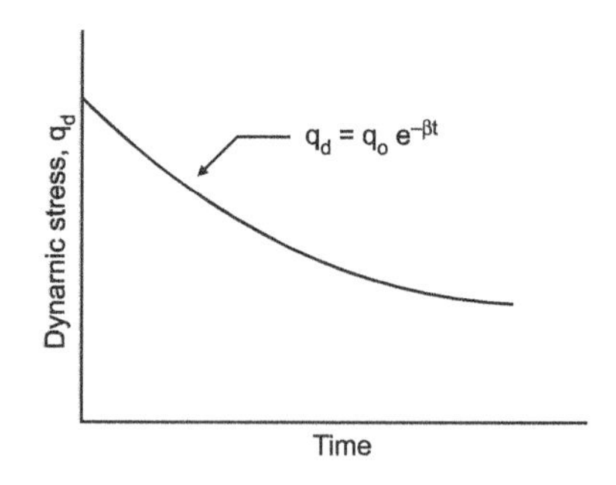

Fig. 6.14. Transient vertical load

Analysis

Let the transient stress pulse be expressed in the form

$$q_d = q_o e^{-\beta t} = \lambda q_u e^{-\beta t} \tag{6.35}$$

where q_d = Stress at time t

β = Decaying function

q_u = Static bearing capacity of continuous footing
q_0 = Instantaneous peak intensity of the stress pulse
λ = Over load factor = $\frac{q_0}{q_u}$

The rupture surface is shown in Fig. 6.12 with centre of rotation at point O located at a height of 0.43 *B* above the ground surface. The equation of motion is written by equating the moment of the disturbing and restoring forces taken about the point O. The only disturbing force is an externally applied dynamic pulse. The restoring forces consist of shearing resistance along the rupture surface, the inertia of the soil mass participating in motion and the resistance caused by the displacement of centre of gravity of soil mass.

Driving moment M_{dp} due to applied dynamic pulse is

$$M_{dp} = \frac{1}{2} q_d B^2 \tag{6.36}$$

where B = Width of footing.

The static bearing capacity of a continuous footing along the failure surface (Fellenius, 1948) is

$$q_u = 5.54\, c_u \tag{6.37a}$$

where c_u = Undrained shear strength.

Resisting moment M_{rs} due to shear strength is

$$M_{rs} = \frac{1}{2} q_u B^2 \tag{6.37b}$$

An applied pulse imparts an acceleration to the soil mass. The resisting moment M_{ri} due to the rigid body motion of the failed soil mass is

$$M_{ri} = J_0 \ddot{\theta} \tag{6.38}$$

where J_o = Polar mass moment of inertia

$$J_o = \frac{WB^2}{1.36\, g} \tag{6.39}$$

W = Weight of the cylinderical soil mass

$$= 0.31\, \gamma\, \pi B^2 \tag{6.40}$$

γ = Unit weight of soil

Therefore

$$M_{ri} = \frac{WB^2}{1.36\, g} \ddot{\theta} \tag{6.41}$$

The displaced position of the soil mass generates a restoring moment M_{rw}, which may be expressed as

$$M_{rw} = W\, \bar{r}\, \sin\theta \tag{6.42a}$$

For small rotations,

$$M_{rw} = W \bar{r}\, \theta \tag{6.42b}$$

where

$$\bar{r} = \frac{2.205\, B}{\pi} \tag{6.43}$$

By equating the moments of driving forces to those of the restoring forces, the following equation of motion is obtained:

$$M_{dp} = M_{rs} + M_{ri} + M_{rw} \tag{6.44}$$

Substituting for moments and rearranging, we get

$$\ddot{\theta} + \frac{3g}{\pi B}\theta = \left[\frac{0.68g}{W}\right] q_u \left[\lambda e^{-\beta t} - 1\right] \tag{6.45}$$

Equation (6.45) is a second order, nonhomogeneous, linear differential equation with constant coefficients. The natural frequency and the time period of the system are given by

$$\omega_n = \sqrt{\frac{3g}{\pi B}} \tag{6.46}$$

$$T = \frac{1}{2\pi}\sqrt{\frac{\pi B}{3g}} \tag{6.47}$$

Solution of Eq. (6.45) gives the following relation:

$$\frac{W}{0.68 g q_u}(\theta) = \frac{T^2}{4\pi^2 + \beta^2 T^2}\left[\left\{1 - \lambda + \frac{\beta^2 T^2}{4\pi^2}\right\}\cos\left(\frac{2\pi t}{T}\right) + \frac{\beta \lambda T}{2\pi}\sin\left(\frac{2\pi t}{T}\right) + \lambda e^{-\beta t} - \frac{\beta^2 T^2}{4\pi^2} - 1\right] \tag{6.48}$$

The above relation can be used to trace the history of motion of the foundation. For determination of the maximum angular deflection θ, Eq. (6.48) can be differentiated with respect to time. Thus

$$\frac{W}{0.68 g q_u}(\theta) = \frac{2\pi T}{4\pi^2 + \beta^2 T^2}\left[\left\{\lambda - 1 - \frac{\beta^2 T^2}{4\pi^2}\right\}\sin\left\{\frac{2\pi t}{T}\right\} + \frac{\beta \lambda T}{2\pi}\cos\left\{\frac{2\pi t}{T}\right\} - \frac{\beta \lambda T}{2\pi}e^{-\beta t}\right] \tag{6.49}$$

For obtaining the critical time $t = t_c$ which corresponds to $\theta = \theta_{max}$, the right-hand side of Eq. (6.49) is equated to zero. Since $2\pi T/(4\pi^2 + \beta^2 T^2) \neq 0$

$$\left[\lambda - 1 - \frac{\beta^2 T^2}{4\pi^2}\right]\sin\left[\frac{2\pi t}{T}\right] + \frac{\beta \lambda T}{2\pi}\cos\left[\frac{2\pi t}{T}\right] - \frac{\beta \lambda T}{2\pi}e^{-\beta t} = 0 \tag{6.50}$$

By using small increments of time t in Eq. (6.50), the value of t_c can be obtained. This value of $t = t_c$ can then be substituted in to Eq. (6.48) with known values of β, λ and B to obtain $(W/0.68\, g\, q_u)\, \theta_{max} = K$, dynamic load factor. Figures 6.15, 6.16 and 6.17 give the values of $K(s^2)$ for $B = 0.6$, 1.5 and 0.3 m, respectively, with $\lambda = 1\text{-}5$ and $\beta = 0\text{-}50\ s^{-1}$.

6.5.2 Wallace's Solution

Analysis presented by Trianadafilidis (1965) is based on rotational mode of failure. However, it is possible that a foundation may fail by vertically punching into the soil mass due to the application of vertical transient load. Wallace (1961) presented a procedure for the estimation of the vertical displacement of continuous footing considering punching mode of failure. The analysis is based on the following assumptions:

(i) The failure surface in the soil mass is assumed to be of similar type as suggested by Terzaghi (1943) for the evaluation of static bearing capacity of strip footings. This is shown in Fig. 6.18.
(ii) The soil behaves as a rigid plastic material (Fig. 6.13)
(iii) The ultimate shear strength is given by

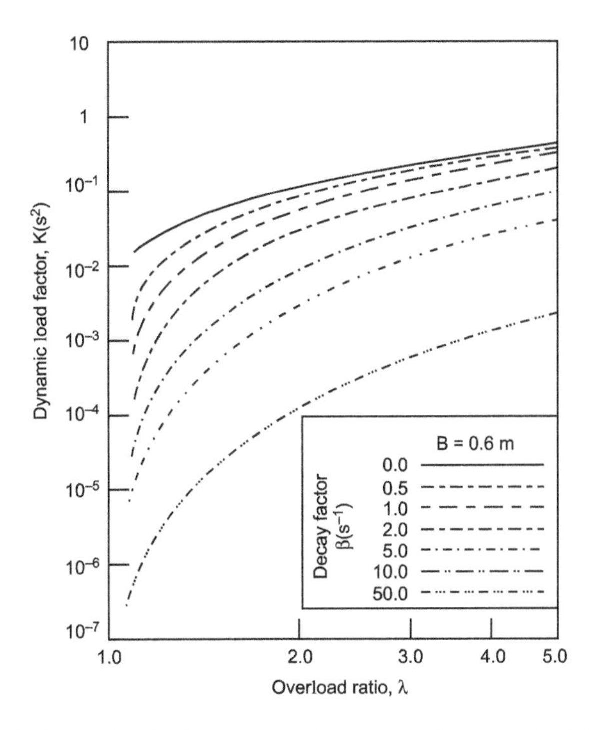

Fig. 6.15. Relationship between overload ratio and dynamic load factor for continuous footings 0.6 m wide

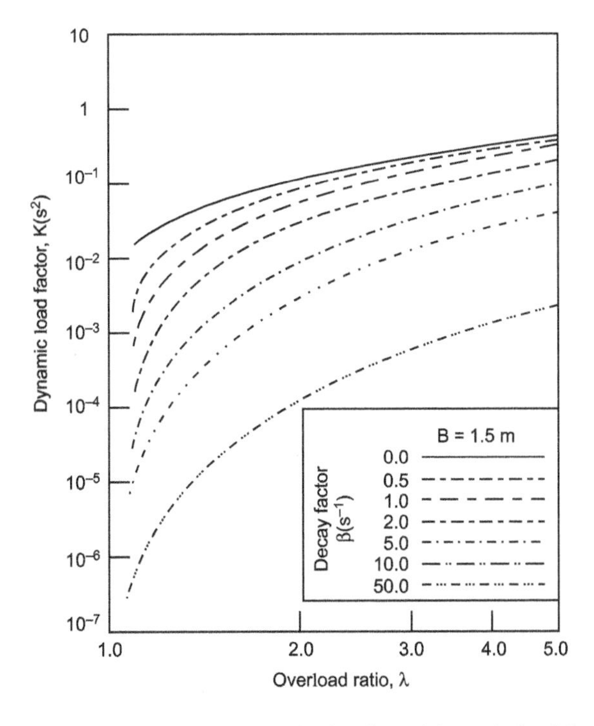

Fig. 6.16. Relationship between overload ratio and dynamic load factor for continuous footings 1.5 m wide

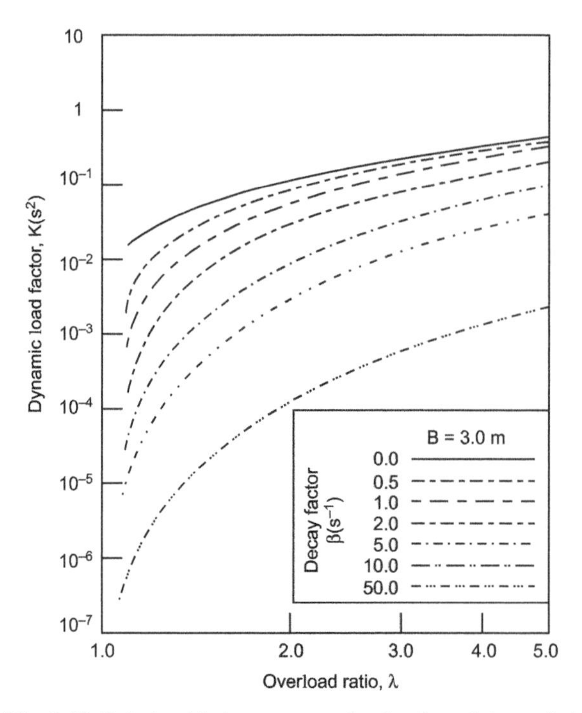

Fig. 6.17. Relationship between overload ratio and dynamic load factor for continuous footings 3.0 m Wide

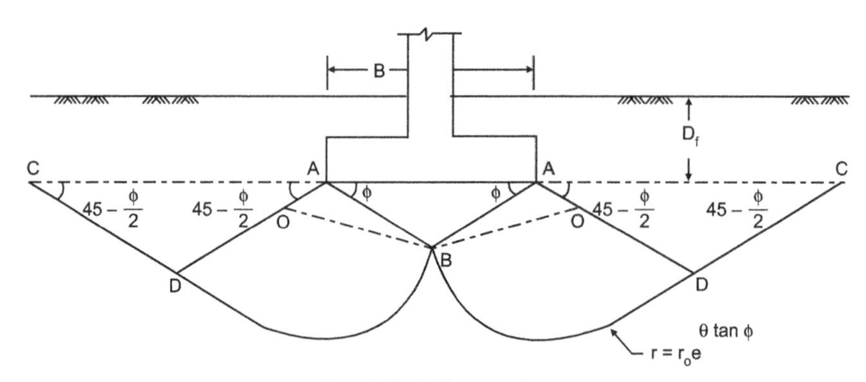

Fig. 6.18. Failure surface

$$S = c + \sigma \tan \phi \tag{6.51}$$

where S = Ultimate shear strength

c = Cohesion

σ = Normal stress

ϕ = Angle of internal friction

(iv) The dynamic load applied to the footing is initially peak triangular force pulse (Fig. 6.19).

(v) The footing is assumed to be weightless and to impart uniform load to the soil surface.

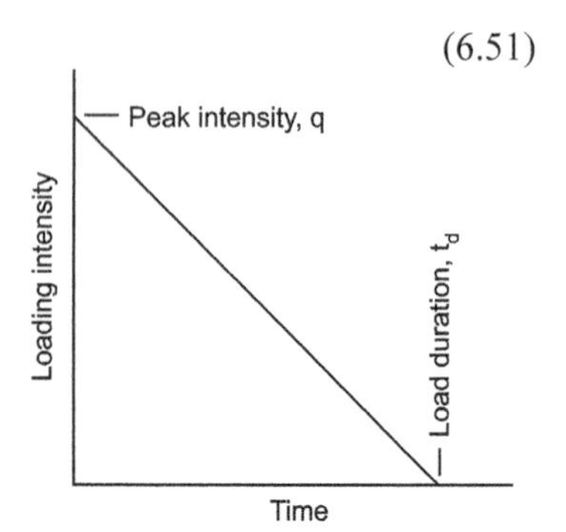

Fig. 6.19. Loading function

Analysis

The applied load is assumed to be an initial-peak triangular force which decays to zero at time t_d (Fig. 6.19). The peak load q is expressed in pressure units. Since the function is discontinuous at time t_d, two equations are necessary.

For $0 \le t \le t_d$, Loading function $= qB\left(1-\frac{t}{t_d}\right)$ (6.52)

For $t \ge t_d$, Loading function $= 0$ (6.53)

In Fig. 6.18, BD is an arc of a logarithmic spiral with its centre at O. It is defined by Eq. (6.54)

$$r = r_o e^{\theta \tan \phi} \tag{6.54}$$

where r_o = Distance OB (Fig. 6.18)

ϕ = Angle of internal friction

The static bearing capacity q_u for such a failure surface is given by

$$q_u = cN_c + qN_q + \frac{1}{2}\gamma B N_\gamma \tag{6.55}$$

where c = Cohesion

$q = \gamma D_f$ (6.56)

γ = Density of soil

D_f = Depth of footing

N_c, N_q, and N_γ = Bearing capacity factors

The bearing capacity factors depend on ϕ and K, K being 2 (Distance OA)/B, Fig. 6.18. The value of K locates the centre of the spiral which is the centre of rotation. Obviously the correct value of K is that which yields the minimum value of the bearing capacity. It is obtained by trial and error for each set of problem parameters. The values of N_γ, N_c and N_q for various values of ϕ and K are given in columns 3, 4 and 5 of Table 6.4.

Any acceleration of the soil mass ACDBA due to the downward movement of the footing will cause inertial forces which will resist such movement. The inertial forces are directly proportional to the acceleration of each individual soil mass and thereby dependent on displacements. The effective total inertial force is obtained by combining the inertial forces on each separate mass using energy considerations.

The inertial force is given by, $I_f = N_I \gamma B \frac{d^2\Delta}{dt^2}$ (6.57)

where Δ = Displacement at any time t

N_I = Coefficient of dynamic inertial shear resistance

The coefficient N_I depends on ϕ and K, and its values are listed in column no. 6 of Table 6.4.

Displacement of the soil mass within the failure surface due to downward movement of the footing will increase the restoring moment about the point O, and the increase in moment will be proportional to the displacement provided the rotation is not excessive. It is expressed as

$$R_f = N_R B \gamma \Delta \tag{6.58}$$

The coefficient N_R also depends on ϕ and K. Its values are listed in column no. 7 of Table 6.4.

The differential equations are set up by equating the four vertical forces to zero. There must be separate equations for before and after time t_d, since the loading function is defined in that manner.

For $0 \le t \le t_d$

$$N_I \gamma B^2 \frac{d^2\Delta}{dt^2} + N_R \gamma B \Delta + q_u B - qB\left(1-\frac{t}{t_d}\right) = 0 \tag{6.59a}$$

Table 6.4. Bearing Capacity Factors (N_γ, N_c, N_q, N_1, N_R)

ϕ (deg.) (1)	K (2)	N_γ (3)	N_c (4)	N_q (5)	N_1 (6)	N_R (7)	$\sqrt{\frac{N_R}{N_1}}$ (8)
0	– 0.05	0.0000	5.7277	1.0	0.0633	2.0125	5.6366
	0.00	0.0000	5.7124	1.0	0.0631	1.9723	5.5887
	+ 0.05	0.0000	5.7258	1.0	0.0633	1.9433	5.5394
5	– 0.65	0.1454	79.6255	7.9664	0.3755	8.9076	4.8709
	– 0.60	0.1445	29.8163	3.6086	0.2280	6.4362	5.3126
	– 0.55	0.1481	18.9958	2.6619	0.1579	5.0332	5.6460
	– 0.50	0.1553	14.3469	2.2552	0.1213	4.1699	5.8636
	– 0.45	0.1655	11.8179	2.0339	0.1011	3.6088	5.9750
	– 0.40	0.1786	10.2699	1.8985	0.0897	3.2299	6.0020
	– 0.35	0.1945	9.2580	1.8100	0.0833	2.9674	5.9698
	– 0.30	0.2131	8.5723	1.7500	0.0799	2.7828	5.9005
	– 0.25	0.2344	8.1007	1.7087	0.0786	2.6523	5.8108
	– 0.20	0.2585	7.7778	1.6805	0.0785	2.5604	5.7116
	– 0.15	0.2855	7.5629	1.6617	0.0793	2.4969	5.6099
	– 0.10	0.3154	7.4291	1.6500	0.0809	2.4547	5.5096
	– 0.05	0.3483	7.3580	1.6437	0.0829	2.4288	5.4128
	0.00	0.3843	7.3366	1.6419	0.0853	2.4155	5.3205
	+ 0.05	0.4233	7.3553	1.6435	0.0881	2.4122	5.2330
10	– 0.60	0.5700	53.9491	10.5127	0.1120	5.7922	7.1922
	– 0.55	0.5588	28.9945	6.1125	0.0935	4.8411	7.1948
	– 0.50	0.5645	20.5266	4.6194	0.0833	4.2238	7.1228
	– 0.45	0.5832	16.3539	3.8837	0.0779	3.8095	6.9932
	– 0.40	0.6127	13.9337	3.4569	0.0757	3.5264	6.8273
	– 0.35	0.6521	12.4031	3.1870	0.0755	3.3323	6.6445
	– 0.30	0.7008	11.3881	3.0080	0.0767	3.2008	6.4587
	– 0.25	0.7586	10.7004	2.8868	0.0790	3.1147	6.2781
	– 0.20	0.8253	10.2345	2.8046	0.0821	3.0625	6.1071
	– 0.15	0.9012	9.9267	2.7503	0.0858	3.0360	5.9474
	– 0.10	0.9863	9.7361	2.7167	0.0901	3.0294	5.7994
	– 0.05	1.0807	9.6352	2.6990	0.0948	3.0386	5.6676
	0.00	1.1848	9.6049	2.6936	0.0999	3.0604	5.5360
	+ 0.05	1.2986	9.6313	2.6983	0.1053	3.0923	3.4187
15	– 0.55	1.5462	46.5473	13.4724	0.0707	5.2677	8.6324
	– 0.50	1.5198	30.2759	9.1124	0.0696	4.7177	8.2310
	– 0.45	1.5342	23.2038	7.2175	0.0707	4.3564	7.8481
	– 0.40	1.5806	19.3483	6.1844	0.0734	4.1189	7.4903
	– 0.35	1.6540	16.9964	5.5542	0.0773	3.9669	7.1622
	– 0.30	1.7520	15.4722	5.1458	0.0823	3.8766	6.8645
	– 0.25	1.8730	14.4550	4.8732	0.0881	3.8322	6.5961
	– 0.20	2.0166	13.7730	4.6905	0.0947	3.8232	6.3542
	– 0.15	2.1825	13.3257	4.5706	0.1020	3.8418	6.1361
	– 0.10	2.3710	13.0501	4.4968	0.1101	3.8825	5.9388

(*Contd.*)

Table 6.4. (*Contd.*)

(1)	(2)	(3)	(4)	(5)	(6)	(7)	(8)
	– 0.05	2.5823	12.9048	4.4579	0.1183	3.9413	5.7596
	0.00	2.8168	12.8613	4.4462	0.1282	4.0149	5.5961
	+ 0.05	3.0750	12.8991	4.4563	0.1383	4.1008	5.4463
20	– 0.50	3.6745	46.2884	17.8477	0.0673	5.6658	9.1768
	– 0.45	3.6419	33.8986	13.3381	0.0728	5.3067	8.5380
	– 0.40	3.6943	27.6099	11.0492	0.0796	5.0886	7.9941
	– 0.35	3.8151	23.9213	9.7067	0.0877	4.9684	7.5267
	– 0.30	3.9952	21.5875	8.8572	0.0970	4.9199	7.1214
	– 0.25	4.2298	20.0542	8.2992	0.1076	4.9258	6.7672
	– 0.20	4.5161	19.0369	7.9289	0.1194	4.9746	6.4552
	– 0.15	5.8533	18.3742	7.6877	0.1325	5.0582	6.1783
	– 0.10	5.2413	17.9678	7.5398	0.1470	5.1704	5.9309
	– 0.05	5.6804	17.7542	7.4620	0.1629	5.3068	5.7084
	0.00	6.1717	17.6903	7.4368	0.1802	5.4638	5.5072
	+ 0.05	6.7161	17.7457	7.4589	0.1989	5.6486	5.3243
25	– 0.50	8.5665	73.8778	35.4499	0.0732	7.2346	9.9384
	– 0.45	8.3599	51.2706	24.9079	0.0835	6.8363	9.0503
	– 0.40	8.3728	40.7056	19.9814	0.0954	6.6214	8.3291
	– 0.35	8.5541	34.7663	17.2119	0.1094	6.5339	7.7297
	– 0.30	8.8760	31.1015	15.5029	0.1254	6.5404	7.2223
	– 0.25	9.3230	28.7315	14.3977	0.1437	6.6199	6.7864
	– 0.20	9.8871	27.1750	13.6720	0.1646	6.7584	6.4075
	– 0.15	10.5646	26.1681	13.2024	0.1882	6.9462	6.0748
	– 0.10	11.3542	25.5533	12.9157	0.2148	7.1761	5.7803
	– 0.05	12.2569	25.2309	12.7654	0.2445	7.4429	5.5178
	0.00	13.2745	25.1345	12.7205	0.2775	7.7423	5.2825
	+ 0.05	14.4095	25.2180	12.7594	0.3139	8.0710	5.0704
30	– 0.45	19.3095	80.8644	47.6872	0.1064	9.3123	9.3540
	– 0.40	19.1315	62.4470	37.0539	0.1267	9.0899	8.4705
	– 0.35	19.3718	52.5548	31.3426	0.1506	9.0494	7.7518
	– 0.30	19.940	46.6067	27.9084	0.1787	9.1446	7.1533
	– 0.25	20.1887	42.8208	25.7226	0.2116	9.3473	6.6458
	– 0.20	21.9566	40.3597	24.3017	0.2500	9.6392	6.2095
	– 0.15	23.3512	38.7778	23.3884	0.2944	10.0081	5.8303
	– 0.10	24.9984	37.8159	22.8330	0.3456	10.4452	5.4979
	– 0.05	26.8993	37.3127	22.5425	0.4041	10.9441	5.2044
	0.00	29.0580	37.1624	22.4558	0.4706	11.4998	4.9436
	+ 0.05	31.4810	37.2926	22.5309	0.5457	12.1084	4.7107
35	– 0.45	46.2942	134.3023	95.0397	0.1527	13.4981	9.4021
	– 0.40	45.4427	100.6609	71.4837	0.1887	13.2639	8.3844
	– 0.35	45.6687	83.4477	59.4308	0.2323	13.3114	7.5703
	– 0.30	46.7356	73.3676	52.3727	0.2849	13.5708	6.9017
	– 0.25	48.5145	67.0529	47.9511	0.3481	14.0015	6.3419
	– 0.20	50.9356	62.9887	45.1052	0.4237	14.5786	5.8661
	– 0.15	53.9640	60.3926	43.2874	0.5133	15.2895	5.4569
	– 0.10	57.8568	58.8199	42.1862	0.6191	16.1127	5.1018
	– 0.05	61.8051	57.9989	41.6113	0.7428	17.0515	4.7911
	0.00	66.6296	57.7539	41.4398	0.8868	18.0970	4.5175
	+ 0.05	72.0773	57.9662	41.5884	1.0529	19.2451	4.2753

(*Contd.*)

40	– 0.40	115.7097	172.8231	146.0161	0.3229	20.8738	8.0404
	– 0.35	115.5504	141.1002	119.3973	0.4107	21.1138	7.1701
	– 0.30	117.6386	123.0124	104.2199	0.5195	21.7125	6.4650
	– 0.25	121.5875	111.8576	94.8599	0.6536	22.6077	5.8817
	– 0.20	127.1879	104.7472	88.8935	0.8175	23.7619	5.3914
	– 0.15	134.3346	100.2323	85.1051	1.0168	25.1570	4.9741
	– 0.10	142.9868	97.5069	82.8181	1.2572	26.7775	4.6152
	– 0.05	153.1451	96.0866	81.6263	1.5450	28.6173	4.3038
	0.00	164.839	95.6630	81.2709	1.8870	30.6724	4.0317
	+ 0.05	178.1176	96.0303	81.5791	2.2904	32.9409	3.7924
45	– 0.40	327.6781	322.2748	323.2752	0.6576	36.2961	7.4295
	– 0.35	325.4943	259.1345	260.1349	0.8611	37.0113	6.5559
	– 0.30	329.9752	224.0769	225.0772	1.1194	38.3965	5.8568
	– 0.25	339.8627	202.7837	203.7840	1.4447	40.3468	5.2846
	– 0.20	354.4804	189.3358	190.3361	1.8515	42.8070	4.8083
	– 0.15	373.4971	180.8450	181.8452	2.3565	45.7496	4.4062
	– 0.10	393.7473	175.7358	176.7361	2.9784	49.1634	4.0628
	– 0.05	424.2605	173.0775	174.0778	3.7386	53.0475	3.7669
	0.00	456.1177	172.2851	173.2853	4.6607	57.4067	3.5096
	+ 0.05	492.4763	172.9729	173.9732	5.7709	62.2499	3.2843

or

$$\frac{d^2\Delta}{dt^2}+\frac{N_R}{N_I B}\Delta = \frac{q-q_u}{N_I\gamma B}-\frac{q}{N_I\gamma B t_d}I \tag{6.59b}$$

For $t \geq t_d$

$$N_I\gamma B^2\frac{d^2\Delta}{dt^2}+N_R\gamma B\Delta+q_u B = 0 \tag{6.59c}$$

or

$$\frac{d^2\Delta}{dt^2}+\frac{N_R}{N_I B}\Delta = \frac{q_u}{N_I\gamma B} \tag{6.59d}$$

The solution of the differential equations will yield equations of footing displacement versus time. The forms of the particular solutions of Eq. (6.59b) and Eq. (6.59d) are found to be

$$\Delta = C_1\cos(K't)+C_2\sin(K't)+\left(\frac{q-q_u}{N_R\gamma}\right)-\left(\frac{q}{N_R\gamma t_d}\right)t \tag{6.60a}$$

and

$$\Delta = C_3\cos(K't)+C_4\sin(K't)-\frac{q_u}{N_R\gamma} \tag{6.60b}$$

respectively, in which $K'=\sqrt{\dfrac{2N_R}{N_I B}}$; and C_1, C_2, C_3 and C_4 are coefficients of integration. The coefficients C_1 and C_2 are evaluated by the initial conditions. The coefficients C_3 and C_4 are evaluated by the conditions of displacement and velocity at t_d as defined by Eq. (6.60a). Solution and substitution of the coefficients yield nondimensional Eqs. 6.60.

For $0 \leq t \geq t_d$

$$\left(\frac{N_R\gamma}{q_u}\right)\Delta = \left(\frac{q}{q_u}-1\right)[1-\cos(K't)]+\frac{\frac{q}{q_u}}{t_d K'}[\sin(K't)-(K't)] \tag{6.60c}$$

For $t \geq t_d$

$$\left(\frac{N_R\gamma}{q_u}\right)\Delta = \left\{\left(1-\frac{q}{q_u}\right)+\frac{\frac{q}{q_u}}{t_dK'}\sin(K't_d)\right\}\cos(K't_d)+\left\{\frac{\frac{q}{q_u}}{t_dK'}\{1-\cos(K't_d)\}\right\}\sin(K't)-1 \quad (6.60d)$$

The coefficients N_γ, N_c, N_q, N_I and N_R are dependent only on values of ϕ and K. Using magnitudes of ϕ from 0° to 45° and of K for the region where the ultimate static shear resistance could be a minimum, these coefficients were evaluated. The values obtained are given in Table 6.4 for every fifth degree.

The maximum displacement from Eq. (6.60a) and Eq. (6.60b) is the predicated permanent footing displacement, since downward motion ceases at the time of maximum displacement and rebound is not considered. These data along with the times of maximum displacement are given in Fig. 6.20.

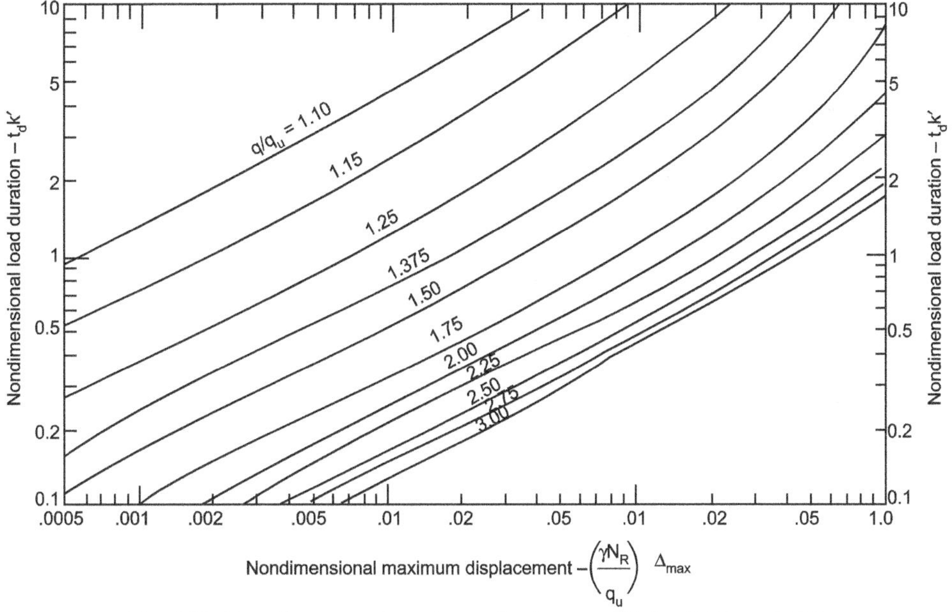

Fig. 6.20. Non-dimensional maximum displacement (After Wallace, 1961)

6.5.3 Chummar's Solution

Chummar (1965) persented a solution for dynamic response of a strip footing supported by c - ϕ soil and subjected to horizontal transient load. The analysis is based on the following assumptions:

(i) The failure of the footing occurs with the application of a horizontal dynamic load acting at a certain height above the base of the footing.

(ii) The resulting motion in the footing is of a rotatory nature. The failure surface is a logarithmic spiral with its centre on the base corner of the footing, which is also the centre of rotation Fig. (6.21).

(iii) The rotating soil mass is considered to be a rigid body rotating about a fixed axis.

(iv) The soil exhibits rigid plastic, stress-strain characteristics.

Analysis

The static bearing capacity of the footing is calculated by assuming that the footing fails when acted upon by a vertical static load, which causes rotation of the logarithmic spiral failure. The ultimate static bearing capacity q_u is given by

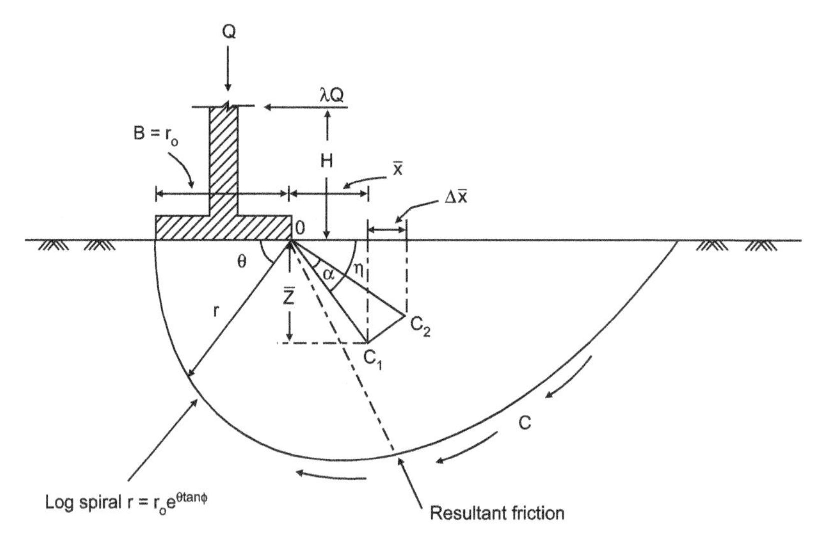

Fig. 6.21. Transient horizontal load on a continuous footing resting on ground surface

$$q_u = c N_c + \frac{1}{2} \gamma B N_\gamma \tag{6.61}$$

where c = Cohesion

B = Footing width and equal to the initial radius of spiral

γ = Unit weight of the soil

N_c and N_γ = Bearing capacity factors for the assumed type of failure

Considering moment of the forces about O, the centre of rotation:

$$\text{Moment due to cohesion } c,\ M_{RC} = \frac{c B^2}{2 \tan \phi}(e^{2\pi \tan \phi} - 1) = \psi c B^2 \tag{6.62a}$$

where

$$\psi = \frac{(e^{2\pi \tan \phi} - 1)}{2 \tan \phi} \tag{6.62b}$$

Moment due to weight W of soil wedge,

$$M_{RW} = \gamma \frac{B^3 \tan \phi (e^{3\pi \tan \phi} + 1)}{9 \tan^2 \phi + 1} = \in \gamma B^3 \tag{6.63a}$$

where

$$\in = \frac{\tan \phi (e^{3\pi \tan \phi} + 1)}{9 \tan^2 \phi + 1} \tag{6.63b}$$

ϕ = Angle of internal friction

$$\text{Moment of } q_u \text{ about O} = q_u \frac{B^2}{2} = M_{RC} + M_{RW} \tag{6.64}$$

It gives

$$q_u = \frac{c}{\tan \phi}(e^{2\pi \tan \phi} - 1) + \frac{2 \gamma B \tan \phi (e^{3\pi \tan \phi} + 1)}{9 \tan^2 \phi + 1} \tag{6.65}$$

Combining Eqs. (6.61) and (6.65), we get

$$N_\gamma = \frac{4\tan\phi(e^{3\pi\tan\phi}+1)}{9\tan^2\phi+1} \tag{6.66}$$

and,
$$N_c = \frac{e^{2\pi\tan\phi}-1}{\tan\phi} \tag{6.67}$$

With a suitable factor of safety F, the static vertical force on the foundation per unit length can be given as

$$Q = \frac{B}{F}\left(cN_c + \frac{1}{2}\gamma B N_\gamma\right) \tag{6.68}$$

The variation of dynamic force considered in the analysis is shown in Fig. 6.22. In this

$$Q_{d\,(max)} = \lambda\, Q \tag{6.69}$$

where

$Q_{d\,(max)}$ = Maximum value of horizontal transient load per unit length acting at height H above base of the footing

λ = Over load factor

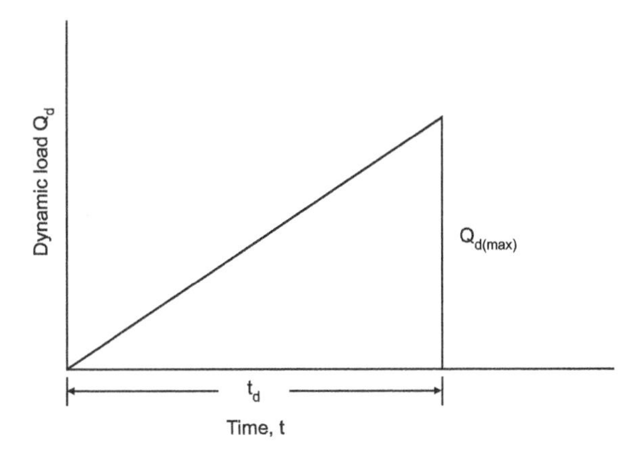

Fig. 6.22. Loading function

For considering the dynamic equilibrium of the foundation with the horizontal transient load, the moment of each of the forces (per unit length) about the centre of the log spiral needs to be considered:

1. Moment due to the vertical force Q

$$M_1 = \frac{1}{2}QB \tag{6.70}$$

2. Moment due to the horizontal force Q_d at any time t

$$M_2 = Q_d H = \frac{Q_{d\,(max)}\, Ht}{t_d} = \frac{M_{d\,(max)}\, t}{t_d} \tag{6.71}$$

where $M_{d\,(max)} = Q_{d\,(max)}\, H$

3. Moment due to the cohesive force acting along the failure surface is given by Eq. (6.62a).
4. Moment due to weight of soil mass in the failure wedge is given by Eq. (6.63a).
5. Moment of the force due to displacement of the centre of gravity of the failure wedge (C' in Fig. 6.21) from its initial position:

$$M_3 = W\Delta\bar{x} \tag{6.72}$$

where W is the weight of the failure wedge, and given by

$$W = \frac{\gamma B^2 (e^{2\pi \tan\phi} - 1)}{(4\tan\theta)} \tag{6.73}$$

$$\Delta\bar{x} = R\cos(\eta - \alpha) - R\cos\eta \tag{6.74}$$

and $R = \overline{OC_1}$ (Fig. 6.21). When α is small, Eq. (6.74) can be written as

$$\Delta\bar{x} = (R\sin\eta).\alpha = \bar{z}.\alpha \tag{6.75}$$

However,

$$R = \sqrt{(\bar{x})^2 + (\bar{z})^2} \tag{6.76}$$

where

$$\bar{x} = \frac{-4B\tan^2\phi(e^{3\pi\tan\phi} + 1)}{(9\tan^2\phi + 1)\,(e^{2\pi\tan\phi} - 1)} \tag{6.77}$$

$$\bar{z} = \frac{4B\tan\phi(e^{3\pi\tan\phi} + 1)}{3\left(\sqrt{9\tan^2\phi + 1}\right)(e^{2\pi\tan\phi} - 1)} \tag{6.78}$$

Combining Eqs. (6.72) - (6.78)

$$M_3 = \gamma\beta B^3 \bar{z}\alpha \tag{6.79}$$

where

$$\beta = \frac{(e^{3\pi\tan\phi} + 1)}{3\left(\sqrt{9\tan^2\phi + 1}\right)} \tag{6.80}$$

6. Moment due to inertia force of soil wedge:

$$M_4 = \left(\frac{d^2\alpha}{dt^2}\right) J \tag{6.81}$$

where J is the mass moment of inertia of the soil wedge about the axis of rotation.

$$J = \left[\frac{\gamma B^4}{16 g \tan\phi}\right](e^{4\pi\tan\phi} - 1) \tag{6.82}$$

and g is the acceleration due to gravity. Substitution of Eq. (6.62) into Eq. (6.81) yields

$$M_4 = \frac{\mu_c \gamma B^4}{g} \cdot \frac{d^2\alpha}{dt^2} \tag{6.83}$$

where

$$\mu_c = \frac{(e^{4\pi\tan\phi} - 1)}{16\tan\phi} \tag{6.84}$$

Moment due to the frictional resistance along the failure surface will be zero as its resultant will pass through the centre of log spiral.

Now for the equation of motion,

$$M_1 + M_2 = M_{RC} + M_{RW} + M_3 + M_4 \tag{6.85}$$

Substitution of the proper terms for the moments in Eq. (6.85) gives

$$\left(\frac{d^2\alpha}{dt^2}\right)+k^2\alpha = A\left[\left(\frac{M_{d(\max)t}}{t_d}\right)+\frac{1}{2}QB-E\right] \tag{6.86}$$

where
$$k = \sqrt{\frac{g\beta\sin\eta}{\mu_c B}} \tag{6.87}$$

$$A = \frac{g}{(\gamma B^4 \mu_c)} \tag{6.88}$$

$$E = \psi c B^2 + \epsilon\gamma B^3 \tag{6.89}$$

Solution of the differential equation of motion [Eq. (6.86)] with proper boundary conditions yields the following results:
For $t \leq t_d$

$$\alpha = \frac{A}{k^2}\left(E-\frac{1}{2}QB\right)\cos(kt)-\frac{A}{k^3}\cdot\frac{M_{d(\max)}}{t_d}\sin(kt)+\frac{A}{k^2}\left(\frac{M_{d(\max)}.t}{t_d}+\frac{1}{2}QB-E\right) \tag{6.90}$$

For $t > t_d$

$$\alpha = \left(\frac{1}{k}\right)[G_1 k\cos(kt_d)-G_2\sin(kt_d)]\cos(kt)+\left(\frac{1}{k}\right)[G_1 k\sin(kt_d)$$

$$-G_2\cos(kt_d)\sin(kt)+\left(\frac{A}{k^2}\right)\left(\frac{1}{2}QB-E\right) \tag{6.91}$$

where
$$G_1 = \frac{A}{k^2}\left(E-\frac{1}{2}QB\right)\cos(kt_d)-\frac{A}{k^3}\cdot\frac{M_{d(\max)}}{t_d}\sin(kt_d)+\frac{AM_{d(\max)}}{k^2} \tag{6.92}$$

and
$$G_2 = -\frac{A}{k}\left(E-\frac{1}{2}QB\right)\sin(kt_d)-\frac{A}{k^2}\cdot\frac{M_{d(\max)}}{t_d}\cos(kt_d)+\frac{AM_{d(\max)}}{k^2 t_d} \tag{6.93}$$

The procedure of comutations have been discussed in Example 6.4.

6.6 Seismic Bearing Capacity and Seismic Settlement

The strength of the soil notably reduces when subjected to seismic excitation, because of the extensive inertial stresses induced during shaking. The decrease in strength of soil leads to a significant reduction in bearing capacity of footing; while increase in the settlement. This reduced bearing capacity is termed as seismic bearing capacity of footing and the increased settlement is called seismic settlement.

6.6.1 Seismic Bearing Capacity

Determination of seismic bearing capacity of footing by pseudo static method has received considerable attention in last two decades. Due to seismic loading bearing capacity of footings reduces significantly as a result of induced stresses; as mentioned above this reduced bearing capacity is termed as seismic bearing capacity of footing.

In comparison with the extensive studies on the static bearing capacity of shallow foundations, only a limited amount of information is available on seismic bearing capacity of footings. Using limit equilibrium approach, various researchers like Sarma and Iossifelis (1990), Richard et al. (1993), Budhu and Al-kami (1993) and Saran and Rangwala (2011) have carried out analysis considering seismic forces both on the structure and the supporting soil mass. Apart from above mentioned studies, Dormieux and Pecker (1995), Paolucci and Pecker (1997) and Soubra (1999) used upper-bound limit analysis to develop seismic bearing capacity factors for strip footings. Kumar and Rao (2002) used method of characteristics to evaluate the bearing capacity of footing subjected horizontal seismic accelerations only. Shafiee and Jahanandish (2010) obtained seismic bearing capacity factors using finite element method (FEM).

Recently, Rangwala et al. (2012) developed an analysis for determination of seismic bearing capacity of a strip footing resting on the flat ground subjected to seismic accelerations. Pseudo-static approach has been used in the analysis considering the inertial forces within the soil mass. The seismic accelerations were considered both in horizontal and vertical directions. One sided rupture surface, as shown in Fig. 6.23, was considered in the direction of horizontal seismic force. Zones I and III are respectively the elastic and passive Rankine zones. These are bounded with radial shear zone II. Zone I is unsymmetrical triangle zone defined by wedge angles α_1 and α_2. On the other side the soil was considered partially mobilized having $\varphi_m = \tan^{-1}(m \tan \varphi)$, and $c_m = m_c$ where m is mobilization factor less than unity.

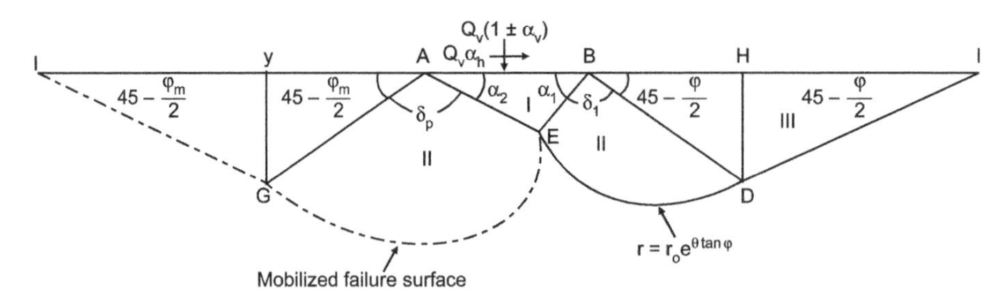

Fig. 6.23. Seismic failures surface (Rangwala et al., 2012)

The bearing capacity expression was then developed by considering the equilibrium of elastic wedge ABE (Fig. 6.24). The forces acting on the wedge are:

(i) Seismic vertical force $Q_v(1 \pm a_v)$, where Q_v is the static load

(ii) Seismic horizontal force $Q_v\alpha_h$

(iii) On face BE

$P_{p\gamma E}$ = Seismic passive earth pressure due to weight only,
P_{pqE} = Seismic passive earth pressure due to surcharge only,
P_{pcE} = Seismic passive earth pressure due to cohesion only,
C = Cohesive force acting on BE. It is equal to c. BE, c being unit cohesion.

On face AE

$P_{m\gamma E}$ = Mobilised seismic passive earth pressure due to weight only,
P_{mqE} = Mobilised seismic passive earth pressure due to surcharge only,
P_{mcE} = Mobilised seismic passive earth pressure due to cohesion only,
C_m = Mobilised cohesive force acting on AE. It is equal to m.c.AE.

(iv) Inertial forces $W(1 \pm \alpha_v)$ and $W\alpha_h$, W being the weight of wedge ABE.

Magnitudes of C, C_m, $W(1 + \alpha_v)$ and $W\alpha_h$ are very small in comparison to passive earth pressures and therefore neglected in further analysis.

Considering the equilibrium of the wedge ABE ($\Sigma V = 0$)

$$Q_v(1 \pm \alpha_v) = (P_{p\gamma E} + P_{pqE} + P_{pcE}) \cos(\alpha_1 - \varphi) + (P_{m\gamma E} + P_{mqE} + P_{mcE}) \cos(\alpha_2 - \varphi_m) \quad (6.94)$$

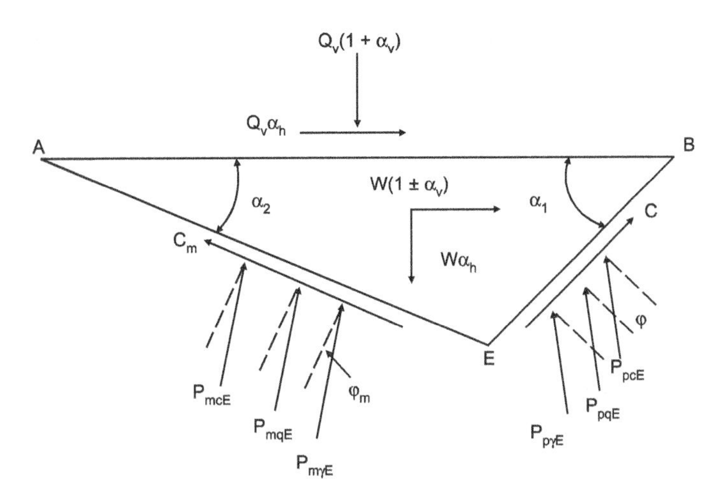

Fig. 6.24. Forces acting on central elastic wedge

By introducing non-dimensional quantities,

$$N_{\gamma E} = \frac{2}{(1 \pm \alpha_v)}\left\{\frac{P_{p\gamma E}}{\gamma B^2}\cos(\alpha_1 - \varphi) + \frac{P_{m\gamma E}}{\gamma B^2}\cos(\alpha_2 - \varphi_m)\right\} \tag{6.95a}$$

$$N_{qE} = \frac{2}{(1 \pm \alpha_v)}\left\{\frac{P_{pqE}}{qB}\cos(\alpha_1 - \varphi) + \frac{P_{mqE}}{qB}\cos(\alpha_2 - \varphi_m)\right\} \tag{6.95b}$$

$$N_{cE} = \frac{2}{(1 \pm \alpha_v)}\left\{\frac{P_{pcE}}{cB}\cos(\alpha_1 - \varphi) + \frac{P_{mcE}}{cB}\cos(\alpha_2 - \varphi_m)\right\} \tag{6.95c}$$

The expression (6.94) may be rewritten as,

$$Q_v = B(cN_{cE} + qN_{qE} + 0.5\, B\gamma N_{\gamma E}) \tag{6.96}$$

Non-dimensional quantities $N_{\gamma E}$, N_{qE}, and N_{cE} are termed as seismic bearing capacity factors and these depend on φ, α_h and α_v.

The problem has been solved considering three cases namely, (1) $c = q = 0$; (2) $c = \gamma = 0$; and (3) $q = \gamma = 0$ separately. These cases enabled the values of seismic bearing capacity factors $N_{\gamma E}$, N_{qE}, and N_{cE} respectively. Principle of superposition was then assumed to hold good. It will lead the results slightly on the conservative side.

For each case, the following steps indicate the manner, in which computations were made for a given value of angle of internal friction, φ, horizontal seismic acceleration coefficient, α_h and horizontal seismic acceleration coefficient, α_v.

1. Establish relationship between wedge angles α_1 and α_2. This is done by writing the three conditions of equilibrium of wedge ABE (i.e. $\Sigma H = 0$, $\Sigma V = 0$ and $\Sigma M = 0$). Solving these conditions and eliminating passive earth pressures, wedge angle relationship is obtained.
2. A particular value of mobilization factor m is assumed.
3. Assume value of wedge angles α_1 and obtain the values of α_2 by the derived wedge angle relationship or by iterations satisfying the moment equilibrium condition.
4. Find a set of wedge angles α_1 and α_2, which satisfies the equilibrium of horizontal forces.
5. Determine the values of passive pressure and their respective point of application.
6. Find the wedge angles and satisfying equilibrium conditions iteratively.
7. Steps 2-5 are repeated for different values of mobilization factor m. The passive pressures for

value of m, satisfying all the equilibrium conditions, $\Sigma H = 0$; $\Sigma V = 0$; and $\Sigma M = 0$, and giving minimum value of the bearing capacity factor is adopted.

The bearing capacity factors ($N_{\gamma S}$, N_{qS} and N_{cS}), in static conditions ($\alpha_h = a_v = 0$), were obtained as:

$$N_{\gamma S} = 0.5(e^{8\tan(0.9\,\varphi)} - 1) \tag{6.97}$$

$$N_{qS} = \frac{e^{2\left(\frac{3\pi}{4} - \frac{\varphi}{2}\right)\tan\phi}}{1 - \sin\varphi} \tag{6.98}$$

$$N_{cS} = (N_{qS} - 1)\cot\phi \tag{6.99}$$

For the seismic bearing capacity, the results are obtained and presented in form of the ratio of seismic bearing capacity factor to its static counterpart, termed as seismic reductions factor and is denoted as SRF.

A numerical analysis has been done varying the values of parameters affecting the bearing capacity factors. A regression analysis has been carried out on the results of the parametric study. The expressions determined are as given below:

$$SRF_\gamma = N_{\gamma E}/N_{\gamma S} = e^{-6.994\tan\psi} \tag{6.100}$$

where $\psi = \tan^{-1}[\alpha_h/(1 - \alpha_v)]$

$$SRF_q = N_{qE}/N_{qs} = e^{-a_q\alpha_h} \tag{6.101}$$

where $a_q = 5(\tan\varphi)^{0.44}$

$$SRF_c = N_{cE}/N_{cS} = e^{-a_c\alpha_h} \tag{6.102}$$

where $a_c = 4.45(\tan\varphi)^{0.63}$

Error in regression analysis was found to be in the range of ± 5%. The details of the analysis and computational procedure are given elsewhere (Rangwala et al., 2012).

6.6.2 Seismic Settlement

Al-Karni (1993) using the concept of critical acceleration, developed an analysis to obtain the seismic settlement of a footing. The critical acceleration is defined as the acceleration at which the footing will start to move and therefore the ultimate seismic bearing capacity q_{uE} reduced to allowable static bearing capacity (q_{uS}/F), F being factor of safety. The method is cumbersome and gives only total uniform settlement of the footing. Under seismic vibrations, footing invariably gets tilted. Keeping this in view, the details of the method are not summarised here.

Illustrative Examples

Example 6.1

Proportion an isolated footing for a column of 500 mm × 500 mm size subjected to a vertical load of 2400 kN. The structure is located in seismic region. The earthquake force results a moment of 400 kN-m and shear load of 360 kN at the base of the footing.

The soil properties are as follows:

$$c = 6 \text{ kN/m}^2,\ \phi = 39^\circ \text{ and } \gamma = 18 \text{ kN/m}^3$$

A plate load test was performed at the anticipated depth of foundation on a plate of size 600 mm × 600 mm and a pressure settlement record as given below was obtained. The permissible values of settlement, tilt and lateral displacement are 50 mm, 1 degree and 25 mm respectively.

Pressure (kN/m²)	0.0	240	480	720	960	1200	1440	1680
Settlement (mm)	0.0	2.0	5.0	7.5	12.0	16.0	23.0	28.0

Solution

1. **Safe bearing capacity**

Eccentricity of load, $e = \dfrac{M}{Q_v} = \dfrac{400}{2400} = 0.1667 \text{ m}$

$$\text{Inclination of load} = i = \tan^{-1}\left(\frac{Q_h}{Q_v}\right) = \tan^{-1}\left(\frac{360}{2400}\right) = 8.53^\circ$$

Let the size of footing is 2.0 m × 2.0 m and is located at 1.0 m depth below ground surface.

Hence, $B' = B - 2e = 2 - 2(0.1667) = 1.667 \text{ m}$

and $L' = L = 2 \text{ m}$

$$q_{nu} = [cN_c S_c d_c i_c + \gamma_1 D_f (N_q - 1) S_q d_q i_q r_w + \frac{1}{2}\gamma_2 B N_\gamma S_\gamma d_\gamma i_\gamma r'_w]$$

Since $\phi = 39^\circ$ it is the case of general shear failure. For $\phi = 39^\circ$, $N_c = 70.79$, $N_q = 59.62$ and $N_\gamma = 100.71$

Depth factors are given as:

$$d_c = 1 + 0.4\frac{D_f}{B'} = 1 + 0.4\frac{1}{1.667} = 1.24$$

$$d_q = 1 + 2\tan\phi(1 - \sin\phi)^2 \frac{D_f}{B'}$$

$$= 1 + 2\tan 39(1 - \sin 39)^2 \frac{1}{1.667} = 1.13$$

and $d_\gamma = 1$

Shape factors are calculated as:

$$S_c = 1 + 0.2\frac{B'}{L} = 1 + 0.2\frac{1.667}{2} = 1.167$$

$$S_q = 1 + 0.2\frac{B'}{L} = 1 + 0.2\frac{1.667}{2} = 1.167$$

$$S_\gamma = 1 - 0.4\frac{B'}{L} = 1 - 0.4\frac{1.667}{2} = 0.667$$

Inclination factors: $Q = \sqrt{Q_h^2 + Q_v^2} = \sqrt{360^2 + 2400^2} = 2426.85 \text{ kN}$

and $m = \left(2 + \dfrac{B'}{L}\right)\left(1 + \dfrac{B'}{L}\right) = \left(2 + \dfrac{1.667}{2}\right)\left(1 + \dfrac{1.667}{2}\right) = 5.195$

Therefore, $$i_q = \left[1 - \frac{Q}{Q + B' L_c \cot\phi}\right]^m$$

$$= \left[1 - \frac{360}{2426.85 + 1.667(2)(6)\cot 39}\right]^{5.195} = 0.438$$

and $$i_c = i_q - \frac{(1 - i_q)}{N_c \tan\phi} = 0.438 - \frac{(1 - 0.438)}{70.79 \tan 39} = 0.428$$

$$i_\gamma = \left[1 - \frac{Q_h}{Q + B\, Lc \cot\phi}\right]^{m+1}$$

$$= \left[1 - \frac{360}{2426.85 + 1.667\,(2)\,(6)\cot 39}\right]^{6.195} = 0.374$$

Assuming water table to be below the ground surface at a depth greater than $(D_f + B)$, hence $r_w = r_w' = 1$, and also say $\gamma_1 = \gamma_2 = \gamma$.

Hence $q_{nu} = cN_c\, S_c\, d_c\, i_c + \gamma_1\, D_f\, (N_q - 1) S_q\, d_q\, i_q\, r_w + \frac{1}{2}\gamma_2\, BN_\gamma\, S_\gamma\, d_\gamma\, i_\gamma\, r'_w$

$$q_{nu} = 6(70.79)\,(1.167)\,(1.24)\,(0.428)$$

$$+ 18\,(1)\,(59.62–1)\,(1.167)\,(1.13)\,(0.438)\,(1)$$

$$+ (0.5)\,(18)\,(1.667)\,(100.7)\,(0.6677)\,(1)\,(0.374)\,(1)$$

$$= 263.06 + 609.45 + 376.92 = 1249.43 \text{ kN/m}^2$$

According to Meyerhof,

$$i_c = i_q = \left(1 - \frac{i}{90}\right)^2 = \left(1 - \frac{8.53}{90}\right)^2 = 0.82$$

$$i_\gamma = \left(1 - \frac{i}{\phi}\right)^2 = \left(1 - \frac{8.53}{39}\right)^2 = 0.61$$

Therefore,

$$q_{nu} = 6(70.79)\,(1.167)\,(1.24)\,(0.82)$$

$$+ 18\,(1)\,(59.62 - 1)\,(1.167)\,(1.13)\,(0.82)$$

$$+ \frac{1}{2}(18)\,(1.667)\,(100.71)\,(0.667)\,(1)\,(0.61)\,(1)$$

$$= 503.99 + 1140.98 + 614.76 = 2259.74 \text{ kN/m}^2$$

From charts (According to Saran and Agarwal, 1991)

$$\phi = 39°,\ \frac{e}{B} = \frac{0.1667}{2} = 0.083, \text{ and } \ i = 8.53°$$

	For $\frac{e}{B}$ = 0.0 (Figs. 6.4 to 6.6)			For $\frac{e}{B}$ = 0.10 (Figs. 6.4 to 6.6)		
	$i = 0°$	$i = 10°$	$i = 8.53°$	$i = 0°$	$i = 10°$	$i = 8.53°$
$N_{\gamma ei}$	144	68	79.64	75	42	46.85
N_{qei}	68	41	44.97	47	30	32.50
N_{cei}	88	48	53.88	58	37	40.09

Hence for $\frac{e}{B} = 0.083$ and $i = 8.53$,

$$N_{cei} = 42.43, N_{qei} = 34.62 \text{ and } N_{\gamma ei} = 52.42$$

In this case [As effect of eccentricity and inclination has been considered already]

$$i_c = i_q = i_\gamma = 1$$

and $B = 2$ m

Depth factors will be: $d_c = 1+0.4\frac{D_f}{B} = 1+0.4\frac{1}{2} = 1.2$

$$d_q = 1+2\tan 39(1-\sin 39)^2\frac{1}{2} = 1.11$$

$$d_\gamma = 1$$

and shape factors will be:

$S_c = 1.3$, $S_q = 1.2$ and $S_\gamma = 0.8$ (As footing is square)

Thus
$$q_{nu} = 6(42.43)(1.3)(1.2)(1)$$
$$+ 18(1)(34.62-1)(1.2)(1.11)(1)(1)$$
$$+ \frac{1}{2}(18)(2)(52.42)(0.8)(1)(1)(1)$$
$$= 397.14 + 806.07 + 754.84$$
$$= 1958.06 \text{ kN/m}^2$$

Therefore, value of q_{nu} from charts lies between values of q_{nu} obtained by Eq. (6.17) and Meyerhof's method.

Hence $Q_{nu} = q_{nu} \times$ Area of footing

$$Q_{nu} = 1958.06 \times 2 \times 2 = 7832.23 \text{ kN}$$

Factor of safety $= \frac{7832.3}{2400} = 3.26 > 3$. Therefore foundation is safe against shear.

2. Settlement computation:

When footing is subjected to a central vertical load only, in that case $e/B = 0$ and $i = 0$

For $\phi = 39°$, $N_c = 88$, $N_\gamma = 144$ and $N_q = 68$

Thus
$$q_{nu} = (6)(88)(1.33)(1.2)(1) + 18(1)(68-1)(1.2)(1.11)(1)(1) + \frac{1}{2}(18)(2)(144)(0.8)(1)(1)(1)$$

$$= 823.68 + 1606.39 + 2703.6 = 4503.67 \text{ kN/m}^2$$

Pressure on footing corresponding to a F.O.S. = 3.26, will be $= \frac{4503.67}{3.26} = 1381.28 \text{ kN/m}^2$

From plate load test data corresponding to pressure 1381.28 kN/m².

$$S_p = 22.2 \text{ mm}$$

Now since
$$\frac{S_f}{S_p} = \left[\frac{B_f(B_p+30)}{B_f(B_f+30}\right]^2$$

Thus settlement of footing when it is subjected to central vertical load will be

$$S_o = S_f = \left[\frac{200(60+30)}{60(20+30)}\right]^2 \times 22.2 = 37.77 \text{ mm}$$

Now,
$$\frac{i}{\phi} = \frac{8.53}{39} = 0.2187$$

Hence
$$A_o = 1 - 0.56\left(\frac{i}{\phi}\right) - 0.82\left(\frac{i}{\phi}\right)^2$$

$$= 1 - 0.56(0.2187) - 0.82(0.2187)^2 = 0.838$$

$$A_1 = -3.51 + 1.47\left(\frac{i}{\phi}\right) + 5.67\left(\frac{i}{\phi}\right)^2$$

$$= -3.51 + 1.47(0.2187) + 5.67(0.2187)^2 = -2.917$$

$$A_2 = 4.74 - 1.38\left(\frac{i}{\phi}\right) - 12.45\left(\frac{i}{\phi}\right)^2$$

$$= 4.47 - 1.38(0.2187) - 12.45(0.2187)^2 = -3.843$$

$$\frac{S_e}{S_o} = A_o + A_1\left(\frac{e}{B}\right) + A_2\left(\frac{e}{B}\right)^2$$

$$= 0.838 - 2.917\left(\frac{0.1667}{2}\right) + 3.843\left(\frac{0.1667}{2}\right)^2 = 0.6216$$

$$S_e = 0.6216\, S_o = 0.6216 \times 37.77 = 23.48 \text{ mm}$$

and
$$B_o = 1 - 0.48\left(\frac{i}{\phi}\right) - 0.82\left(\frac{i}{\phi}\right)^2 = 1 - 0.48(0.2187) - 0.82(0.2187)^2 = 0.856$$

$$B_1 = -1.80+0.94\left(\frac{i}{\phi}\right)+1.63\left(\frac{i}{\phi}\right)^2$$

$$= -1.80 + 0.94(0.2187) + 1.63(0.2187)^2 = -1.516$$

$$\frac{S_m}{S_o} = B_o + B_1\left(\frac{e}{B}\right) = 0.856 + (-1.516)\left(\frac{0.1667}{2}\right) = 0.7296$$

$$S_m = 0.7296\,S_o = 0.7296\,(37.77) = 27.56 \text{ mm} < 50 \text{ mm (safe)}$$

$$\sin t = \frac{S_m - S_e}{\frac{B}{2}-e} = \frac{27.56 - 23.48}{\frac{2000}{2} - 166.7} = 0.0049$$

$$t = 0.28° < 1° \text{ (safe)}$$

and

$$\frac{H_D}{B} = 0.12(0.2187) - 0.682(0.2187)^2$$

$$+ 1.99(0.2187)^3 - 2.01(0.2187)^4 = 0.0101$$

$$H_D = 0.0101\,B = 0.0101(2000) = 21.21 \text{ mm} < 25 \text{ mm (safe)}$$

Example 6.2

A 1.5 m wide strip foundation is subjected to a vertical transient stress pulse which can be given as $q_d = 650\, e^{-10\,t}$ kN/m². The soil supporting the foundation is saturated clay with $c_u = 60$ kN/m². The unit weight of soil is 19 kN/m³. Determine the maximum angular rotation of the footing might undergo.

Solution

1. $q_u = 5.54\, c_u = 5.54 \times 60 = 332.4$ kN/m²

$$\lambda = \frac{650}{332.4} = 1.955$$

2. Refering Fig. 6.16. For $\lambda = 1.955$ and $\beta = 10$ s^{-1}

$$\left(\frac{W}{0.68\, g\, q_u}\right)\theta_{\max} = 0.00298$$

$$W = 0.31\, \pi\, \gamma\, B^2$$

$$= 0.31\, \pi\, 19 \times (1.5)^2 = 41.61 \text{ kN}$$

Therefore

$$\theta_{max} = (0.00298)\frac{(0.68 \times 9.81 \times 332.4)}{41.61}$$

$$= 0.158 \text{ rad} = 9.1°$$

Example 6.3

A 2.5 m wide continuous footing located at 1.5 m below the ground surface is subjected to a vertical transient load ($q_{d\,(max)} = 3000$ kN/m², $t_d = 0.3$ s). The properties of the soil are $\gamma = 18$ kN/m³, $\phi = 30°$ and $c = 50$ kN/m². Calculate the maximum vertical movement of the foundation.

Solution

1.
$$q_u = cN_c + \gamma D_f N_q + \frac{1}{2}\gamma B N_\gamma$$
$$= 50N_c + 18\times1.5N_q + \frac{1}{2}\times18\times2.5N_\gamma$$
$$= 50N_c + 27N_q + 22.5N_\gamma$$

The computations of q_u are done for $\phi = 30°$ and different values of k by taking N_c, N_q and N_γ factors from Table 6.4. These are given below in Table 6.5.

Table 6.5. Computations of q_u for Different Values of k

k	q_u (kN/m^2)
– 0.45	5765
– 0.40	4553
– 0.35	3910
– 0.30	3533
– 0.25	3304
– 0.20	2512
– 0.15	2204
– 0.10	3070
– 0.05	3080
– 0.00	3118
+ 0.05	3181

The minimum value of q_u is obtained at $k = -0.15$ as 2204 kN/m^2.

2. For $k = -0.15$ and $\phi = 30°$, from Table 6.4.

$$N_R = 10.0081 \text{ and } \sqrt{\frac{N_R}{N_I}} = 5.8303$$

$$\frac{N_R\gamma}{q_u} = \frac{10.0081\times18}{2204} = 0.08174$$

$$k' = \sqrt{\frac{2N_R}{N_I B}} = \sqrt{\frac{N_R}{N_I}}\cdot\sqrt{\frac{2}{B}}$$

$$= 5.8303\times\sqrt{\frac{2}{2.5}} = 5.22$$

3.
$$\frac{q_{d(\max)}}{q_u} = \frac{3000}{2204} = 1.36$$

$$t_d K' = 0.3\times5.22 = 1.566$$

For $\frac{q_{d(\max)}}{q_u} = 1.36$ and $t_d k' = 1.566$, Fig. 6.20 gives

$$\frac{\gamma N_R}{q_u}\cdot\Delta_{\max} = 0.034$$

Therefore,
$$\Delta_{max} = \frac{0.034}{0.08174} = 0.416 \text{ m.}$$

Example 6.4

A 2.5 m wide continuous surface footing is subjected to a horizontal transient load of duration 0.4 s applied at a height of 4.0 m from the base of footing. The properties of the soil are $\gamma = 17$ kN/m^3, $c = 30$ kN/m^2 and $\phi = 32°$. Determine the value of the maximum horizontal load that can be applied on the footing. Also compute the rotation at time equal to 0.6 s.

Solution

(i) Determine Q using a suitable factor of safety (= 2.0). For $c = 30$ kN/m^2, $\phi = 32°$, $\gamma = 17$ kN/m^3 and $B = 2.5$ m

$$N_c = \frac{e^{2\pi \tan\phi} - 1}{\tan\phi} = \frac{e^{2\pi \tan 32°} - 1}{\tan 32°} = 79.4$$

$$N_\gamma = \frac{4\tan\phi(e^{3\pi\tan\phi} + 1)}{1 + 9\tan^2\phi} = \frac{4\tan 32°(e^{3\pi\tan 32°} + 1)}{1 + 9\tan^2 32°} = 200$$

$$Q = \frac{1}{2}B\left(cN_c + \frac{1}{2}\gamma BN_\gamma\right)$$

$$= \frac{1}{2} \times 2.5\left(30 \times 79.4 + \frac{1}{2} \times 17 \times 2.5 \times 200\right) = 8290 \text{ kN}$$

(ii) Determine ψ, ε, μ_c, β and $\sin\eta$

$$H = 4.0 \text{ m}$$

$$t_d = 0.4 \text{ s}$$

$$\psi = \frac{e^{2\pi\tan\phi} - 1}{2\tan\phi} = \frac{79.4}{2} = 39.7$$

$$\varepsilon = \frac{\tan\phi(e^{3\pi\tan\phi} + 1)}{(1 + 9\tan^2\phi)} = \frac{200}{4} = 50$$

$$\mu_c = \frac{e^{4\pi\tan\phi} - 1}{16\tan\phi} = 256$$

$$\beta = \frac{e^{3\pi\tan\phi} + 1}{3\left[\sqrt{9\tan^2\phi + 1}\right]} = 56.6$$

$$\bar{x} = \frac{-4B\tan^2\phi(e^{3\pi\tan\phi} + 1)}{(1 + 9\tan^2\phi)(e^{2\pi\tan\phi} - 1)} = (-2B)\frac{50}{39.7} = -2.52B$$

$$\bar{z} = \frac{4B\tan\phi(e^{3\pi\tan\phi} + 1)}{3\left[\sqrt{9\tan^2\phi + 1}\right](e^{2\pi\tan\phi} - 1)} = (2B)\frac{56.6}{39.7} = 2.85B$$

$$\sin\eta = \frac{\bar{z}}{\sqrt{\bar{x}^2 + \bar{z}^2}} = \frac{2.85B}{\sqrt{(-2.52B)^2 + (2.85B)^2}} = 0.75$$

(iii) Determine k, A and E,

$$k = \sqrt{g\beta \frac{\sin\eta}{u_c B}} = \sqrt{\frac{9.81 \times 56.6 \times 0.75}{(256 \times 2.5)}} = 0.807$$

$$A = \frac{g}{\gamma B^4 \mu} = \frac{9.81}{17 \times 2.5^4 \times 256} = 0.0000577$$

$$E = \psi c B^2 + \in \gamma B^3$$

$$= 39.7 \times 30 \times 2.5^2 + 50 \times 17 \times 2.5^3 = 20700 \text{ kN}$$

(iv) Determine $M_{d(max)}$ in terms of λ

$$M_{d(max)} = H.\ Q_{d(max)} = H.\ \lambda.\ Q$$

$$= 4 \times 8290\ \lambda = 33160\ \lambda$$

(v) Determine λ_{cr} which corresponds to $\alpha = 0$

$$\alpha = \frac{A}{k^2}\left(E - \frac{1}{2}QB\right)\cos(kt) - \frac{A}{k^3} \cdot \frac{M_{d(\max)}}{t_d}\sin(kt) + \frac{A}{k^2}\left[\frac{M_{d(\max)\cdot t}}{t_d} + \frac{1}{2}QB - E\right]$$

For $t = t_d$

$$\alpha = \frac{0.0000577}{0.807^2}\left(20700 - \frac{1}{2} \times 8290 \times 2.5\right)\cos(0.8070 \times 0.4)$$

$$- \frac{0.0000577}{0.807^3} \cdot \frac{33160\lambda}{0.40}\sin(0.8070 \times 0.4)$$

$$+ \frac{0.0000577}{0.807^2} \cdot \left[\frac{33160\lambda \times 0.4}{0.40} + \frac{1}{2} \times 8290 \times 2.5 - 20700\right]$$

$$= 0.9159 \cos(0.3228) - 9.10\ \lambda \sin(0.3228) + 2.94\ \lambda$$
$$+ 0.9181 - 1.834 = 0.05\ \lambda - 0.0474$$

For $\alpha = 0$, $\lambda = 0.948 = \lambda_{cr}$

(vi) Determine $M_{d(\max)}$ for $\lambda = \lambda_{cr}$

$$M_{d(\max)} = 33160\ \lambda_{cr} = 33160 \times 0.948 = 31436 \text{ kN-m}$$

(vii) Determine G_1 and G_2

$$G_1 = \frac{A}{k^2}\left(E - \frac{1}{2}Q.B\right)\cos(kt_d)$$

$$- \frac{A}{k^3} \cdot \frac{M_{d(\max)}}{t_d}\sin(kt_d) + \frac{A}{k^2} M_{d(\max)}$$

$$= 0.8685 - 2.89 \times 0.948 + 2.94 \times 0.948$$

$$= 0.9159$$

$$G_2 = -\frac{A}{k}\left(E - \frac{1}{2}Q.B\right)\sin(kt_d) - \frac{A}{k^2}\cdot\frac{M_{d(\max)}}{t_d}\cos(kt_d) + \frac{A}{k^2}\cdot\frac{M_{d(\max)}}{t_d}$$

$$= -0.9159 \times 0.807 \sin(0.3228) - 9.1 \times 0.948 \times 0.807 \cos(0.3228) + 2.94 \times 0.948 = -0.2346 - 6.60 + 2.79 = -4.05$$

(viii) Determine α for $t = 0.6$ *s*

$$\alpha = \left(\frac{1}{k}\right)[G_1 k\cos(kt_d) - G_2 \sin(kt_d)]\cos(kt_d)$$

$$+\left(\frac{1}{k}\right)[G_1 k\sin(kt_d) - G_2\cos(kt_d)]\sin(kt_d) + \frac{A}{K^2}\left(\frac{1}{2}QB - E\right)$$

$$= \frac{1}{0.807}[0.9159\times0.807\cos(0.807\times0.4)$$

$$+4.05\sin(0.807\times0.4)]\cos(0.807\times0.6)$$

$$+\frac{1}{0.807}[0.9159\times0.807\sin(0.807\times0.4)$$

$$-4.05\cos(0.807\times0.4)]\sin(0.807\times0.6)$$

$$+\frac{0.0000577}{0.807^2}\left[\frac{1}{2}\times8290\times2.5 - 20700\right]$$

$$= \frac{1}{0.807}[0.701+1.285]\times884$$

$$+\frac{1}{0.807}[0.2346-3.841]\times0.466-0.9159$$

$$= 2.175 - 2.082 - 0.9159 = -0.8229 \text{ rad.}$$

Example 6.5

A rectangular footing of size 1.5 m × 3.0 m is provided at the depth of 1.0 m below the ground surface. The soil properties are

$$c = 10 \text{ kN/m}^2, \varphi = 30° \text{ and } \gamma = 17.5 \text{ kN/m}^2$$

Determine (i) Ultimate static bearing capacity and (ii) Ultimate seismic bearing capacity for horizontal and vertical accelerations as $0.3g$ and $0.2g$ respectively, using Meyerhof' method and (ii) Rangwala et al. method.

Solution

Shape factors:

$$S_c = 1+0.2\frac{B}{L} = 1 + 0.2 \times \frac{1.5}{3.0} = 1.1$$

$$S_c = 1+0.2\frac{B}{L} = 1+0.2\times\frac{1.5}{3.0} = 1.1$$

$$S_\gamma = 1-0.4\frac{B}{L} = 1-0.4\times\frac{1.5}{3.0} = 0.80$$

Depth factors:

$$d_c = 1-0.4\frac{D_f}{B} = 1+4\times\frac{1.0}{1.5} = 1.27$$

$$d_q = 1+2\tan\phi(1-\sin\phi)^2\frac{D_f}{B} = 1+2\tan 30(1-\sin 30)^2\times\frac{1.0}{1.5} = 1.192$$

$$d_\gamma = 1.0$$

From Table 6.2, for $\varphi = 30^\circ$

$$N_{cS} = 30.14,\ N_{qS} = 18.4 \text{ and } N_{\gamma S} = 22.40$$

$$q_{uS} = cN_{cS}s_c d_c + qN_{qS}s_q d_q + {}^1\!/_2\,\gamma BN_{\gamma S}s_\gamma d_\gamma$$

$$= 10 \times 30.14 \times 1.1 \times 1.27 + 17.5 \times 1.0 \times 18.4 \times 1.1 \times 1.192$$

$$+ 1/2 \times 17.5 \times 1.5 \times 22.4 \times 0.8 \times 1.0$$

$$= 421.05 + 422.21 + 235.20 = 1078.45 \text{ kN/m}^2$$

From Meyerhof (1953) approach
Load will act at inclination

$$i = \tan^{-1}\left(\frac{\alpha_h}{1-\alpha_v}\right) = \tan^{-1}\left(\frac{0.3}{1-0.2}\right) = 20.5^\circ$$

$$i_c = i_q = \left(1-\frac{i}{90}\right)^2 = \left(1-\frac{20.5}{90}\right)^2 = 0.595$$

$$i_q = \left(1-\frac{i}{\varphi}\right)^2 = \left(1-\frac{20.5}{30}\right)^2 = 0.1$$

$$q_{uE} = (421.05 + 422.21) \times 0.595 + 235.20 \times 0.1$$

$$= 501.74 + 23.52 = 525.26 \text{ kN/m}^2$$

$$\frac{q_{uE}}{q_{uS}} = \frac{525.26}{1078.45} = 0.487$$

From Rangwala et al. (2012) approach

$$N_{\gamma S} = 0.5\,(e^{8\tan(0.9\varphi)} - 1) = 0.5(e^{8\tan(0.9\times 30)} - 1) = 28.96$$

$$N_{qS} = \frac{e^{2\left(\frac{3\pi}{4}-\frac{\varphi}{2}\right)\tan\phi}}{1-\sin\varphi} = \frac{e^{2\left(\frac{3\pi}{4}-\frac{30}{2}\times\frac{\pi}{180}\right)\tan 30}}{1-\sin 30} = 22.43$$

$$N_{cS} = (N_{qS} - 1)\cot\phi = (22.43 - 1)\cot 30 = 37.11$$

$$q_{uS} = cN_{cS}s_c d_c + qN_{qS}s_q d_q + \frac{1}{2}\gamma BN_{\gamma S}s_\gamma d_\gamma$$

$$= 10 \times 37.11 \times 1.1 \times 1.27 + 17.5 \times 10 \times 2243 \times 1.1 \times 1.192$$

$$+ 1/2 \times 17.5 \times 1.5 \times 28.96 \times 0.8 \times 1.0$$

$$q_{uS} = 518.43 + 514.68 + 314.58 = 1347.69 \text{ kN/m}^2$$

$$\psi = \tan^{-1}\left(\frac{\alpha_h}{1-\alpha_v}\right) = \tan^{-1}\left(\frac{0.3}{1-0.2}\right) = 20.5$$

$$SRF_\gamma = e^{-6.994 \tan \psi} = e^{-6.994 \tan 20.5} = 0.073$$

$$a_q = 5(\tan \phi)^{0.44} = 5(\tan 30)^{0.44} = 3.926$$

and $$SRF_q = e^{-3.926 \times 0.3} = 0.3079$$

$$a_c = 4.45 (\tan \varphi)^{0.63} = 4.45(\tan 30)^{0.63} = 3.148$$

and $$SRF_c = e^{-a_c A_h} = e^{-3.148 \times 0.3} = 0.3889$$

$$q_{uE} = 518.43 \times 0.3889 + 514.68 \times 0.3079 + 314.58 \times 0.073$$

$$= 201.62 + 158.47 + 22.96 = 383.05 \text{ kN/m}^2$$

$$\frac{q_{uE}}{q_{uS}} = \frac{383.05}{1347.69} = 0.284$$

Using the static bearing capacity factors as given in Table 6.2 and seismic reduction factors as proposed by Rangwala et al. (2012).

$$q_{uE} = 421.05 \times 0.3889 + 422.21 \times 0.3079 + 235.19 \times 0.073$$

$$= 136.75 + 130.00 + 17.17 = 310.92 \text{ kN/m}^2$$

$$\frac{q_{uE}}{q_{uS}} = \frac{310.92}{1078.45} = 0.289$$

Using the static bearing capacity factors proposed by Rangwala et al. (2012) and inclination factors given by Meyerhof (1953),

$$q_{uE} = 518.43 \times 0.595 + 514.68 \times 0.595 + 314.58 \times 0.1$$

$$= 309.46 + 306.22 + 31.46 = 647.14 \text{ kN/m}^2$$

$$\frac{q_{uE}}{q_{uS}} = \frac{647.14}{1347.69} = 0.48$$

From the above exercise, following points may be noted:

(i) Ratio q_{uE}/q_{uS} depends on (a) inclination factors and (b) seismic reduction factors.
(ii) Values of q_{uE}/q_{uS} is less when obtained using seismic reduction factor.

Therefore it may be concluded that use of seismic reduction factors (or obtaining the value of seismic bearing capacity using analysis considering seismic acceleration in soil mass) should be adopted for safe design. Static bearing capacity factors proposed by Rangwala et al. (2012) matches very well with the bearing capacity factor given by Saran (1971). Readers may adopt these factors for design.

References

Agarwal, R.K. (1986), "Behaviour of shallow foundations subjected to eccentric-inclined loads". Ph. D. Thesis, University of Roorkee, India.

Al-Karni, A. (1993), "Seismic Settlement and Bearing Capacity of Shallow Footings on Cohesionless Soil." Ph.D. Thesis, The University of Arizona, USA.

Budhu, M. and Al-Karni, A. (1993), "Seismic bearing capacity of soils." Geotechnique, 43(1), 181-187.

Caquot. A. (1934), "Equilibre des massifs a frottement interene", Ganthiv Villars, Paris.

Chummar, A.V. (1965), "Dynamic bearing capacity of footings", Master of Engineering Dissertation, University of Roorkee. India.

Cunny, R.W. and Sloan, R.C. (1961), "Dynamic loading machine and results of preliminary small-scale footing tests", A. S. T. M. Symposium on Soil Dynamics, Special Technical Publications. Nc. 3.5, pp 65-77.

De Beer, E. and Vesic, A. (1958), "Etude experimental de la capacite portante du sable sons des foundations directed ctablies en surface", Annales des Travaux Public de Delgigue, 59 (3), pp 5-58.

Dormieux, L. and Pecker, A. (1995), "Seismic Bearing Capacity of Foundation on Cohesionless Soil." Journal of Geotechnical Engineering, 121(3), 300.

Fellenius, W. (1948), "Erdstatische berchnugen", 4th ed., W. Ernst Und Sohn, Berlin.

Fisher, W.E. (1962), "Experimental studies of dynamically loaded footings on sand", Report to U.S. Army Engineer Waterways Experiment Station. University of Illinois, Soil Mechanics Series No. 6.

Hansen, J.B. (1970), "A revised and extended formula for bearing capacity", Bull. No. 28. Danish Geotechnical Institute, Copenhegen.

IS : 6403 (1981), "Code of practice for determination of bearing capacity of shallow foundations", ISI, New Delhi.

Johnson, T.D. and Ireland H.O. (1963), "Tests on clay subsoils beneath statically and dynamically loaded spread footings", Report to U.S. Army Engineer Waterways Experiment Station, University of Illinois, Soil Mechanics Series no. 7.

McKee, K.E. and Shenkman, S. (1962), "Design and analysis of foundations for protective structures", Final Report to Armour Research Foundation, Illinois Institute of Technology.

Meyerhof, G.G. (1951), "The ultimate bearing capacity of foundations", Geotechnique, Vol. 2, No. 4, pp. 301-331.

Meyerhof, G.G. (1953), "The bearing capacity of footings under eccentric and inclined loads", Proc. Third Int. Conf. Soil Mech. Foun Engg., Zurich, vol. 1. pp. 440-445.

Paolucci, R., and Pecker, A. (1997). "Seismic bearing capacity of shallow strip foundations on dry soils." Soils and Foundations, 37(3), 95-105.

Rangwala, H., Saran, S. and Mukerjee, S. (2012). "Seismic bearing capacity of footings adjacent to a stable slope." Ph. D. Research Proposal Report strip, IIT Roorkee, Roorkee.

Richards, R., Elms, D.G., and Budhu, M. (1993). "Seismic Bearing Capacity and Settlements of Foundations."Journal of Geotechnical Engineering, 119(4), 662.

Saran, S.K. and Rangwala, H. (2011), "Seismic bearing capacity of footings." Geotechnical Engineering. 5(4). 475-483.

Saran, S.K. (1975), "Seismic stability of earth dams and embankments." Geotechnique, 25(4), 743-761.

Sarma, S.K. and Iossifelis, I.S. (1990), "Seismic bearing capacity factors of shallow strip footing." Geotechnique, 40(2), 265-273.

Shafiee, A.H. and Jahanandish, M. (2010), "Seismic bearing capacity factors for strip footings." 5th National Congress on Civil Engineering, Mashhad, Iran.

Soubra, A.H. (1999), "Upper-Bound Solutions for Bearing Capacity of Foundations." Journal of Geotechnical and Geoenvironmental Engineering, 125(1), 59.

Saran, S. and Agarwal, R.K. (1989), "Eccentrically-obliquely loaded footings", ASCE, Journal of Geot. Engs. Vol. 115, No. 11, pp. 1673-1680.

Saran, S. and Agarwal, R.K. (1991), "Bearing capacity of eccentrically-obliquely loaded footings", ASCE. Journal of Geot. Engs., Vol. 117, No. 11, p. 1669-1690.

Terzaghi, K. (1943). "Theoretical soil mechanics", John Wiley and Sons, New York.

Terzaghi, K. and Peck, R.B. (1967), "Soil mechanics in engineering practice", Ist Ed. John Wiley and Sons, New York.

Triandafilidis, G.E. (1961), "Analytical study of dynamic bearing capacity of foundations", Ph. D. Thesis, University of Illinois, Urbana, Illinois.

Triandafilidis, G.E. (1965), "Dynamic response of continuous footings supported on cohesive soils", Proc. sixth Int. Conf. Soil Mech. Found. Engin., Montreal, Vol. 2, pp. 205-208.

Vesic, A.S. (1973), "Analysis of ultimate loads of shallow foundation", J SMFD, ASCE, Vol, 99, SMI. pp. 45-73.

Wallace, W.L. (1961), "Displacement of long footings by dynamic loads", ASCE, Journal of the Soil Mechanics and Foundation Division, 87, SM5, pp. 45-68.

White, C.R. (1964), "Static and dynamic plate bearing tests on dry sand without overburden", Report R 277, U. S. Naval Civil Engineering Laboratory.

Practice Problems

6.1 Describe stepwise pseudo-static analysis of designing footing subjected to earthquake loading.

6.2 Differentiate between Triandafilidis and Wallace analyses of dynamic bearing capacity of footing subjected to transient vertical load. Give the salient features of any one.

6.3 Describe the method of obtaining the maximum horizontal dynamic load that can be applied on the footing. Give the expression of determining the rotation of the footing.

6.4 A 2.0 m wide strip footing is subjected to a vertical transient pulse ($q_u = 600\ e^{-8\,t}$). The soil supporting the foundation is clay with c_u = 50 kN/m^2. The unit weight of soil is 18 kN/m^3. Determine the maximum angular rotation of footing that it might undergo.

6.5 A 2.0 m wide footing located at 1.0 m below ground surface is subjected to a vertical transient load ($q_{d\,(max)}$ = 2000 kN/m^2, t_d = 2 s). The properties of the soil are γ = 17 kN/m^3, ϕ = 32° and c = 30 kN/m^2. Using Wallace's approach, determine maximum vertical movement of the foundation.

6.6 A 3.0 m wide surface footing is subjected to a horizontal dynamic load having duration 0.3 s. The properties of soil are γ = 18 kN/m^3. ϕ = 35° and c = 20 kN/m^2. Using Chummar's approach, determine the value of maximum horizontal load that can be applied on the footing. Also determine the rotation of footing after 0.2 s and 0.4 s.

CHAPTER

7

Pile Foundations under Dynamic Loads

7.1 Introduction

Piles are relatively long and slender members used to transfer loads through weak soil or water to deeper soil or rock strata having a high bearing capacity. When a pile passes through poor material and its tip penetrates a small distance into a stratum of good bearing capacity, it is called a bearing pile. When a pile is installed in deep stratum of limited supporting ability and the pile develops its carrying capacity by friction on its sides, it is called friction pile. Many times, the load carrying capacity of a pile results from the combination of point resistance and skin friction.

The introduction of piles in a soil stratum makes the system stiff. Due to this both natural frequency and amplitudes of motion are affected. In all vibration problems, resonance needs to be avoided. Therefore, the natural frequency of the structure-soil-pile system is required for analysis and design.

Basically there are two types of approaches namely (i) pseudo-static analysis, and (ii) dynamic analysis. In this chapter, pseudo-static analysis is discussed first, and it is followed by dynamic analysis.

7.2 Pseudo-Static Analysis

Pseudo-static analysis is sometimes used for designing the pile foundation for the structures located in seismic regions. Adopting appropriate values of horizontal and vertical seismic coefficients, equivalent seismic forces are evaluated. These forces in combination of static forces make the foundation subjected to eccentric-inclined load.

7.2.1 Distribution of Load in Piles of a Pile Group Having only Vertical Piles and Subjected to Eccentric-Inclined Load

Considering that at the base of pile cap, the loads transferred from superstructure are (i) Vertical load (V), (ii) Moment (M), and (iii) Horizontal load (H) as shown in Fig. 7.1 (a). The load in a pile can be obtained using Eq. (7.1) (Saran, 1998, Fig. 7.1b):

$$V_i = \frac{V}{n} + \frac{V.e.x_1}{\sum x_i^2} \tag{7.1a}$$

$$e = \frac{M}{V} \tag{7.1b}$$

where
V_i = Load on i^{th} pile
V = Total vertical load acting on the pile group
n = Total number of piles in the pile group
e = Amount of eccentricity with respect to the centre of the pile group
x_i = Distance of the centre of the ith pile from the centre of pile group, measured parallel to e.

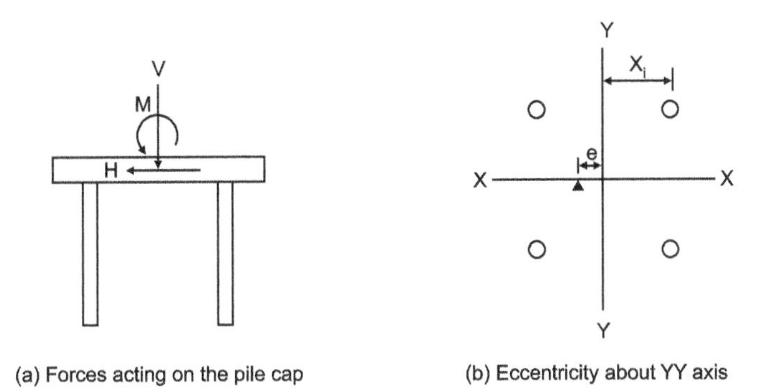

Fig. 7.1. Typical arrangement of vertical piles in a group subjected to vertical load, moment and horizontal force

The ith pile is then analysed considering that it is subjected to a lateral load of magnitude $(1.2\ H/n)$ at the top of the pile in addition to an axial load of V_i. The procedure of analyzing the single pile subjected to lateral load has been discussed subsequently.

7.2.2 Distribution of Load between Vertical and Batter Piles of a Pile Group

Figure 7.2 (a) shows a pile group having vertical positive battered and negative battered piles. Usually the inclination of piles are decided in such a way that all the piles are subjected to axial loads only. For example, let the piles 3 and 4 are inclined with vertical by an angle θ_1, and piles 5 and 6 are inclined with vertical by an angle θ_2 (Fig. 7.2a). The forces in each pile can be obtained by the following procedure:

(i) Resolve resultant force R into a vertical component V and a horizontal component H.

(ii) Ignore the horizontal component, and treat the pile group as if all piles were vertical. Pile load is then determined using Eq. (7.1a).

(iii) Each pile is assumed to be subjected to an axial, R_1, R_2,... where

$$R_{pi} = V_{pi}/\cos\theta_i \tag{7.2}$$

where θ_i is the angle of ith pile with vertical.

(iv) Usually the inclination of either of the positive or negative battered piles is decided first, then the inclination of other is determined satisfying the following equation:

$$(R_{p3} + R_{p4})\sin\theta_1 = (R_{p5} + R_{p6})\sin\theta_2 \tag{7.3a}$$

or

$$(V_{p3} + V_{p4})\tan\theta_1 = (V_{p5} + V_{p6})\tan\theta_2 \tag{7.3b}$$

The values of V_{p3}, V_{p4}, V_{p5} and V_{p6} are known in step no. (ii). If value of θ_1 is selected, then θ_2 can be obtained solving Eq. (7.3b). If the arrangement of the piles in a group is done as illustrated above, then the piles will be designed only for their axial forces i.e. R_{p1}, R_{p2}, ..., R_{pn}. Figure 7.2b shows polygon of forces giving axial loads in various piles. If θ_1 and θ_2 are properly selected then $H' = 0$. However it is considered acceptable if H' is less than 4 kN per pile.

7.2.3 Axial Capacity of Single Pile

The axial capacity of a single pile mainly depends on two factors namely (i) soil properties along the shaft and below the tip of the pile, and (ii) method of installation of pile, i.e. whether the pile is precast-driven pile or cast-in-situ bored pile.

In this section, approaches common in use to determine the bearing capacity of pile have been discussed, keeping in view the method of installation and type of soil.

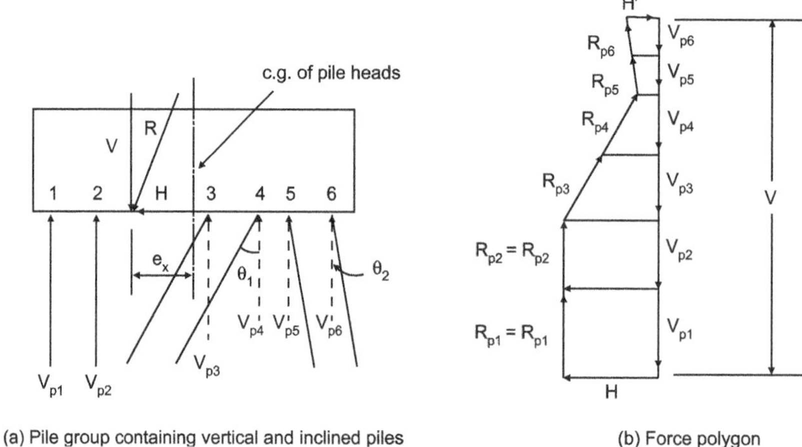

Fig. 7.2. Procedure of determining loads in piles of a pile group having vertical and inclined piles

Ultimate Bearing Capacity of a Driven Pile in Cohesionless Soil

Standard penetration test and static cone penetration test are the two important tests by using which the bearing capacity of driven pile in cohesionless soil can be predicted with reasonable accuracy.

Using standard penetration test, the value of ultimate capacity of a driven pile in cohesionless soil can be obtained by the procedure described below:

(i) Perform standard penetration tests up to the expected depth of penetration of pile. Let N_1 represents the average of observed N-values along the length of pile and N_2 is the observed N-value near the pile tip. It may be noted that the N-values are not corrected for effective overburden pressure.

(ii) The point bearing load (or base resistance), Q_p is given by:

$$Q_p = A_b \cdot \sigma'_v (N_q - 1) \tag{7.4}$$

where A_b = Area of cross-section of pile tip, m²

σ'_v = Effective overburden pressure at the pile tip, kN/m² and

N_q = Bearing capacity factor

For determining N_q, firstly using Fig. 7.3 the value of ϕ is obtained corresponding to SPT value N_2. The value of N_q is then read from Fig. 7.4 in case the pile is penetrated into the bearing stratum greater than five times the pile diameter. For lesser penetration, it is recommended to get the value of N_q factor using Fig. 7.5.

As per Tomlinson (1986), the maximum unit base resistance ($q_p = Q_p/A_b$) is normally limited to 11000 kN/m², whatever might be the penetration depth.

(iii) The side friction resistance, Q_s is given by:

$$Q_s = f \cdot A_s = K \cdot \sigma'_v \tan\delta \cdot A_s \tag{7.5}$$

where f = Unit skin friction, kN/m²

K = Earth pressure coefficient

σ'_v = Average effective overburden pressure over embedded depth of pile, kN/m²

δ = Angle of wall friction and

A_S = Embedded surface area of pile, m²

For determining K and δ, firstly using Fig. 7.3 the value of ϕ is obtained for SPT value N_1. Broms (1966) has suggested the values of K and δ as shown in Table 7.1.

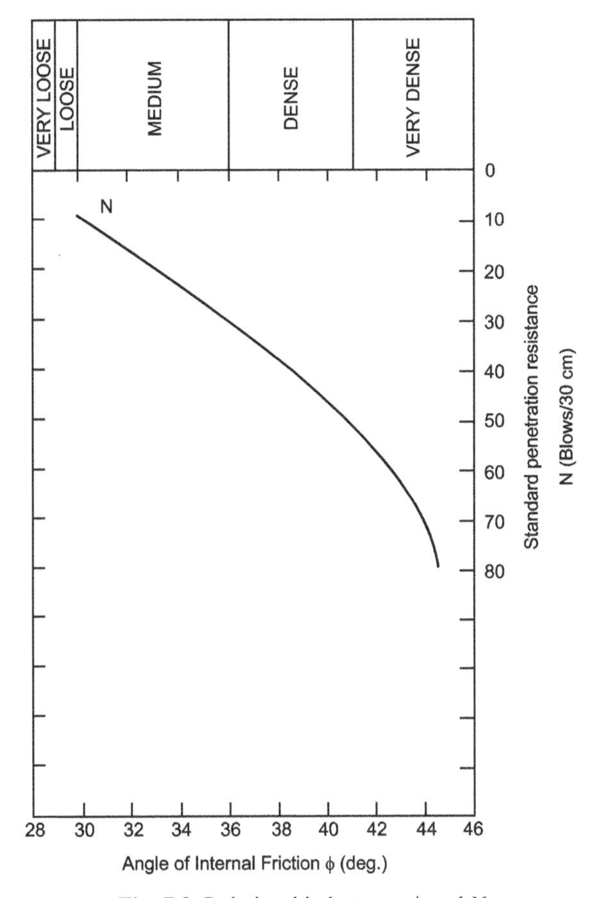

Fig. 7.3. Relationship between ϕ and N

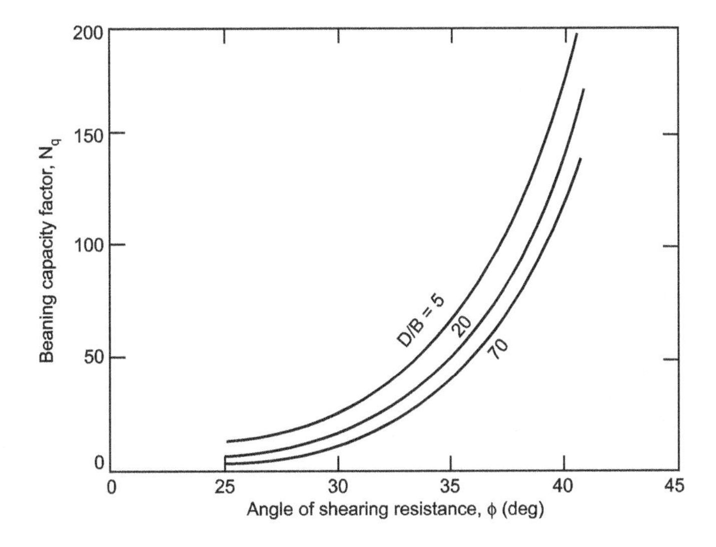

Fig. 7.4. Bearing capacity factor N_q (Berezantesv, 1961)

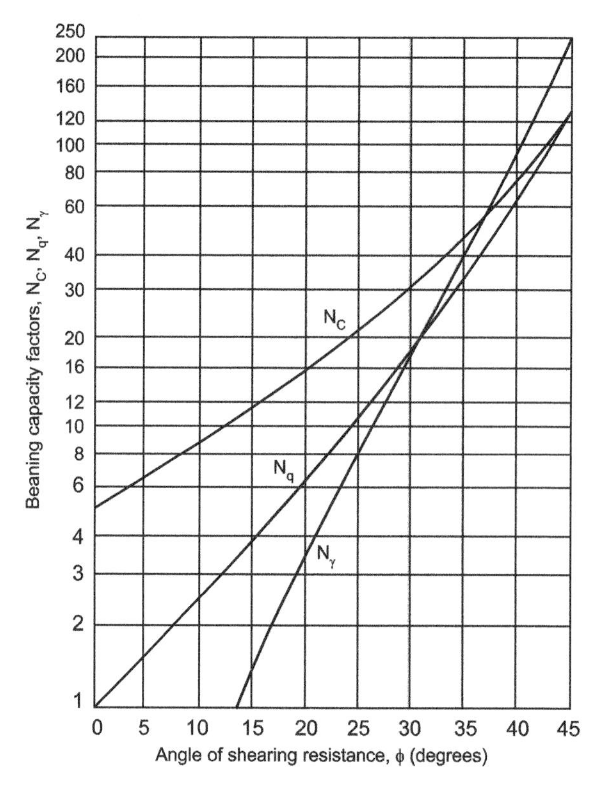

Fig. 7.5. Bearing capacity factor N_q (Hansen, 1961)

Table 7.1. Values of K and δ

Pile material	*δ (Deg.)*	*Values of K*	
		Low relative density	*High relative density*
Steel	20	0.5	1.0
Concrete	3/4 φ	1.0	2.0
Wood	2/3 φ	1.5	4.0

Equation (7.5) implies that the unit-skin-friction (f) increases continuously with increasing depth of embedment of pile. Vesic (1970) showed that at some penetration depth between 10 and 20 pile diameters, a peak value of unit skin friction is reached which is not exceeded at greater penetration depths. Therefore, the use of Eq. (7.5) is recommended if the depth of pile penetration is limited to 20 diameter. For pile driven deeper than 20 pile diameter, Tomlinson (1986) suggested the value of unit skin friction as given in Table 7.2.

Table 7.2. Values of Unit Skin Friction ($L_f/B > 20$)*

Relative density (%)	*Unit skin friction, f (kN/m²)*
Less than 35 (loose)	10
35-65 (medium dense)	10-25
65-85 (dense)	25-70

* L_f = Embedded length of pile, B = Diameter of pile

Using static cone penetration test data, the procedure of obtaining bearing capacity of a pile is described in the following steps:

(i) Make the plot of cone resistance (q_c) versus depth

(ii) Determine the average cone resistance over the range of depth ($3D + D = 4D$) as shown in Fig. 7.6. Let it be q_{c1} kN/m^2.
The ultimate point resistance is given by

$$Q_p = q_{c1}.A_b \text{ (kN)} \tag{7.6}$$

(iii) Determine the average cone resistance along the pile shaft. Let it be q_{c2}, kN/m^2
The ultimate side resistance of a driven concrete pile is given by

$$Q_s = (q_{c2}/200).A_s \text{ (kN)} \tag{7.7}$$

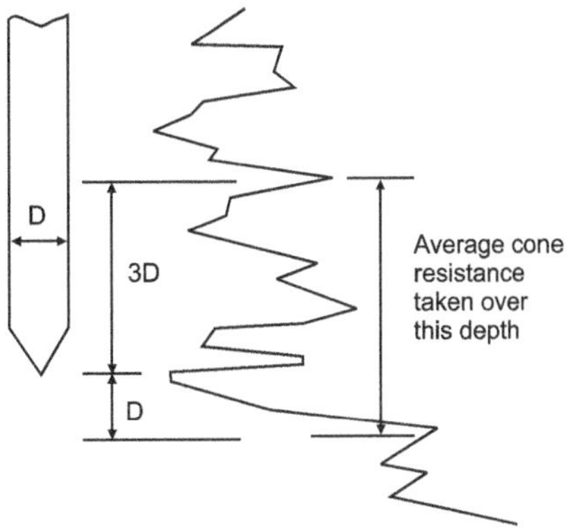

Fig. 7.6. Zone considered for computing end resistance of pile from SCPT.

For driven steep *H* piles,

$$Q_s = (q_{c2}/400).A_s \text{ (kN)} \tag{7.8}$$

Working load of a pile can be used using a factor of safety of 2.5. Therefore

$$\text{Working load} = (Q_p + Q_s)/2.5 \tag{7.9}$$

Ultimate Bearing Capacity of a Bored and Cast-in-Situ Pile in Cohesionless Soil

In all cases of bored piles formed in cohesionless soils, the soil will get loosened due to boring even though it may be initially in dense or medium dense state. Equations (7.5) and (7.6) may be used for computing point bearing and skin friction resistance adopting the value of ϕ corresponding to loose condition ($\phi \leq 30°$). However, if the piles are installed by rotary drilling under a bentonite slurry, computations of both the skin friction and point bearing resistance may be done for the ϕ corresponding to undisturbed conditions (Fleming and Sliwinski, 1977).

Ultimate Bearing Capacity of Driven and Cast-in-Situ Piles in Cohesionless Soils

Driven and cast-in-situ piles are formed by driving a tube into the ground. The bearing capacity of pile in which the tube is left into the ground can be obtained by the procedure given for driven piles. In the case the tube is withdrawn with the placing of concrete, the procedure given in for bored and cast-in-situ piles may be adopted for computing bearing capacity.

Ultimate Bearing Capacity of Driven Piles in Cohesive Soils

The ultimate bearing capacity of a driven pile in cohesive soil is given by :

$$Q_u = N_c \cdot c_b \cdot A_b + \alpha \bar{c}_u A_s \tag{7.10}$$

where
N_c = Bearing capacity factor = 9.0
c_b = Undisturbed shear strength of clay at the pile tip
A_b = Area of cross-section of pile tip
α = Adhesion factor
$\bar{c}_u$ = Average undrained shear strength of clay adjacent to shaft, and
A_s = Embedded surface area of pile.

Driven piles cause displacement of soil. Due to this pore pressures get increased which reduces the effective stress and therefore skin friction capacity will be smaller. Keeping this in view, Tomlinson (1986) recommended the charts shown in Fig. 7.7 to get the value of adhesion factor α. These charts are essentially applicable to piles carrying light to moderate loading driven to a relatively shallow penetration into the bearing stratum. Where heavy loads are carried, the piles may be driven very deeply into the bearing stratum. For such cases the ultimate bearing capacity of pile is given by:

$$Q_u = N_c \cdot c_b \cdot A_b + F \alpha_p \bar{c}_u A_s \tag{7.11}$$

where N_c, c_b, A_b, $\bar{c}_u$ and A_s have the same meaning as in Eq. (7.10). α_p is peak adhesion factor depending on the ratio of average undrained shear strength and average effective overburden pressure $(\bar{c}_u / \sigma_v)$. This is shown in Fig. 7.8a. F is a factor depending on the slenderness ratio of pile (L/B). This is shown in Fig. 7.8b.

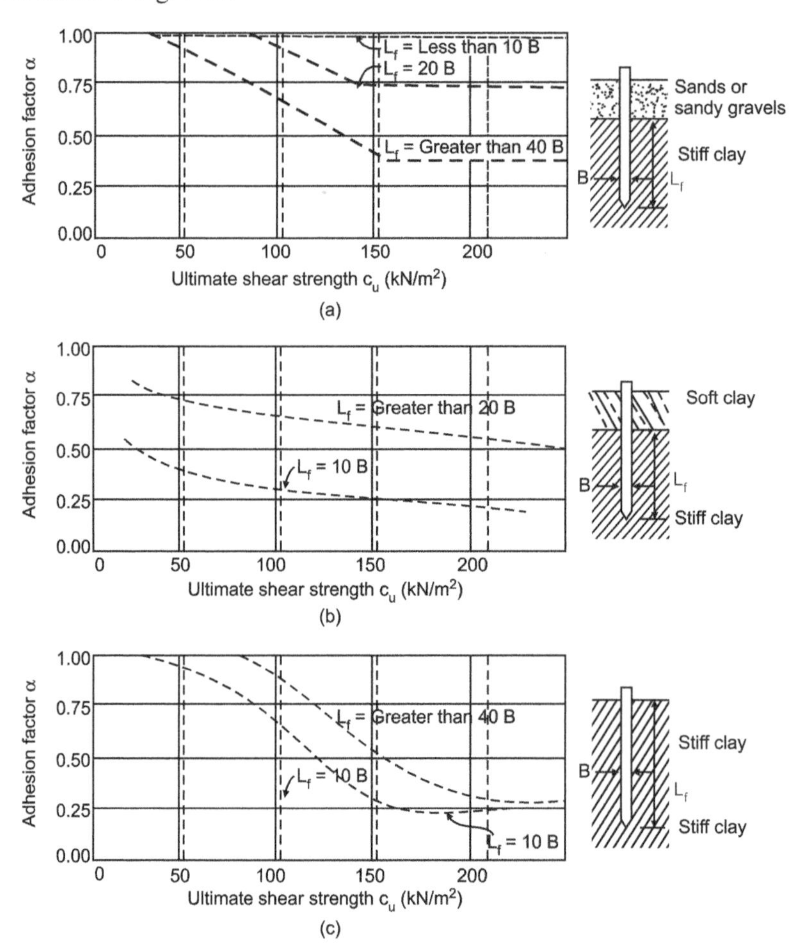

Fig. 7.7. Adhesion factors for driven piles in clay (Tomlinson, 1986)

Ultimate Bearing Capacity of Bored and Cast-in-Situ Pile in Cohesive Soil

Ultimate bearing capacity of bored and cast-in-situ piles can be obtained using Eq. (7.10) by taking the value of adhesion factor α as 0.45 in medium clays and 0.3 in highly fissured clays.

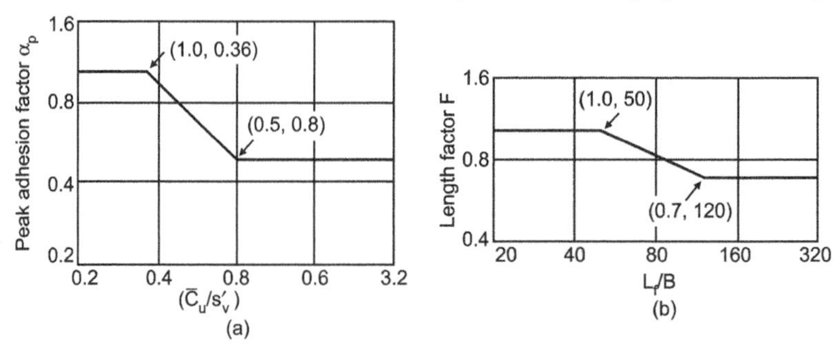

Fig. 7.8. Adhesion factors for heavily-loaded piles driven to deep penetration

Ultimate Bearing Capacity of Piles in c–ϕ Soils

The point bearing capacity of a pile in c–ϕ soil is given by:

$$Q_p = A_b[1.2\, cN_c + \gamma L_f(N_q - 1) + 0.4\, \gamma\, BN_\gamma] \tag{7.12}$$

where
A_b = Area of cross-section of pile tip
c = Unit cohesion
γ = Unit weight of soil
B = Diameter of pile
N_c, N_q, N_γ = Bearing capacity factors (Fig. 7.5) and
L_f = Embedded length of pile.

The side soil resistance can be computed as the sum of adhesion and skin friction given by the following equation:

$$Q_s = A_s[K\bar{\sigma}_v \tan\delta + \alpha\bar{c}_u] \tag{7.13}$$

Values of K and α should be appropriately selected keeping in view the method of installation of pile and density/stiffness of the soil.

If the pile toe is terminated in a layer of stiff clay or dense sand underlain by soft lay or loose sand there is a risk of pile punching through the weak layer. Meyerhof (1976) suggested the following equation to compute the base resistance of the pile in the strong layer where the thickness H between the pile tip, and the top of the weak layer is less than the critical thickness of about 10 B (Fig. 7.9):

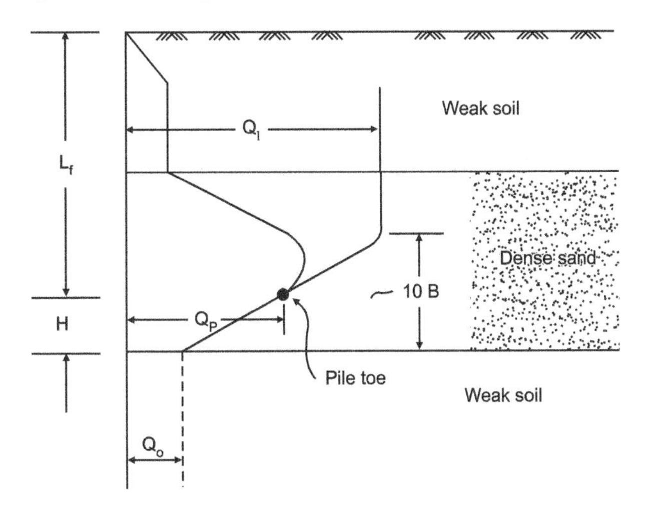

Fig. 7.9. End bearing resistance of piles in layered soils (Meyerhof, 1976)

$$Q_p = Q_0\left(\frac{Q_l - Q_0}{10B}\right)H < Q_l \tag{7.14}$$

where Q_o and Q_l are the ultimate base resistance of the pile in lower weak and upper strong layers respectively.

Pile Load Test

Pile load test is the most reliable and acceptable method of determining pile load capacity. It may be performed on a driven pile or a cast-in-situ pile. The number of tests to be conducted at a project site may vary upto 2 percent or more, depending on the nature of the soil strata and importance of the project. The details of conducting a pile load test are given by Saran (1998). IS:2911 Part IV-2010 suggested the procedure of getting the allowable load capactiy of pile.

7.2.4 Vertical Pile Subjected to Lateral Load

Lateral load and moments may act on piles in addition to axial loads. Two types of piles are encountered in practice, namely (i) long piles, and (ii) short piles. When a pile length is greater than a particular length, the length loses its significance. The behaviour of the pile will not be affected if the length is greater than this particular length. Such piles are called as long flexible piles. In case of short piles, the flexural stiffness (EI) on the material of pile loses its significance. The pile behaves as a rigid member and rotates as a unit.

Three types of boundary conditions occur in practice, namely (i) free-head pile, (ii) fixed-head pile, and (iii) partially-restrained head pile. In the case of free-head pile, the lateral load may act at or above ground level and the pile head is free to rotate without any restraint. A fixed-head pile is free to move only laterally but rotation is prevented completely, whereas a pile with partially restrained head moves and rotates under restraint.

Methods of calculating lateral resistance of vertical piles can broadly be divided into two categories:

(i) Methods of calculating ultimate lateral resistance
   (a) Brinch Hanson's method (1961): This method is based on earth pressure theory and is applicable only to short piles.
   (b) Brom's method (1964a, b): This method is also based on earth pressure theory with the simplifying assumption for distribution of ultimate soil resistance along the pile length. This method is applicable for both short and long piles.

(ii) Methods of calculating acceptable deflection at working load
   (a) Modulus of subgrade reaction approach (Reese and Matlock, 1956): In this method it is assumed that the soil acts as a series of independent linearly elastic springs.
   (b) Elastic approach (Poulos, 1971a and b): In this method, the soil is assumed as an ideal elastic continuam.

Modulus of subgrade reaction approach is relatively simple and has been used in practice for a long time. This method can incorporate factors such as nonlinearity, variation of subgrade reaction with depth and layer system. In this chapter only this method has been dealt with.

Subgrade Reaction Approach

In this approach a laterally loaded pile is treated as a beam on elastic foundation. It is assumed that the pile is supported by a series of infinity closed spaced independent and elastic springs as shown in Fig. 7.10b. The stiffness of these springs K_h (also called the modulus of horizontal subgrade reaction) is expressed as below:

$$K_h = p/y \tag{7.15}$$

where p = Soil reaction per unit length of pile, and

y = Pile deformation.

Palmer and Thompson (1948) employed the following form to express the modulus of subgrade reaction:

$$K_x = K_h(x/L)^n \tag{7.16}$$

where K_h = Value of K_x at tip of pile, i.e. at x equal to L

x = Depth of any point along pile length measured from ground surface

n = Coefficient equal to or greater than zero

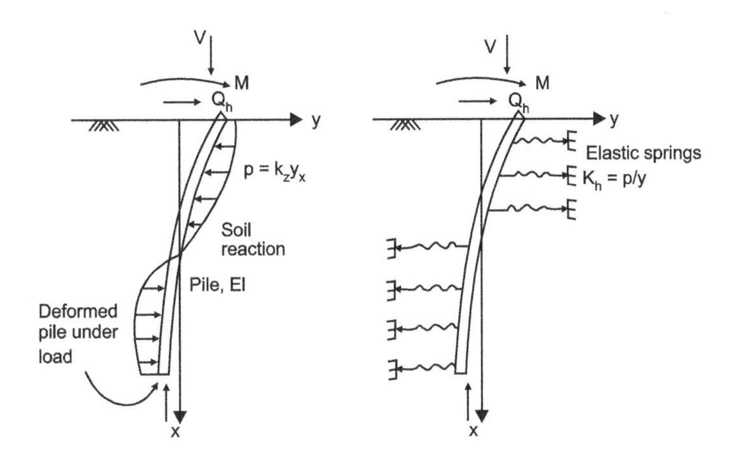

Fig. 7.10. Behaviour of laterally loaded pile as beam on elastic foundation

The most commonly used value of n for sands and normally consolidated clays is unity. For over consolidation clay, n is taken as zero. For the value of $n = 1$, the variation of K_h with depth is expressed by the following relationship:

$$K_h = \eta_h \cdot x \tag{7.17}$$

where η_h is the constant of modulus of subgrade reaction expressed in kN/m^3. For the value of $n = 0$, the modulus will be constant with depth. Value of η_h and K_h can depend on soil properties (density for granular soils and shear strength for clayey soils), the width or diameter of pile, the flexural stiffness of the pile material and the deflection of the pile. Values of η_h and K_h can be obtained from Tables 7.3 and 7.4 respectively (IS : 2911 - Part I).

The behaviour of a pile can be analysed by using the equation of an elastic beam supported on elastic foundation and is given by the following equation:

$$EI\frac{d^4 y}{dx^4} + p = 0 \tag{7.18}$$

where E = Modulus of elasticity of pile

I = Moment of inertia of pile section

p = Soil reaction = $K_h \cdot y$

Table 7.3. Typical Values of η_h (IS : 2911 - Part I)

Soil type	η_h *in kN/m*^{3}	
	Dry	*Submerged*
Loose sand	2.6×10^3	1.5×10^3
Medium sand	7.7×10^3	5.2×10^3
Dense sand	20×10^3	12.5×10^3
Very loose sand under repeated loading	—	0.41×10^3
Very soft organic soil	—	0.15×10^3
Very soft clay		
Static loads	—	0.45×10^3
Repeated loads	—	0.27×10^3

Table 7.4. Typical Values of K_h for Preloaded Clays (IS : 2911 - Part I)

Unconfined compressive strength, kN/m²	*Range of values of K_h, kN/m²*	*Probable value of K_h, kN/m²*
20 to 40	700 to 4200	773
100 to 200	3200 to 6500	4879
200 to 400	6500 to 13000	9773
More than 400	—	19546

Equation (7.18) can be written as

$$\frac{d^4 y}{dx^4} + \frac{K_h y}{EI} = 0 \tag{7.19}$$

Solution of Eq. (7.19) is obtained to get the values of deflection, shear force and moment for the pile in cohesionless soil or in cohesive soil.

Laterally Loaded Pile in Cohesionless Soil

Free-head Pile

Figure 7.11 shows the distribution of pile deflection y, pile slope variation, moment, shear, and soil reaction along the pile length due to a lateral load Q_{hg} and moment M_g applied at the pile head. The behaviour of this pile can be expressed by Eq. (7.19). In general, solution for this equation can be expressed by the following formulation:

$$y = f(x, T, L, K_h, EI, Q_{hg}, M_g) \tag{7.20}$$

where
- x = Depth below ground level
- T = Relative stiffness factor
- L = Pile length
- K_h = $\eta_h x$ = Modulus of horizontal subgrade reaction
- η_h = Constant of subgrade reaction
- EI = Pile stiffness
- Q_{hg} = Lateral load applied at the pile head
- M_g = Moment applied at the pile head

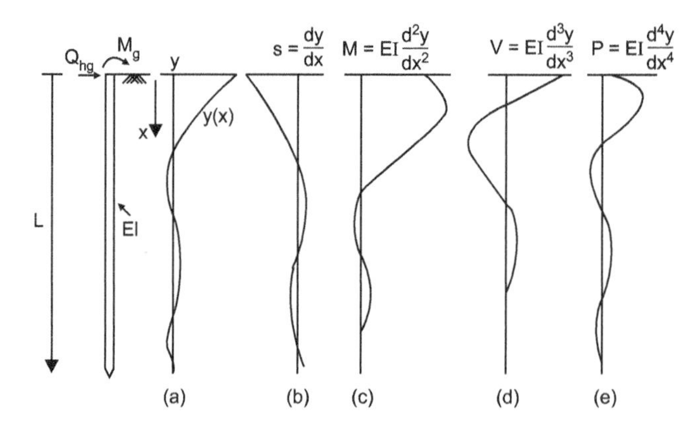

Fig. 7.11. Pile subjected to lateral load and moment

Elastic behaviour can be assumed for small deflections relative to the pile dimensions. For such a behaviour the principle of superposition may be applied. By utilizing the principle of superposition, the effects of lateral load Q_{hg} on deformation y_A and the effect of moment M_g on deformation y_B can be considered separately. Then the total deflection y_x at depth x can be given by the following:

$$y_x = y_A + y_B \tag{7.21}$$

where $$y_A/Q_{hg} = f_1(x, T, L, K_h, EI) \tag{7.22a}$$

and $$y_B / M_g = f_2(x, T, L, K_h, EI) \tag{7.22b}$$

f_1 and f_2 are two different functions of the same terms. In Eqs. (7.22a) and (7.22b) there are six terms and two dimensions; force and length are involved. Therefore, following four independent nondimensional terms can be determined (Matlock and Reese, 1962).

$$\frac{y_A EI}{Q_{hg} T^3}, \frac{x}{T}, \frac{L}{T}, \frac{K_h T^4}{EI} \tag{7.23a}$$

$$\frac{y_B EI}{M_g T^2}, \frac{x}{T}, \frac{L}{T}, \frac{K_h T^4}{EI} \tag{7.23b}$$

Furthermore, the following symbols can be assigned to these non-dimensional terms:

$$\frac{y_A EI}{Q_{hg} T^3} = A_y \text{ (Deflection coefficient for lateral load)} \tag{7.24}$$

$$\frac{y_B EI}{M_g T^2} = B_y \text{ (Deflection coefficient for moment)} \tag{7.25}$$

$$x/T = Z \text{ (Depth coefficient)} \tag{7.26}$$

$$L/T = Z_{max} \text{ (Maximum depth coefficient)} \tag{7.27}$$

$$K_h T^4/EI = \phi(x) \text{ (Soil modulus function)} \tag{7.28}$$

From equations (7.24) and (7.25) one can obtain:

$$y_x = y_A + y_B = A_y \frac{Q_{hg} T^3}{EI} + B_y \frac{M_g T^2}{EI} \tag{7.29}$$

Similarly, one can obtain expressions for moment M_x, slope S_x, shear V_x, and soil reaction p_x as follows:

$$M_x = M_A + M_B = A_m Q_{hg} T + B_m M_g \tag{7.30}$$

$$S_x = S_A + S_B = A_s \frac{Q_{hg} T^2}{EI} + B_s \frac{M_g T}{EI} \tag{7.31}$$

$$V_x = V_A + V_B = A_v Q_{hg} + B_v (M_g / T) \tag{7.32}$$

$$p_x = p_A + p_B = A_p \frac{Q_{hg}}{T} + B_p \frac{M_g}{T^2} \tag{7.33}$$

Referring to the basic differential equation (7.19) of beam on elastic foundation and utilizing the principle of superposition, we get :

$$\frac{d^4 y_A}{dx^4} + \frac{k_h y_A}{EI} = 0 \tag{7.34}$$

$$\frac{d^4 y_B}{dx^4} + \frac{k_h}{EI} y_B = 0 \tag{7.35}$$

Substituting for y_A and y_B from equations (7.24) and (7.25), k_h/EI from Eq. (7.28) and x/T from Eq. (7.26) :

$$\frac{d^4 A_y}{dz^4} + \phi(x) A_y = 0 \tag{7.36}$$

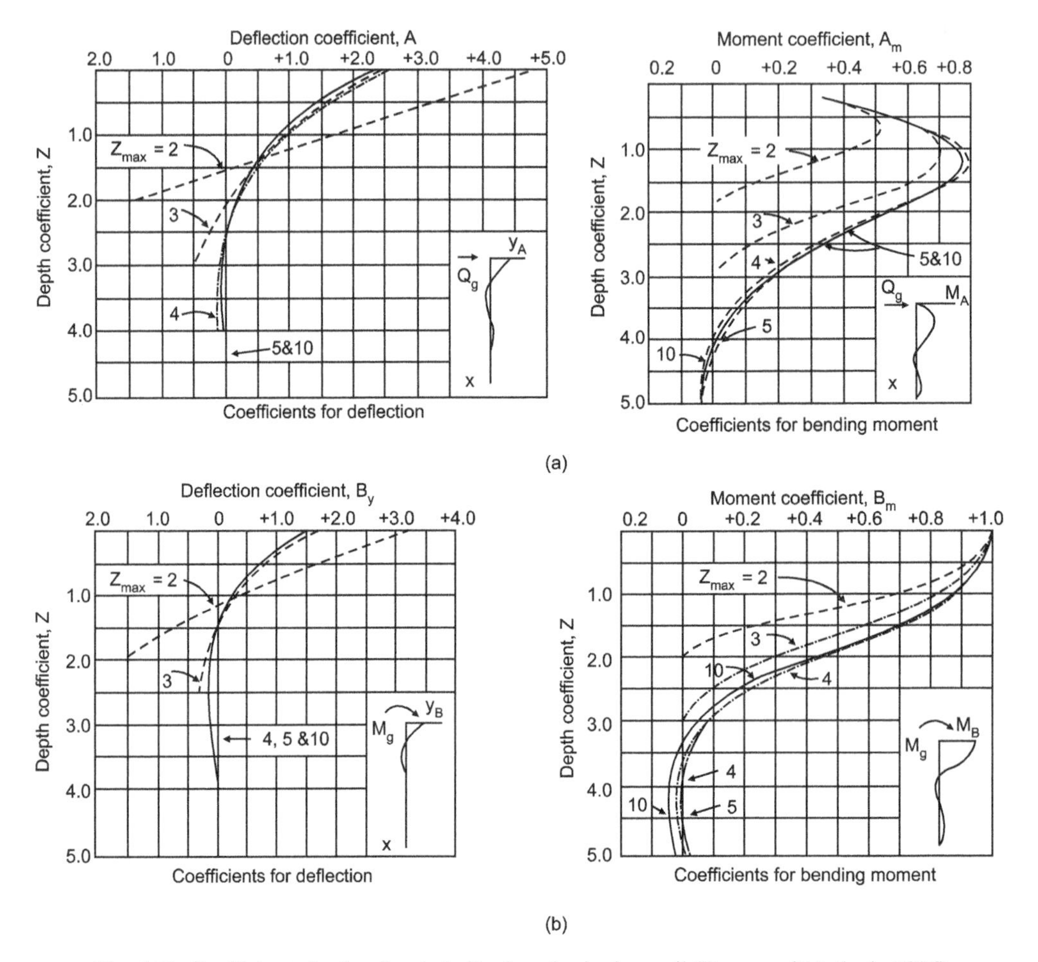

Fig. 7.12. Coefficients for free-headed piles in cohesionless soil (Reese and Matlock, 1956)

$$\frac{d^4 B_y}{dz^4} + \phi(x) B_y = 0 \tag{7.37}$$

For cohesionless soil where soil modulus is assumed to increase with depth $K_h = \eta_h x$, $\phi(x)$ may be equated to $Z = x/T$. Therefore, equation (7.28) becomes:

$$\frac{\eta_h x T^4}{EI} = \frac{x}{T} \tag{7.38}$$

This gives

$$T = \left(\frac{EI}{\eta_h}\right)^{1/5} \tag{7.39}$$

Solutions for equations (7.36) and (7.37), by using finite-difference methods, were obtained by

Reese and Matlock (1956) for values of A_y, A_s, A_m, A_v, A_p, B_y, B_s, B_m, B_v and B_p for various $Z = x/T$.

It has been found that pile deformation is like a rigid body (small curvature) for $Z_{max} \leq 2$. Therefore, piles with $Z_{max} \leq 2$ will behave as rigid piles or poles. Also, deflection coefficients are same for $Z_{max} = 5$ to 10. Therefore, pile length beyond $Z_{max} = 5$ does not change the deflection. In practice, in most cases pile length is greater than $5T$; therefore, coefficients given in Tables 7.5 and 7.6 can be used. Figure 7.12 provides values of A_y, A_m, B_y and B_m for different values of Z_{max}.

Table 7.5. Coefficient A for Long Piles ($Z_{max} > 5$) : Free Head (After Matlock and Reese, 1961, 1962)

Z	A_y	A_s	A_m	A_v	A_p
0.0	2.435	– 1.623	0.000	1.000	0.000
0.1	2.273	– 1.618	0.100	0.989	– 0.227
0.2	2.112	– 1.603	0.198	0.956	– 0.422
0.3	1.952	– 1.578	0.291	0.906	– 0.586
0.4	1.796	– 1.545	0.379	0.840	– 0.718
0.5	1.644	– 1.503	0.459	0.764	– 0.822
0.6	1.496	– 1.454	0.532	0.677	– 0.897
0.7	1.353	– 1.397	0.595	0.585	– 0.947
0.8	1.216	– 1.335	0.649	0.489	– 0.973
0.9	1.086	– 1.268	0.693	0.392	– 0.977
1.0	0.962	– 1.197	0.727	0.295	– 0.962
1.2	0.738	– 1.047	0.767	0.109	– 0.885
1.4	0.544	– 0.893	0.772	– 0.056	– 0.761
1.6	0.381	– 0.741	0.746	– 0.193	– 0.609
1.8	0.247	– 0.596	0.696	– 0.298	– 0.445
2.0	0.142	– 0.464	0.628	– 0.371	– 0.283
3.0	– 0.075	– 0.040	0.225	– 0.349	0.226
4.0	– 0.050	– 0.052	0.000	– 0.106	0.201
5.0	– 0.009	– 0.025	– 0.033	0.013	0.046

Fixed-head Pile

For a fixed-head pile, the slope (S) at the ground surface is zero. Therefore, from equation (7.31).

Table 7.6. Coefficient B for Long Piles ($Z_{max} > 5$) : Free Head (After Matlock and Reese, 1961, 1962)

Z	B_y	B_s	B_m	B_v	B_p
0.0	1.623	– 1.750	1.000	0.000	0.000
0.1	1.453	– 1.650	1.000	– 0.007	– 0.145
0.2	1.293	– 1.550	0.999	– 0.028	– 0.259
0.3	1.143	– 1.450	0.994	– 0.058	– 0.343
0.4	1.003	– 1.351	0.987	– 0.095	– 0.401
0.5	0.873	– 1.253	0.976	– 0.137	– 0.436
0.6	0.752	– 1.156	0.960	– 0.181	– 0.451
0.7	0.642	– 1.061	0.939	– 0.226	– 0.449
0.8	0.540	– 0.968	0.914	– 0.270	– 0.432
0.9	0.448	– 0.878	0.885	– 0.312	– 0.403
1.0	0.364	– 0.792	0.852	– 0.350	– 0.364
1.2	0.223	– 0.629	0.775	– 0.414	– 0.268
1.4	0.112	– 0.482	0.688	– 0.456	– 0.157
1.6	0.029	– 0.354	0.594	– 0.477	– 0.047
1.8	– 0.030	– 0.245	0.498	– 0.476	– 0.054
2.0	– 0.070	– 0.155	0.404	– 0.456	0.140
3.0	– 0.089	0.057	0.059	– 0.213	0.268
4.0	– 0.028	0.049	– 0.042	0.017	0.112
5.0	0.000	0.011	– 0.026	0.029	– 0.002

Therefore,

$$\frac{M_g}{Q_{hg}T} = \frac{A_s}{B_s} \text{ at } x = 0$$

From Tables 7.5 and 7.6 for $Z = x/t = 0$

$$\frac{A_s}{B_s} = -\frac{-1.623}{-1.75} = -0.93$$

Therefore, $M_g/Q_{hg}\,T = -0.93$. The term $M_g/Q_{hg}\,T$ has been defined as the non-dimensional fixity factor by Prakash (1962). Then the equations for deflection and moment for fixed head can be modified as follows:

$$y_x = A_y \frac{Q_{hg}T^3}{EI} + B_y \frac{M_g T^2}{EI} \tag{7.40}$$

Substituting $M_g = -0.93\,Q_{hg}\,T$ for fixed head, we get

$$y_x = (A_y - 0.93B_y)\frac{Q_{hg}T^3}{EI} \tag{7.41a}$$

or

$$y_x = C_y \frac{Q_{hg}T^3}{EI} \tag{7.41b}$$

Similarly,

$$M_x = C_m\,Q_{hg}\,T \tag{7.42}$$

Values of C_y and C_m can be obtained from Fig. 7.13.

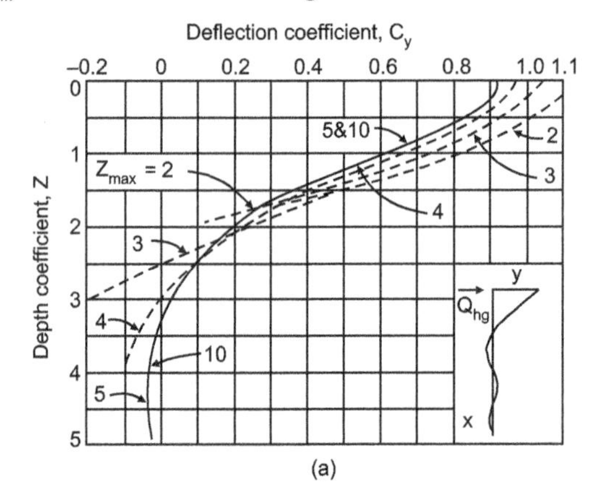

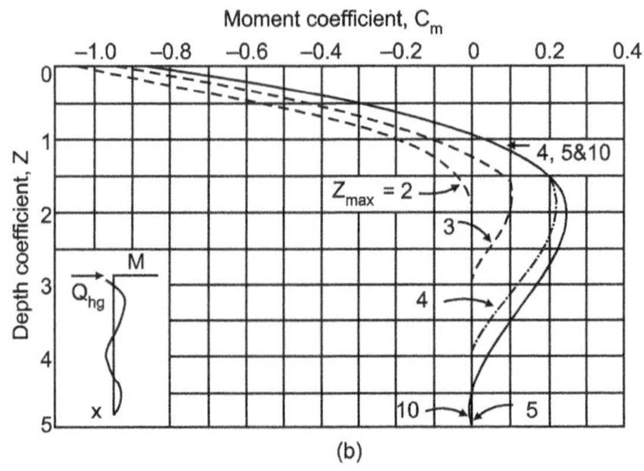

Fig. 7.13. Coefficient for fixed-head pile in cohesionless soils (Reese and Matlock, 1956)

Partially fixed head pile

In cases where the piles undergo some rotation at the joints of their head and the cap, these are called partially fixed piles. In such a situation, the coefficient C needs modification as follows:

$$C_y = (A_y - 0.93\ \lambda\ B_y) \tag{7.43}$$

$$C_m = (A_m - 0.93\ \lambda\ B_m) \tag{7.44}$$

where λ is percent fixity (i.e. $\lambda = 1$ for 100 percent fixity or fully restrained pile head and $\lambda = 0$ for fully free pile head). At intermediate fixity levels, proper λ can be taken.

Laterally Loaded Pile in Cohesive Soil

Free-head Pile

For normally consolidated clays, the modulus of subgrade reaction increases linearly with depth. Therefore, for such clays the above analysis for cohesionless soil shall apply.

For over consolidated clays, subgrade modulus is constant with depth. For such clays, deflection coefficients are defined as

$$\frac{y_A EI}{Q_{hg} R^3} = A_{yc} \tag{7.45}$$

$$\frac{y_B EI}{M_g R^2} = B_{yc} \tag{7.46}$$

where A_{yc} and B_{yc} are deflection coefficients in clay for Q_{hg} and M_g respectively. Letting $y = y_A + y_B$ as in equation (7.29), we get deflection y at any depth.

$$y = A_{yc}\frac{Q_{hg}R^3}{EI} + B_{yc}\frac{M_g R^2}{EI} \tag{7.47}$$

Similarly, moment M at any depth is

$$M = A_{mc}\, Q_{hg}\, R + B_{mc}\, M_g \tag{7.48}$$

Solutions for A and B coefficients similar to those presented for cohesionless soil had been obtained by Davisson and Gill (1963). In equation (7.36), by replacing A_y with A_{yc}, we get

$$\frac{d^4 A_{yc}}{dZ^2} + \phi(x)\, A_{yc} = 0 \tag{7.49}$$

Now putting $\phi(x) = 1$, and replacing T with R, equation (7.28) becomes :

$$\frac{K_h R^4}{EI} = 1 \tag{7.50}$$

$$R = \left(\frac{EI}{K_h}\right)^{1/4} \tag{7.51}$$

and $$Z = x/R \tag{7.52}$$

Substituting the above equations in equation (7.49), the solutions for A and B coefficients can be obtained in a similar manner as for cohesionless soils.

The solutions for A_{yc} and A_{mc} have been plotted with nondimensional depth coefficient Z in Fig. 7.14a and B_{yc} and B_{mc} in Fig. 7.14b. It will be seen in Fig. 7.14 that if Z_{max} (= L/R) < 2, the pile behaves as a rigid pile or a pole. And for Z_{max} (= L/R) $\geq$ 4, the pile behaves as an infinitely long pile.

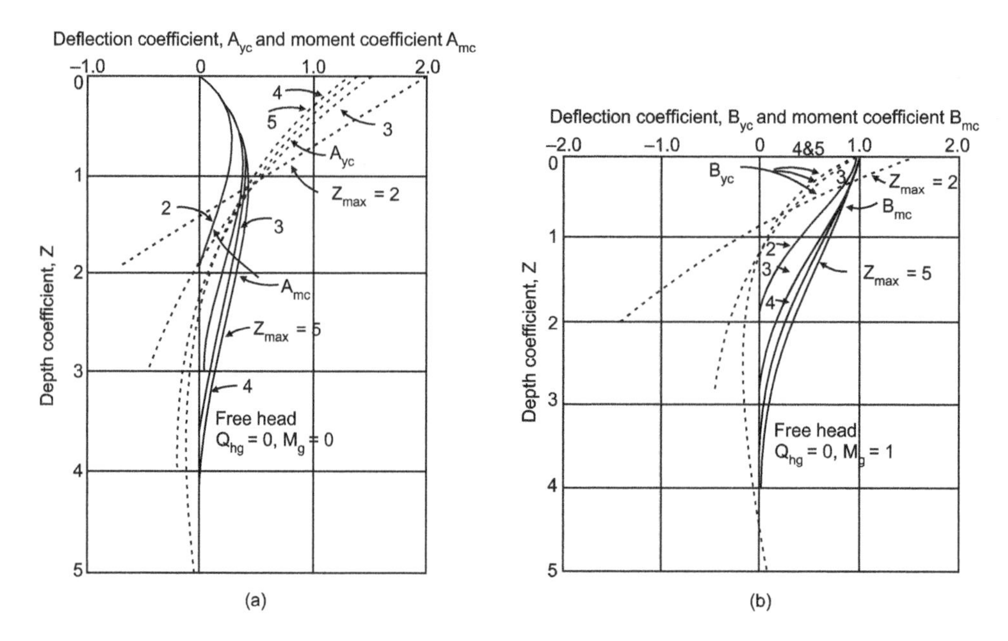

Fig. 7.14. Coefficients for free-headed piles in cohesive soils (Davission and Gill, 1963)

Lateral Load Capacity of Single Pile

Jain (1983) developed a method of computing lateral load capacity of a pile with a known value of pile head deflection. The method is based on analyzing the pile as an equivalent cantilever (Fig. 7.15). The length of the equivalent cantilever can be obtained using chart given in Fig. 7.16. The results presented in Fig. 7.16 are obtained from an analysis based on modulus of subgrade reaction approach, Values of relative stiffness (T or R) may be obtained using Eqs. (7.39) and (7.51). For a given limiting pile head deflection y_o, the lateral load capacity can be obtained using following expressions (Ramasamy et al., 1987).

(a) For free-head pile

$$Q_h = (3\, y_o\, EI) / (h + L_f)^3 \tag{7.53}$$

(b) For fixed-head pile

$$Q_h = (12\, y_o\, EI) / (h + L_f)^3 \tag{7.54}$$

Lateral Pile Load Test

Lateral load capacity of a single pile can be obtained by performing a lateral pile load test. In this, the lateral load is applied to the test pile by using a hydraulic jack and a suitable reaction system. The amount of the lateral load applied is measured either by a calibrated load cell or a pressure gauge. The pile is tested by applying the lateral load in 10 steps with a maximum of twice the design lateral load. For each increment of lateral load, the lateral movement of the pile head is measured by dial gauges.

Lateral load versus deflection curve is then plotted. Allowable lateral load capacity will be the least from the following criteria:

(i) Half the final lateral load at which lateral movement of pile is 12 mm.
(ii) Lateral load corresponding to 5 mm lateral movement.
(iii) Lateral load corresponding to allowable lateral movement.

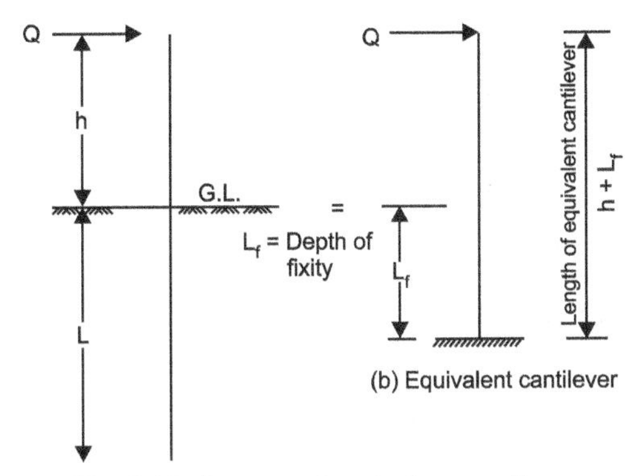

Fig. 7.15. Concept of equivalent cantilever

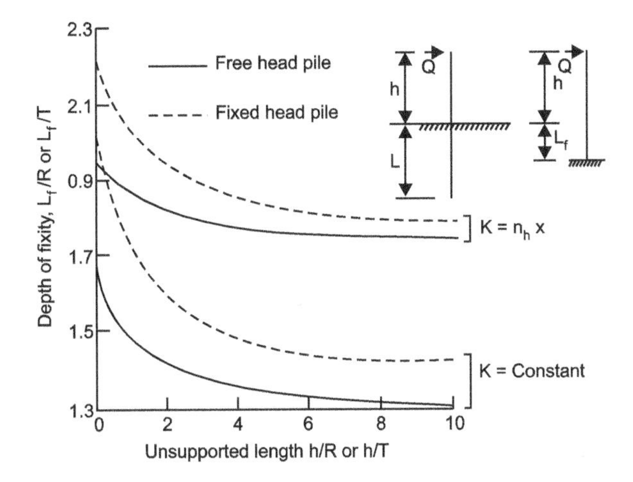

Fig. 7.16. Charts for getting depth of fixity, L_f (Ramasamy et al. 1987).

7.3 Single Pile under Vertical Vibrations

A pile subjected to sinusoidally varying dynamic force is shown in Fig. 7.17a, and its single degree freedom model is shown in Fig. 7.17b. With appropriate values of mass m, spring constant K and dampig constant k, the foundation response can be predicted. In this model, mass m represent the mass of the structure above the ground, mass of pile cap and pile. Novak (1974, 1977) and Sheta and Novak (1982) have suggested a method for obtaining stiffness and damping of the soil-pile system. In this method, the soil has been assumed as composed of a set of independent infinitesimally thin horizontal layers of infinite extent. This concept could be thought of as a generalised Winkler material that possesses inertia and dissipates energy.

Using the above concept, Novak and Sharnouby (1983) have developed following solutions of getting the stiffness and damping constant of single pile subjected to vertical vibration :

$$K_{zp} = \frac{E_p \cdot A}{r_0} f_{w1} \tag{7.55}$$

$$C_{zp} = \frac{E_p \cdot A}{v_s} f_{w2} \tag{7.56}$$

where K_{zp} = Stiffness of single pile

E_{zp} = Modulus of elasticity of pile material
A = Area of cross section pile
r_o = Equivalent radius of pile
v_s = Shear wave velocity = $\sqrt{\dfrac{G_s}{\rho}}$
G_s = Shear modulus of soil
ρ = Mass density of soil
C_{zp} = Damping constant for single pile
l = Embedded length of pile

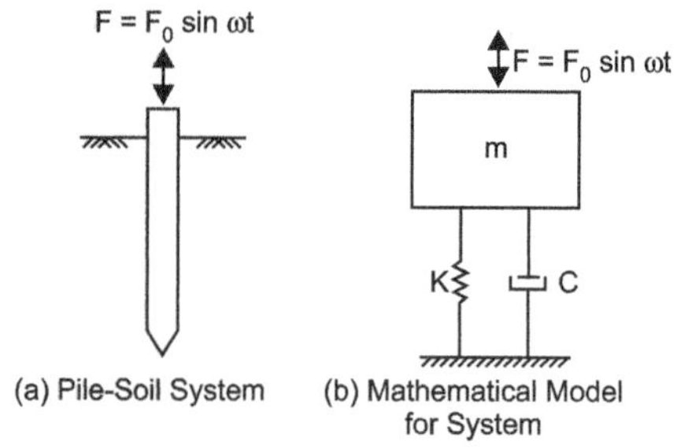

F = F_0 sin ωt

(a) Pile-Soil System (b) Mathematical Model for System

Fig. 7.17. Single pile subjected to vertical vibrations

f_{w1} and f_{w2} are the parameters which depend on (i) type of pile whether end bearing or friction, (ii) pile slenderness (l/r_o), (iii) E_p/G_s ratio and (iv) variation G_s with depth. Considering these factors, Novak and Sharnouby (1983) have developed charts as shown in Figs 7.18 and 7.19.

Knowing the values of K_{zp} and C_{zp}, response of the pile can be obtained.

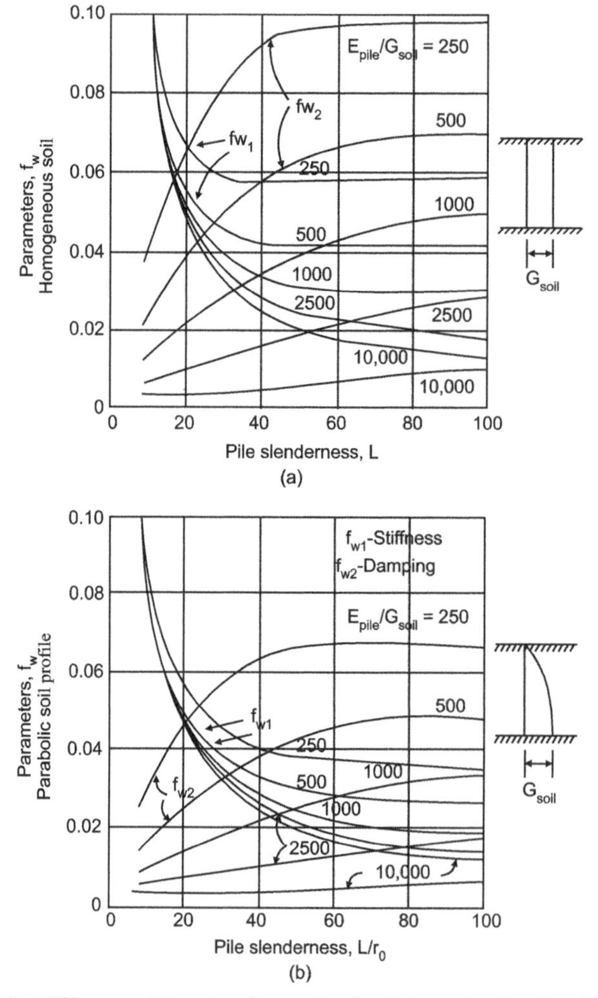

Fig. 7.18. Stiffness and damping factors for fixed tip vertically vibrating piles (Novak and El-Sharnouby, 1983)

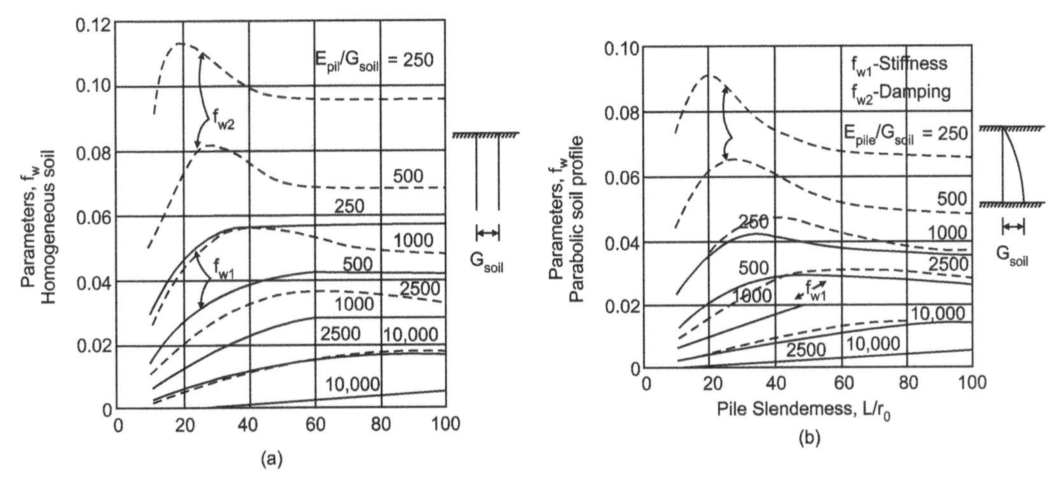

Fig. 7.19. Stiffness and damping parameters of vertical response of friction piles (Novak and EI-Shamouby, 1983)

7.4 Single Pile under Lateral Vibrations

A simple analysis of pile under lateral dynamic loads has been proposed by Chandrasekaran (1974). The soil pile system has been idealized as a discretized mathematical model (Fig. 7.20) where M_t is the lumped mass at the pile top which is the super-imposed load per pile. The remaining mass is obtained by dividing the pile into a convenient number of segments. The mass of each segment is assumed to be lumped at the centre of the segment. The soil-pile interaction affects in the idealized model are considered as Winkler reactions and are discretized as springs (Prakash and Chandrasekaran, 1977). The springs are connected at the mass points at one end and immovable support at the other. The concept of subgrade modulus has been utilized in this process. The adopted end conditions of the model have also been identified in the figure.

$$Y = Y(x) \sin \omega t \tag{7.57}$$

where ω is circular natural frequency.

Using this idealization, Chandrasekaran (1974) obtained solutions for piles embedded in two different soil types, namely, one in which the soil modulus remains constant with depth as in precompressed clay and the other in which the soil modulus increase linearly with depth as in sands and silts. For the above cases, the influence of important factors, e.g. soil stiffness, pile stiffness, sustained vertical load and pile length have been considered. Based on the parametric study, Chandrasekaran (1974) developed following non-dimensional frequency factor.

(i) For soil modulus constant with depth

$$F_{cL1} = \omega_{n1}\sqrt{\frac{W}{g.K.R}} \tag{7.58}$$

where $M_t = W/g$ = Lumped mass at pile top (7.59)

K = Soil modulus constant with depth

R = Relative stiffness factor

$$= \sqrt[4]{\frac{EL}{K}} \tag{7.60}$$

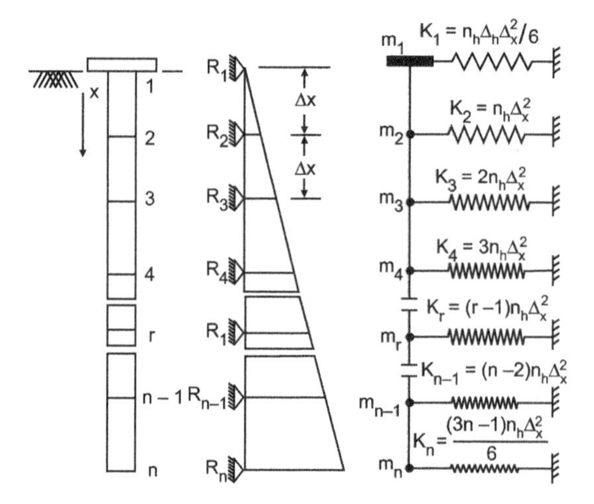

(a) Assuming soil modulus proportional with depth $K_x = \eta_h x$

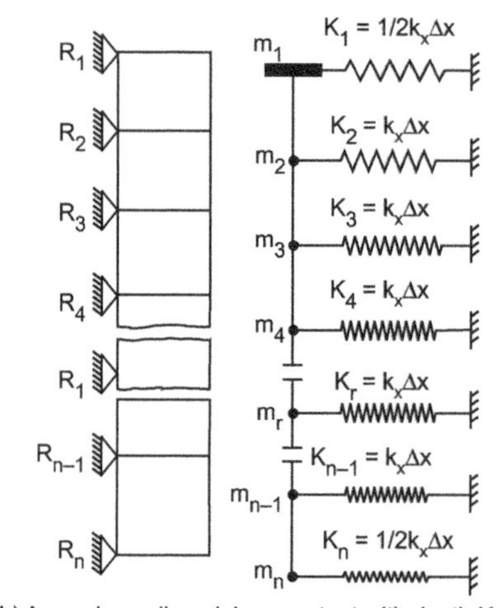

(b) Assuming soil modulus constant with depth K = k

Fig. 7.20. Discretization of soil-pile interaction effects. (a) Soil modulus linearly varying with depth. (b) Soil modulus constant with depth (After Chandrasekaran, 1974)

Figure 7.21a shows the variation of frequency factor F_{cL1} with maximum depth factor, Z_{max} where $Z_{max} = L_s/R$, L_s being the length of embedment of pile. It is reported that for any soil pile characteristics such as flexural stiffness EI, soil stiffness in terms of K and sustained vertical loads and for same Z_{max} unique values of frequency factors exist.

(ii) For soil modulus linearly varying with depth

The variation of frequency factor F_{sL1}. and F'_{sL1} in soils in which modulus vary linearly with depth have been presented in Figs. 7.21b and 7.21c. The definition of F_{sL1} for F'_{sL1} for pile top free to rotate and fixed against rotation conditions are identical and given by Eq. (7.61).

$$F_{sL1} \text{ and } F'_{sL1} = \omega_{n1}\sqrt{\frac{W}{g}\cdot\frac{1}{n_h T^2}} \tag{7.61}$$

where n_h is the constant of horizontal subgrade reaction for

$$K = n_h \cdot x \text{ and } T = \sqrt[5]{\frac{\text{EI}}{n_h}} \tag{7.62}$$

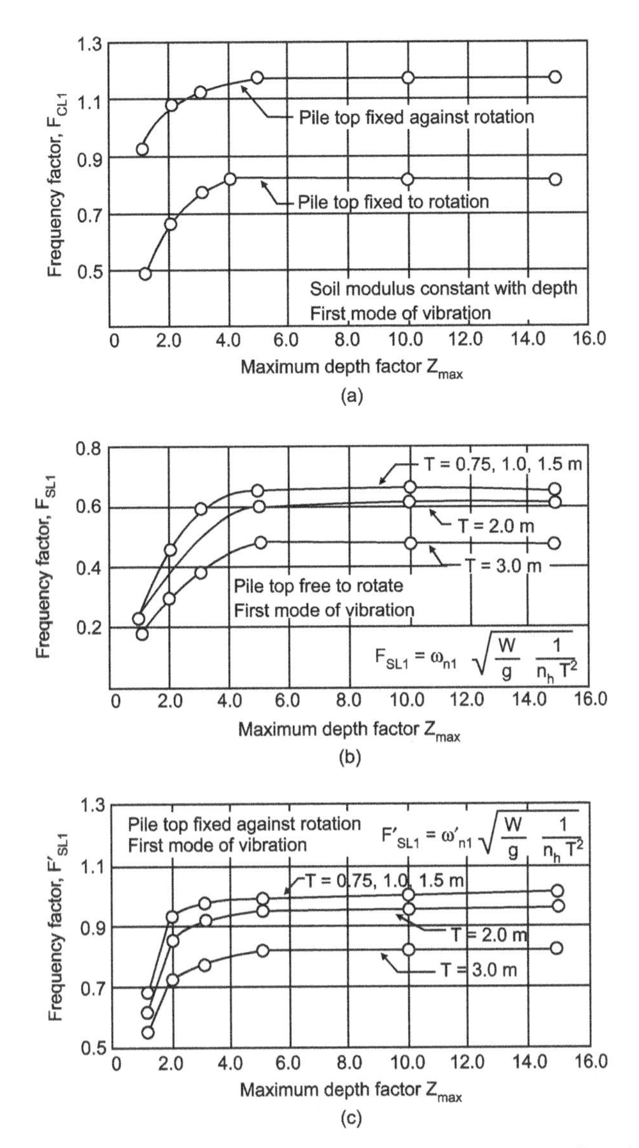

Fig. 7.21. Nondimensional frequency factors in first mode of vibrations. (a) Soil modulus constant with depth. (b) Soil modulus linearly varying with depth and pile top free. (c) Soil modulus linearly varying with depth and pile top restrained against rotation (After Prakash and Chandrasekaran, 1977)

With these two sets of curves having known the soil pile charateristics, length and the fixity conditions, the designer could predict the time period for first mode of vibrations.

Method of Analysis

On the basis of dynamic analysis discussed above and non-dimensional solutions for moments, shear and soil reactions have been obtained by Chandrasekaran (1974). The following procedure for dynamic analysis of piles may be adopted:

(1) Estimate the dynamic soil modulus (K or n_h). In the absence of realistic data, values from static lateral load test can be modified based upon engineering judgement.
(2) Using the soil modulus value, compute the relative stiffness factor.
(3) Calculate the maximum depth factor Z_{max} for pile.

(4) For the computed value of the maximum depth factor and the pile end condition read the frequency factor (Fig. 7.21).

(5) Estimate the dead load on the pile. This provides mass at the pile top.

(6) Using Eqs. (7.63) and (7.64) determine the natural frequency ω_{n1}.
Soil modulus constant with depth :

$$\omega_{n1} = \frac{F_{cL1}}{\sqrt{\frac{W}{gKR}}} \tag{7.63}$$

Soil modulus proportional to depth :

$$\omega_{n1} = \frac{F_{sL1} \text{ or } F_{sL1}}{\sqrt{\frac{W}{g}\frac{1}{n_h T^2}}} \tag{7.64}$$

(7) Compute the natural time period $t = \dfrac{2\pi}{\omega_n}$ (7.65)

(8) For the computed natural time period read the average acceleration coefficient for a given value of damping (Fig. 7.22) and compute spectral displacement.

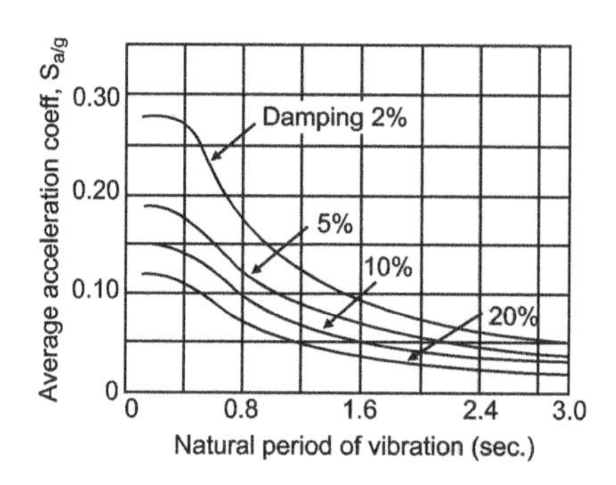

Fig. 7.22. Average acceleration spectra

(9) Estimate the maximum bending moment in the pile section.
(i) Soil modulus constant with depth
Bending moment = $A_m \cdot K R^2 \cdot S_d$ (7.66)
The maximum values of A_m are given in Table 7.7.
(ii) Soil modulus increasing linearly with depth
Bending moment = $B_m \, n_h \, T^3 . S_d$ (7.67)
The maximum values of B_m are given in Table 7.8.
The pile section need be checked against this moment.

Table 7.7. Maximum Values of Coefficient A_m

Maximum depth factor Z_{max}	*Pile free to rotate*	*Coefficient* A_m	
		Pile fixed at top against rotation	
		– ve	*+ ve*
2	0.13	– 0.9	+ 0
3	0.24	– 0.9	+ 0.04
5– 15	0.32	– 0.9	+ 0.18

Table 7.8. Maximum Values of Coefficient B_m

Maximum depth factor Z_{max}	*Pile free to rotate*	*Coefficient* B_m	
		Pile fixed at top against rotation	
		– *ve*	+ *ve*
2	0.100	– 0.93	+ 0
3	0.255	– 0.93	+ 0.10
5– 15	0.315	– 0.90	+ 0.28

(10) For the maximum displacement at the ground computed above the displacement all along the length of pile may be determined approximately assuming the deflections as per the solutions of Davisson and Gill (1963) and Reese and Matlock (1956) for the two cases of soil modulus. The soil reaction all along the pile length is then computed as follows:
(i) For soil modulus constant with depth

$$p = K \cdot y_x \tag{7.68a}$$

(ii) For soil modulus linearly varying with depth

$$p = \eta_h \cdot x \cdot y_x \tag{7.68b}$$

The allowable soil reaction may be taken as that corresponding to Rankine passive pressure at all depth.

It is recommended that the solution be obtained for the two cases of pile restraint, i.e. free to rotate and pile top restrained against rotation. Once the fixity condition of the actual pile in the group is estimated, the solution is obtained for the case by linear interpolation.

Novak (1974) developed solutions for obtaining lateral stiffness and damping constants for single pile with soil modulus constant with depth in three modes of vibration namely (i) Translation alone, (ii) Rotation alone, and (iii) Coupled translation and rotation. Novak and El-Sharnouby (1983) extended these solutions considering parabolic variation of soil shear modulus. The solutions can be summarised in the following equations:

Translational stiffness and damping constants:

$$K_{xp} = \frac{E_p I_p}{r_o^3} \cdot f_{x1} \tag{7.69}$$

$$C_{xp} = \frac{E_p I_p}{r_o^2 v_s} \cdot f_{x2} \tag{7.70}$$

Rotational stiffness and damping constants :

$$K_{\phi p} = \frac{E_p I_p}{r_o} \cdot f_{\phi 1} \tag{7.71}$$

$$C_{\phi p} = \frac{E_p I_p}{v_s} \cdot f_{\phi 2} \tag{7.72}$$

Coupled vibration stiffness and damping constants :

$$K_{x\phi p} = \frac{E_p I_p}{r_o^2} \cdot f_{x\phi 1} \tag{7.73}$$

$$C_{x\phi p} = \frac{E_p I_p}{r_o v_s} \cdot f_{x\phi 2} \tag{7.74}$$

where I_p = Moment of inertia of pile cross-section

E_p = Young's modulus of pile
v_s = Shear wave velocity in soil
v_c = Longitudinal wave velocity in pile
r_o = Equivalent radius of pile
f = Constants (Table 7.9)

Table 7.9. Stiffness and Damping Parameters of Horizontal Response for Piles with $L/r_o > 25$ for Homogeneous Soil Profile and $L/r_o > 30$ for Parabolic Soil Profile (After Novak and El-Sharnouby, (1983)

v	*Stiffness parameters*				*Damping parameters*				
	E_{pile}/G_{soil}	$(f_{\phi 1})$	$(f_{x\phi 1})$	(f^*_{x1})	(f^P_{x1})	$(f_{x\phi 2})$	$(f_{x\phi 2})$	(f^*_{x2})	(f^P_{x2})
(1)	(2)	(3)	(4)	(5)	(6)	(7)	(8)	(9)	(10)
Homogeneous Soil Profile									
0.25	10.000	0.2135	–0.0217	0.0042	0.0021	0.1577	–0.0333	0.0107	0.0054
	2.500	0.2998	–0.0429	0.0119	0.0061	0.2152	–0.0646	0.0297	0.0154
	1000	0.3741	–0.0668	0.0236	0.0123	0.2598	–0.0985	0.0579	0.0306
	500	0.4411	–0.0929	0.0395	0.0210	0.2953	–0.1337	0.0953	0.0514
	250	0.5186	–0.1281	0.0659	0.0358	0.3299	–0.1786	0.1556	0.0864
0.40	10,000	0.2207	–0.0232	0.0047	0.0024	0.1634	–0.0358	0.0119	0.0060
	2,500	0.3097	–0.0459	0.0132	0.0068	0.2224	–0.0692	0.0329	0.0171
	1000	0.3860	–0.0714	0.0261	0.0136	0.2677	–0.1052	0.0641	0.0339
	500	0.4547	–0.0991	0.0436	0.0231	0.3034	–0.1425	1.1054	0.0570
	250	0.5336	–0.1365	0.0726	0.0394	0.3377	–0.1896	0.1717	0.0957
Parabolic Soil Profile									
0.25	10,000	0.1800	–0.0144	0.0019	0.0008	0.1450	–0.0252	0.0060	0.0028
	2,500	0.2452	–0.0267	0.0047	0.0020	0.2025	–0.0484	0.0159	0.0076
	1000	0.3000	–0.0400	0.0086	0.0037	0.2499	–0.0737	0.0303	0.0147
	500	0.3489	–0.0543	0.0136	0.0059	0.2910	–0.1008	0.0491	0.0241
	250	0.4049	–0.0734	0.0215	0.0094	0.3361	–0.1370	0.0793	0.0398
0.40	10,000	0.1857	–0.0153	0.0020	0.0009	0.1508	–0.0271	0.0067	0.0031
	2,500	0.2529	–0.0284	0.0051	0.0022	0.2101	–0.0519	0.0177	0.0084
	1000	0.3094	–0.0426	0.0094	0.0041	0.2589	–0.0790	0.0336	0.0163
	500	0.3596	–0.0577	0.0149	0.0065	0.3009	–0.1079	0.0544	0.0269
	250	0.4170	–0.0780	0.0236	0.0103	0.3468	–0.1461	0.0880	0.0443

f^P_{x1} and f^P_{x2} are parameters for pinned head.

* Fixed-translating head

7.5 Group Action under Dynamic Loading

7.5.1 Vertical Vibrations

Novak and Grigg (1976) proposed following expressions for piles in a group for computing stiffness and damping

$$K_{zg} = \frac{\sum_{I}^{n} K_{zp}}{\sum_{I}^{n} \alpha_A} \tag{7.75}$$

$$C_{zg} = \frac{\sum_{I}^{n} C_{zp}}{\sum_{I}^{n} \alpha_A} \tag{7.76}$$

where n = Number of piles

α_A = Axial displacement interaction factor for a typical reference pile in the group relative to itself and to all other piles in the group, assuming the reference pile and all other piles carry the same load (Fig. 7.23).

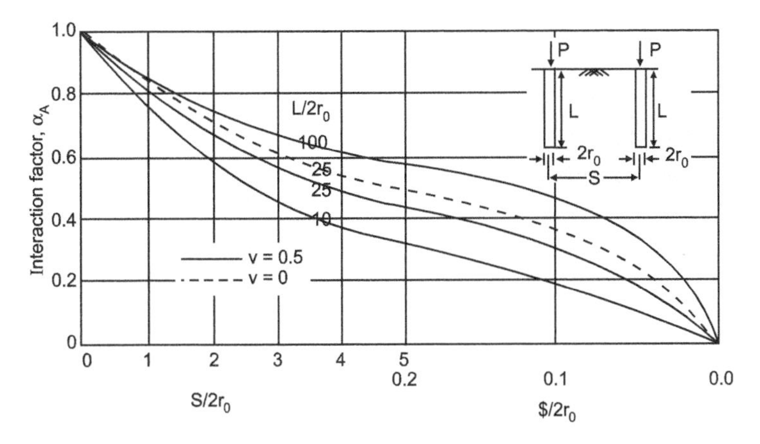

Fig. 7.23. α_A as a function of pile length and spacing (Poulos, 1968)

Novak and Beredugo (1972) have developed following expressions for getting additional stiffness and damping constants due to side friction in embedded pile caps:

$$K_{zf} = G_s D \bar{S}_{z1} \tag{7.77}$$

$$C_{zf} = D r_{oc} \bar{S}_{z2} \sqrt{G_s \rho} \tag{7.78}$$

where D = Depth of embedment of pile cap

r_{oc} = Equivalent radius of pile cap

G_s = Shear modulus of backfill soil

ρ_s = Mass density of backfill soil

$\bar{S}_{z1}$ and $\bar{S}_{z2}$ are constants and are 2.70 and 6.70 respectively.

Then, total spring stiffness and damping constants for the group will be:

$$(K_z)_g = K_{zg} + K_{zf} \tag{7.79}$$

$$(C_z)_g = C_{zg} + C_{zf} \tag{7.80}$$

7.5.2 Lateral Vibrations

The spring stiffness and damping constants of the total piles in a group are given by:

$$K_{xg} = \frac{\sum_{1}^{n} K_{xp}}{\sum_{1}^{n} \alpha_L} \tag{7.81}$$

$$C_{xg} = \frac{\sum_{1}^{n} C_{xp}}{\sum_{1}^{n} \alpha_L} \tag{7.82}$$

Value of α_L is taken from Fig. 7.24.

Stiffness and damping constants for the pile cap are:

$$K_{xf} = G_s D \overline{S}_{x1} \tag{7.83}$$

$$C_{xf} = D r_{oc} \sqrt{G_s \rho} \times \overline{S}_{x2} \tag{7.84}$$

Values of S_{x1} and S_{x2} are listed in Table 7.10.

Table 7.10. Stiffness and Damping Constants for Half-Space and Side Layers for Sliding (After Beredugo and Novak, 1972)

Poisson's ratio v	*Validity range*	*Constant parameter*
0.0	$0 < a_o < 2$	$\overline{S}_{x1} = 3.6$ $\overline{S}_{x2} = 8.2$
0.25	$0 < a_o < 2$	$\overline{S}_{x1} = 4.0$ $\overline{S}_{x2} = 9.1$
0.4	$0 < a_o < 2.0$	$\overline{S}_{x1} = 4.1$ $\overline{S}_{x2} = 10.6$

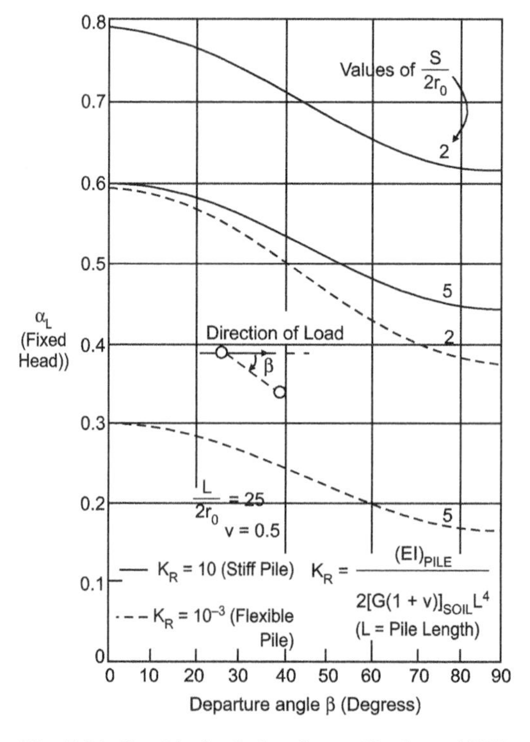

Fig. 7.24. Graphical solution for α_L (Poulous, 1972)

Total stiffness and damping constants are:

$$(K_x)_g = K_{xg} + K_{xf} \tag{7.85}$$

$$(C_x)g = C_{xg} + C_{xf} \tag{7.86}$$

7.5.3 Rocking Vibrations

Stiffness and damping constants for total piles in the group are:

$$K_{\phi g} = \sum_{1}^{n}(K_{\phi p} + K_{zp}x_r^2 + K_{xp}Z_c^2 - 2Z_c K_{x\phi p}) \tag{7.87}$$

$$C_{\phi g} = \sum_{1}^{n}(C_{\phi p} + C_{zp}x_r^2 + C_{xp}Z_c^2 - 2Z_c C_{x\phi p}) \tag{7.88}$$

where x_r = Distance of each pile from c.g.
Z_c = Height of centre of gravity of the pile cap above its base

Stiffness and damping constants for the pile cap are:

$$K_{\phi f} = G_s r_{oc}^2 D \times \bar{S}_{\phi 1} + G_s r_{oc}^2 D\left[\left(\frac{\delta^2}{3}\right) + \left(\frac{Z_c}{r_0}\right)^2 - \left(\frac{Z_c}{r_0}\right)\right]\bar{S}_{x1} \tag{7.89}$$

$$C_{\phi f} = \delta r_{oc}^4 \sqrt{\frac{G_s \gamma}{g}}\left\{\bar{S}_{\phi 2} + \left[\frac{\delta^2}{3} + \left(\frac{Z_c}{r_0}\right)^2 - \delta\left(\frac{Z_c}{r_0}\right)\right]\bar{S}_{x2}\right\} \tag{7.90}$$

where $$\delta = D/r_0 \tag{7.91}$$

In the case of rotation about a horizontal axis (rocking), for any of Poisson's ratio (v), values of constant parameters $\bar{S}_{\phi 1}$ and $\bar{S}_{\phi 2}$ are respectively 2.5 and 1.8.

Total stiffness and damping constants of pile group are:

$$(K_\phi)_g = K_{\phi g} + K_{\phi f} \tag{7.92}$$

$$(C_\phi)_g = C_{\phi g} + C_{\phi f} \tag{7.93}$$

Illustrative Examples

Example 7.1

It is plan to construct a seven storeyed residential building. The building frame has three spans of 7 m each in both directions. The bore logs indicate that the soil at the site consists of medium sand to great depth. It was found that interior columns were heaviest loaded. An interior column had:

Axial static load = 4000 kN
Static moment = 8 kN-m
Static shear force = 30 kN
Additional moment due to earthquake = ±500 kN-m
Additional shear force due to earthquake = ±180 kN

It was decided to rest the column on piles of 500 mm diameter and length 16.0 m.

Vertical pile load test performed on a test pile indicated that the capacity of the pile may be adopted as 1100 kN. A lateral load test was also carried out on this pile. Table 7.11 gives the data obtained from static lateral load test.

Table 7.11. Lateral Load Test Data

Lateral load (kN)	5	10	15	20	25	30
Horizontal deflect (mm)	0.03	0.06	0.13	0.34	0.53	0.80

Determine the maximum moment in a pile by (i) Pseudo static analysis, (ii) Dynamic analysis.

Solution:

(i) Length of pile = 12.0 m
Diameter of pile = 0.5 m
Capacity of pile = 1100 kN

$$\text{Moment of inertia of pile} = \frac{\pi \times d^4}{64} = \frac{\pi \times 0.5^4}{64} = 3.066 \times 10^{-3}\,\text{m}^4$$

Let the pile is constructed in M-25 grade concrete

$$E = 5700\sqrt{25} = 28500\ \text{N/mm}^2 = 2.85 \times 10^7\ \text{kN/m}^2$$

$$\text{EI of a pile} = 2.85 \times 10^7 \times 3.066 \times 10^{-3} = 8.74 \times 10^4\ \text{kN-m}^2$$

(ii) Let us use 5 piles group as shown in Fig. 7.25.

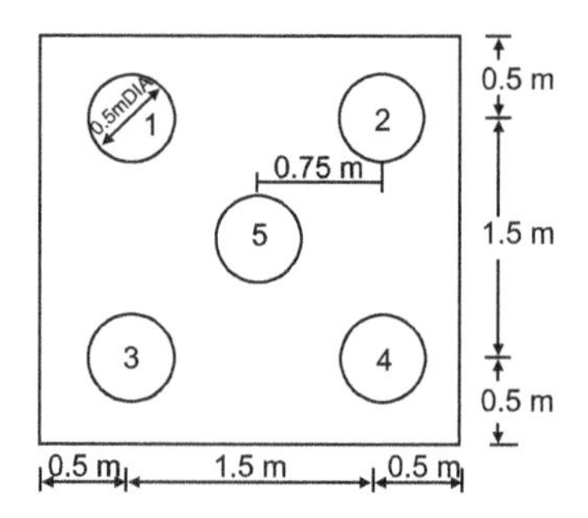

Fig. 7.25. Arrangement of piles (Example 7.1)

Total vertical load on the pile cap = 4000 kN

Add 10% as the weight of pile cap = $\frac{400\ \text{kN}}{4400\ \text{kN}}$

Total moment acting on the pile cap = 8 + 500 = 508 kN-m
Total shear force at the pile cap = 30 + 180 = 210 kN

$$e = M/V = 508/4400 = 0.1155\ \text{m}$$

$$V_{pi} = \frac{V}{n} \pm \frac{V \cdot e \cdot x_1}{Ex_i^2}$$

$$\text{Load in the central pile} = \frac{4400}{5} = 880\ \text{kN}$$

$$\text{Load in the end pile} = \frac{4400}{5} + \frac{508 \times 0.75}{4 \times 0.75^2}$$

$$= 880 + 169 = 1049\ \text{kN} \quad [< 1100\ \text{kN, safe}]$$

(iii) Using the data of lateral load test:

$$y = \frac{Q_{hg}T^3}{EI} \times A_y$$

From Table 7.5, for $Z = 0, A_y = 2.435$.

For $Q_{hg} = 30$ kN, $M_g = 0.0, y = 0.080$ mm $= 0.80 \times 10^{-3}$ m,

Eq. (7.19) gives

$$0.80 \times 10^{-3} = \frac{30 \times T^3}{8.75 \times 10^4} \times 2.435$$

$$T = 3.06 \text{ m} = (EI/\eta_h)^{1/5}$$

or $$\eta_h = EI/3.06^5 = 8.74 \times 10^4/3.06^5 = 326 \text{ kN/m}^3$$

(iv) Analysis by Reese & Matlock (1956) method

Each pile will be subjected to:

Vertical load, $V = 4400/5 = 880$ kN

Moment, $M_g = 508/5 = 102$ kN-m

Horizontal load, $Q_{hg} = 1.2 \times (210/5) = 50.4$ kN

$$Z_{max} = L_s/T = 16/3.06 = 5.23 \text{ (> 5.0, long pile)}$$

$$M_x = A_m Q_{hg} T + B_m M_g$$

$$= A_m \times 50.4 \times 3.06 + 102\, B_m$$

$$= 154.22\, A_m + 102\, B_m$$

Considering pile top free to rotate, from Tables 7.5 and 7.6.

Z	A_m	B_m	M_x (kN-m)
0.0	0.0	1.0	102.0
0.2	0.198	0.999	132.5
0.4	0.379	0.987	159.2
0.6	0.532	0.960	179.9
0.8	0.649	0.914	193.3
1.0	0.727	0.852	199.1
1.2	0.767	0.775	197.3
1.4	0.772	0.688	189.2
1.6	0.746	0.594	175.6
1.8	0.696	0.498	153.1
2.0	0.628	0.404	138.1

Therefore, maximum moment = 199.10 kN-m

It occurs at $Z = 1.0 = x/T$ or $x = 3.06$ m

Hence, maximum moment of magnitude 180.4 kN-m occurs at a depth of 3.06 m below ground level.

(v) Analysis by Chandrasekaran (1974) method

Considering the dynamic soil modulus equal to static soil modulus, then for $T = 3.06$ m, $Z_{max} = 5.2$ and pile top free to rotate condition, Fig. 7.21b gives :

$$F_{SL1} = 0.483$$

$$\omega_{n1} = \frac{F_{SL1}}{\sqrt{\frac{W}{g} \cdot \frac{1}{\eta_h \cdot T^2}}}$$

$$W = 880 \text{ kN} = 88000 \text{ kg}$$

$$\eta_h = 326 \text{ kN/m}^2 = 32600 \text{ kg/m}^2$$

$T = 3.06$ m

$$\omega_{n1} = \frac{0.983}{\sqrt{\frac{88000}{9.81} \cdot \frac{1}{32600 \times 3.06^2}}} = 2.817 \text{ rad./s}$$

Therefore, $f_{n1} = 2.817/2\pi = 0.4485$ cps

Period of vibration in first mode = $T_1 = 1/f_{n1} = 1/0.4485 = 2.23$ s

Assuming damping = 5%

For 5% damping and $T_1 = 2.23$ s, Fig. 7.22 gives

$$S_a/g = 0.051$$

$$\text{Spectral displacement} = \frac{0.051 \times 9.81}{2.817^2} = 0.063 \text{ m}$$

From Table 7.8, for pile free to rotate and $Z_{max} > 5$.

$B_m = 0.315$

Bending moment $= B_m \cdot \eta_h \cdot T^3 \cdot S_d$

$= 0.315 \times 326 \times 3.06^3 \times 0.063$

$= 185.34$ kN-m

Example 7.2

A reciprocating machine is symmetrically mounted on a pile cap of size 3.5 m × 3.5 m × 1.5 m thick which rested on nine piles arranged as shown in Fig. 7.26a.

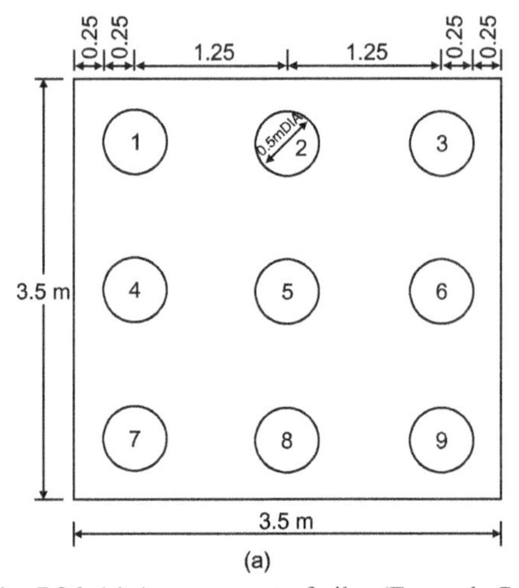

Fig. 7.26. (a) Arrangement of piles (Example 7.2)

Each pile is of 500 mm diameter and length 12.0 m. The axial capacity of each pile is 800 kN. The soil at the site is sandy in nature having $\phi = 35°$ and shear modulus 1.10×10^4 kN/m^2. The moist density of soil is 17 kN/m^3. The machine vibrating at a speed of 180 rpm generates:

Maximum vertical unbalanced force per pile = 2.5 kN

Maximum horizontal unbalanced force per pile = 2.0 kN at a height of 0.4 m above the pile cap.

Determine the natural frequencies and amplitudes of the system in vertical vibration and coupled sliding and rocking vibration.

Solution

(i) Operating frequency of machine $\omega = 180 \times \frac{2\pi}{60} = 18.84$ rad/sec

$$F_z = 2.5 \times 9 \sin \omega t \text{ kN} = 22.5 \sin \omega t \text{ kN}$$

$$F_x = 2.0 \times 9 \sin \omega t \text{ kN} = 18 \sin \omega t \text{ kN}$$

Length of pile = 12.0 m

Diameter of pile = 0.5 m

$$G_s = 1.1 \times 10^4 \text{ kN/m}^2,\ \gamma_s = 17.0 \text{ kN/m}^3,\ \varphi = 35°$$

$$\gamma_p = 24.0 \text{ kN/m}^3,\ E_p = 2.85 \times 10^7 \text{ kN/m}^2,\ A_p = 0.19625 \text{ m}^2$$

$$I_p = 3.066 \times 10^{-3} \text{ m}^4,\ E_pI_p = 8.74 \times 10^4 \text{ kN-m}^2$$

(ii) Vertical vibration

Estimation of stiffness and damping values:

Single Pile

$$r_o = 0.25 \text{ m};\ E_p/G_s = 2.85 \times 10^7/(1.1 \times 10^4) = 2590;$$

$$L/r_o = 12/0.25 = 48$$

Considering the pile as friction pile and parabolic variation of G_s, Fig. 7.19b gives $f_{w1} = 0.0085$. $f_{w2} = 0.029$.

$$v_s = \sqrt{\frac{G_s}{\rho}} = \sqrt{\frac{1.1 \times 10^4 \times 9.81}{17.0}} = 79.67 \text{ m/s}$$

$$K_{zp} = \frac{E_pA_p}{r_o} f_{w1} = \frac{2.85 \times 10^7 \times 0.19625}{0.25} \times 0.0085 = 19.02 \times 10^4 \text{ kN/m}$$

$$C_{zp} = \frac{E_pA_p}{v_s} f_{w2} = \frac{2.85 \times 10^7 \times 0.19625}{79.67} \times 0.029 = 2.036 \times 10^3 \text{ kN-s/m}$$

3 × 3 Pile Group

Cap thickness = 1.5 m (Fig. 7.26b)

Selecting one of a pile (No. 5) as reference pile (Fig. 7.26b)

Computation of α_A

$$L/2r_o = 12/0.5 = 24$$

$$\mu = 0.4 \text{ (assumed)}$$

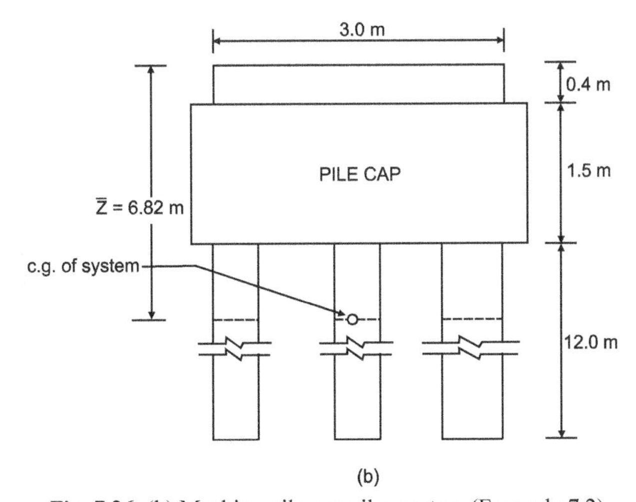

Fig. 7.26. (b) Machine-pile cap piles system (Example 7.2)

For adjacent piles 2 and 4

$$S/2\,r_o = 1.25/0.5 = 2.5$$

For diagonal piles 3

$$S/2\,r_o = (1.25\sqrt{2})/0.5 = 3.54$$

From Fig. 7.23

α_A for reference pile (No. 5) = 1.00
α_A for adjacent pile (No. 2, 4, 6 and 8) = 0.60
a_A for diagonal pile (No. 1, 3, 7 and 9) = 0.53
$\Sigma\alpha_A = 1.00 + 4 \times 0.60 + 4 \times 0.53 = 5.52$

$$K_{zg} = \frac{\sum_1^n K_{zp}}{\sum \alpha_A} = \frac{9 \times 19.02 \times 10^4}{5.52} = 31.01 \times 10^4 \text{ kN/m}$$

$$C_{zg} = \frac{\sum_1^n C_{zp}}{\sum \alpha_A} = \frac{9 \times 2.036 \times 10^3}{5.52} = 3.32 \times 10^3 \text{ kN-s/m}$$

$$K_{zf} = G_s D\bar{S}_{z1} = 1.1 \times 10^4 \times 1.25 \times 2.7 = 3.71 \times 10^4 \text{ kN/m}$$

$$r_{oc} = \sqrt{\frac{3.5 \times 3.5}{\pi}} = 1.98 \text{ m}$$

$$C_{zf} = Dr_{oc}\bar{S}_{z2}\sqrt{G_s\rho} = 1.25 \times 1.98 \times 6.7 \times \sqrt{1.1 \times 10^4 \times 17.0/9.81} = 2.29 \times 10^3 \text{ kN-s/m}$$

Therefore,

$$(K_z)_g = K_{zg} + K_{zf} = (31.01 + 3.71) \times 10^4 = 34.72 \times 10^4 \text{ kN/m}$$

$$(C_z)g = C_{zg} + C_{zf} = (3.32 + 2.29) \times 10^3 = 5.61 \times 10^3 \text{ kN-s/m}$$

Refer Fig. 7.26b

Weight of pile cap = 3.5 × 3.5 × 1.5 ×24 = 441 kN

Weight of piles = 9 × (π/4) × 0.5^2 × 12 × 24 = 508.68 kN

Weight of soil = (3.5 × 3.5 × 12 – (π/4) × 0.5^2 × 12 × 9) × 17.0 kN

= 2138.685 kN

Weight of machine = 50 kN

Total weight = 441 + 508.68 + 2138.685 + 50 = 3138.365 kN

m = 3138.365/9.81 = 319.91 kg

$$(C_c)_g = 2\sqrt{(K_z)_g m} = 2\sqrt{34.72 \times 10^4 \times 319.91} = 21.07 \times 10^3 \text{ kN-s/m}$$

$$(\xi)_g = \frac{5.61 \times 10^3}{21.07 \times 10^3} = 0.266$$

$$(\omega_{nz})_g = \sqrt{\frac{34.72 \times 10^4}{319.91}} = 32.94\,\text{rad/s}$$

$$(\omega_{ndz})_g = \frac{(\omega_{nz})_g}{\sqrt{1-2\xi^2}} = \frac{32.94}{\sqrt{1-2\times 0.266^2}} = 35.55 \text{ rad/s}$$

$$\frac{\omega}{\omega_{ndz}} = \frac{18.84}{35.55} = 0.53 \text{ (hence, safe)}$$

$$2m_e e\omega^2 = 22.5 \times 9.81$$

$$2m_e e = \frac{220.725}{18.84^2} = 0.622\,\text{kN-m}$$

$$\eta = \frac{\omega}{\omega_{nz}} = \frac{18.84}{32.94} = 0.572$$

$$\frac{A_z}{\frac{2m_e e}{m}} = \frac{\eta^2}{\sqrt{(1-\eta^2)^2+(2\eta\xi)^2}} = \frac{0.572^2}{\sqrt{(1-0.572^2)^2+(2\times 0.572\times 0.266)^2}} = 0.443$$

$$A_2 = [(0.443 \times 0.622)/3138.365] \times 1000 = 0.0878 \text{ mm}$$

(iii) Combined sliding and rocking
From Table 7.9, for parabolic soil profile, $(L/r_o) > 30$,

$$\mu = 0.40 \text{ and } (E_p/G_s) = 2590, \text{ we get}$$

$$f_{x1} = 0.0022,\ f_{\varphi 1} = 0.2529;\quad f_{x\varphi 1} = -0.0284$$

$$f_{x2} = 0.0084,\ f_{\varphi 2} = 0.2109;\quad f_{x\varphi 2} = -0.0519$$

Values of coefficients for $\mu = 0.5$ are almost same as for $\mu = 0.4$. The difference between the two values may be neglected.

$$K_{xp} = \frac{E_p I_p}{r_0^3} f_{x1} = \frac{8.74\times 10^4}{0.25^3}\times 0.0022 = 1.23\times 10^4 \text{ kN/m}$$

$$C_{xp} = \frac{E_p I_p}{r_0^2 v_s} f_{x2} = \frac{8.74\times 10^4}{0.25^2\times 79.67}\times 0.0084 = 147.41 \text{ kN-s/m}$$

$$K_{\varphi p} = \frac{E_p I_p}{r_o} f_{\varphi 1} = \frac{8.74\times 10^4}{0.25}\times 0.2529 = 8.84\times 10^4 \text{ kN-m/rad}$$

$$C_{\varphi p} = \frac{E_p I_p}{v_s} f_{\varphi 2} = \frac{8.74\times 10^4}{79.67}\times 0.2109 = 231.31 \text{ kN-s-m}$$

$$K_{x\varphi p} = \frac{E_p I_p}{r_0^2} f_{x\varphi 1} = \frac{8.74\times 10^4}{0.25^2}\times(-0.0284) = -3.97\times 10^4 \text{ kN/rad}$$

$$C_{x\varphi p} = \frac{E_p I_p}{r_o v_s} f_{x\varphi 1} = \frac{8.74\times 10^4}{0.25\times 79.67}\times(-0.0519) = -227.69 \text{ kN-s}$$

$$K_R = \frac{E_p I_p}{2G(1+\mu)L^4} = \frac{8.74\times10^4}{2\times1.1\times10^4\times(1+0.4)\times12^4} = 13.7\times10^{-5}$$

Therefore, pile behaves as a flexible pile.
Refer Fig. 7.26 and considering pile No. 1 as reference pile.

Pile No.	*S/2 r_o*	β	α_A
1	0.0	0°	1.00
2	2.5	0°	0.58
3	5.0	0°	0.30
4	2.5	90°	0.36
5	3.53	45°	0.38
6	5.59	26.5°	0.26
7	5.0	90°	0.17
8	5.59	63.4°	0.18
9	7.07	45°	0.04
			$\Sigma\alpha_L = 3.27$

$$K_{xg} = \frac{\sum_1^n K_{xp}}{\sum \alpha_L} = \frac{9\times1.23\times10^4}{3.27} = 3.385\times10^4 \text{ kN/m}$$

$$C_{xg} = \frac{\sum_1^n C_{xp}}{\sum \alpha_L} = \frac{9\times147.41}{3.27} = 405.71 \text{ kN-s/m}$$

$$a_o = \omega r_o\sqrt{\frac{\rho}{G}} = 18.84\times1.98\sqrt{\frac{17.0/9.81}{1.1\times10^4}} = 0.47$$

From Table 7.10 for $a_o = 0.29$ and $\mu = 0.4$, $S_{x1} = 4.1$ and $S_{x2} = 10.6$. Therefore,

$$K_{xf} = G_s D\bar{S}_{x1} = 1.1\times10^4\times1.25\times4.1 = 5.638\times10^4 \text{ kN/m}$$

$$C_{zf} = Dr_{oc}\bar{S}_{x2}\sqrt{G_s\rho} = 1.25\times1.98\times10.6\times\sqrt{1.1\times10^4\times17.0/9.81} = 3.62\times10^3 \text{ kN-s/m}$$

$$(K_x)_g = (3.385 + 5.638)\times10^4 = 9.023\times10^4 \text{ kN/m}$$

$$(C_x)_g = (0.405 + 3.62)\times10^3 = 4.028\times10^3 \text{ kN-s/m}$$

$$K_{\phi g} = \sum_1^n (K_{\varphi p} + K_{zp}x_r^2 + K_{xp}Z_c^2 - 2Z_c K_{x\varphi p})$$

$x_r = 1.25$ m for piles 1, 3, 4, 6, 7 and 9

$x_r = 0$ for piles 2, 5 and 8

$Z_c = 1.5/2 = 0.75$ m

$$K_{\varphi g} = 9\times8.84\times10^4 + 6\times19.02\times10^4\times1.25^2 + 9\times1.23\times10^4\times0.75^2 - 2\times0.75\times9\times(-3.97\times10^4)$$

$$K_{\varphi g} = 3.177\times10^6 \text{ kN-m/rad}$$

$$C_{\phi g} = \sum_{1}^{n}(C_{\varphi p} + C_{zp}x_r^2 + C_{xp}Z_c^2 - 2Z_c C_{x\varphi p})$$

$$= 9 \times 231.31 + 6 \times 2.036 \times 10^3 \times 1.25^2$$

$$+ 9 \times 147.41 \times 0.75^2 - 2 \times 0.75 \times 9 \times (-227.69)$$

$$= 24.98 \times 10^3 \text{ kN-s-m}$$

For rocking vibration, $r_{oc} = \left(\dfrac{ab^3}{3\pi}\right)^{1/4} = \left(\dfrac{3.5^4}{3\pi}\right)^{1/4} = 2 \text{ m}$

$$\delta = \frac{D}{r_{oc}} = \frac{1.25}{2} = 0.63$$

From Table 7.10, $\bar{S}_{x1} = 4.1$, $\bar{S}_{x2} = 10.6$, $\bar{S}_{\phi 1} = 2.5$ and $\bar{S}_{\phi 2} = 1.8$

$$K_{\phi f} = G_s r_{oc}^2 D \times \bar{S}_{\phi 1} + G_s r_{oc}^2 h\left[\left(\frac{\delta^2}{3}\right) + \left(\frac{Z_c}{r_0}\right)^2 - \delta\left(\frac{Z_c}{r_0}\right)\right]\bar{S}_{x1}$$

$$= 1.1 \times 10^4 \times 2^2 \times 1.25 \times 2.5 + 1.1 \times 10^4 \times 2^2$$

$$\times 1.25\left[\frac{0.63^2}{3} + \left(\frac{0.75}{2}\right)^2 - 0.63 \times \frac{0.75}{2}\right] \times 4.1$$

$$= 145.77 \times 10^3 \text{ kN-m/rad}$$

$$C_{\phi f} = \delta r_{oc}^4 \sqrt{\frac{G_s \gamma}{g}}\left\{\bar{S}_{\phi 2} + \left[\left(\frac{\delta^2}{3}\right) + \left(\frac{Z_c}{r_0}\right)^2 - \delta\left(\frac{Z_c}{r_0}\right)\right]\bar{S}_{x2}\right\}$$

$$= 0.63 \times 2^4 \sqrt{\frac{1.1 \times 10^4 \times 17}{9.81}}\left\{1.8 + \left[\frac{0.63^2}{3} + \left(\frac{0.75}{2}\right)^2 - 0.63 \times \frac{0.75}{2}\right] \times 10.6\right\}$$

$$= 3.046 \times 10^3 \text{ kN-s-m}$$

$$(K_\varphi)_g = 3.177 \times 10^6 + 145.77 \times 10^3 = 3.322 \times 10^6 \text{ kN-m/rad}$$

$$(C_\varphi)_g = 24.98 \times 10^3 + 3.046 \times 10^3 = 28.03 \times 10^3 \text{ kN-s-m}$$

Refer Fig. 7.26b

$$\bar{Z} = \frac{50 \times 0.2 + 441(0.4 + 0.75) + (506.68 + 2138.685)(0.4 + 1.5 + 6.0)}{3138.365}$$

$$= \frac{21431.334}{3138.365} = 6.83 \text{ m}$$

$$M_m = \frac{50(3^2 + 0.4^2)}{12} + 50 \times 6.63^2 + \frac{441(3.5^2 + 1.5^2)}{12} + 441 \times 5.68^2$$

$$+ \frac{(508.68 + 2138.685)(3.5^2 + 12^2)}{12} + (508.68 + 2138.685)(-1.07)^2$$

$$= 54498.54 \text{ kN-m}^2$$

$$M_{mo} = M_m + mL^2 = 54498.54 + 3138.365 \times 7.07^2 = 211369.4 \text{ kN-m}^2$$

$$r = \frac{M_m}{M_{mo}} = \frac{54498.54}{211369.4} = 0.258$$

$$M_y = 18 \times 6.83 \sin \omega t = 122.94 \sin \omega t \text{ kN-m}$$

$$\omega_{nx} = \sqrt{\frac{K_x}{m}} = \sqrt{\frac{9.023 \times 10^4}{319.91}} = 16.79 \text{ rad/s}$$

$$\omega_{n\varphi} = \sqrt{\frac{K_\phi}{M_{mo}}} = \sqrt{\frac{3.322 \times 10^6 \times 9.81}{211369.4}} = 12.43 \text{ rad/s}$$

Undamped natural frequency in coupled rocking and sliding are given by:

$$\omega_{n1,2}^2 = \left[(\omega_{nx}^2 + \omega_{n\varphi}^2) \pm \sqrt{(\omega_{nx}^2 + \omega_{n\varphi}^2)^2 - 4r\omega_{nx}^2\omega_{n\varphi}^2}\right]$$

$$= \frac{1}{2 \times 0.258}\left[(16.79^2 + 12.43^2) \pm \sqrt{(16.79^2 + 12.43^2)^2 - 4 \times 0.258 \times 16.79^2 \times 12.43^2}\right]$$

$$\omega_{n1} = 39.81 \text{ rad/s}, \ \omega_{n2} = 10.32 \text{ rad/s}$$

$$(C_{cx})_g = 2\sqrt{K_x m} = 2\sqrt{9.023 \times 10^4 \times 319.91} = 10.745 \times 10^3 \text{ kN-s/m}$$

$$(C_{c\varphi})_g = 2\sqrt{K_\varphi M_{mo}} = 2\sqrt{3.322 \times 10^6 \times (211369.4/9.81)} = 535.08 \times 10^3 \text{ kN-s-m}$$

$$(\xi_x)_g = \frac{4.028 \times 10^3}{10.745 \times 10^3} = 0.375$$

$$(\xi_\varphi)_g = \frac{28.03 \times 10^3}{535.08 \times 10^3} = 0.052$$

$$\Delta(\omega^2) = \left[\left\{\omega^4 - \omega^2\left[\frac{\omega_{nx}^2 + \omega_{n\varphi}^2}{r} - \frac{4\xi_x\xi_\varphi\omega_{nx}\omega_{n\varphi}}{r}\right] + \frac{\omega_{nx}^2\,\omega_{n\varphi}^2}{r}\right\}^2\right.$$

$$\left. + 4\left\{\xi_x\frac{\omega_{nx}\omega}{r}(\omega_{n\phi}^2 - \omega^2) + \frac{\xi_\varphi\omega_{n\phi}\omega}{r}(\omega_{nx}^2 - \omega^2\right\}^2\right]^{1/2}$$

$$= \left[\left\{18.84^4 - 18.84^2\left[\frac{(16.79^2 + 12.43)^2}{0.258} - \frac{4 \times 0.375 \times 0.052 \times 16.79 \times 12.43}{0.258}\right] + \frac{16.79^2 \times 12.43^2}{0.258}\right\}^2\right.$$

$$\left. + 4\left\{0.375 \times \frac{16.79 \times 18.84}{0.258}(12.43^2 - 18.84^2) + \frac{0.052 \times 12.43 \times 18.84}{0.258}(16.79^2 - 18.84^2)\right\}^2\right]^{1/2}$$

$$= 3.414 \times 10^5$$

$$A_{x1} = \frac{F_x\left[(-M_m\omega^2 + K_\phi + K_x L^2)^2 + \omega^2 (C_\phi + C_x L^2)^2\right]^{1/2}}{mM_m \Delta\omega^2}$$

$$= \frac{22.5\left[(-54498.54\times18.84^2 + 3.322\times10^6 + 9.023\times10^4\times7.07^2)^2 + 18.84^2(28.03\times10^3 + 4.028\times10^3\times7.07^2)^2\right]^{1/2}}{319.91\times\left(\frac{54498.54}{9.81}\right)\times3.414\times10^5}$$

$$= 215.975\times10^{-6}\text{ m}$$

$$A_{\phi 1} = \frac{F_x L\omega_{nx}\left[\omega_{nx}^2 + 4\xi_x\omega^2\right]^{1/2}}{M_m\Delta\omega^2} = \frac{22.5\times7.07\times16.79\times[16.79^2 + 4\times0.375\times18.84^2]^{1/2}}{\left(\frac{54498.54}{9.81}\right)\times3.414\times10^5}$$

$$= 32.148\times10^{-6}\text{ rad}$$

$$A_{x2} = \frac{M_y L[(\omega_{nx}^2)^2 + (2\xi_x\omega_{nx}\omega)^2]^{1/2}}{M_m\Delta\omega^2} = \frac{122.94\times7.07[(16.79^2)^2 + (2\times0.375\times16.79\times18.84)^2]^{1/2}}{\left(\frac{54498.54}{9.81}\right)3.414\times10^5}$$

$$= 168.82\times10^{-6}\text{ m}$$

$$A_{\phi 2} = \frac{M_y[(\omega_{nx}^2 - \omega^2)^2 + (2\xi_x\omega_{nx}\omega)^2]^{1/2}}{M_m\Delta\omega^2} = \frac{122.94[(16.79^2 - 18.84^2)^2 + (2\times0.375\times16.79\times18.84)^2]^{1/2}}{\left(\frac{54498.54}{9.81}\right)3.414\times10^5}$$

$$A_x = A_{x1} + A_{x2} = 2.96\times10^{-4}\text{ m}$$

$$A_\varphi = A_{\varphi1} + A_{\varphi2} = 4.82\times10\text{–}5\text{ rad}$$

References

Berezantsev, V.G., V. Khristoforov and Gulubkov, V. (1961), "Load Bearing Capacity and Deformation of Piles Foundations", Proc. 5th Int. Conf., SM & FE, Vol. 2.

Broms, B. (1964 a), "The lateral resistance of piles in cohesive soils", J. Soil Mech. Found. Div., ASCE, Vol. 90, No. SM2, pp. 27-63.

Broms, B. (1964 b), "The lateral resistance of piles in cohesive soils", J. Soil Mech. Found. Div., ASCE, Vol. 90, No. SM2, pp. 123-156.

Chandrasekaran, V. (1974), "Analysis of pile foundations under static and dynamic loads", Ph.D. Thesis I.I.T., Roorkee, Roorkee.

Davisson, M.T. and Gill, H.L. (1963), "Laterally-loaded piles in a layered soil system", J. Soil Mech. Found, Div., ASCE, Vol. 89, No. SM3, pp. 63-94.

Fleming, W.G.K. and Sliwinski, Z.J. (1977), "The use and influence of bentonite in bored pile construction", Construction Industry Research and Information Association Report, No. PDP2.

Hansen, J.B., (1961), "The ultimate resistance of rigid piles against transversal forces", Danish Geotechnical Institute (Geoteknisk Instiut) Bull No. 12, Copenhagen, p. 5-9.

IS : 2911 (Part I and Part IV), (2010). "Code of Practice for Design and Construction of Pile Foundations", ISI, New Delhi.

Jain, N.K. (1983), "Flexural Behaviour of Partially Embedded Pile Foundations", Ph.D. Thesis, Deptt. of Civil Engg., University of Roorkee.

Matlock, H. and Reese, L.C. (1962), "Generalized solutions for laterally loaded piles", Transactions of the American Society of Civil Engineers, Vol. 127, Part-I, pp. 1220-1247.

Meyerhof, G.G. (1976), "Bearing capacity and settlement of pile foundations", J. Geotech. Div. ASCE, Vol. 102, No. GT 3, pp. 197-228.

Novak, M. (1974), "Dynamic stiffness and damping of piles", Can Geotech. J., Vol. 11, No. 4, pp. 574-598.

Novak, M. and Beredugo, Y.O. (1972), "Vertical vibration of embedded footings", J. Soil Mech. Found. Div., ASCE, Vol. 98, No. SM12, pp. 1291-1310.

Novak, M. and El-Sharnouby, B. (1983), "Stiffness and damping constants of single piles", J. Geotech. Engg. Div., ASCE, Vol. 109, No. 8, pp. 961-974.

Novak, M. and Grigg, R.F. (1976), "Dynamic experiments with small pile foundation", Can. Geot. J. Vol., 13, No. 4, pp. 372-395.

Palmer, L.A. and Thompson, J.B. (1948), "The earth pressure and deflection along the embedded lengths of piles subjected to lateral thrust", Proceedings Second International Conferene on Soil Mechanics and Foundation Engineering, Rotterdam, Holland, Vol. V, pp. 156-161.

Prakash, S. and Chandrasekaran, V. (1977), "Free vibration characteristics of piles", Proceedings Ninth International Conference on Soil Mechanics and Foundation Engineering, Tokyo, Vol. 2. pp. 333-336.

Poulos, H.G. (1971 a), "Behaviour of laterally loaded piles: I-single piles", J. Soil Mech. Found. Div., ASCE, Vol. 97, No. SM5, pp. 711-731.

Poulos, H.G. (1971 b), "Behaviour of pile groups subjected to lateral loads piles: II-piles groups", J. Soil Mech. Found. Div., ASCE, Vol. 97, No. SM5, pp. 733-751.

Prakash, S. (1962), "Behaviour of pile groups subjected to lateral loads", Ph.D. Thesis, University of Illinois, Urbana, p. 397.

Ramasamay, G., Ranjan, G. and Jain, N.K. (1987), "Modification to Indian Standard Code Procedure on Lateral Capacity of Piles", Indian Geotechnical Journal, Vol. 17, No. 3, pp. 249-259.

Tomlinson, M.J. (1986), "Foundation Design Construction", Longman, Singapore Publishers, Singapore.

Reese, L.C. and Matlocck, H. (1956), "Non-dimensiontal solutions for laterally loaded piles with soil modulus assumed proportional to depth", Proceedings 8th Texas Conference on Soil Mechanics and Foundation Engineering, Austin, TX, pp. 1-41.

Saran, S. (1998), "Analysis and design of substructures", Oxford & IBH Publishing Co. Pvt. Ltd., New Delhi.

Sheta, M. and Novak, M. (1982), "Vertical vibrations of pile groups", J. Geot. Engg., ASCE, Vol. 108, No. GT4, pp. 570-590.

Practice Problems

7.1 Describe the subgrade reaction method of obtaining the variation of deflection and moment along the length of pile subjected to lateral load. Give the analysis considering that the pile is embedded in (i) sandy strata, and (ii) clayey strata. Discuss the effect of fixity of pile head on deflection and bending moment.

7.2 Give the methods of determining the loads in the various piles of a pile group containing both vertical and batter piles. The pile group is subjected to a vertical load, a lateral load and a moment.

7.3 A 400 mm × 400 mm size concrete pile is driven through a deposit of looses and ($\phi = 25°$) 8.0 m thick into a deposit of dense sand ($\phi = 41°$). The pile penetrates to a depth of 2.5 m in the dense sand deposit which extends to a great depth. If the unit weights of loose sand and dense sand deposits are 16 kN/m^3 and 19 kN/m^3 respectively, estimate the ultimate pile capacity.

7.4 Estimate the required length of 500 mm diameter concrete pile driven in a deposit of clay having an unconfined compressive strength of 120 kN/m^2. The pile is required to carry a load of 280 kN. If the cast-in-situ bored pile of the same length is used, what will be its capacity?

7.5 Proportion a friction pile group to carry a load of 3000 kN including the weight of pile cap on a site where the subsoil consists of uniform clay to a depth of 20 m underlain by rock. The undrained cohesion of clay is 50 kN/m^2. The clay is normally loaded and has a void ratio of 0.95 and the liquid limit of 55%. Compute the settlement of the pile group.

7.6 It is plan to construct a 60 m high tower near Delhi. The soil stratum is such that it is required to be supported on pile foundation. It was decided to rest the foundation of the tower on piles of 500 mm diameter and length 15.0 m. The pile load test on a test pile gave the allowable pile capacity as 1500 kN. The base of pile cap is subjected to:
Axial vertical load = 6000 kN
Total horizontal force including due to earthquake = 250 kN
Total moment including due to earthquake = 600 kN-m
A lateral pile load test was also performed at the site on a test pile. It was observed that a horizontal load of 25 kN applied very close to ground surface gave a horizontal deflection of 0.5 mm.
Determine the maximum moment in the pile by (i) Pseudo-static analysis and (ii) Dynamic analysis.

7.7 A reciprocating machine supported on pile foundation and vibrating at speed of 200 rpm generates:
Maximum vertial unbalanced force = 20 kN
Maximum horizontal unbalanced force = 15 kN
The horizontal force was acting at a height of 0.5 m from the top of the pile cap (4.0 m × 4.0 m × 2.0 m thick). It was decided to use 16 piles each of 12.0 m length. The soil at the site is over-consolidated clay having shear modulus (G) as 2.5 × 10^4 kN/m^2. Considering suitable arrangement of piles, determine natural frequencies and amplitudes of the system.

CHAPTER

8

Seismic Stability Analysis of Slopes

8.1 General

The most commonly used methods of stability analysis are based on the assumption of potential sliding surfaces. In the past a number of different sliding surfaces have been considered for stability analysis. The oldest one was plane surface postulated by Coulomb (1776) and Culmann (1886), and this assumptions led to fairly accurate results for vertical slopes. The circular arc method of analysis was first proposed by Petterson (1916), based on his field studies and it was developed by Fellenius (1927) by use of method of slices. This concept of 'method of slice' is widely used in most of the present day methods. Taylor (1937) introduced a more mathematically sound stability analysis based on total stresses, known as 'φ-circle method', which led to the preparation of simple stability charts for homogeneous slopes. These charts are valid only for analysis based upon total stresses. Analyses based on logarithmic spiral shaped rupture surfaces were also available (Taylor, 1937; Rendulic, 1935; Frohlich, 1953), but it was pointed out by Taylor that these methods did not differ from that of φ-circle method in giving the factor of safety or the position of critical rupture surface. Bishop (1955) introduced the concept of 'Limit design' into stability analysis and used method of slices with effective stresses. His method is too complicated and has been approximated to a simplified method with a little sacrifice of accuracy, by neglecting the effect of shear forces acting on the vertical faces of the slices. Based on this simplified method, non-dimensional charts were obtained for constant pore pressure ratios (Bishop and Morgenstern, 1960) and for sudden-draw down conditions (Morgenstern, 1963). Spencer (1967) has provided stability charts which are quite useful for selecting safe slope for an embankment in a soil of known properties. This method of analysis was based on the assumption that the inter-slice forces are parallel to each other. Apart from these, stability charts were also provided by different authors Janbu (1954a), Meyer (1958), Isao Minami (1963) based on the circular arc method.

Bell (1963) has reduced the number of dimensionless variables used by others in earlier investigations and presented a replot of charts of Taylor and Bishop in terms of modified factor of safety with respect to strength and modified stability number.

Nature never provides homogeneous and ideal conditions of slope configuration, pore pressure etc. and thus the soil may slip along a non-circular surface and the minimum factor of safety may be associated with such a surface. Janbu (1954b) presented a generalized method of slices considering general shape for the rupture surface. His method necessitates a reasonable assumption for the position of line of thrust. His solution may be applied safely to elongated shallow slip surfaces, but is reported by Nonveiller (1965) to be in error when applied to deep slip surfaces. A method of slices procedure for non-circular slip surfaces with an arbitrary point as movement center was proposed by Nonveiller (1965). He claimed that in many cases the inter-slice forces may be neglected without large errors. However, a more comprehensive investigation is necessary to find the limits within which this approximation is acceptable for practical computations. Morgenstern and Price's (1965) general slip surface analysis has considered the limiting equilibrium of a potential sliding body composed of a series of slices. They assumed a more general functional relationship between the normal and shear forces at the interface of a slice. In their opinion, this functional relationship may be estimated from elastic theory or from field observations. This method is complete in all respects if the functional relationship is evaluated correctly.

None of the above mentioned solutions attempt to solve the problem of stability of slopes subjected to earthquake. This problem is of great importance to engineers.

Stability analysis of a slope subjected to earthquake force is carried out adopting three approaches, namely (i) pseudo-static, (ii) assessing non-recoverable i.e. plastic deformation, and (iii) dynamic analysis using finite element technique. In pseudo-static analysis, a surface of sliding is assumed and a quantitative estimate of the factor of safety is obtained by examining the equilibrium conditions at the time of incipient failure and compare the strength necessary to maintain limiting equilibrium with the available strength of soil. The earthquake forces are considered as equivalent static forces which are obtained by multiplying the weight of the sliding wedge with seismic coefficients. In deformation analysis, displacement of the slope is estimated for the amount of acceleration which is in excess to yield acceleration. In finite element method, dynamic analysis of the slope is carried out by discritizing the slope into small elements and imposing on it more realistic estimated ground motion. In this method stresses and strains in the slope section are evaluated. This method is more general as dynamic strain dependent material properties can be utilized in the analysis. However the accuracy of the results depend on the precision of input parameters. Software packages are available commercially to analyse the slope by finite element technique.

In this chapter, the pseudo static methods, and procedures of determining non-recoverable deformation are discussed subsequently.

8.2 Pseudo-Static Methods

8.2.1 Modified Swedish Circle Method

The method of slices was first introduced by Fellinius (1936) and is also known as Swedish circle method. The sliding surface is assumed circular. The analysis is described in following steps:

(i) The sliding mass is divided into a number of vertical slices, usually 6 to 12 being consistent with the general accuracy of the method (Fig. 8.1a). The width of the slices need not be the same and is usually adjusted such that the base of any one slice is located in a single material.

(ii) The forces acting on the ith slice of width b are shown in (Fig. 8.1b). These are:
Weight of the slice, W_i.
Shear forces X_i and X_{i+1} on the vertical sides of the slice.
Normal forces E_i and E_{i+1} on the vertical sides of the slice.
Cohesion force c, ΔS, ΔS being the curved length of the element base and c is unit cohesion.
Resultant friction force, P acting on the base of the slice making an angle ϕ with normal.
Horizontal seismic force, $W_i \cdot \alpha_h$ acting horizontally at the c.g. of the slice. α_h being the horizontal seismic coefficient.
Vertical seismic force, $\pm W_i \alpha_v$ acting in the vertical diretion at the c.g. of the slice. α_v being the vertical seismic coefficient.
Pore pressure u acting normally at the base of slice

(iii) The analysis is based on the assumption that for each individula slice.

$$E_i - E_{i+1} = 0 \tag{8.1a}$$

$$X_i - X_{i+1} = 0 \tag{8.1b}$$

It may be noted that if the sliding wedge is considered as one unit, then

$$\sum_{i=1}^{n}(E_i - E_{i+1}) = 0 \tag{8.2a}$$

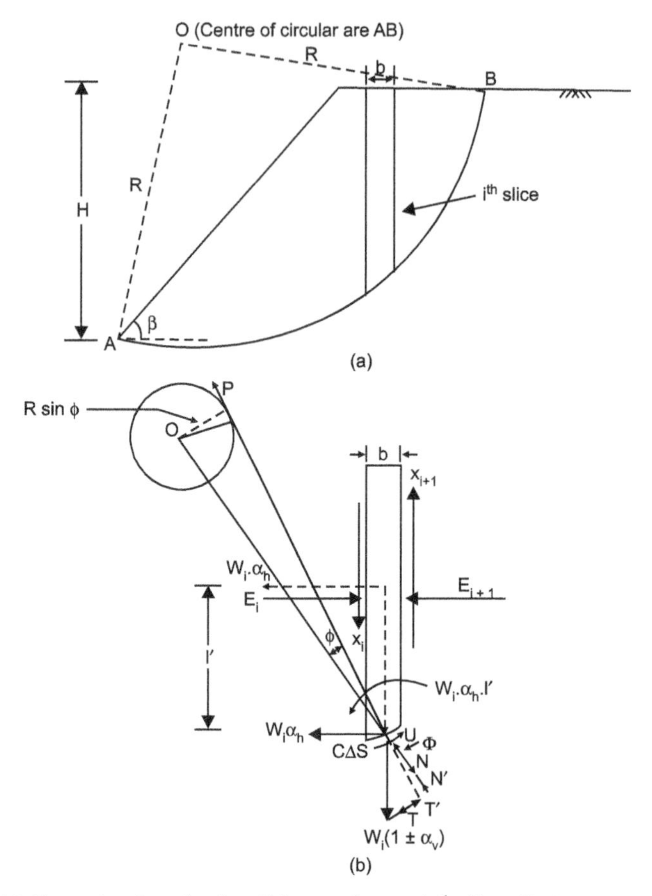

Fig. 8.1. (a) Slope showing circular sliding surface and i^{th} slice, (b) Forces on the ith Slice

$$\sum_{i=1}^{n}(X_i - X_{i+1}) = 0 \tag{8.2b}$$

(iv) Let N *and* N' represent respectively the normal component of $W_i(1 \pm \alpha_v)$ and $W_i \cdot \alpha_h$; and T and T' represent their tangential components, then resolving the force P in normal direction, it can be written as:

$$P \cos \phi = (N - N' - u) \tag{8.3a}$$

or

$$P = \frac{N - N' - u}{\cos \phi} \tag{8.3b}$$

(v) As shown in Fig. 8.1, 'O' represents the center of circular wedge and R being the radius. Further the force $W_i\, \alpha_h$ has been shifted at the base the slice with an moment equal to $W_i\, \alpha_h \cdot l'$. Taking the moments of all the forces about the center 'O'.

$$\text{Resisting moments} = c \cdot \Delta S \cdot R + P \cdot R \sin \phi + W_i\, \alpha_h \cdot l' \tag{8.4a}$$

$$\text{Overturning moment} = (T + T')R \tag{8.4b}$$

Adding the moments for all the n slices, and substituting value of P from Eq. (8.3b), the expression of factor of safety will be:

$$F = \frac{\Sigma[c \cdot \Delta S \cdot R + (N - N' - u)\tan\phi \cdot R + W_i\, \sigma_h \cdot l']}{\Sigma[(T + T')R]} \tag{8.4c}$$

The value of $\Sigma W_i \alpha_h \cdot l'$ is small in comparison to other terms, and usually neglected. It understimates the factor of safety by little amount. Then the expression of factor of safety can be written as:

$$F = \frac{cS + \tan\phi \cdot \Sigma(N - N' - u)}{\Sigma[(T + T')]} \tag{8.4d}$$

S is the total length the base of circular wedge i.e. AB (Fig. 8.1).

The role of pore water pressure is important in analyzing the stability of a slope. Pore water is present if the soil mass is located below water table or under steady seepage. In the later case, the flow net is constructed, and this makes it possible to evaluate the pore pressure at the desired point.

This method has been incorporated in IS:1893-2016.

8.2.2 Modified Taylor's Method

Prakah, Saran and Raj (1968) extended the 'Friction Circle Method' proposed by Taylor (1937) considering horizontal seismic force. The limitation of this method is that it is applicable to homogeneous slopes. In this method also, sliding wedge is considered circular. Figure (8.2) shows the circular wedge ABM with its center at point 'O' and radius R. The slope is of height H and inclined with horizontal by an angle β. The chord AB makes an angle x with horizontal and the two radii OA and OB make angle y with the perpendicular OP.

The wedge ABM is in equilibrium under the following forces:

Weight of mass ABM, W acting at it c.g.

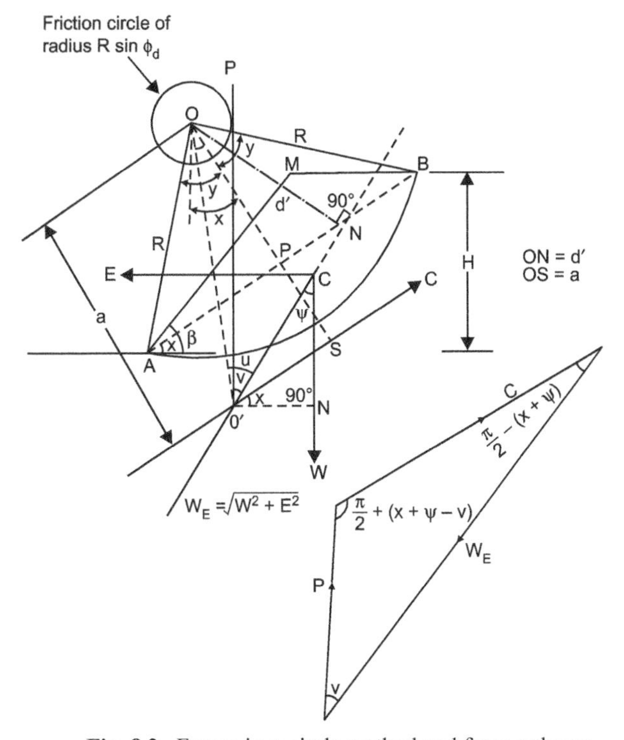

Fig. 8.2. Forces in φ circle method and force polygon

Cohesion force C acting such that (Fig. 8.2)

$$a = RS/L = R.R \cdot 2y/(2R \sin y) = Ry \text{ cosec } y \tag{8.5}$$

where S = Length of circular arc AB

L = Length of chord AB

Inertia force, $E = W \cdot \alpha_h$ acting horizontally at the c.g. of mass *ABM*. α_h. being the horizontal seismic coefficient.

Resultant frictional force, *P*. It will be tangential to ϕ circle of radius $R \sin \phi_d$ (Fig. 8.2). ϕ_d is developed angle of internal friction needed to keep the wedge in equilibrium.

Cohesion force *C* given by

$$C = c_d \cdot S = c \cdot S \cdot /F_c = 2c \cdot R \sin y/F_c \tag{8.6}$$

where

c_d = Value of unit cohesion required to keep the wedge in equilibrium

F_c = Factor of safety with respect to cohesion

Let W_E is the resultant of the forces *W* and *E*, i.e.

$$W_E = \sqrt{W^2 + E^2} \tag{8.7}$$

The resultant W_E is inclined to vertical at an angle ψ such that

$$\psi = \tan^{-1} (\alpha_h) \tag{8.8}$$

Refer Fig. (8.2),

$$R = H/2 \operatorname{cosec} x \operatorname{cosec} y \tag{8.9}$$

$$W = \gamma R^2 y - \gamma R^2 \sin y \cos y + \frac{\gamma H^2}{2}(\cot x - \cot \beta) \tag{8.10}$$

$$\begin{aligned} W_E d' = {} & \gamma R^2 y\left(\frac{2R}{3}\frac{\sin y}{y}\sin x\right) - \gamma R^2 \sin y \cos y \times \left(\frac{2R}{3}\cos y \sin x\right) + \frac{\gamma H^2}{2}\cot x \\ & \times\left[\frac{H}{3}\cot x + R\sin(x-y)\right] - \frac{\gamma H^2}{2}\cot\beta \times \left[\frac{H}{3}\cot\beta + R\sin(x-y)\right] - \gamma H^2 y \alpha_h \\ & \times\left[\frac{2R}{3}\frac{\sin y}{y}\cos x\right] - \gamma R^2 \sin y \cos y \alpha_h \times\left[\frac{2R}{3}\cos y \cos x\right] + \frac{\gamma H^2}{2}\cos x \alpha_h \\ & \times\left[R\cos(x-y) - \frac{2H}{3}\right] - \frac{\gamma H^2}{2}\cot\beta\,\alpha_h \times\left[R\cos(x-y) - \frac{2H}{3}\right] \end{aligned}$$

or

$$\begin{aligned} W_E \cdot d' = {} & \frac{\gamma H^3}{12}(1 - 2\cot^2\beta + 3\cot\beta\cot x - 3\cot\beta\cot y + 3\cot x\cot y) \\ & + \frac{\gamma H^3}{12}\alpha_h\left[\cot x\{\operatorname{cosec}^2 x + 3\cot x\cot y - 3\cot\beta\cot y - 1\} + \cot\beta\right] \end{aligned} \tag{8.11}$$

$$\begin{aligned} OO' &= d' \operatorname{cosec} u = a \sec(x + \psi - u) = R \cdot y \cdot \operatorname{cosec} y \sec(x + \psi - u) \\ &= R \sin \phi_d \operatorname{cosec}(u - v) \end{aligned} \tag{8.12}$$

From the force polygon

$$\frac{W_E}{C} = \frac{\cos(x+\psi-v)}{\sin v} = \cos(x+\psi)\cot v + \sin(x+\psi) \tag{8.13}$$

Substituting (8.9) in (8.10)

$$W = \frac{\gamma H^2}{4}(y \operatorname{cosec}^2 x \operatorname{cosec}^2 y - \operatorname{cosec}^2 x \cot y + 2\cot x - 2\cot\beta) \tag{8.14}$$

Hence, $W_E = W\sqrt{1+\alpha_h^2}$ (8.15)

From equations (8.9) and (8.12)

$$\cot u = \frac{H}{2d'} y \sec(x+y) \operatorname{cosec} x \operatorname{cosec}^2 y - \tan(x+y) \tag{8.16}$$

From equations (8.9) and (8.12) again

$$\sin(u-v) = \frac{H}{2d'} \sin u \operatorname{cosec} x \operatorname{cosec} y \sin \phi_d \tag{8.17}$$

From equations (8.6), (8.9) and (8.13)

$$\frac{c}{F_c \gamma H} = \frac{\sqrt{1+\alpha_h^2}\sin x\left[\frac{1}{2}\operatorname{cosec}^2 x(y \operatorname{cosec}^2 y - \cot y) + \cot x - \cot\beta\right]}{2\cos(x+\psi)\cot v + 2\sin(x+\psi)} \tag{8.18a}$$

Rewriting the above equation

$$F_c = \frac{c}{\gamma H}\left[\frac{2\cos(x+\psi)\cot y + 2\sin(x+\psi)}{\sqrt{1+\alpha_h^2}\sin x\left[\frac{1}{2}\operatorname{cosec}^2 x(y \operatorname{cosec}^2 y - \cot y) + \cot x - \cot\beta\right]}\right] \tag{8.18b}$$

Computations

The stability charts of Taylor were prepared by finding the maximum value of $c/(F_c\gamma H)$ or minimum F_c by plotting the contours with the values of $c/(F_c\gamma H)$ at the appropriate trial centers. This is quite laborious. With the advent of fast digital computers and with the availability of numerical methods it is quite possible to obtain the minimum value of factor of safety without such tedious procedure. To obtain minimum value of F_c, the partial derivatives of F_c (Eq. 8.18b) with respect to x and y should be zero.

$$\left.\begin{aligned}\frac{\delta F_c}{\delta x} &= 0\\ \frac{\delta F_c}{\delta y} &= 0\end{aligned}\right\} \tag{8.19}$$

Satisfying the above two conditions one obtain the values of x and y for the critical circle. Substitution of these values of x and y in Eq. 8.18b, will yield the maximum $c/(F_c\gamma H)$. The minimization procedure used for the computations was 'steepest descent method'. In this method, the minimum value of a function is determined by starting from an arbitrary point and traveling along steepest descend path. The convergence of this method is inevitable provided the chosen values of x and y are close to their actual values. For $\alpha_h = 0$ case the values of x, y and $c_d/(\gamma H)$ were computed and these values of x and y were used as initial values for further computation of stability numbers for different values of α_h and slope angles.

Figure 8.3 shows the curves between stability number and slope angle β for different values of angle of internal friction φ_d for static case. The curves are similar to those of Taylor (1937).

Similar curves are shown in Figs. 8.4, 8.5 and 8.6 for values of α_h equal to 0.05, 0.10 and 0.15 respectively. It may be seen from the curves that as α_h increases the value of slope stability number $(c_d/\gamma H)$ increases for the same value of slope angle β and angle of internal friction φ_d.

In Figs. 8.7, 8.8, 8.9 and 8.10 are plotted angles x and y respectively, to determine the location of the failure surfaces. Angle of internal friction is plotted as the abscisae while the chord angle and semicentral angle are plotted on ordinate. It is seen from these figures that the chord angle x decreases with slope angle β and increases with φ_d. The semi-central angle y shows the reverse trend. Both the chord angle x and semi-central angle y decrease with α_h.

Figures 8.11, 8.12, 8.13 and 8.14 show the plots for determining the factor of safety with respect to strength. These charts were obtained in a similar manner as produced by Bell (1966) from Taylor's data. The procedure used for reduction of the computed data is briefly described below:

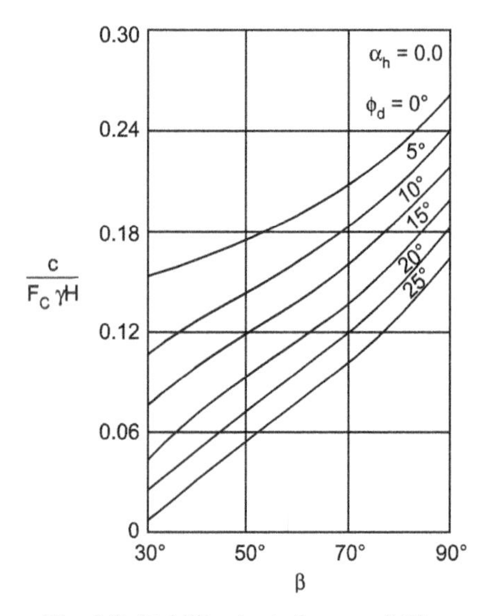

Fig. 8.3. Stability charts for $\alpha_h = 0.00$

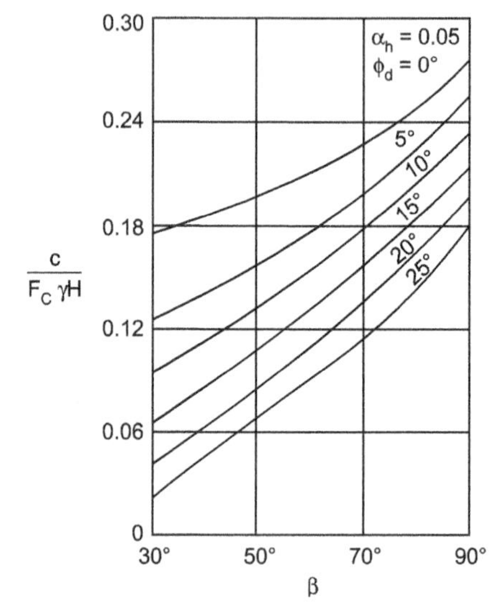

Fig. 8.4. Stability charts for $\alpha_h = 0.05$

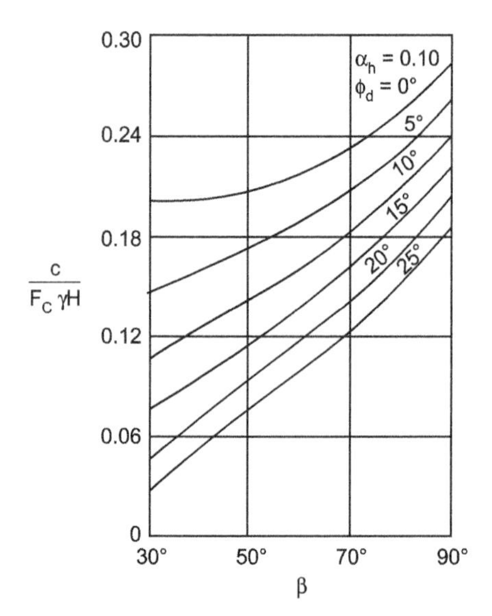

Fig. 8.5. Stability charts for $\alpha_h = 0.10$

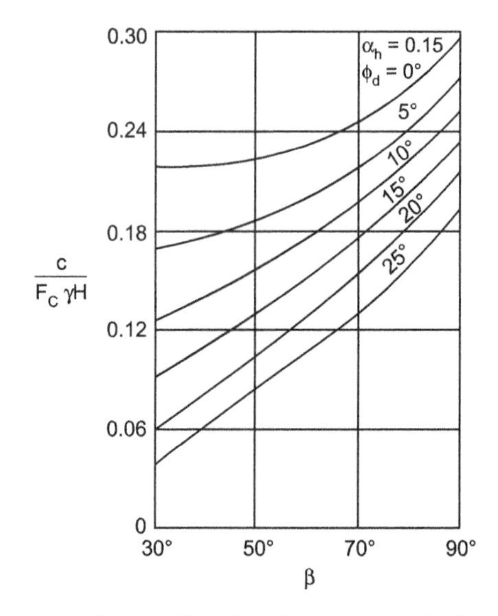

Fig. 8.6. Stability charts for $\alpha_h = 0.15$

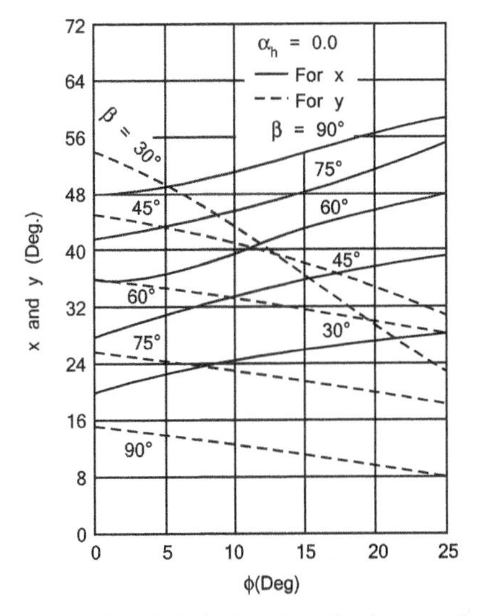

Fig. 8.7. Critical circle location plot for $\alpha_h = 0.00$

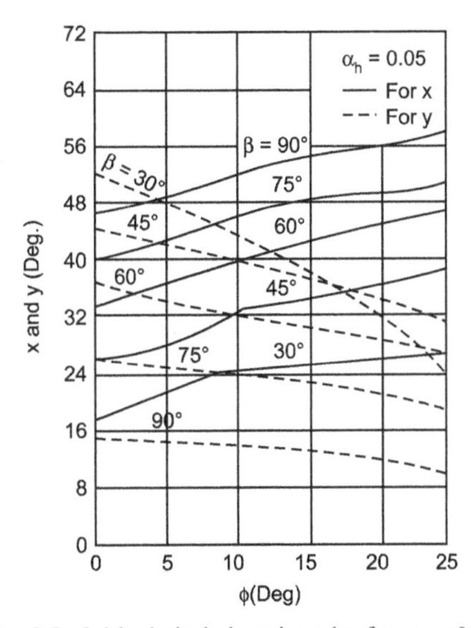

Fig. 8.8. Critical circle location plot for $\alpha_h = 0.05$

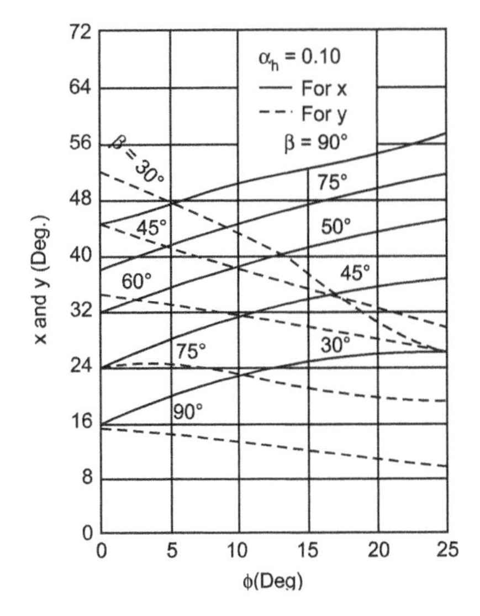

Fig. 8.9. Critical circle location plot for $\alpha_h = 0.10$

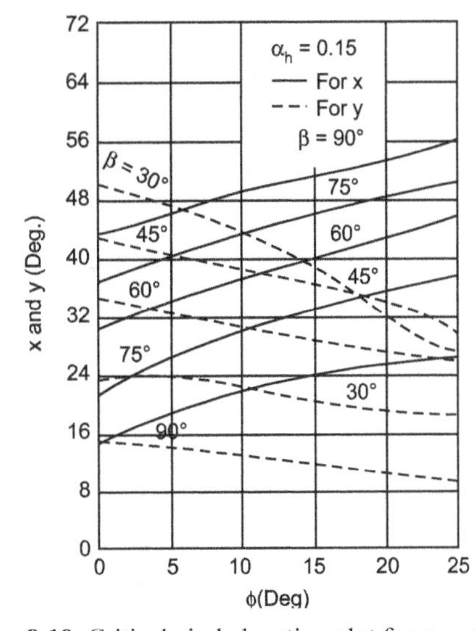

Fig. 8.10. Critical circle location plot for $\alpha_h = 0.15$

The relationship between effective and developed parameters is expressed by $\tan \varphi / F_s = \tan \varphi_d$ and $c/F_s = c_d$. The modified dimensionless factor of safety is expressed by

$$F^* = \frac{F_s}{\tan \varphi} = \frac{1}{\tan \varphi_d} \tag{8.20}$$

and the modified dimensionless stability number is given by

$$N^* = \frac{c}{\gamma H \tan \varphi} = \frac{\dfrac{c_d}{\gamma H}}{\tan \varphi_d} \tag{8.21}$$

Prakash, Saran and Raj (1970) extended the above presented analysis by considering the parabolic variation in the acceleration due to earthquake as shown in Fig. 8.15. The details of

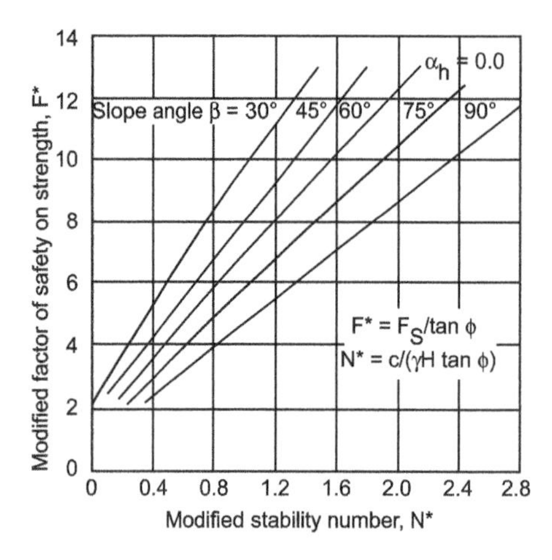

Fig. 8.11. Modified dimensionless parameters on strength for $\alpha_h = 0.00$

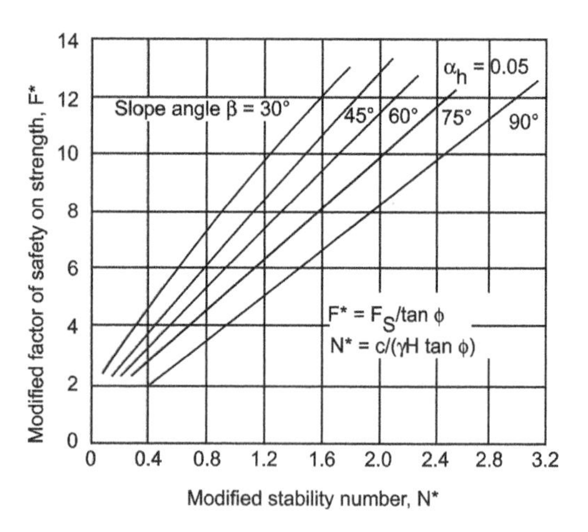

Fig. 8.12. Modified dimensionless parameters on strength for $\alpha_h = 0.05$

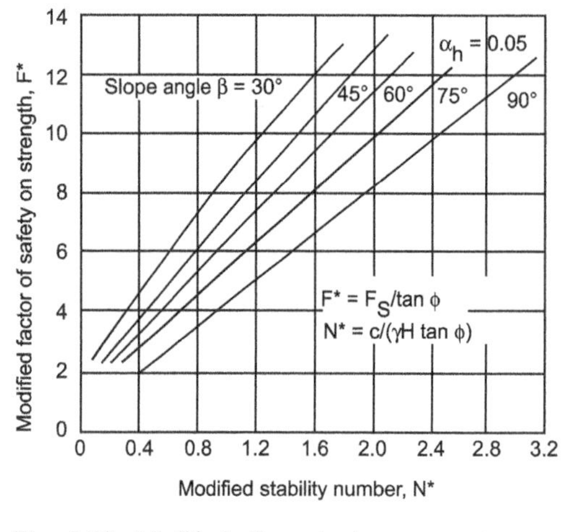

Fig. 8.13. Modified dimensionless parameters on strength for $\alpha_h = 0.10$

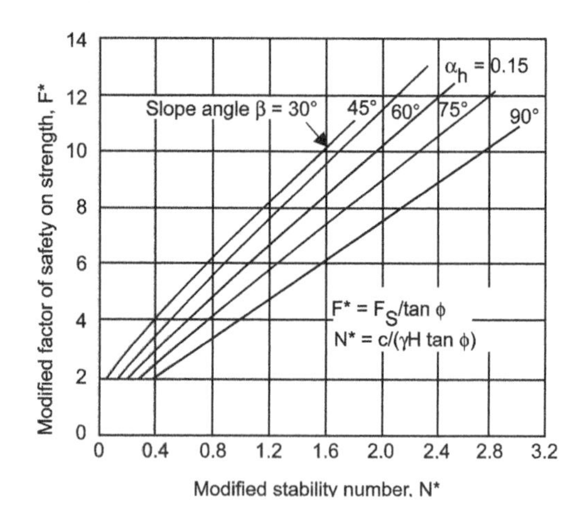

Fig. 8.14. Modified dimensionless parameters on strength for $\alpha_h = 0.15$

analysis and computations are given in above cited paper. Figures 8.16, 8.17 and 8.18 show the typical plots of stability number, chord angles x and y, and modified factor of safety for $\alpha_h = 0.10$.

As discussed later, by taking the parabolic variation of acceleration, the factor of safety with respect to strength is more than the factor of safety with respet to strength for constant acceleration.

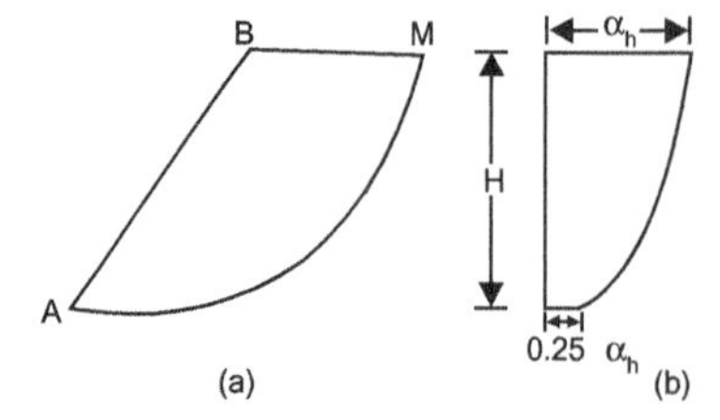

Fig. 8.15. (a) Slope; (b) Variation of seismic coefficient

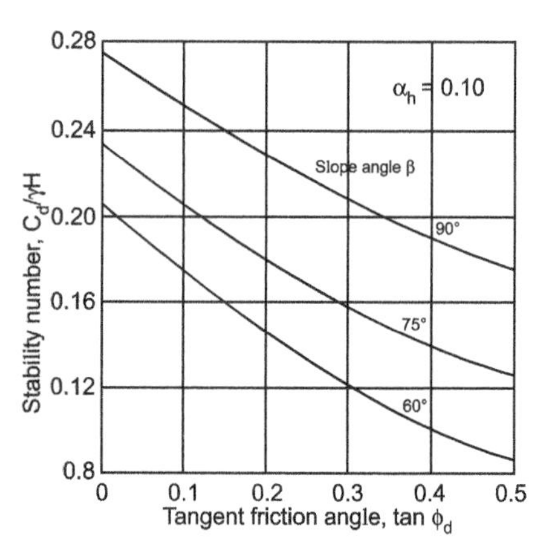

Fig. 8.16. Stability chart

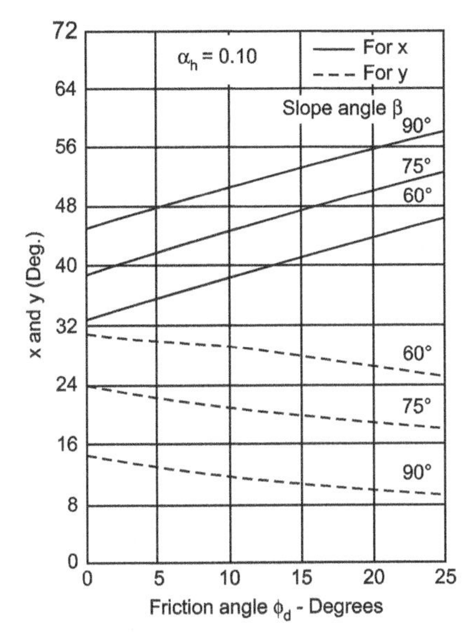

Fig. 8.17. Critical circle location plot

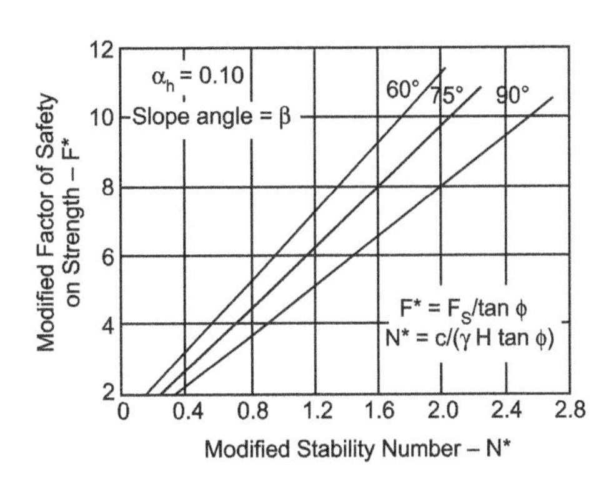

Fig. 8.18. Modified dimensionless parameters on strength

8.3 Displacement Analysis

The adverse earthquake forces usually consists of a few cycles lasting for a very short duration less than half a minute in most cases. If the average shear resistance falls below the shear stress on a surface through the slope, then only small displacement can take place in such small duration force. Further the reversal of the acceleration will arrest the displacement till the next adverse cycle. As discussed in chapter 4, the strength of soil under short duration dynamic load is not effected significantly. Keeping this in view, according to the new concept the criterion for design should be based on the allowable displacement rather than the limiting equilibrium.

8.3.1 Concept of Yield Acceleration

Newmark (1965), defined the yield acceleration as the acceleration which drops the dynamic factor of safety of the slope to unity. Figure 8.19 shows a slope of height H and inclined to horizontal by an angle β. ABM represents a circular sliding wedge with its center at O. In static condition (Fig. 8.19a), this wedge is in equilibrium under three forces, namely

(i) Weight W of the wedge acting at its c.g. in vertical downward direction

(ii) $\Sigma\tau_s \cdot ds$ along the arc $AB \cdot \tau_s$ is the shear strength of the soil given by:

$$\tau_s = c + \sigma_n \tan\phi \tag{8.22}$$

where c = Unit cohesion of slope material

ϕ = Angle of internal friction of slope material

σ_n = Normal stress on the portion ds of the arc

(iii) $\Sigma\sigma_n \cdot ds$ may be noted that $\sigma_n \cdot ds$ will act in the direction normal to arc ds. Therefore $\sigma_n \cdot ds$ and hence $\Sigma\sigma_n \cdot ds$ will pass through the center of circle.

If F_s represents the static factor of safety, then

$$F_s = \frac{\text{Stabilising moment}}{\text{Overturning moment}} \tag{8.23a}$$

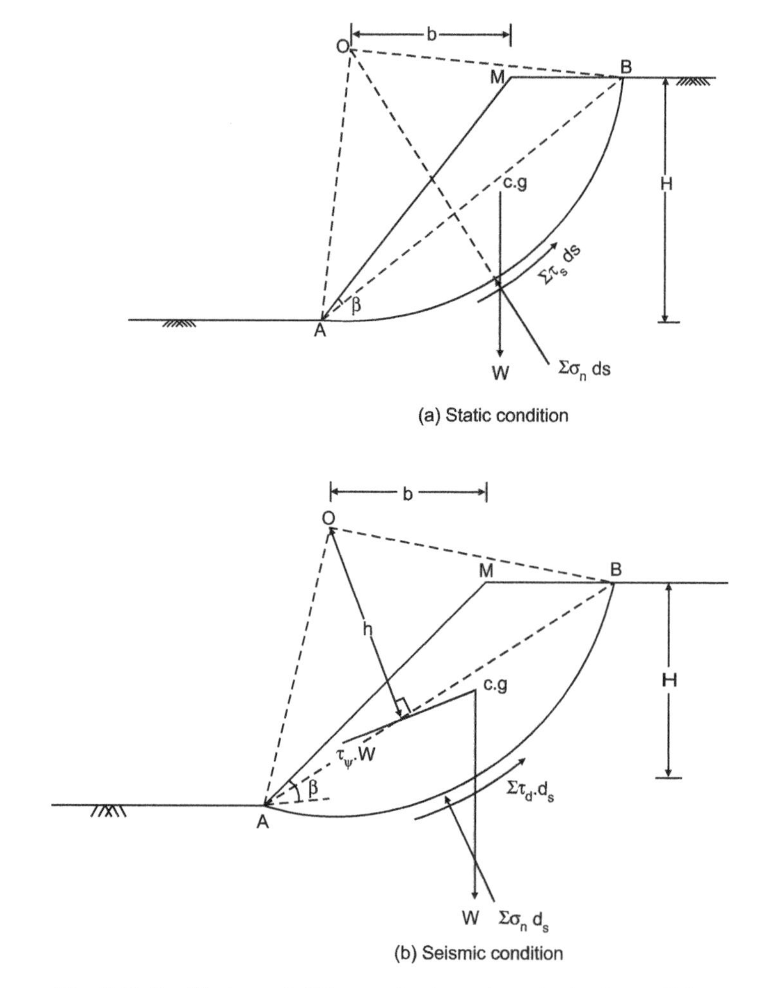

Fig. 8.19. Equilibrium of sliding wedge in static and seismic conditions

or
$$F_s = \frac{R\sum\tau_s\,ds}{W\cdot b} \tag{8.23b}$$

Let during earthquake, acceleration of magnitude $\alpha_\psi.g$ is generated in the direction making an angle ψ with vertical. α_ψ and g are respectively the seismic coefficient and acceleration due to gravity. As shown in Fig. 8.19b, the stability of the slope, in this case can be analysed by taking an additional force $\alpha_\psi.W$ (Fig. 8.19b). If F_d is the dynamic factor of safety, then,

$$F_d = \frac{R\sum\tau_d\,ds}{W.b+\alpha_\psi.W.h} \tag{8.24}$$

τ_d is dynamic strength of soil. Assuming that $\tau_d \approx \tau_s$.

$$\frac{F_s}{F_d} = \frac{W.b+\alpha_\psi.W.h}{W.b} = \left(1+\frac{h}{b}\alpha_\psi\right) \tag{8.25}$$

or
$$\alpha_\psi.g = \left(\frac{F_s}{F_d}-1\right)\frac{b}{h}\cdot g \tag{8.26a}$$

According to the definition of yield acceleration, $F_d = 1$, i.e.

$$\alpha_{\psi y} \cdot g = (F_s - 1)\frac{b}{h} \cdot g \tag{8.26b}$$

where $\alpha_{\psi y} \cdot g$ is yield acceleration.

8.3.2 Rigid Plastic Approach for Determining Displacement

Let a rigid body placed on the horizontal surface (Fig. 8.20a) has horizontal yield acceleration of magnitude $\alpha_{hy} \cdot g$. If this body is subjected to horizontal acceleration $\alpha_{ha} \cdot g$ for time t_o, then displacement of this body will occur due to the acceleration equal to $(\alpha_{ha} - \alpha_{hy})\, g$. The velocity V_m of the body after time t_o with acceleration equal to $\alpha_{ha} \cdot g$ will be $\alpha_{ha} \cdot g \cdot t_o$. The same velocity will be achieved in time t_m for the acceleration $\alpha_{hy} \cdot g$ (Fig. 8.20b) Therefore,

$$V_m = \alpha_{ha} \cdot g \cdot t_o = \alpha_{hy} \cdot g \cdot t_m \tag{8.27}$$

or

$$t_o = \frac{V_m}{\alpha_{ha} \cdot g} \tag{8.28a}$$

$$t_m = \frac{V_m}{\alpha_{hy} \cdot g} \tag{8.28b}$$

Displacement of the body U for acceleration $(\alpha_{ha} - \alpha_{hy})\, g$ will be given by (Fig. 8.20b)

$$U = \text{Area of triangle ADE} - \text{Area of triangle ABC} \tag{8.29}$$

$$= (1/2)\, V_m \cdot t_m - (1/2)\, V_m \cdot t_o$$

Substituting the values of t_o and t_m from Eqs. (8.28a) and (8.28b)

$$U = \frac{V_m^2}{2\alpha_{hy} \cdot g}\left[1 - \frac{\alpha_{hy}}{\alpha_{ha}}\right] \tag{8.30}$$

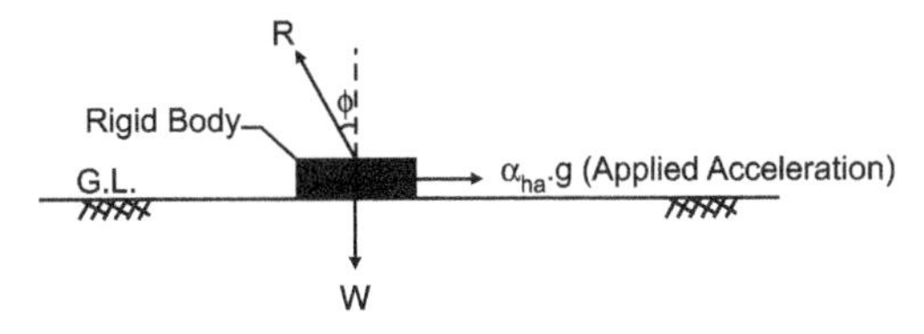

(a) Movement of rigid body

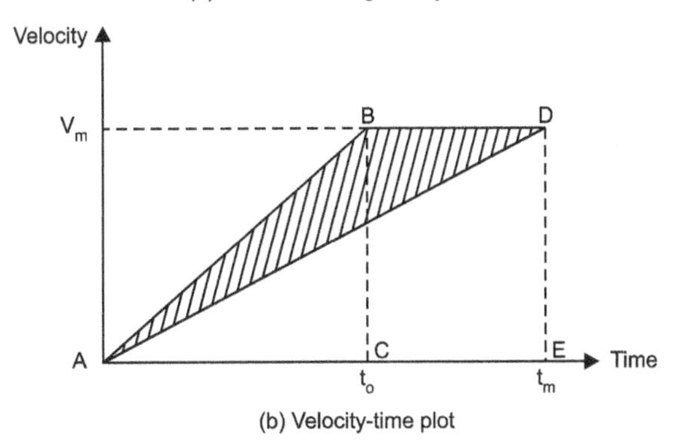

(b) Velocity-time plot

Fig. 8.20. Rigid plastic approach for determining displacement

The displacement given by Eq. (8.30) will be due to the single pulse, and assuming that the displacement cannot be reversed by the opposite pulses, each adverse pulse will increase the displacement. Further, the displacement varies as the square of ground velocity for the same value of the resistance coefficient α_{hy}/α_{ha}.

8.3.3 Goodman and Seed's Approach

The approach developed by Goodman and Seed (1996) is applicable to slopes consisted of dry cohesionless materials or materials having high permeability so that no change of pore pressures takes place during earthquake.

Yield Acceleration

The situation is considered analogus to sliding of a block on frictional surface. The block is in equilibrium under three forces, namely (i) Weight of the block, W; (ii) Inertial force $\alpha_{\psi y}$. W at an angle ψ to the horizontal; and (iii) Resultant friction force, R at an angle ϕ with the normal to the slope surface (Fig. 8.21). Applying sine rule,

$$\frac{\alpha_{\psi y}.W}{\sin(\phi-\beta)} = \frac{W}{\sin[90-(\psi+\beta-\phi)} \tag{8.31}$$

or

$$\alpha_{\psi y}\cdot g = \frac{\sin(\phi-\beta)}{\cos(\phi-\psi-\beta)}\cdot g \tag{8.32}$$

The yield acceleration $\alpha_{\psi y}\cdot g$ will be maximum when denominator becomes unity i.e.

$$\phi-\psi-\beta = 0 \tag{8.33a}$$

or

$$\psi = \phi-\beta \tag{8.33b}$$

Therefore maximum yield acceleration will occur when it is applied at an angle of $(\phi - \beta)$ with horizontal. Its magnitude will be $\sin(\phi - \beta) \cdot g$.

If the acceleration is acting in the horizontal direction, i.e. $\psi = 0$, the yield acceleration will be

$$\alpha_{hy}\cdot g = \tan(\phi-\beta).\, g \tag{8.34}$$

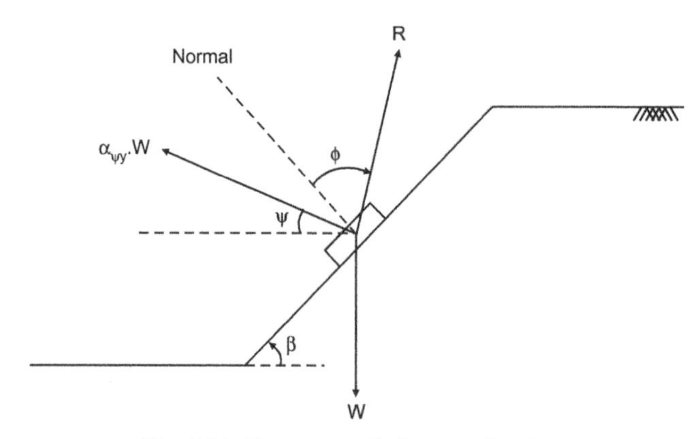

Fig. 8.21. Forces on soil element of a slope

Displacement

The mechanism of obtaining the displacement is illustrated in Fig. 8.22. $\alpha_h(t)\cdot g$ is the applied acceleration. No displacement would occur till the induced acceleration exceeds the yield acceleration for the first cycle. Assuming the yield acceleration to remain constant, it is marked off by a horizontal line. The variation in velocity of the sliding mass is obtaining by integrating

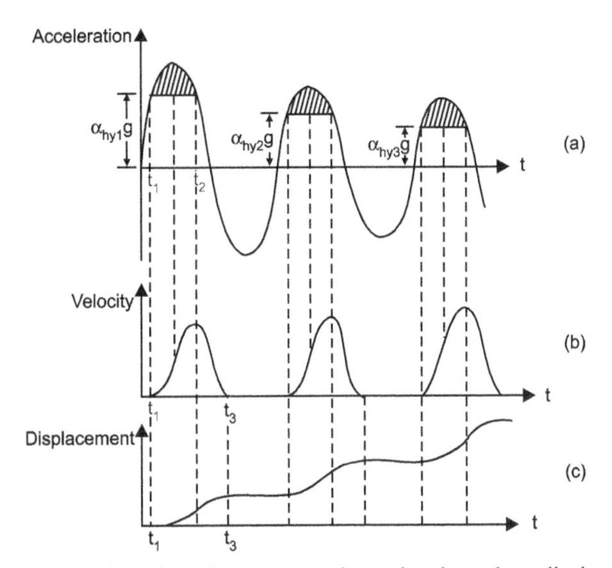

Fig. 8.22. Integration of accelerograms to determine downslope displacements

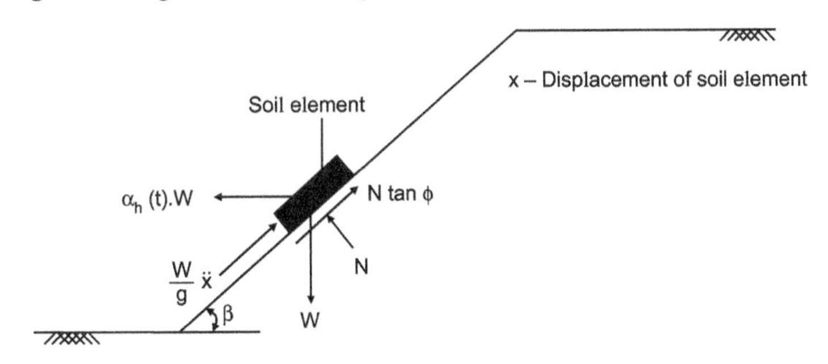

8.23. Forces on the soil element for determining displacement

the shaded area. The velocity will continue to increase until time t_2 when the acceleration again drops below the yield value. The velocity is finally brought to zero at time t_3 as the direction of acceleration is reversed (8.22b). The integration of velocity-time relationship gives displacement as shown in Fig. 8.22c.

It is not always convenient to use accelerograms and if the ground motion is assumed harmonic, a somewhat less accurate but more convenient method of computation may be used. Consider a block element of soil on the slope subjected to seismic force $\alpha_h(t)$. W, where $\alpha_h(t)$ is time dependent seismic coefficient. N represents the normal component of resultant frictional force (Fig. 8.23).

Resolving the forces along the slope and normal to the slope.

$$\frac{W}{g}\ddot{x} = W \sin\beta + \alpha_h(t) W \cos\beta - N \tan\phi \tag{8.35}$$

$$N = W \cos\beta - \alpha_h(t) W \sin\beta \tag{8.36}$$

Solving Eqs. (8.35) and (8.36), the following expression works out:

$$X = B[\alpha_h(t) - \alpha_{hy}] \tag{8.37}$$

where

$$B = g(\sin\beta \tan\phi + \cos\beta) \tag{8.38}$$

$$\alpha_{hy} = \tan(\phi - \beta) \tag{8.39a}$$

If,

$$\alpha_h(t) = P \sin(\omega t + \theta) \tag{8.39b}$$

where P = Maximum value of acceleration coefficient

ω = Frequency

θ = Phase angle = $\sin^{-1}(\alpha_{hy}/P)$

Equation (8.36) can be written as

$$x = B[P \sin(\omega t + \theta) - \alpha_{hy}] \tag{8.40}$$

Using two boundary conditions: (i) At $t = 0$, $x = 0$; (ii) At $t = 0$, $\dot{x} = 0$, the solution of Eq. (8.40) comes:

$$x = -\frac{1}{2}B\alpha_{hy}t^2 - \frac{BP}{w^2}\sin(\omega t + \theta) + \frac{BP}{\omega}\left[\frac{\sin\theta}{\omega} + t\cos\theta\right] \tag{8.41}$$

The Eq. (8.41) gives the downward displacement of the slope in first cycle of harmonic motion. By adjusting the values of B and α_{hy} for subsequent cycles, displacement after n cycles can be obtained. However conservatively the displacement after n cycles may be taken as n times the displacement obtained in the first cycle.

Using Eq. (8.41), Goodman and Seed (1966) developed charts (Fig. 8.24) for determining the displacement of a surface layer of cohesionless slope for different values of slope angles, angle of internal friction (ϕ_{eq}) and maximum acceleration coefficient P. These curves are for a frequency of 1 cps. However these can be used for determining the displacement for any other desired frequency using the following expression:

$$x_f = x_1/f^2 \tag{8.42}$$

where x_f = Displacement for the frequency of f *c/s*

x_1 = Displacement for the frequency of 1 *c/s*

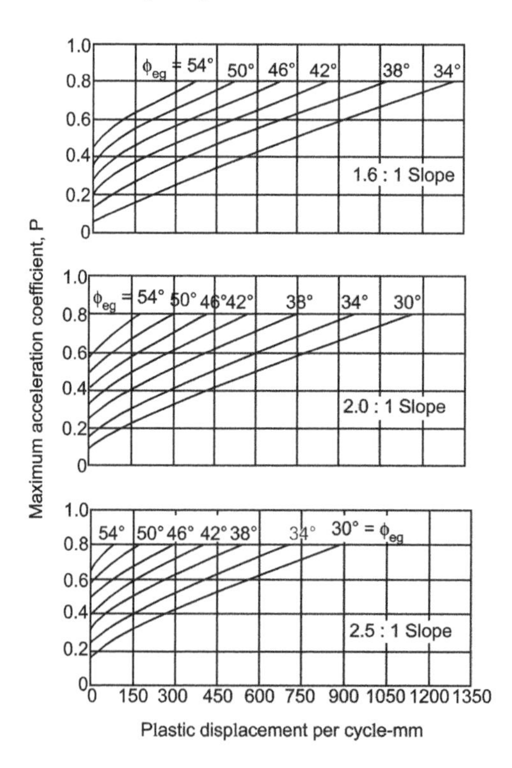

Fig. 8.24. Downslope displacement of a block slide during horizontal simple harmonic shaking: Frequency = 1 cycle per second

The charts shown in Fig. 8.24 are prepared for particular slope angles. However, these charts can be used for different slope angles by adjusting ϕ_{eq} value in the opposite direction i.e. of the actual slope is flatter by 2° than that in the chart, the ϕ_{eq} angle should be increased by 2°, and vice-versa.

Illustrative Examples

Example 8.1

A slope of height 7.0 m is inclined to horizontal by an angle 50°. The slope material has unit cohesion of 25 kN/m^2 and angle of internal friction 25°. The unit weight of soil is 17.0 kN/m^2. Determine the factor of safety of the slope by method of slices for the rupture surface shown in Fig. 8.25 for following cases:

(i) Static case
(ii) Seismic case, $\alpha_h = 0.1$
(iii) Seismic case, $\alpha_h = 0.1$, $\alpha_v = +0.05$
(iv) Seismic case, $\alpha_h = 0.1$, $\alpha_v = -0.05$

Solution

(a) Let the sliding wedge *ABM* is divided in five slices (Fig. 8.25)

$$\text{Length of arc } AB = \frac{\pi}{180} \times 100 \times 8.0 = 13.95\,\text{m}$$

$$c = 25\ \text{kN/m}^2,\ \phi = 25°,\ \gamma = 17\ \text{kN/m}^3$$

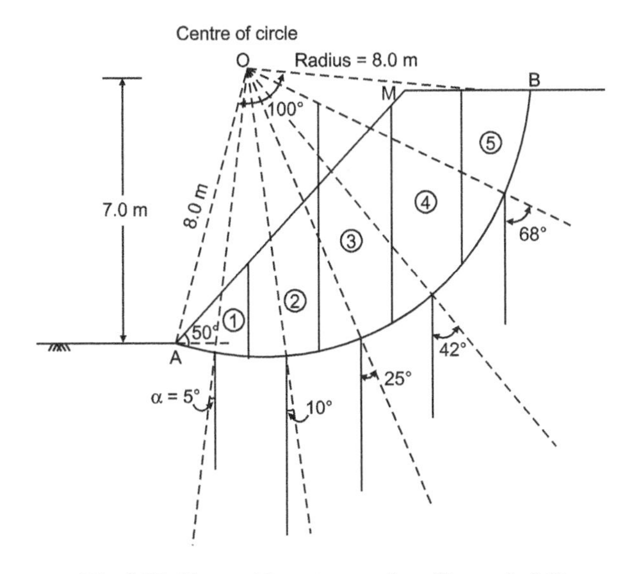

Fig. 8.25. Slope with rupture surface (Example 8.1)

(b) Computation of weight of each slice, W_i, N and T are given below:

Slice No.	α *(Deg.)*	W_i *(kN)*	*N (kN)*	*T (kN)*
1	– 5	43.2	43.0	– 3.8
2	10	119.0	117.2	20.7
3	25	176.0	159.5	74.4
4	42	179.5	133.4	120.1
5	68	119.8	44.9	111.2
			498.0	**322.6**

Factor of safety under static condition

$$F_s = \frac{c \cdot S + \tan\phi \Sigma N}{\Sigma T} = \frac{25 \times 13.95 + \tan 25^\circ \times 498.0}{322.56} = 1.80$$

(c) Seismic condition, $\alpha_h = 0.1$, $\alpha_v = 0.0$

Slice No.	$W_i \alpha_h$ *(kN)*	*N'(kN)*	*T'(kN)*
1	4.32	– 0.38	4.30
2	11.90	2.06	11.72
3	17.60	7.44	15.95
4	17.90	11.97	13.30
5	12.00	11.12	4.49
		32.21	**49.76**

$$F_s = \frac{c.S.+\tan\phi(N-N')}{\Sigma(T+T')}$$

$$= \frac{25 \times 13.95 + \tan 25^\circ \times (498-32.21)}{(322.56 + 49.76)}$$

$$= 1.52$$

(d) Seismic condition, $\alpha_h = 0.1$, $\alpha_v = +0.5$

Slice No.	$W_i(1+\alpha_v)$ *(kN)*	*N (kN)*	*T (kN)*
1	45.34	45.2	– 3.95
2	124.90	123.0	21.68
3	184.70	167.90	78.05
4	188.50	140.10	126.13
5	125.80	40.90	116.39
		521.10	**338.55**

$$F_s = \frac{25 \times 13.95 + \tan 25^\circ \times (521.10-32.21)}{(338.55 + 49.76)}$$

$$= 1.48$$

(e) Seismic condition $\alpha_h = 0.1$, $\alpha_v = -0.05$

Slice No.	$W_i(1-\alpha_v)$ *(kN)*	*N (kN)*	*T (kN)*
1	41.02	40.86	– 3.57
2	113.05	111.33	19.63
3	167.15	151.49	70.64
4	170.52	126.72	114.10
5	113.86	42.65	105.57
		473.05	**306.37**

$$F_s = \frac{25 \times 13.95 + \tan 25^\circ \times (473.05-32.21)}{(306.37+49.76)}$$

$$= 1.55$$

Summary:

Seismic coefficients		*F.O.S.*
α_h	α_v	
0.0	0.0	1.80
0.1	0.0	1.52
0.1	+0.05	1.48
0.1	– 0.05	1.55

Example 8.2

Using modified friction circle method determine the factor of safety for the slope given in Example 8.1.

Solution

$$c = 25 \text{ kN/m}^2, \phi = 25°, \beta = 50°, H = 7.0 \text{ m}, \gamma = 17.0 \text{ kN/m}^3$$

(a) Using Fig. 8.3, for $\phi = 25°$, $\beta = 50°$

$$c_d/\gamma H = 0.06, c_d = 0.06 \times 17 \times 7.0 = 7.14 \text{ kN/m}^2$$

$$F_c = c/c_d = 25/7.14 = 3.50$$

$$N^* = \frac{c}{\gamma H \tan\phi} = \frac{25}{17.0 \times 7.0 \times \tan 25°} = 0.4505$$

From Fig. 8.11, for $N^* = 0.4505$, $\alpha_h = 0.0$ and $\beta = 50°$, $F^* = 4.15$, $F_s = F^* \tan\phi = 4.15 \tan 25° = 1.88$

(b) Seismic case, for $\phi = 25°$, $\beta = 50°$ and $\alpha_h = 0.10$

$$\frac{c_d}{\gamma H} = 0.0767, c_d = 0.0767 \times 17 \times 7 = 9.12 \text{ kN/m}^2$$

$$F_c = 25/9.12 = 2.73$$

From Fig. 8.13, for $N^* = 0.4505$, $\alpha_h = 0.0$ and $\beta = 50°$

$$F^* = 3.8$$

$$F_s = 3.8 \tan 25° = 1.77$$

(c) Seismic case ($\alpha_h = 0.10$, variable acceleration)

From Fig. 8.16, for $\phi = 25°$, $\beta = 50°$ and $\alpha_h = 0.10$

$$\frac{c_d}{\gamma H} = 0.092, c_d = 0.092 \times 17 \times 7 = 10.94 \text{ kN/m}^2$$

$$F_c = c/c_d = 25/10.94 = 2.28$$

From Fig. 8.18, for $N^* = 0.4505$, $\alpha_h = 0.10$ and $\beta = 50°$

$$F^* = 3.5$$

$$F_s = 3.5 \tan 25° = 1.63$$

Example 8.3

Determine the yield acceleration and value of horizontal displacement of a cohesionless slope (2 H : IV) subjected to a harmonic dynamic acceleration of $0.5\, g \text{ Sin } (4\pi\, t + \theta)$ magnitude. The slope material has angle of internal friction of 40°.

Solution

Yield acceleration coefficient:

$$\alpha_{hy} = \tan(\phi - \beta);\ \beta = \tan^{-1} ½ = 26.5°$$

$$= \tan(40° - 26.5°) = 0.24$$

$$\theta = \sin^{-1}\frac{0.24}{0.5} = 28.7°$$

$$B = g(\sin\beta\tan\phi + \cos\beta)$$

$$= 9.81\ (\sin 26.5 \tan 40 + \cos 26.5)$$

$$= 12.45\ \text{m/s}^2$$

$$x = -\frac{1}{2}B\alpha_{hy}t^2 - \frac{BP}{\omega^2}\sin(\omega t + \theta) + \frac{BP}{\omega}\left(\frac{\sin\theta}{\omega} + t\cos\theta\right)$$

Time for one cycle = 1/2 s; $\omega = 2\pi f = 4\pi$

$$f = 2\text{c/s}$$

Time for half cycle = 1/4 s

In the expression of x, t will be put as 1/4 s, because the acceleration in one cycle is higher than yield acceleration only in the half cycle.

Therefore,

$$x_t = -\frac{1}{2}\times 12.45\times 0.24\times\left(\frac{1}{4}\right)^2 - \frac{12.45\times 0.5}{(4\pi)^2}\sin\left(45\times\frac{1}{4} + \frac{28.7\times\pi}{180}\right)$$

$$+\frac{12.45\times 0.5}{4\times\pi}\left(\frac{\sin 28.7}{4\pi} + \frac{1}{4}\cos 28.7\right) = 0.05320\ \text{m}$$

or $$x = 53.2\ \text{mm}$$

From Fig. 8.24, for slope 2 : 1, $\phi = 40°$ and $P = 0.5$.

Plastic displacement = 219 mm

$$\therefore \qquad x = \frac{x_1}{f^2} = \frac{219}{2^2} = \frac{219}{4} = 54.7\ \text{mm}$$

Therefore, yield acceleration = 0.24 × 9.81 = 2.35 m/s^2

Horizontal displacement per cycle = 54 mm.

References

Bell, J.M. (1966), "Dimensionless Parameters for Homogeneous Earth Slopes", Proc. ASCE, Vol. 92, SM-5, pp. 51-65.

Bishop, A.W. (1955), "The Use of the Slip Circle in the Stability Analysis of Slopes", Geotechnique, Vol. 5. No. 1, pp. 7-17.

Bishop, A.W. and Morgenstern, N. (1960), "Stability Coefficient for Earth Slopes", Geotechnique, Vol. 10, No. 4, pp. 129-150.

Coulomb, C.A. (1776), "Essai Sur une Application des Rigles des Maximis et Minimis a eueiques Problems Sta-ique Relatifs a 1, 9 Architecture, Mem. Acad Roy. Pres. Divers savants, Vol. 7, Paris.

Culmann, K. (1885), Die graphische Statik, Zurich.
Fellenius, W. (1927), "Erdtische Berechung Unit Reibung and Kohaesion (Adhaesion) und unter Annahme Kreiszyhindeisrher Gleitflaechen, Ernst, Berlin.
Fellenius, W. (1936), "Calculation of Stability of Earth Dams", Transactions, 2nd Congress Large Dams,Vol. 4, p. 445.
Frohlich, O.K. (1953), "The factor of Safety with Respect to Sliding of a Mass of Soil along the Arc of a Logarithmic Spiral", Proceedings 3rd Int. Conf., SMFE, Vol. 2, pp. 230-238.
Goodman, R.E. and Seed, H.B. (1966), "Earthquake Induced Displacements in Sand Embankments", Journal of Soil Mech. and Founds. Division, ASCE, Vol. 92, No. SM2, pp. 125-141.
IS : 1893-2016 : Criteria for Earthquake Resistant Design of Structures, ISI, New Delhi.
Isao, Minami (1963), "The Stability Table of Slope of Earth Dam", Proc. 2nd Asian Regional Conference on SMFE, Vol. 1, pp. 286, 293.
Janbu, N. (1954a), "Stability Analysis of Slopes with Dimensionless Parameters", Harvard Soil Mechanics Series No. 46.
Janbu, N. (1954b), "Applications for Composite Slip Surfaces for Stability Analysis", European Conf. of Stability of Earth Slopes, Discussion, Vol. 3, Stockholm, pp. 43-49.
Meyer, O.H. (1958), "Computation of the Stability of Slopes", Proc. ASCE, Vol. 84, SM-4, pp. 1-12.
Morgenstern, N. (1963), "Stability Charts for Earth Slopes During Rapid Drawdown", Geotechnique, Vol. 13. No. 2, pp. 121-131.
Morgenstern, N. and Price, V.E. (1965), "The Analysis of the Stability of General Slip Surfaces", Geotechnique, Vol. 15, No. 1, pp. 79-93.
Newmark, N.M. (1965), "Effects of Earthquakes on Dams and Dam Embankments", Geotechnique, Vol. XV, No. 2, pp. 129-160.
Nonveiller, E. (1965), "The Stability Analysis of Slopes with a Slip Surface of General Shape", Proc. 6th Int. Conf. SMFE, Vol. 2, pp. 522-535.
Petterson, K.E. (1916), "Meddelande angaende den av Goteborgs Hamnstyrelse tillsatta kajkammissioneus arbeteu (Information about the work of the quoy commission of the Harbour Board). Tekniska Samfundets Hand-Linear, Vol. 6, No. 1.
Prakash, S., Saran, S. and P. Raj (1968), "Seismic Stability Analysis of Homogeneous Slopes", Symposium on Earth and Rockfill Dams", November, Beas Dam Site, pp. 10-20.
Prakash, S., Saran, S. and P. Raj (1970), "Seismic Analysis of Stability of Slopes", VII International Conference on Soil Mechanics and Foundation Engg., Vol. 1, pp. 653-657.
Rendulic, L. (1935), "Ein Beitrag Zur Bestimmung der Gleitsicherheit", Der Banin Genieur, No. 19/20.
Spencer, E. (1967), "Analysis of Embankments Assuming Parallel Inter-Slice Forces", Geotechnique, Vol. 17, No. 1, pp. 11-26.
Taylor, D.W. (1937), "Stability of Earth Slopes", Contributions to Soil Mechanics (1925-1940): Boston Society of Civil Engineers, pp. 337-386.

Practice Problems

8.1 Assuming a suitable trial rupture surface, determine the factor of safety of a slope ($\beta = 60°$; $H = 10.0$ m; $c = 30$ kN/m^2; $\phi = 28°$; $\gamma = 16.5$ kN/m^3) using both method of slices and modified friction circle method in the following cases:
(a) Static case
(b) Seismic case
(i) $\alpha_h = 0.05$
(ii) $\alpha_h = 0.15$

8.2 Determine the yield acceleration and horizontal displacement of a cohesionless slope (1.6 H : 1V, $\phi = 38°$, $\gamma = 17.0$ kN/m^3) which is subjected to a harmonic acceleration 0.6 g($3\pi\, t + \theta$).

CHAPTER

9

Liquefaction of Soils

9.1 General

Many failures of earth structures, slopes and foundations on saturated sands have been attributed in the literature to liquefaction of the sands. The best known cases of foundation failures due to liquefaction are those that occurred during the 1964 earthquake in Nigata, Japan (Kishida, 1965). Classical examples of liquefaction are the flow slides that have occurred in the province of Zealand in Holland (Geuze, 1948; Koppejan et al., 1948) and in the point bar deposits along the Mississippi river (Waterways experiment station, 1967). The failures of Fort Peck Dam in Montana in 1938 (Casagrande, 1965; Corps of Engineers, 1939; Middlebrooks, 1942), the Calaveras Dam in California in 1920 (Hazen, 1920) and the Lower Lan Norman Dam during the 1971 San Fernando Earthquake (Seed et al., 1975) in California provide typical examples of liquefaction failures of hydraulic-fill dams.

Liquefaction often appears in the form of sand fountains, and a large number of such fountains have been observed during Dhubri Earthquake in Assam in 1930 and Bihar Earthquake in 1934 (Housner, 1958; Dunn et al., 1939). When soil fails in this manner, a structure resting on it simply sinks into it. The most recent Koyna earthquake of 1995 is an illustration of liquefaction phenomenon causing catastrophic damages to structures and resulting in loss of life and property.

9.2 Definitions

9.2.1 Liquefaction

It denotes a condition where a soil will undergo continued deformation at a constant low residual stress or with no residual resistance, due to the build-up and maintenance of high pore water pressure which reduces the effective confining pressure to a very low value; pore pressure build-up leading to true liquefaction of this type may be due either to static or cyclic stress applications.

9.2.2 Initial Liquefaction

It denotes a condition where, during the course of cyclic stress applications, the residual pore water pressure on completion of any full stress cycle becomes equal to the applied confining pressure, the development of initial liquefaction has no implications concerning the magnitude of the deformations which the soil might subsequently undergo; however, it defines a condition which is a useful basis for assessing various possible forms of subsequent soil behaviour.

9.2.3 Initial Liquefaction with Limited Strain Potential, Cyclic Mobility or Cyclic Liquefaction

It denotes a condition in which cyclic stress applications develop a condition of initial liquefaction and subsequent cyclic stress applications cause limited strains to develop either because of the remaining resistance of the soil to deformation or because the soil dilates, the pore pressure drops, and the soil stabilizes under the applied loads.

In laboratory undrained cyclic tests (triaxial, direct simple shear and gyratory shear) on saturated sands, cyclic mobility has been observed to develop and to result in large strains (Lee and Seed, 1967; Seed and Lee, 1966). It is controversial whether cyclic mobility occurs in dilative sands in situ during earthquakes to the same extreme degree as has been observed in the laboratory. A simple means for understanding the difference between liquefaction and cyclic mobility as observed in the laboratory is through the use of the state diagram, which is shown in Fig. 9.1 (Castro and Poulos, 1976). The axes are void ratio and effective minor principal stress. The steady state line shown represents the locus of states in which a soil can flow at constant void ratio, constant effective minor principal stress $\bar{\sigma}_3$ and constant shear stress. The void ratio at the steady state is the same as the critical void ratio.

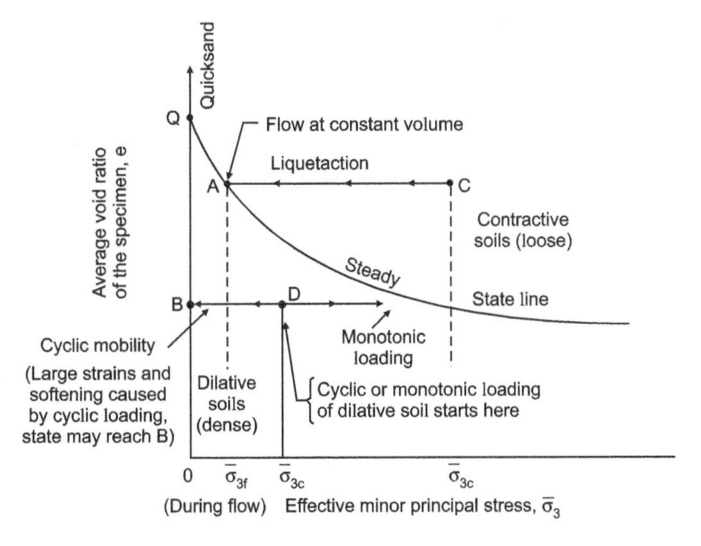

Fig. 9.1. Undrained tests on fully saturated sands depicted on state diagram (Castro and Poulos, 1976)

Liquefaction is the result of undrained failure of a fully saturated, highly contractive (loose) sand, for example starting at point *C* and ending with steady state flow at constant volume and constant $\bar{\sigma}_3$ at point *A*. During undrained flow, the soil remains at point *A* in the state diagram.

The quicksand condition that is so familiar through the use of quicksand devices for instruction in soil mechanics is depicted by points on the zero effective stress axis at void ratios above *Q*. In this state, sand has zero strength and is also neither dilative nor contractive. At void ratios above *Q* the sand grains are not in close contact at all times.

The mechanics of cyclic mobility may also be illustrated with the aid of Fig. 9.1. Consider first the behaviour in Fig. 9.1 when a fully saturated dilative sand starting, for example, at point *D* is loaded monotonically (statically) in the undrained condition. In that case the point on the state diagram may move slightly to the left of point *D* but then it will move horizontally towards the steady state line as load is applied. If one now starts a new test at point *D*, but this time applies cyclic loading, one can follow the behaviour by plotting the average void ratio and the effective stress each time the applied cyclic load passes through zero. In this case the state point moves horizontally to the left, because the average void ratio is held constant and the pore pressure rises due to cyclic loading.

The magnitude of pore pressure build up in the cyclic test will depend on the magnitude of the cyclic load, the number of cycles, the type of test, and the soil type, to name a few variables. In particular, it has been observed in the laboratory that in triaxial tests for which the hydrostatic stress condition is passed during cycling, and if a large enough number of cycles of sufficient size are applied, the state point for the average conditions in the specimen eventually reaches zero effective stress at point *B* each time the hydrostatic stress state is reached. Subsequent application of undrained monotonic loading moves the state point to the right toward the steady state line, and the resistance of the specimen increases.

During cycling in the test described above, strains develop and the specimen becomes softer. If these strains are large enough, one can say that the specimen has developed cyclic mobility. Adequate evidence has been presented to show that most of the strains measured in cyclic load tests in the laboratory are due to internal redistribution of void ratio in the laboratory specimens. For example, at the completion of such tests the void ratio at the top of the specimen is much higher than at the bottom (Castro, 1969). Thus the horizontal line *D-B* in Fig. 9.1 is fictitious in the sense that it represents average conditions. Near the top of the specimen, the void ratio increases, and near the bottom the void ratio decreases. The pore pressures that build up and the strains measured in the laboratory are due to the formation of such loose zones (Castro and Poulos, 1976).

In summary then, specimens that lie above the steady state line on Fig. 9.1 can liquefy if the load applied is large enough. Such liquefaction can be triggered by monotonic or cyclic undrained loading. The further to the right of the steady state line that the starting point is, the greater will be the deformation associated with the liquefaction. If the initial point is above *Q*, the strength after liquefaction will be zero. If the starting point is below *Q* the strength after liquefaction will be small but finite. Saturated sands starting at points on or below the steady state line, will be dilative during undrained monotonic loading in the triaxial cell and the state point will move to the right. If cyclically loaded the state points will shift to the left as strains occur and the specimen softens. If enough cycles are applied, if they are large enough, and if the hydrostatic stress condition is passed during each cycle, then the zero effective stress condition (*i.e.* initial liquefaction) can ultimately be reached in the laboratory.

9.3 Mechanism of Liquefaction

The strength of sand is primarily due to internal friction. In saturated state it may be expressed as (Fig. 9.2).

$$S = \bar{\sigma}_n \tan \phi \tag{9.1}$$

where, S = Shear strength of sand

$\bar{\sigma}_n$ = Effective normal pressure on any plane xx at depth $z = \gamma\, h_w + \gamma_{sub}\,(z - h_w)$

ϕ = Angle of internal friction

γ = Unit weight of soil above water table

γ_{sub} = Submerged unit weight of soil

If a saturated sand is subjected to ground vibrations, it tends to compact and decrease in volume,

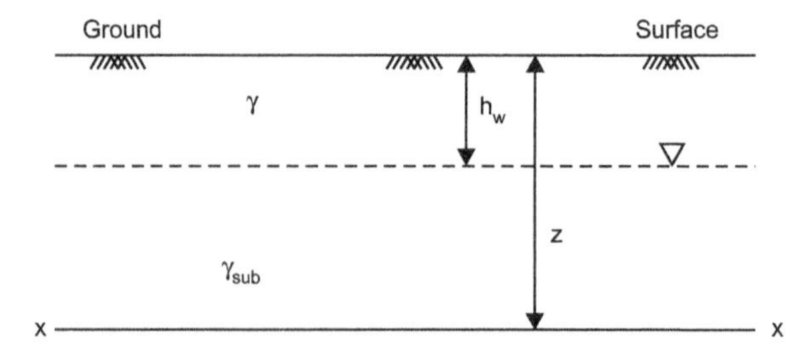

Fig. 9.2. Section of ground showing the position of water table

if drainage is restrained the tendency to decrease in volume results in an increase in pore pressure. The strength may now be expressed as:

$$S_{dyn} = (\bar{\sigma}_n - u_{dyn}) \tan \phi_{dyn} \tag{9.2}$$

where S_{dyn} = Shear strength of soil under vibrations
u_{dyn} = Excess pore water pressure due to ground vibrations
ϕ_{dyn} = Angle of internal friction under dynamic conditions

It is seen that with development of additional positive pore pressure, the strength of sand is reduced. In sands. ϕ_{dyn} is almost equal to ϕ, i.e. angle of internal friction in static condition.

For complete loss of strength i.e. S_{dyn} is zero

Thus, $$\bar{\sigma}_n - u_{dyn} = 0$$

or $$\bar{\sigma}_n = u_{dyn}$$

or $$\frac{u_{dyn}}{\bar{\sigma}_n} = 1 \tag{9.3}$$

Expressing u_{dyn} in terms of rise in water head, h_w and γ_{sub} as $(G - 1/1 + e)\gamma_w$, Eq. (9.3) can be written as:

$$\frac{\gamma_w \cdot h_w}{\frac{G-1}{1+e} \cdot \gamma_w \cdot z} = 1$$

or $$\frac{h_w}{z} = \frac{G-1}{1+e} = i_{cr} \tag{9.4}$$

where G = Specific gravity of soil particles
e = Void ratio
i_{cr} = Critical hydraulic gradient.

It is seen that, because of increase in pore water pressure the effective stress reduces, resulting in loss of strength. Transfer of intergranular stress takes place from soil grains to pore water. Thus if this transfer is complete, there is complete loss of strength, resulting in what is known as complete liquefaction. However, if only partial transfer of stress from the grains to the pore water occurs, there is partial loss of strength resulting in partial liquefaction.

In case of complete liquefaction, the effective stress is lost and the sand-water mixture behaves as a viscous material and process of consolidation starts, followed by surface settlement, resulting in closer packing of sand grains. Thus the structures resting on such a material start sinking. The rate of sinking of structures depends upon the time for which the sand remains in liquefied state.

Liquefaction of sand may develop at any zone of a deposit, where the necessary combination of in-situ density, surcharge conditions and vibration characteristics occur. Such a zone may be at the surface or at some depth below the ground surface, depending only on the state of sand and the induced motion.

However, liquefaction of the upper layers of a deposit may also occur, not as a direct result of the ground motion to which they are subjected, but because of the development of liquefaction in an underlying zone of the deposit. Once liquefaction develops at some depth in a mass of sand, the excess pore water pressure in the liquefied zone will dissipate by flow of water in an upward direction. If the hydraulic gradient becomes sufficiently large, the upward flow of water will induce a quick or liquefied condition in the surface layers of the deposit.

Thus, an important feature of the phenomenon of liquefaction is the fact that, its onset in one zone of deposit may lead to liquefaction of other zones, which would have remained stable otherwise.

9.4 Laboratory Studies

9.4.1 Field Conditions for Soil Liquefaction

An element of soil located at depth z below the horizontal ground surface will be subjected to vertical effective stress $\bar{\sigma}$. which is equal to σ_{vi}, and horizontal effective stress $K_o\ \bar{\sigma}$, where K_o is the coefficient of earth pressure at rest (Fig. 9.3a). There is no initial shear stress acting on the element. Due to ground shaking during an earthquake, a cyclic shear stress τ_h will be imposed on the soil element (Fig. 9.3b). In the case of sloping ground, an element of soil will also have initial shear stress. τ_{hi} (Fig. 9.4a). During earthquake, the stresses on the element will be as shown in Fig. 9.4b. The presence of the initial shear stresses can have major effect on the response of the soil to a superimposed cyclic stress condition and in general, the presence of initial stresses tends to reduce the rate of pore pressure generation due to cyclic stress applications. Since the most critical conditions are likely to be those associated with no initial shear stresses on horizontal planes, a condition analogus in earthquake problems to soil response under essentially level ground is considered. Hence, soil elements can be considered to undergo a series of cyclic stress conditions as illustrated in Fig. 9.5. The actual stress series are somewhat random in pattern (Fig. 9.6) but nevertheless cyclic in nature as illustrated in Fig. 9.5.

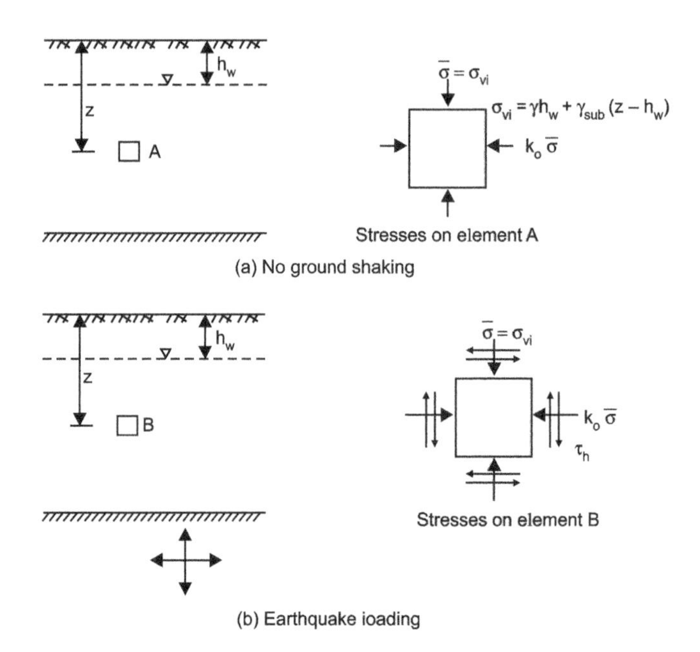

Fig. 9.3. Stress conditions for soil element below horizontal ground in cyclic loading conditions

9.4.2 Different Laboratory Tests

Simulating cyclic shear stress conditions, following types of test procedures have been adopted for liquefaction studies:

(1) Dynamic triaxial test (Seed and Lee, 1966; Lee and Seed, 1967).
(2) Cyclic simple shear test (Peacock and Seed, 1968, Finn et al., 1970, Seed and Peacock, 1971).

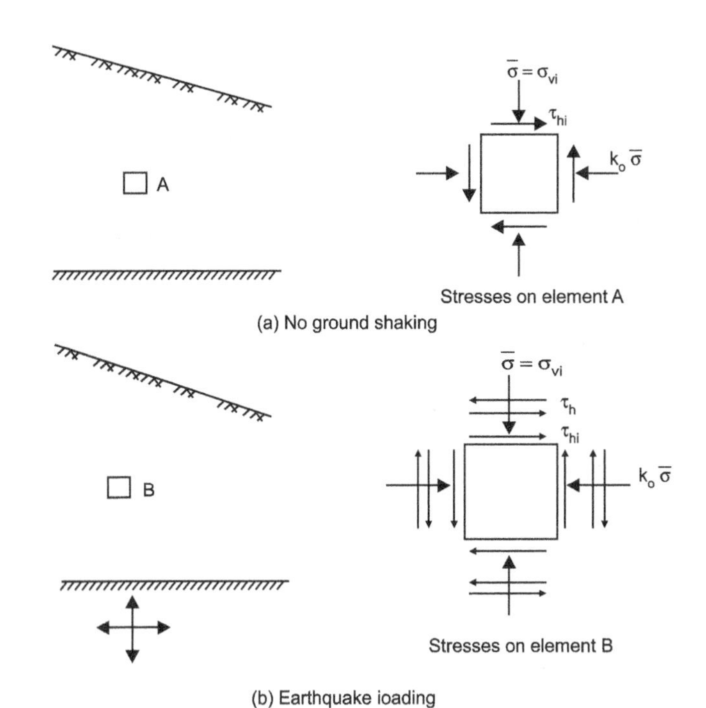

Fig. 9.4. Stress conditions for soil element below sloping ground in cyclic loading conditions

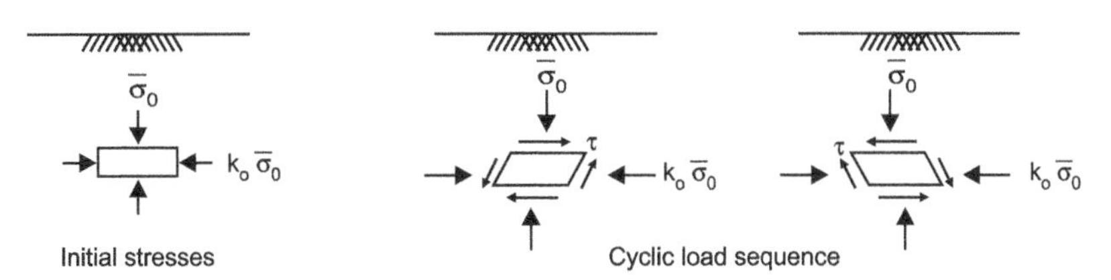

Fig. 9.5. Idealised field loading conditions

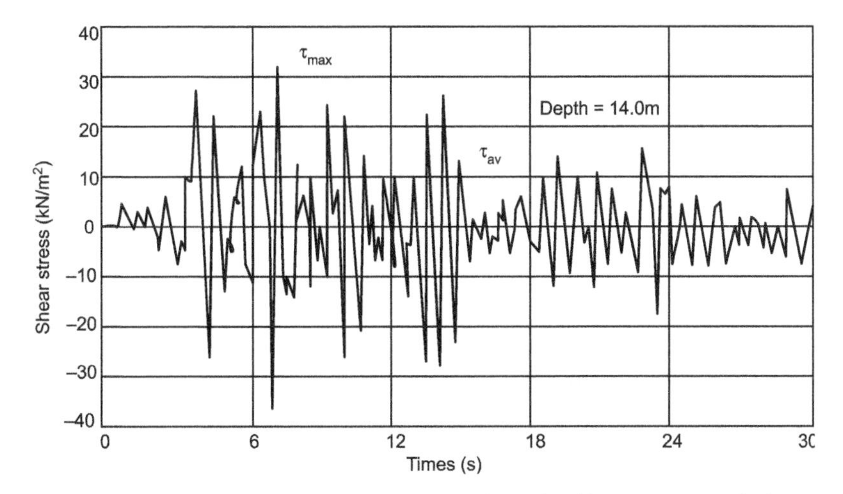

Fig. 9.6. Typical shear stress variation determined by response analysis

(3) Cyclic torsional shear test (Yoshimi and Oh-Oka, 1973; Ishibashi and Sherif, 1974).
(4) Shaking table test (Prakash and Mathur, 1965; Yoshimi, 1967; Finn et al., 1970)

Typical studies on above mentioned laboratory tests are described herein.

9.5 Dynamic Trixial Test

Seed and Lee (1996) reported the first set of comprehensive data on liquefaction characteristics of sand studied by dynamic triaxial test. They used the concept of developing cyclic shear stress on the soil sample as shown by the stress conditions I and II on the sample given in Col. 1 of Fig. 9.7. The stress condition I is achieved by increasing axial stress on the specimen by an amount σ_d, keeping lateral stress σ_3 constant, and simultaneously reducing the all round pressure on the specimen by an amount $\sigma_d/2$ (Cols. 3 and 4 of Condition I). Similarly the desired stress condition II can be induced by reducing the vertical stress by σ_d, and simultaneously applying an increase in all round pressure equal to $\sigma_d/2$ (Cols. 3 and 4 of Condition II). It may be noted that during testing the pore pressure should be corrected by reducing it by $\sigma_d/2$ in condition 1, and increasing by $\sigma_d/2$ in condition II.

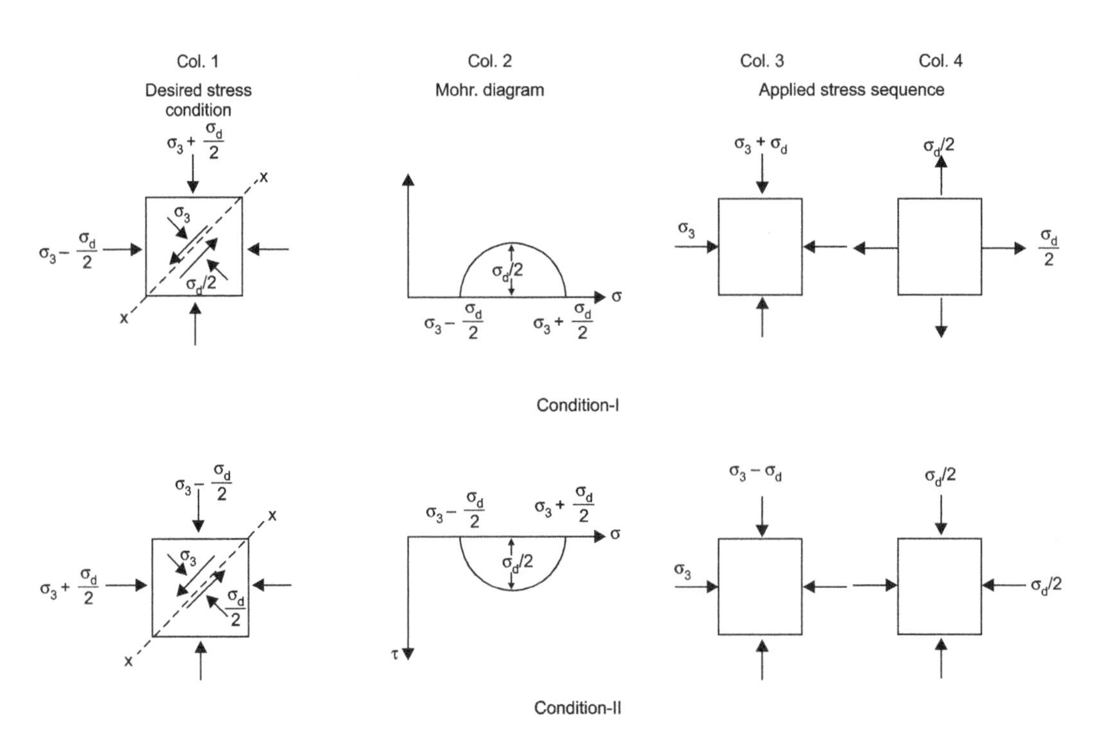

Fig. 9.7. Simulation of cyclic shear stress on a plane of triaxial test specimen (Seed and Lee, 1966)

Seed and Lee (1966) performed several undrained triaxial tests on Sacramento river sand (e_{max} = 1.03, e_{min} = 0.61). The grain size ranged between 0.149 mm and 0.297 mm. The results of a typical test in loose sand (e = 0.87, D_R = 38%) are shown in Fig. 9.8. In this test the initial all round pressure and the initial pore water pressure were 196 kN/m^2 and 98 kN/m^2 respectively, thus giving the value of effective confining pressure as 98 kN/m^2. The cyclic deviator stress σ_d of magnitude 38.2 kN/m^2 was applied with a frequency of 2 cps. The test data in Fig. 9.8 show the variation of load, deformation and pore-water pressure with time.

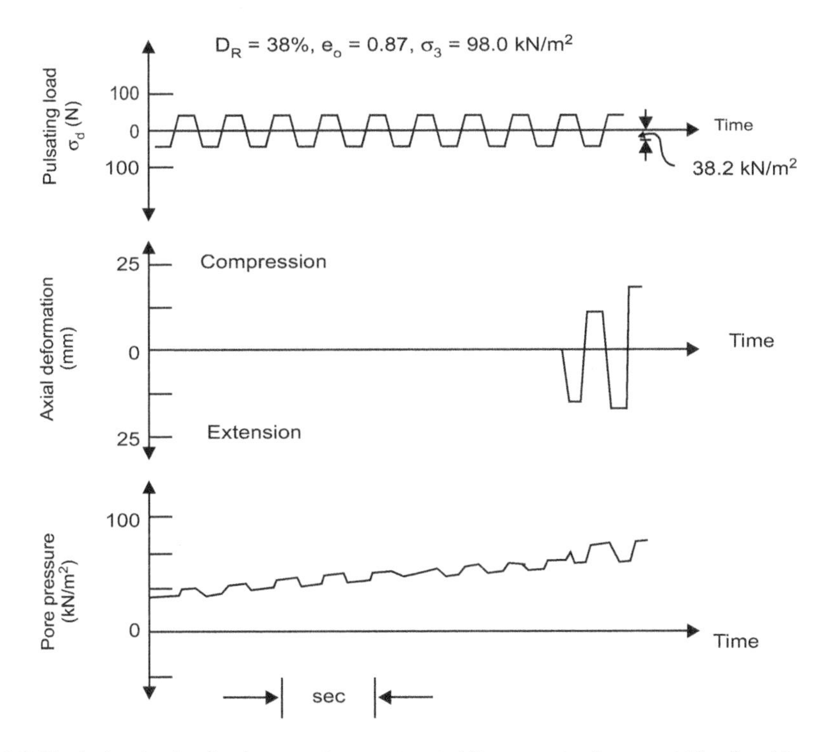

Fig. 9.8. Typical pulsating load test on loose saturated Sacramento river sand (Seed and Lee, 1966)

From this data, variation of axial strain amplitude, the observed changes in pore-water pressures, and the pore water pressure changes corrected to mean extreme principal stress conditions with number of stress cycles have been plotted in Fig. 9.9. It was observed that, during the first eight cycles of stress application, the sample showed no noticeable deformation although pore-pressure increased gradually. The pore pressure became equal to σ_3 during the ninth cycle, indicating zero effective confining pressure. During the tenth cycle, the axial strain exceeded 20% and the soil liquefied.

Similar tests as described above were performed by Seed and Lee (1966) for different values of σ_d. The relationship between σ_d against the number of cycles of pulsating load applications is shown in Fig. 9.10. It is evident from this figure that number of cycles of pulsating load application increases with the decrease of the value of σ_d.

The data obtained on dense Sacramento river sand is shown in Fig. 9.11. It may be noted that the change in pore water pressure become equal to σ_3 after about 13 cycles; however the axial strain

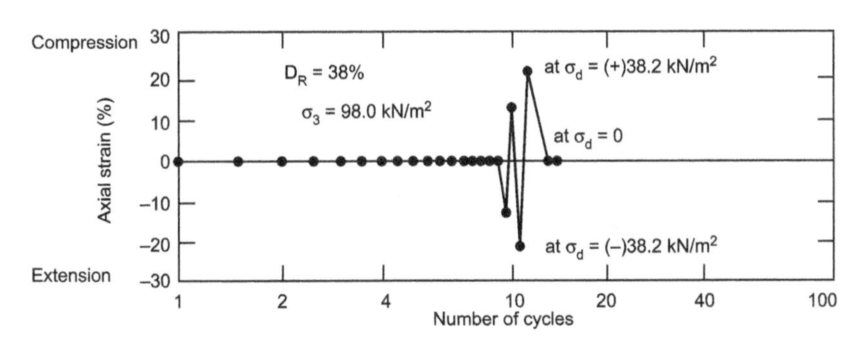

(a) Axial strain versus number of cycles

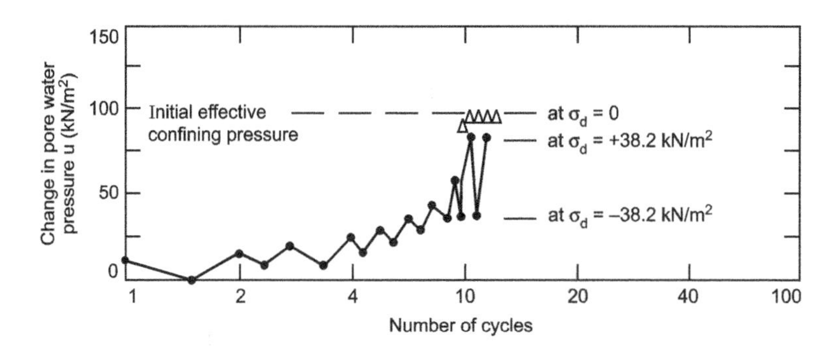

(b) Observed change in pore water pressure and number of cycles

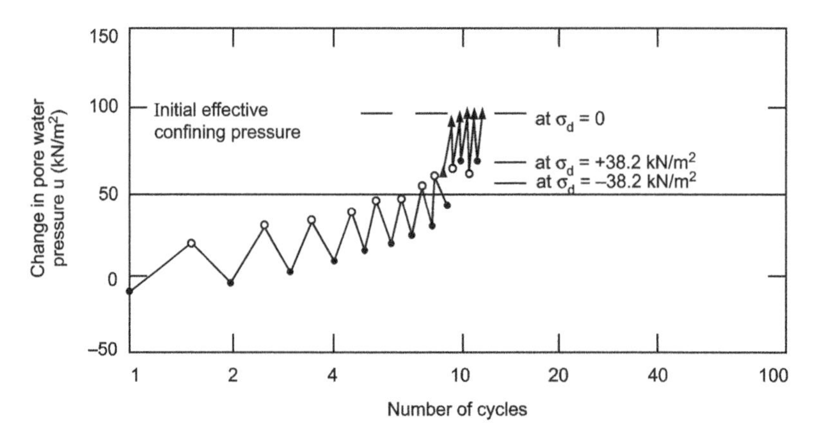

(c) Change in pore pressure and number of cycles

Fig. 9.9. Typical pulsating load test on loose Sacramento river sand (Seed and Lee,1966)

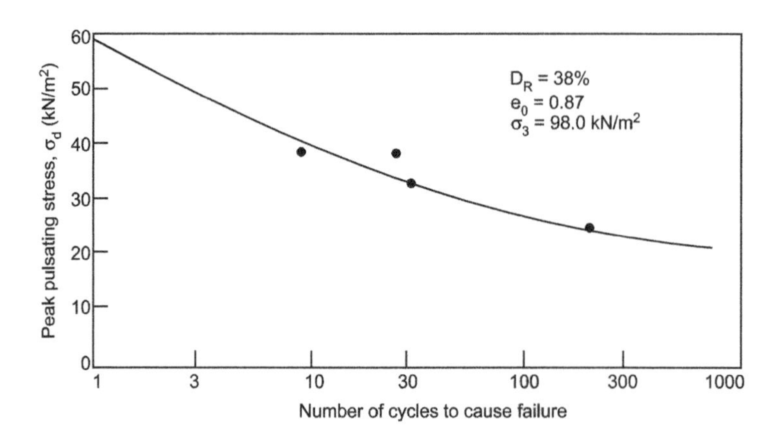

Fig. 9.10. Relationship between pulsating deviator stress and number of cycles required to cause failure in Sacramento river sand (Seed and Lee, 1966)

amplitude did not exceed 10% even after 30 cycles. This is due to the fact that in dense condition soil dilates, and the pore water pressure reduces which in turn stabilises the soil under load. As discussed earlier this corresponds to cyclic mobility (Seed, 1976; Castro, 1976).

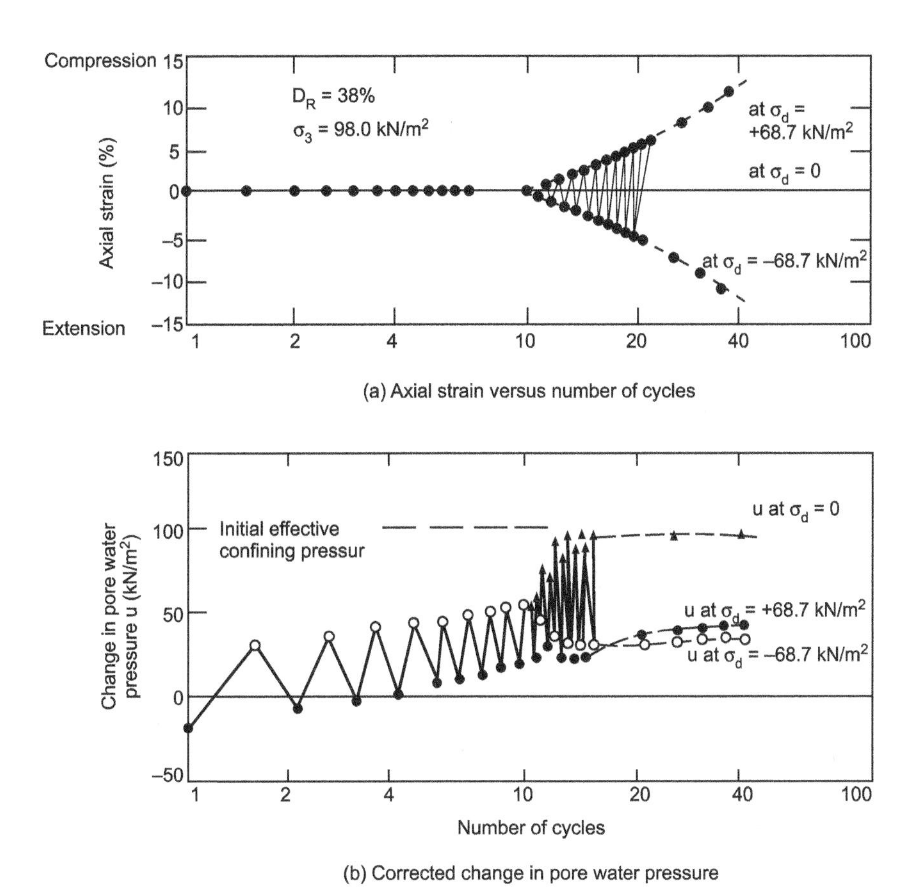

Fig. 9.11. Typical pulsating load test on dense Sacramento river sand (Seed and Lee, 1966)

Lee and Seed (1967) have extended this work for studying the various factors affecting liquefaction and identified the following:

Relative density. Figure 9.12 shows the plots of peak pulsating stress (i.e. the stress causing liquefacation) against number of cycles of stress in loose and dense sands. The initial liquefaction corresponds to the condition when the pore water pressure becomes equal to the confining pressure σ_3. Criterion for complete liquefaction is taken corresponding to 20% double amplitude strain. The figure indicates that in loose sand, initial liquefaction and failure occur simultaneously (Fig. 9.12a). With the increase in relative density, the difference between the number of cycles to cause initial liquefaction and failure increases.

Confining pressure. Figure 9.13 shows the influence of confining pressure on initial liquefaction and failure conditions. At all relative densities for a given peak pulsating stress, the number of cycles to cause initial liquefaction (Figs. 9.13 a and b) or failure (i.e. 20% strain, Figs. 9.13 c and d) increased with the increase in confining pressure.

Peak pulsating stress. Figures 9.14 a and b show respectively the variation of peak pulsating stress σ_d with confining pressure for initial liquefaction and 20% axial strain in 100 cycles. It may be noted that for a given relative density and number of cycles of load application, the σ_d increases linearly with σ_3 for initial liquefaction, while for 20% axial strain condition, similar linear trend exists only in loose sands.

Number of cycles of pulsating stress. From Figs. 9.12 and 9.13, one can conclude that for a given pulsating stress, number of cycles needed for causing initial liquefaction and failure increases with the increase in relative density and confining pressure.

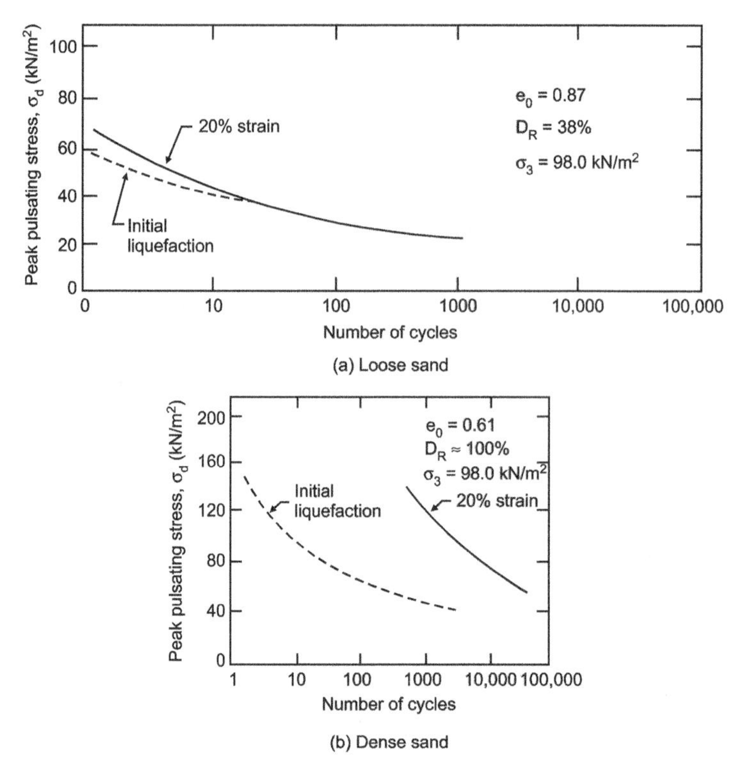

Fig. 9.12. Peak pulsatiing stress versus number of cycles (Lee and Seed, 1967)

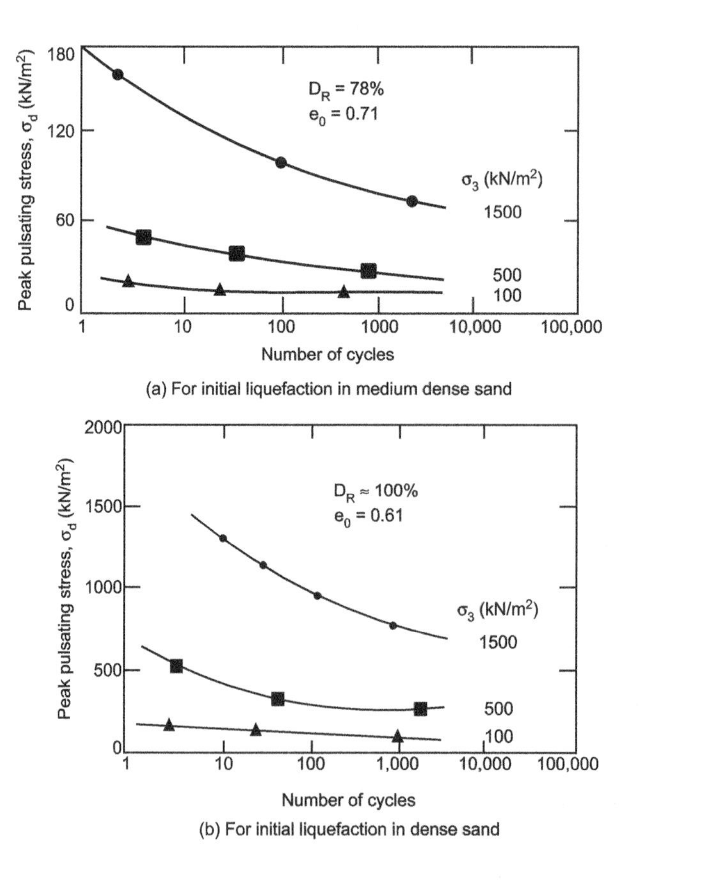

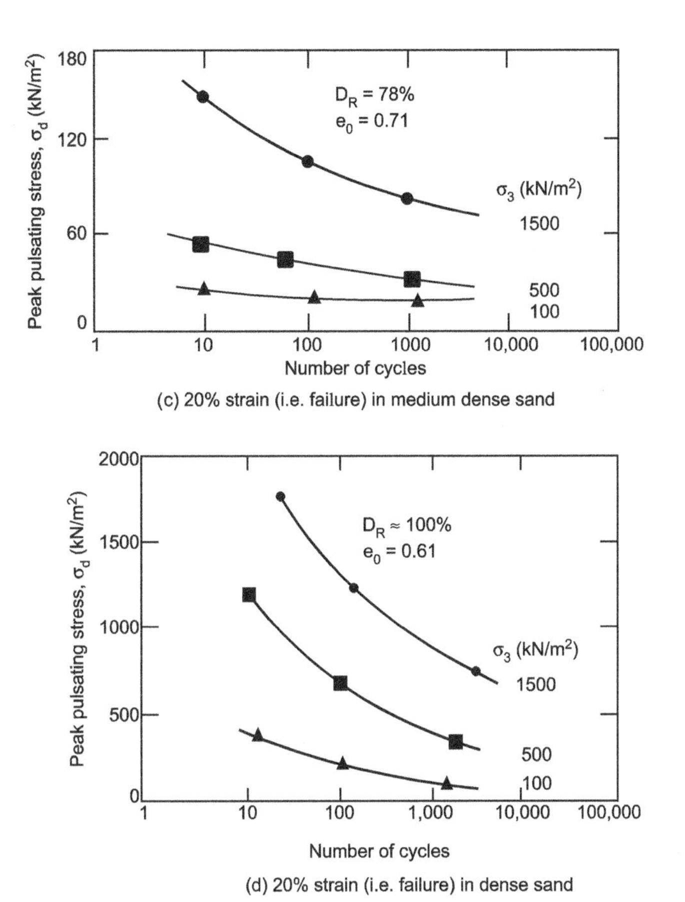

(c) 20% strain (i.e. failure) in medium dense sand

(d) 20% strain (i.e. failure) in dense sand

Fig. 9.13. Peak pulsatiing stress versus number of cycles (Lee and Seed, 1967)

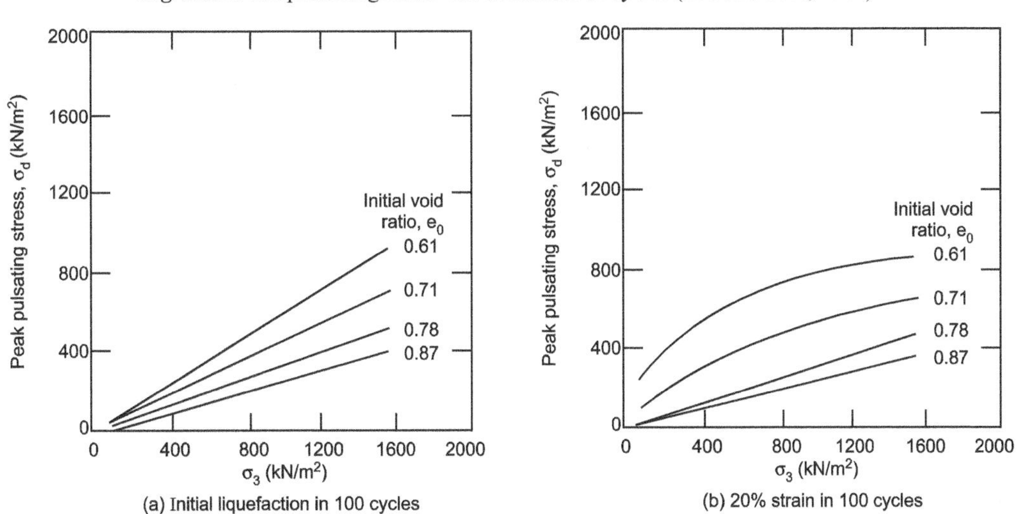

(a) Initial liquefaction in 100 cycles

(b) 20% strain in 100 cycles

Fig. 9.14. Influence of pulsating stress on the liquefaction of Sacramento river sand (Lee and Seed, 1967)

9.6 Cyclic Simple Shear Test

The cyclic simple shear test device is already described in Chapter 4 with a mention that it simulates earthquake condition in a better way. Peacock and Seed (1968) reported the first set of comprehensive data on liquefaction studies by using this test. They performed tests on Monterey sand (*SP*, e_{max} =

0.83, e_{min} = 0.53, D_{10} = 0.54 mm, D_{60} = 0.66, G = 2.64). The sample size was 60 mm square and 20 mm thick. The samples were tested at relative densities (D_R) of 50%, 80% and 90% giving the sand in loose, medium dense and dense states respectively. The oscillatory shear stress was applied at a frequency of 1 cps or 2 cps keeping the normal stress constant.

The typical test data in Fig. 9.15 (D_R = 50%, e_0 = 0.68, σ_v = 500 kN/m^2, f = 1 Hz) show the variation of shear stress, shear strain and pore water pressure with time. As evident from Fig. 9.15 b, there was no significant shear strain of the sample during the application of the first 24 cycles of stress. During the twenty-fifth stress cycle, the shear strain suddenly increased to a value of about 15% and become 23% in the next cycle. Pore pressure increased gradually until the effective confining pressure is reduced to zero (Fig. 9.15c). At this point the resulting deformations became extremely large, and the soil had essentially liquefied. Similar trend was also observed in triaxial test.

Peackock and Seed (1968) have also studied the effects of following factors on liquefaction:

Relative density. Figure 9.16 shows a plot of peak pulsating stress (τ_h) causing initial liquefaction with number of cycles of application for different relative densities and confining pressures. From this figure it can be concluded that for a given value of confining pressure and number of cycles of stress application, τ_h increases with the increase of relative density. A more clear presentation is shown in Fig. 9.17.

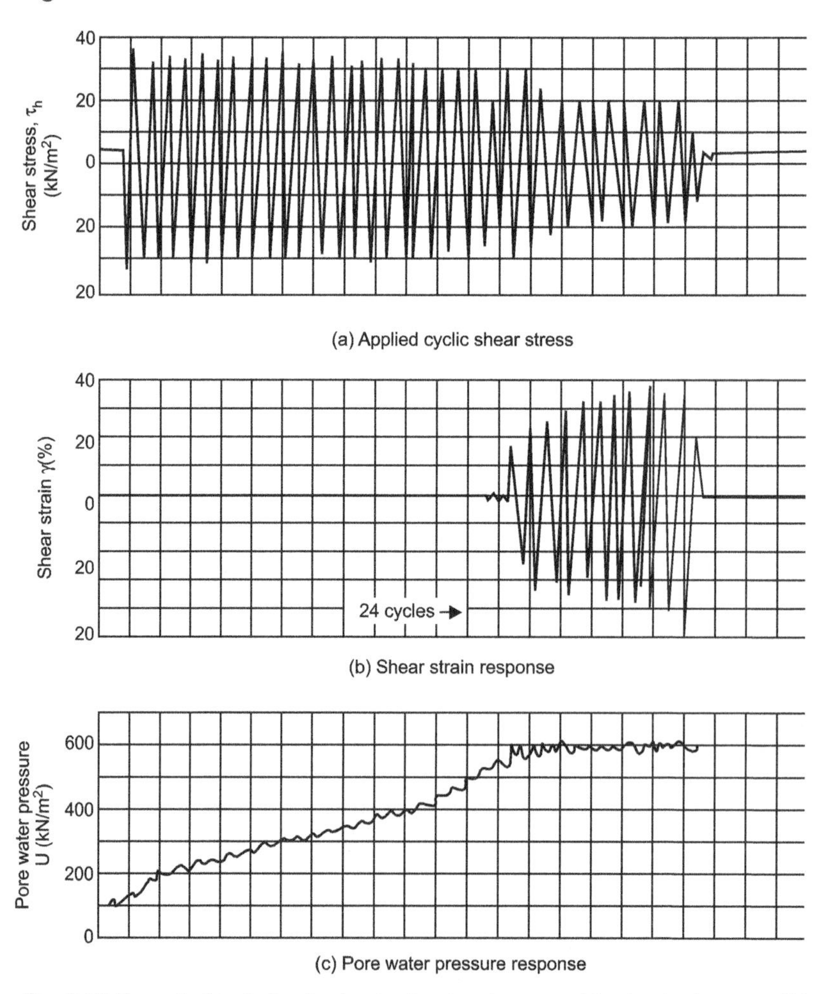

Fig. 9.15. Record of typical pulsating load test on loose sand in simple shear conditions (Peacock and Seed, 1968)

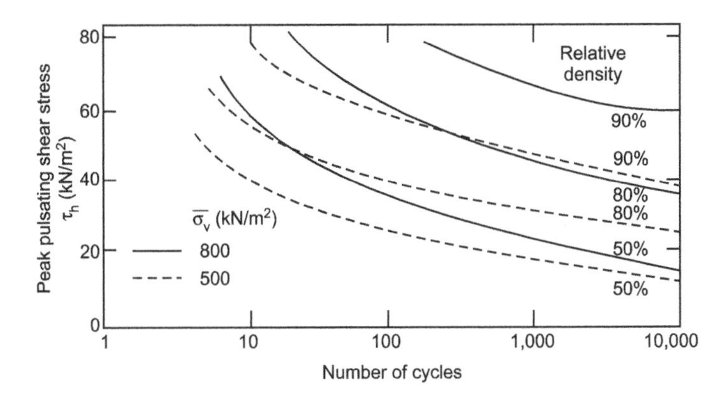

Fig. 9.16. Initial liquefaction of cyclic simple shear test on Monterery sand (Peacock and Seed, 1968)

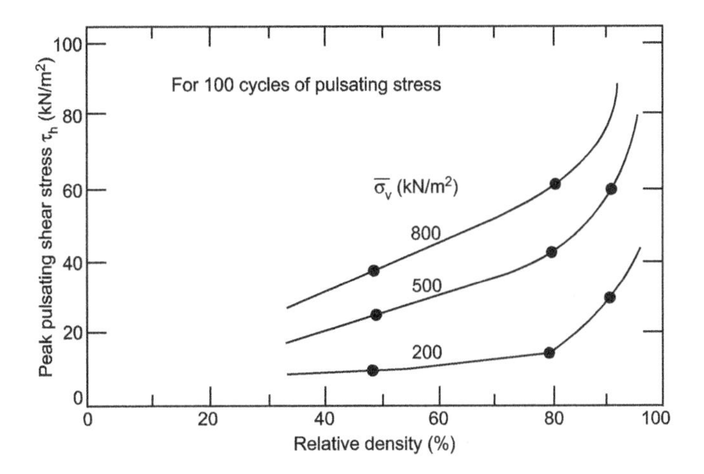

Fig. 9.17. Effect of relative density on cyclic stress causing initial liquefaction (Peacock and Seed, 1968)

Confining pressure. From the data presented in Figs. 9.16 and 9.17, plot of τ_h versus $\overline{\sigma}_v$ was prepared as shown in Fig. 9.18. For a given value of D_R and number of cycles of stress application, τ_h increases linearly with the increase in $\overline{\sigma}_v$.

Peak pulsating stress and number of cycles of stress application. From Fig. 9.16, it can be seen that for a given value of $\overline{\sigma}_v$ and relative density D_R, a decrease of τ_h requires an increase

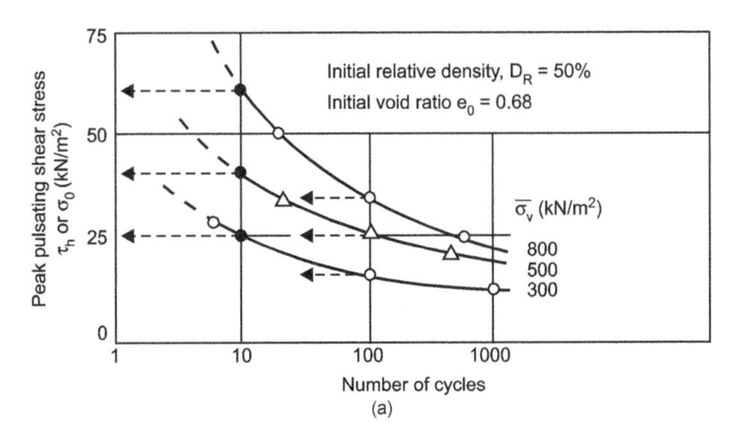

Fig. 9.18. (a) Cyclic stresses required to cause initial liquefaction at different confining pressure

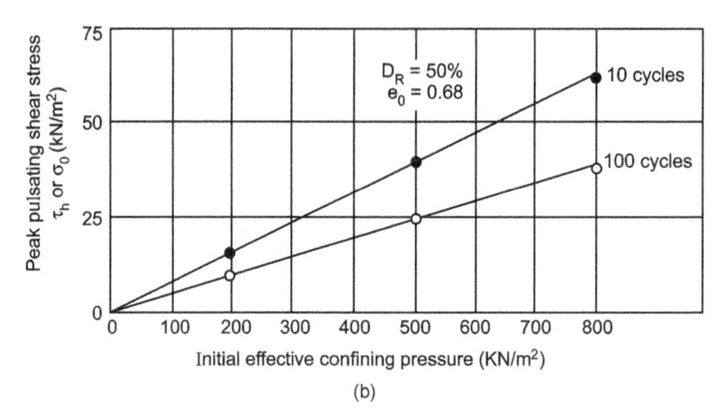

Fig. 9.18 (b) Effect of confining pressure on cyclic stress to cause failure in 10 cycles and 100 cycles (Peacock and Seed, 1968)

of number of cycles to cause liquefaction. Further for a given value τ_h, number of cycles of stress application required to cause liquefaction increases with the increase in relative density D_R and confining pressure $\bar{\sigma}_v$.

Frequency of load application. Tests were performed at frequencies of 1 Hz, 2 Hz, and 4 Hz, and the effect of frequency on the stress causing liquefaction was found negligible.

Seed and Peacock (1971) have studied the effect of coefficient of earth pressure (K_o) on the peak pulsating shear stress τ_h causing liquefaction in cyclic simple shear test. The value of K_o depends on the overconsolidation ratio (*OCR*).

Figure 9.19 shows a plot of stress ratio $(\tau_h / \bar{\sigma}_v)$ with number of cycles of stress application for different values of K_o. For a given relative density and number of cycles of stress application, the value of $(\tau_h / \bar{\sigma}_v)$ decreases with the decrease of K_o.

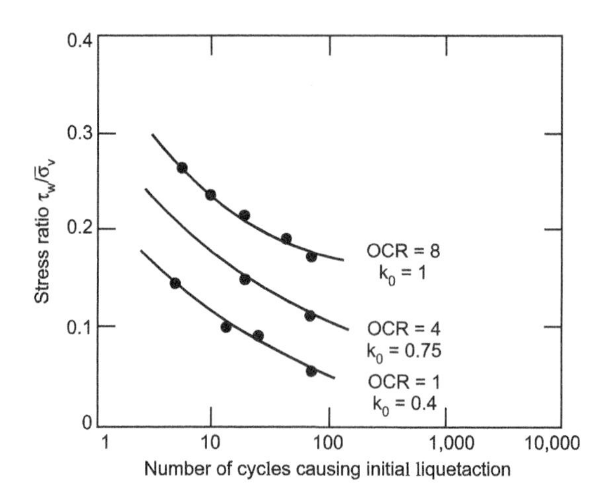

Fig. 9.19. Influence of overconsolidation ratio (*OCR*) on stress causing liquefaction in simple shear tests (Seed and Peacock, 1971)

9.7 Comparison of Cyclic Stresses Causing Liquefaction under Triaxial and Simple Shear Conditions

Peacock and Seed (1968) performed both cyclic triaxial and cyclic simple shear tests for liquefaction studies on Monterey sand with a relative density of 50% and confining pressures $(\sigma_3$ or $\bar{\sigma}_v)$ of 300, 500 and 800 kN/m². Results are plotted in Figs. 9.20 and 9.21. It may be seen from these figures that the cyclic stress required to cause liquefaction of loose sands under simple shear condition (τ_h) is about 35 percent of the cyclic stress required to cause liquefaction in triaxial condition ($\sigma_d/2$).

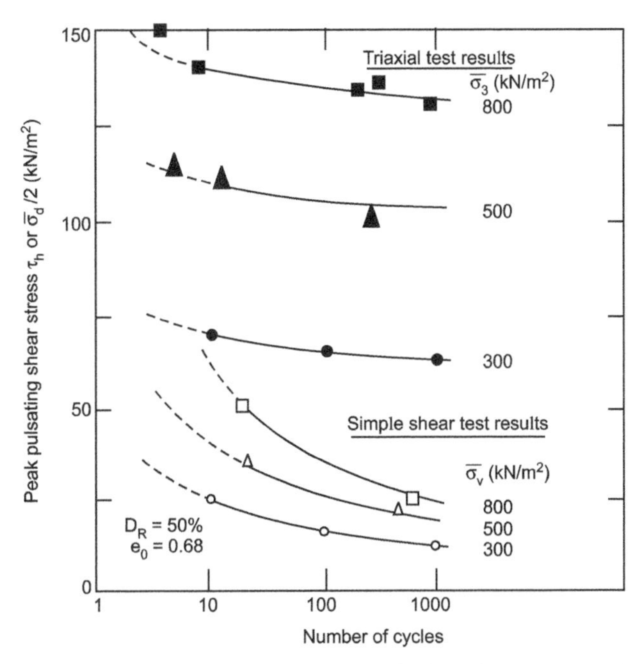

Fig. 9.20. Cyclic stress required to cause liquefaction of Monterey sand at different confining pressures in triaxial and simple shear tests (Peacock and Seed, 1968)

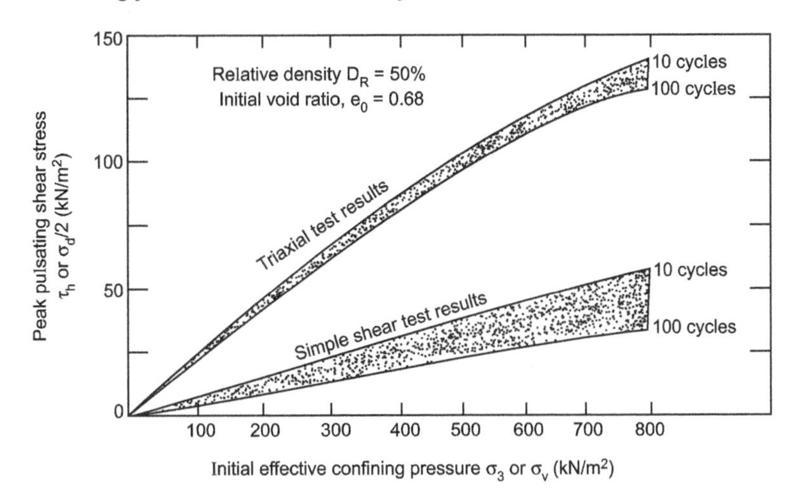

Fig. 9.21. Comparison of pulsating shear strength of loose Monterey sand under cyclic loading-simple shear and triaxial condition (Peacock and Seed, 1968)

9.8 Standard Curves and Correlations for Liquefaction

For evaluation of liquefaction potential, Seed and Idriss (1971) developed standard curves between cyclic stress ratio ($\sigma_d/2/\sigma_3$) versus mean grain size (D_{50}) for 10 and 30 number of cycles of stress application for an initial relative density of compaction of 50% (Figs. 9.22 a and b). These curves were prepared by compiling the results of various tests conducted by several investigators on various types of sand.

The values of stress ratio $(\tau_h / \overline{\sigma}_v)$ causing liquefaction, estimated from the result of simple shear tests, have shown that the value of $(\tau_h / \overline{\sigma}_v)$ is less than the corresponding value of $\sigma_d/2\sigma_3$ (Figs. 9.20, 9.21 and 9.22). The two stress ratios may be expressed by the relation:

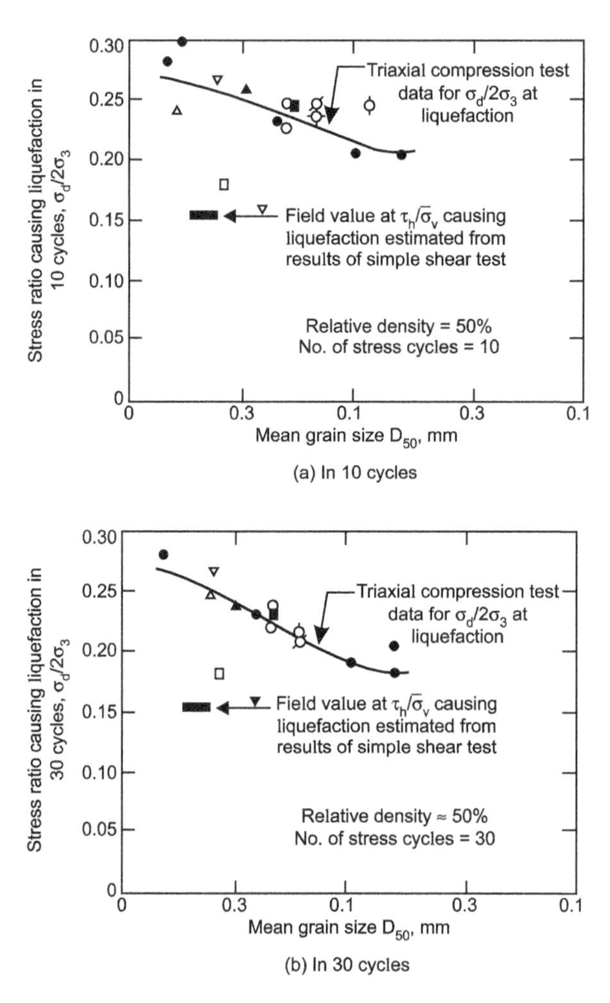

Fig. 9.22. Stress conditions causing liquefaction of sands (Seed and Idriss, 1971)

$$\left(\frac{\tau_h}{\bar{\sigma}_v}\right)_{\text{simple shear}} = \left(\frac{\sigma_d}{2\sigma_3}\right)_{triax} \cdot C_1 \tag{9.5}$$

where C_1 = Correction factor to be applied to laboratory triaxial test data to obtain stress conditions causing liquefaction in the field.

Seed and Peacock (1971) have proposed the following three alternative criteria of obtaining C_1:

(i) **Maximum ratio of shear stress developed during cyclic loading to the normal stress:** The initial stress conditions of a specimen in simple shear device are shown in Fig. 9.23a, the corresponding Mohr's circle is shown in Fig. 9.23b. Figures 9.23 c and d show respectively the stress conditions on the soil specimen during cyclic simple shear test and corresponding Mohr's circle. It can be seen from Fig. 9.23d, that the maximum ratio of shear stress to normal stress in cyclic simple shear stress is $\tau_h / K_o \bar{\sigma}_v$. This ratio in triaxial test is $\sigma_d/2\sigma_3$ (Fig. 9.7). Therefore,

$$\frac{\tau_h}{K_o \bar{\sigma}_v} = \frac{\sigma_d}{2\sigma_3} \tag{9.6}$$

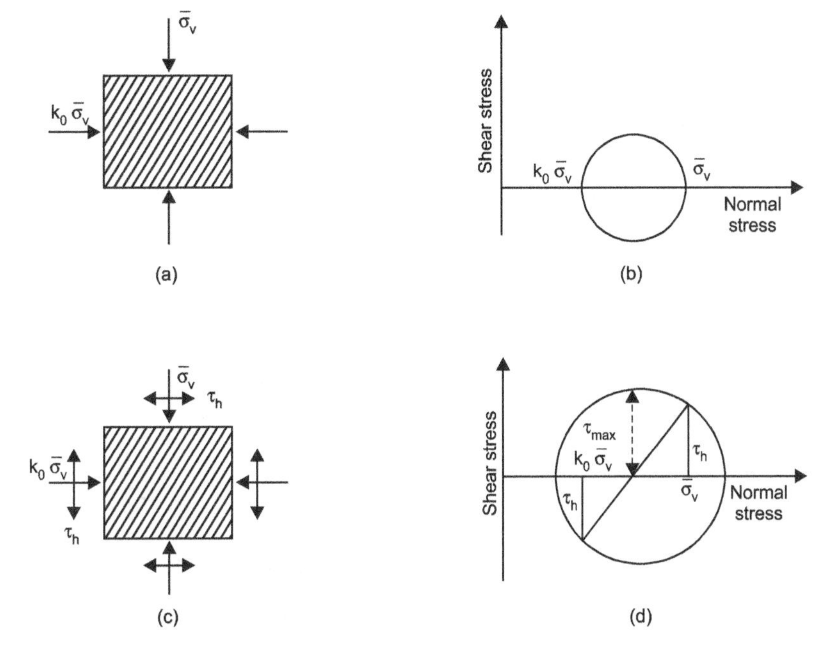

Fig. 9.23. Maximum shear stress for cyclic simple shear tests

Now $$C_1 = \frac{\tau_h / \bar{\sigma}_v}{\sigma_d / 2\sigma_3} = \frac{\tau_h / \bar{\sigma}_v}{\tau_h / K_o \bar{\sigma}_v} = K_o \tag{9.7}$$

(ii) **Ratio of maximum shear stress to the mean principal stress.** In simple shear test (Fig. 9.23d)

$$\text{Maximum shear stress, } \tau_{max} = \sqrt{\tau_h^2 + \left[\frac{1}{2}\bar{\sigma}_v (1 - K_o)\right]^2} \tag{9.8}$$

Mean principal stress during consolidation (Fig. 9.23a)

$$= \frac{1}{3}[\bar{\sigma}_v + K_o\,\bar{\sigma}_v + K_o\,\bar{\sigma}_v]$$

$$= \frac{1}{3}\bar{\sigma}_v\,(1+2K_o) \tag{9.9}$$

In Triaxial test

$$\text{Maximum shear stress} = \frac{\sigma_d}{2} \tag{9.10}$$

$$\text{Minor principal stress} = \sigma_3 \tag{9.11}$$

Therefore, $$\frac{\sqrt{\tau_h^2 + \frac{1}{2}\{\bar{\sigma}_v(1-K_o)\}^2}}{\left[\frac{1}{3}\bar{\sigma}_v(1+2K_o)\right]} = \frac{\sigma_d}{2\sigma_3} \tag{9.12}$$

It gives $$\left(\frac{\tau_h}{\bar{\sigma}_v}\right) = \left(\frac{\sigma_d}{2\sigma_3}\right)\sqrt{\frac{1}{9}(1 + 2K_o)^2 - \frac{1}{4}(1 - K_o)^2 / (\sigma_d / 2\sigma_3)^2} \tag{9.13}$$

Hence, $$C_1 = \sqrt{\frac{1}{9}(1+2K_o)^2 - \frac{1}{4}(1-K_o)^2/(\sigma_d/2\sigma_3)^2} \tag{9.14}$$

(iii) **Ratio of maximum change in shear stress to the mean principal stress during consolidation**

$$\frac{\tau_h}{\left[\dfrac{\bar{\sigma}_v(1+2K_o)}{3}\right]} = \frac{\left(\dfrac{\sigma_d}{2}\right)}{\sigma_3} \tag{9.15}$$

It gives $$C_1 = \frac{(1+2K_o)}{3} \tag{9.16}$$

Finn et al. (1970) have shown that for initial liquefaction of normally consolidated sands

$$C_1 = \frac{(1+K_o)}{2} \tag{9.17}$$

Castro (1975) has proposed that the initial liquefaction is better represented by the criteria of the ratio of the octahedral shear stress during cyclic loading to the effective octahedral normal stress during consolidation. It gives the value of C_1 as

$$C_1 = \frac{2(1+2K_o)}{(3\sqrt{3})} \tag{9.18}$$

Values of C_1 computed from the above equations are given in Table 9.1. Weighted average values of C_1 are given in the last column of the table. In normally consolidated sands, value of K_o ranges from 0.3 to 0.5 which in turn gives the value of C_1 varying from 0.45 to 0.55.

Table 9.1. Values of C_1

K_o	*Value of C_1 using*					
	Equation (9.7)	*Equation (9.14)**	*Equation (9.16)*	*Equation (9.17)*	*Equation (9.18)*	*Average value*
0.3	0.3	Negative value	0.53	0.65	0.61	0.45
0.4	0.4	Negative value	0.60	0.70	0.69	0.53
0.5	0.5	0.25	0.67	0.75	0.77	0.55
0.6	0.6	0.54	0.73	0.80	0.85	0.68
0.7	0.7	0.71	0.80	0.85	0.92	0.78
0.8	0.8	0.83	0.87	0.90	1.00	0.87

* For $\dfrac{\sigma_d}{2\sigma_3} = 0.4$

In simple shear test equipment, there is always some nonuniformity of stress condition. This causes specimens to develop liquefaction under lower horizontal cyclic stresses as compared to that in the field. Seed and Peacock (1971) demonstrated this fact for a uniform medium sand (D_R = 50%) in which the field values were about 1.2 times the laboratory values. It can be expressed by the following relation:

$$\left(\frac{\tau_h}{\bar{\sigma}_v}\right)_{\text{field}} = C_2\left(\frac{\tau_h}{\bar{\sigma}_v}\right)_{\text{simple shear}} \tag{9.19}$$

where C_2 = Constant to account the non-uniformity of stress conditions in simple shear test.

Combining Eqs. (9.5) and (9.19), we get

$$\left(\frac{\tau_h}{\bar{\sigma}_v}\right)_{field} = C_1 C_2 \left(\frac{\sigma_d}{2\sigma_3}\right)_{triax} = C_r \left(\frac{\sigma_d}{2\sigma_3}\right)_{triax} \tag{9.20}$$

where $$C_r = C_1 C_2 \tag{9.21}$$

Seed and Idriss (1971) suggested the values of C_r as given in Table 9.2.

Table 9.2. Values of C_r

Relative density D_R (%)	C_r
0-50	0.57
60	0.60
80	0.68

As evident from Fig. 9.17, upto a relative density of 80%, the peak pulsating shear stress causing liquefaction increases almost linearly with the increase in relative density. Keeping this fact in view, the following general relation is suggested:

$$\left(\frac{\tau_h}{\bar{\sigma}_v}\right)_{\substack{field\\D_R}} = \left(\frac{\sigma_d}{2\sigma_3}\right)_{\substack{triax\\50}} \cdot C_r \cdot \frac{D_R}{50} \tag{9.22}$$

where

$\left(\frac{\tau_h}{\bar{\sigma}_v}\right)_{\substack{field\\D_R}}$ = Cyclic shear stress ratio in the field at relative density of D_R percent

$\left(\frac{\sigma_d}{2\sigma_3}\right)_{\substack{triax\\50}}$ = Stress ratio obtained from triaxial test at relative density of 50%. It can be determined using Fig. 9.22.

9.9 Evaluation of Zone of Liquefaction in Field

At a depth below the ground surface, liquefaction will occur if shear stress induced by earthquake is more than the shear stress predicted by Eq. (9.22). By comparing the induced and predicted shear stresses at various depths, liquefaction zone can be obtained.

In a sand deposit, consider a column of soil of height h and unit area of cross section subjected as maximum ground acceleration a_{max} (Fig. 9.24). Assuming the soil column to behave as a rigid body, the maximum shear stress τ_{max} at a depth h is given by

$$\tau_{max} = \left(\frac{\gamma h}{g}\right) \cdot a_{\max} \tag{9.23}$$

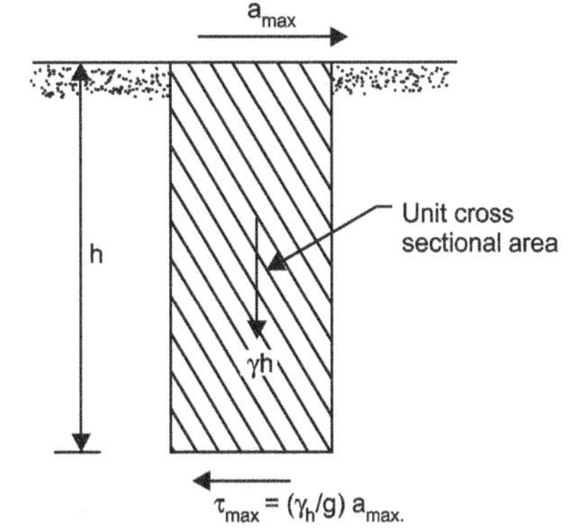

Fig. 9.24. Maximum shear stress at a depth for a rigid soil column

where g = Acceleration due to gravity

γ = Unit weight of soil

Since the soil column behaves as a deformable body, the actual shear stress at depth h, $(\tau_{max})_{act}$ is taken as

$$(\tau_{max})_{act} = r_d \cdot \tau_{\max} = r_d \left(\frac{\gamma h}{g}\right) \cdot a_{\max} \tag{9.24}$$

where r_d = Stress reduction factor

Seed and Idriss (1971) recommended the use of charts shown in Fig. 9.25 for obtaining the values of r_d at various depths. In this figure the range of r_d for different soil profiles alongwith the average value upto depth of 12 m is shown. The critical depth for development of liquefaction is usually less than 12 m.

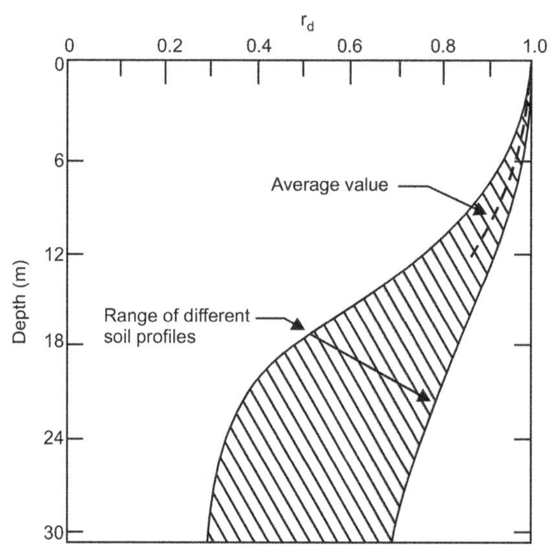

Fig. 9.25. Reduction factor r_d versus depths (Seed and Idriss, 1971)

The actual time history of shear stress at any point in a soil deposit during an earthquake will be as shown in Fig. 9.6. According to Seed and Idriss (1971), the average equivalent uniform shear stress τ_{av} is about 65 percent of the maximum shear stress τ_{max}. Therefore

$$\tau_{av} = 0.65\frac{\gamma h}{g} \cdot a_{\max} \cdot r_d \tag{9.25}$$

The corresponding number of significant cycles N_S for τ_{av} is given in Table 9.3.

Table 9.3. Significant Number of Cycles N_S Corresponding to τ_{av}

Earthquake magnitude, M on Richter's scale	N_S
7	10
7.5	20
8	30

The procedure of locating liquefaction zone can be summarised in following steps:

(i) Establish the design earthquake, and obtain peak ground acceleration a_{max}. Also obtain number of significant cycles N_S corresponding to earthquake magnitude using Table 9.3.

(ii) Using Eq. (9.25), determine τ_{av} at depth h below ground surface.

(iii) Using Fig. (9.22), determine the value of $(\sigma_d/2\ \sigma_3)$ for given value D_{50} of soil and number of equivalent cycles N_S for the relative density of 50%.

(iv) Using Eq. (9.25), determine the value of $\left(\dfrac{\tau_h}{\bar{\sigma}_v}\right)_{\substack{\text{field}\\ D_R}}$ for the relative density D_R of the soil at site. Multiplying $\left(\dfrac{\tau_h}{\bar{\sigma}_v}\right)_{\substack{\text{field}\\ D_R}}$ with effective stress at depth h, one can obtain the value of shear stress τ_h required for causing liquefaction.

(v) At depth h, liquefaction will occur if $\tau_{av} > \tau_h$

(vi) Repeat steps (ii) to (iv) for other values of h to locate the zone of liquefaction. τ_{av} and τ_h can be plotted as shown in Fig. 9.26.

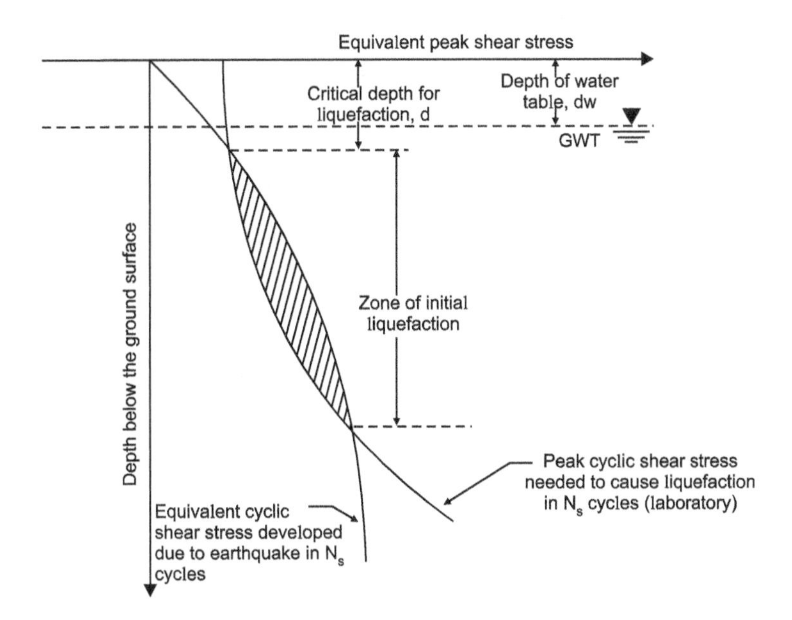

Fig. 9.26. Zone of initial liquefaction in field

9.10 Vibration Table Studies

In vibration table studies, a large specimen of saturated sand is prepared in a tank which is placed on a vibration table. Figure 9.27 shows a typical setup of a horizontal shake table available at Indian Institute of Technology Roorkee. It consists of a rigid platform on which the test tank (1.05 m × 0.6 m × 0.6 m high) is mounted. The platform with wheels rests on four knife edges being rigidly fixed on two pairs of rails anchored to the foundation. The platform is connected with motor and brake assembly for imparting vibrations. Some important characteristics of the table are:

Amplitude – 0-10 mm
Frequency – 0-20 Hz
Acceleration – 0-20 g

Facilities are available for measuring pore pressures at different depths in the sample placed in tank. The procedure of carrying out test is simple. Firstly the sand is placed in the tank under saturated condition. The table is then excited with the desired amplitude and acceleration. Variation of pore pressure with time and number of cycles are then noted.

The main advantages of vibration table studies are:

(i) It simulates field conditions in better way as the size of sample is large, prepared and consolidated under anisotropic conditions.

(ii) It is possible to trace the actual pore-water pressure distribution during liquefaction.
(iii) Deformation occurs under plane strain conditions.
(iv) Visual examination of sample during vibration is possible.

Fig. 9.27. General view of horizontal shake table

Since 1957, many investigators have studied liquefaction characteristics of sand using vibration table on different sizes of soil samples and dynamic characteristics of load. The effect of the following aspects have been studied:

(i) Grain size characteristics of soil.
(ii) Relative density.
(iii) Initial stress condition i.e. overburden pressure.
(iv) Intensity and character of excitation force.
(v) Entrapped air.

The important conclusions drawn from vibration table studies are:

1. For a given sand placed at a particular density, there is sudden increase in pore pressure at a definite acceleration. This is termed as 'critical acceleration'. Critical acceleration is not unique property of sand. It depends on the type of sand, its density, the amplitude and frequency of oscillation and the overburden pressure (Maslov, 1957; Matsuo and Ohara, 1960; Florin and Ivanov, 1961).
2. If sand is subjected to shock loading, the whole stratum is liquefied at the same time, while under steady-state vibrations, the liquefaction starts from the top and proceeds downward (Florin and Ivanov, 1961).
3. As the surcharge pressure increased, the number of cycles required to cause liquefaction increased (Fig. 9.28; Fin, 1970). Tests have shown that, even small drainage surcharge will reduce the time of the liquefied state tenfold (Fig. 9.29; Florin and Ivanov, 1961).

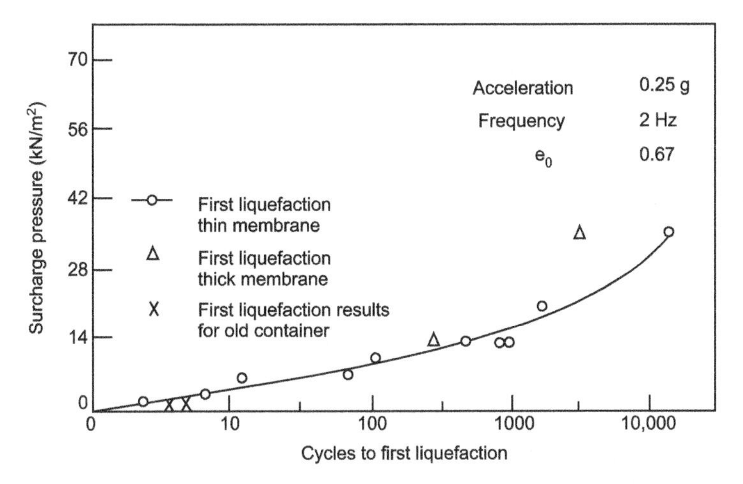

Fig. 9.28. Effect of surcharge pressure on resistance to initial liquefaction in vibration table studies (Finn, 1972)

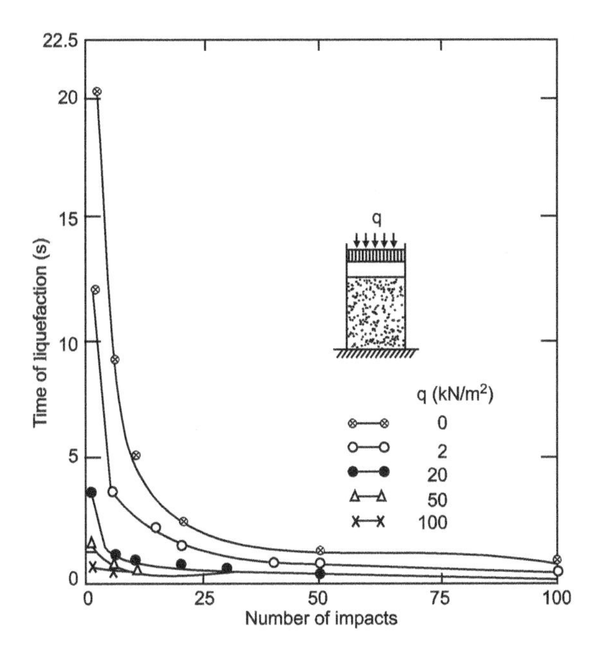

Fig. 9.29. Influence of the intensity of draining surcharge on the period of time within which the sand remain liquid (Florin and Ivanov, 1961)

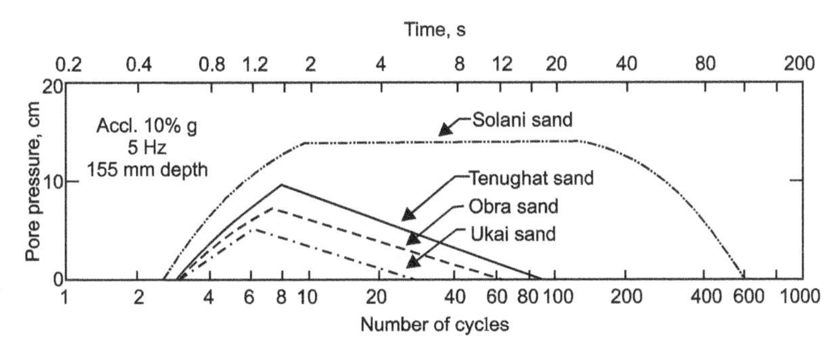

Fig. 9.30. Pore pressure versus number of cycles for different sands (Gupta, 1979)

4. The time during which the liquefied state lasts is much less for coarser grained soils than for fine grained soils (Fig. 9.30; Gupta, 1979). He carried out liquefaction studies on four sands namely (i) Ukai sand (D_{50} = 1.8 mm), (ii) Obra sand (D_{50} = 1.0 mm), (iii) Tenughat sand (D_{50} = 0.47 mm) and Solani sand (D_{50} = 0.15 mm). The maximum pore water pressure is developed in about 6 to 10 cycles. It started dissipating immediately after attaining maximum value. The total time required for dissipation was about 6s for Ukai sand and 20s for Obra and Tenught sands. The corresponding value for Solani sand was 120s. It remained constant for about 35s. Thus the time required for dissipation decreases with the increase in coarseness.

   Since the liquid state lasts for only a short time, the liquefied masses of soil have no time for displacements, so that there is practically no indication that the phenomenon of liquefaction occurs in coarse-grained soils.
5. The excess pore-water pressures decrease with the increase in initial relative density (Maslov, 1957; Gupta, 1979). Figure 9.31 shows a typical test data indicating the effect of relative density on an increase in pore pressure at 10 percent *g* for Solani sand (Gupta 1979). In this case, no pore-water pressure increase was observed when initial density became 62 percent. Tests performed on other types of sand with different accelerations gave the values of relative densities as listed in Table. 9.4 beyond which no pore-water pressure was observed.

Table 9.4. Initial Relative Density Beyond which no Excess Pore Pressure Develops (Gupta, 1979)

Acceleration (g) Percent	*Initial Relative Density*			
	Solani sand D_{50} – (0.15 mm)	*Tenughat sand (0.47 mm)*	*Obra sand (1.0 mm)*	*Ukai sand (1.8 mm)*
10	62.5	52.0	51.5	50.5
20	62.5	61.5	60.0	59.5
40	66.0	64.0	62.5	62.0
50	66.5	65.0	64.0	63.0

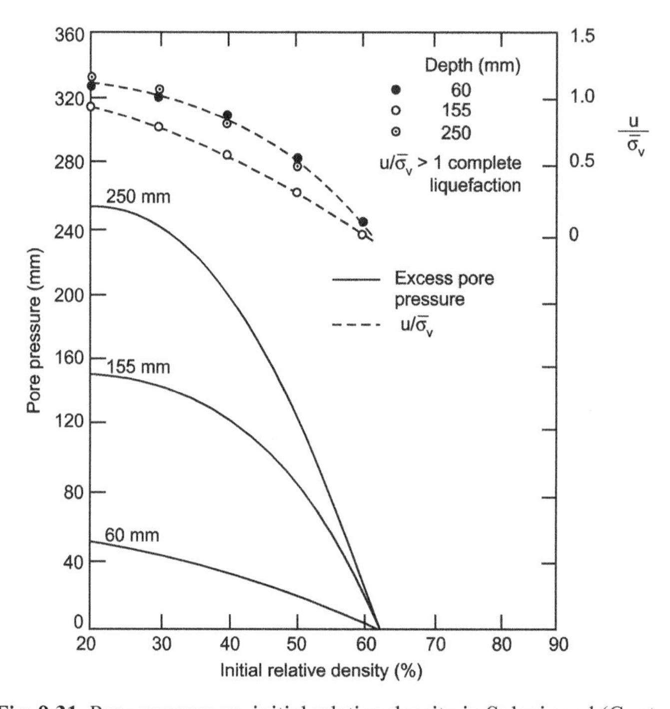

Fig. 9.31. Pore pressure vs. initial relative density in Solani sand (Gupta, 1979)

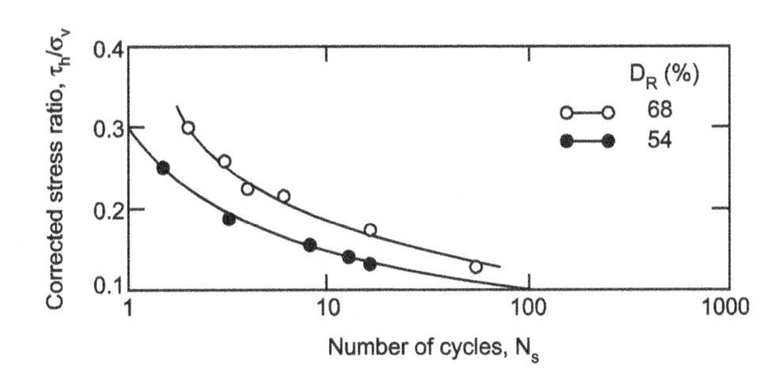

Fig. 9.32. Corrected τ_h/σ_v versus N_s for initial liquefaction from shake table studies (DeAlba. Seed and Chan, 1976)

DeAlba, Seed and Chan (1976) presented the results of shake table tests in the form of stress ratio (τ_h/σ_v) and number of cycles required for causing liquefaction as shown in Fig. 9.32. It indicates that for a given value of τ_h, more number of cycles are required for liquefying a sand having more relative density. This is a similar conclusion as obtained by Seed and Peacock (1971) by cyclic simple tests. In shake table tests, the value of τ_h is given by:

$$\tau_h = \frac{W}{g} \cdot a_m \tag{9.26}$$

where W = Total pressure exerted on the base of tank placed on the shake table
a_m = Peak acceleration of the uniform cyclic motion

9.11 Field Blast Studies

In blast tests, predetermined charge (like ammonia, gelatin etc.) with electric detonators is installed at predetermined depth in a cased bore hole. The hole is later filled with sand and the casing is withdrawn. Lead wires from detonators are connected with blaster so that the charges may be fired at any desired moment. Acceleration pickups are placed at regular intervals from the blast point to record horizontal and vertical acceleration at the time of blasting. Similarly porewater pickups and settlement gauges are placed at certain distances from the blast point to record the increase in porewater pressure and ground settlements. Accelerations, porewater pressure and ground settlement at the blast point are then obtained by extrapolation. The data is then interpreted to obtain the liquefaction potential.

One of the main purpose of carrying out blast tests is to ascertain whether the soil at the site will liquefy under simulated earthquake loading. Field data using small explosives at some depth at the site along with pore pressure and settlement observations for predicting liquefaction potential are available in literature from few investigations (Florin and Ivanov, 1961; Kummeneja and Eide, 1961; Krishna and Prakash, 1968; Prakash and Gupta, 1970; Arya et al., 1978; Gupta and Mukherjee, 1979).

For examining the chances of liquefaction at a barrage site, Gupta and Mukherjee (1979) performed blast tests in a river bed having the soil profile as given below:

Depth	*Description of soil*	*Average N-value*	*Remarks*
0-4 m	Fine sand	5	C_u = 2.47 m, D_{50} = 0.12 mm
4 m-7 m	Clay	–	Position of water table near the surface
7 m-20 m	Silty sand mixed with kankars	21-60 (increases with depth)	C_u = Uniformity coefficient

The critical hydraulic gradient of top loose sandy deposit works out to be 0.8. Special gelatin (60 percent. 2 kg) was installed at 4 m depth in 150 mm diameter cased bore holes. Blasting was done with the help of an electric exploder. Horizontal surface accelerations were measured using acceleration pickups placed at various distance from the source point. The depth of each acceleration

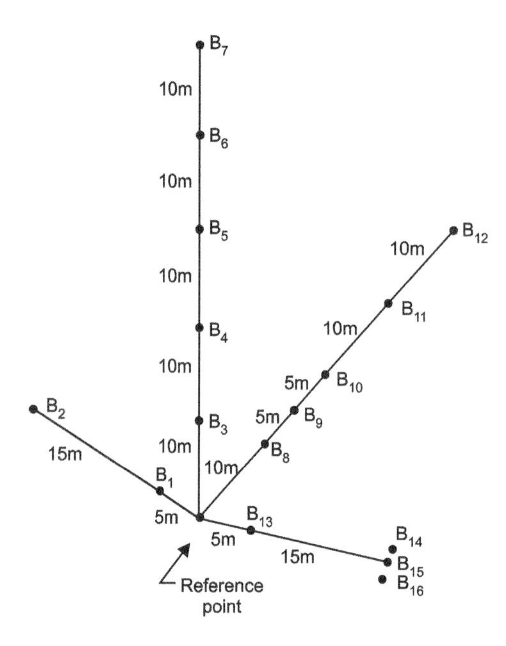

Fig. 9.33. Site layout for field blasting tests (Gupta and Mukherjee, 1979)

pickup was 200 mm below the ground surface. The porewater pressures at various distances from source point were measured at 2.5 m depth from the ground surface.

Figure 9.33 shows the sketch of layout of the tests at the site. A typical acceleration record obtained at 35 m distance from the blast point is shown in Fig. 9.34. Variation of maximum horizontal acceleration and porewater pressure with distance from blast point are shown respectively in Figs. 9.35 and 9.36. On the basis of past earthquakes, the maximum possible acceleration record at the site is assumed as shown in Fig. 9.37. Using the method of Lee and Chan (1972) this earthquake record is worked out to be equivalent to 19.6 cycles of 0.075 g acceleration (Table 9.5)

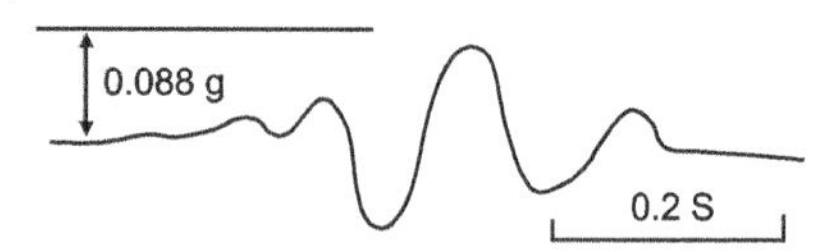

Fig. 9.34. Surface acceleration (Gupta and Mukherjee, 1979)

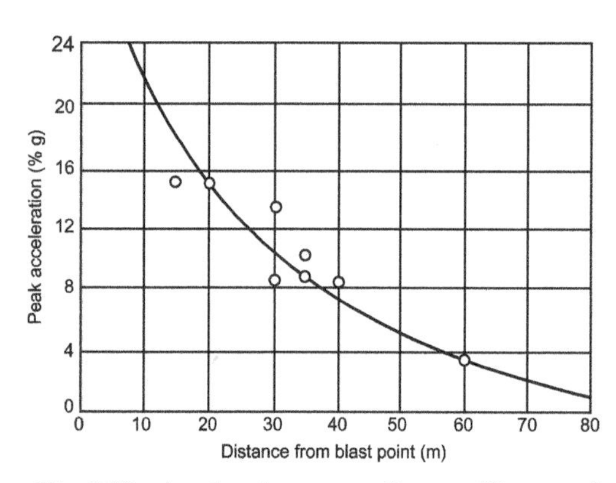

Fig. 9.35. Acceleration versus distance (Gupta and Mukherjee, 1979)

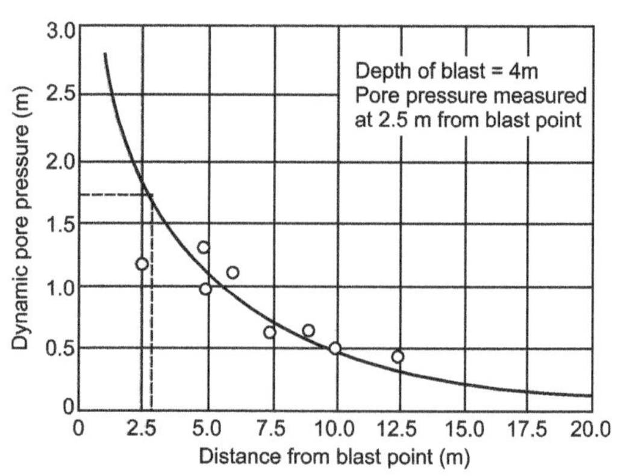

Fig. 9.36. Pore pressure vs. distance (Gupta and Mukherjee, 1979)

Table 9.5. Equivalent Number of Cycles for Anticipated Earthquake

Accl. level in percent of peak accl.	*Average accl. in percent*	*Number of cycles*	*Conversion factor (Fig. 1.6)*	*Equivalent number of cycles at 0.65 τ_{max}*
(1)	(2)	(3)	(4)	(5)
100-80	90	17/2 = 8.5	2.6	22.1
80-60	70	8/2 = 4.0	1.2	4.8
60-40	50	25/2 = 12.5	0.20	2.5
40-00	20	> 1000	Negligible	0.0
			Total	29.4
Total number of cycles for 0.75 τ_{max} = 29.4/1.5 = 19.6				

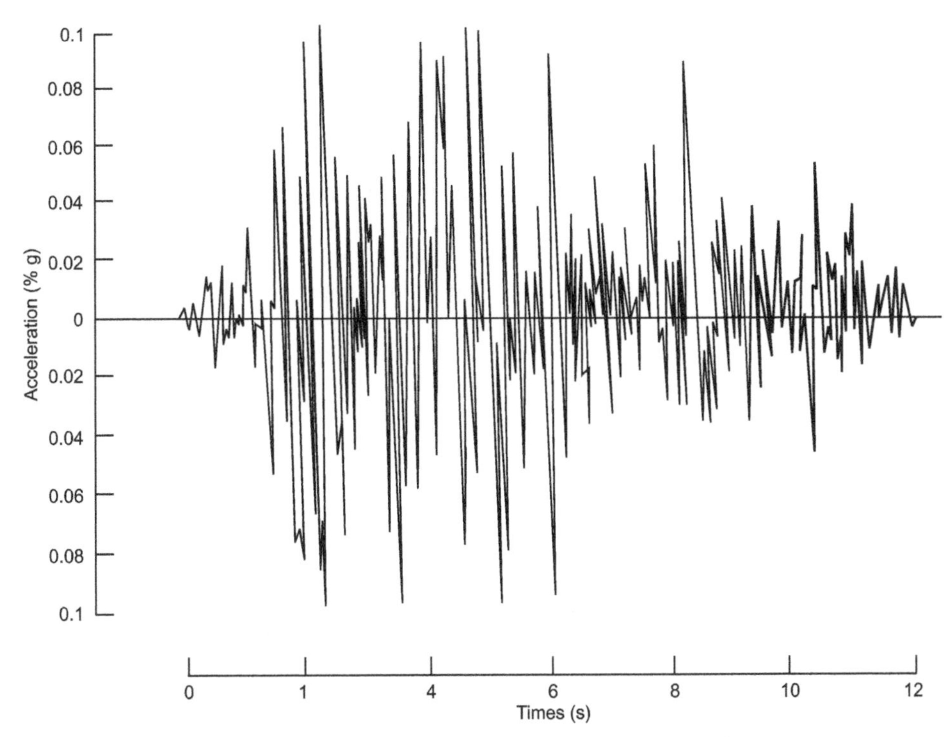

Fig. 9.37. Assumed acceleration for site (Gupta and Mukherjee, 1979)

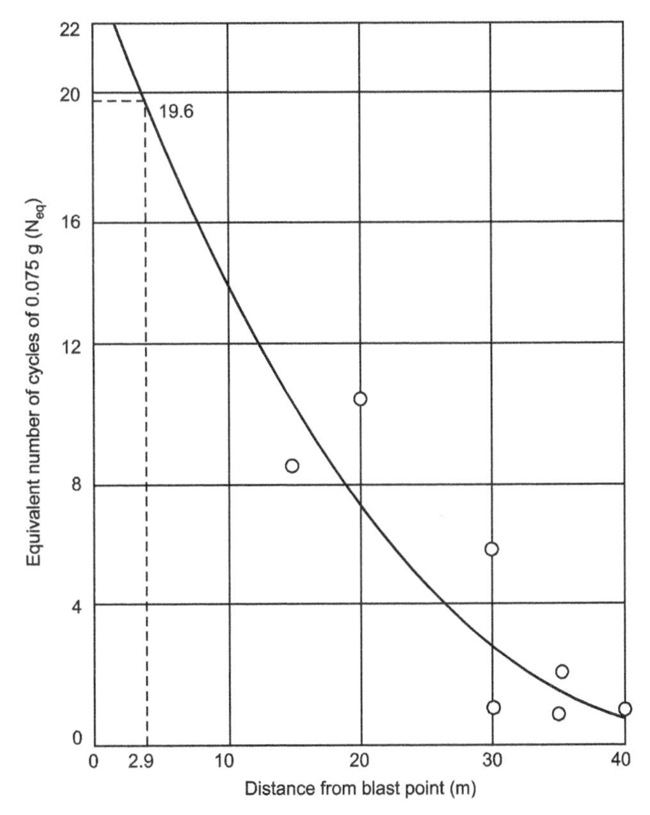

Fig. 9.38. Equivalent cycles versus distance (Gupta and Mukherjee, 1979)

Similarly, the blast records at different distances are also converted into equivalent number of cycles of 0.075 g acceleration (Fig. 9.38). From this figure it can be observed that vibrations generated due to blast at a distance of 2.9 m are equivalent to 19.6 cycles of 0.075 g, the expected earthquake i.e. the blast has the same severity as the design earthquake at a distance of 2.9 m from the blast hole.

From Fig. 9.36, the pore pressure developed at a distance of 2.9 m, and at a depth of 2.5 m is 1.74 m of water column. The critical hydraulic gradient for this site is 0.8; therefore at a depth of 2.5 m the critical porewater pressure or hydraulic head is $0.8 \times 2.5 = 2.0$ m. The pore pressure developed is 1.74 m. The actual porewater pressures developed will be larger than the measured value of 1.74 m, because there will be a time lag in rise of water level in piezometer pipe. Hence under the above conditions, a larger pore pressure is expected to be developed and complete liquefaction of site is expected during the earthquake.

9.12 Procedures Based on In-Situ Tests

In-situ testing procedures are more widely used, since these are not subjected to the difficulties experienced in obtaining truly undisturbed samples that retain their in-situ liquefaction characteristics. Of the in-situ tests available, standard penetration test (SPT) and cone penetration test (CPT) are the most commonly used for liquefaction assessment. Techniques using data from the SPT include those developed by Seed and Idriss (1971), Seed et al. (1985), Iwaski et al. (1978), Tokimatsu and Yoshimi (1983), Iai et al. (1989) and the Japan Road Association (1980, 1991). Methods using the CPT include those developed by Seed and DeAlba (1986), Ishihara (1985), Shibata and Teparaksa (1988) and Robertson and Companella (1985).

9.12.1 Evaluation Based on Standard Penetration Test

The standard penetration test is most commonly used in-situ test in a bore hole to have fairly good estimation of relative density of cohesionless soil. Since liquefaction primarily depends on the initial relative density of saturated sand, many reaserchers have made attempt to develop correlation in liquefaction potential and standard penetration resistance.

After the occurrence of Niigata earthquake, Kishida (1966), Kuizumi (1966), and Ohasaki (1966) studied the areas in Niigata where liquefaction had not occurred and developed criteria for differentiating between liquefaction and non-liquefaction conditions in that city, based on N-values of the sand deposits (Seed, 1979). The results of these studies for Niigata are shown in Fig. 9.39. Ohasaki (1970) have a useful rule of thumb that says liquefaction is not a problem if the blow count from a standard penetration test exceeds twice the depth in metres.

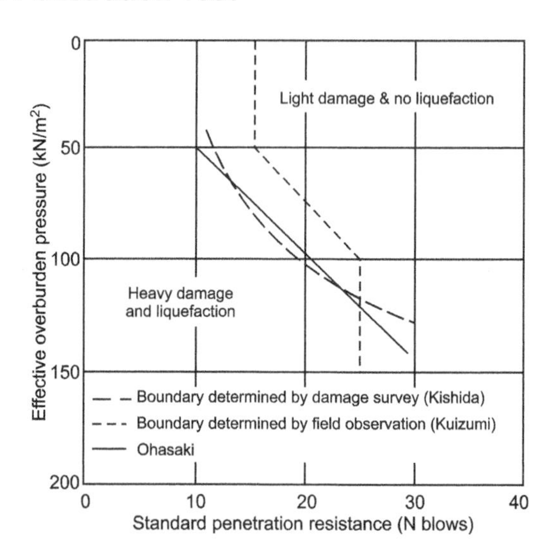

Fig. 9.39. Analysis of liquefaction potential at Niigata for earthquake of June 16, 1994 (Seed, 1979)

(a) Correction of SPT N-value for Energy Efficiency: The blowcount or N-value derived from the SPT is measured in different parts of the world using different delivery systems. This has led to a variety of different test procedures, each different in their energy delivery (Seed et al., 1985;

Skempton,1986). A recommended procedure is given in Table 9.6. It is important to understand and correct for this when using the SPT. This can be achieved by standardising the measured blowcount against available empirical correlations.

Table 9.6. Recommended SPT Procedure for Use in Liquefaction Correlations (Seed et al., 1985)

1.	Borehole : 4 to 5 inch diameter rotary borehole with bentonite drilling mud for borehole stability
2.	Drill Bit : Upward deflection of drilling mud (tricone of baffled bit).
3.	Sampler : O.D. = 2.00 inch; I.D. = 1.38 inch - constant (i.e. no room for liners in barrel).
4.	Drill Rods : A for AW for depth less than 50 feet; or NW for greater depths.
5.	Energy delivered to Sampler: 2,520 in-Ibs. (60% of theoretical maximum).
6.	Blowcount rate : 30-40 blows per minute
7.	Penetration Resistance Count: Measures over range of 6 to 18 inch of penetration into the ground.

Table 9.7. Summary of Energy Ratios for SPT Procedures (Seed et al., 1985; Skempton, 1986)

Country	*Hammer type*	*Hammer release*	*Estimated rod energy ER_m(%)*	*Correction factor for 60% rod energy*
Japan[a]	Donut Donut[b]	Tombi Rope and Pulley with special throw release	78 67	1.30 1.12
U.S.A.	Safety[b]	Rope and Pulley	60	1.00
	Donut	Rope and Pulley	45	0.75
Argentina	Donut[b]	Rope and Pulley	45	0.75
China	Donut[b]	Free fall[c]	60	1.00
	Donut	Rope and Pulley	50	0.85
U.K.	Pilcon	Trip	60	1.00
	Old Standard	Rope and Pulley	60	1.00

[a] Japanese SPT results have additional corrections for borehole diameter and frequency effects
[b] Prevalent method in each country today
[c] Pilcon type hammers develop an energy ratio of about 60%

If the energy efficiency of a particular procedure is different from that adopted in the proposed empirical correlation, the measured blowcount N_m should be scaled as follows:

$$N_{ER} = (ER_m/ER)N_m \tag{9.27}$$

where ER_m denotes the energy efficiency used for obtaining the N_m value, and ER and N_{ER} denote the energy efficiency and blowcount respectively used for establishing the correlation between the cyclic strength and the SPT value. The energy efficiencies for typical SPT test procedures are given in Table 9.7.

Simplified Procedure by Seed and Idriss

The cyclic stress ratio, induced at particular depth beneath a level ground surface may be estimated by using the relation develped by Seed and Idriss (1971).

$$\frac{\tau_{av}}{\bar{\sigma}_v} = 0.65\frac{a_{\max}}{g}\frac{\sigma_v}{\bar{\sigma}_v}r_d \tag{9.28}$$

where τ_{av} is the average cyclic shear stress during a particular time history, $\bar{\sigma}_v$ is the effective overburden stress at the depth in question, σ_v is the total overburden stress at that depth, a_{max} is the peak horizontal ground acceleration induced by the earthquake at the ground surface, g is the acceleration due to gravity and r_d is a stress reduction factor which is a function of depth and the rigidity of the soil column (Fig. 9.25). The second part of the Seed and Idriss procedure requires the determination of the cyclic strength of the soil deposit. This is estimated based on either empirical correlations with the SPT N_m value, Seed et al.(1985) or from cone penetration resistance, q_c, allowing for the effects of the soil fines content. Empirical charts, Figs. 9.40 and 9.41, have been prepared to determine the cyclic strength based on corrected SPT blowcount. $(N_1)_{60}$, calculated as follows:

$$(N_1)_{60} = C_N \frac{ER_m}{60\%} N_m \tag{9.29}$$

where C_N is a correction coefficient for overburden pressure as shown in Fig. 9.42, and ER_m is the actual energy efficiency delivered to the drill rod from Table 9.7.

Based on $(N_1)_{60}$, then, the cyclic stress ratio required to induce liquefaction for a magnitude 7.5 earthquake. $(\tau_h / \bar{\sigma}_v)_{LM=7.5}$ is given by the relationship drawn in Figs. 9.40 and 9.41. For earthquakes of other magnitudes, the appropriate cyclic strength is obtained by multiplying by a magnitude scaling factor as given in Table 9.8.

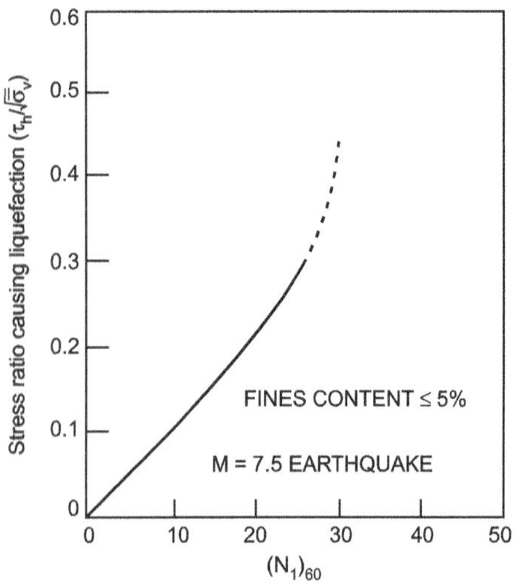

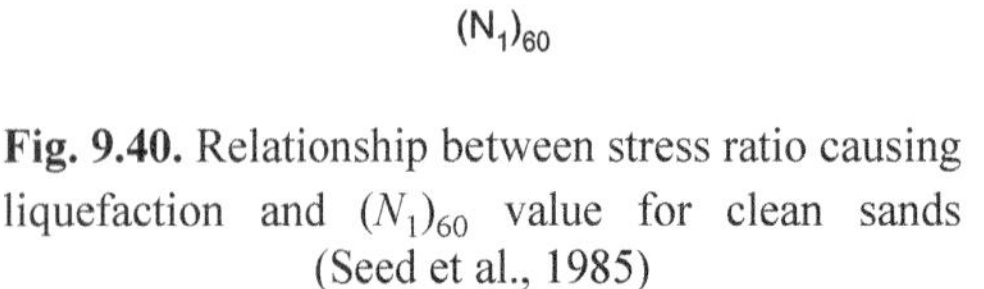

Fig. 9.40. Relationship between stress ratio causing liquefaction and $(N_1)_{60}$ value for clean sands (Seed et al., 1985)

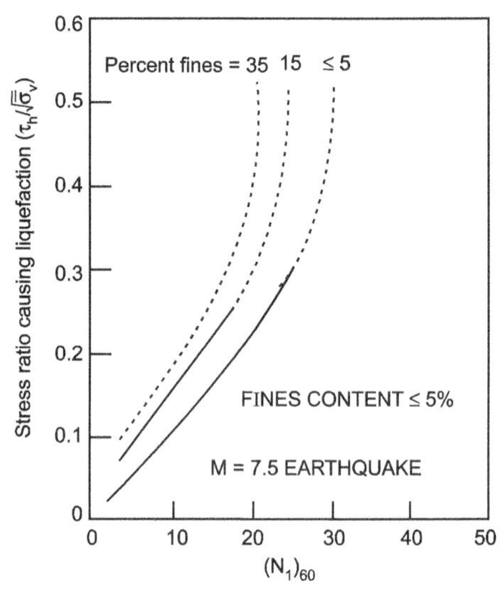

Fig. 9.41. Relationship between stress causing liquefaction and $(N_1)_{60}$, value for silty sands (Seed et al., 1985)

Table 9.8. Correction Factors for the Influence of Earthquake Magnitude on Liquefaction Resistance (Seed et al., 1985)

Earthquake magnitude, M	*Number of representative cycles at 0.6* τ_{max}	$\frac{(\tau_h / \bar{\sigma}_v)_{LM=M}}{(\tau_h / \bar{\sigma}_v)_{LM=7.5}}$
8.50	26	0.89
7.50	15	1.00
6.75	10	1.13
6.00	5-6	1.32
5.25	2-3	1.50

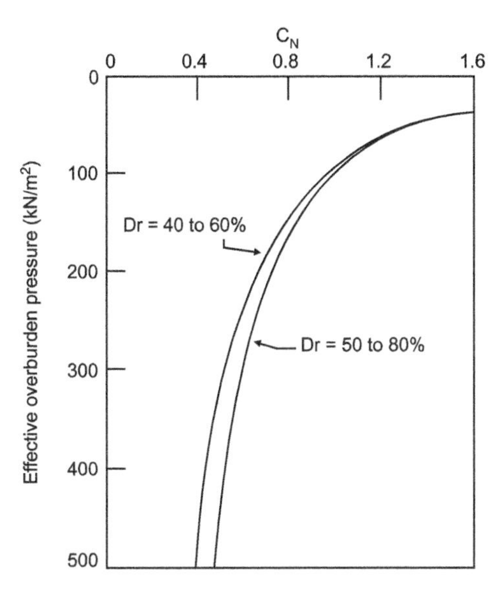

Fig. 9.42. Recommended curves for C_N values (Seed et al., 1985)

The factor of safety against liquefaction, F_1, is then estimated as:

$$F_L = \frac{(\tau_h / \bar{\sigma}_v)_{LM=M}}{(\tau_{av} / \bar{\sigma}_v)} \tag{9.30}$$

(b) Simplified procedure adopted in the Japanese Highway Bridge Code: Iwasaki (1986) introduced the concept of liquefaction resistance factor F_L which is defined as:

$$F_L = \frac{R}{L} \tag{9.31}$$

R is the ratio of in-situ cyclic strength of soil and effective overburden pressure. It depends on relative density, effective overburden pressure and mean particle size. It is given by:

For 0.02 mm < D_{50} < 0.6 mm.

$$R = 0.882\sqrt{\frac{N}{\bar{\sigma}_v + 70}} + 0.225 \log_{10}\left(\frac{0.35}{D_{50}}\right) \tag{9.32a}$$

For 0.6 mm < D_{50} < 2.0 mm

$$R = 0.882\sqrt{\frac{N}{\bar{\sigma}_v + 70}} - 0.05 \tag{9.32b}$$

where N = Observed value of standard penetration resistance

$\bar{\sigma}_v$ = Effective overburden pressure at the depth under consideration for liquefaction examination in kN/m^2.

D_{50} = Mean grain size in mm.

L = Ratio of dynamic load induced by seismic motion and effective overburden pressure. It is given by:

$$L = \frac{a_{max}}{g}\frac{\sigma_v}{\bar{\sigma}_v} r_d \tag{9.33}$$

$$a_{max} = \text{Peak ground acceleration due to earthquake}$$
$$= 0.184 \times 10^{0.320M} (D)^{-0.8} \tag{9.34}$$

where M = Magnitude of earthquake on Richter's scale
D = Maximum epicentral distance in km. (Fig. 9.43)
σ_v = Total overburden pressure
r_d = Reduction factor to account the flexibility of the ground = 1 – 0.015 h
h = Depth of plane below ground surface in m
g = Acceleration due to gravity, m/s^2

For the soil not to liquefy, F_L should be greater than unity.

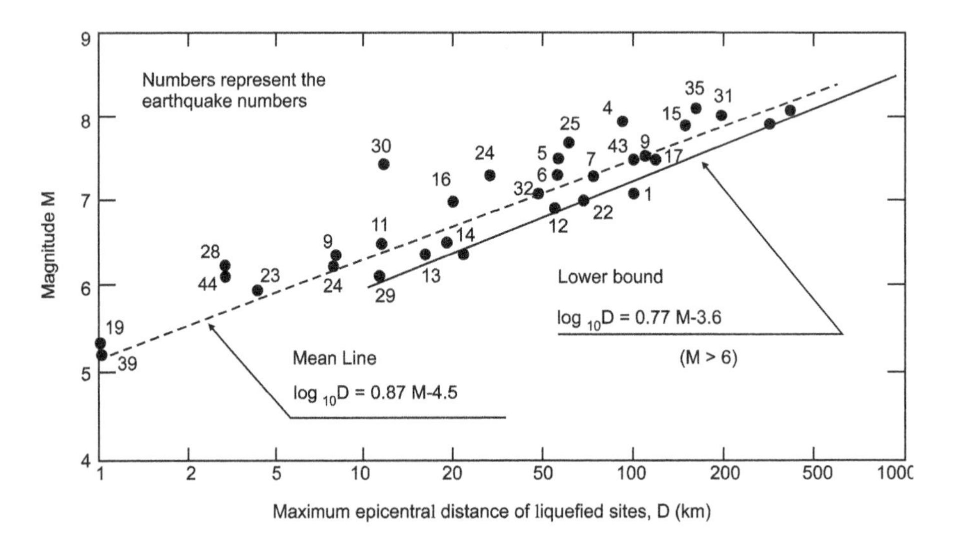

Fig. 9.43. Relationship between the maximum epicentral distance of liquefied site (D) and earthquake magnitude (M) (Kuribayashi, Tatsuoka and Yoshida, 1977)

(c) Chinese Criteria for Liquefaction Evaluation: Correlations between the liquefaction resistance of sand deposits and SPT N-values have also been developed in China (Taiping et al., 1984; Ishihara 1986). These are presented in the form of codal requirements for determining liquefiable sites, expressed in terms of critical penetration resistance, N_{crit}.

$$N_{crit} = \overline{N}[1+0.125(d_s-3)-0.05(d_w-2)] \tag{9.35}$$

where d_s is the depth in metres to the sand layer under consideration, d_w is the depth to the water table in metres, and $\overline{N}$ is a function of the earthquake shaking intensity as summarized in Table 9.9.

Table 9.9. $\overline{N}$ Value as a Function of Earthquake Shaking Intensity (Seed et al., 1985)

Earthquake intensity	*N in blow / ft.*	*Peak ground acceleration*
VII	6	0.10 g
VIII	10	0.20 g
IX	16	0.40 g

In recent years, it has been proposed that this original equation be supplemented by an additional term reflecting the influence of the fines content of the soil, defined in terms of the percent clay content P_c (Taiping et al., 1984).

$$N_{crit} = \overline{N}[1+0.125(d_s-3)-0.05(d_w-2)-0.07P_c] \tag{9.36}$$

The SPT N-values (denoted by $\overline{N}$) associated with this equation are those determined using a "free-fall" type of hammer which is presumed to deliver an energy ratio of about 60%. The field performance data from which Eq. (9.36) has been developed appears to relate to earthquakes with a magnitude of about 7.5.

9.12.2 Evaluation Based on Cone Penetration Resistance

In recent years, there has been considerable interest in using the CPT to evaluate liquefaction susceptibility. Procedures which use CPT resistance are very similar to those using SPT resistance. A boundary can be defined separating liquefiable from non-liquefiable conditions, for example as shown in Fig. 9.44 (Seed and DeAlba, 1986), in which the normalised cone penetration resistance q_{c1} is defined as:

$$q_{c1} = C_q q_c \tag{9.37}$$

where q_{c1} denotes the cone tip resistance corresponding to an overburden pressure of 1 kgf/cm^2, q_c is the cone tip resistance, in the same units, and C_q is a factor obtained from the chart shown in Fig. 9.45. Thus, once the cone penetration resistance q_c for a deposit is known, its liquefaction resistance can be estimated using the charts shown in Figs. 9.44 and 9.45. Several limiting criteria are compared with those proposed by Seed and DeAlba in Fig. 9.46, for clean sands, and Fig. 9.47, for silty sands.

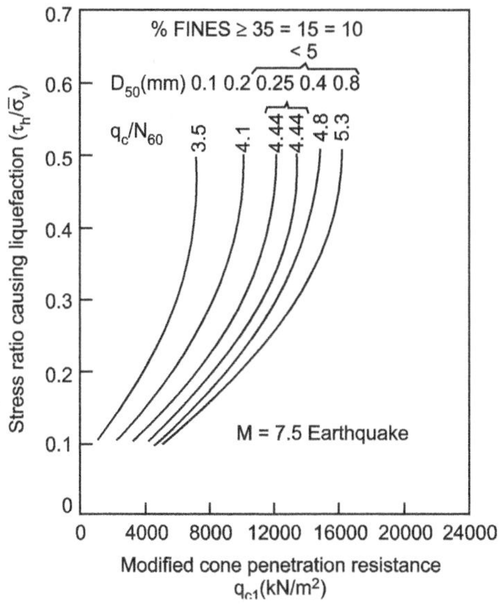

Fig. 9.44. Relationship between stress ratio causing liquefaction and cone tip resistance for sands and silty sands (Seed and DeAlba, 1986)

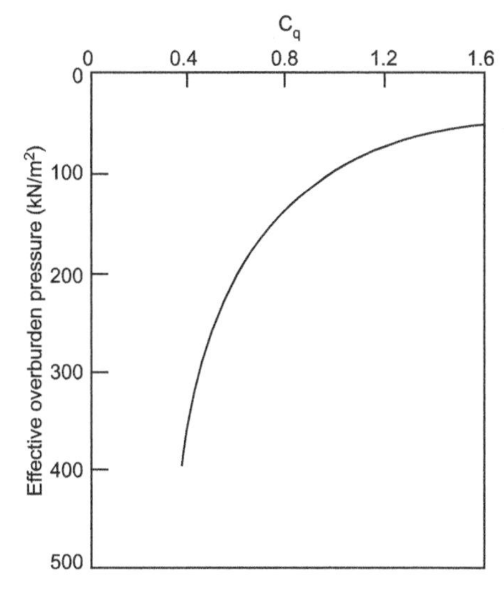

Fig. 9.45. Recommended curve for C_q values (Seed et al., 1983)

9.13 Factors Affecting Liquefaction

Laboratory investigations and observations of field performance during past earthquakes have shown that the liquefaction potential of a soil deposit subjected to earthquake motions depends basically upon the characteristics of the soil, the initial stresses acting on the soil and the characteristics of the earthquake ground motion. The significant factors and their influence on the liquefaction characteristics of sand are discussed below.

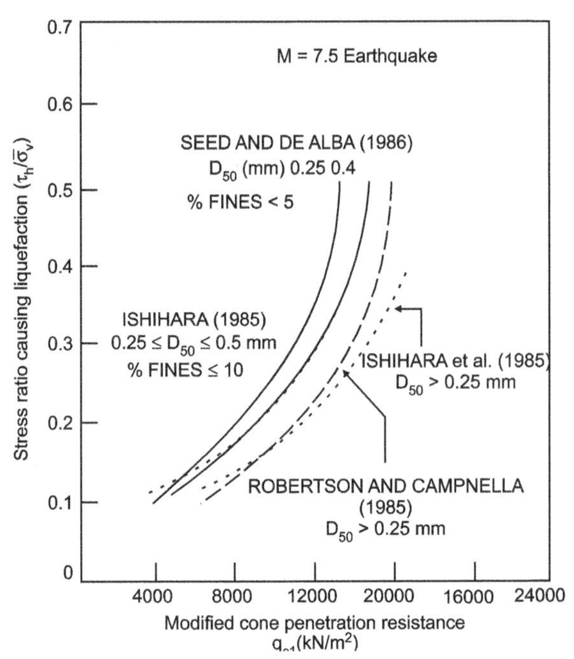

Fig. 9.46. Comparison of proposed boundary curves for liquefaction resistance of clean sands in terms of CPT (modified from Tokimatsu. 1988)

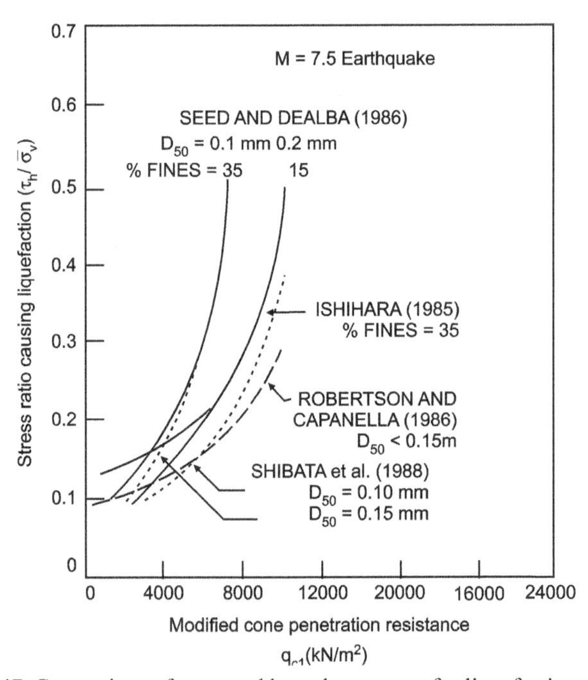

Fig. 9.47. Comparison of proposed boundary curves for liquefaction resistance of silty sands in terms of CPT (modified from Tokimatsu. 1988)

9.13.1 Soil Type

For cohesionless soils, the soil type is perhaps most easily characterised by the grain size distribution. There is some evidence to show that uniformly graded materials are more susceptible to liquefaction than well graded materials (Ross et al., 1969; Lee and Fitton, 1969), and that for uniformly graded soils, fine sands tend to liquefy more easily than do coarse sands, gravelly soils, silts or clays (Lee and Fitton, 1969). In a study of bridge foundation displacements in the Alaska earthquake, Ross et al. (1969) reported that there were no cases of bridge damage due to this cause for structures supported on gravels but there were many cases of damage for bridges supported on sands. Again in the Fukui earthquake, Kishida (1966) reported that liquefaction occurred at a site where the upper 5 m consisted of medium sand but there was no liquefaction at an adjacent site where the soil in this depth range was a sandy silt. These field observations are supported by the results of laboratory cyclic load tests on a wide range of materials (Lee and Fitton, 1969).

9.13.2 Relative Density or Void Ratio

Since the classical work of Casagrande (1936) on the volume changes accompanying shear deformations in cohesionless soils, it has been generally recognised that the susceptibility of a given soil to liquefaction will be determined to a high degree by its void ratio or relative density. In any given earthquake loose sands may liquefy but the same materials in a denser condition may not.

In the Niigata earthquake of 1964, Ohsaki (1966) reported that liquefaction was extensive where the relative density of the sand was about 50 percent but it did not develop in areas where the relative density exceeded about 70 percent. Laboratory test data of all types shows the important influence of this factor on soil behaviour.

9.13.3 Initial Confining Pressure

There is considerable evidence to show that under earthquake loading conditions, in contrast to flow-slide susceptibility under static load conditions, the liquefaction potential of a soil is reduced by an increase in confining pressure. Laboratory tests by numerous investigators (Maslov, 1957; Florin and Ivanov, 1961; Seed and Lee, 1966; Lee and Seed, 1967; Peacock and Seed, 1968; Lee and Fitton, 1969; and Finn et al., 1969) have shown that for a given initial density, the stress required to initiate liquefaction under cyclic load conditions increases with the initial confining pressure. The effect was also shown in the field during the Niigata earthquake where soil under a 9 ft fill remained stable but similar soils surrounding the fill liquefied extensively (Seed and Idriss, 1967).

9.13.4 Intensity of Ground Shaking

For a soil in a given condition and under a given confining pressure, the vulnerability to liquefaction during an earthquake depends on the magnitude of the stresses or strains induced in it by the earthquake; these in turn are related to the intensity of ground shaking. The significance of the applied stresses has been shown by many laboratory investigations but the important effect of ground shaking intensity in the field is very well illustrated by the soil behaviour at Niigata in Japan. The relatively loose sands in Niigata have been shaken by 25 earthquakes in the past 370 years, historical records show only three occasions in which liquefaction has been reported in or near Niigata itself; on these occasions the estimated ground accelerations were in excess of 0.13 g, culminating with extensive liquefaction in 1964 when ground accelerations had their probable maximum value of 0.16 g. Of special significance, however, is the fact that in 22 other earthquakes producing estimated ground accelerations ranging from 0.005 g to 0.12 g, there was no indication of any soil liquefaction in the city (Kawasumi, 1968). The intensity of ground motions must thus be considered an important factor in evaluating soil liquefaction potential.

9.13.5 Duration of Ground Shaking

The duration of ground shaking is a significant factor in determining liquefaction potential since it determines in a general way the number of significant stress or strain cycles to which a soil is subjected. All laboratory studies of soil liquefaction under cyclic loading conditions show that for any given stress or strain level, the onset of liquefaction depends on the application of a requisite number of stress or strain cycles. In the field, the importance of this is perhaps best illustrated by the landslides which were triggered by liquefaction in anchorage during the Alaska earthquake of 1964. These slides did not occur until about 90 seconds after the earthquake motions started indicating the need for development of a sufficient number of stress cycles to induce liquefaction and instability. Clearly if the duration of ground shaking had been only 45 seconds, no liquefaction or soil instability would have developed.

9.13.6 Location of Drainage and Dimensions of Deposit

Sands are generally more pervious than fine-grained soils. However, if a pervious deposit has large dimensions the drainage path increases and under quick loading conditions during an earthquake, the deposit may behave as if it were undrained. Therefore the chances of liquefaction are increased in such a deposit. The introduction of gravel drains to stabilise a potentially liquefiable sand deposit has been proposed by Yoshimi and Kuwabara (1973). Seed and Booker (1976) and Seed (1976) have proposed an analytical procedure for designing such drainage. The drains are considered fully effective if the material of which they are constructed is about 200 times more permeable than the soil in which they are installed. The drainage path is reduced by the introduction of these gravel drains.

9.13.7 Method of Soil Formation

Sands are generally known not to display a characteristic structure as do fine-grained soils, such as clays. But investigations carried out by Ladd (1976) and others have demonstrated that liquefaction characteristics of saturated sands under cyclic loading are significantly influenced by the method of sample preparation and by soil structure. It is shown by Seed (1976) that, depending on the method of sample preparation, the stress condition required to cause liquefaction in a given number of stress cycles for samples of the same sand at the same density may vary as much as 200 percent. It will, therefore, be necessary to simulate the orientation of soil particles and the soil fabric in the laboratory. More research is needed on the definition of soil fabric in quantitative terms and on the methods of reproducing it in the laboratory.

9.13.8 Period under Sustained Load

The age of a sand deposit may influence its liquefaction characteristics. A study of liquefaction in an undisturbed sand and its freshly prepared sample indicates that the liquefaction resistance may increase by 75 percent (Seed, 1976). Lee (1975) explains this strength increase as being due to some form of cementation or welding which may occur at contact points between sand particles, and as being associated with secondary compression of the soil. This effect must be recognised as different from that due to orientation of soil particles in the soil fabric.

9.13.9 Previous Strain History

Sands may be subjected to some strains due to earthquakes. To determine the effect of previous strain history, studies of the liquefaction characteristics of freshly deposited sand and of a similar deposit previously subjected to some strain history were conducted in simple shear by Finn et al. (1970). It was found that liquefaction characteristics were influenced by the strain undergone previously. Seed (1976) showed that although the prior strain history caused no significant change in the density of the sand, it increased the stress that causes liquefaction by a factor of about 1.5. Much larger increases have been shown to result from more severe restrain conditions (Bjerrum, 1973; Lee and Focht, 1975).

9.13.10 Entrapped Air

If air is entrapped in water in which pore pressures develop, part of it is dissipated in compression of the entrapped air. Hence, entrapped air helps to reduce the possibility of liquefaction.

9.14 Anti-liquefaction Measures

A comprehensive study is required to find out various possible measures to prevent liquefaction. Though it depends on a number of factors, however, few can be controlled in field. Based on these, certain methods have been suggested (Lew, 1984). Liquefaction resistance to some extent can be improved by:

9.14.1 Compaction of Loose Sands

As has been indicated earlier, loose saturated sands are more prone to liquefaction than dense saturated sands. Therefore, the liquefaction potential can be reduced by compacting the loose sand deposit before any structure is constructed. The various methods suggested for compaction of loose sands in situ are:

9.14.1.1 Rolling with Rubber Tyre Rollers

It may be accomplished by excavating some depth, then carefully backfilling in controlled lift thickness and compacting the soil. When rubber tyres are used, lifts are commonly 150 mm to 200 mm. This method, however, cannot be used for compacting deep sand deposits.

9.14.1.2 Compaction with Vibratory Plates and Vibratory Rollers

Compaction of cohesionless soils can be achieved using smooth wheel rollers commonly with a vibratory device inside. Lift depths upto about 1.5 m to 2 m can be compacted with this equipment (Bowles, 1982). Also plates mounted with vibratory assembly can be used; however, small thickness of soils can be compacted by these methods and they cannot be used for large deposits.

9.14.1.3 Driving of Piles

Piles when driven in loose deposits, compacts the sand within an area covered by eight times around it. This concept may be utilized in compacting the site having loose sand deposits. As pile remains in the sand, the overall stiffness of the soil stratum increases substantially.

9.14.1.4 Vibrofloatation

The method is most commonly used to densify cohesionless deposits of sands and gravel with having not more than 20% silt or 10% clay. Vibrofloatation utilizes a cylindrical penetrator. It is an equipment of about 4 m long and 400 mm in diameter. The lower half is vibrator and upper half is stationary part. The device has water jets at top and bottom. Vibrofloat is lowered under its own weight with bottom jet on which induces the quick sand condition; when it reaches the desired depth, the flow of water is diverted to upper jet and vibrofloat is pulled out slowly. Top jet aids the compaction process. As the vibrofloat is pulled out a crater is formed. Sand or gravel is added to the crater formed.

9.14.1.5 Blasting

The explosion of buried charges induces liquefaction of the soil mass followed by escape of excess pore water pressure which acts as a lubricant to facilitate re-arrangement and thus leading the sand to a more compacted state.

The earliest use of detonating buried charges of explosive for compacting loose cohesionless soils in their natural state has been reported by Lyman (1942). He concluded that:

(i) Lateral distribution of charges should be based on results obtained from a series of single shots.
(ii) Where loose sands greater than 10 m thick are to be compacted, two or more tiers of small charges are preferred.
(iii) For deposits less than 10 m thick, charges placed at 2/3rd depth from surface will generally suffice.
(iv) There is no apparent limit of depth that can be compacted by means of explosive.

Later Hall (1962) reported that

(i) Repeated blasts are more effective than a single blast of several small changes detonated simultaneously.
(ii) Very little compaction can be achieved in top 1 m.
(iii) Small charges are more effective than large charges for compacting upper 1.5 m of sand.
(iv) The compaction gained by repeating the blasts more than three times is small.
(v) The relative densities can be increased to 80%.

9.14.2 Grouting and Chemical Stabilization

Grouting is a technique of inserting some kind of stabilizing agent into the soil mass under pressure. The pressure forces the agent into the soil voids in a limit space around the injection tube. The agent reacts with the soil and/or itself to form a stable mass. The most common grout is a mixture of cement and water, with or without sand. Generally grout can be used if the permeability of the deposit is greater than 10^{-5} m/s. Chemical stabilization is in the form of lime, cement, flyash or combination of these.

9.14.3 Application of Surcharge

Application of surcharge over the deposit liable to liquefy can also be used as an effective measure against liquefaction. Figure 9.48 shows a plot between rise in pore pressure and effective over burden pressure at an acceleration of ten percent of *g*. It indicates that pore pressure increases with increase in overburden pressure till a maximum value of pore pressure is reached, after which it starts decreasing with further increase in surcharge. Thus an overburden pressure above this value, depending upon the situation, makes the deposit safe against liquefaction.

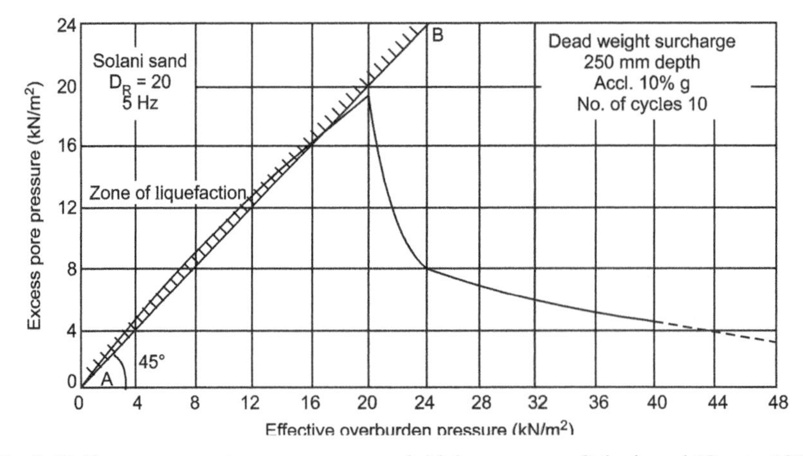

Fig. 9.48. Excess pore water pressure versus initial pressure on Solani sand (Gupta, 1979)

9.14.4 Drainage Using Coarse Material Blanket and Drains

Blankets and drains of material with higher permeability reduces the length of drainage path and also due to higher coefficient of permeability, speed up the drainage process (Katsumi et al., 1988; Susumu et al., 1988).

A comparative study of the various soil improvement methods, their effective depths and applications, as reported by Suematsu et al. (1982), is presented in Table 9.10.

Table 9.10. Soil Improvement Methods and their Applications (Suematsu et al., 1982)

Principle	*Counter measure method*	*Depth approximate applicable*	*Effectiveness*	*Associated problems*	*Remarks*
Densification	Vibro-floatation	upto 20 m	upto N-value of 15 to 20	less troublesome because of horizontal vibration	permeability may increase in some cases
	Sand compaction piles	upto 35 m	upto N-value of 25 to 30	troublesome because of vertical vibration	permeability may increase in some cases
	Blasting	upto 20 m	upto Relative Density of 70 to 80 %	large shock	execution management is difficult
	Rod compaction	upto 20 m	upto N-value of 15 to 20	troublesome because of vertical vibration	in-situ sands are often used
	Dynamic consolidation (heavy weight dropping)	upto 10 m	apt to execution practices	large shock	compaction for shallow layers
	Vibro-tamper	upto 3 m	apt to execution practices	less troublesome	compaction for shallow layers, other methods may be used jointly
	Roller compaction	upto 0.2 to 0.3 m, Shallower than water table	upto relative Density of 95 %	less troublesome	water proofing measures necessary for depth below water table
	Pile driving	upto 10 to 12 m	apt to execution practices	vibration and noise from diesel pile hammer	
	Chemical piles	upto 20 m	apt to execution practices	less troublesome	

Particle improvement of hardening	Substitution	generally upto 5 m	effective by using gravels	less troublesome	
	Injection hardening	upto the depth of boring	apt to execution management	influences neighbouring structures	execution management is difficult in case of cement grouting
	Surface mixing treatment	upto 5 m	apt to mixing contents	less	
	Deep mixing treatment	upto 30 m	apt to mixing contents	less	
Decrease in degree of saturation	Well point	lowers water table by 5 to 6 m	uncertainty of permeability	lower water table at neighbouring sites	long time operation is necessary
	Deep point	lowers water table by 15 to 20 m	uncertainty of permeability	lowers water table at neighbouring sites	long time operation is necessary
Dissipation of pore water pressures	Gravel drain	upto 20 m		less	retrofitting of existing structures
Restraint of shear deformation	Sheet pile	upto 10 m	difficult, to assess the effects of restraint	vibration from driving of sheet piles	underground continuous walls may be used instead of sheet piles

9.15 Studies on Use of Gravel Drains

Yoshimi and Kuwabara (1973) were first to introduce gravel drains to stabilize a potentially liquefiable sand deposit. Seed and Brooker (1976) have proposed an analytical procedure for designing such drains (Fig. 9.49). These drains are considered fully effective if the permeability of material of drains, k_d is about 200 times the permeability, k_s, of the soil in which they are installed i.e. $(k_d/k_s) > 200$. The effective drainage path is reduced by the introduction of number of artificial drains. Seed and Brooker (1976) developed nondimensional charts as shown in Fig. 9.50 for determining the spacing of drains. Various terms shown on this figure are as below:

$$r_g = \frac{\text{Limiting value of } u_g \text{ chosen for design}}{\sigma_v}$$

u_g = Excess porewater pressure build up in a cyclic simple shear test (Fig. 9.46)
σ_v = Initial consolidation pressure
N_S = Number of cyclic stress applications
N_1 = Number of stress cycles needed for liquefaction
R_d = Radius of rock or gravel drains
R_e = Effective radius of the rock or gravel drains

$$T_{ad} = \frac{K_h}{\gamma_\omega} \cdot \left(\frac{t_d}{m_v R_d^2} \right)$$

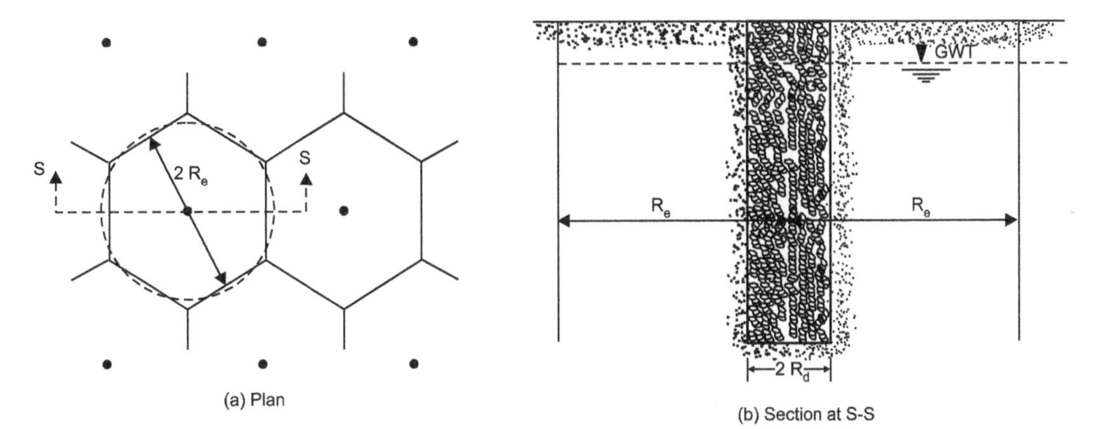

Fig. 9.49. Gravel drains

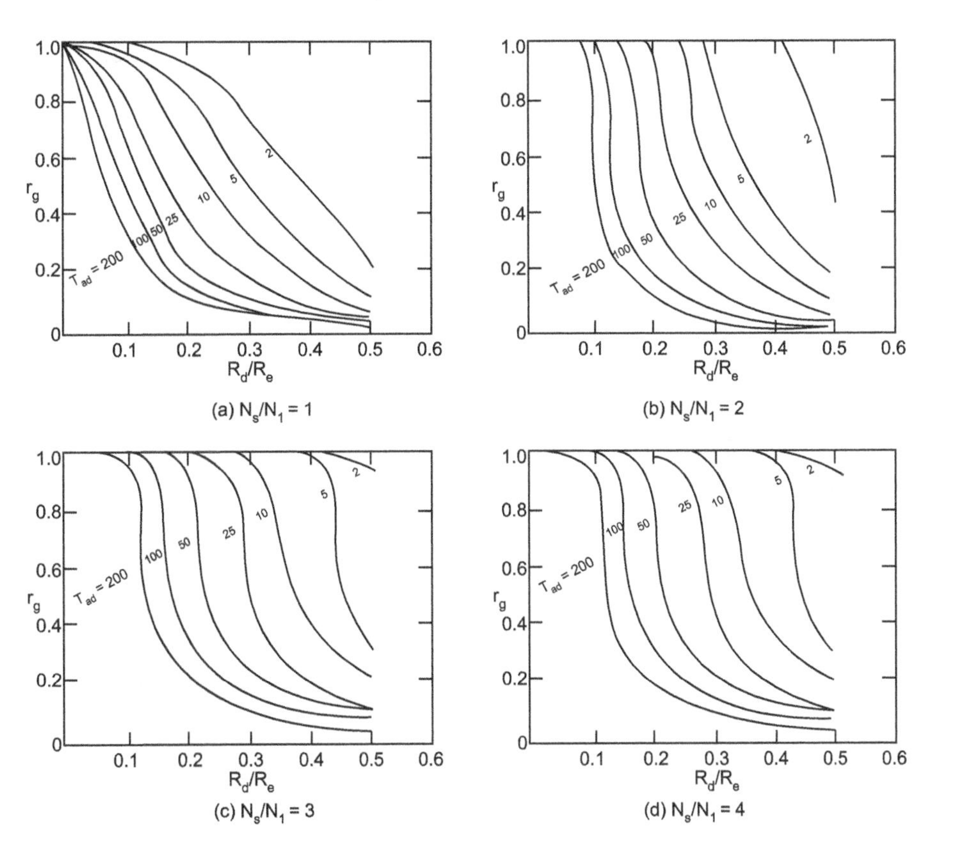

Fig. 9.50. Relation between greatest porewater pressure ratio and drain system parameters

K_h = Coefficient of permeability of sand in horizontal direction
γ_ω = Unit weight of water
m_v = Coefficient of volume compressibility of sand
t_d = Duration of earthquake

Yosufumi et al. (1984) had developed a method to evaluate liquefaction resistance under partially drained condition. They assumed that the dissipation of excess pore water pressure induced by an earthquake will occur according to Darcy's law. Dynamic triaxial apparatus was used to conduct tests under perfectly undrained and perfectly drained conditions. In case of perfectly

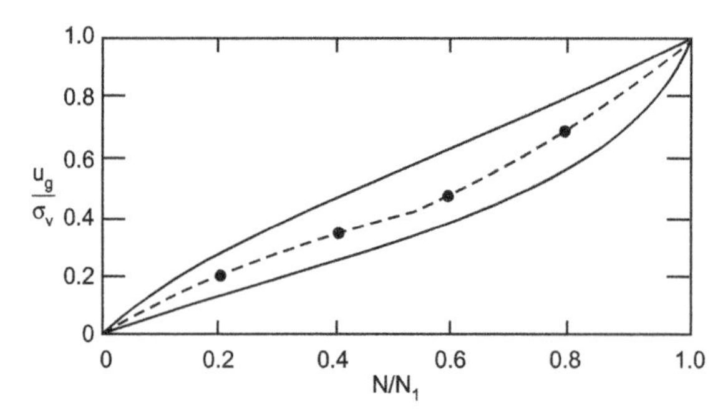

Fig. 9.51. Rate of pore water pressure buildup in cyclic simple shear test

undrained condition, liquefaction resistance was neither influenced by permeability of sample nor by the frequency of loading. This is reasonable, because, no drainage of water is involved and hence permeability of soil does not have a role to play. It was also concluded that in case of partially drained condition, the effect of drainage and frequency is remarkable for soils with relatively larger relative density. D_R. and not so significant for soils with smaller relative density. But for loose soils rate of generation of pore pressure depends on number of cycles of stresses which in turn depends on frequency. Rate of pore pressure built up depends on the rate of dissipation of pore pressure, which is based on drainage.

Yasushi and Taniguchi (1982) carried out large scale model tests to confirm the effectiveness of gravel drains for preventing liquefaction of sand deposits. The purpose of the tests, as stated by them, was:

(i) to know the generation and dissipation characteristics of pore water pressure,

(ii) to clarify the effective area of the gravel drain from the view point of preventing liquefaction and

(iii) to know whether the gravel drain is effective in preventing the liquefaction of subsoils under a road that is partially buried.

They performed tests on shaking table of 12 m × 12 m × 3 m (deep) size filled with cohesionless soil. The acceleration of loading was 200 gals, the duration of shaking was one minute and the frequency was 5 cps. They concluded that pore water pressure within 500 mm from the edge of a gravel drain is much smaller than that, for away from the gravel drain.

Wang (1984) made experimental study on liquefaction inhibiting effect of gravel drains. A shaking box of size 1.5 m × 0.28 m × 0.5 m was used. He used gravel drains wells under the foundation and it was assumed that under plain strain condition the wells are referred as drains. It was noted by him that the section of non-liquefied zone of deposit was basically a trapezoid in which pore pressure ratio (i.e. the ratio of excess pore water pressure to the effective overburden pressure) was generally below 0.6. Basically this zone did not reduce with increasing vibration time.

As the number of drains installed is increased, the non-liquefied zone increases. As the acceleration increases, the zone reduces gradually but the increase in time does not reduce the non-liquefied zone. The angle of trapezoid was found to be 15° to 17° in the direction of depth. The zone is about 40 mm outside the drains. It was also observed by him that the surface drains may effectively prevent foundation settlement. In order to obtain good effect in reducing foundation settlement it must be ensured that adequate depth and width of drains be installed when installing shallow drains and outside drains.

O-hara and Tamamoto (1987) presented a fundamental study on gravel pile for preventing liquefaction. They used a shaking box of size 1.0 m × 0.35 m × 0.65 m (deep). Radii of gravel piles

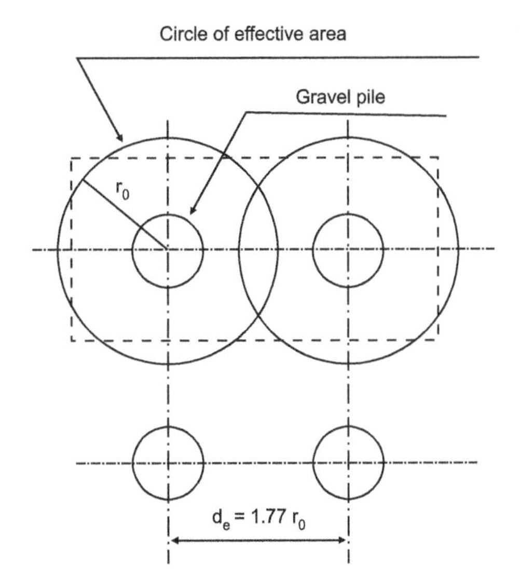

Fig. 9.52. Radius and distance (or spacing) of gravel piles (O-hara, 1987)

were 0.75 m, 0.15 m etc. The flow of pore water was assumed to be horizontal. They measured the pore water pressure at points near the drains and away from the drains. They concluded that liquefaction occurred at points too far from the drain and that at points close to the gravel drain, liquefaction did not occur. Results obtained by them presented in form of optimum radius of pile and optimum spacing between gravel piles. Figure 9.52 shows the effective circle which is defined as the circle with area equal to area of square with sides equal to the line joining mid points of the spacing between adjacent gravel piles. The sides of the squares are taken as optimum spacing between gravel piles. They found that the effective area of gravel pile increases in proportion to the diameter of the gravel pile and the permeability. For a fixed diameter of pile and permeability of soil, as the optimum distance decreases, the pore pressure ratio decreases. As the permeability increases, pore pressure ratio decreases very sharply. Highly permeable gravel drains are much more effective even at higher optimum distance and smaller diameter of drains.

A flexible vertical drain formed by using organic fibres like jute or coir has been used in several projects. The most important properties of such drains are permeability and tensile strength. The jute filter cover has permeability better than 10^{-5} m/sec. This facilitates the flow of water from pervious lenses present in the seams and layer of sand and speeds up the pore water pressure dissipation. They have the advantage of decaying and getting mixed with the soil without harming the environment. When the filter permeability is large, the clogging of the drain has to be considered.

Geotextiles are used fairly widely in surface and subsurface and installation (Krishnaswamy and Issac, 1995). Crushed stone wrapped in geotextiles have often been used as surface and subsurface drains. Perforated plastic pipes too may be used for this purpose. They may be filled with crushed stones, if necessary.

9.16 Evaluation of the Effects of Liquefaction

For engineering purposes, it is not usually the occurrence of liquefaction itself that is important but its consequences for damage to the ground or adjacent structures. Two approaches are available for identifying the effects of liquefaction, based on data from site specific geotechnical investigations.

(a) **Damage in the Presence of an Unliquefiable Surface Layer or Crust:** To decide whether liquefaction will or will not cause damage on the ground surface, the thickness of the liquefiable layer can be compared with the thickness of the surface crust using criteria such as that shown in Fig. 9.48 (Ishihara. 1985). If the thickness of the surface layer, H_1 , is larger than that of the underlying liquefied layer, resulting damage on the ground surface may be insignificant. If the water table is below the ground surface the definition of H_1 depends on the nature of the superficial deposits, as shown in Fig, 9.53. For a deposit of sandy soil, the thickness H_1 can be taken to be equal to the depth of the water table.

(b) Liquefaction Potential Index: Iwasaki et al. (1982) quantified the severity of possible liquefaction at any site by introducing a factor called the liquefaction potential index. P_L, defined as

$$P_L = \int_0^{20} F(z)w(z)dz \tag{9.38}$$

where z is the depth below the ground surface, measured in metres, $F(z)$ is a function of the liquefaction resistance factor. F_L, where $F(z) = 1 - F_L$, but if $F_L > 1.0$, $F(z) - 0$, and $w(z) = 10 - 0.5\ z$. Equation (9.38) gives values of P_L ranging from 0 to 100. By calculating this index for 63 liquefied and 22 nonliquefied sites in Japan Iwasaki et al. concluded that sites with P_L values greater than about 15 suffer severe liquefaction damage, whereas damage is minor at sites with a value of P_L less than about 5.

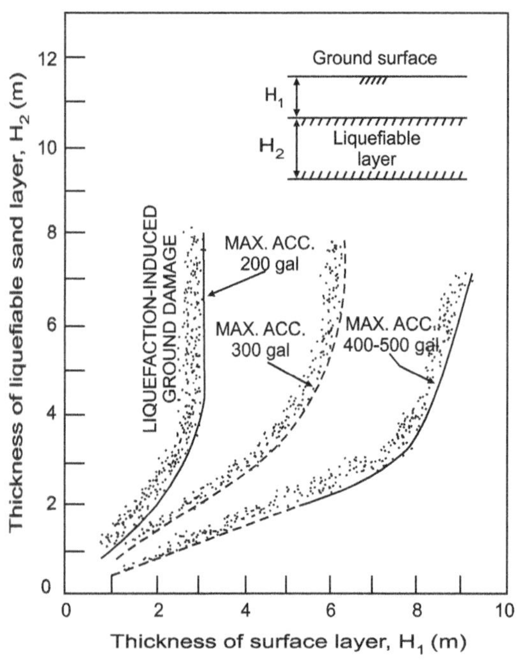

Fig. 9.53. Proposed boundary curves for surface manifestation of liquefaction induced damage. (Ishihara, 1985)

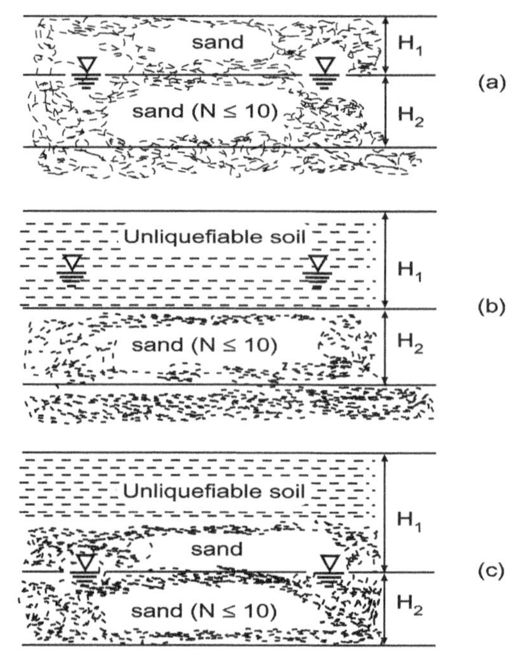

Fig. 9.54. Definitions of the surface unliquefiable layer and the underlying liquefiable sand layer. (Ishihara, 1985)

Illustrative Examples

Example 9.1

At a given site, a boring supplemented with standard penetration tests was done upto 15.0 m depth. The results of the boring are as given below:

Depth (m)	*Classification of soils*	D_{50} *(mm)*	*N-Value*	D_R *(%)*	*Remarks*
1.5	SP	0.18	3	19	(i) Position of water table lies 1.5 m below the ground surface
3.0	SP	0.20	5	30	
4.5	SM	0.12	6	35	
6.0	SM	0.14	9	40	
7.5	SM	0.13	12	45	(ii) γ_{moist} = 19 kN/m^2
9.0	SP	0.16	17	52	γ_{sub} = 10 kN/m^3
10.5	SW	0.20	20	52	
12.0	SW	0.22	18	46	
13.0	SW	0.22	24	60	(iii) SW – < 5%times
15.0	SW	0.24	30	65	SP – 10% times
					SM – 15% times

The site is located in seismically active region, and is likely to be subjected by an earthquake of magnitude 7.5. Determine the zone of liquefaction using

(a) Seed and Idriss (1971) method
(b) Seed et al. (1985) method
(c) Iwasaki (1986) method

Solution

(a) Seed and Idriss (1971) method

(i) From Fig. 9.43, For $M = 7.5$, D = 106 km

From Eq. 9.31, $\frac{a_{\max}}{g} = \frac{0.184}{9.81} \times 10^{(0.302 \times 7.5)}\ 106^{-0.8} = 0.083$

Number of significant cycles (Table 9.3), $N_s = 20$ (For $M = 7.5$)

(ii) From Eq. (9.25), $\tau_{av} = 0.65\gamma h \cdot \frac{a_{\max}}{g} \cdot r_d$

$= 0.65 \times \gamma h \times (0.083) r_d = 0.054\ \gamma h\ r_d$

It may be noted that $\gamma\ h$ represents the total stress at depth h below ground surface. Value of r_d are read from Fig. 9.25. Values of total stresses and τ_{av} at different depths are given in Cols, 3 and 5 of Table 9.11.

(iii) For 50% relative density, the stress ratio $\sigma_d/2\sigma_3$ is read from Figs. 9.22 *a* and *b* for given values of D_{50}. Average of the two values is the stress ratio for number of significant cycles equal to 20. The stress causing liquefaction at any depth is then computed using Eq. 9.22.

$$\tau_h = \left(\frac{\sigma_d}{2\bar{\sigma}_3}\right)_{50\%} \cdot \frac{D_R}{50} \cdot C_r \cdot \bar{\sigma}_v$$

Values of C_r are obtained using Table 9.2. The details of computations of τ_h are summarised in Table 9.12.

Table 9.11. Computation of τ_{av}

S. No.	*Depth (m)*	*Total stress (kN/m²)*	r_d	τ_{av} *(kN/m²)*
(1)	(2)	(3)	(4)	(5)
1.	1.5	28.50	0.99	1.52
2.	3.0	28.50	0.98	3.10
3.	4.5	88.50	0.97	4.64
4.	6.0	118.50	0.97	6.21
5.	7.5	148.50	0.96	7.69
6.	9.0	178.50	0.95	9.14
7.	10.5	208.50	0.94	10.54
8.	12.0	238.50	0.92	11.90
9.	13.5	268.50	0.91	13.23
10.	15.0	298.50	0.90	14.51

Table 9.12. Computations of τ_h from Seed and Idriss (1971) method

S. No.	*Depth (m)*	*Effective stress* $\bar{\sigma}_v$ *(kN/m²)*	D_R	C_r	$\tau_h / \bar{\sigma}_v$	τ_h *(kN/m²)*
1.	1.5	28.50	19	0.55	0.2198	1.31
2.	3.0	43.50	30	0.55	0.2260	3.24
3.	4.5	58.50	35	0.55	0.2012	4.53
4.	6.0	73.50	40	0.55	0.2074	6.71
5.	7.5	88.50	45	0.555	0.2043	9.03
6.	9.0	103.50	52	0.573	0.2136	13.17
7.	10.5	118.50	52	0.573	0.2260	15.96
8.	12.0	133.50	46	0.556	0.2298	15.69
9.	13.5	148.50	60	0.60	0.2298	24.57
10.	15.0	163.50	65	0.61	0.3336	30.29

(b) Seed et al. (1985) Method

(i) In this method, the value of shear stress at any depth induced by the earthquake is obtained exactly in the same manner as illustrated in Seed and Idriss (1971) method (Table 9.10)

(ii) To determine τ_h firstly *N*-values are corrected for effective overburden pressure using Fig. 9.42. The stress ratio $\tau_h / \bar{\sigma}_v$ is then obtained using the relevant curve of Fig. 9.41 for the given value of corrected N. The details of computations are given in Table 9.13.

Table 9.13. Detail of Computations of τ_h by Seed (1979) method

S. No.	*Depth (m)*	*N-Value*	*Corrected N*	*Effective stress*	$\tau_h / \bar{\sigma}_v$	τ_h
(1)	(2)	(3)	(4)	(5)	(6)	(7)
1.	1.5	3	4	28.50	0.062	1.77
2.	3.0	5	6	43.50	0.088	3.83
3.	4.5	6	7	58.50	0.012	7.02
4.	6.0	9	10	73.50	0.150	11.02
5.	7.5	12	13	88.50	0.185	16.37
6.	9.0	17	17	103.50	0.223	23.08
7.	10.5	20	18	118.50	0.194	22.09
8.	12.0	18	17	133.5+0	0.183	24.43
9.	13.5	24	20	148.50	0.225	31.93
10.	15.0	30	25	163.50	0.269	43.98

(c) Iwasaki's Method

(i) Firstly the value of factor R is obtained using Eq. 9.32a.

$$R = 0.882\sqrt{\frac{N}{\bar{\sigma}_v + 70}} + 0.225 \log_{10}\left(\frac{0.35}{D_{50}}\right)$$

The details of computations of factor R are given in Table 9.14.

Table 9.14. Details of Computations for Obtaining Factor R in Iwasaki's Method

S. No.	*Depth (m)*	$\bar{\sigma}_v$ *(kN/m²)*	D_{50} *(mm)*	N	R
(1)	(2)	(3)	(4)	(5)	(6)
1.	1.5	28.50	0.18	3	0.2189
2.	3.0	43.50	0.20	5	0.2398
3.	4.5	58.50	0.12	6	0.2950
4.	6.0	73.50	0.14	9	0.3104
5.	7.5	88.50	0.13	12	0.3394
6.	9.0	103.50	0.16	17	0.3525
7.	10.5	118.50	0.20	20	0.3419
8.	12.0	133.50	0.22	18	0.3077
9.	13.5	148.50	0.22	24	0.3377
10.	15.0	163.50	0.24	30	0.3464

(ii) The factor L is then obtained using Eq. 9.33:

$$L = \frac{a_{max}}{g} \cdot \frac{\sigma_v}{\bar{\sigma}_v} \cdot r_d$$

The details of computations are given in Table 9.15. The ratio of factor of safety R/L is listed in the last column of table 9.15.

Table 9.15. Details of Computations of Obtaining Liquefaction Potential by Iwasaki (1986) Method

S. No.	*Depth* (m)	$\sigma_v / \bar{\sigma}_v$	r_d	L	F_L
(1)	(2)	(3)	(4)	(5)	(6)
1.	1.5	1.000	0.99	0.0820	2.670
2.	3.0	1.345	0.98	0.1092	2.196
3.	4.5	1.513	0.97	0.1215	2.427
4.	6.0	1.612	0.97	0.1295	2.397
5.	7.5	1.678	0.96	0.1334	2.544
6.	9.0	1.725	0.95	0.1357	2.597
7.	10.5	1.758	0.94	0.1369	2.498
8.	12.0	1.787	0.92	0.1362	2.260
9.	13.5	1.808	0.91	0.1363	2.478
10.	15.0	1.826	0.90	0.1361	2.545

In Table 9.16, summary of different methods are given. It is evident from this table that liquefaction occurs only upto 1.5 m depth according to Seed and Idriss (1971) method, Seed et al. (1985) method; and no liquefaction according to Iwasaki's method.

Table 9.16. Summary of Different Methods

S. No.	*Depth (m)*	τ_{av} *(kN/m²)*	τ_h *Seed and Idriss method (1971) (kN/m²)*	τ_h *Seed et al. (1985) (kN/m²)*	*Iwasaki (1986) method* F_L (R/L)
1.	1.5	1.52	1.31	1.28	2.670
2.	3.0	3.10	3.24	2.91	2.196
3.	4.5	4.64	4.53	4.62	2.427
4.	6.0	6.21	6.71	8.30	2.397
5.	7.5	7.69	9.03	12.92	2.544
6.	9.0	9.14	13.17	19.77	2.597
7.	10.5	10.54	15.96	22.99	2.498
8.	12.0	11.90	15.69	24.43	2.260
9.	13.5	13.23	24.57	31.93	2.478
10.	15.0	14.51	30.29	43.98	2.545

References

Arya. A.S., Nandakumaran, P., Puri, V.K. and S. Mukherjee (1978), "Verification of liquefaction potential by field blast tests", Proc. 2nd International Conference on Microzonation, Seattle, USA, Vol. II, p. 865.

Bjerrum, L. (1973), "Geotechnical Problems Involved in Foundation of Structures in the North Sea", Geotechnique, Vol 23, No. 3.

Bowles. J.E. (1982), "Foundation analysis and design", McGraw Hill Book Co. Singapore.

Casagrande, A. (1965), "The role of the calculated risk in earthwork and foundation engineering", J. Geotech, Engg. Div., ASCE. Vol. 91. No. SM4, pp. 1-40, Proc. Paper 4390.

Castro, G. (1969), "Liquefaction of sands", Harvard Soil Mechanics Series No. 81, Harvard University, Cambridge, Massachusetts.

Castro, G. (1975), "Liquefaction and cyclic mobility of saturated sands", Journal of the Geotechnical Engineering Division, ASCE, 101 (GTS), pp. 551-569.

Castro, G. and S.J. Poulos (1976), "Factors affecting liquefaction and cyclic mobility ". Symposium on Soil Liquefaction, ASCE, National Connvention, Philadelphia, pp. 105-138.

Christian, J.T. and W.F. Swiger (1976), "Statistics of liquefaction and S.P.T. results", J. Geotech. Engg. Div., ASCE, Vol. 101, No. G T 11, pp. 1135-1150.

Corps of Engineers, U.S. Dept. of the Army (1939), "Report of the slide of a portion of the upstream face at fort peck dam", U.S. Government Printing Office, Washington, D. C.

DeAlba. P., H-B. Seed and C.K. Chan (1976), "Sand liquefaction in large scale simple shear tests", J. Geotech. Engin. Div., ASCE. Vol. 102, No. GT9. pp. 909-927.

Dunn, J.A., J.B. Auden and A. M. N. Ghosh (1939). "The Bihar Nepal earthquake of 1934", Mem. Geol. Surv., India. Vol. 73, p. 32.

Finn, W.D.L., J.J. Emery and Y.P. Gupta (1970), "A shaking table study of the liquefaction of saturated sands during earthquakes", Proc. Third Europ. Symp., Earthquake Engin., pp. 253-262.

Finn. W.D.L., P.L. Bransby and D.J. Pickering (1970), "Effect of strain history on liquefaction of sands". J. Soil Mech. Found. Div., ASCE, Vol. 96, No. SM 6, pp. 1971-1934.

Finn, W.D. Liam., Pickering, D.J. and Bransby, P.L. (1969), "Sand Liquefaction in Triaxial and Simple Shear Tests", Soil Mechanics Research Report Series No. 11, University of British Columbia, Vancouver. Canada.

Finn, W.D.L., Bransby, P.L. and Pickering, D.J. (1970), "Effect of Strain History on Liquefaction of Sands", Journal Soil Mech. and Found. Division, ASCE, Vol. 96, No. SM6.

Florin. V.A. and P.L. Ivanov (1961), "Liquefaction of saturated sandy soils", Proc. Fifth Int. Conf. Soil Mech. Found. Engin., Paris, Vol. l, pp. 107-111.

Geuze, E. (1948), "Critical density of some dutch sands", Proc. 2nd ICSMFE, Vol. III, pp. 125-130.

Gupta, M.K. (1979), "Liquefaction of sands during earthquakes", Ph.D. Thesis, University of Roorkee. India.

Gupta, M.K. and S. Mukherjee (1979), "Blast tests for liquefaction studies", Proc. International Symposium on Insitu Testing of Soil and Rock and Performance of Structures, Roorkee, India, Vol. 1, p. 253.

Hall, E.C. (1962), "Compacting of a dam foundation by blasting", ASCE Journal, Vol. 80.

Hazen, A. (1920), "Hydraulic fill dams", ASCE Transactions, Vol. 83, pp. 1713-1745.

Housner. G.W. (1958), "Mechanics of sand blows". Bull. Seismol. Soc. Am., Vol. 48, No. 2, pp. 155-168.

Iai, S., Tsuchida, H. and Koizumi, K. (1989), "A Liquefaction Criterion Based on Field Performance around Seismograph Stations". Soils and Foundations, Vol. 29, No. 2.

Ishibashi, 1. and Sherif. M.A. (1974), "Soil liquefaction by torsional simple shear device". Journal of the Geotechnical Engineering Division. ASCE, 100, G T 8, pp. 871-888.

Ishihara, K. and Yasuda, S. (1975), "Sand Liquefaction in Hollow Cylinder Torsion Under Irregular Excitation". Soils and Foundations, Vol. 15, No. 1.

Ishihara, K. (1985), "Stability of Natural Deposits during Earthquakes", Proceedings 11th International Conf. on Soil Mech. and Found. Engg., San Francisco, USA.

Iwasaki, T. (1986), "Soil Liquefaction Studies in Japan", Soil Dynamics and Earthquake Engineering, Vol. 5. No. 1.

Iwasaki, T. Tokida, K., Tatsuoka, F., Watanabe, S., Yasuda, S. and Sato, H. (1982), "Microzonation for Soil Liquefaction Potential Using Simplified Methods". Proceedings 3rd International Conf. on Microzonation, Seattle, USA.

Japan Road Association (1991), "Specifications for Highway Bridges", Part V, Earthquake Resistant Design, 1900.

Katsumi. M., M. Maraya and T. Miteuru (1988), "Analysis of gravel drain against liquefaction and its application to design", IXth WCEE, Tokyo, Vol. III, pp. 249-254.

Kawasumi, Hirosi. (1968), "Historical Earthquakes in the Disturbed Area and Vicinity", General Report on the Niigata Earthquake of 1964, Electrical Engineering College Press, University of Tokyo, Japan.

Kishida, H. (1966), "Damage of reinforced concrete buildings in Niigata city with special reference to foundation engineering", Soil Found. Engin. (Tokyo), Vol. 9, No. 1, pp. 75-92.

Koppejan, A.W., Wamelen, B.M. and L.J. Weinberg (1948), "Coastal flow slides in the Dutch province of Seeland", Proc. 2nd ICSMFE, Vol. 5, pp. 89-96, Rotterdam.

Krishna, J. and S. Prakash (1968), "Blast tests at Obra dam site", J. Inst. Engin. (India), Vol. 47. No. 9, pt. C15, pp. 1273-1284.

Krishnamaswamy, N.R. and N.T. Issac (1995), "Liquefaction analysis of saturated reinforced granular soil", ASCE. Vol. 121, No. 9. pp. 645-652.

Kuizumi, Y. (1966), "Change in density of sand subsoil caused by the Niigata earthquake", Soil Found. Engin. (Tokyo), Vol. 8, No. 2, pp. 38-44.

Kuribayshi, E., Tatsuoka, F. and Yoshida, S. (1977), "History of earthquake induced soil liquefaction in Japan", Bulletion of PWRI, 31.

Kuwabara, F. and Yoshumi, Y. (1973), "Effect of subsurface liquefaction on strength of surface soil", ASCE, JGE, Vol. 19, No. 2.

Ladd, R.S. (1976), "Effects of Specimen Preparation on the Cyclic Structural Stability of Sands", Symp. on Soil Liquefaction, ASCE National Convention, Philadelphia, USA.

Lee, K.L. and C.K. Chan (1972), "Number of equivalent significant cycles in strong motion earthquakes", Proc. First Int. Conf., Microzonation, Seattle, Vol. 2, pp. 609-627.

Lee, K.L. and H.B. Seed (1967), "Cyclic stress conditions causing liquefaction of sands", J. Soil Mech. Found. Div., ASCE, Vol. 93, No. SMI, pp. 47-70.

Lee, K.L. and Fitton, J.A. (1969), "Factors Affecting the Cyclic Loading Strength of Soil", Vibration Effects of Earthquakes on Soils and Foundations. ASTM STP 450, American Society for Testing and Materials.

Lee, K.L. and Focht, Jr. J.A. (1975), "Cyclic Testing of Soil to Ocean Wave Loading Problems", 7th Annual Offshore Technology Conference, Houston, Texas, USA.

Lee, K.L. (1975), "Formation of Adhesion Bonds in Sands at High Pressures", Report No. UCLA-ENG-7586. UCLA School of Engineering and Applied Science, USA.

Lew. M. (1984), "Risk and mitigation of liquefaction hazard", Proc. VIIIth WCEE, Vol. I, pp. 183-190.

Lyman. A.R.N. (1942), "Compaction of cohesionless foundation soil by explosive", ASCE Trans., Vol. 107.

Maslov. N.N. (1957), "Questions of seismic stability of submerged sandy foundations and structures", Proc. Fourth Int. Conf. Soil Mech. Found. Engin., London, Vol. 1, pp. 368-372.

Matsuo. H. and S. Ohara (1960), "Lateral earth pressure and stability of quay walls", Proc. Second World Conference on Earthquake Engineering, Tokyo, Vol. 1, pp. 165-182.

Middlebrooks, T.A. (1942), "Fort peck slide". ASCE Transactions, Vol. 107, pp. 723-764.

Ohara.S. and T. Tamamoto (1987), "Fundamental study on gravel pile method for preventing liquefaction", ECEE 87, pp. 5.3/41-48.

Ohasaki. Y. (1966), "Niigata earthquake 1964, building damage and soil conditions", Soil Found. (Tokyo), Vol. 6, No. 2, pp. 14-37.

Ohasaki, Y. (1970), "Effects of sand compaction on liquefaction during the Tokachioki earthquake", Soil Found. (Tokyo), Vol. 10, No. 2, pp. 112-128.

Peacock, W.H. and H.B. Seed (1968), "Sand liquefaction under cyclic loading simple shear conditions", J. Soil Mech. Found., Div., ASCE, Vol. 94, No. SM 3, pp. 689-708.

Prakash, S. and M.K. Gupta (1970a), "Final report on liquefaction and settlement characteristics of loose sand under vibrations", Proc. International Conference on Dynamic Waves in Civil Engineering, Swansea, pp. 323-328.

Prakash. S. (1981), "Soil dynamics", McGraw Hill Book Co.

Prakash, S. and M.K. Gupta (1970b), "Blast tests at Tenughat dam site", J. Southeast Asian Soc. Soil Mech. Found Engin. (Bangkok), Vol. 1, No. 1, pp. 41-50.

Prakash, S. and M.K. Gupta (1970c), "Liquefaction and settlement characteristics of Ukai dam sand", Bull Indian Soc. Earthquake Tech. (Roorkee), Vol. 7, No. 3, pp. 123-132.

Prakash, S. and Mathur, J.N. (1965), "Liquefaction of fine sand under dynamic loads", Proc. 5th Symposium of the Civil and Hydraulic Engineering Department, Indian Institute of Sciene, Bangalore., India.

Robertson, P.K. and R.G. Companella (1985), "Liquefaction Potential of Sands Using the CPT", Journal Geotechnical Engg. Division, ASCE, Vol. III, No. 3.

Seed, H.B. (1976a), "Some aspects of sand liquefaction under cyclic loading", Conference on Behaviour of Offshore Structures, The Norwegian Institute of Technology, Norway.

Seed, H.B. (1976 b), "Evaluation of soil liquefaction effects on level ground during earthquakes". State-of-the-Art Paper, Symposium on Soil Liquefaction, ASCE National Convention, Philadelphia, pp. 1-104.

Seed, H.B. (1979), "Soil liquefaction and cyclic mobility evaluation for level ground during earthquakes", J. Geotech. Engin. Div., ASCE, Vol. 105, No. GT2, pp. 201-255.

Seed, H.B., K. Mori and C.K. Chan (1977), "Influence of seismic history on liquefaction of sands", J. Geotech. Engin. Div., ASCE, Vol. 103, No. GT 4, pp. 246-270.

Seed, H.B. and I.M. Idriss (1967), "Analysis of soil liquefaction Niigata earthquake", J. Soil Mech. Found. Div. ASCE, Vol. 93, No. SM 3, pp. 83-108.

Seed, H.B. and Booker, J.R. (1976), "Stabilisation of potentially liquefiable sand deposits using Gravel drain system", Report No. EERC 76-10, Earthquake Engineering Research Centre, University of California, Berkeley.

Seed, H.B. and I.M. Idriss (1971), "Simplified procedure for evaluating soil liquefaction potential", Journal of Soil Mechanics and Foundations Division, ASCE, 97, SM9, pp. 1249-1273.

Seed, H.B. and K.L. Lee (1966), "Liquefaction of saturated sands during cyclic loading", ASCE, JGE, Vol. 92, No. Sm 6. pp. 105-134.

Seed. H.B. and W.H. Peacock (1971), "Test procedures for measuring soil liquefaction characteristics". J. Soil Mech. Found. Div. ASCE, Vol. 97, No. SM 8, pp. 1099-1199.

Seed. H.B., I. Arango and C.K. Chan (1975), "Evaluation of soil liquefaction potential during earthquakes" Report on EERC. 75-28, Earthquake Engineering Research Center, University of California. Berkeley.

Seed, H.B., K.L. Lee, I.M. Idriss and F.I. Makdisi (1971), "The slides in the San Fernando dams during the earthquake of Feb.7, 1971", Journal of the Geotechnical Engineering Division, Proceedings. ASCE. Vol. 101. No.CT 7.

Seed, H. Bolton and I.M. Idriss (1967), "Analysis of Soil Liquefaction: Niigata Earthquake", Journal Soil Mech. and Found. Division, ASCE, Vol. 93, No. SM3.

Seed, H.B. and P. DeAlba (1986), "Use of SPT and CPT Tests for Evaluating the Liquefaction Resistance of Sands", Proceedings In-Situ 86, ASCE.

Seed, H.B., Tokimatsu, K., Harder, L.F. and Chung, R.M. (1985), "Influence of SPT Procedures in Soil Liquefaction Resistance Evaluation", Journal Geotechnical Engg. Division, ASCE, Vol. III, No. 12.

Shibata, T. and Teparaksa, W. (1988), "Evaluation of Liquefaction Potential of Soils using Cone Penetration Tests", Soils and Foundations, Vol. 28, No. 2.

Skempton, A.W. (1956), "Standard Penetration Test Procedures and the Effects in Sands of Overburden Pressure, Relative Density. Particle Size, Aging and Over consolidation", Geotechnique, Vol. 36, No. 3.

Suematsu, N., Yoshimi, Y. and Sasaki, Y. (1982), "Countermeasures for Reducing Damages Due to Liquefaction'", Soils and Foundations, (in Japanese).

Susumu, I.A.I., K. Koizimi, S. Node and H. Ysuchia (1988), "Large scale model tests and analysis of Gravel drains", IXth WCEE, Tokyo, Vol. III, pp. 261-266.

Taiping Q., W. Chenchun, W. Lunian and L. Hoishan (1984), "Liquefaction Risk Evaluation During Earthquakes", Proceedings International Conf. on Case Histories in Geotechnical Engineering, St. Louis, USA.

Terzaghi, K. and R.B. Peck (1967), "Soil mechanics in engineering practice", John Wiley and Sons, Inc., New York.

Tokimatsu. K. and Y. Yoshimi (1983), "Empirical Correlation of Soil Liquefaction Based on SPT N-value and Fines Content", Soils and Foundations, Vol. 30, No. 3.

Wang, S. (1984), "Experimental study on liquefaction inhibiting effect of gravel drains", Proc. VIII WCEE. California, Vol. 1, pp. 207-214.

Waterways Experiment Station U. S. Corps of Engineers (1967), "Potamology investigations, report 12-18, verification of empirical methods for determining river bank stability", 1965 Data, Vicksburg, Mississippi.

Yasushi, S. and E. Taniquchi (1982), "Large scale shaking table tests on the effectiveness of Gravel drains". Earthquake Engg. Conference, Southampton, pp. 843-847.

Yoshimi, Y. (1967), "Experimental study of liquefaction of saturated sands". Soil Found. (Tokyo). Vol. 7, No. 2, pp. 20-32.

Yoshimi. Y. and H. Ohaka (1975), "Influence of degree of shear stress reversal on the liquefaction potential of saturated sands". Soil Found. (Tokyo), Vol. 15, No. 3, pp. 27-40.

Yoshimi. Y. and H. Oh-Oka (1973), "A ring torsion apparatus for simple shear tests", Proc. 8th

International Conference on Soil Mechanics and Foundation Engineering, Vol. 12, Moscow. USSR.

Yoshimi, Y. and F. Kuwabara (1973), "Effect of Subsurface Liquefaction on the Strength of Surface Soil", Soils and Foundations, Vol. 13, No. 2.

Yosufumi. T., G. Kokusho and Matsui (1984), "On preventing liquefaction of level ground using gravel piles", Proc. JASCE, No. 352, pp. 89-98.

Practice Problems

9.1 Explain the terms 'initial liquefaction', 'liquefaction' and 'cyclic mobility'. Illustrate your answer with neat sketches.

9.2 List and discuss the factors on which liquefaction of saturated sand depends.

9.3 Give the salient features of the liquefaction studies made by (a) triaxial tests, (b) shake table tests, and (c) blast tests.

9.4 Describe briefly the following methods of predicting liquefaction potential:
(a) Seed and Idriss (1971) method
(b) Seed (1979) method
(c) Iwasaki (1986) method

9.5 At a given site, boring supplement with SPT was done upto 20.0 m depth. The results of the boring are as given below:

Depth (m)	*Classification of soils*	D_{50} (mm)	*N-Value*	D_R (%)	*Remarks*
2.0	SM	0.20	4	25	(i) Position of water table lies 2.0 m below the ground surface
4.0	SM	0.18	4	30	
6.0	SP	0.16	5	33	
8.0	SP	0.18	7	40	
10.0	SP	0.19	9	43	(ii) γ_{moist} = 18 kN/m^3
12.0	SP	0.19	10	45	γ_{sat} = 20 kN/m^3
14.0	SP	0.19	12	52	γ_{sub} = 10 kN/m^3
16.0	SW	0.30	14	56	
18.0	SW	0.32	16	60	
20.0	SW	0.20	18	63	

This site is located in seismically active region, and likely to be subjected by an earthquake of magnitude 7.5. Determine the zone of liquefaction by the three methods mentioned in Problem 9.4.

CHAPTER

10

General Principles of Machine Foundation Design

10.1 General

For machine foundations which are subjected to dynamic loads in addition to static loads, the conventional considerations of bearing capacity and allowable settlement are insufficient to ensure a safe design. In general, a foundation weighs several times as much as machine (Cozens, 1938; Rausch, 1959). Also the dynamic loads produced by the moving parts of the machine are small in comparison to the static weight of the machine and foundation. But the dynamic load acts repetitively on the foundation-soil system over long periods of time. Therefore, it is necessary that the soil behaviour be elastic under the vibration levels produced by the machine, otherwise deformation will increase with each cycle of loading and excessive settlement may occur. The most important parameters for the design of a machine foundation are: (i) natural frequency of the machine-foundation-soil system; and (ii) amplitude of motion of the machine at its operating frequency.

10.2 Types of Machines and Foundations

There are various types of machines that generate different periodic forces. The main categories are:

10.2.1 Reciprocating Machines

These include steam, diesel and gas engines, compressors and pumps. The basic mechanism of a reciprocating machine consists of a piston that moves within a cylinder, a connecting rod, a piston rod and a crank. The crank rotates with a constant angular velocity. Figure 10.1 shows the outline of a typical gangsaw in which the out of balance forces may lead to vibration problems.

The operating speeds of reciprocating machines are usually smaller than 1000 rpm. Large reciprocating engines, compressors and blowers generally operate at frequencies ranging within 50-250 rpm. Reciprocating engines such as diesel and gas engines usually operate within 300-1000 rpm.

The magnitude of the unbalanced forces and moments depend upon the number of cylinders in the machine, their size, piston displacement and the direction of mounting. The mechanism developing out of balance inertia forces for a single crank is shown in Fig. 10.2. It consists of a piston of mass m_p moving within a cylinder, a connecting rod AB of mass m_r and crank AO of mass m which rotates about point O at a frequency ω. The centre of gravity of the connecting rod is located at a distance L_1 from point A. If the rotating masses are to be partially or fully balanced, counterweights of mass m_w may be located with their centre of gravity at point C.

In order to simply the analysis of the motion of the connecting rod, the mass m_r is replaced by two equivalent masses: one rotating with the crank pin A, the other translating with the wrist pin B. The inertia forces can then be expressed in terms of the total rotating mass (m_{rot}) and the

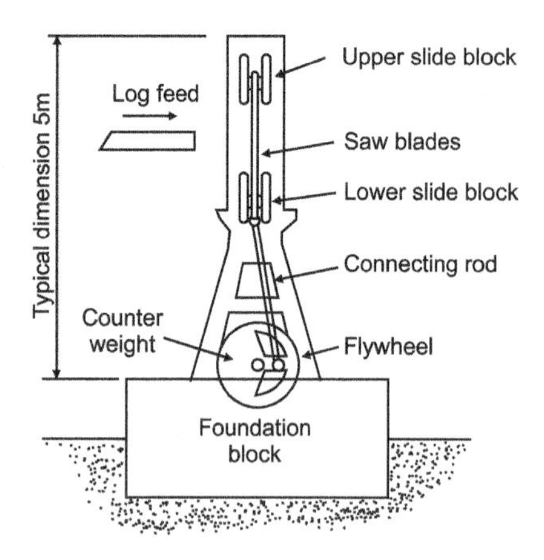

Fig. 10.1. Outline of a typical gangsaw machine

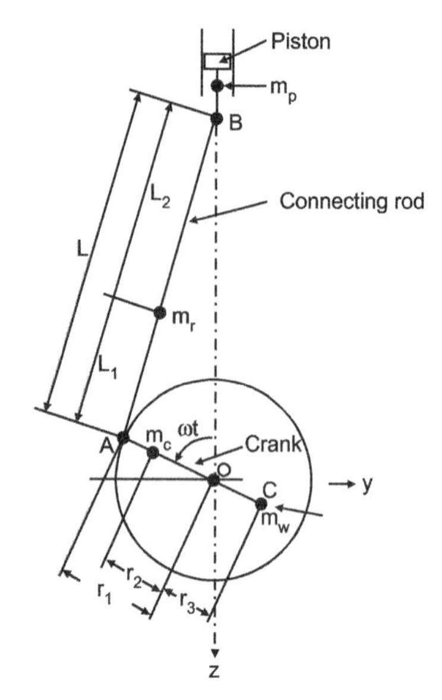

Fig 10.2 Crank mechanism

total reciprocating mass (m_{rec}). The total rotating mass is assumed to be concentrated at the crank in A.

$$m_{rot} = \frac{r_2}{r_1} m_c + \frac{L_2}{L} m_r - \frac{r_3}{r_1} m_w \tag{10.1}$$

$$m_{rec} = m_p + \frac{L_1}{L} m_r \tag{10.2}$$

The inertia force (F_z) in the z direction may be shown to be

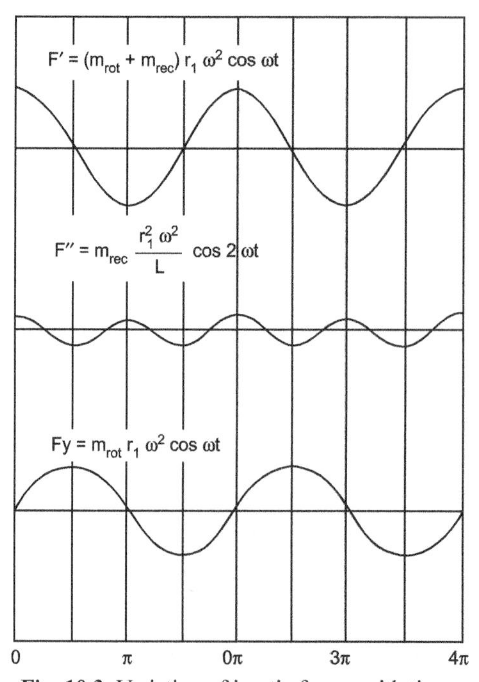

Fig. 10.3. Variation of inertia forces with time

$$F_z = (m_{rot} + m_{rec}) r_1 \omega^2 \cos \omega t + m_{rec} \frac{r_1^2}{L} \omega^2 \cos 2\omega t \tag{10.3}$$

which has a primary component (F') acting at the frequency of rotation, and a secondary component (F'') acting at twice the rotation frequency.

$$F_z = F' + F'' \tag{10.4}$$

And in the y direction

$$F_y = m_{rot} \cdot r_1 \omega^2 \sin \omega t \tag{10.5}$$

The time variations of these inertia forces are illustrated in Fig. 10.3.

If the rotating mass is balanced, the inertia force in the y direction disappears and that in the z direction becomes

$$F_z = m_{rec} r_1 \omega^2 \left(\cos \omega t + \frac{r_1}{L} \cos 2\omega t \right) \tag{10.6}$$

The amplitude of the primary (F'_{max}) and secondary (F''_{max}) inertia forces are then related as follows.

$$F''_{\max} = \frac{r_1}{L} F'_{max} \tag{10.7}$$

The preceding development relates to a single cylinder machine, which posseses unbalanced primary and secondary forces. As more cylinders are added the unbalanced forces and couples are modified as shown in Table 10.1 (Newcomb, 1951). With a six-cylinder machine complete balance is achieved.

Different crank arrangements pertaining to Table 10.1 are shown in Fig. 10.4.

Reciprocating machines are very frequently encountered in practice. Usually the following two types of foundations are used for such machines:

(a) Block type foundation consisting of a pedestal of concrete on which the machine rests (Fig. 10.5).
(b) Box or Caisson type foundation consisting of a hollow concrete block supporting the machinery on its top (Fig. 10.6).

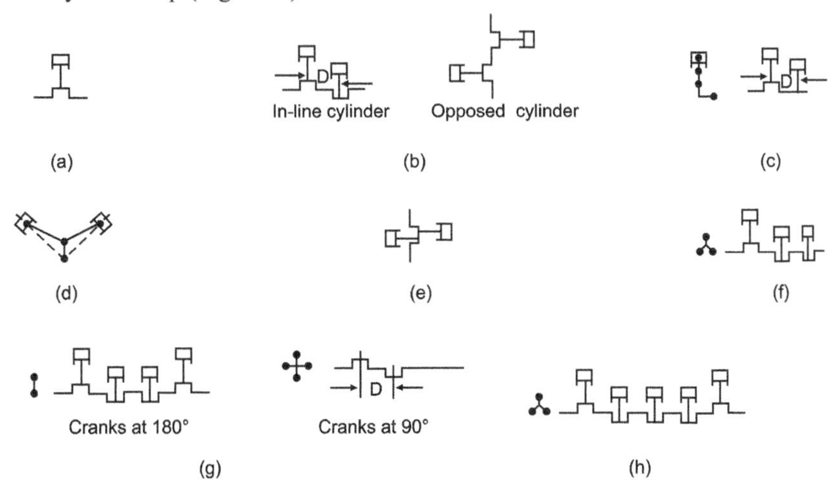

Fig. 10.4. Different crank arrangements: (a) Single crank, (b) Two cranks at 180°, (c) Two cranks at 90°, (d) Two cylinders at 90° on one crank, (e) Two opposed cylinders on one crank, (f) Three cranks at 120°, (g) Four cylinders and (h) Six cylinders

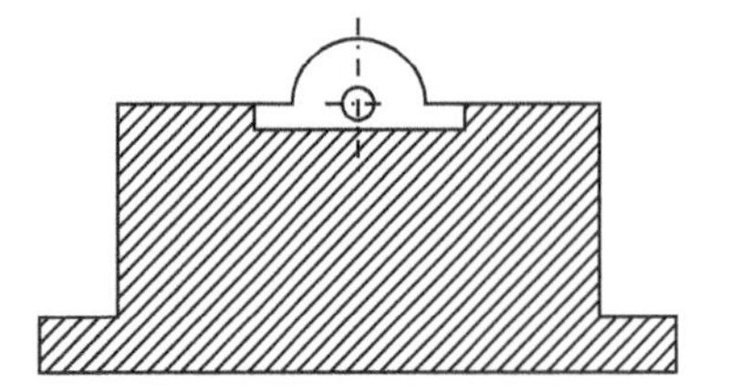

Fig. 10.5. Block type foundation

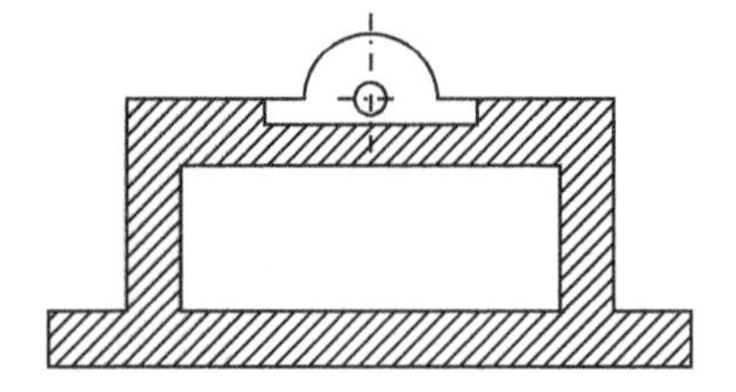

Fig. 10.6. Box type foundation

10.2.2 Impact Machines

These include machines like forging hammers, punch presses, and stamping machines which produce impact loads. Forge hammers are divided into two groups: drop hammers for die stamping and forge hammers proper. These machines consist of falling ram, an anvil, and a frame (Fig. 10.7). The speeds of operation usually range from 50 to 150 blows per minute. The dynamic loads attain a peak in a very short interval and then practically die out.

Table 10.1. Unbalanced Forces and Couples for Different Crank Arrangements (Newcomb, 1951)

Crank arrangements (Fig. 10.4)		*Forces*		*Couples*	
		Primary	*Secondary*	*Primary*	*Secondary*
a.	Single crank	F' without counter wts. (0.5) F' with counter wts.	F''	0	0
b.	Two cranks at 180° In-line cylinders Opposed cylinders	 0 0	2 F'' 0	$F'D$ without counter wts. (0.5) $F'D$ with counter wts.	0 0 0
c.	Two cranks at 90°	(1.41) F' without counter wts. (0.707)F' with counter wts.	0	(1.41) $F'D$ without counter wts. (0.707)$F'D$ with counter wts.	$F''D$
d.	Two cylinders on one crank, cylinders at 90°	F' without counter wts. 0 with counter wts.	1.41 F''	0	0
e.	Two cylinders on one crank, opposed cylinders	2 F' without counter wts. F' with counter wts.	0	0	0
f.	Three cranks at 120°	0	0	(3.46) $F'D$ without counter wts. (1.73) $F'D$ without counter wts.	(3.46) $F''D$
g.	Four cylinders: Cranks at 180° Cranks at 90°	 0 0	 0 0	 (1.41) $F'D$ without counter wts. (0.707) $F'D$ with counter wts.	 O (4.0) $F''D$
h.	Six cylinders	0	0	0	0

F' = primary force; F'' = secondary force; D = cylinder-centre distance

Impact machines may also be mounted on block foundations, but their details would be quite different from those of reciprocating machines.

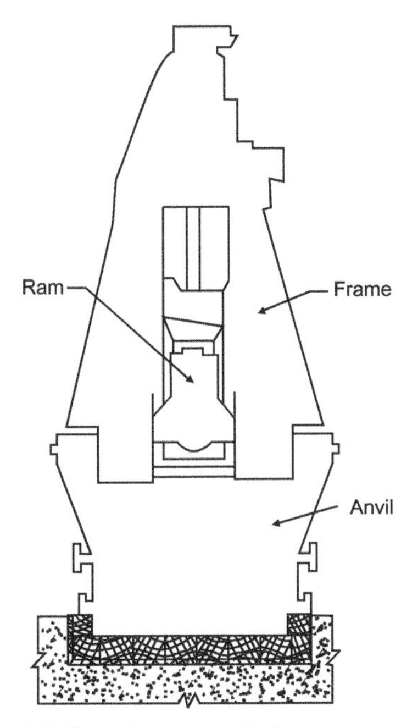

Fig. 10.7. Drop hammer with frame mounted on anvil

10.2.3 Rotary Machines

These include high speed machines such as turbo generators, turbines, and rotary compressors which operate at frequencies of the order of 3000 rpm to 10000 rpm. Associated with these machines there may be a considerable amount of auxiliary equipment such as condensers, coolers and pumps with connecting pipework and ducting. To accommodate these auxiliary equipments a common foundation arrangement is a two-storey frame structure with the turbine located on the upper slab and the auxiliary equipment placed beneath, the upper slab being flush with the floor level of machine hall (Fig. 10.8).

Rotating machinery is balanced before erection. However, in actual operation some unbalance always exists. It means that the axis of rotation lies at certain eccentricity with respect to principal axis of inertia of the whole unit. Although the amount of eccentricity is small in rotary machines the

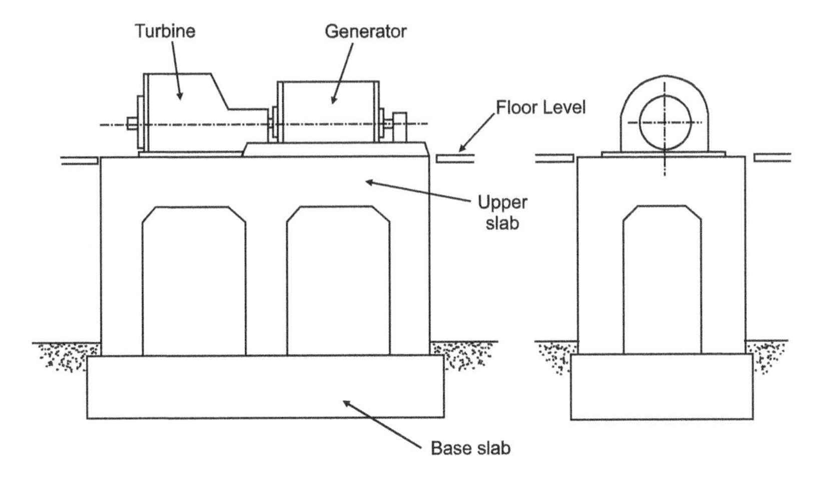

Fig. 10.8. Concrete frame turbogenerator foundation

unbalanced force may be large due to their high speed. Figure 10.9a shows a typical rotating mass type oscillator in which a single mass m_e is placed on a rotating shaft at an eccentricity e from axis of rotation. The unbalanced forces produced by such a system in vertical and horizontal directions are given by

$$F_V = m_e e\omega^2 \sin\omega t \tag{10.8a}$$

$$F_H = m_e e\omega^2 \cos\omega t \tag{10.8b}$$

Figure 10.9b shows two equal masses mounted on two parallel shafts at the same eccentricity, the shafts rotating in opposite directions with the same angular velocity. Such an arrangement produces an oscillating force with a controlled direction. For the arrangement shown in Fig. 10.9b, horizontal force components cancel and the vertical components are added to give

$$F = 2m_e\omega^2 \sin\omega t \tag{10.9}$$

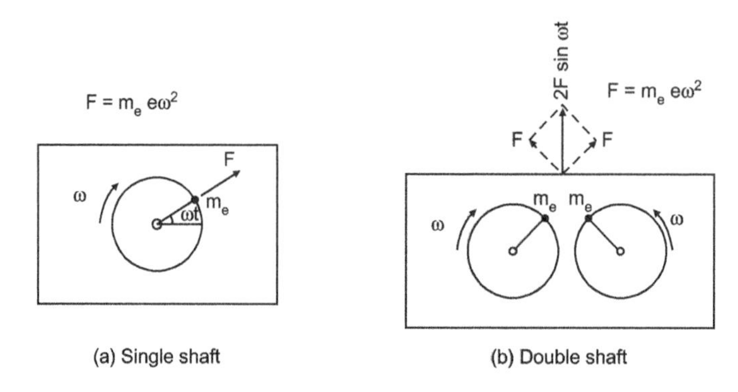

Fig. 10.9. Rotating mass type oscillator

10.3 General Requirements of Machine Foundations

For the satisfactory design of a machine foundation, the following requirements are met:

1. The combined centre of gravity of the machine and foundation should as far as possible be in the same vertical line as the centre of gravity of the base plane.
2. The foundation should be safe against shear failure.
3. The settlement and tilt of the foundation should be within permissible limits.
4. No resonance should occur; i.e., the natural frequency of the machine-foundation-soil system should not coincide with the operating frequency of the machine. Generally, a zone of resonance is defined and the natural frequency of the system should lie outside this zone.

   If ω represents the operating frequency of the machine and ω_n as the natural frequency of the system, then

   (a) In reciprocating machines (IS: 2974 pt I-2018)

   For important machines : $0.5 > \dfrac{\omega}{\omega_n} > 2.0$

   For ordinary machines : $0.6 > \dfrac{\omega}{\omega_n} > 1.5$

   (b) In impact machines (IS: 2974 Pt II-2018)

   $0.4 > \dfrac{\omega}{\omega_n} > 1.5$

(c) In rotary machines (IS: 2974 Pt III-2015)

$$0.8 > \frac{\omega}{\omega_n} > 1.25$$

It may be noted that where natural frequency of system ω_n is below the operating frequency of machine ω, the amplitudes during the transient resonance should be considered. For low speed machines, the natural frequency should be high, and vice versa. When natural frequency is lower than the operating speed, the foundation is said to be low tuned or under tuned; when the natural frequency is higher than the operating speed, it is high tuned or over tuned.

5. The amplitude of motion at operating frequencies should not exceed the permissible amplitude. In no case the permissible amplitude should exceed the limiting amplitude of the machine which is prescribed by the manufacturer.
6. The vibrations must not be annoying to the persons working in the factory or be damaging to other precision machines. The nature of vibrations that are perceptible, annoying, or harmful depends on the frequency of the vibrations and the amplitudes of motion. Richart (1962) developed a plot for vibrations (Fig. 10.10) that gives various limits of frequency and amplitude for different purposes. In this figure, the envelop described by the shaded line indicates only a limit for safety and not a limit for satisfactory operation of machines.

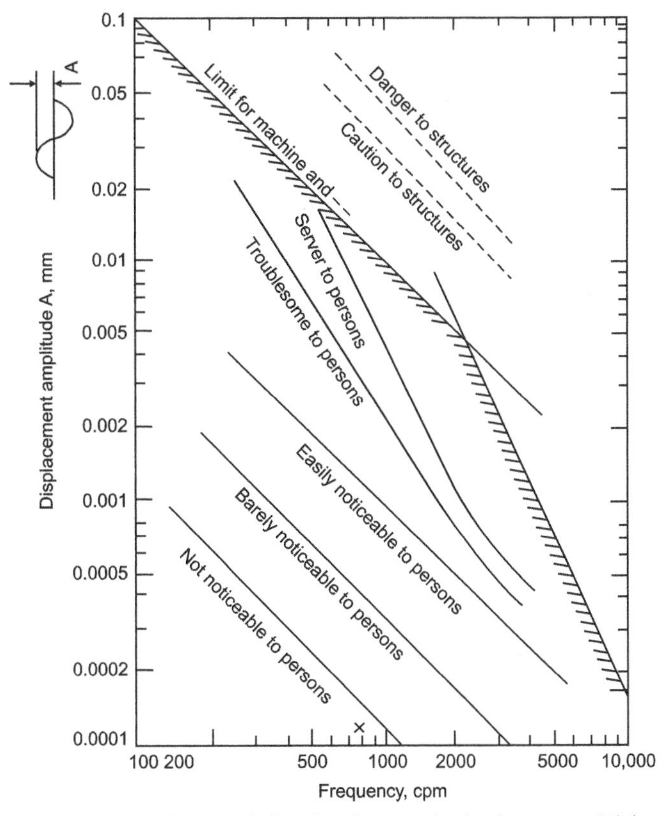

Fig. 10.10. Limiting amplitudes of vibration for a particular frequency (Richart, 1962).

10.4 Permissible Amplitude

For the design of machine foundation, the values of permissible amplitudes suggested by Bureau of Indian Standards for the foundations of different types of machines are given in Table 10.2.

Table 10.2. Values of Permissible Amplitudes for Foundations of Different Machines

S. No.	*Type of machine*	*Permissible amplitude, mm*	*Reference*
1.	Reciprocating machines	0.2	IS: 2974(Pt-I)
2.	Hammer* (a) For foundation block (b) For anvil	 1.0 to 2.0 1.0 to 3.0	 IS: 2974 (Pt-II)
3.	Rotary machines (a) Speed < 1500 rpm (b) Speed 1500 to 3000 rpm (c) Speed > 3000 rpm	 0.2 0.4 to 0.6 Vertical vibration 0.7 to 0.9 Horizontal vibration 0.2 to 0.3 Vertical vibration 0.4 to 0.5 Horizontal vibration	 IS:2974(Pt-IV) IS:2974 (Pt - III): 1992

* Permissible amplitude depends on the weight of tup, lower value for 10 kN tup and higher value for the tup weight equal to 30 kN or higher.

10.5 Allowable Soil Pressure

The allowable soil pressure should be evaluated by adequate sub-soil exploration and testing in accordance with IS: 1904 -1986, The soil stress below the foundation shall not exceed 80 percent of the allowable soil pressure. When seismic forces are considered, the allowable soil pressure may be increased as specified in IS: 1893-2016.

10.6 Permissible Stresses of Concrete and Steel

For the construction of the foundation of a machine M20 or M25 concrete in accordance with IS: 456-2005 shall be used. The allowable stresses of concrete and steel shall be reduced to 40 percent for concrete and 55% for steel, if the detailed design of foundation and components is limited to static load of foundation and machine. Considering temperature and all other loadings together, these assumed stresses may be exceeded by 33.5 percent. Alternatively, full value of stresses for concrete and steel as specified in IS: 456-2005 may be used if dynamic loads are separately considered in detailed design by applying suitable creep and fatigue factors.

The following dynamic moduli of concrete may be used in the design:

Grade of concrete	*Dynamic elastic modulus (kN/m^2)*
M20	3.0×10^7
M25	3.4×10^7
M30	3.7×10^7

10.7 Permissible Stresses of Timber

The timber is generally used under the anvil of hammer foundations. Grade of timber is specified according to the size of defects like knots, checks etc. in the timber. Timber is thus classified into three grades: Select, Grade I and Grade II IS: 3629 (1986). The best quality timber having minimum or no defects at all is of the select grade Grade I timber is one having defects not larger than the specified ones. Grade II timber is poorer in quality than grade I. The permissible values of stresses are given in Table 10.3 for species of timber of grade I. In machine foundations timber of select grade I is used. The permissible stresses of timber given in Table 10.3 may be multiplied by 1.16 to get the permissible stress of timber of select grade.

Table 10.3. Minimum Permissible Stress Limits (N/mm^2) in Three Groups of Structural Timbers (For Grade I Material)

S. No.	*Strength character*	*Location of use*	*Group A*	*Group B*	*Group C*
(i)	Bending and tension along grain	Inside[2]	18.0	12.0	8.5
(ii)	Shear[1]	1.05	0.64	0.49	
	Horizontal All locations				
	Along grain All locations				
(iii)	Compression parallel to grain	Inside[2]	11.7	7.8	4.9
(iv)	Compression perpendicular to grain	Inside[2]	4.0	2.5	1.1
(v)	Modulus of elasticity ($\times 10^3$ N/mm^2)	All locations and grade	12.6	9.8	5.6

[1] The values of horizontal shear to be used only for beams. In all other cases shear along grain to be used.
[2] For working stresses for other locations of use, i.e., outside and wet, generally factors of 5/6 and 2/3 are applied

The permissible bearing pressures on other elastic materials such as felt, cork and rubber are generally given by the manufacturers of these materials. No specific values are recommended here since they vary in wide limits.

References

Barkan, D.D. (1962), “Dynamics of bases and foundations”, McGraw Hill, New York

Cozens, W.J. (1938), “Machinery foundations”, J. I. E. E., Vol. 82, pp. 512-523.

IS: 2974 (Part I) - (2018), “Foundations for reciprocating machines”, I. S. I., New Delhi.

IS: 2974 (Part II) - (2018), “Foundations for impact type machines (Hammer foundations)”, I.S.I., New Delhi.

IS-2974 (Part III) - (2015), “Foundations for rotary type machines (medium and high frequency), I.S.I., New Delhi.

IS: 2974 (Part IV) - (2015), “Foundations for rotary type machines (low frequency)”. I.S.I., New Delhi.

IS: 2974 (Part V) - (1987), “Foundations for impact type machines other than hammers”, I.S.I., New Delhi.

IS: 456-2005, “Code of practice for plain and reinforced concrete”, I.S.I., New Delhi.

IS: 3629 (1986), “Design of structural timber in building: Code of Practice”, I.S.I., New Delhi.

IS: 1893-2016, “Earthquake resistant design of structures”, I.S.I., New Delhi.

IS: 1904-1986, “Code of practice for structural safety of buildings: Shallow foundation”, I.S.I., New Delhi.

Newcomb, W.K. (1951), “Principles of foundation design for engines and compressors”, trans. A.S.M.E, Vol. 73, pp. 307-312.

Raush, E. (1959), “Maschinen fundamente und andere dynamische beanspruchte Baukonstructionen”, Dusseldorf: VDI, Verlag.

Richart, F.E. (1962), “Foundation vibrations”, Trans A.S.C.E., Vol. 127, Part I, pp. 863-898.

CHAPTER

11

Foundations of Reciprocating Machines

11.1 General

Reciprocating machines are common in use. Steam engines, internal combustion engines (e.g. diesel and gas engines), pumps and compressors fall in this category of machines. Block type or box type foundations are used for reciprocating machines.

For the satisfactory performance of the machine-foundation system, the requirements given in Sec. 10.3 should be fulfilled. For this, one has to obtain (i) the natural frequency of the system, and (ii) the amplitude o f foundation during machine operation. In this chapter, methods have been presented to obtain these two parameters in different modes of vibration. The basic assumptions made in the analyses are: (i) the foundation block is considered to have only inertial properties and to lack elastic properties, and (ii) the soil is considered to have only elastic properties and to lack properties of inertia. Design steps and illustrative examples are given at the end of the chapter.

11.2 Modes of Vibration of a Rigid Foundation Block

A rigid block has, in general, six degrees of freedom. Three of them are translations along the three principal axes and the other three are rotations about these axes. Thus, under the action of unbalanced forces, the rigid block may undergo vibrations as follows (Fig. 11.1):

1. Translation along Z axis – Vertical vibration
2. Translation along Y axis – Longitudinal or sliding vibration
3. Translation along X axis – Lateral or sliding vibration
4. Rotation about Z axis – Yawing motion
5. Rotation about Y axis – Rocking vibration
6. Rotation about X axis – Pitching or rocking vibration

The vibratory modes may be 'decoupled' or 'coupled'. Of the six modes, translation along Z axis and rotation around the Z axis can occur independently of any other motion and are called decoupled modes. However. translation along the X or Y axis and the corresponding rotation about the Y or X axis, respectively, always occur together and are called coupled modes. Therefore the dynamic analysis of a block foundation should be carried out for the following cases:

(i) Uncoupled translatory motion along Z axis i.e. vertical vibration.
(ii) Coupled sliding and rocking motion of the foundation in *X–Z* and *Y–Z* planes passing through the common centre of gravity of machine and foundation.
(iii) Uncoupled twisting motion about Z axis.

A rigid block being a problem of six degrees of freedom has six natural frequencies. The natural frequency is determined in a particular mode (decoupled or coupled) and compared with the operating frequency. Similarly, amplitude is worked out in a particular mode and compared with the permissible value.

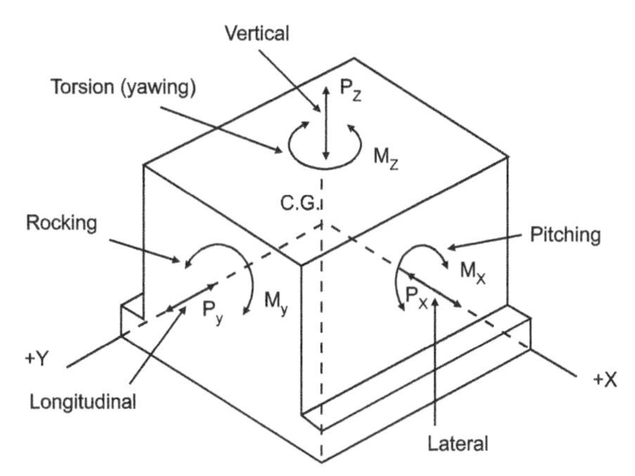

Fig. 11.1. Modes of vibration of a rigid block foundation

11.3 Methods of Analysis

The following two methods are commonly used for analysing a machine foundation:

(i) Linear elastic weightless spring method (Barkan, 1962)
(ii) Elastic half-space method (Richart, 1962)

In the first method which is proposed by Barkan (1962), soil is replaced by elastic springs. In developing the analysis the effects of damping and participating soil mass are neglected. Damping does not affect the natural frequency of the system appreciably, but it affects resonant amplitudes considerably. Since the zone of resonance is avoided in designing machine foundations, the effect of damping on amplitudes computed at operating frequency is also small. Hence neglecting damping may not affect the design appreciably, and if any that on the conservative side. Empirical methods have been suggested to obtain the soil mass participating in vibration.

In the elastic half-space method, the machine foundation is idealised as a vibrating mechanical oscillator with a circular base resting on the surface of ground. The ground is assumed to be an elastic, homogeneous, isotropic, semi-infinite body, which is referred to as an elastic half-space. This approach is apparently more rational, but relatively more complicated.

In the above two methods, the effect of side soil resistance is not considered i. e., the foundation is assumed to rest on the ground surface.

11.4 Linear Elastic Weightless Spring Method

Barkan (1962) has given the analysis of block foundation in following modes of vibration:

(i) Vertical vibration
(ii) Pure sliding vibration
(iii) Pure rocking vibration
(iv) Coupled sliding and rocking vibration
(v) Yawing motion

Let us consider a block foundation of base contact area A placed at a depth D_f below the ground level (Fig. 11.2). Neglecting the effect of side soil resistance and considering soil as weightless elastic material, the machine-foundation soil systems can be idealised to mass-spring systems shown in Fig. 11.3a to 11.3e for different modes of vibration. Barkan (1962) had introduced the following soil parameters which yield the spring stiffnesses of soil in various modes:

(a) Coefficient of elastic uniform compression (C_u): It is defined as the ratio of compressive stress applied to a rigid foundation block to the 'elastic' part of the settlement induced consequently. Thus

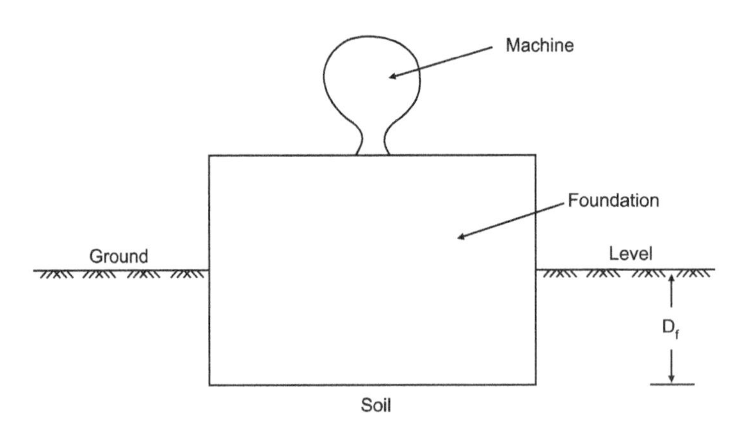

Fig. 11.2. Block foundation

$$C_u = \frac{q_z}{S_{ez}} \quad \text{(Fig.11.3a)} \tag{11.1}$$

where
$$q_z = \frac{\text{Load at the base of foundation}}{\text{Base contact area}} \tag{11.2}$$

It is used in vertical vibration mode. From definition, spring constant K_z

$$K_z = \frac{\text{Load}}{\text{Elastic deformation}} = \frac{q_z A}{s_{ez}} C_u \cdot A \tag{11.3}$$

(b) ***Coefficient of elastic uniform shear*** **(C_τ):** It is defined as the ratio of the average shear stress at the foundation contact area to the elastic part of the sliding movement of the foundation.

$$C_\tau = \frac{q_x}{S_{ex}} \quad \text{(Fig.11.3b)} \tag{11.4}$$

It is used in analysing sliding vibration mode. The spring constant K_x is given by

$$K_x = \frac{\text{Shear load}}{S_{ex}} = \frac{q_x \cdot A}{S_{ex}} = C_\tau \cdot A \tag{11.5}$$

(c) *Coefficient of elastic non-uniform compression* **(C_ϕ):** It is used in rocking vibration (Fig. 11.3c). In this case the elastic settlement of the block is not uniform over the base. It is defined as the ratio of intensity of pressure at certain location from the centre of the base of block to the corresponding elastic settlement. If ϕ is the angle of rotation of block, then at a distance l from the centre of the base of block, the elastic deformation will be $l\,\phi$. Taking the intensity of pressure at this location as q, C_ϕ, is given by

$$C_\phi = \frac{q}{l\phi} \tag{11.6}$$

The stiffness K_ϕ is defined as the moment per unit rotation, and is given by

$$K_\phi = \frac{M}{\phi} = C_\phi \cdot I \tag{11.7}$$

where I = Moment of inertia of the base of block about the axis of rotation
M = Moment caused due to soil reaction

(d) *Coefficient of elastic non-uniform shear* (C_ψ): It is used in yawing motion. If a foundation is acted upon by a moment with respect to vertical axis, it will rotate about this axis

(Fig 11.3d). Tests have shown that the angle of rotation ψ of the block is proportional to the external moment.

Therefore, $$M_z = K_\psi \psi \qquad (11.8)$$

where $$K_\psi = C_\psi \cdot J_z \qquad (11.9)$$

J_z = Polar moment of the inertia of contact base area of foundation.

In the rotation of a foundation around a vertical axis, the base of the foundation undergoes non uniform sliding, hence the term "Coefficient of elastic non uniform shear" is applied to the Coefficient C_ψ.

Barkan (1962) derived the Eq. (11.10) for determining the value of C_u. It is based on theory of elasticity.

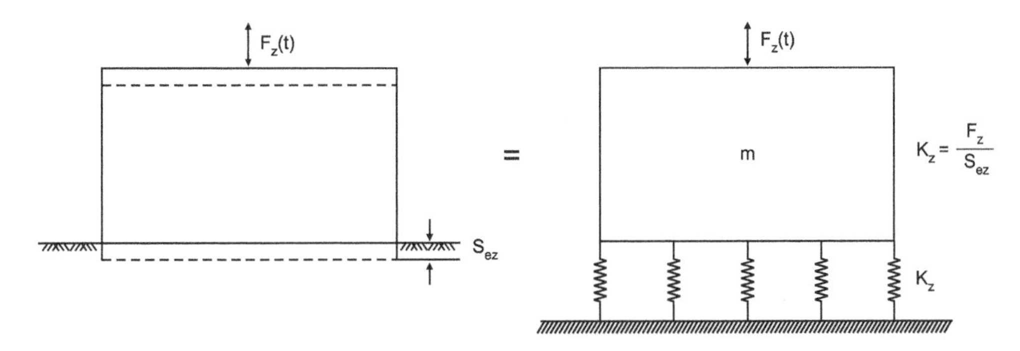

(a) Vertical vibration

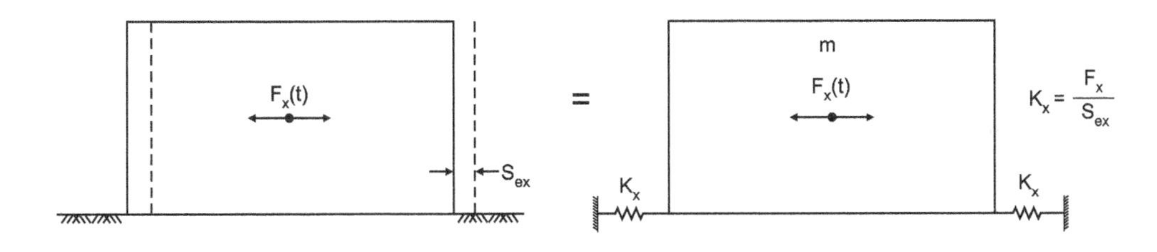

(b) Pure sliding vibration

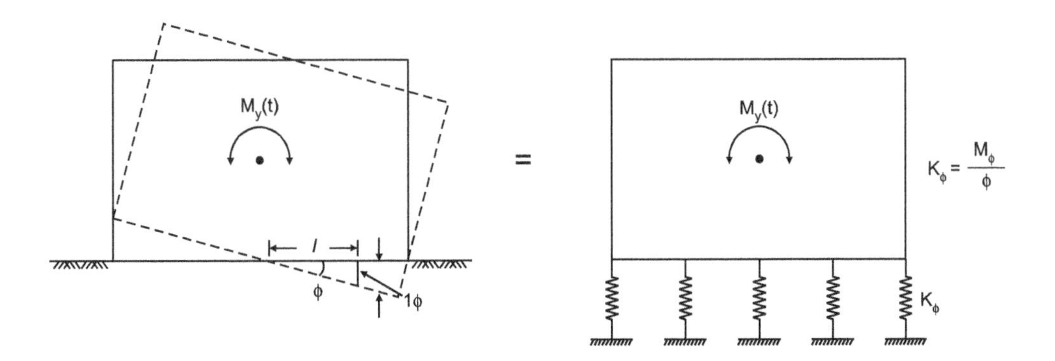

(c) Pure rocking vibration

Fig. 11.3. Types of motion of a rigid foundation (... Contd.)

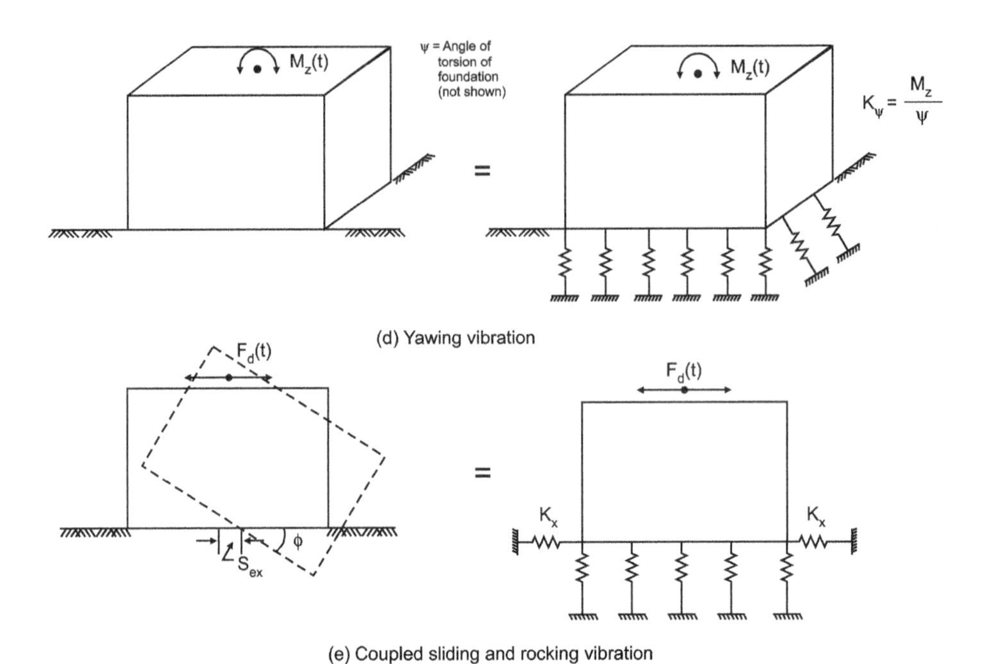

Fig. 11.3. Types of motion of a rigid foundation.

$$C_u = \frac{1.13\,E}{1-\mu^2}\cdot\frac{1}{\sqrt{A}} \tag{11.10}$$

where, E = Young's modulus of soil
μ = Poisson's ratio
A = Area of base of the foundation.

He also developed the relationships between C_u, C_ϕ, C_τ and C_ψ. For the analysis and design of machine foundation, he recommended that

$$C_u = 2\,C_\tau \tag{11.11}$$

$$C_\phi = 2\,C_u \tag{11.12}$$

$$C_\tau = 1.5\,C_\psi \tag{11.13}$$

For preliminary design, Barkan (1962) recommended the values of C_u as listed in Table 11.1.

Table 11.1. Recommended Design Values for the Coefficient of Elastic Uniform Compression C_u for $A = 10\ m^2$*

Soil	*Soil group*	*Permissible static load*, kN/m^2	C_u, kN/m^3
1	2	3	4
I	Weak soils (clays and silty clays with sand in plastic state; clayey and silty sands; also soils of categories II and III with laminae of organic silt and of peat)	upto 150	upto 3×10^4
II	Soil of medium strength (clays and silty clays with sand close to the plastic limit; sand)	150 to 350	$(3–5) \times 10^4$
III	Strong soils (clays and silty clays with sand of hard consistency; gravels and gravelly sands; loess and loessial soils)	350 to 500	$(5–10) \times 10^4$
IV	Rocks	> 500	$> 10 \times 10^4$

The procedure of determining the values of C_u, C_τ E and G have been given in detail in Sec 4.3 of Chapter 4. As discussed in that section, dynamic elastic constants depend on (i) base area of foundation, (ii) confining pressure and (iii) strain level. The method of converting the value of a dynamic elastic constant obtained from a field test for using in the design of actual foundation has been illustrated in examples 4.2 and 4.3.

11.4.1 Vertical Vibrations

For the purpose of analysis, the machine-foundaton-soil system shown in Fig. 11.3 *a* is represented by the idealised mass-spring system shown in Fig 11.4. Let the unbalanced force is representd by

$$F_z(t) = F_z \sin \omega\, t \tag{11.14}$$

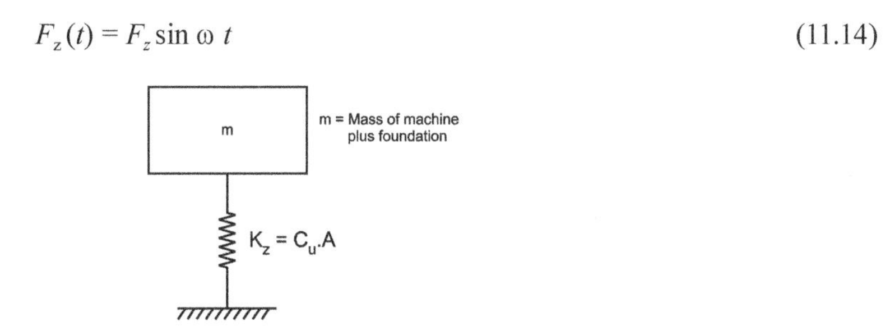

Fig. 11.4. Equivalent model for vertical vibration

If the centre of gravity of the foundation and machine and the centroid of the base area of the foundation in contact with the soil lie on a vertical line that coincides with the line of action of the exciting force F_z, then foundation will vibrate vertically only.

The equation of motion of the system is

$$m\ddot{z} + K_z \cdot Z = F_z \sin \omega\, t \tag{11.15}$$

where M = Mass of machine and foundation

K_z = Equivalent spring constant of the soil in vertical direction for base area A of the foundation = $C_u A$

C_u = Coefficient of elastic uniform compression

Therefore, the natural frequency ω_{nz} of the system is

$$\omega_{nz} = \sqrt{\frac{K_z}{m}} = \sqrt{\frac{C_u\, A}{m}} \tag{11.16}$$

The amplitude of motion A_z is given by

$$A_z = \frac{F_z \sin \omega t}{K_z - m\omega^2} = \frac{F_z \sin \omega t}{C_u\, A - m\omega^2} \tag{11.17a}$$

or

$$A_z = \frac{F_z \sin \omega t}{m(\omega_{nz}^2 - \omega^2)} \tag{11.17b}$$

Maximum amplitude of motion A_z is given by

$$A_z = \frac{F_z}{m(\omega_{nz}^2 - \omega^2)} \tag{11.18}$$

11.4.2. Sliding Vibrations of a Block

In practice, rocking and sliding occur simultneously. But if the vibration in rocking can be neglected, then only horizontal displacement of the foundation would occur under an exciting force F_x (t) on the block of area A (Fig. 11.3b). This system can be idealised as shown in Fig. 11.5.

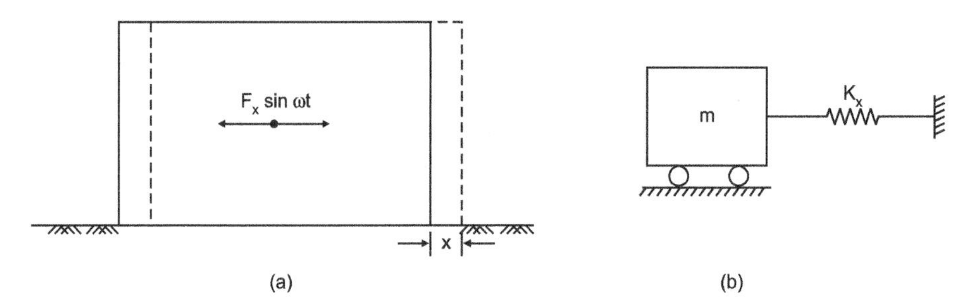

Fig. 11.5 (a) Block foundation in pure sliding vibration, (b) Equivalent model

The equation of motion of the system is

$$m\ddot{x}+K_x x = F_x \sin \omega t \tag{11.19}$$

where x = Sliding displacement of the foundation
K_x = Equivalent spring constant of the soil in sliding for base area A foundation = $C_\tau \cdot A$
C_τ = Coefficient of elastic uniform shear

Therefore, the natural frequency ω_{nx} of the system is

$$\omega_{nx} = \sqrt{\frac{K_x}{m}} = \sqrt{\frac{C_\tau A}{m}} \tag{11.20}$$

Maximum amplitude of motion A_x is given by

$$A_x = \frac{F_x}{m(\omega_{nx}^2 - \omega^2)} \tag{11.21}$$

11.4.3 Pure Rocking Vibrations of a Block

Consider only the rocking vibrations induced in a foundation block by an externally exciting moment M_y (t) (Fig. 11.3 *c*). This is also a hypothetical case as rocking vibrations are coupled with sliding vibrations. Let the unbalanced moment be given by

$$M_y(t) = M_y \sin \omega t \tag{11.22}$$

where M_y = Moment acting in the x-z plane.

At any time t, considering that the applied moment is acting in clockwise direction the displaced position of the block will be as shown in Fig. 11.6. In machine foundations, as the rotation ϕ is small $\tan \phi = \phi$ in radians. The equation of motion can be obtained by applying Newton's second law of motion.

The various moments acting on the foundation about the centre of rotation are obtained as described below:

(i) *Moment M_R due to soil reaction*: Consider an element dA of the foundation area in contact with the soil and located at distance l from the axis of rotation (Fig. 11.6 *b*). At any time, the soil will be compressed nonuniformly. From the definition of coefficient of elastic uniform compression, C_ϕ is

$$C_\phi = \frac{dR/dA}{l\phi} \tag{11.23a}$$

where dR = Soil reaction force acting on element dA
ϕ = Angle of rotation

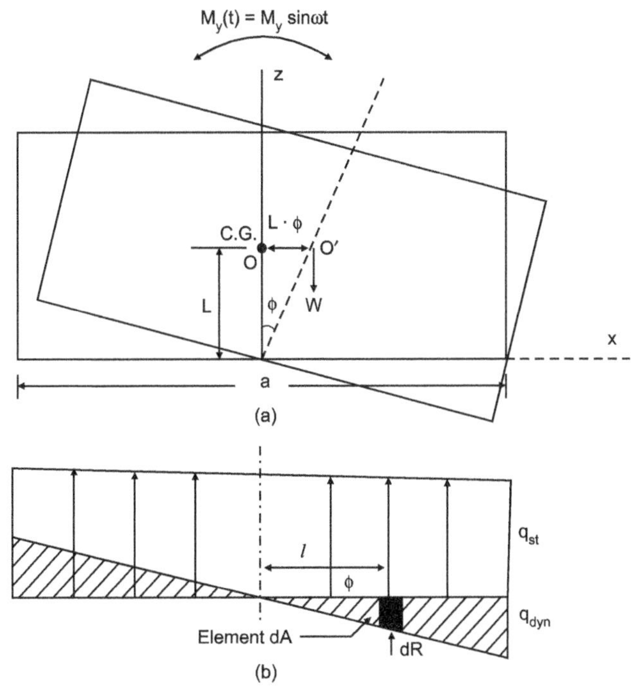

Fig. 11.6. Block foundation under pure rocking vibrations

If the foundation does not lose contact with soil, then the soil reaction will be as shown in Fig. 11.6 *b*. The total reactive moment M_R against the foundation area in contact with soil is given by

$$M_R = \int^A C_\phi \cdot l\phi \, dAl \;=\; C_\phi \, \phi \int l^2 \, dA \;=\; C_\phi \, \phi I \tag{11.24}$$

where I = Moment of inertia of the foundation area in contact with the soil with respect to the axis of rotation.

This moment acts in the anticlockwise direction.

(ii) *Moment M_w due to the displaced position of centre of gravity of the block:* As shown in Fig. 11.6 *a*, the centre of gravity of the block is shifted from point O to O'. As angle of rotation ϕ is small, the moment M_w of weight W will be

$$M_w = W L \phi \tag{11.25}$$

where L = Distance between the centre of gravity of block and axis of rotation.
This moment 11 acts in the clockwise direction.

(iii) *Moment M_i, caused by inertia of foundation*:

It is given by $$M_i = M_{mo} \ddot{\phi} \tag{11.26}$$

where M_{mo} = Moment of inertia of the mass of the foundation and machine with respect of axis of rotation.

This moment acts in the anticlockwise direction.

The equation of motion can be written by equating clockwise moments to anticlockwise moments

Therefore, $$M_y \sin \omega t + WL \phi = C_\phi I\phi + M_{mo} \ddot{\phi}$$

or $$M_{mo} \ddot{\phi} + (C_\phi I - WL)\phi = M_y \sin \omega t \tag{11.27}$$

The natural frequency $\omega_{n\phi}$ of this system is given by

$$\omega_{n\phi} = \sqrt{\frac{C_\phi I - WL}{M_{mo}}} \tag{11.28}$$

and maximum displacement A_ϕ is given by

$$A_\phi = \frac{M_y}{M_{mo}(\omega_{n\phi}^2 - \omega^2)} \tag{11.29}$$

In practice, $C_\phi I$ is many times WL; hence Eq. (11.28) may be written:

$$\omega_{n\phi} = \sqrt{\frac{C_\phi I}{M_{mo}}} \tag{11.30}$$

If the dimensions of the footing at the base are a and b in the X and Y directions, respectively,

$$I = \frac{ba^3}{12} \tag{11.31}$$

$$\omega_{n\phi} = \sqrt{\frac{C_\phi ba^3}{M_{mo} 12}} \tag{11.32}$$

It is seen from Eq. (11.32) that the linear dimension of the contact area perpendicular to the axis of rotation exercises a considerably greater effect on the natural frequency of rocking vibrations than the other dimension. This principle is sometimes used in proportioning the sides of the machine foundation undergoing predominantly rocking vibrations.

The amplitude of the vertical motion of the edge of the footing is

$$A_{zr} = \frac{a}{2} \times A_\phi$$

$$= \frac{M_y a/2}{M_{mo}(\omega_{n\phi}^2 - \omega^2)} \tag{11.33}$$

Similarly, the contribution of rocking towards the horizontal amplitude is

$$A_{xr} = h \cdot A_\phi \tag{11.34}$$

where h = Height of the point above the base where amplitude is to be determined.

A_{zr} and A_{xr} are added to A_z (Eq. 11.18) and A_x (Eq. 11.21) respectively to obtain total vertical and sliding amplitudes when rocking is combined with vertical and sliding vibrations.

11.4.4 Yawing Vibrations of a Block

A foundation is subjected to yawing motion if it is subjected to a torsional moment M_z (t) about Z-axis (Fig. 11.7 a). The position of the foundation at any time t may be defined in terms of angle of rotation ψ.

Let the unbalanced moment is given by

$$M_z(t) = M_z \sin \omega t \tag{11.35}$$

As explained in Sec. 11.4, the resistive moment due to soil is $C_\psi . J_z . \psi$.

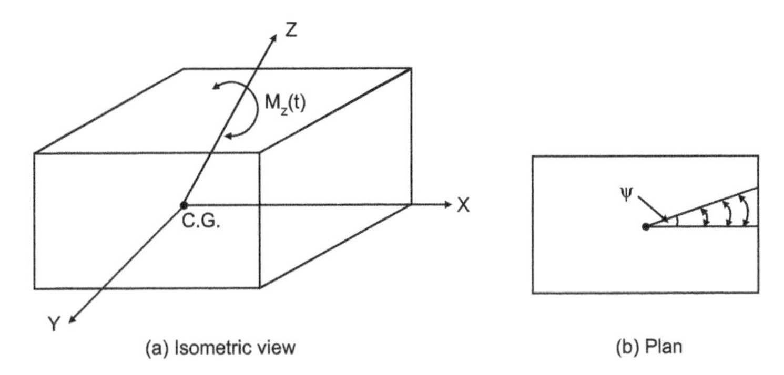

Fig. 11.7. (a) Yawing motion of a rigid block and (b) Development of non-uniform shear below the base

The equation of motion is written by taking moment about *Z*-axis. It gives

$$M_{mz}\ddot{\psi} + C_{\psi} J_z \psi = M_z \sin \omega t \tag{11.36}$$

where M_{mz} = Mass moment of inertia of the machine and foundation about the axis of rotation (Z-axis)

J_z = Polar moment of inertia of foundation contact area

C_{ψ} = Coefficient of elastic nonuniform shear

The expressions for natural frequency and maximum angular displacements are as follows

$$\omega_{n\psi} = \sqrt{\frac{C_{\psi} J_z}{M_{mz}}} \tag{11.37}$$

$$A_{\psi} = -\frac{M_z}{M_{mz}(\omega_{n\psi}^2 - \omega^2)} \tag{11.38}$$

The horizontal displacement $A_{h\psi}$ caused by torsion is

$$A_{h\psi} = rA_{\psi} \tag{11.39}$$

where r = Horizontal distance of the point on the foundation from the axis of motion (*Z*-axis).

11.4.5 Simultaneous Vertical, Sliding and Rocking Vibrations

In general, a machine foundation is subjected to time dependent vertical force, horizontal force and moment, and therefore it simultaneously slide, rock and vibrate vertically. In Fig. 11.8, a foundation block subjected to a vertical force ($F_z \sin \omega t$), a horizontal force ($F_x \sin \omega t$) and an oscillatory moment ($M_y \sin \omega t$) is shown. These forces and moment are considered to act at the combined centre of gravity O of the machine and the foundation, which is also taken as the origin of coordinates. At any time *t*, considering the vertical force acting in downward direction, horizontal force in right-hand side direction, and moment in the clockwise direction, the foundation block will be displaced as shown in Fig. 11.8. It is therefore subjected to (*i*) displacement *z*, in the vertical direction, (*ii*) displacement x_o in the horizontal direction at the base and (*iii*) rotation ϕ of the base.

The equations of motion can be written by evaluating the resisting and actuating forces and moments acting on the foundation in the displaced position. These forces and moments are obtained as given below:

(*i*) Upward soil reaction R_v due to vertical displacement *z*:

$$R_v = C_u Az \tag{11.40}$$

(*ii*) Horizontal soil reaction R_x due to horizontal displacement x_o:

$$R_x = C_{\tau} Ax_o \tag{11.41}$$

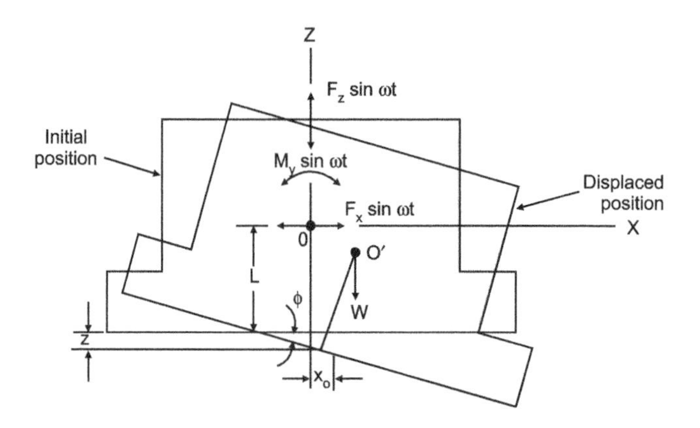

Fig. 11.8. Block foundation subjected to simultaneous vertical, sliding and rocking vibrations

As the origin is at O, x_o can be expressed in terms of x and ϕ as below:

$$x_o = x - L\,\phi \tag{11.42}$$

where L = Height of centre of gravity O from base of the block.

(*iii*) Moment M_R due to resistance of soil induced by rotation of the foundation by ϕ:
The M_R about point O is given by

$$M_R = C_\phi\, \phi\, I \tag{11.43}$$

(iv) Moment M_w due to displaced position of the centre of gravity of block:
The moment M_w about point O is given by

$$M_w = W\,L\,\phi \tag{11.44}$$

(*v*) Moment M_{xR} due to horizontal resisting force R_x:
Moment M_{xR} about point O is given by

$$M_{xR} = R_x \cdot L = C_\tau A\,(x - L\phi)\cdot L \tag{11.45}$$

(*vi*) Intertial forces and moment:

(a) In the Z - direction $F_{iz} = m\ddot{z}$ (11.46)

(b) In the X - direction $F_{ix} = m\ddot{x}$ (11.47)

(c) In the rotational mode $M_{i\phi} = M_m\,\ddot{\phi}$ (11.48)

where M_m = Mass moment of inertia of the machine and foundation about an axis passing through combined centre of gravity O and in the direction of Y-axis

The equations of the motion can now be written as below:

In the Z - direction: $m\ddot{z} + C_u\,Az = F_z \sin \omega\, t$ (11.49)

In the X - Direction: $m\ddot{x} + C_\tau\,Ax_o = F_x \sin \omega\, t$ (11.50)

Substituting the value of x_o from Eq. (11.42) in Eq. (11.50), we get

$$m\ddot{x} + C_\tau\,A(x - L\phi) = F_x \sin \omega\, t \tag{11.51}$$

In the rotational mode:

$$M_m\ddot{\phi} + C_\phi\,\phi I - W\,L\phi - C_\tau\,A(x - L\phi) = M_y \sin \omega\, t$$

or

$$M_m\ddot{\phi} - C_\tau\,AL_x + (C_\phi\,I - W\,L + C_\tau\,AL^2)\phi = M_y \sin \omega\, t \tag{11.52}$$

Equation (11.49) contains only the terms of z, therefore the motion in Z - direction is independent of any other motion. The solution of this equation is already given in Eqs. (11.16) to (11.18). Equations (11.51) and (11.52) contain both x and ϕ and are interdependent. Therefore, sliding and rocking are coupled modes. A solution for simultaneous rocking and sliding vibrations is presented below.

11.4.5.1 Natural frequencies of coupled rocking and sliding

The system represented by Eqs. (11.51) and (11.52) is a two-degree-of-freedom system. The solutions for natural frequencies are obtained by considering the free vibrations of the system.

Hence,
$$m\ddot{x} + C_\tau Ax - C_\tau AL\phi = 0 \tag{11.53}$$

and
$$M_m\ddot{\phi} - C_\tau ALx + (C_\phi I - WL + C_\tau AL^2)\phi = 0 \tag{11.54}$$

Particular solutions of these equations may be assumed as

$$x = x_1 \sin(\omega_n t + \alpha) \tag{11.55}$$

and
$$\phi = \phi_1 \sin(\omega_n t + \alpha) \tag{11.56}$$

in which x_1, ϕ_1 and α are arbitrary constants whose values depend upon the initial conditions of motion.

By substituting Eqs.(11.55) and (11.56) into Eqs. (11.53) and (11.54) and dividing by $\sin(\omega_n t + \alpha)$ we get

$$-m\omega_n^2 x_1 + C_\tau Ax_1 - C_\tau AL\phi_1 = 0$$

or
$$x_1(C_\tau A - m\omega_n^2) - C_\tau AL\phi_1 = 0 \tag{11.57}$$

and
$$-M_m\omega_n^2\phi_1 + \phi_1(C_\tau AL^2 + C_\phi I - WL) - C_\tau ALx_1 = 0 \tag{11.58}$$

From Eq. (11.57)
$$x_1 = \frac{C_\tau AL\phi_1}{C_\tau A - m\omega_n^2} \tag{11.59}$$

By substituting the value f x_1 from Eq. (11.51) into Eq. (11.58), we get

$$\phi_1[-C_\tau^2 A^2 L^2 + (C_\phi I - WL + C_\tau AL^2 - M_m\omega_n^2)(C_\tau A - m\omega_n^2)] = 0 \tag{11.60}$$

For a nontrivial solution, ϕ_1 cannot be zero. Hence the expression within the parentheses must be zero.

This gives

$$-C_\tau^2 A^2 L^2 + (C_\phi I - WL + C_\tau AL^2 - M_m\omega_n^2)(C_\tau A - m\omega_n^2) = 0 \tag{11.61}$$

The term ω_n, which represents the natural frequency in combined sliding and rocking, is the only unknown in Eq. (11.61), which can now be solved. Equation (11.61) may be rewritten as follows:

$$\begin{aligned} -C_\tau^2 A^2 L^2 + C_\tau^2 A^2 L^2 + C_\tau A(C_\phi I - WL) - C_\tau AM_m\omega_n^2 \\ -C_\tau AL^2 m\omega_n^2 - (C_\phi I - WL)m\omega_n^2 + M_m m\omega_n^4 = 0 \end{aligned} \tag{11.62}$$

By dividing by mM_m and rearranging, one obtains

$$\omega_n^4 - \omega_n^2\left[\left(\frac{C_\phi I - WL}{M_m}\right) + \frac{C_\tau A(M_m + mL^2)}{mM_m}\right] + \frac{C_\tau A}{m}\left(\frac{C_\phi I - WL}{M_m}\right) = 0 \tag{11.63}$$

By definition, the quantity $(M_m + mL^2)$ is the mass moment of inertia of the foundation and machine about an axis that passes through the centroid of the base contact area and is perpendicular to the plane of vibrations. This is denoted by M_{mo}. Thus,

$$M_{mo} = M_m + mL^2 \tag{11.64}$$

Further, by denoting $\dfrac{M_m}{M_{mo}} = r$ where $1 > r > 0$ (11.65)

Equation (11.63) may be rewritten as

$$\omega_n^4 - \frac{\omega_n^2}{r}\left(\frac{C_\phi I - WL}{M_{mo}} + \frac{C_\tau A}{m}\right) + \frac{C_\tau A}{m}\frac{C_\phi I - WL}{rM_{mo}} = 0 \tag{11.66}$$

Now

$$\frac{C_\tau A}{m} = \omega_{nx}^2 \tag{11.67}$$

$$\frac{C_\phi I - WL}{M_{mo}} = \omega_{n\phi}^2 \tag{11.68}$$

By writing Eq. (11.66) in terms of $\omega_{n\phi}$ and $\omega_{n\phi}$, we get

$$\omega_n^4 - \left(\frac{\omega_{nx}^2 + \omega_{n\phi}^2}{r}\right)\omega_n^2 + \frac{\omega_{nx}^2\,\omega_{n\phi}^2}{r} = 0 \tag{11.69}$$

Equation (11.69) has two positive roots, ω_{n1} and ω_{n2}, which correspond to two natural frequencies of the system. The roots of Eq. (11.69) are:

$$\omega_{n1,2}^2 = \frac{1}{2}\left[\left(\frac{\omega_{nx}^2 + \omega_{n\phi}^2}{r}\right) \pm \sqrt{\left(\frac{\omega_{nx}^2 + \omega_{n\phi}^2}{r}\right) - \frac{4\omega_{nx}^2\,\omega_{n\phi}^2}{r}}\right] \tag{11.70}$$

Equation (11.70) may be rewritten as

$$\omega_{n1,2}^2 = \frac{1}{2r}\left[(\omega_{nx}^2 + \omega_{n\phi}^2) \pm \sqrt{(\omega_{nx}^2 + \omega_{n\phi}^2)^2 - 4r\omega_{nx}^2\,\omega_{n\phi}^2}\right] \tag{11.71}$$

From the property of quardratic equation:

$$\omega_{n1}^2 + \omega_{n2}^2 = \frac{\omega_{nx}^2 + \omega_{n\phi}^2}{r} \tag{11.72}$$

$$\omega_{n1}^2 \times \omega_{n2}^2 = \frac{\omega_{nx}^2\,\omega_{n\phi}^2}{r} \tag{11.73}$$

and

$$\omega_{n1}^2 - \omega_{n2}^2 = \frac{1}{r}\left[(\omega_{nx}^2 + \omega_{n\phi}^2) - 4r\,\omega_{n\phi}^2\,\omega_{n\phi}^2\right]^{1/2} \tag{11.74}$$

It can be proved that ω_{nx} and $\omega_{n\phi}$ will always lie between limiting natural frequencies ω_{n1} and ω_{n2}.

11.4.5.2 Amplitudes of coupled rocking and sliding

The amplitudes of vibration are determined in the following three cases:

Case I. If only the horizontal force $F_x \sin \omega t$ is acting: Equations (11.51) and (11.52) may be rewritten as follows:

$$m\ddot{x}+C_\tau Ax-C_\tau AL\phi = F_x \sin \omega t \tag{11.75}$$

$$M_m\ddot{\phi}+\phi(C_\tau AL^2+C_\phi I-WL)-C_\tau AL_x = 0 \tag{11.76}$$

Assume that the particular solution to these equations are

$$x = A_x \sin \omega t$$

$$\phi = A_\phi \sin \omega t$$

in which A_x and A_ϕ are the maximum sliding and rocking amplitudes respectively. By substituting these solutions into the above equations, we get

$$A_x(C_\tau A-m\omega^2)-C_\tau ALA_\phi = F_x \tag{11.77}$$

$$-C_\tau ALA_x+A_\phi(C_\tau AL^2+C_\phi I-WL-M_m\omega^2) = 0 \tag{11.78}$$

or

$$A_x = \frac{(C_\tau AL^2+C_\phi I-WL-M_m\omega^2)}{C_\tau AL}\cdot A_\phi \tag{11.79}$$

By substituting for A_x in Eq. (11.77), we get

$$\frac{(C_\tau AL^2+C_\phi I-WL-M_m\omega^2)(C_\tau A-m\omega^2)}{C_\tau AL}A_\phi - C_\tau ALA_\phi = 0$$

$$A_\phi = \frac{C_\tau AL}{mM_m\left[\frac{C_\tau A+(C_\phi I-WL)}{mM_m}-\omega^2\left[\frac{C_\tau A(mL^2+M_m)}{mM_m}+\frac{(C_\phi I-WL)}{M_m}\right]+\omega^4\right]}\times F_x$$

$$= \frac{C_\tau AL}{mM_m\left[\frac{\omega_{nx}^2\,\omega_{n\phi}^2}{r}-\frac{\omega^2}{r}(\omega_{nx}^2+\omega_{n\phi}^2)+\omega^4\right]}F_x \tag{11.80}$$

By using the relations given in (11.72) and (11.73) into Eq. (11.80), we get

$$A_\phi = \frac{C_\tau AL}{mM_m\left[\omega_{n1}^2\,\omega_{n2}^2-\omega^2(\omega_{n1}^2+\omega_{n2}^2)+\omega^4\right]}\times F_x$$

$$= \frac{C_\tau AL}{mM_m(\omega_{n1}^2-\omega^2)(\omega_{n2}^2-\omega^2)}F_x \tag{11.81}$$

Let, $$mM_m(\omega_{n1}^2-\omega^2)(\omega_{n2}^2-\omega^2) = \Delta(\omega^2) \tag{11.82}$$

$$A_\phi = \frac{C_\tau AL}{\Delta(\omega)^2}F_x \tag{11.83}$$

By substituting for A_ϕ in Eq. (11.79). we get

$$A_x = \frac{C_\tau AL^2+C_\phi I-WL-M_m\omega^2}{\Delta(\omega^2)}F_x \tag{11.84}$$

Case II. If only moment $M_y \sin \omega\, t$ is acting: Equations (11.51) and (11.52) may be rewritten as

$$m\ddot{x}+C_\tau Ax-C_\tau AL\phi = 0 \tag{11.85}$$

and $$M_m\ddot{\phi}-\phi(C_\tau AL^2+C_\phi I-WL)-C_\tau ALx = M_y \sin \omega t \tag{11.86}$$

By assuming solutions as for Eqs. (11.75) and (11.76), it can be shown that the following expressions hold:

$$A_x = \frac{C_\tau AL}{\Delta(\omega^2)} M_y \tag{11.87}$$

and $$A_\phi = \frac{C_\tau A - m\omega^2}{\Delta(\omega^2)} M_y \tag{11.88}$$

Case III. If both the unbalanced force F_x and moment M_y are acting, the amplitudes of motion are determined as follows:

$$A_x = \frac{(C_\tau AL^2 + C_\phi I - WL - M_m\omega^2)F_x + (C_\tau AL)M_y}{\Delta(\omega^2)} \tag{11.89}$$

and $$A_\phi = \frac{(C_\tau AL)F_x + (C_\tau A - m\omega^2)M_y}{\Delta(\omega^2)} \tag{11.90}$$

The total amplitude of the vertical and horizontal vibrations are given by

$$A_v = A_z + \frac{a}{2} A_\phi \tag{11.91}$$

and $$A_h = A_x + hA_\phi \tag{11.92}$$

where h = Height of the top of the foundation above the combined centre of gravity.

In foundations with two degrees of freedom, specific forms of vibrations correspond to the frequencies ω_{n1} and ω_{n2}. These vibrations are characterised by a certain interrelationship between the amplitudes A_x and A_ϕ which depends on the foundation size and the soil properties, but does not depend on the initial conditions of foundation motion.

Let us examine the case when the foundation is subjected to exciting moment M_y only. The ratio of amplitudes A_x and A_ϕ obtained using Eqs. (11.87) and (11.88) is given by

$$\rho = \frac{A_x}{A_\phi} = \frac{C_\tau AL}{C_\tau A - m\omega^2} = \frac{\omega_{nx}^2}{\omega_{nx}^2 - \omega^2} L \tag{11.93}$$

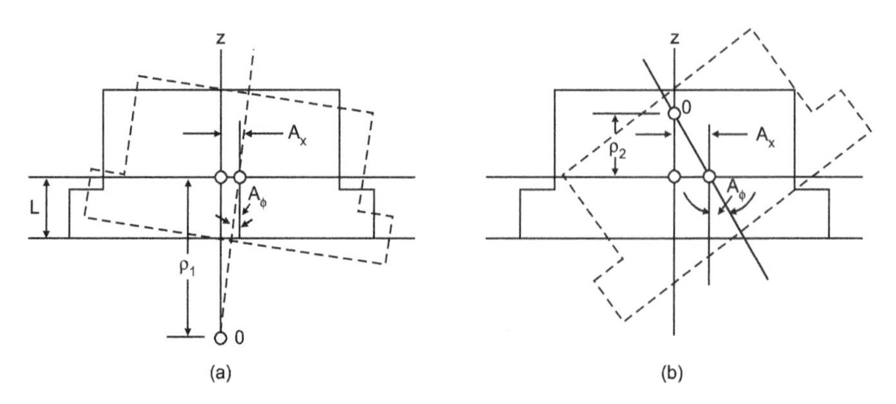

Fig. 11.9. (a) Rocking and sliding in phase with each other and (b) Rocking and sliding in opposite phase

The following cases are important for consideration of form of vibration:

(i) If $\omega << \omega_{nx}$, then $\rho \cong L$. It means that the foundation rotates about an axis that passes through the centroid of the base contact area and sliding is absent.

(ii) If $\omega = \omega_{n2}$, ω_{n2} being the lower limiting natural frequency, then $\omega_{nx}^2 - \omega_{n2}^2 > 0$. It means that during vibration at frequency ω_{n2}, when the centre of gravity deviates from the equilibrium position, for example, the positive direction of the X - axis, the rotation of the foundation will also be positive, and changes of amplitudes A_x and A_ϕ will be in phase. The form of vibration will be as shown in Fig. ll.9 *a*, i.e. the foundation will undergo rocking vibrations with respect to a point situated at a distance ρ_1 from the centre of gravity of foundation. The value of ρ_1 is determined by the absolute value of expression (Eq. 11.93) if ω_{n2}, is substituted for ω_n.

(iii) If $\omega = \omega_{n1}$ then $\omega_{nx}^2 - \omega_{n1}^2 < 0$, ρ will be negative, and A_x and A_ϕ will be out of phase. Figure 11.9 *b* illustrates the form of vibrations around a point which lies higher than the centre of gravity and at a distance ρ_2 determined from expression (Eq. 11.93) if ω_{n1} is substituted for ω_n.

11.5 Elastic Half-Space Method

11.5.1 Vertical Vibrations

Lamb (1904) studied the problem of vibration of a single oscillating force (Vertical or horizontal. Fig. 11.10) acting at a point on the surface of an elastic half space. Reissner (1936) developed the analysis for the problem of vibration of a uniformly loaded flexible circular area (Fig. 11.11) by integration of Lamb's solution for a point load. Based on his work, the vertical displacement at the centre of the circular area is given by

$$Z_o = \frac{F_o e^{i\omega t}}{G r_o}(f_1 + i f_2) \tag{11.94}$$

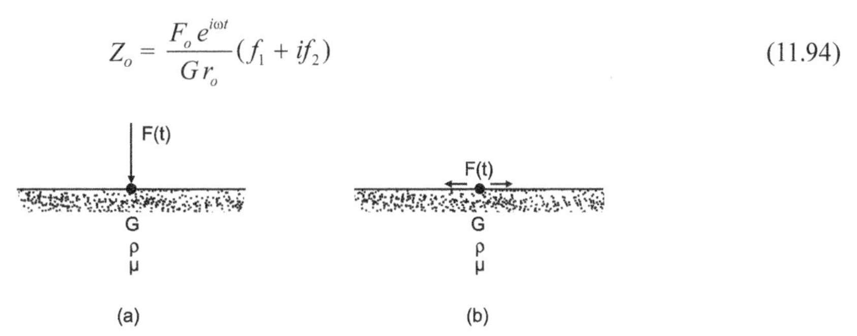

Fig. 11.10. Oscillating force on the surface of elastic half-space

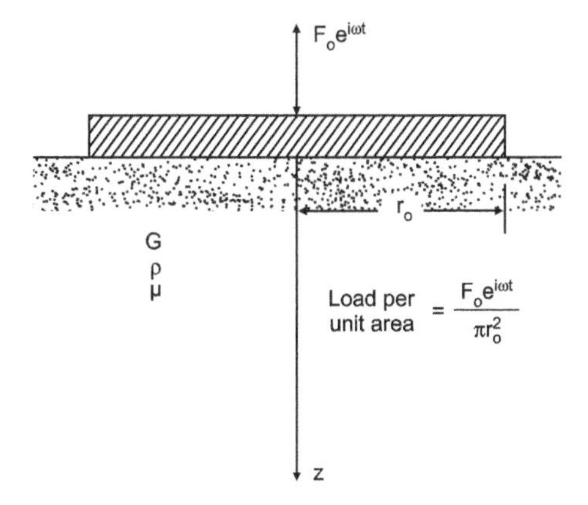

Fig. 11.11. Vibration of a uniformly loaded circular flexible area

where F_o = Magnitude of oscillatory force

ω = Forcing frequency, rad/s

G = Dynamic shear modulus of the medium

r_o = Radius of the footing

f_1, f_2 = Displacement functions

$i = \sqrt{-1}$

Reissner (1936) introduced following two non-dimensional terms:

(a) *Mass ratio, b*: It is given by

$$b = \frac{m}{\rho r_o^3} = \frac{W}{\gamma r_o^3} \tag{11.95}$$

It describes the relation between the mass of vibrating footing and a certain mass of the elastic half-space.

(b) *Dimensionless frequency, a_o*: It is given by

$$a_o = \omega r_o \sqrt{\frac{\rho}{G}} = \omega r_o / v_s \tag{11.96}$$

where m = Mass of the machine and foundation

v_s = Shear wave velocity

γ = Unit weight of soil

ρ = Mass density of soil

Using the displacement Eq. (11.94), and solving the equation of equilibrium forces, Reissner obtained the following expression for the amplitude of foundation having circular base:

$$A_{zn} = \frac{A_z}{(F_o / Gr_o)} = \sqrt{\frac{f_1^2 + f_2^2}{(1 - ba_o^2 f_1)^2 + (ba_o^2 f_2)^2}} \tag{11.97}$$

where A_z = Amplitude of foundation

A_{zn} = Dimensionless amplitude

Values of displacement functions f_1 and f_2 are found dependent on Poisson's ratio μ and dimensionless frequency factor a_o. Their values for flexible circular foundation are given in Table 11.2.

Table 11.2 Values of Displacement Function of Flexible Foundations (Bowles, 1977)

Poisson's Ratio μ	*Values of* (f_1)
0	$0.318310 - 0.092841\ a_o^2 + 0.007405\ a_o^4$
0.25	$0.238733 - 0.059683\ a_o^2 + 0.004163\ a_o^4$
0.5	$0.159155 - 0.039789\ a_o^2 + 0.002432\ a_o^4$
	Values of $(-f_2)$
0	$0.214474\ a_o - 0.029561\ a_o^3 + 0.001528\ a_o^5$
0.25	$0.148594\ a_o - 0.017757\ a_o^3 + 0.000808\ a_o^5$
0.5	$0.104547\ a_o - 0.011038\ a_o^3 + 0.000444\ a_o^5$

The classical work of Reissner (1936) for circular loaded area was extended by Quinlan (1953) and Sung (1953) for the following three contact pressure distributions:

(i) Rigid base (Fig. 11.12 a), $f_z = \dfrac{F_o e^{i\omega t}}{2\pi r_o \sqrt{r_o^2 - r^2}}$ for $r \le r_o$ (11.98)

(ii) Uniform (Fig. 11.12b), $$f_z = \frac{F_o e^{i\omega t}}{\pi r_o^2} \text{ for } r \leq r_o \tag{11.99}$$

It is the same as considered by Reissner *i.e.* for flexible foundations with circular base.

(iii) Parabolic (Fig. 11.12c), $$f_z = \frac{2(r_o^2 - r^2)F_o e^{i\omega t}}{\pi r_o^4} \text{ for } r \leq r_o \tag{11.100}$$

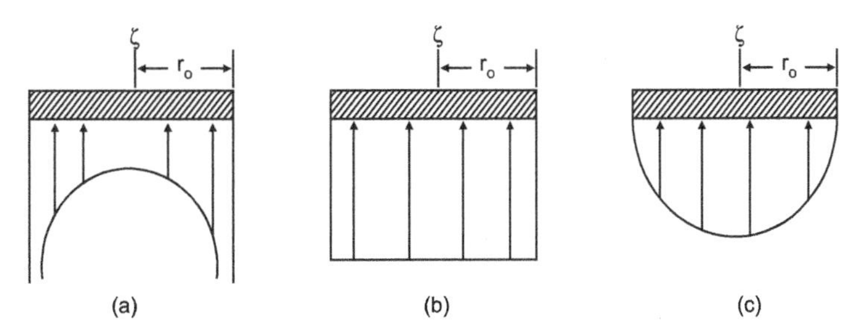

Fig. 11.12. Contact pressure distribution under a circular foundation

In the above equations f_z is the contact pressure at a distance r measured from the centre of foundation. Equation (11.97) holds good for all the three types of contact pressure distributions with changed values of f_1 and f_2. The values of f_1 and f_2 for rigid base foundations were computed by Sung (1953) on the assumption that the pressure distribution remain unchanged with frequency. Their values are given in Table 11.3.

Table. 11.3. Values of Displacement Functios for Rigid Foundations (Bowles, 1977)

Poisson's ratio μ	*Values of* f_1
0	$0.250000 - 0.1011375\, a_o^2 + 0.0101105\, a_o^4$
0.25	$0.187500 - 0.070313\, a_o^2 + 0.006131\, a_o^4$
0.5	$0.125000 - 0.046875\, a_o^2 + 0.003581\, a_o^4$
	Values of $-f_2$
0	$0.214470\, a_o - 0.311416\, a_o^3 + 0.002444\, a_o^5$
0.25	$0.148594\, a_o - 0.023677\, a_o^3 + 0.0012114\, a_o^5$
0.5	$0.104547\, a_o - 0.014717\, a_o^3 + 0.007170\, a_o^5$

Figure 11.13 shows a typical plot of A_{zn} versus a_o for various values of mass ratio b for a rigid base circular footing subjected to a constant force excitation $F_o e^{i\omega t}$. A high mass ratio (greater height of footing and smaller contact radius) implies a large amplitude of vibration for a given set of conditions.

Manytimes foundations are subjected to a frequency dependent excitation (Fig. 2.15). The amplitude of the external oscillating force is given by

$$F_o = 2m_e e\omega^2 \tag{11.101}$$

where $2m_e$ = Total rotating mass.

For this condition, the amplitude of vibration A_{ze} is given by

$$A_{ze} = \frac{2m_e e\omega^2}{Gr_o}\sqrt{\frac{f_1^2 + f_2^2}{(1-ba_o^2 f_1)^2 + (ba_o^2 f_2)^2}} \tag{11.102}$$

or $$A_{zen} = \frac{A_{ze}}{(2m_e e/\rho r_o^3)} = \sqrt{\frac{f_1^2 + f_2^2}{(1-ba_o^2 f_1)^2 + (ba_o^2 f_2)^2}} \cdot a_o^2 \tag{11.103}$$

where A_{zen} = Non-dimensional amplitude in frequency dependent excitation.

Figure 11.14 shows a typical plot of A_{zen} versus a_o for various values of mass ratio b for rigid base circular footing subjected to frequency dependent excitation.

It may be noted that the curves shown in Figs. 11.13 and 11.14 are similar to the frequency-amplitude curves, shown in Figs. 2.13 and 2.16 respectively.

Richart and Whitman (1967) have studied the effect of the shape of contact pressure distribution and Poisson's ratio on amplitude-frequency response of rigid circular footing subjected to frequency dependent excitation. Figure 11.15 demonstrates the nature of variation of A_{zen} with a_o for three types of contact pressure distribution; *i.e.* uniform, rigid and parabolic. Parabolic and uniform pressure distributions produced higher displacement than a rigid base. The effect of Poisson's ratio on the variation of A_{zen} can be seen in Fig. 11.16. The peak value of A_{zen} decreases with the increases in the value of μ, but the corresponding value of a_o increases with increase in μ.

From the amplitude-frequency curves in Figs. 11.13 and 11.14, one can obtain: (i) maximum amplitude, and (ii) value of a_o for maximum amplitude (*i.e.* resonance condition) for different values of b. Richart (1962) have plotted this data in graphical form as shown in Figs. 11.17 and 11.18 which are convenient for design use.

Lysmer and Richart (1966) proposed a simplified mass-spring-dashpot analog for calculating the response of a rigid circular footing subjected to vertical oscillations. The values of spring constant K_z and damping constant c_z were taken as given below:

$$K_z = \frac{4Gr_o}{1-\mu} \tag{11.104}$$

and

$$C_z = \frac{3.4r_o^2}{1-\mu}\sqrt{\rho G} \tag{11.105}$$

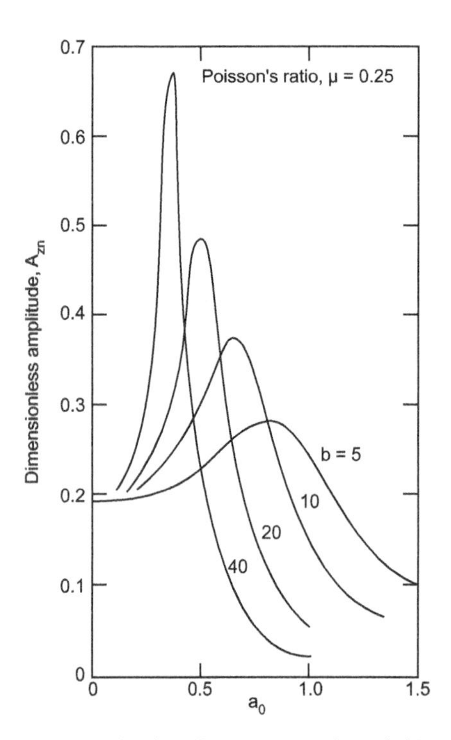

Fig. 11.13. Plot of A_{zn} versus a_0 for a rigid circular foundation subjected to constant excitation force (Richart, 1962)

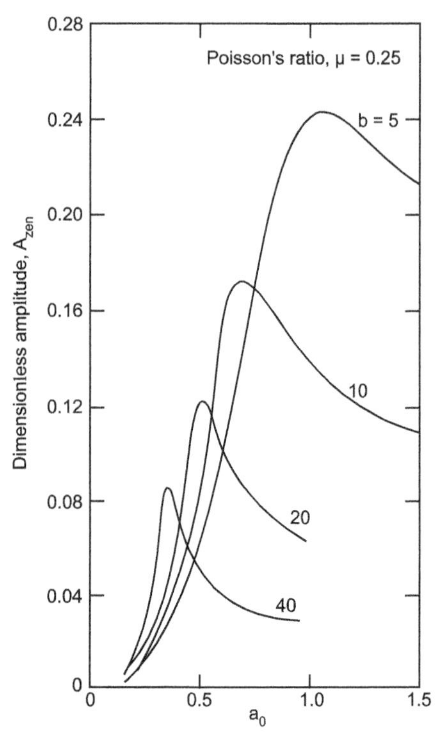

Fig. 11.14. Plot of A_{zen}, versus a_0 for a rigid circular foundation subjected to frequency dependent excitation (Richart, 1962)

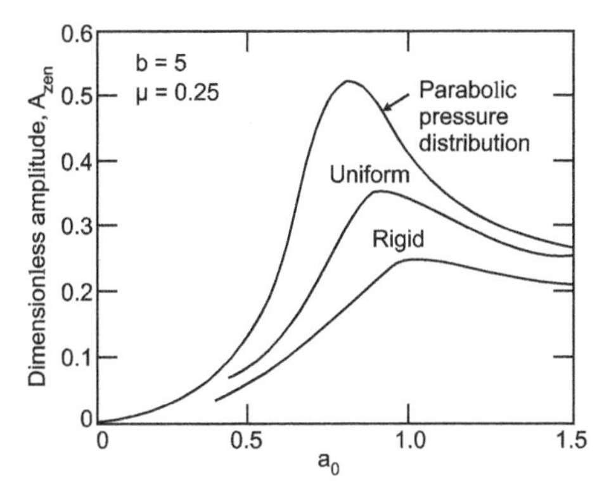

Fig. 11.15. Effect of contact pressure distribution on the variation of A_{zen} with a_0 (Richart and whitman, 1967)

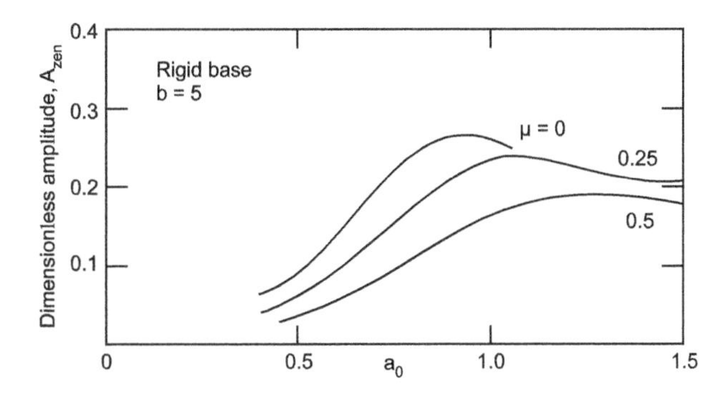

Fig. 11.16. Effect of Poisson's ratio on the variation of A_{zen} with a_0 (Richart and Whitman, 1967)

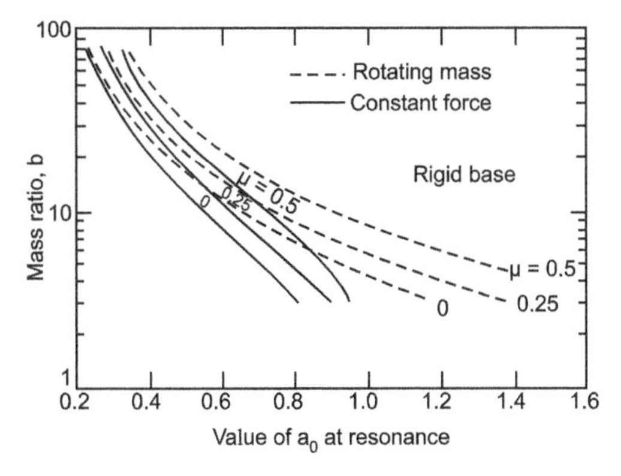

Fig. 11.17. Plot of mass ratio b versus a_0 for resonance condition for vertical vibrations (Richart, 1962)

The equation of motion may thus be written as

$$m\ddot{z}+\frac{3.4r_o^2}{1-\mu}\sqrt{\rho G}\cdot\dot{z}+\frac{4Gr_o}{1-\mu}\cdot z = F_z e^{i\omega t} \tag{11.106}$$

Lysmer and Richart (1966) also suggested the modified mass ratio as

$$B_z = \frac{1-\mu}{4}b = \frac{1-\mu}{4}\cdot\frac{m}{\rho r_o^3} \tag{11.107}$$

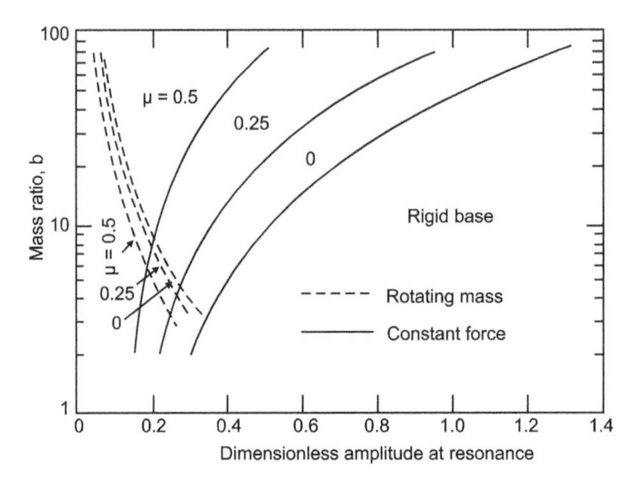

Fig. 11.18. Plot of mass ratio b versus dimensionless amplitude at resonance (Richart, 1962)

The damping ratio ξ_z is obtained as

$$\xi_z = \frac{C_z}{C_c} = \frac{C_z}{2\sqrt{K_z m}} = \frac{3.4r_o^2}{2(1-\mu)} \cdot \frac{\sqrt{\rho G}}{\sqrt{\dfrac{4Gr_o m}{1-\mu}}} \tag{11.108}$$

Putting the value of m in terms of B_z from Eq. (11.107), Eq. (11.108) becomes

$$\xi_z = \frac{0.425}{\sqrt{B_z}} \tag{11.109}$$

The response of the system can be studied using Eq. (2.58).

The dashed curves in Fig. 11.19 illustrate how well the response curves for the analog agree with the response curves for half space method. The derivation of magnification factor, M_z is given below.

The natural frequency of the system is described by Eq. (11.106), *i.e.*

$$\omega_{nz} = \sqrt{\frac{K_z}{m}} \cdot \sqrt{1-\xi_z^2} \tag{11.110}$$

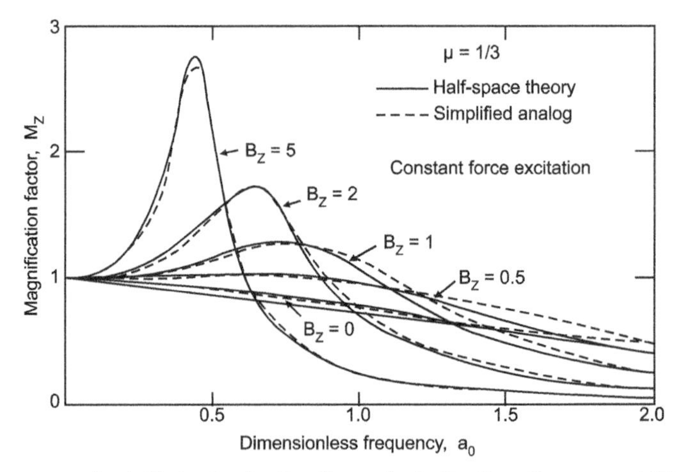

Fig. 11.19. Response of a rigid circular footing for vertical vibrations (Lysmer and Richart, 1966)

Putting the values of K_z, m and ξ_z from Eqs. (11.104), (11.107) and (11.109) into Eq.(11.110), we get

$$\omega_{nz} = \sqrt{\frac{(B_z - 0.36)G}{\rho}} \cdot \frac{1}{B_z r_o} \tag{11.111}$$

and at resonance the amplitude is given by Eq. (2.61). *i.e.*

$$(A_z)_{max} = \frac{F_o / K_z}{2\xi_z \sqrt{1-\xi_z^2}} \tag{11.112}$$

Putting the values of ξ_z, from Eq. (11.109) into Eq. (11.112), we get

$$(A_z)_{max} = \frac{F_o}{K_z\, 0.85} \frac{B_z}{\sqrt{B_z - 0.18}} \tag{11.113a}$$

or

$$(A_z)_{max} = \frac{F_o \cdot (1-\mu)}{4Gr_o} \cdot \frac{B_z}{0.85\sqrt{B_z - 0.18}} \tag{11.113b}$$

$$\text{Magnification factor,} \quad M_z = \frac{(A_z)_{\max} \cdot 4Gr_o}{F_o(1-\mu)} = \frac{B_z}{0.85\sqrt{B_z - 0.18}}$$

For a frequency-dependent excitation the resonant frequency and the maximum amplitude are given by:

$$\omega_{nz} = \sqrt{\frac{0.9G}{(B_z - 0.45)\rho}} \cdot \frac{1}{r_o} \tag{11.114}$$

and

$$(A_z)_{max} = \frac{2m_e e}{m} \cdot \frac{B_z}{0.85\sqrt{B_z - 0.18}} \tag{11.115a}$$

$$\text{Magnification factor,} \quad M_{ze} = \frac{(A_z)_{\max}}{\dfrac{2m_e e}{m}} = \frac{B_z}{0.85\sqrt{(B_z - 0.18)}} \tag{11.115b}$$

where $2m_e$ = Unbalanced rotating mass

e = Eccentricity of mass from the axis of rotation.

11.5.2 Pure Sliding Vibrations

Arnold et al. (1955) have obtained theoretical solutions for sliding vibrations of rigid circular foundations (Fig. 11.20) subjected to an oscillatory horizontal force $F_o e^{i\omega t}$. They have presented the solutions for two cases namely (i) constant force excitation, and (ii) frequency dependent excitation.

In constant force excitation (F_o = constant), the amplitude A_x is given by

$$A_x = \frac{F_o}{Gr_o} A_{xn} \tag{11.116}$$

where A_{xn} = Non-dimensional amplitude factor.

The variation of amplitude versus frequency is shown in Fig. 11.21 by dotted lines. The envelop drawn to these curves is used to define the frequency at maximum amplitude. The definition of mass ratio b is same as given in Eq. (11.95). The plot of b versus a_o for resonant amplitude is given in Fig. 11.22.

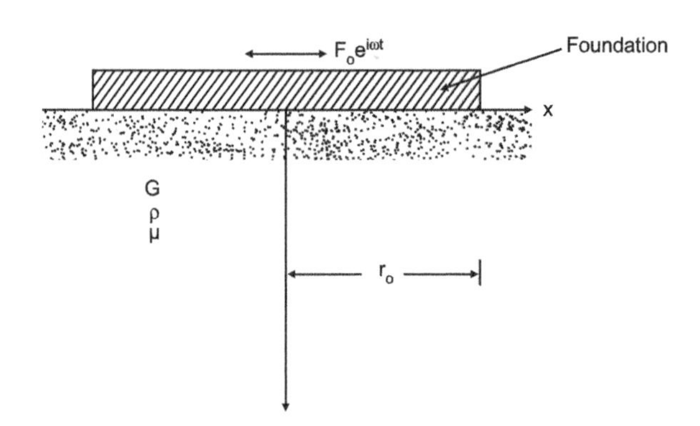

Fig. 11.20. Rigid circular foundation subjected to sliding oscillations

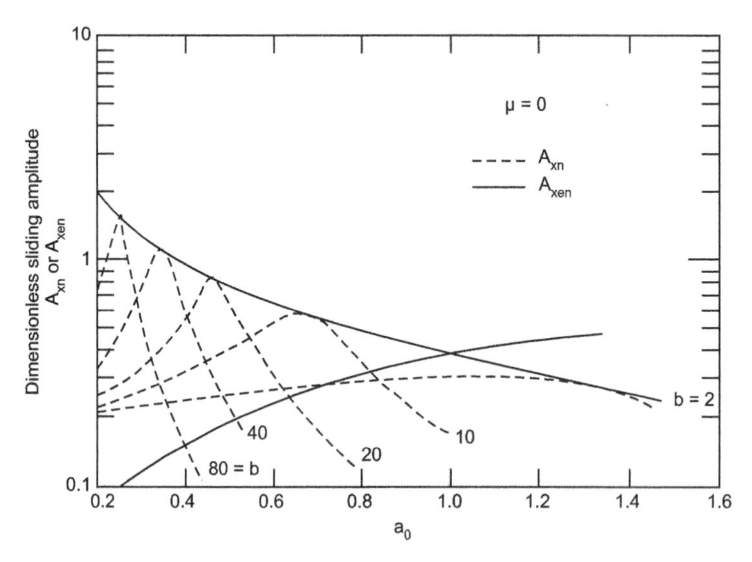

Fig. 11.21. Plots of A_{xn} and A_{xen} versus a_0 for sliding vibrations (Richart, 1962)

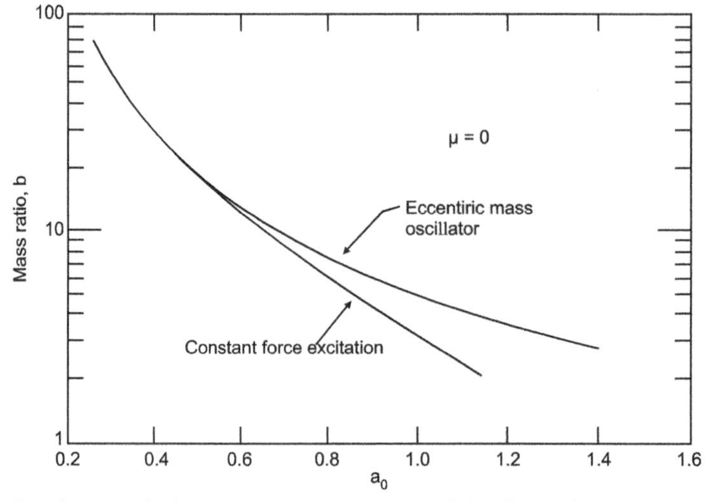

Fig. 11.22. Plot of mass ratio b versus a_0 for resonance condition for sliding vibrations (Richart, 1962)

For the case $F_o = 2\, m_e\, e\, \omega^2$, the value of A_x is given by:

$$A_x = \frac{2 m_e e}{\rho r_o^3} A_{xen} \tag{11.117}$$

The firm line in Fig. 11.21 shows the envelop of A_{xen} versus a_o for resonant condition. The variation of the mass ratio b versus a_o for resonant condition is given in Fig. 11.22.

Hall (1967) proposed a simplified mass-spring dashpot analog for calculating the response of a rigid circular footing subjected to sliding vibrations. The values of the spring constant K_x and damping constant C_x were taken as given below.

$$K_x = \frac{32(1-\mu)}{7-8\mu} G r_o \tag{11.118}$$

and

$$C_x = \frac{18.4(1-\mu)}{7-8\mu} r_o^2 \sqrt{\rho G} \tag{11.119}$$

The equation of motion thus can be written as

$$m\ddot{x} + \frac{18.4(1-\mu)}{7-8\mu} r_o^2 \sqrt{\rho G} \cdot \dot{x} + \frac{32(1-\mu)}{7-8\mu} G r_o \cdot x = F_o e^{i\omega x} \tag{11.120a}$$

He also suggested the modified mass ratio B_x, for sliding vibrations as

$$B_x = \frac{7-8\mu}{32(1-\mu)} \frac{m}{\rho r_o^3} \tag{11.120b}$$

The damping ratio ξ_x, is given by

$$\xi_x = \frac{C_x}{C_c} = \frac{C_x}{2\sqrt{K_x m}} \tag{11.121}$$

Putting the values of C_x and K_x frorn Eqs. (11.118) and (11.119) in Eq. (11.121) we get

$$\xi_x = \frac{0.2875}{\sqrt{B_x}} \tag{11.122}$$

Figure 11.23 illustrates how well the response curves for the analog agree with the response curves for the half space model.

The natural frequency ω_{nx} and maximum value of amplitude $(A_x)_{max}$ can be computed using Eqs. (11.123) and (11.124) respectively.

$$\omega_{nx} = \sqrt{\frac{K_x}{m}} \cdot \sqrt{1-\xi_x^2} \tag{11.123}$$

and

$$(A_x)_{max} = \frac{F_x / K_x}{2\xi_x \sqrt{1-\xi_x^2}} \tag{11.124}$$

11.5.3 Pure Rocking Vibrations

Arnold et al. (1955) and Bycroft (1956) have obtained theoretical solutions for rigid circular foundations subjected to pure rocking vibrations (Fig. 11.24). The contact pressure below the foundation is varied according to

$$q = \frac{3 M_y\, r \cos\alpha}{2\pi r_o^3 \sqrt{r_o^2 - r^2}} e^{i\omega t} \quad (\text{for } r \le r_o) \tag{11.125}$$

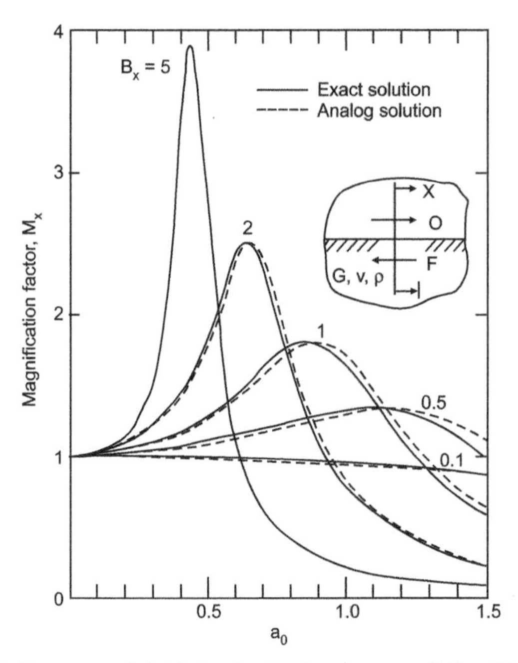

Fig. 11.23. Response of rigid circular footing for pure sliding (Hall, 1967)

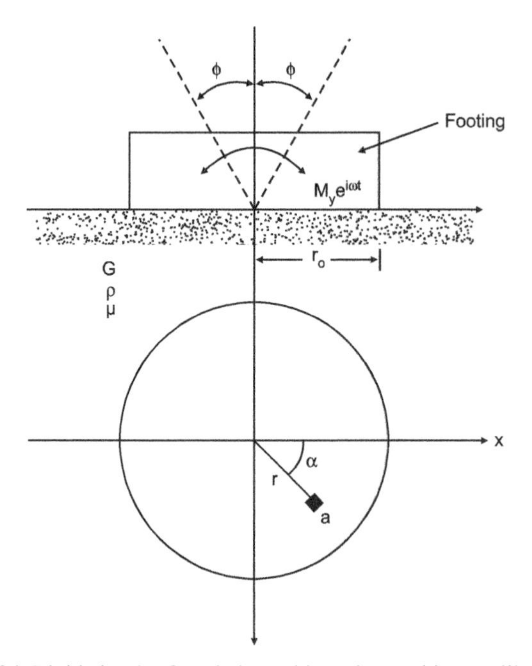

Fig. 11.24. Rigid circular foundation subjected to rocking oscillations

where M_y = Exciting moment about Y-Y axis.
α = Angle of rotation

Borowicka (11143) gave the following equation for computing static rotation of the foundation under the static application of moment M_y.

$$A_{\phi s} = \frac{3(1-\mu)}{8} \frac{M_{mo}}{\rho r_o^3} \tag{11.126}$$

Under dynamic conditions the amplitude of rocking is a function of the inertia ratio B_ϕ which is given by

$$B_\phi = \frac{3(1-\mu)}{8}\frac{M_{mo}}{\rho r_o^5} \tag{11.127}$$

where M_{mo} = Mass moment of inertia of machine and foundation about the axis of rotation.

For the dynamic moment M_y, the amplitude of angular rotation A_ϕ. can be expressed as

$$A_\phi = \left(\frac{M_y}{G r_o^3}\right)\cdot A_{\phi n} \tag{11.128}$$

where $A_{\phi n}$ = Non-dimensional rotational amplitude.

Figure 11.25 *a* shows the variation of $A_{\phi n}$ with dimensionless frequency a_0 for various values of inertia ratio B_ϕ. The envelop curve shown by the firm line can be used to define the relation between a_o at maximum amplitude (resonant condition). The plot of inertia ratio B_ϕ versus a_o for resonant amplitude is shown in Fig. 11.25*b*.

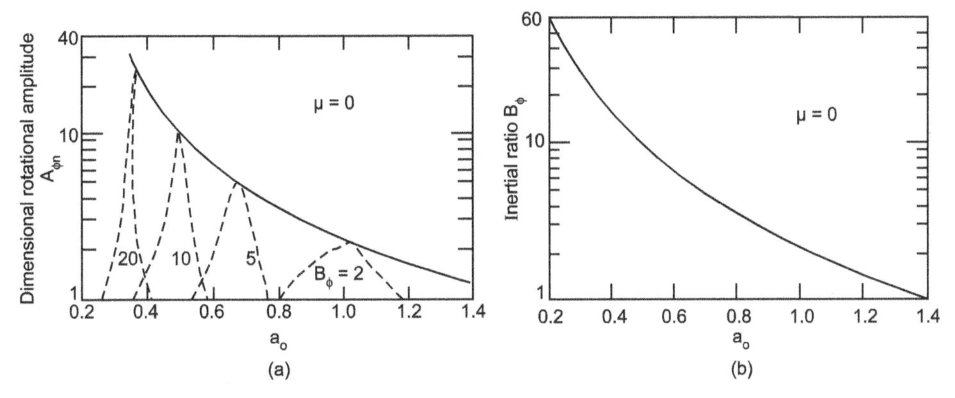

Fig. 11.25. (a) Plot of A_{fn} versus a_o (Richart, 1962) and (b) Plot of B_f versus a_o for resonance condition for rocking oscillations of rigid circular foundation (Richart, 1962)

Hall (1967) proposed an equivalent mass-spring-dashpot analog for calculating the response of a rigid circular footing subjected to rocking vibrations. The values of spring constant K_ϕ and damping constant C_ϕ were taken as given below:

$$K_\phi = \frac{8 G r_o^3}{3(1-\mu)} \tag{11.129}$$

$$C_\phi = \frac{0.8 r_o^4 \sqrt{G\rho}}{(1-\mu)(1+B_\phi)} \tag{11.130}$$

The equation of motion can be written as

$$M_{m\phi}\,\ddot{\phi} + \frac{0.8 r_o^4\sqrt{G\rho}}{(1-\mu)(1+B_\phi)}\cdot\dot{\phi} + \frac{8 G r_o^3}{3(1-\mu)}\cdot\phi = M_y\, e^{i\omega t} \tag{11.131}$$

For critical damping,

$$C_{\phi c} = 2\sqrt{K_\phi M_{mo}} \tag{11.132}$$

Therefore damping ratio ξ_ϕ will be

$$\xi_\phi = \frac{C_\phi}{C_{\phi c}} = \frac{0.15}{(1+B_\phi)\sqrt{B_\phi}} \tag{11.133}$$

Figure 11.26 illustrates how well the response curves for the analog agree with the response curves for the half space model. The undamped natural frequency $\omega_{n\phi}$ and amplitude A_ϕ in rocking vibrations are given by

$$\omega_{n\phi} = \sqrt{\frac{K_\phi}{M_{mo}}} \tag{11.134}$$

$$A_\phi = \frac{M_y}{K_\phi \left[\left\{ 1 - \frac{\omega^2}{\omega_{n\phi}^2} \right\}^2 + \left\{ 2\xi_\phi \frac{\omega}{\omega_{n\phi}} \right\}^2 \right]^{1/2}} \tag{11.135}$$

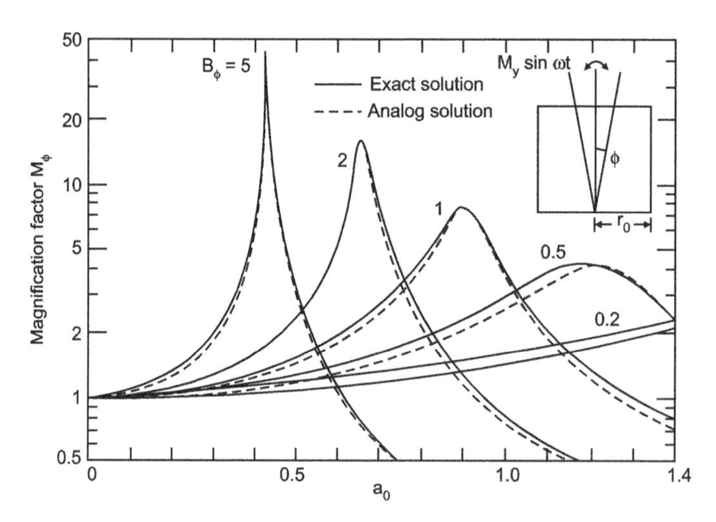

Fig. 11.26. Response of rigid circular foundations subjected to pure rocking vibrations (Hall, 1967)

11.5.4 Torsional Vibrations

Ressner and Sagoci (1944) have obtained theoretical solutions for rigid circular foundations subjected to torsional vibrations (Fig. 11.27). The variation of tangential shear stress is given by

$$\tau_{z\theta} = \frac{3}{4\pi} \cdot \frac{Mr}{r_o^3\sqrt{r_o^2 - r^2}} \quad \text{for } 0 < r < r_o \tag{11.136}$$

where $\tau_{z\theta}$ = Tangential shear stress
$M = M_z e^{i\omega t}$ = Moment at any time t
M_z = Maximum moment about Z-axis

For a static moment M_z the angle of rotation $A_{\psi s}$ is given by

$$A_{\psi s} = \left[\frac{3}{16 G r_o^3} \right] M_z \tag{11.137}$$

The amplitude of the angle of rotation A_ψ can be expressed as

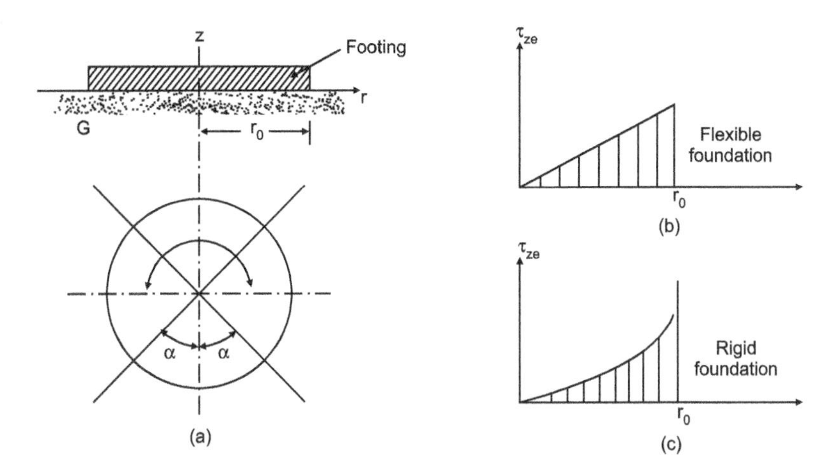

Fig. 11.27. Rigid circular foundation subjected to torsional vibrations

$$A_\psi = \frac{M_z}{G r_o^3} A_{\psi n} \tag{11.138}$$

where $A_{\psi n}$ = Non-dimensional amplitude factor.

Under dynamic condition the amplitude of torsion is a function of inertia ratio B_ψ which is given by

$$B_\psi = \frac{M_{mz}}{\rho r_o^5} \tag{11.139}$$

where M_{mz} = Polar mass moment of inertia of the machine and foundation about the axis of rotation.

Figure 11.28 a shows the variation of $A_{\psi n}$ with dimensionless frequency a_o for various values of inertia ratio B_ψ. The envelop curve shown by the firm line can be used to define the relation between a_o at maximum amplitude (resonant condition) and the values of inertia ratio B_ψ (Fig. 11.28 b).

Richart and Whitman (1967) proposed an equivalent mass-spring-dashpot analog for calculating the response of a rigid circular footing subjected to torsional vibrations. The values of spring constant K_ψ and damping constant C_ψ were taken as below:

$$K_\psi = \frac{16}{3} G r_o^3 \tag{11.140}$$

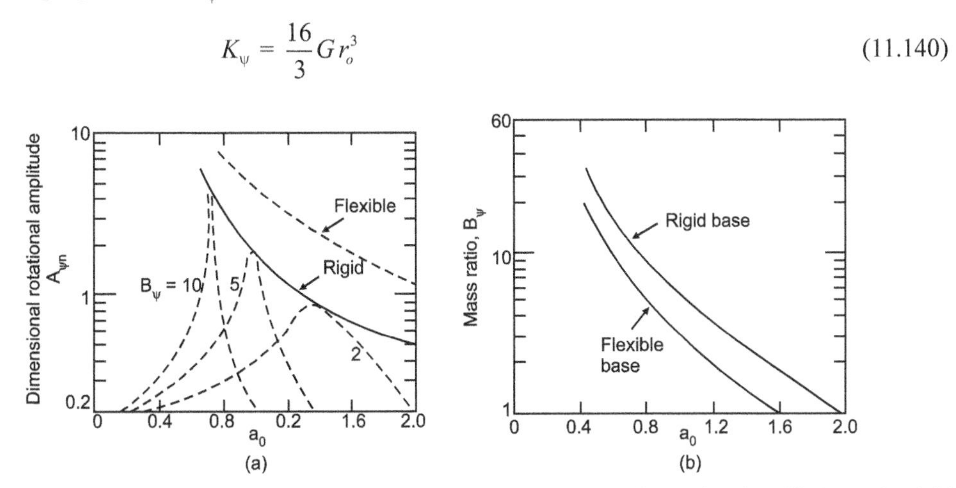

Fig. 11.28. (a) Plot of $A_{\psi n}$ versus a_o; (b) Plot of B_ψ, at resonance versus a_o for torsional oscillations of a rigid circular foundation (Richart, 1962)

and $$C_\psi = \frac{1.6 r_o^4 \sqrt{\rho G}}{1+B_\psi} \tag{11.141}$$

The equation of motion can be written as

$$M_{mz}\,\ddot{\psi} + \frac{1.6 r_o^4 \sqrt{\rho G}}{1+B_\psi}\cdot\dot{\psi} + \frac{16}{3} G r_o^3 \cdot \psi = M_z e^{i\omega t} \tag{11.142}$$

The damping ratio ξ_ψ will be

$$\xi_\psi = \frac{C_\psi}{2\sqrt{K_\psi \cdot m}} = \frac{0.5}{(1+2B_\psi)} \tag{11.143}$$

The undamped natural frequency $\omega_{n\psi}$ and amplitude A_ψ of the torsional vibration are given by

$$\omega_{n\psi} = \sqrt{\frac{K_\psi}{M_{m\psi}}} \tag{11.144}$$

$$A_\psi = \frac{M_z}{K_\psi \left[\left\{1 - \frac{\omega^2}{\omega_{n\psi}^2}\right\}^2 + \left\{2\xi_\psi \frac{\omega}{\omega_{n\psi}}\right\}^2\right]^{1/2}} \tag{11.145}$$

11.5.5 Coupled Rocking and Sliding Vibrations

Figure 11.29a shows a rigid circular foundation resting on the surface of elastic half-space and subjected to an oscillatory moment $M_y\, e^{i\omega t}$ and an oscillatory horizontal force $F_x\, e^{i\omega t}$. The sign convention chosen is illustrated in Fig. 11.29b, which indicates that $+x$ and $+F$ act to the right and that $+\phi$ and $+M$ are clockwise. Figure 11.29c shows the motion of the footing when both translation of the c.g. and rotation about c.g. are positive (i.e. in phase). In this case, the centre of rotation lies below the base of the footing, and the motion is termed *as first mode of vibration.* If the translation is positive while the rotation is negative, the centre of rotation lies above the c.g. and the motion is designated as *the second mode of vibration.*

From Fig. 11.29a

$$x_o = x - L\phi \tag{11.146}$$

where, x_o = Displacement of base
x = Displacement of c.g.
ϕ = Angle of rotation in radians

Therefore, $$\dot{x}_o = \dot{x} - L\dot{\phi} \tag{11.147}$$

The equations of motion are written in terms of x and ϕ. The equation of motion for sliding is

$$m\ddot{x} + C_x \dot{x}_o + K_x x_o = F_x e^{i\omega t}$$

or $$m\ddot{x} + C_x(\dot{x} - L\dot{\phi}) + K_x(x - L\phi) = F_x e^{i\omega t}$$

or $$m\ddot{x} + C_x\dot{x} + K_x x - LC_x\dot{\phi} - LK_x\phi = F_x e^{i\omega t} \tag{11.148}$$

The equations of motion for rocking is

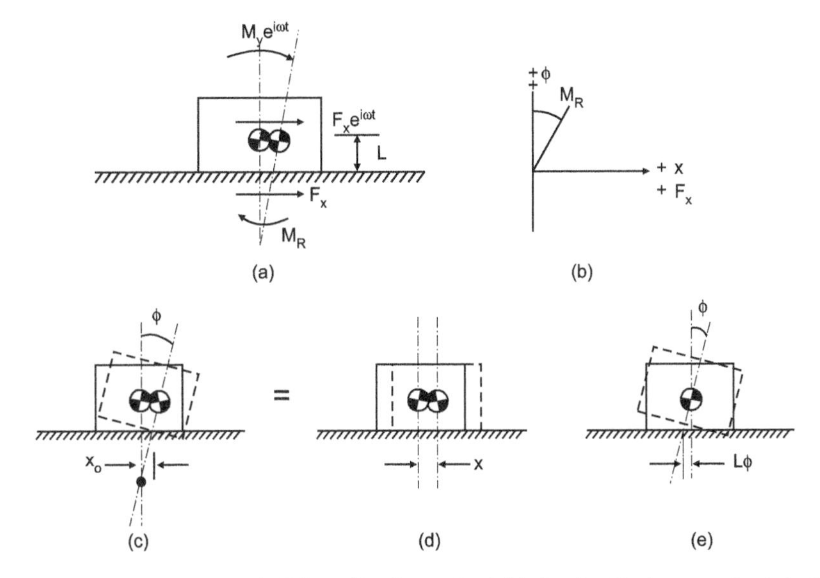

Fig. 11.29. Coupled rocking and sliding vibrations of a rigid circular block on an elastic half space (After Richart and Whitman, 1967)

$$M_m\ddot{\phi} + C_\phi\,\dot{\phi} + K_\phi\phi + C_x\cdot\dot{x}_o\cdot L + K_x\cdot x_o\cdot L = M_y\,e^{i\omega t}$$

Substituting the value of x_o from Eq. (11.147), the above equation simplifies to

$$M_m\ddot{\phi} + (C_\phi + L^2C_x)\dot{\phi} + (K_\phi + L^2K_x)\phi - LC_x\,\dot{x} - LK_x\cdot x = M_y\,e^{i\omega t} \tag{11.149}$$

The natural frequencies of coupled rocking and sliding are obtained by putting forcing functions in Eqs. (11.148) and (11.149) equal to zero. Thus

$$m\ddot{x} + C_x\,\dot{x} + K_x\,x - LC_x\dot{\phi} - LK_x\cdot\phi = 0 \tag{11.150}$$

and
$$M_m\ddot{\phi} + (C_\phi + L^2C_x)\dot{\phi} + (K_\phi + LK_x^2)\phi - LC_x\,x - LK_x\cdot x = 0 \tag{11.151}$$

The solutions of above equations can be obtained by substituting

$$x = Ae^{i\omega_{nd}t} \tag{11.152}$$

$$\phi = Be^{i\omega_{nd}t} \tag{11.153}$$

in which A and B are arbitrary constants. By doing this Eqs. (11.150) and (11.151) become

$$\frac{A}{B} = \frac{LK_x + iLC_x\cdot\omega_{nd}}{-m\omega_{nd}^2 + K_x + iC_x\,\omega_{nd}} \tag{11.154}$$

and
$$\frac{A}{B} = \frac{-M_m\,\omega_{nd}^2 + (K_\phi + L^2K_x) + i(C_\phi + L^2C_x)\omega_{nd}}{LK_x - iC_x\,\omega_{nd}} \tag{11.155}$$

Equating Eqs. (11.154) and (11.155). and substituting

$$\omega_{nx} = \sqrt{\frac{K_x}{m}},\ \omega_{n\phi} = \sqrt{\frac{K_\phi}{M_{mo}}},\ r = \frac{M_m}{M_{mo}},\ \xi_x = \frac{C_x}{2\sqrt{K_x m}}$$

and $$\xi_\phi = \frac{C_\phi}{2\sqrt{K_\phi M_{mo}}}, \text{ on simplification we get}$$

$$\left[\omega_{nd}^4 - \left(\frac{\omega_{n\phi}^2 + \omega_{nx}^2}{r} - \frac{4\xi_x \xi_\phi \omega_{nx} \omega_{n\phi}}{r}\right)\omega_{nd}^2 + \frac{\omega_{n\phi}^2 \omega_{nx}^2}{r}\right]^2$$

$$+4\left[\frac{\xi_x \omega_{nx} \omega_{nd}}{r}(\omega_{n\phi}^2 - \omega_{nd}^2) + \frac{\xi_\phi \omega_{n\phi} \omega_{nd}}{r}(\omega_{nx}^2 - \omega_{nd}^2)\right]^2 = 0 \quad (11.156)$$

It may be noted that Eq. (11.156) reduces to Eq. (11.69) when $\xi_x = \xi_\phi = 0$. As the effect of damping on natural frequency is small, the undamped natural frequencies for coupled sliding and rocking vibrations can be computed using Eq. (11.70).

The damped amplitudes of rocking and sliding of a foundation subjected to a horizontal force $F_x e^{i\omega t}$ are given by

$$A_{x1} = \frac{F_x}{m M_m} \frac{\left[(-M_m \omega^2 + K_\phi + K_x L^2) + 4\omega^2 (\xi_\phi \sqrt{K_x M_{mo}} + L^2 \xi_x \sqrt{K_x m})^2\right]^{1/2}}{\Delta(\omega^2)} \quad (11.157)$$

and $$A_{\phi 1} = \frac{F_x L}{M_m} \frac{\omega_{nx}(\omega_{nx}^2 + 4\xi_x \omega^2)^{1/2}}{\Delta(\omega^2)} \quad (11.158)$$

where $$\Delta(\omega^2) = \left[\omega^4 - \omega^2\left\{\frac{\omega_{n\phi}^2 + \omega_{nx}^2}{r} - \frac{4\xi_x \xi_\phi \omega_{nx} \omega_{n\phi}}{r}\right\} + \frac{\omega_{n\phi}^2 \omega_{nx}^2}{r}\right]$$

$$+\left[4\left\{\xi_x \frac{\omega_{nx}\omega}{r}(\omega_{n\phi}^2 - \omega^2) + \frac{\xi_\phi \omega_{n\phi}\omega}{r}(\omega_{n\phi}^2 - \omega^2)\right\}^2\right]^{1/2} \quad (11.159)$$

The damped amplitudes of rocking and sliding of the foundation subjected to an exciting moment $M_y e^{i\omega t}$ are given by

$$A_{x2} = \frac{M_y \cdot L}{M_m} \frac{[(\omega_{nx}^2)^2 + (2\xi_x \omega_{nx} \cdot \omega)^2]^{1/2}}{\Delta(\omega^2)} \quad (11.160)$$

and $$A_{\phi 2} = \frac{M_y}{M_m} \frac{[(\omega_{nx}^2 - \omega^2)^2 + (2\xi_x \omega_{nx} \omega)^2]^{1/2}}{\Delta(\omega^2)} \quad (11.161)$$

When a footing is subjected to an oscillatory moment $M_y e^{i\omega t}$ and a horizontal force $F_x e^{i\omega t}$ simultaneously, the resulting amplitudes of sliding and rocking are

$$A_x = A_{x1} + A_{x2} \quad (11.162)$$

$$A_\phi = A_{\phi 1} + A_{\phi 2} \quad (11.163)$$

11.6 Effect of Footing Shape on Vibratory Response

Elastic half-space theory was developed for a footing with circular contact area. Response of a footing is influenced by the shape of contact area. The usual practice is to transform area of any shape to an equivalent circle of same area (for translational modes) or equivalent moment of inertia (for rocking and torsional modes) (Richart and Whitman, 1967; Whitman and Richart, 1967). Thus

For translatory vibrations: $$r_o = \sqrt{\frac{ab}{\pi}} \tag{11.164}$$

For rocking vibrations: $$r_o = \left(\frac{ba^3}{3\pi}\right)^{1/4} \tag{11.165}$$

For torsional vibrations: $$r_o = \left[\frac{ab(a^2+b^2)}{6\pi}\right]^{1/2} \tag{11.166}$$

where r_o = Equivalent radius
a = Width of foundation (parallel to the axis of rotation for rocking)
b = Length of foundation (perpendicular to the axis of rotation for rocking).

11.7 Dynamic Response of Embedded Block Foundations

For an embedded foundation, the soil resistances are mobilised both below the base and on the sides. The additional soil reaction that comes into play on the sides may have significant influence on the dynamic response of embedded foundations. Typical response curves showing the effect of embedment are presented in Fig. 11.30. It gives that as a result of embedment, the natural frequency of the foundation-soil system increases and the amplitude of vibration decreases (Novak, 1970, 1985; Beredugo, 1971; Beredugo and Novak, 1972; Fry, 1963; Stokoe, 1972; Stokoeand Richart, 1974; Chae, 1971; Gupta, 1972; Vijayvergiya, 1981).

The problem of embedded foundations has been analysed by both linear elastic weightless spring approach (Prakash and Puri, 1971, 1972; Vijayvergiya, 1981) and elastic half-space theory (Anandkrishan and Krishnaswamy, 1973; Baranov, 1967; Berdugo and Novak, 1972; Novak and Beredugo, 1971). The analysis developed by Vijayvergiya (1981) is simple and logical, and therefore selected for presentation here. On the basis of theoretical analysis, he had recommended the equivalent spring stiffnesses in different types of motion as given below:

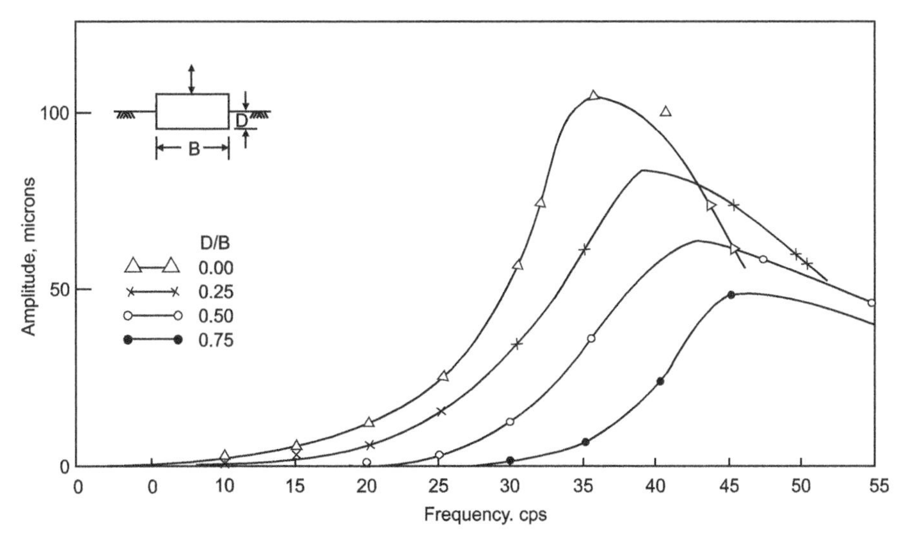

Fig. 11.30. Amplitude-frequency curves for vertical tests at various D/B ratios ($\theta = 70°$ and no air gap) (Vijayvergiya, 1981)

11.7.1 Vertical Vibrations

Refer Fig. 11.31,

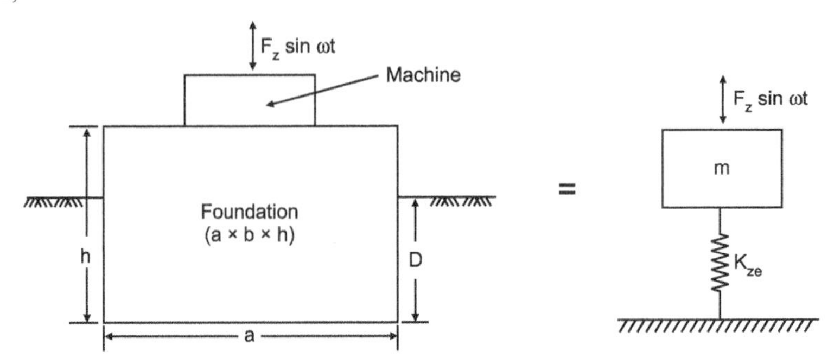

Fig. 11.31. Embedded block foundation subjected to pure vertical vibration

$$K_{ze} = C_{uD}A + 2\,C_{\tau av}\,(b\,D + a\,D) \tag{11.167}$$

where K_{ze} = Equivalent spring stiffness of the embedded foundation

C_{uD} = Coefficient of elastic uniform compression obtained at the base of foundation

$C_{\tau av}$ = Average value of coefficient of elastic uniform shear = $\dfrac{C_\tau + C_{\tau D}}{2}$

C_τ = Coefficient of elastic uniform shear at the ground surface

$C_{\tau D}$ = Coefficient of elastic uniform shear at the base of foundation

D = Embedment depth

b = Width of base of foundation

a = Length of base of foundation

The equation of motion will be

$$m\ddot{z} + K_{ze} \cdot z = F_z \sin \omega\, t \tag{11.168}$$

The natural frequency ω_{nze} and maximum amplitude A_{ze} of motion are given by

$$\omega_{nze} = \sqrt{\frac{K_{ze}}{m}} \tag{11.169a}$$

$$A_{ze} = \frac{F_z}{m(\omega_{nze}^2 - \omega^2)} \tag{11.169b}$$

11.7.2 Pure Sliding Vibrations

Refer Fig. 11.32,

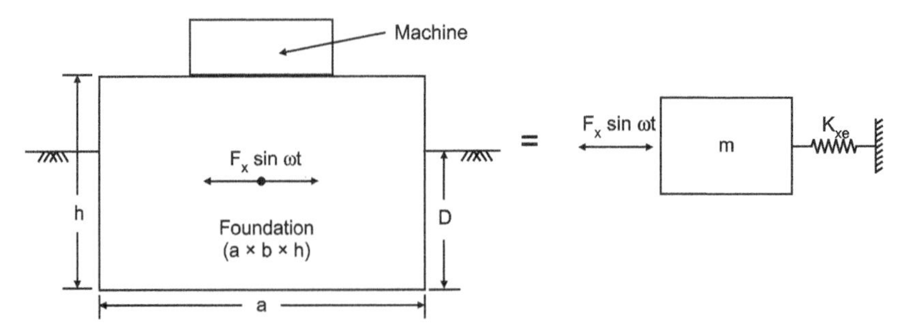

Fig. 11.32. Embedded block foundation subjected to pure sliding vibration

$$K_{xe} = C_{\tau D} A + C_{uav}\, b\, D + 2\, C_{\tau av}\, aD \tag{11.170}$$

where, K_{xe} = Equivalent spring stiffness of the embedded foundation

C_{uav} = Average value of coefficient of elastic uniform compression = $\dfrac{C_u + C_{uD}}{2}$

C_u = Coefficient of elastic uniform compression at the ground surface.

Similarly, the equation of motion will be

$$m\ddot{x} + K_{xe} \cdot \dot{x} = F_x \sin \omega\, t \tag{11.171}$$

The natural frequency and maximum amplitude of vibration are given by

$$\omega_{nxe} = \sqrt{\frac{K_{xe}}{m}} \tag{11.172}$$

$$A_{xe} = \frac{F_x}{m(\omega_{nxe}^2 - \omega^2)} \tag{11.173}$$

11.7.3 Pure Rocking Vibrations

Refer Fig. 11.33,

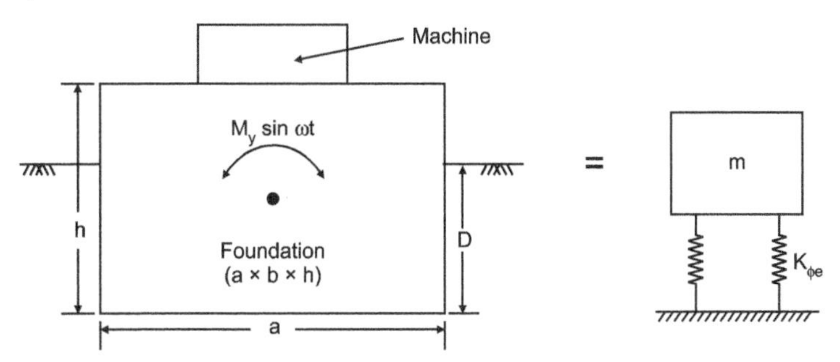

Fig. 11.33. Embedded block foundation subjected to rocking vibration

$$K_{\phi e} = C_{\phi D} \cdot I - W \cdot L + \frac{C_{\phi av} b}{24}(16D^3 - 12hD^2) + 2C_{\phi av} I_o + C_{\tau av}\frac{Dba^2}{2} \tag{11.174}$$

where $K_{\phi e}$ = Equivalent spring stiffness of the embedded foundation

$C_{\phi D}$ = Coefficient of elastic non-uniform compression at the base level of foundation

$C_{\phi av}$ = Average value of coefficient of elastic non-uniform compression $\dfrac{C_\phi + C_{\phi D}}{2}$

C_ϕ = Coefficient of elastic non-uniform compression at the ground surface

L = Height of the combined *c.g.* of machine and foundation from the centre of the base

W = Weight of foundation

$I = \dfrac{ba^3}{12}$

$I_o = \dfrac{aD^3}{3}$

The equation of motion will be

$$M_{mo}\ddot{\phi} + K_{\phi e} \cdot \phi = M_y \sin \omega\, t \tag{11.175}$$

The natural frequency and maximum amplitude of vibration are given by

$$\omega_{n\phi e} = \sqrt{\frac{K_{\phi e}}{M_{mo}}} \tag{11.176}$$

$$A_{\phi e} = \frac{M_y}{M_{mo}(\omega_{n\phi}^2 - \omega^2)} \tag{11.177}$$

11.7.4 Coupled Sliding and Rocking Vibrations

The equations of motion in coupled vibrations are given as

$$m\ddot{x} + K_{xx} \cdot x + K_{x\phi} \cdot \phi = F_x \sin \omega t \tag{11.178}$$

and $$M_m\ddot{\phi} + K_{\phi\phi} \cdot \phi + K_{\phi x} \cdot x = M_y \sin \omega t \tag{11.179}$$

where $$K_{xx} = C_{\tau D}A + 2C_{uav}\, bD + 2C_{\tau av}\, aD \tag{11.180}$$

$$K_{x\phi} = C_{\phi av}\, b(D^2 - 2DL) - C_{\tau D} \cdot AL \tag{11.181}$$

$$K_{\phi\phi} = C_{\phi D}\, I + C_{\tau D}AL^2 - WL + 2C_{\psi av}I_y + C_{\tau av}bD\frac{a^2}{2} + \frac{2}{3}C_{\phi av}[L^3 + (D - L^3)] \tag{11.182}$$

$$K_{\phi x} = -\left[C_{\tau D}AL + 2C_{uav}\, bD\left(L - \frac{D}{3}\right) + 2C_{\tau av}\left(L - \frac{D}{3}\right)aD\right] \tag{11.183}$$

where $C_{\psi av}$ = Average value of coefficient of elastic nonuniform shear
I_y = Moment of inertia of area $a \times D$ lying in the plane of vibration about axis of rotation
$= \dfrac{Da^3}{12} + \dfrac{aDb^2}{4}$

The natural frequencies of the system can be obtained by solving Eq. (11.184)

$$mM_m\omega_{ne}^4 - (mK_{\phi\phi} + M_m\, K_{xx})\omega_{ne}^2 + (K_{\phi\phi}\, K_{xx} - K_{\phi x} \times K_{x\phi}) = 0 \tag{11.184}$$

The amplitudes of vibration of the system can be obtained as below:

(a) Only the horizontal force $F_x \sin \omega t$ is acting: The equations of motion will be:

$$m\ddot{x} + K_{xx} \cdot x + K_{x\phi} \cdot \phi = F_x \sin \omega t \tag{11.185}$$

$$M_m\ddot{\phi} + K_{\phi\phi} \cdot \phi + K_{\phi x}x = 0 \tag{11.186}$$

The solution of the above equations can be represented by

$$x = A_{x1} \sin \omega t \tag{11.187}$$

$$\phi = A_{\phi 1} \sin \omega t \tag{11.188}$$

By substituting x and ϕ from Eqs. (11.187) and (11.188) in Eqs. (11.185) and (11.186), we get

$$(K_{xx} - m\omega^2)A_{x1} + K_{x\phi}\, A_{\phi 1} = F_x \sin \omega t \tag{11.189}$$

$$K_{\phi x}\, A_{x1} + (K_{\phi\phi} - M_m\, \omega^2)A_{\phi 1} = 0 \tag{11.190}$$

By solving Eqs. (11.189) and (11.190), we get

$$A_{x1} = \frac{(K_{\phi\phi} - M_m\, \omega^2)}{(K_{xx} - m\omega^2)(K_{\phi\phi} - M_m\, \omega^2) - K_{x\phi}\, K_{\phi x}} F_x \tag{11.191}$$

$$A_{\phi 1} = \frac{-K_{\phi x}}{(K_{xx} - m\omega^2)(K_{\phi\phi} - M_m\,\omega^2) - K_{x\phi}\,K_{\phi x}} F_x \tag{11.192}$$

(b) Only the moment $M_y \sin \omega\, t$ is acting: The equations of motion will be:

$$m\ddot{x} + K_{xx} \cdot x + K_{\phi x}\,\phi = 0 \tag{11.193}$$

$$M_m\,\ddot{\phi} + K_{\phi\phi} \cdot \phi + K_{\phi x}\,x = M_y \sin \omega\, t \tag{11.194}$$

The solutions of the above equations can be represented as:

$$x = A_{x2} \sin \omega\, t \tag{11.195}$$

$$\phi = A_{\phi 2} \sin \omega\, t \tag{11.196}$$

Substituting the values of x and ϕ from Eqs. (11.195) and (11.196) in Eqs. (11.193) and (11.194), we get

$$(K_{xx} - m\,\omega^2)A_{x2} + K_{x\phi}\,A_{\phi 2} = 0 \tag{11.197}$$

$$K_{\phi x} \cdot A_{x2} + (K_{\phi\phi} - M_m\,\omega^2)A_{\phi 2} = M_y \tag{11.198}$$

By solving Eqs. (11.197) and (11.198), we get

$$A_{x2} = \frac{-K_{x\phi}}{(K_{xx} - m\omega^2)(K_{\phi\phi} - M_m\,\omega^2) - K_{\phi x}\,K_{x\phi}} M_y \tag{11.199}$$

and

$$A_{\phi 2} = \frac{(K_{xx} - m\omega^2)}{(K_{xx} - m\omega^2)(K_{\phi\phi} - M_m\,\omega^2) - K_{\phi x}\,K_{x\phi}} M_y \tag{11.200}$$

If both $F_x \sin \omega t$ and $M_y \sin \omega t$ are acting simultaneously,

$$A_x = A_{x1} + A_{x2} \tag{11.201}$$

$$A_\phi = A_{\phi 1} + A_{\phi 2} \tag{11.202}$$

Sometimes to screen the vibrations, some air gap is left between the pit and the foundation block (Fig. 11.34). Figure 11.35 shows the comparison between the response of embedded foundation with air gap and without air gap. From this it can be concluded that if air gap is provided around the foundation the amplitude of vibration increases whereas the natural frequency decreases when compared with corresponding foundation with no air gap around it. The response of embedded foundation with air gap can be obtained by analysis given in Sec. 11.4 by using C_{uD}, $C_{\tau D}$, $C_{\phi D}$ and $C_{\psi D}$ in place of C_u, C_τ, C_ϕ and C_ψ respectively.

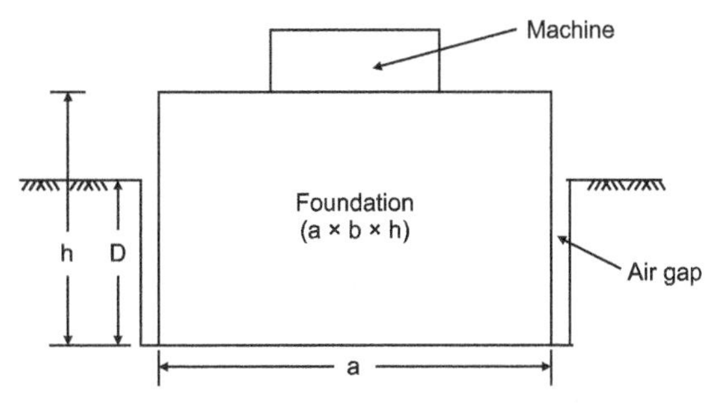

Fig. 11.34. Embedded block foundation with air gap

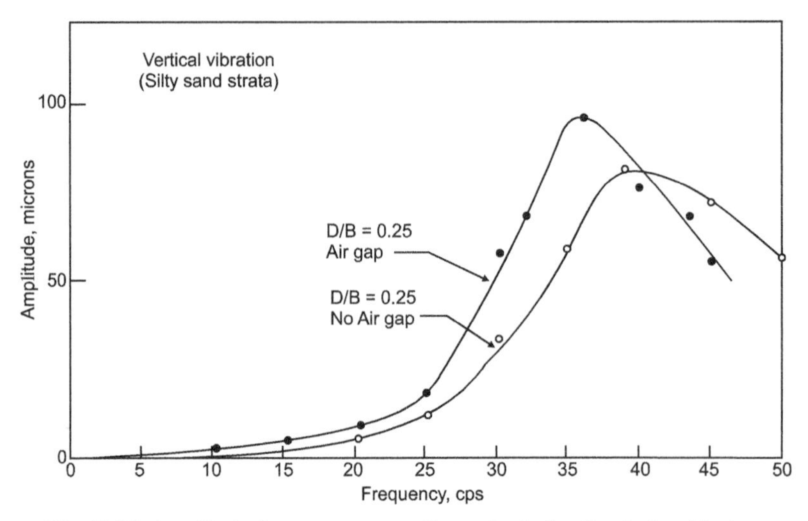

Fig. 11.35. Amplitude-frequency curves for vertical vibration tests with air gap and without air gap (Vijavergiya. 1981)

11.8 Soil Mass Participating in Vibrations

In both the methods of analysis and design of foundations of reciprocating machines described in Sec. 11.4 and 11.5, the effect of soil mass participating in vibrations has not been considered.

Pauw (1953) developed equations for the apparent soil mass by equating the kinetic energy of the affected zone to the kinetic energy of a mass assumed to be concentrated at the base of the foundation. He gave the following expression for apparent soil mass m_s for translatory modes of vibration:

$$m_s = \frac{\gamma b^3}{g\alpha} C_m \tag{11.203}$$

where α = Factor which defines the slope of truncated pyramid (Fig. 11.36). It is generally taken unity.

C_m = Function which depends on s and r

$s = \dfrac{\alpha h_e}{b}$

$r = \dfrac{a}{b}$

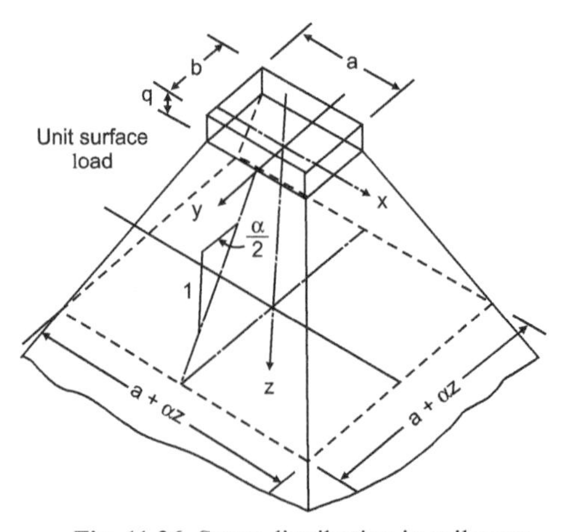

Fig. 11.36. Stress distribution in soil mass

h_e = Equivalent surcharge defined by the ratio of foundation pressure to unit weight of soil.

For non-cohesive soils, C_m is obtained from Fig. 11.37a. No graphical data is suggested by Pauw for cohesive soils.

The expression for mass moment of inertia of soil in rotational vibrations is given by

$$M_{ms} = \frac{\gamma b^5 C_i}{12 g \alpha} \tag{11.204}$$

where C_i can be obtained from Fig. 11.37b to *e*. In these figures, C_i^x, C_i^y, and C_i^z denote the factors of mass moments of inertia about *X*, *Y* and *Z* axes respectively. These factors can be obtained from Figs. 11.37b to d for cohesionless soils and from Fig. 11.37e for cohesive soils.

Balkrishna (1961) has developed the following expression for the apparent soil mass in vertical vibrations:

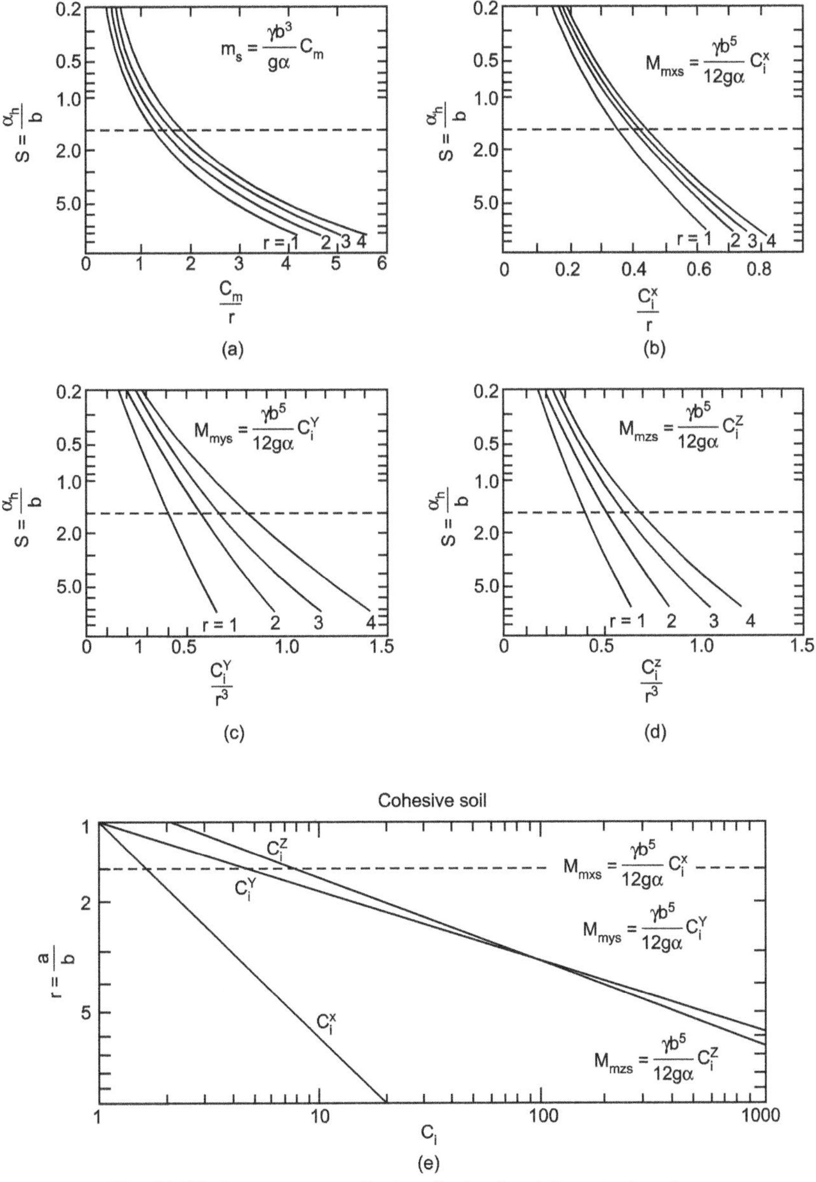

Fig. 11.37. Apparent mass factors for horizontal contact surfaces

$$m_s = \frac{4}{3}\left(\frac{0.4775\,Q}{4\,\rho}\right)^{3/2} \pi\rho \tag{11.205}$$

where Q = Sum of the static and dynamic load.

Barkan (1962) has suggested that the apparent soil mass may be taken as 23% of the mass of machine plus foundation.

Hsieh (1962) gave the expressions for getting apparent soil mass as given in Table 11.4.

Table 11.4. Effective Mass and Mass Moment of Inertia for Soil below a Vibrating Footing (Hsieh, 1962)

Mode of vibration	*Effective mass or mass moment of inertia of soil*		
	$\mu = 0$	$\mu = 0.25$	$\mu = 0.5$
Vertical translation	$0.5\,\rho r_o^3$	$1.0\,\rho r_o^3$	$2.0\,\rho r_o^3$
Horizontal translation	$0.2\,\rho r_o^3$	$0.2\,\rho r_o^3$	$1.0\,\rho r_o^3$
Rocking	$0.4\,\rho r_o^5$	—	—
Torsion (about vertical axis)	$0.3\,\rho r_o^5$	$0.3\,\rho r_o^5$	$0.3\,\rho r_o^5$

The apparent soil mass/mass moment of inertia is added to the mass m/mass moment of inertia M_m or J_o to get the natural frequency and amplitude of vibration.

11.9 Design Procedure for a Block Foundation

The design of a block foundation provided for a reciprocating machine may be carried out in following steps.

11.9.1 Machine Data

The following information shall be obtained from the manufacturer of the machine for guidance in designing:

(a) A detailed loading diagram comprising the plan, elevation and section showing details of connections and point of application of all loads on foundation;
(b) Distance between axis of the main shaft of the machine and the top face of foundation;
(c) Capacity or rated output of machine;
(d) Operating speed of machine; and
(e) Exciting forces of the machine and short circuit moment of motor, if any.

11.9.2 Soil Data

The following information about the subsurface soil should be known:

(a) Soil profile and data (including soil properties generally for depth equal to twice the width of the proposed foundation or up to hard stratum).
(b) Soil investigation to ascertain allowable soil pressures and to determine the dynamic properties of the soil.
(c) The relative position of the water table below ground at different times of the year.

The minimum distance to any important foundation in the vicinity of the machine foundation should also be ascertained.

11.9.3 Trial Size of the Foundation

Area of Block: The size of the foundation block (in plan) should be larger than the bed plate of the machine it supports, with a minimum all-round clearance of 150 mm.

Depth: In all cases, the depth of foundation should be such as to rest the foundation on good bearing strata and to ensure stability against rotations in vertical plane.

Centre of Gravity: The combined centre of gravity of the machine and the block shall be as much below the top of foundation as possible, but in no case it shall be above the top of foundation.

Eccentricity: The eccentricity shall not exceed 5 percent of the least width in any horizontal section.

Sharp corners shall be avoided, whenever possible, particularly in the openings.

11.9.4 Selecting Soil Constants

The values of dynamic elastic constants (C_u, C_ϕ, C_τ, C_ψ, G, E and μ) are obtained from relevant tests and corresponding strain levels are noted. These values are reduced to 10 m^2 contact area and 10 kN/m^2 confining pressure. A plot is then prepared between dynamic elastic constants and strain level. The value of dynamic elastic constants are picked up corresponding to the strain level expected in the actual foundation. These values of dynamic elastic constants are then corrected for the actual area of the foundation (if $<$ 10 m^2), and confining pressure. The details of this has been discussed in illustrative example 4.2.

11.9.5 Centering the Foundation Area in Contact with Soil

Determine the combined centre of gravity (Table 11.5) for the machine and the foundation in X, Y and Z planes and check to see that the eccentricity along X or Y axis is not over 5 percent. This is the upper limit for this type of analysis. If eccentricity exceeds 5 percent, the additional rocking due to vertical eccentric loading must be considered in the analysis (Barkan, 1962).

The static pressure should be checked; it should be less than 80 percent of the allowable soil pressure under static conditions. This condition is met in most practical foundations.

Table 11.5. Determination of CG of the System

Element of system	*Dimensions* a_x a_y a_z	*Weight of element*	*Mass of element*	*Coordinate c.g. of the element* x_i y_i z_i	*Static moments of mass of element* $m_i\ x_i\ m_i\ y_i\ m_i$
1. 2. 3.					

11.9.6 Design Values of Exciting Loads and Moments

The final values of force and resulting moments are now obtained with respect to the combined centre of gravity of the system. The relative magnitudes of the unbalanced forces and moments will decide the nature of vibrations in the block foundation.

11.9.7 Determination of Moments of Inertia and Mass Moments of Inertia

The moments of inertia and mass moments of inertia may be obtained using the formulae given in Tables 11.6 and 11.7.

Table 11.6. Moments of Inertia

Shape of the area	*Figure*	*Formula for*		
		I_x	I_y	I_z
Rectangle	y, C.G., x, x, b, y, a	$\frac{ab^3}{12}$	$\frac{ba^3}{12}$	$\frac{ab(a^2+b^2)}{12}$
Circle	y, C.G., x, x, d = 2r, y	$\frac{\pi}{64}a^4$	$\frac{\pi d^4}{64}$	$\frac{\pi d^4}{32}$

Table 11.7. Mass Moments of Inertia

Shape of elements	*Figure*	*Formula for*		
		M_{mx}	M_{my}	M_{mz}
Rectangular block	z, x, y, C.G., h, a, b	$\frac{m}{12}(b^2+h^2)$	$\frac{m}{12}(a^2+h^2)$	$\frac{m}{12}(a^2+b^2)$
Circular block	z, x, y, h, d	$\frac{m}{12}\left(\frac{3d^2}{4}+h^2\right)$	$\frac{m}{12}\left(\frac{3d^2}{4}+h^2\right)$	$\frac{md^2}{8}$

$$M_{mo} = M_m + m\,L^2$$

where M_{mo} = Mass moment of inertia of machine and foundation about the axis of rotation passing through base.

L = Distance of combined centre of gravity above base.

$$r = \frac{M_m}{M_{mo}}$$

11.9.8 Determination of Natural Frequencies and Amplitudes

Linear weightless spring approach

(i) ***Vertical Vibration***

$$\omega_{nz} = \sqrt{\frac{C_u \cdot A}{m}}$$

$$A_z = \frac{F_z}{m(\omega_{nz}^2 - \omega^2)}$$

(ii) *Torsional Vibration*

$$\omega_{n\psi} = \sqrt{\frac{C_\psi J_z}{M_{mz}}}$$

and

$$A_\psi = \frac{M_z}{M_{mz}(\omega_{n\psi}^2 - \omega^2)}$$

(iii) *Combined Rocking and Sliding*

Sliding and rocking are coupled modes of vibration. The natural frequencies are determined as follows:

$$\omega_{nx} = \sqrt{\frac{C_\tau A}{m}}$$

$$\omega_{n\phi} = \sqrt{\frac{C_\phi I - WL}{M_{mo}}}$$

and

$$\omega_{n1,2}^2 = \frac{1}{2r}\left[(\omega_{nx}^2 + \omega_{n\phi}^2) \pm \sqrt{(\omega_{nx}^2 + \omega_{n\phi}^2) - 4r\,\omega_{nx}^2\,\omega_{n\phi}^2}\right]$$

The amplitudes of vibration can be computed with the following equations:

$$A_x = \frac{(C_\tau AL^2 + C_\phi I - WL - M_m\omega^2)F_x + (C_\tau AL)M_y}{\Delta(\omega)^2}$$

$$A_\phi = \frac{(C_\tau AL)F_x + (C_\tau A - m\omega^2)M_y}{\Delta(\omega^2)}$$

where A_x = Linear horizontal amplitude of the combined centre of gravity

A_ϕ = Rotational amplitude in radians around the combined centre of gravity.

$\Delta(\omega^2) = mM_m(\omega_{n1}^2 - \omega^2)(\omega_{n2}^2 - \omega^2)$

The amplitude of the block should be determined at the bearing level of the foundation as

$$A_v = A_z + \frac{a}{2}A_\phi$$

$$A_h = A_x + h\,A_\phi$$

where A_h = Horizontal amplitude at bearing level

h = Height of the bearing above the combined centre of gravity of the system

A_v = Maximum vertical amplitude

Elastic half-space approach

Equivalent radius, mass ratio, spring constants and damping factors are listed in Table 11.8.

Table 11.8. Values of Equivalent Radius, Mass Ratio, Spring Constants and Damping Factor

Mode of vibration (1)	*Equivalent radius* (2)	*Mass (or inertia) ratio* (3)	*Damping factor* (4)	*Spring constant* (5)
Vertical	$r_{oz}=\sqrt{\frac{ab}{\pi}}$	$B_z=\frac{(1-\mu)}{4}\frac{m}{\rho r_o^3}$	$\xi_z=\frac{0.425}{B_z}$	$K_z=\frac{4Gr_o}{1-\mu}$
Sliding	$r_{ox}=\sqrt{\frac{ab}{\pi}}$	$B_x=\frac{(7-8\mu)}{32(1-\mu)}\frac{m}{\rho r_o^3}$	$\xi_x=\frac{0.2875}{B_x}$	$K_x=\frac{32(1-\mu)Gr_o}{7-8\mu}$
Rocking	$r_{o\phi}=\left(\frac{ba^3}{3\pi}\right)^{1/4}$	$B_\phi=\frac{3(1-\mu)}{8}\frac{M_{mo}}{\rho r_o^5}$	$\xi_\phi=\frac{0.15}{(1+B_\phi)\sqrt{B_\phi}}$	$K_\phi=\frac{8Gr_o^3}{3(1-\mu)}$
Torsional	$r_{o\psi}=\left(\frac{ba(a^2+b^2)}{6\pi}\right)^{1/4}$	$B_y=\frac{M_{mz}}{\rho r_o^5}$	$\xi_\psi=\frac{0.5}{1+2B_\psi}$	$K_\psi=\frac{16}{3}Gr_o^3$

Natural frequencies and amplitude of vibrations

(i) Vertical Vibration

$$\omega_{nz}=\sqrt{\frac{K_z}{m}}$$

$$A_z=\frac{F_z}{K_z\left[\left\{1-\left(\frac{\omega}{\omega_{nz}}\right)^2\right\}^2+\left(2\xi_z\frac{\omega}{\omega_{nz}}\right)^2\right]^{1/2}}$$

(ii) Torsional Vibration

$$\omega_{n\psi}=\sqrt{\frac{K_\psi}{M_{m\psi}}}$$

and

$$A_\psi=\frac{M_z}{K_\psi\left[\left\{1-\left(\frac{\omega}{\omega_{n\psi}}\right)^2\right\}^2+\left(2\xi_\psi\frac{\omega}{\omega_{n\psi}}\right)^2\right]^{1/2}}$$

(iii) Coupled Sliding and Rocking Vibration

$$\omega_{nx}=\sqrt{\frac{K_x}{m}}$$

$$\omega_{n\phi}=\sqrt{\frac{K_\phi}{M_{mo}}}$$

Damped natural frequencies are obtained as the roots of the following equation:

$$\left[\omega_{nd}^4-\omega_{nd}^2\left\{\frac{(\omega_{n\phi}^2+\omega_{nx}^2)}{r}-\frac{4\xi_x\,\xi_\phi\,\omega_{nx}\,\omega_{n\phi}}{r}\right\}+\frac{\omega_{nx}^2\,\omega_{n\phi}^2}{r}\right]^2$$

$$+4\left[\frac{\xi_x\,\omega_{nx}\,\omega_{nd}}{r}(\omega_{n\phi}^2-\omega_{nd}^2)+\frac{\xi_\phi\,\omega_{nd}\,\omega_{n\phi}}{r}(\omega_{nx}^2-\omega_{nd}^2)\right]^2=0$$

Undamped natural frequencies can be obtained by using following equation:

$$\omega_{n1,2}^2=\frac{1}{2r}\left[(\omega_{nx}^2+\omega_{n\phi}^2)\pm\sqrt{(\omega_{n\phi}^2+\omega_{nx}^2)^2-4r\omega_{n\phi}^2\,\omega_{nx}^2}\right]$$

Damped amplitudes for motion occasioned by the applied moment, can be obtained as below:

$$A_x=\frac{M_y\cdot L}{M_m}\frac{\left[(\omega_{nx}^2)^2+(2\xi_x\,\omega_{nx}.\omega)^2\right]^{1/2}}{\Delta(\omega^2)}$$

$$A_\phi=\frac{M_y}{M_m}\frac{\left[(\omega_{nx}^2-\omega^2)^2+(2\xi_x\,\omega_{nx})^2\right]^{1/2}}{\Delta(\omega^2)}$$

where $\Delta(\omega^2)$ is given by

$$\Delta(\omega^2)=\left[\left\{\omega^4-\omega^2\left(\frac{(\omega_{n\phi}^2+\omega_{nx}^2)}{r}-\frac{4\xi_x\,\xi_\phi\,\omega_{nx}\,\omega_{n\phi}}{r}\right)-\frac{\omega_{nx}^2\,\omega_{n\phi}^2}{r}\right\}^2+4\left\{\xi_x\frac{\omega_{nx}\,\omega}{r}(\omega_{n\phi}^2-\omega^2)+\frac{\xi_\phi\,\omega_{n\phi}\,\omega}{r}(\omega_{nx}^2-\omega^2)\right\}^2\right]^{1/2}$$

Damped amplitudes for motion occasioned by an applied force F_x acting at the centre of gravity of the foundation may be obtained as below:

$$A_x=\frac{F_x}{mM_m}\frac{\left[(-M_m\,\omega^2+K_\phi+L^2K_x)^2+4\omega^2(\xi_\phi\sqrt{K_\phi\,M_{mo}}+L^2\xi_x\sqrt{K_xm})^2\right]^{1/2}}{\Delta(\omega^2)}$$

and
$$A_\phi=\frac{F_x\,L\,\omega_{nx}(\omega_{nx}^2+4\xi_x\,\omega^2)^{1/2}}{M_m\,\Delta(\omega^2)}$$

11.9.9 Check for Adequate Foundation

The natural frequencies computed in Art. 11.9.8 should be away from the resonance zone i.e.

$$\frac{\omega}{\omega_n}<0.5 \text{ or } \frac{\omega}{\omega_n}>1.5$$

The amplitudes computed in Art. 11.9.8 should be less than the limiting amplitudes of the machine which are usually specified by the manufacturer of the machine.

Illustrative Examples

Example 11.1

A reciprocating machine is symmetrically mounted on a block of size 4.0 m × 3.0 m × 3.5 m high. The soil at the site is sandy in nature having ϕ = 35° and γ_{sat} = kN/m³. The water table lies at a depth of 3.0 m below the ground surface. The block is embedded in the ground by 2.0 m depth. The machine vibrating at a speed of 250 rpm generates

Maximum vertical unbalanced force = 2.5 kN

Torque about Z-axis = 4.0 kN-m

Maximum horizontal unbalanced force = 2.0 kN at a height of 0.2 m above the top of the block.

The machine weight is small in comparison to the weight of foundation. Limiting amplitude of the machine is 150 microns. A block resonance test was conducted at the site to evaluate the dynamic elastic constants. The data obtained from the test is the same as given in Example 4.2 (Chapter 4).

Determine the natural frequencies and amplitudes by (*a*) Weightless spring method, and (*b*) Elastic half-space approach.

Solution

1. Machine data (Fig. 11.38)

Operating speed of machine = 500 rpm

Weight of machine is small and can be neglected.

$$F_z = 2.5 \sin \omega t \text{ kN}$$

$$M_z = 4.0 \sin \omega t \text{ kN-m}$$

$$F_z = 2.0 \sin \omega t \text{ kN acting at a height of 0.2 m above the top of block}$$

$$\omega = 250 \text{ rpm} = 250 \times \frac{2\pi}{60} = 26.2 \text{ rad/s}$$

$$M_y = (0.2 + 1.75) \times 2.0 \sin \omega t = 3.9 \sin \omega t \text{ kN-m}$$

2. Dynamic elastic constants

Refer example 4.2. The soil data and the size of the actual block are same as in that example. The procedure of determining the values of dynamic elastic constants for analysing the block foundation is illustrated there. Therefore, following values of dynamic elastic constants may be adopted:

$$C_u = 3.62 \times 10^4 \text{ kN/m}^3$$

$$G = 3.29 \times 10^4 \text{ kN/m}^2$$

$$E = 8.89 \times 10^4 \text{ kN/m}^2$$

$$\mu = 0.35$$

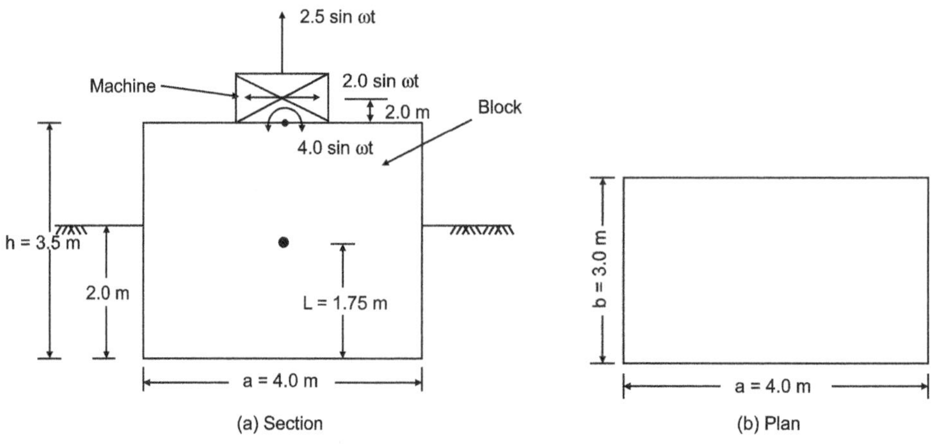

Fig. 11.38. Details of foundation

$$C_\tau = \frac{C_u}{2} = \frac{3.62}{2}\times 10^4 = 1.81\times 10^4 \text{ kN/m}^3$$

$$C_\phi = 3.46\ C_\tau = 3.46 \times 1.81 \times 10^4 = 6.26 \times 10^4 \text{ kN/m}^3$$

$$C_\psi = 0.75\ C_u = 0.75 \times 3.62 \times 10^4 = 2.71 \times 10^4 \text{ kN/m}^3$$

3. Foundation data

Let the block is casted in M20 concrete. The unit weight material is taken as 24.0 kN/m^3.

$$\text{Weight of block} = 24.0 \times 4.0 \times 3.0 \times 3.5$$

$$= 1008 \text{ kN}$$

$$m = \frac{1008 \times 10^3}{9.81} = 102.8\times 10^3 \text{ kg}$$

$$\text{Area of foundation base} = 4.0 \times 3.0 = 12.0 \text{ m}^2$$

In further analysis and design of foundation, the depth of embedment is neglected.

4. Linear weightless spring approach

(a) Vertical vibration

$$\omega_{nz} = \sqrt{\frac{C_u A}{m}} = \sqrt{\frac{3.62 \times 10^7 \times 12.0}{102.8 \times 10^3}} = 65 \text{ rad/s}$$

$$(A_z)_{max} = \frac{F_z}{m(\omega_{nz}^2 - \omega^2)}$$

$$= \frac{2.5}{102.8(65^2 - 26.2^2)} = 6.8\times 10^{-6} \text{m} = 6.8 \text{microns}$$

(b) Torsional vibration

$$J_z = \frac{ab}{12}(a^2 + b^2) = \frac{4\times 3}{12}\times(4^2 + 3^2) = 25 \text{m}^4$$

$$M_{mz} = \frac{\text{m}}{12}(a^2 + b^2) = \frac{102.8\times 10^3}{12}(4^2 + 3^2) = 214\times 10^3 \text{ kg-m}^2$$

$$\omega_{n\psi} = \sqrt{\frac{C_\psi J_z}{M_{mz}}} = \sqrt{\frac{2.71\times 10^7 \times 25}{214 \times 10^3}} = 56.2 \text{ rad/s}$$

(c) Coupled rocking and sliding

$$\omega_{nx} = \sqrt{\frac{C_\tau A}{m}} = \sqrt{\frac{1.81\times 10^7 \times 12}{102.8\times 10^3}} = 46 \text{ rad/s}$$

$$\omega_{n\phi} = \sqrt{\frac{C_\phi I - WL}{M_{mo}}}$$

$$I = \frac{ba^3}{12} = \frac{3\times 4^3}{12} = 16 \text{ m}^4$$

$$W = 1008 \text{ kN}$$

$$L = 1.75 \text{ m}$$

$$M_{mo} = M_m + m\,L^2$$

$$M_m = \frac{m}{12}(a^2 + h^2) = \frac{102.8\times10^3}{12}(3^2 + 3.5^2) = 182\times10^3 \text{ kg-m}^2$$

Therefore, $$M_{mo} = 182 \times 10^3 + 102.8 \times 10^3 \times 1.75^2 = 496.8 \times 10^3 \text{ kg-m}^2$$

$$\omega_{n\phi} = \sqrt{\frac{6.26\times10^7 \times 16 - 1008\times10^3\times1.75}{496.8\times10^3}} = 44.8 \text{ rad/s}$$

$$r = \frac{M_m}{M_{mo}} = \frac{182\times10^3}{496.8\times10^3} = 0.366$$

$$\omega_{n1,2}^2 = \frac{1}{2r}\left[(\omega_{nx}^2 + \omega_{n\phi}^2) \pm \sqrt{(\omega_{n\phi}^2 + \omega_{nx}^2)^2 - 4r\omega_{n\phi}^2\,\omega_{nx}^2}\right]$$

$$= \frac{1}{2\times0.366}\left[(46^2 + 44.8^2) \pm \sqrt{(46^2 + 44.8^2)^2 - 4\times0.366\times46^2\times44.8^2}\right]$$

$$= \frac{1}{0.732}[4123 \pm 3283]$$

$$\omega_{n1} = 33.8 \text{ rad/s and } \omega_{n2} = 100.6 \text{ rad/s}$$

$$\Delta(\omega^2) = mM_m(\omega_{n1}^2 - \omega^2)(\omega_{n2}^2 - \omega^2)$$

$$= 102.8 \times 10^3 \times 182 \times 10^3\,(33.8^2 - 26.2^2)\,(100.6^2 - 26.2^2)$$

$$= 102.8 \times 10^3 \times 182 \times 10^3 \times 456 \times 4934 = 8.05 \times 10^{16} \text{ kg}^2\text{m}^2$$

$$M_y = 2(0.2 + 1.75) = 3.9 \text{ kN/m}$$

$$A_x = \frac{(C_\phi I - WL + C_\tau AL^2 - M_m\omega^2)F_x + (C_\tau AL)M_y}{\Delta(\omega^2)}$$

$$= \frac{(6.26\times10^4\times16 - 1008\times1.75 + 1.81\times10^4\times12\times1.75^2 - 182\times26.2^2)\times2.0\times10^6}{8.05\times10^{16}}$$

$$+\frac{1.81\times10^4\times12\times1.75\times(3.9)\times10^6}{8.05\times10^{16}} = \frac{3080158 + 1482390}{8.05\times10^{10}} = 56.6\times10^{-6} \text{ m}$$

$$A_\phi = \frac{(C_\tau AL)F_x + (C_\tau A - m\omega^2)M_y}{\Delta(\omega^2)}$$

$$= \frac{(1.81\times10^4\times12\times1.75)\times2.0\times10^6 + (1.81\times10^4\times12 - 102.8\times26.2^2)3.9\times10^6}{8.05\times10^{16}}$$

$$= \frac{760200+571876}{8.05\times10^{10}} = 16.5\times10^{-6} \text{ rad}$$

Hence,

$$A_v = A_z + \frac{a}{2}A_\phi = 6.8\times10^{-6} + \frac{4.0}{2}\times16.5\times10^{-6}$$

$$= 39.8\times10^{-6} \text{ m} = 39.8 \text{ microns}$$

$$A_h = A_x + h'A_\phi = 56.6\times10^{-6} + 1.75\times16.5\times10^{-6}$$

$$= 85.47\times10^{-6} \text{ m} = 85.47 \text{ microns}$$

5. Elastic half-approach

(a) Vertical vibration

$$r_o = \sqrt{\frac{A}{\pi}} = \sqrt{\frac{12}{\pi}} = 1.95 \text{ m}$$

Average effective unit weight of soil

$$= \frac{20+10}{2} = 15 \text{ kN/m}^3$$

$$B_z = \frac{1-\mu}{4}\cdot\frac{m}{\rho r_o^3} = \frac{1-0.35}{4}\times\frac{\frac{1008}{9.81}}{\left(\frac{15}{9.81}\right)\times1.95^3} = 1.47$$

$$K_z = \frac{4Gr_o}{1-\mu} = \frac{4\times3.29\times10^4\times1.95}{1-0.35} = 39.48\times10^4 \text{ kN/m}$$

$$\xi_z = \frac{0.425}{\sqrt{B_z}} = \frac{0.425}{\sqrt{1.95}} = 0.304$$

$$\omega_{nz} = \sqrt{\frac{K_z}{m}} = \sqrt{\frac{39.48\times10^4}{102.8}} = 61.92 \text{ rad/s}$$

$$A_z = \frac{F_z}{K_z\left[\left\{1-\left(\frac{\omega}{\omega_{nz}}\right)^2\right\}^2 + \left(2\xi_z\frac{\omega}{\omega_{nz}}\right)^2\right]^{1/2}}$$

$$= \frac{2.5}{39.48\times10^4\left[\left\{1-\left(\frac{26.2}{61.92}\right)^2\right\}^2 + \left(2\times0.304\times\frac{26.2}{61.92}\right)^2\right]^{1/2}}$$

$$= 7.36\times10^{-6} \text{ m}$$

(b) Torsional vibration

$$r_o = \left[\frac{ab(a^2+b^2)}{6\pi}\right]^{1/4} = \left[\frac{4\times3(4^2+3^2)}{6\pi}\right]^{1/4} = 1.9973 \text{ m}$$

$$M_{mz} = 214 \times 10^3 \text{ kg-m}^2$$

$$B_\psi = \frac{M_{mz}}{\rho r_o^5} = \frac{214\times9.81}{15\times1.9973^5} = 4.40$$

$$K_\psi = \frac{16}{3}G r_o^3 = \frac{16}{3}\times3.29\times10^4\times1.9973^3 = 139.82\times10^4 \text{ kNm/rad}$$

$$\xi_\psi = \frac{0.5}{1+2B_\psi} = \frac{0.5}{1+2\times4.4} = 0.051$$

$$\omega_{n\psi} = \sqrt{\frac{K_\psi}{M_{mz}}} = \sqrt{\frac{139.82\times10^4}{214}} = 80.83 \text{ rad/s}$$

(c) Coupled sliding and rocking vibration

Sliding

$$r_o = \sqrt{\frac{A}{\pi}} = \sqrt{\frac{4\times3}{\pi}} = 1.95 \text{ m}$$

$$K_x = \frac{32(1-\mu)G r_o}{7-8\mu} = \frac{32(1-0.35)\times3.29\times10^4\times1.95}{7-8\times0.35}$$

$$= 31.76\times10^4 \text{ kN/m}$$

$$B_x = \frac{7-8\mu}{32(1-\mu)}\cdot\frac{m}{\rho r_o^3} = \frac{7-8\times0.35}{32(1-0.35)}\times\frac{1008}{15\times1.95^3} = 1.83$$

$$\xi_x = \frac{0.2875}{\sqrt{B_x}} = \frac{0.2875}{\sqrt{1.83}} = 0.212$$

$$\omega_{nx} = \sqrt{\frac{K_x}{m}} = \sqrt{\frac{31.76\times10^4}{102.8}} = 55.58 \text{ rad/s}$$

Rocking

$$r_o = \left(\frac{ab^3}{3\pi}\right)^{1/4} = \left(\frac{4.0\times3^3}{3\pi}\right)^{1/4} = 1.84 \text{ m}$$

$$K_\phi = \frac{8G r_o^3}{3(1-\mu)} = \frac{8\times3.29\times10^4\times1.84^3}{3(1-0.35)} = 84.04\times10^4 \text{ kN-m/rad}$$

$$B_\phi = \frac{3(1-\mu)}{8}\times\frac{M_{mo}}{\rho r_o^5} = \frac{3(1-0.35)}{8}\times\frac{496.8\times9.81}{15\times1.84^5} = 3.75$$

$$\xi_\phi = \frac{0.15}{(1+B_\phi)\sqrt{B_\phi}} = \frac{0.15}{(1+3.75)\sqrt{3.75}} = 0.016$$

$$\omega_{n\phi} = \sqrt{\frac{K_\phi}{M_{mo}}} = \sqrt{\frac{84.04\times10^4}{496.8}} = 41.13 \text{ rad/s}$$

(d) Coupled Vibration.

Undamped natural frequencies in couples rocking and sliding are given by

$$\omega_{n1,2}^2 = \frac{1}{2r}\left[(\omega_{nx}^2+\omega_{n\phi}^2) \pm \sqrt{(\omega_{nx}^2+\omega_{n\phi}^2)^2 - 4r\,\omega_{nx}^2\,\omega_{n\phi}^2}\right]$$

$$= \frac{1}{2\times0.366}\left[(55.58^2+41.13^2) \pm \sqrt{(55.58^2+41.13^2)^2 - 4\times0.366\times55.58^2\times51.13^2}\right]$$

$$= \frac{1}{0.732}[4781 \pm 3899]$$

$$\omega_{n1} = 34.70 \text{ rad/s} \text{ and } \omega_{n2} = 108.9 \text{ rad/s}$$

$$\Delta(\omega^2) = \left[\left\{\omega^4 - \omega^2\left(\frac{(\omega_{n\phi}^2+\omega_{nx}^2)}{r} - \frac{4\xi_x\,\xi_\phi\,\omega_{nx}\,\omega_{n\phi}}{r}\right) + \frac{\omega_{nx}^2\,\omega_{n\phi}^2}{r}\right\}^2 + 4\left\{\xi_x\frac{\omega_{nx}\,\omega}{r}(\omega_{n\phi}^2-\omega^2) + \frac{\xi_\phi\,\omega_{n\phi}\,\omega}{r}(\omega_{nx}^2-\omega^2)\right\}^2\right]^{1/2}$$

$$= \left[\left\{26.2^4 - 26.2^2\left(\frac{4781}{0.366} - \frac{4\times0.212\times0.016\times55.58\times41.13}{0.366}\right) + \frac{55.58^2\times41.13^2}{0.366}\right\}^2 + 4\left\{0.212\times\frac{55.58\times26.2}{0.366}(41.13^2-26.2^2) + \frac{0.016\times41.13\times26.2}{0.366}(55.58^2-26.2^2)\right\}^2\right]^{1/2}$$

$$= \left[(471200 - 8908687 + 14278199)^2 + 4(847897+113187)^2\right]^{1/2} = 61488873$$

$$A_{x1} = \frac{F_x}{mM_m}\frac{\{(-M_m\,\omega^2 + K_\phi + K_x\,L^2)^2 + 4\omega^2(\xi_\phi\sqrt{K_\phi\,M_{mo}} + L^2\,\xi_x\sqrt{K_x m})^2\}^{1/2}}{\Delta(\omega^2)}$$

$$= \frac{2.0}{102.8\times182\times919697}\left[(-182\times26.2^2 + 28.1\times10^4 + 10.62\times10^4\times1.75^2)^2 + 4\times26.2^2\left\{(0.016\sqrt{28.1\times10^4\times496.8} + 1.75^2\times0.212\sqrt{10.62\times10^4\times102.8})\right\}^2\right]^{1/2}$$

$$= \frac{2.0}{102.8\times182}\,\frac{(2.3165\times10^{11} + 0.1496\times10^{11})^{1/2}}{61488873} = 0.87\times10^{-6}\text{ m}$$

$$A_{\phi 1} = \frac{F_x L}{M_m} \frac{\omega_{nx}(\omega_{nx}^2 + 4\xi_x \omega^2)^{1/2}}{\Delta(\omega^2)}$$

$$= \frac{2.0 \times 1.75}{182}\left[\frac{55.58(55.58^2 + 4 \times 0.212 \times 26.2^2)^{1/2}}{61488873}\right] = 1.05 \times 10^{-6} \text{ rad}$$

$$A_{x2} = \frac{M_y \cdot L}{M_m} \frac{\left[(\omega_{nx}^2)^2 + (2\xi_x \omega_{nx} \cdot \omega)^2\right]^{1/2}}{\Delta(\omega^2)}$$

$$= \frac{3.9 \times 1.75}{182} \frac{\left[(55.58^2)^2 + (2 \times 0.212 \times 55.58 \times 26.2)^2\right]^{1/2}}{61488873} = 0.38 \times 10^{-6} \text{ m}$$

$$A_{\phi 2} = \frac{M_y}{M_m} \frac{\left[(\omega_{nx}^2 - \omega^2)^2 + (2\xi_x \omega_{nx} \omega)^2\right]^{1/2}}{\Delta(\omega^2)}$$

$$= \frac{3.9}{182} \frac{\left[(55.58^2 - 26.2^2)^2 + (2 \times 0.212 \times 55.58 \times 26.2)^2\right]^{1/2}}{61488873} = 0.22 \times 10^{-6} \text{ rad}$$

$$A_x = A_{x1} + A_{x2}$$

$$= 0.87 \times 10^{-6} + 0.38 \times 10^{-6} = 1.25 \times 10^{-6} \text{ m}$$

$$A_\phi = A_{\phi 1} + A_{\phi 2}$$

$$= 1.05 \times 10^{-6} + 0.22 \times 10^{-6} = 1.27 \times 10^{-6} \text{ rad}$$

Hence, $$A_v = A_z + \frac{a}{2} A_\phi$$

$$= 7.36 \times 10^{-6} + \frac{4.0}{2} \times 1.27 \times 10^{-6} = 9.9 \times 10^{-6} \text{ m}$$

$$A_h = A_x + h A_\phi$$

$$= 1.25 \times 10^{-6} + 1.75 \times 1.275 \times 10^{-6} = 3.47 \times 10^{-6} \text{ m}$$

It may be noted the there are significant differences in the magnitudes of natural frequencies and amplitudes computed by the two approaches. It may be due to the reason that the value of shear modulus G is computed from the block resonance test data using the relation between C_u, E and G. Actually in elastic half-space theory, it is desirable that the value of G is obtained from wave propagation test. Author's experience indicates that the value of G obtained from wave propagation test is much higher than computed from block resonance test data. Use of apropriate value of G may bring the results of the two approahes closer.

Example 11.2

Determine the natural frequencies and amplitudes of motion of foundation (Example 11.1) taking into account the embedment effect and apparent soil mass.

Solution

(i) Assume that the values of dynamic elastic constants at a depth of 2.0 m are 20% higher than the values at the surface of ground. Therefore

$$C_{uD} = 1.2 \times 3.62 \times 10^4 = 4.35 \times 10^4 \text{ kN/m}^3$$

$$C_{\tau D} = 1.2 \times 1.81 \times 10^4 = 2.18 \times 10^4 \text{ kN/m}^3$$

$$C_{\phi D} = 1.2 \times 6.26 \times 10^4 = 7.51 \times 10^4 \text{ kN/m}^3$$

$$C_{\psi D} = 1.2 \times 2.71 \times 10^4 = 3.25 \times 10^4 \text{ kN/m}^3$$

The average values of dynamic elastic constants will be

$$C_{uav} = 4.0 \times 10^4 \text{ kN/m}^3$$

$$C_{\tau av} = 2.0 \times 10^4 \text{ kN/m}^3$$

$$C_{\phi av} = 6.89 \times 10^4 \text{ kN/m}^3$$

$$C_{\psi av} = 3.0 \times 10^4 \text{ kN/m}^3$$

$$D = 20 \text{ m}$$

(ii) Vertical vibration

$$K_{ze} = C_{uD} \times A + 2\, C_{\tau av}\,(bD + aD)$$

$$= 4.35 \times 10^4 \times 12 + 2 \times 2.0 \times 10^4 (3 \times 2 + 4 \times 2)$$

$$= 108.2 \times 10^4 \text{ kN/m}$$

$$h_e = \frac{24 \times 3.5}{15} = 5.6 \text{ m}$$

$$S = \frac{\alpha h_e}{b} = \frac{1.0 \times 5.6}{3.0} = 1.867; \quad r = \frac{a}{b} = \frac{4}{3} = 1.334$$

From Fig. 11.36*a* for $S = 1.867$ and $r = 1.334$; $\frac{C_m}{r} = 1.45$

or $$C_m = 1.45 \times 1.334 = 1.93$$

Therefore, $$m_s = \frac{\gamma b^3}{\alpha g} C_m = \frac{15 \times 3^3}{1.0 \times 9.81} \times 1.93 = 79.7 \times 10^3 \text{ kg}$$

$$\omega_{nze} = \sqrt{\frac{108.2 \times 10^4}{102.8 + 79.7}} = 75.8 \text{ rad/s}$$

$$A_{ze} = \frac{F_e}{(m + m_s)(\omega_{nze}^2 - \omega^2)}$$

$$= \frac{2.5}{(102.8 + 79.7)(75.8^2 - 26.2^2)} = 2.70 \times 10^{-6} \text{ m}$$

(iii) Coupled vibration

$$K_{xe} = C_{\tau D} \cdot A + 2\, C_{uav}\, bD + 2\, C_{\tau av}\, aD$$

$$= 2.18 \times 10^4 \times 12 + 2 \times 4 \times 10^4 \times 3 \times 2 + 2 \times 2 \times 10^4 \times 4 \times 2$$

$$= 106.16 \times 10^4 \text{ kN/m}$$

From Fig. 11.36*b*, for $S = 1.867$ and $r = 1.334$, $\frac{C_1^x}{r} = 0.43$

$$M_{mxs} = \frac{\gamma b^5}{12 g\alpha} C_i^x = \frac{15\times 3^5}{12\times 9.81\times 1} 0.43 \times 1.334 = 17.72 \text{ kNms}^2$$

$$M_{mos} = M_{mxs} + m_s \times L^2 = 17.72 + 79.7 \times 1.75^2 = 261.80 \text{ kNms}^2$$

$$\omega_{nxe} = \sqrt{\frac{K_{xe}}{m+m_s}} = \sqrt{\frac{106.16\times 10^4}{102.8+79.7}} = 76.2 \text{ rad/s}$$

$$K_{\phi e} = C_{\phi D} I - WL + \frac{C_{\phi av}\cdot b}{24}(16D^3 - 12hD^2) + 2C_{\phi av}\times I_o + C_{\tau av}\times \frac{Dba^2}{2}$$

$$I = \frac{ba^3}{12} = \frac{3\times 4^3}{12} = 16\,\text{m}^4$$

$$W = 24 \times 3 \times 4 \times 3.5 = 1008 \text{ kN}$$

$$I_o = \frac{aD^3}{3} = \frac{4\times 2^3}{3} = 10.67\,\text{m}^4$$

$$K_{\phi e} = 7.51\times 10^4 \times 16 - 1008\times 1.75 + \frac{6.89\times 10^4\times 3}{24}(16\times 2^3 - 12\times 3.5\times 2^2)$$

$$+2\times 6.89\times 10.67 + 2.0\times 10^4 \times \frac{2\times 3\times 4^2}{2}$$

$$= 120.16 \times 10^4 - 0.1764 \times 10^4 - 34.45 \times 10^4 + 147 \times 10^4 + 48 \times 10^4$$

$$= 280.53 \times 10^4 \text{ kNm/rad}$$

$$\omega_{n\phi e} = \sqrt{\frac{K_{\phi e}}{M_{mo}+M_{mos}}} = \sqrt{\frac{280.53\times 10^4}{182+261.80}} = 79.5 \text{ rad/s}$$

$$K_{xx} = C_{\tau D} \times A + 2C_{uav} bD + 2C_{\tau av}\, aD$$

$$= 2.18 \times 10^4 \times 12 + 2 \times 4 \times 10^4 \times 3 \times 2 + 2\times 2\times 10^4 \times 4 \times 2$$

$$= 90.16 \times 10^4 \text{ kN/m}$$

$$K_{x\phi} = C_{\phi av}\, b(D^2 - 2DL) - C_{\tau D}\, AL$$

$$= 6.89 \times 10^4 \times 3(2^2 - 2 \times 2 \times 1.75) - 2.18\times 10^4 \times 12\times 1.75$$

$$= -231.81 \times 10^4 \text{ kN}$$

$$I_y = \frac{Da^3}{12} + \frac{aDb^2}{4} = \frac{2\times 4^3}{12} + \frac{4\times 2\times 3^2}{4} = 28.67\,\text{m}^4$$

$$K_{\phi\phi} = C_{\phi D} I + C_{\tau D} AL^2 - WL + 2C_{\psi av} I_y + C_{\tau av}\frac{bDa^2}{2} + \frac{2}{3} C_{\phi av}[L^3 + (D-L)^3]$$

$$= 7.51 \times 10^4 \times 16 + 2.18 \times 10^4 \times 12 \times 1.75^2 - 1008 \times 1.75 + 2 \times 3 \times 10^4 \times 28.67$$

$$+2\times10^4\times\frac{3\times2\times4^2}{2}+\frac{2}{3}\times6.89\times10^4[1.75^3+(2.0-1.75)^3]$$

$$=(120.16+80.11-0.17+172.02+96+24.69)\times10^4$$

$$=512.81\times10^4\text{ kNm}$$

$$K_{\phi x}=-\left[C_{\tau D}\,AL+2C_{uav}\,bD\left(L-\frac{D}{3}\right)+2C_{\tau av}\left(L-\frac{D}{3}\right)aD\right]$$

$$=-\left[2.18\times10^4\times12\times1.75+2\times4\times10^4\times3\times2\left(1.75-\frac{2}{3}\right)+2\times2\times10^4\left(1.75-\frac{2}{3}\right)4\times2\right]$$

$$=-[45.78+51.84+34.56]10^4=-132.18\times10^4\text{ kN}$$

$$A_{x1}=\frac{\{K_{\phi\phi}-(M_m+M_{mxs})\omega^2\}F_x}{\{K_{xx}-(m+m_s)\omega^2\}\{K_{\phi\phi}-(M_m+M_{mxs})\omega^2\}-K_{x\phi}\cdot K_{\phi x}}$$

$$=\frac{\{512.81\times10^4-(182+17.72)26.2^2\}2.0}{\left[\begin{array}{c}\{90.16\times10^4-(102.8+79.7)26.2^2\}\{512.81\times10^4-(182+17.72)26.2^2\}\\-(-231.81\times10^4)(-132.18\times10^4)\end{array}\right]}$$

$$=\frac{998.18\times10^4}{77.64\times10^4\times498.3\times10^4-30639.32\times10^8}$$

$$=\frac{998.18\times10^4}{8048.69\times10^8}=12.40\times10^{-6}\text{ m}$$

$$A_{\phi1}=\frac{-K_{\phi x}\cdot F_x}{\{K_{xx}-(m+m_s)\omega^2\}\{K_{\phi\phi}-(M_m+M_{mxs})\omega^2\}-K_{x\phi}\cdot K_{\phi x}}$$

$$=\frac{132.18\times10^4\times2.0}{8048.69\times10^8}=3.2\times10^{-6}\text{ rad}$$

$$A_{x2}=\frac{-K_{x\phi}\cdot M_y}{\{K_{xx}-(m+m_s)\omega^2\}\{K_{\phi\phi}-(M_m+M_{mxs})\omega^2\}-K_{x\phi}\cdot K_{\phi x}}$$

$$=\frac{231.81\times10^4\times3.9}{8048.69\times10^8}=11.23\times10^{-6}\text{ m}$$

$$A_{\phi2}=\frac{\{K_{xx}-(m+m_s)\omega^2\}M_y}{\{K_{xx}-(m+m_s)\omega^2\}\{K_{\phi\phi}-(M_m+M_{xs})\omega^2\}-K_{x\phi}\cdot K_{\phi x}}$$

$$=\frac{[90.16\times10^4-(102.8+79.7)\times26.2^2]\times3.9}{8048.69\times10^8}=3.7\times10^{-6}\text{ rad}$$

$$A_x=A_{x1}+A_{x2}=(12.40+11.23)\times10^{-6}=23.63\times10^{-6}\text{ m}$$

$$A_\phi = A_{\phi 1} + A_{\phi 2} = (3.2 + 3.7) \times 10^{-6} = 6.9 \times 10^{-6} \text{ rad}$$

Displacement of the top of the block

$$= A_x + (h - L)A_\phi$$

$$= 23.63 \times 10^{-6} + (3.5 - 1.75) \times 6.9 \times 1^{0-6} = 35.70 \times 10^{-6} \text{ m}$$

References

Anandakrishnan, M. and Krishnaswamy, N. R. (1973 a), "Response of embedded footings to vertical vibrations", J. Soil Mech. Found. Div., Am. Soc. Civ. Engg. 99, pp. 863-883.

Arnold, R.N. , Bycroft, G.N. and Warburton, G.B. (1955), "Forced vibrations of a body on an infinite elastic solid", Trans. ASME, 77, pp. 391-401.

Balkrishna. R. H. A. (1961) "The design of machine foundations related to the bulb of pressure", Proc. Int. Conf. Soil Mech. Found. Eng., 5th, Paris, Vol. 1, pp. 563-568.

Baranov, V.A. (1967), "On the calculation of excited vibrations of an embedded foundation (in Russian)". Vopr. Dyn. Prochn., 14 pp. 195-209.

Barkan, D.D. (1962), "Dynamics of base and foundations", McGraw-Hill, New York.

Bereduge, Y.O. and Novak, M. (1972), "Coupled horizontal and rocking vibration of embedded footings", Can. Geotech. J., 9(4), pp. 477-497.

Beredugo, Y.O. (1971), "Vibrations of embedded symmetric footings". Ph. D. Thesis, University of Western Ontario. London, Canada.

Bowles, J.E. (1982), "Foundation analysis and design" , McGraw-Hill, New York.

Bycroft, G.N. (1956), "Forced vibrations of a rigid circular plate on a semi-infinite elastic space and on an elastic stratum", Philos. Trans. R. Soc. London, Ser. A, 248, pp. 327-368.

Chae, Y. S. (1971), "Dynamic behaviour of embedded foundation-soil system", Highw. Res. Rec, 323, pp. 49-59.

Fry, Z. B. (1973), "Development and evaluation of soil bearing capacity. Foundation of Structures", Waterways Exp. Sta., Tech. Rep. No. 3, p. 632.

Gupta. B. N. (1972), "Effect of foundation embedment on the dynamic behaviour of the foundation-soil system", Geotechnique, 22 (1), pp. 129-137.

Hall, J. R. (1967), "Coupled rocking and sliding oscillations of rigid circular footing", Proc. Int Symp. Wave Propag. Dyn. Prop. Earth Mater, Albuquerque, NM, pp. 139-148.

Hsieh, T. K. (1962), "Foundation vibrations", Proc. Inst. Civ. Eng., 22, pp. 211-226.

Lamb, H. (1904), "On the propagation of tremors over the surface of an elastic solid", Philos. Trans. R. Soc. London, Ser. A 203, pp. 1-72.

Lysmer J. and Richart, F. E., Jr. (1966), "Dynamic response of footing to vertical loading", J. Soil Mech. Found. Div., Am. Soc. Civ. Eng., 92 (SM-1), pp. 65-91.

Novak, M. (1970). "Prediction of footing vibrations", J. Soil Mech. Found. Div., Am. Soc. Civ. Eng., 96 (SM-3), pp. 836-861.

Novak, M., (1985), "Experiments with shallow and deep foundations". Proc. Symp. Vib. Probl. Geotech. Eng., Am. Soc. Civ. Eng., Annu. Conv., pp. 1-26.

Novak, M. and Beredugo, Y. O. (1971), "Effect of embedment on footing vibration", Proc. Can. Conf. Earthquake Eng., 1st, Vancouver, pp. 111-125.

Pauw, A. (1953), "A dynamic analogy for foundation soils system", ASTM Spec. Tech. Publ., STP, pp. 3-34.

Reissner, E. (1936), "Stationare Axialymmeterische durch eine Schuttelnde Masse Erregte Schwingungen Homogenen Elastichen Halbraumes", Ing. Arch. 7(6), pp. 381-396.

Reissner, E. (1937), "Freie und erzwungene Torsionschwing-ungen des Elastichen Halbraumes", Ing.-Arch, 8 (4), pp. 229-245.

Reissner. E. and Sagoci, H. F. (1944), "Forced torsional oscillations of an elastic half space", J. Appl. Phys., 15, pp. 652-662.

Richart. F. E. Jr. (1962), "Foundation vibrations". Trans. Am. Soc. Civ. Eng., 127, Part, I, pp. 863-898.

Richart, F. E., Jr., and Whitman, R. V. (1967), "Comparison of footing vibrations tests with theory", J. Soil Mech. Found. Div., Am. Soc. Civ. Eng., 93 (SM - 6), pp. 143-168.

Stokoe, K. H II (1972). "Dynamic response of embedded foundations", Ph. D. thesis presented, University of Michigan, Ann Arbor, Michigan.

Stokoe, K. H., II. and Richart, F. E. Jr. (1974), "Dynamic response of embedded machine foundation", J. Geotech. Eng. Div., Am. Soc. Civ. Eng., 100 (GT-4), pp. 427-447.

Sung, T. Y. (1953), "Vibrations in semi-infinite solids due to periodic surface loading", Ph. D. Thesis, Harvard University, Cambridge, Massachusetts.

Vijayvergiya, R. C. (1981). "Response of embedded foundations", Ph. D. Thesis, Univeristy of Roorkee, Roorkee, India.

Whitman. R. C. and Richart, F. E., Jr. (1967), "Design procedures for dynamically loaded foundations", J. Soil Mech. Found. Div., Am. Soc. Civ. Eng., 93 (SM - 6), pp. 169-193.

Practice Problems

11.1 (a) List the basic differences in analysing a reciprocating machine foundation by the two approaches namely (*i*) Linear weightless spring-mass system, and (*ii*) Elastic half-space theory.

(b) Derive the expressions of natural frequency and amplitude of a block foundation subjected to vertical vibration.

11.2 Starting from fundamentals, derive the expressions of natural frequencies and amplitudes of block foundation subjected to a horizontal force $F_x \sin \omega t$ and a moment $M_y \sin \omega t$ at the combined c.g. of machine and foundation.

11.3 A concrete block shown in Fig. 11.39 is to be used as a foundation for a reciprocating engine operating at 500 rpm and mounted symmetrically with respect to foundation. The weight of the engine is 10 kN. It is likely that the operation of machine exerts the following:

Unbalanced vertical force = 1.8 sin ωt kN

Unbalanced torsional moment = 6 sin ωt kN

The values of the dynamic elastic constants for the design of the foundation may be adopted as given below:

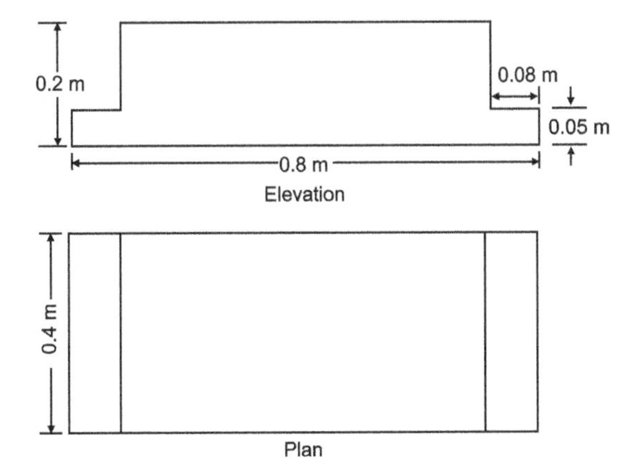

Fig. 11.39. Details of foundation

$C_u = 4.5 \times 10^4$ kN/m^3
$G = 4.8 \times 10^4$ kN/m^3
$\mu = 0.34$
$\gamma_{soil} = 17$ kN/m^3
$\gamma_{conc.} = 24$ kN/m^3
Determine the natural frequencies and amplitudes of the block by (*i*) Linear elastic spring-mass approach and (*ii*) Elastic half space theory.

11.4 Design a suitable foundation for a horizontal compressor driven by an electric motor. The following data are available:
Weight of compressor = 160 kN
Weight of motor = 60 kN
Speed of compressor = 250 rpm
Horizontal unbalanced force = 75 kN acting at a height of 1.0 m above top of pedestal.
The following tests were performed at the site to determine the values of dynamic elastic constants:

(a) A vertical vibration test was conducted on an M-20 concrete block 1.5 × 0.75 × 0.7 m high, using different eccentricities. The data obtained is given in Table 11.9.

Table 11.9

Sl. No.	*Angle of setting of eccentric masses*	f_{nz} *Hz*	*Amplitude of resonance* mm
1	15	35.0	0.06375
2	30	32.0	0.150
3	45	31.0	0.210
4	60	29.5	0.300
5	120	28.0	0.525
6	140	27.0	0.620

(b) A cyclic-plate load test was done on plate 300 mm × 300 mm. The elastic settlement corresponding to a load intensity of 250 kN/m^2 was 6.00 mm.

(c) A wave-propagation test gave an average value of travel time of compression waves as 0.02*s*, corresponding to a distance between geophones of 6 m.

The water table at the site is 2.0 m below the proposed depth of the foundation 3.0 m. The soil at the site is sandy in nature.

CHAPTER

12

Foundations of Impact Type Machines

12.1 General

Impact type machines produce transient dynamic loads of short duration. Hammers are most typical of impact machines. A hammer-foundation-soil system consists of a frame, a falling weight known as 'tup' the anvil and the foundation block. Figure 12.6 shows a typical foundation for a hammer with its frame mounted on the anvil. In Fig. 12.1, a foundation for a hammer with its frame mounted on the foundation is shown.

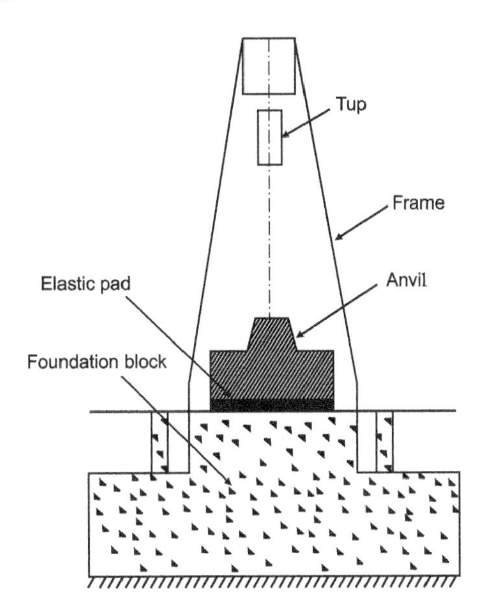

Fig 12.1. Typical arrangement of a hammer foundation with A-frame mounted on foundation

The foundation of a hammer generally consists of a reinforced concrete block. The followmg arrangements are used depending upon the size of the hammer:

(i) For small hammers, the anvil may be directly mounted on the foundation block (Fig. 12.2a). This system can be modelled as single degree of freedom system as shown in Fig. 12.2b.

(ii) In medium capacity hammer, a vibration isolation layer is placed between the anvil and the foundation block (Fig. 12.3a). Usually the isolation layer is an elastic pad consisting of rubber, felt, cork, or timber adequately protected against water and oil. In case of high capacity hammers, special elements such as coil springs and dampers may be used in place of elastic pads (Fig. 12.3b). The systems shown in Fig. 12.3a and 12.3b can be modelled as two degrees of freedom system as shown in Fig. 12.3c.

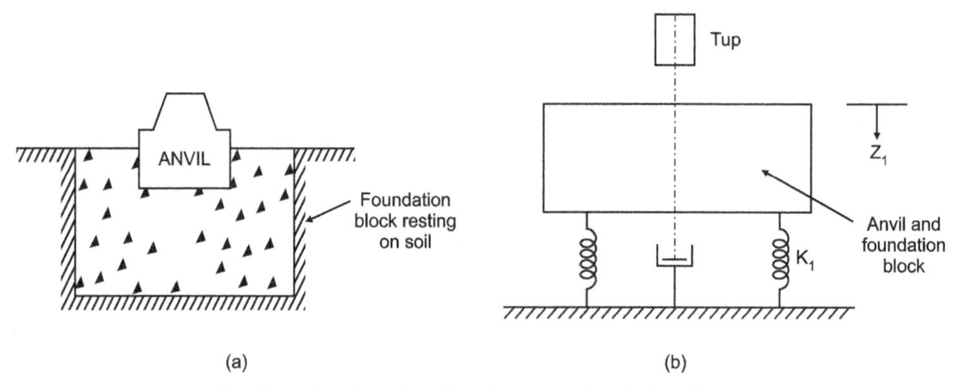

Fig. 12.2. Anvil resting directly on the foundation block

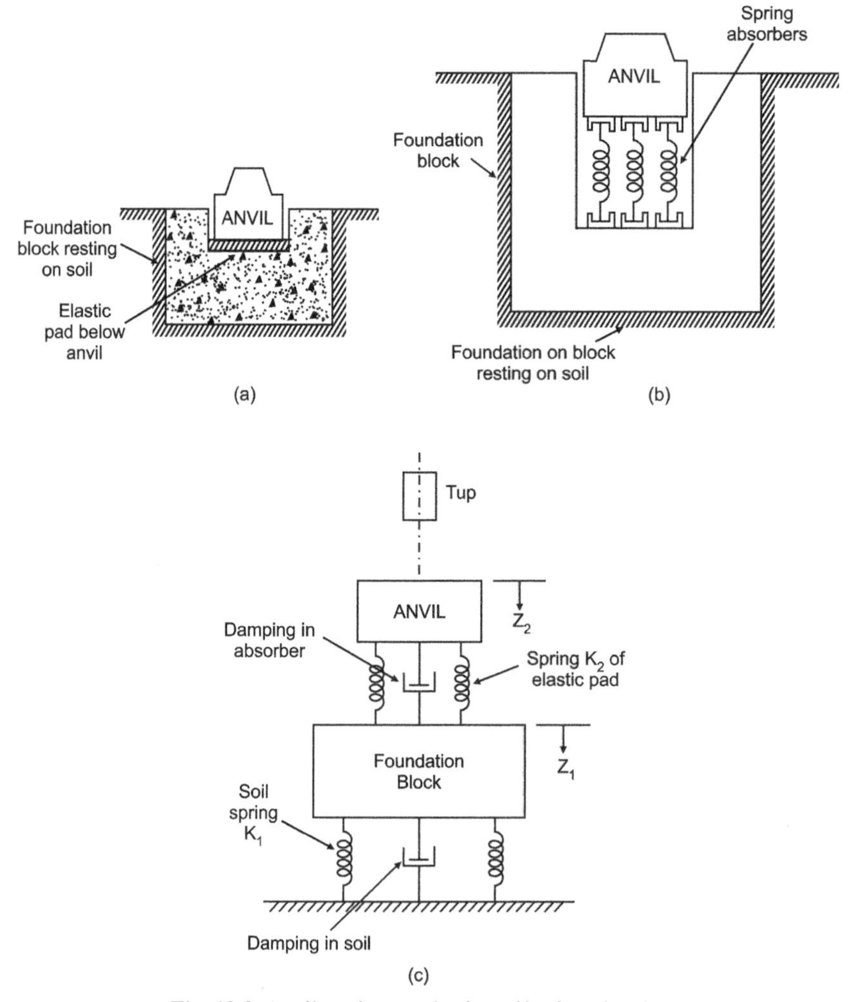

Fig. 12.3. Anvil resting on elastic pad/spring absorbers

(iii) For reducing the transmission of vibrations to the adjoining machines or structures, the foundation block may also be supported on elastic pads (Fig. 12.4a) or on spring absorbers (Fig. 12.4b). In such a case, the foundation is placed in a reinforced concrete trough. The space between the foundation and side of trough is filled up with some soft materials or an air gap is left. The systems shown in Fig. 12.4a and 12.4b can be modelled as three degrees freedom system as shown in Fig. 12.4c. The stiffness of trough (Fig. 12.4a) is very high compared to that of the pad below the foundation block, the trough may be assumed to be rigidly supported on the soil (Novak, 1983), and therefore a two degree freedom model (Fig. 12.3c) may give sufficiently accurate results for all practical purposes.

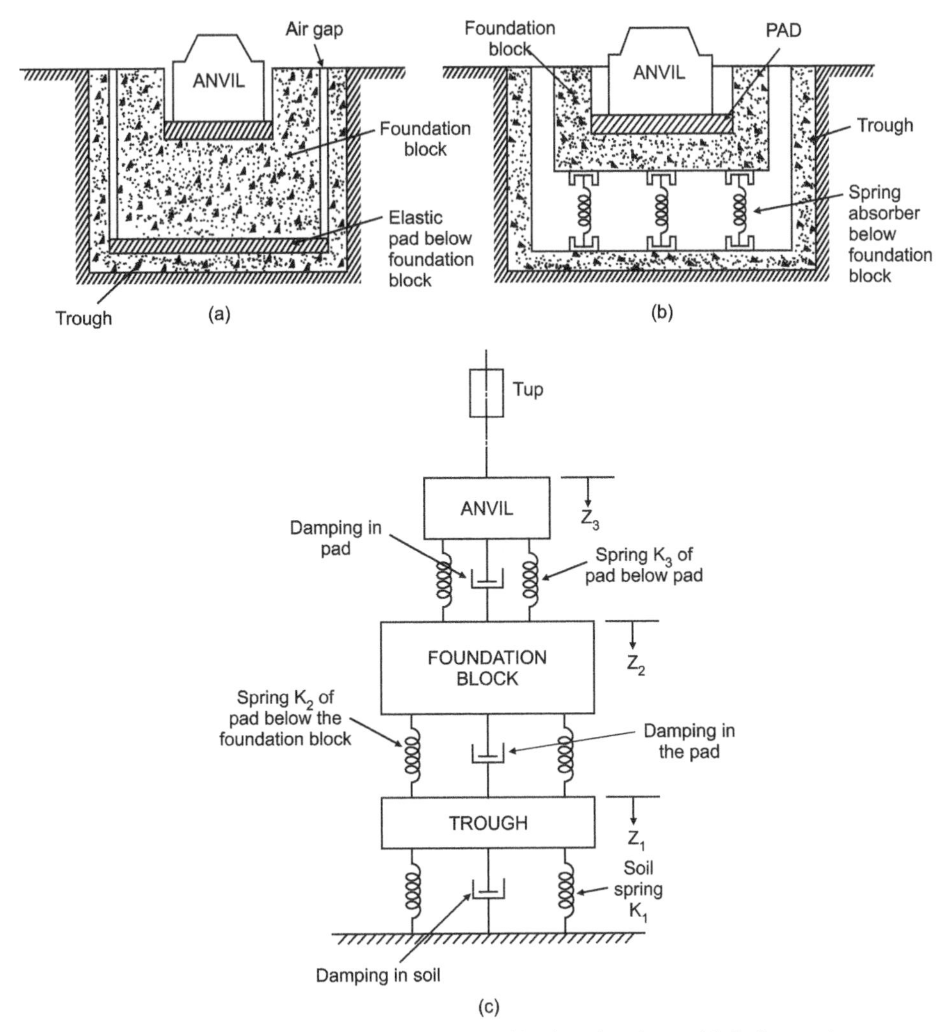

Fig. 12.4. Foundation block on elastic pad/spring absorbers with R.C. trough

In hammer foundations, tup, anvil and foundation are geometrically so aligned that their centres fall on one vertical axis. This will ensure that the loads act on the anvil and foundation without any eccentricity.

12.2 Dynamic Analysis

12.2.1 Two Degrees Freedom System

In general the anvil, pad, foundation, and soil constitute a two degrees system as shown in Fig. 12.5. This model is based on the following assumptions:

(i) The anvil, foundation block, frame, and tup are rigid bodies.

(ii) The pad and the soil can be simulated by equivalent weightless, elastic springs.
(iii) The damping of the elastic pad and soil is neglected.
(iv) The time of impact is short compared to the period of natural vibrations of the system.
(v) Embedment effects are neglected.

The notations used in the model have the following meanings:

m_1 = Mass of foundation and frame if the latter is mounted on the foundation, as in Fig. 12.1

m_2 = Mass of anvil with frame if the latter is mounted on the anvil as in Fig. 10.7

K_1 = $C_u' \cdot A_1$ = Equivalent spring constant of soil under consideration

C_u' = $\lambda\, C_u$ = Modified coefficient of elastic uniform compression, to take into account impact condition which is different from periodic loading

λ = Multiplying factor that governs the relationship between C_u and C_u' usually 1-2, for impact depending upon the soil type

C_u = Coefficient of elastic uniform compression

K_2 = $(E/b) \times A_2$ = Equivalent spring constant of the pad under the anvil

E = Young's modulus of the pad material

b = Thickness of the pad

A_1 = Area of foundation in contact with soil

A_2 = Area of the pad

Z_1 = Displacement of foundation from the equilibrium position

Z_2 = Displacement of anvil from the equilibrium position

12.2.1.1 Natural Frequencies

The equations of motion in free vibration are

$$m_1 \ddot{Z}_1 + K_1 Z_1 + K_2(Z_1 - Z_2) = 0 \quad (12.1)$$

$$m_2 \ddot{Z}_2 + K_2(Z_2 - Z_1) = 0 \quad (12.2)$$

The solution of the above equation can be written as

Let, $$Z_1 = A \sin \omega_n t \quad (12.3)$$

$$Z_2 = B \sin \omega_n t \quad (12.4)$$

where A and B are arbitrary constants. Substituting the values of Z_1 and Z_2 from Eqs. (12.3) and (12.4) in Eqs. (12.1) and (12.2) respectively, we get

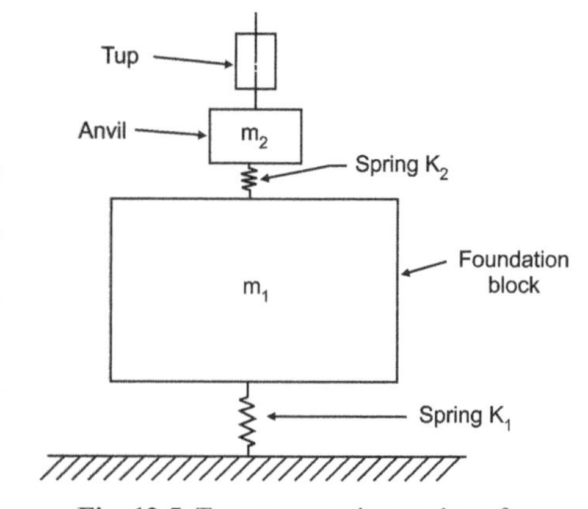

Fig. 12.5. Two-mass-spring analogy for hammer foundation

$$\frac{B}{A} = \frac{K_2 + K_1 - m_1 \omega_n^2}{K_2} \quad (12.5)$$

and $$\frac{B}{A} = \frac{K_2}{K_2 - m_2 \omega_n^2} \quad (12.6)$$

Equating Eqs. (12.5) and (12.6), on simplification we get

$$\omega_n^4 - \left\{ \frac{K_2(m_1 + m_2)}{m_1 m_2} + \frac{K_1}{m_1} \right\} \omega_n^2 + \frac{K_1 K_2}{m_1 m_2} = 0$$

or
$$\omega_n^4 - \left(1+\frac{m_2}{m_1}\right)\left(\frac{K_2}{m_2}+\frac{K_1}{m_1+m_2}\right)\omega_n^2 + \frac{K_2}{m_2}\times\frac{K_1}{m_1+m_2}\left(1+\frac{m_2}{m_1}\right) = 0 \quad (12.7)$$

Let ω_{na} = Circular natural frequency of the foundation of the anvil on the pad

$$\omega_{na} = \sqrt{\frac{K_2}{m_2}} \quad (12.8)$$

ω_{nl} = Limiting natural frequency of the foundation and anvil on soil

$$\omega_{nl} = \sqrt{\frac{K_1}{m_1+m_2}} \quad (12.9)$$

$$\mu_m = \frac{m_2}{m_1}$$

Substituting the values of ω_{na}, ω_{nl} and μ_m in Eq. (12.7), we get

$$\omega_n^4 - (1+\mu_m)(\omega_{na}^2+\omega_{nl}^2)\omega_n^2 + (1+\mu_m)\,\omega_{nl}^2\,\omega_{na}^2 = 0 \quad (12.10a)$$

$$\omega_{n1,2}^2 = \frac{1}{2}\left[(1+\mu_m)(\omega_{na}^2+\omega_{nl}^2) \pm \sqrt{\{(1+\mu_m)(\omega_{na}^2+\omega_{nl}^2)\}^2 - 4(1+\mu_m)\omega_{nl}^2\omega_{na}^2}\,\right] \quad (12.10b)$$

The two natural frequencies of the hammer foundation may be determined by solving the above equations.

12.2.1.2 Amplitude of Vibration

The general solution of the Eqs. (12.1) and (12.2) is given by

$$Z_1 = A_1 \sin\omega_{n1} t + A_2 \cos\omega_{n1} t + A_3 \sin\omega_{n2} t + A_4 \cos\omega_{n2} t \quad (12.11)$$

$$Z_2 = B_1 \sin\omega_{n1} t + B_2 \cos\omega_{n1} t + B_3 \sin\omega_{n2} t + B_4 \cos\omega_{n2} t \quad (12.12)$$

where A_1, A_2, A_3, A_4, B_1, B_2, B_3, B_4 are arbitrary constants.
If system is vibrating at frequency ω_{n1}, then from Eq. (12.6)

$$\frac{B}{A} = \frac{K_2}{K_2 - m_2\,\omega_{n1}^2} = \frac{\omega_{na}^2}{\omega_{na}^2-\omega_{n1}^2} = a_1 \text{ (Say)} \quad (12.13a)$$

Similarly when $\omega_n = \omega_{n2}$

$$\frac{B}{A} = \frac{\omega_{na}^2}{\omega_{na}^2-\omega_{n2}^2} = a_2 \text{ (Say)} \quad (12.13b)$$

It may be noted that values of a_1 and a_2 are known from Eqs. (12.13a) and (12.13b) respectively. Subtracting a_2 from a_1 we get

$$a_1 - a_2 = \omega_{na}^2\left[\frac{1}{\omega_{na}^2-\omega_{n1}^2} - \frac{1}{\omega_{na}^2-\omega_{n2}^2}\right]$$

or
$$a_1 - a_2 = \frac{(\omega_{n1}^2-\omega_{n2}^2)\omega_{na}^2}{(\omega_{na}^2-\omega_{n1}^2)(\omega_{na}^2-\omega_{n2}^2)} \quad (12.14)$$

Equation (12.12) can therefore be written as:

$$Z_2 = a_1 A_1 \sin \omega_{n1} t + a_1 A_2 \cos \omega_{n1} t + a_2 A_3 \sin \omega_{n2} t + a_2 A_4 \cos \omega_{n2} t \tag{12.15}$$

Boundary condition:

(i) At $t = 0$, $Z_1 = Z_2 = 0$

$$A_2 + A_4 = 0 \tag{12.16a}$$

and $$a_1 A_2 + a_2 A_4 = 0 \tag{12.16b}$$

It gives $$A_2 = A_4 = 0 \tag{12.17}$$

(ii) At $t = 0$, $\dot{Z}_1 = 0$; $\dot{Z}_2 = V_a$ (Velocity of anvil)

$$\dot{Z}_1 = A_1 \omega_{n1} \cos \omega_{n1} t - A_2 \omega_{n1} \sin \omega_{n1} t + A_3 \omega_{n2} \cos \omega_{n2} t - A_4 \omega_{n2} \sin \omega_{n2} t$$

or $$A_1 \omega_{n1} + A_3 \omega_{n2} = 0$$

or $$A_3 = -A_1 \frac{\omega_{n1}}{\omega_{n2}} \tag{12.18}$$

$$\dot{Z}_2 = a_1 A_1 \omega_{n1} \cos \omega_{n1} t - a_1 A_2 \omega_{n1} \sin \omega_{n1} t + a_2 A_3 \omega_{n2} \cos \omega_{n2} t - a_2 A_4 \omega_{n2} \sin \omega_{n2} t$$

or $$V_a = a_1 A_1 \omega_{n1} + a_2 A_3 \omega_{n2}$$

or $$A_1 = \frac{V_a}{(a_1 - a_2)\omega_{n1}} \tag{12.19}$$

and $$A_3 = \frac{-V_a}{(a_1 - a_2)\omega_{n2}} \tag{12.20}$$

Therefore, $$Z_1 = \frac{V_a}{(a_1 - a_2)\omega_{n1}} \sin \omega_{n1} - \frac{V_a \sin \omega_{n2} t}{(a_1 - a_2)\omega_{n2}} \tag{12.21}$$

Putting the value of $(a_1 - a_2)$ from Eq. (12. 14) in Eq. (12.21), we get

$$Z_1 = \frac{(\omega_{na}^2 - \omega_{n1}^2)(\omega_{na}^2 - \omega_{n2}^2)}{(\omega_{n1}^2 - \omega_{n2}^2)\omega_{na}^2} \left[\frac{\sin \omega_{n1} t}{\omega_{n1}} - \frac{\sin \omega_{n2} t}{\omega_{n2}} \right] V_a \tag{12.22}$$

Similarly, $$Z_2 = \frac{1}{(\omega_{n1}^2 - \omega_{n2}^2)} \left[\frac{(\omega_{na}^2 - \omega_{n2}^2)\sin \omega_{n1} t}{\omega_{n1}} - \frac{(\omega_{na}^2 - \omega_{n1}^2)\sin \omega_{n2} t}{\omega_{n2}} \right] V_a \tag{12.23}$$

Field observations (Barkan, 1962) on the amplitudes of the anvil and the foundation showed that the vibrations occurred at the lower frequency only. Therefore, it may be assumed that the amplitude of motion for $\sin \omega_{n1} t = 0$ $(\omega_{n1} > \omega_{n2})$.

Hence approximate expressions for maximum displacement will be as follows ($\sin \omega_{n2} t = 1$):

$$Z_{1m} = \frac{(\omega_{na}^2 - \omega_{n1}^2)(\omega_{na}^2 - \omega_{n2}^2)}{\omega_{na}^2(\omega_{n1}^2 - \omega_{n2}^2)} \left[-\frac{1}{\omega_{n2}} \right] V_a \tag{12.24}$$

Similarly, $$Z_{2m} = \frac{1}{(\omega_{n1}^2 - \omega_{n2}^2)} \left[-\frac{(\omega_{na}^2 - \omega_{n1}^2)}{\omega_{n2}} \right] V_a \tag{12.25}$$

12.2.1.3 Stress in the Pad

Maximum compressive stress in the elastic pad below the anvil depends upon the relative displacements of the anvil and the foundation block. The worst case of compression in the pad develops when the anvil moves downward, and at same instant of time, the foundation block moves upward. The maximum compressive stress in the pad is thus expressed by

$$\sigma_p = \frac{Z_{1m} + Z_{2m}}{A_2} \tag{12.26}$$

12.2.1.4 Stress in the Soil

Stresses transmitted to the soil q through the combined static and dynamic loads are expressed by

$$q = \frac{W_a + W_f + Z_{2m} K_2}{A_1} \tag{12.27}$$

where W_a = Weight of the anvil, and

W_f = Weight of the foundation

12.2.2 Single Degree Freedom System

Sometimes in the case of light hammers, no pad is used between anvil and foundation. The system can then be represented as single degree freedom system (Fig. 12.6). In this case the equation of vertical free vibrations of the foundation will be

$$m\ddot{Z} + K_1 Z_1 = 0 \tag{12.28}$$

where Z = Vertical displacement of centre of mass of foundation and anvil, measured from equilibrium position,

m_1 = Total vibrating mass i.e. mass of the anvil plus foundation, and

$K_1 = C_u' A_1$ = equivalent spring constant of soil

The natural frequency of the system will be given by

$$\omega_{nz} = \sqrt{\frac{K_1}{m_1}}$$

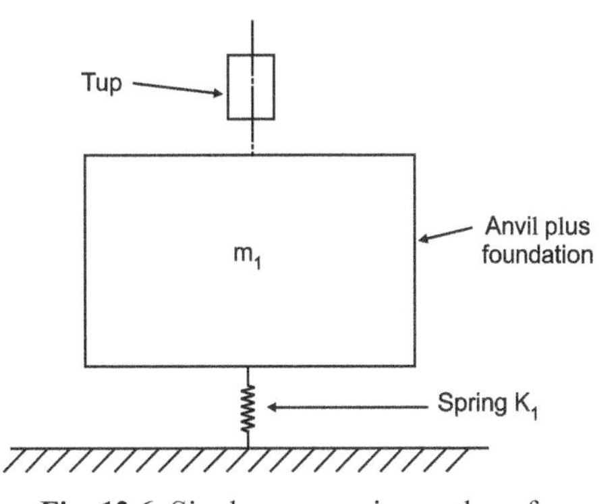

Fig. 12.6. Single-mass-spring analogy for hammer foundation

The Eq. (12.28) is the equation of free vibrations of the foundation without damping. The general solution of this equation is

$$Z_1 = A \sin \omega_n t + B \cos \omega_n t \tag{12.29}$$

The constants A and B, as usual, are determined form initial conditions of motion. Substituting the boundary conditions

At $t = 0$, $Z_1 = 0$, and $\dot{Z}_1 = V_{af}$ = Velocity of combined anvil and foundation

Using these initial conditions, we get

$$A = \frac{V_{af}}{\omega_n} \text{ and } B = 0$$

Therefore,

$$A_z = \frac{V_{af}}{\omega_n} \sin \omega_n t \tag{12.30}$$

The maximum displacement will be

$$Z_{1m} = \frac{V_{af}}{\omega_n} \tag{12.31}$$

Stresses transmitted to the soil q through the combined static and dynamic loads may be computed using Eq. (12.27) using the value of Z_{1m} obtained from Eq. (12.31).

12.2.3 Three Degree of Freedom System

As mentioned in Sec. 12.1, for reducing the transmission of vibrations to the adjoining machines or structures, the foundation block may also be supported on elastic pad or on spring absorbers. In such a case, the foundation is placed in a reinforced concrete trough. The system may be mathematically modelled as shown in Fig. 12.7. m_1, m_2, and m_3 represent respectively the masses of the trough, the foundation block and the anvil respectively. K_1 is the equivalent spring constant of the soil and equal to $C_u' \cdot A_1$, where A_1 is the base contact area of trough. K_2 and K_3 are stiffnesses of elastic pads.

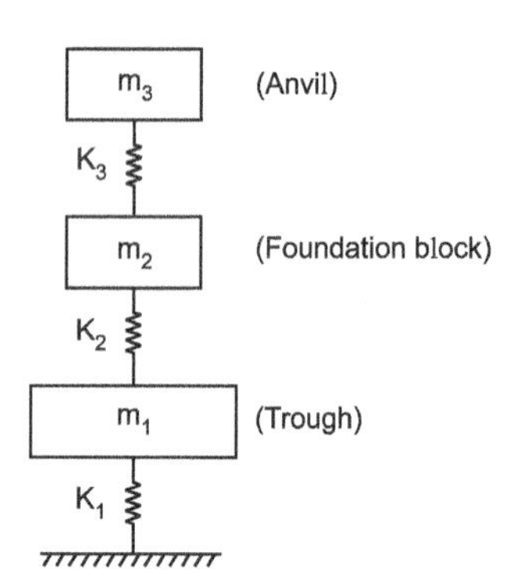

Fig. 12.7. Two mass spring analogy for hammer foundation

12.2.3.1 Natural Frequencies

The equations of motion in free vibration are

$$m_1\ddot{Z}_1 + K_1Z_1 + K_2(Z_1 - Z_2) = 0 \tag{12.32}$$

$$m_2\ddot{Z}_2 + K_2(Z_2 - Z_1) + K_3(Z_2 - Z_3) = 0 \tag{12.33}$$

$$m_3\ddot{Z}_3 + K_3(Z_3 - Z_2) = 0 \tag{12.34}$$

The solution of the above equations can be written as

$$Z_1 = A \sin \omega_n t \tag{12.35}$$

$$Z_2 = B \sin \omega_n t \tag{12.36}$$

$$Z_3 = C \sin \omega_n t \tag{12.37}$$

where A, B and C are arbitrary constants. Substituting the values of Z_1, Z_2 and Z_3 from Eqns. (12.35), (12.36) and (12.37) in Eqns. (12.32), (12.33) and (12.34) respectively, we get

$$-m_1 A\omega_n^2 + K_1A + K_2(A - B) = 0$$

or

$$(-m_1\omega_n^2 + K_1 + K_2)A - K_2B = 0 \tag{12.38}$$

$$-m_2 B\omega_n^2 + K_2(B - A) + K_3(B - C) = 0$$

or

$$-K_2A + (-m_2\omega_n^2 + K_2 + K_3)B - K_3C = 0 \tag{12.39}$$

$$-m_3C\omega_n^2 + K_3(C - B) = 0$$

or

$$-K_3B + (-m_3\omega_n^2 + K_3)C = 0 \tag{12.40}$$

Solution of Eqs. (12.38) (12.39) (12.40) can be expressed in matrix form as below:

$$\begin{vmatrix} (-m_1\omega_n^2 + K_1 + K_2) & -K_2 & 0 \\ -K_2 & (-m_1\omega_n^2 + K_2 + K_3) & -K_3 \\ 0 & -K_3 & (-m_3\omega_n^2 + K_3) \end{vmatrix} = 0 \tag{12.41}$$

On simplification, equation (12.41) gives

$$m_1 m_2 m_3 \omega_n^6 - [m_2 m_3 (K_1 + K_2) + m_1 m_3 (K_2 + K_3) + m_1 m_2 K_3]\omega_n^4$$

$$+[K_1 K_2 m_3 + K_1 K_3 (m_2 + m_3) + K_2 K_3 (m_1 + m_2 + m_3)]\omega_n^2 - K_1 K_2 K_3 = 0 \tag{12.42}$$

The Eq. (12.42) is a cubic in ω_n^2. The three natural frequencies of the system ω_{n1}, ω_{n2} and ω_{n3} are obtained solving this equation.

12.2.3.2. Amplitude of Vibration

The general solutions of Eqns. (12.32), (12.33) and (12.34) are given by:

$$Z_1 = A_1 \sin \omega_{n1} t + A_2 \cos \omega_{n1} t + A_3 \sin \omega_{n2} t + A_4 \cos \omega_{n2} t + A_5 \sin \omega_{n3} t + A_6 \cos \omega_{n3} t \tag{12.43}$$

$$Z_2 = B_1 \sin \omega_{n1} t + B_2 \cos \omega_{n1} t + B_3 \sin \omega_{n2} t + B_4 \cos \omega_{n2} t + B_5 \sin \omega_{n3} t + B_6 \cos \omega_{n3} t \tag{12.44}$$

$$Z_3 = C_1 \sin \omega_{n1} t + C_2 \cos \omega_{n1} t + C_3 \sin \omega_{n2} t + C_4 \cos \omega_{n2} t + C_5 \sin \omega_{n3} t + C_6 \cos \omega_{n3} t \tag{12.45}$$

where $A_1, A_2, A_3, A_4, A_5, A_6, B_1, B_2, B_3, B_4, B_5, B_6, C_1, C_2, C_3, C_4, C_5, C_6$ are arbitrary constants.
From Eq. (12.40)

$$\frac{C}{B} = \frac{K_3}{K_3 - m_3\omega_n^2} = \frac{K_3/m_3}{(K_3/m_3) - \omega_n^2} = \frac{\omega_{na}^2}{\omega_{na}^2 - \omega_n^2}$$

where $\omega_{na}^2 = \dfrac{K_3}{m_3}$ = Natural frequency of anvil and elastic pad system only

If the system is vibrating at $\omega_n = \omega_{n1}$, then

$$\frac{C}{B} = \frac{\omega_{na}^2}{\omega_{na}^2 - \omega_{n1}^2} = a_1 \text{ (say)} \tag{12.46}$$

Similarly for $\omega_n = \omega_{n2}$

$$\frac{C}{B} = \frac{\omega_{na}^2}{\omega_{na}^2 - \omega_{n2}^2} = a_2 \text{ (say)} \tag{12.47}$$

and for $\omega_n = \omega_{n3}$

$$\frac{C}{B} = \frac{\omega_{na}^2}{\omega_{na}^2 - \omega_{n3}^2} = a_3 \text{ (say)} \tag{12.48}$$

Eqn. (12.38) gives:

$$\frac{B}{A} = \frac{K_1 + K_2 - m_1\omega_n^2}{K_2} \tag{12.48}$$

As done above, we may obtain the three values of $\dfrac{B}{A}$ corresponding to the system vibrating at natural frequencies ω_{n1}, ω_{n2} and ω_{n3}

$$\frac{B}{A} = \frac{K_1 + K_2 - m_1\omega_{n1}^2}{K_2} = b_1 \text{ (say)} \tag{12.49}$$

$$\frac{B}{A} = \frac{K_2 + K_2 - m_1\omega_{n2}^2}{K_2} = b_2 \text{ (say)} \tag{12.50}$$

$$\frac{B}{A} = \frac{K_1 + K_2 - m_1\omega_{n3}^2}{K_2} = b_3 \text{ (say)} \tag{12.51}$$

Substituting the boundary conditions in Eqns. (12.43), (12.44) and (12.45) and using the relations (12.46) to (12.51), values of maximum displacements are as follows:

$$Z_{1m} = \frac{V_a(b_1 - b_2)}{\omega_{n3}[a_3b_3(b_1 - b_2) - a_1b_1(b_3 - b_2) - a_2b_2(b_1 - b_3)]} \tag{12.52}$$

$$Z_{2m} = \frac{V_a(b_1 - b_2)b_3}{\omega_{n3}[a_3b_3(b_1 - b_2) - a_1b_1(b_3 - b_2) - a_2b_2(b_1 - b_3)]} \tag{12.53}$$

$$Z_{3m} = \frac{V_a(b_1 - b_2)a_3b_3}{\omega_{n3}[a_3b_3(b_1 - b_2) - a_1b_1(b_3 - b_2) - a_2b_2(b_1 - b_3)]} \tag{12.54}$$

where V_a = Velocity of anvil after impact

12.2.4 Determination of Initial Velocity of Anvil, V_a

For a single acting drop hammer, the initial velocity of the tup V_{Ti} at the time of impact is given by

$$V_{Ti} = \eta\sqrt{2gh} \tag{12.55}$$

where h = Drop of tup in meters

g = Acceleration due to gravity, m/s²

η = Efficiency of drop (It lies between 0.45 to 0.80. An average value equal to 0.65 may be adopted)

For double-acting hammers, operated by pneumatic or steam pressure, V_{Ti} is given by

$$V_{Ti} = \eta\sqrt{2gh\frac{(W_T + pA_c)}{W_T}} \tag{12.56}$$

where W_T = Gross weight of the dropping parts, including upper half of the die in kN.

p = Pneumatic (i.e. steam or air) pressure in kN/m²

A_c = Net area of cylinder in m².

For two degree and three degree freedom systems:

The initial velocity of anvil just after the tup's impact can be determined by using the law of conservation of momentum. Since the anvil is stationary:

$$\text{Momentum of tup and anvil before impact} = \frac{W_T}{g}V_{Ti} \tag{12.57a}$$

and

$$\text{Momentum of tup and anvil after impact} = \frac{W_T}{g}V_{Ta} + \frac{W_a}{g}V_a \tag{12.57b}$$

where W_a = Weight of anvil (plus frame if it mounted on the anvil)

V_a = Velocity of anvil after impact

V_{Ta} = Velocity of tup after impact

Therefore,

$$\frac{W_T}{g}V_{Ti} = \frac{W_T}{g}V_{Ta} + \frac{W_a}{g}V_a \tag{12.58}$$

According to Newton's law, the coefficient of elastic restitution, *e*, is given by

$$e = \frac{\text{Relative velocity after impact}}{\text{Relative velocity before impact}}$$

or

$$e = \frac{V_a - V_{Ta}}{V_{Ti}} \tag{12.59}$$

The value of *e* depends upon the material of the bodies involved in impact. Theoretically value of *e* lies between 0 and 1. In forge hammer, usually the value of *e* does not exceed 0.5 (Barkan, 1962). Since a larger *e* gives larger amplitudes of motion, the value of *e* equal to 0.5 is adopted in designing hammer foundation. On solving Eqs. (12.58) and (12.59) we get

$$V_a = \frac{1+e}{1+\dfrac{W_a}{W_T}} \cdot V_{Ti} \tag{12.60}$$

For single degree freedom system:

Momentum of tup and anvil plus foundation after impact $= \dfrac{W_T}{g}V_{Ta} + \dfrac{W_a + W_f}{g}V_{af}$ (12.61)

where W_f = Weight of the foundation (plus frame if mounted on it), and
V_{af} = Velocity of anvil plus foundation after impact.

Proceeding exactly in the way as described above for two degree freedom system, the expression of V_{af} will be as given below

$$V_{af} = \frac{1+e}{1+\dfrac{W_a + W_f}{W_T}} V_{Ti} \tag{12.62}$$

12.3 Design Procedure for Hammer Foundation

The design of a hammer foundation may be carried out in following steps.

12.3.1 Machine Data

The following information about the hammer is required for the design:

(a) Type and weight of striking part of hammer;
(b) Dimensions of base area of anvil and its weight;
(c) Maximum stroke or fall of hammer, mean effective pressure on piston and effective area of piston;
(d) Arrangement and size of anchor bolts; and
(e) Permissible amplitudes of the anvil motion and the foundation on block. If this information is not available, the amplitudes of motion given in Table 10.2 may be considered as limiting values.

12.3.2 Soil Data

The following information about the sub-surface soil should be known:

(a) Soil Profile: For drop hammers of upto 10 kN capacity, soil investigations should generally be done to a depth of 6 m. For heavier impact machines, it is preferable to investigate soil conditions to a depth of 12 m or to a hard stratum. If piles are used, the investigation should be conducted to a suitable depth.

(b) Soil investigation to ascertain allowable soil pressure and to determine the dynamic properties of the soil specifically the value of C_u.

(c) The relative position of the water table below the ground at different time of the year.

12.3.3 Trial Size of the Foundation

(a) Weight and Area: The weight of the foundation for a hammer and the size of its area in contact with the soil should be selected in such a way that

(i) the static pressure on the soil does not exceed the reduced allowable soil pressure, and

(ii) the foundation does not bounce on the soil.

These conditions may be written as

$$p_{st} \leq \alpha q_a \tag{12.63a}$$

$$A_z < A_p \tag{12.63b}$$

where p_{st} = Static pressure intensity,

α = reduction factor,

q_a = allowable soil pressure,

A_z = amplitude of motion, and

A_p = permissible value of amplitude.

From Eq. (12.63a)

$$\frac{W}{A_1} < \alpha q_a \tag{12.64}$$

Considering an average value of A_p as 1 mm = (10^{-3} m) and assuming the system as single degree freedom system, the Eq. (12.63b) can be written as

$$\frac{(1+e)W_T V_{Ti}}{\sqrt{C_u' W A_1 g}} < 10^{-3} \tag{12.65}$$

where W_T = Weight of tup, kN

W = Weight of foundation, anvil and frame, kN

A_1 = Base area of foundation in contact with soil, m^2

V_{Ti} = Initial velocity of tup, m/s

g = Acceleration due to gravity, m/s^2

C_u' = Coefficient of elastic uniform compression for hammer foundation, kN/m^3

Substituting the value of W from Eq. (12.64) into Eq. (12.65) we get

$$A_1 \geq \frac{(1+e)W_T V_{Ti}}{C_u' \alpha q_a g} \times 10^3 \text{ m}^2 \tag{12.66}$$

Substituting the value of A_1 from Eq. (12.64) into (12.65) we get

$$W = \frac{(1+e)W_T V_{Ti}\sqrt{\alpha q_a}}{\sqrt{C'_u g}} 10^3 \text{ kN} \tag{12.67}$$

From Eq. (12.67), total weight of the anvil and foundation can be obtained. Knowing the weight of anvil, using Eq. (12.68) weight of the foundation can be worked out. On the basis of experience (Barkan, 1962), the weight of the anvil is kept generally 20 times the weight of the tup. Further it is recommended that the weight of the foundation block should be at least 3 to 5 times that of the anvil.

(b) Thickness: The thickness of foundation block shall be so designed that the block is safe both in punching shear and bending. For the calculations the inertia forces developed shall also be included. However, the following minimum thickness of foundation block below the anvil shall be provided:

Mass of tup (kN)	*Thickness of foundation block, Min (m)*
Upto 10	1.0
10 to 20	1.25
20 to 40	1.75
40 to 60	2.25
Over 60	2.50

12.3.4 Selecting the Dynamic Elastic Constant C_u'

The procedure of obtaining C_u has already been discussed in Chapter 4 for relevant strain level. The value of C_u' may be taken as $\lambda C_u'$ where λ varies between l and 2. The value of C_u may also be obtained from the following relation.

$$C_u = \frac{4Gr_o}{1-\mu} \tag{12.69}$$

where G = Shear modulus,

r_o = Equivalent radius $= \sqrt{\frac{A}{\pi}}$ and

μ = Poisson's ratio

12.3.5 Natural Frequencies

Compute

$$\omega_{na} = \sqrt{\frac{K_2}{m_2}}$$

$$\omega_{nl} = \sqrt{\frac{K_1}{m_1 + m_2}}$$

in which $K_2 = \frac{E}{b} A_2$

where E = Young's modulus of pad material

b = Thickness of the pad

A_2 = Area of the pad

$K_1 = C_u' \cdot A = \lambda\, C_u A$

Natural frequencies of the combined system are given by

$$\omega_{n1,2}^2 = \frac{1}{2}[(1+\mu_m)(\omega_{na}^2+\omega_{nl}^2) \pm \sqrt{\{(1+\mu_m)(\omega_{na}^2+\omega_{nl}^2)\}^2 - 4(1+\mu_m)(\omega_{na}^2 \cdot \omega_{nl}^2)}$$

12.3.6 Velocity of Dropping Parts of Anvil

Compute the velocity V_{Ti} of the tup before impact

$$V_{Ti} = \sqrt{\frac{2gh(W_T + pA_c)}{W_T}} \cdot \eta$$

in which, W_T = Gross weight of dropping parts in kN

p = Pneumatic (i.e. steam or air) pressure in kN/m^2

A_c = Net area of cylinder in m^2

h = Drop of the tup

η = Efficiency of drop, usually 0.65.

Compute the velocity of the anvil V_a after impact by

$$V_a = \frac{1+e}{1+\dfrac{W_a}{W_T}} \cdot V_{Ti}$$

in which W_a = Weight of anvil (plus frame if it mounted on the anvil)

V_a = Velocity of anvil after impact

e = Coefficient of elastic restitution. The value of e may be adopted as 0.6

12.3.7 Motion Amplitudes of the Foundation and Anvil

Compute the maximum foundation and anvil amplitudes with following equations

$$Z_{1m} = \frac{(\omega_{na}^2 - \omega_{n1}^2)(\omega_{na}^2 - \omega_{n2}^2)}{\omega_{na}^2(\omega_{n1}^2 - \omega_{n2}^2)}\left[-\frac{1}{\omega_{n2}}\right]V_a$$

$$Z_{2m} = \frac{1}{(\omega_{n1}^2 - \omega_{n2}^2)}\left[-\frac{(\omega_{na}^2 - \omega_{n1}^2}{\omega_{n2}}\right]V_a$$

where ω_{n2} is the smaller natural frequency.

12.3.8 Dynamic Stress in the Pad σ_p

Compute dynamic stress in the pad by

$$\sigma_p = K_2 \frac{Z_{1m} + Z_{2m}}{A_2}$$

12.3.9 Stress in Soil

Compute stress in soil by

$$q = \frac{W_a + W_f + Z_{2m}K_2}{A_1}$$

Computed values of natural frequencies should satisfy the criteria for the frequency of operation of the hammer. Also, motion amplitudes should be smaller than permissible values, and the stress in the elastic pad and in soil should be smaller than the permissible compressive stress of the pad material, and allowable soil pressure respectively.

Illustrative Examples

Example 12.1

A 15 kN forging hammer is proposed to install in an industrial complex. The hammer has the following specifications:

Weight of tup without die	= 11 kN
Maximum tup stroke	= 800 mm
Weight of the upper half of the die	= 4.0 kN
Area of piston	= 0.12 m^2
Supply steam pressure	= 600 kN/m^2
Weight of anvil block	= 300 kN
Total weight of hammer	= 400 kN
Bearing area of anvil	= 1.8 m × 1.8 m
Permissible vibration amplitude for anvil	= 1.5 mm
Permissible amplitude for foundation	= 1.2 mm

It is proposed to use a pine wood pad of thickness 0.5 m below the anvil. The modulus of elasticity of pad material is 5×10^5 kN/m^2, and allowable compressive stress in pad is 3500 kN/m^2.

A vertical resonance test was conducted on a 1.5 m × 0.75 m × 0.70 m high concrete block at the proposed depth of foundation. The data obtained are given below:

S. No.	*θ (Deg)*	f_{nz} *(Hz)*	*Amplitude at resonance (microns)*
1.	36	41	13
2.	72	40	24
3.	108	34	32
4.	144	31	40

The soil at the site is sandy in nature and water table lies at a depth of 3.0 m below ground surface. Allowable soil pressure is 225 kN/m^2. Design a suitable foundation for the hammer.

Solution

(i) *Trial dimensions of foundation*: Let the weight of the block is kept about 5 times the weight of anvil. The details of the suggested foundation are shown in Fig. 12.8.

Weight of foundation = 24 × 7 × 5 × 2 = 1680 kN

(ii) Evaluation of C_u

Area of test block = 1.5 × 0.75 = 1.125 m^2

Weight of test block = (1.125 × 0.75) × 24 = 18.9 kN.

Weight of oscillator and motor = 1.0 kN (Assumed)

Total weight of test block, oscillator and motor = 18.9 + 1.0 = 19.9 kN

$$C_u = \frac{4\pi^2 f_{nz}^2 m}{A} = \frac{4\pi^2 \times 19.9 \times f_{nz}^2}{1.125 \times 9.81} = 71.1 f_{nz}^2 \text{ kN/m}^3$$

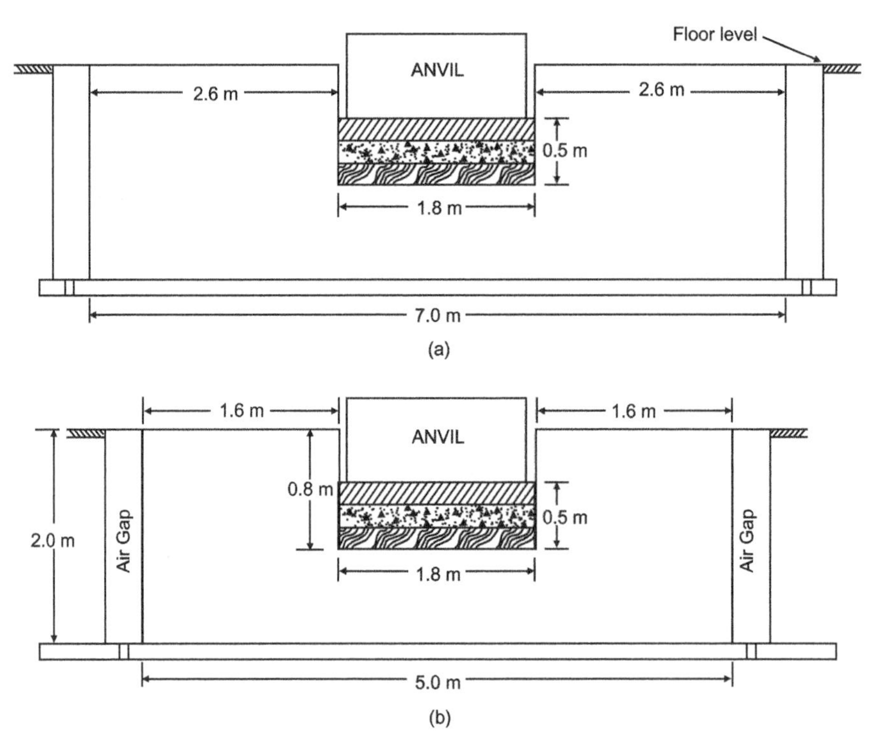

Fig. 12.8. Details of hammer foundation

$$\text{Strain level} = \frac{\text{Amplitude at resonance}}{\text{Width of block}}$$

Putting the given values of f_{nz} and amplitude at resonance, C_u values and corresponding strain levels are computed and listed in columns 2 and 3 of Table 12.1 respectively.

Table 12.1. Values of C_u and Strain Levels

S. No.	$C_{u1} \times 10^4$ *kN/m*^{3}	*Strain Level* $\times 10^{-4}$	$C_{u2} \times 10^4$ *kN/m*^{3}
1.	11.95	0.173	2.96
2.	11.38	0.320	2.82
3.	8.22	0.427	2.04
4.	6.83	0.533	1.69

Corrections for confining pressure and area:

The mean effective confining pressure at a depth of one-half of the width of block is given by

$$\bar{\sigma}_{o1} = \bar{\sigma}_v \frac{(1 + 2K_o)}{3}$$

where, $\bar{\sigma}_v = \bar{\sigma}_{v1} + \bar{\sigma}_{v2}$

$\bar{\sigma}_{v1}$ = Effective overburden pressure at the depth under consideration

$\bar{\sigma}_{v2}$ = Increase in vertical pressure due to the weight of block

Assuming that the top 2.0 m soil has a moist unit weight of 18 kN/m^3, and the next 1.0 m soil i.e. upto water table is saturated having saturated unit weight equal to 20 kN/m^3.

$$\bar{\sigma}_{v1} = 18 \times 2.0 + 20 \frac{0.70}{2} = 43 \text{kN/m}^3$$

$$\bar{\sigma}_{v2} = \frac{4q}{4\pi}\left[\frac{2mn\sqrt{m^2+n^2+1}}{m^2+n^2+1+m^2n^2}\cdot\frac{m^2+n^2+2}{m^2+n^2+1}+\sin^{-1}\frac{2mn\sqrt{m^2+n^2+1}}{m^2+n^2+1+m^2n^2}\right]$$

$$m = \frac{\frac{L}{2}}{z} = \frac{\frac{1.5}{2}}{\frac{0.70}{2}} = 2.14$$

$$n = \frac{\frac{B}{2}}{z} = \frac{\frac{0.75}{2}}{\frac{0.70}{2}} = 1.07$$

$$q = 24 \times 0.70 = 16.8 \text{ kN/m}^2$$

Substituting the above values of m, n and q in the expression $\bar{\sigma}_{v2}$, we get

$$\bar{\sigma}_{v2} = 16.8 \times 0.82 = 13.73 \text{ kN/m}^2$$

$$\bar{\sigma}_{v} = 43.5 + 13.73 = 57.23 \text{ kN/m}^2$$

Assuming $\phi = 35°$, $K_o = 1 - \sin 35° = 0.426$

$$\bar{\sigma}_{o1} = 57.23 \times \frac{(1+2(0.426))}{3} = 35.30 \text{ kN/m}^2$$

For the actual foundation

$$\bar{\sigma}_{v1} = 18 \times 2.0 + 20 \times 1.0 + (20 - 9.81) \times 1.5 = 71.29 \text{ kN/m}^2$$

$$m = \frac{\frac{L}{2}}{Z} = \frac{\frac{7}{2}}{\frac{5}{2}} = 1.4$$

$$n = \frac{\frac{B}{2}}{Z} = \frac{\frac{5}{2}}{\frac{5}{2}} = 1.0$$

$$q = 24 \times 2.0 = 48 \text{ kN/m}^3$$

$$\bar{\sigma}_{v2} = 48 \times 0.76 = 36.58 \text{ kN/m}^2$$

$$\bar{\sigma}_{v} = 71.29 + 36.58 = 107.87 \text{ kN/m}^2$$

$$\bar{\sigma}_{o2} = 107.87 \times \frac{(1+2(0.426))}{3} = 66.88 \text{ kN/m}^2$$

$$\frac{C_{u2}}{C_{u1}} = \sqrt{\frac{\bar{\sigma}_{o2}}{\bar{\sigma}_{o1}} \times \frac{A_1}{A_2}}$$

$$= \sqrt{\frac{66.88}{35.30} \times \frac{1.125}{10}} = 0.462$$

The values of C_u of the actual foundation for different strain levels are listed in column 4 of Table 12.1.

$$\text{Strain level in actual foundation} = \frac{1.0}{5.0 \times 1000} = 2 \times 10^{-4}$$

The strain level in actual foundation is higher than the strain level observed in the tests. Seeing the variation of C_u with respect to strain level, the value of C_u equal to 1.3×10^4 kN/m^3 may be adopted in design.

(iii) Computations of K_1, K_2, m_1 and m_2

$$C_u' = \lambda \cdot C_u$$

$$= 2.0 \times 1.3 \times 10^4 = 2.6 \times 10^4 \text{ kN/m}^3 \ (\lambda = 2.0, \text{Assumed})$$

$$K_1 = C_u' \cdot A_1 = 2.6 \times 10^4 \times 7 \times 5 = 91 \times 10^4 \text{ kN/m}$$

$$K_2 = \frac{E}{b} \cdot A_2 = \frac{5 \times 10^5}{0.5} 1.8 \times 1.8 = 324 \times 10^4 \text{ kN/m}$$

Weight of the foundation block, W_1

$$= [(7 \times 5 \times 2) - (1.8 \times 1.8 \times 0.8)] \times 24$$

$$= 1617.79 \text{ kN}$$

$$m_1 = \frac{1617.79 \times 10^3}{9.81} = 164.91 \times 10^3 \text{ kg}$$

$$W_2 = 400 \text{ kN}$$

$$m_2 = \frac{400 \times 10^3}{9.81} = 40.77 \times 10^3 \text{ kg}$$

(iv) Natural frequencies of soil-foundation system

$$\omega_{na} = \sqrt{\frac{324 \times 10^4}{40.77}} = 281.9 \text{ rad/s}$$

$$\omega_{nl} = \sqrt{\frac{91 \times 10^4}{164.91 + 40.77}} = 66.52 \text{ rad/sec}$$

$$\mu_m = \frac{m_2}{m_1} = \frac{40.77}{164.91} = 0.247$$

The natural frequencies are given by

$$\omega_{n1,2}^2 = \frac{1}{2}\left[(1 + \mu_m)(\omega_{na}^2 + \omega_{nl}^2) \pm \sqrt{\{(1 + \mu_m)(\omega_{na}^2 + \omega_{nl}^2)\}^2 - 4(1 + \mu_m)(\omega_{na}^2 + \omega_{nl}^2)}\right]$$

Substituting the values of ω_{na}, ω_{nl} and μ_m in the above expression, we get

$$\omega_{n1} = 316.61 \text{ rad/sec, and } \omega_{n2} = 66.14 \text{ rad/sec}$$

(v) Velocity of dropping parts

$$V_{Ti} = \eta\sqrt{\frac{2gh(W_T + pA_c)}{W_T}}$$

$\eta = 0.70$ (assumed); $W_T = 15$ kN, $p = 600$ kN/m², $A_c = 0.12$ m², $h = 0.8$

$$V_{Ti} = 0.70\sqrt{\frac{2\times 9.81\times 0.8(15+600\times 0.12)}{15}} = 6.68\,\text{m/sec}$$

$$V_a = \frac{1+e}{1+\frac{W_a}{W_T}}\cdot V_{Ti}$$

$$= \frac{1+0.5}{1+\frac{400}{15}}\cdot 6.68$$

$$= 0.362 \text{ m/sec}$$

(vi) Amplitudes of vibration

$$Z_{1m} = \frac{(\omega_{na}^2 - \omega_{n1}^2)(\omega_{na}^2 - \omega_{n2}^2)}{(\omega_{n1}^2 - \omega_{n2}^2)\omega_{na}^2}\left[-\frac{1}{\omega_{n2}}\right]V_a$$

$$= \frac{(281.9^2 - 316.61^2)(281.9^2 - 66.14^2)}{(316.61^2 - 66.14^2)281.9^2}\left[-\frac{0.362}{66.14}\right]$$

$$= 1.12\times 10^{-3}\text{ m} \qquad (< 1.5 \text{ mm, Safe})$$

$$Z_{2m} = \frac{1}{(\omega_{n1}^2 - \omega_{n2}^2)}\left[-\frac{(\omega_{na}^2 - \omega_{n1}^2}{\omega_{n2}}\right]V_a$$

$$= \frac{(281.9^2 - 316.61^2)}{(316.61^2 - 66.14^2)}\left[-\frac{0.362}{66.14}\right]$$

$$= 1.186\times 10^{-3}\text{ m} \qquad (<1.2 \text{ mm, Safe})$$

(vii) Dynamic stress in pad

$$\sigma_p = K_2\frac{Z_{1m} + Z_{2m}}{A_2}$$

$$= 324\times 10^4 \times \frac{(1.12\times 1.186)\times 10^{-8}}{1.8\times 1.8}$$

$$\sigma_p = 2306 \text{ kN/m}^2 \qquad [< 3500 \text{ kN/m}^2\text{, Safe}]$$

(viii) Stress in the soil

$$q = \frac{W_a + W_f + Z_{2m}K_2}{A_1}$$

$$W_f = 24 \times 7 \times 5 \times 2 + 19.9 = 1699.9 \text{ kN}$$

$$q = \frac{1699.9 + 400 + (1.12 \times 10^{-3} \times 91 \times 10^4)}{7 \times 5}$$

$$q = 89.1 \text{ kN/m}^2 (< 0.8 \times 225 = 180 \text{ kN/m}^2, \text{Safe})$$

Example 12.2
Determine the natural frequency/frequencies and amplitude/amplitudes of anvil-foundation system for a forge hammer having weight of 40 kN. The value of coefficient of elastic uniform compression (C_u) of the soil at the site is 2.5×10^7 N/m^3. Solve the problem considering

(i) Single degree freedom system
(ii) Two degrees freedom system
(iii) Three degrees freedom system and compare the results. Assume the stiffness of elastic pad/pads equal to 1.0×10^9 N/m.

Solution
Single degree freedom system:

(i)

$$W_a = 20 \times W_T = 800 \text{ kN}$$

$$W_f = 80 \times W_T = 3200 \text{ kN}$$

$$W_f = 3200 \times 10^3 = 25 \times 10^3 \times \text{Volume of foundation block}$$

Volume of foundation block = 28 m^3

Assuming the thickness of foundation block as 2.25 m

Base area of foundation block, $A_1 = 128/2.25 = 56.88$ m^2

$$C_u' = 2 \times 2.5 \times 10^7 = 5 \times 10^7 \text{ N/m}^3$$

$$\omega_n = \sqrt{\frac{C_u' A_1 g}{(W_a + W_f)}} = \sqrt{\frac{5 \times 10^7 \times 56.88 \times 9.81}{(800 + 3200) \times 10^3}}$$

$$= 83.515 \text{ rad/sec}$$

$$V_{Ti} = \eta\sqrt{2gh} = 0.65\sqrt{2 \times 9.81 \times 0.8} = 2.575 \text{ m/s}$$

$$V_a = \frac{1+e}{1+\frac{W_a+W_f}{W_T}} V_{Ti}$$

$$= \frac{1+0.5}{1+\frac{800+3200}{40}} \times 2.575$$

$$= 0.0382 \text{ m/sec}$$

$$Z_{1m} = \frac{V_{af}}{\omega_n}$$

$$= 0.382/83.515$$

$$Z_{1m} = 0.000457 \text{ m} = 0.457 \text{ mm}$$

Two degree freedom system:

(i) $W_a = 20 \times W_T = 800$ kN

Volume of anvil block = 10.20 m^3

Assuming the thickness of anvil block as 1.0 m

Base area of anvil = 10.20 m^2

$$W_f = 80 \times W_T = 3200 \text{ kN}$$

$$W_f = 3200 \times 10^3 = 25 \times 10^3 \times \text{Volume of foundation block}$$

Volume of foundation block = 128 m^3

Assuming the thickness of foundation block as 2.25 m

Base area of foundation block, A_1 = 128/2.25 = 256.88 m^2

(ii) $C_u' = 2 \times 2.5 \times 10^7 = 5 \times 10^7$ N/m^3

$$K_1 = C_u' A_1$$

$$= 5 \times 10^7 \times 56.88$$

$$K_1 = 2.84 \times 10^9 \text{ N/m}$$

$$K_2 = 1.0 \times 10^9 \text{ N/m (Assumed)}$$

$$m_1 = (3200 \times 10^3)/9.81 = 326.19 \times 10^3 \text{ kg}$$

$$m_2 = (800 \times 10^3)/9.81 = 81.54 \times 10^3 \text{ kg}$$

$$\mu_m = m_2/m_1 = 0.25$$

$$\omega_{na}^2 = \frac{K_2}{m_2} = (10^9/(81.54 \times 10^3))$$

$$\omega_{na} = 110.74 \text{ rad/sec}$$

$$\omega_{nl}^2 = \frac{K_2}{m_1 + m_2} = 2.84 \times 10^9/(326.19 + 81.54) \times 10^3)$$

$$\omega_{nl} = 83.45 \text{ rad/sec}$$

(iii) Natural frequencies

$$\omega_{n1,2}^2 = \frac{1}{2}\left[(1+\mu_m)(\omega_{na}^2 + \omega_{nl}^2) \pm \sqrt{\{(1+\mu_m)(\omega_{na}^2 + \omega_{nl}^2)\}^2 - 4(1+\mu_m)(\omega_{na}^2\, \omega_{nl}^2)}\right.$$

Substituting the values of ω_{na}, ω_{nl} and μ_m in the above expression, we get

$$\omega_{n1} = 134.74 \text{ rad/sec}$$

$$\omega_{n2} = 76.68 \text{ rad/sec}$$

Note: ω_{n2} is the smaller natural frequency.

(iv) Amplitudes

$$V_{Ti} = \eta\sqrt{2gh} = 0.65\sqrt{2 \times 9.81 \times 0.8} = 2.575\,\text{m/s}$$

$$V_a = \frac{1+e}{1+\dfrac{W_a}{W_T}} V_{Ti} = (1.5/21) \times 2.575 = 0.184\,\text{m/s}$$

$$Z_{1m} = \frac{(\omega_{na}^2 - \omega_{n1}^2)(\omega_{na}^2 - \omega_{n2}^2)}{(\omega_{n1}^2 - \omega_{n2}^2)\omega_{na}^2}\left[-\frac{1}{\omega_{n2}}\right]V_a$$

$$Z_{2m} = \frac{1}{(\omega_{n1}^2 - \omega_{n2}^2)}\left[-\frac{(\omega_{na}^2 - \omega_{n1}^2)}{\omega_{n2}}\right]V_a$$

Substituting the values of ω_{na}, ω_{n1}, ω_{n2} and V_a in the expressions of Z_{1m}, Z_{2m}, we get,

$$Z_{1m} = \frac{(110.74^2 - 134.74^2)(110.74^2 - 76.68^2)}{(134.74^2 - 76.68^2)110.74^2}\left[-\frac{1}{76.68}\right] \times 0.184 = 5.99 \times 10^{-4}\,\text{m}$$

$$Z_{2m} = \frac{(110.74^2 - 134.74^2)}{(134.74^2 - 76.68^2)}\left[-\frac{1}{76.68}\right] \times 0.184 = 1.152 \times 10^{-3}\,\text{m}$$

Three degree freedom system:

(i) $W_a = W_3 = 20 \times W_T = 800$ kN

$W_f = W_2 = 80 \times W_T = 3200$ kN

$W_{tr} = W_1 = 40 \times W_T = 1600$ kN (assumed)

(ii) $W_a = 800 \times 10^3 = 78.5 \times 10^3 \times$ Volume of anvil block
Volume of anvil block = 10.20 m^3
Assuming the thickness of anvil block as 1.0 m
Base area of anvil = 10.20 m^2
$W_f = 3200 \times 10^3 = 25 \times 10^3 \times$ Volume of foundation block
Volume of foundation block = 128 m^3
Assuming the thickness of foundation block as 2.25 m

Base area of foundation block $A_2 = 128/2.25 = 56.88$ m^2

Let the base area of trough, A_1 = 1.25 times the base area of foundation block

$= 1.25 \times 56.88$

$= 71.11$ m^2

(iii) $K_1 = C_u' A_1$

$= 2 \times 2.5 \times 10^7 \times 71.11$

$K_1 = 3.55 \times 10^9$ N/m

$K_2 = K_3 = 1.0 \times 10^9$ N/m (Assumed)

$$m_1 = (1600 \times 10^3)/9.81 = 163.09 \times 10^3 \text{ kg}$$

$$m_2 = (3200 \times 10^3)/9.81 = 326.20 \times 10^3 \text{ kg}$$

$$m_3 = (800 \times 10^3)/9.81 = 81.55 \times 10^3 \text{ kg}$$

(iv) Natural frequencies

$$m_1 m_2 m_3 \omega_n^6 - [m_2 m_3 (K_1 + K_2) + m_1 m_3 (K_2 + K_3) + m_1 m_2 K_3]\omega_n^4 + [K_1 K_2 m_3 + K_1 K_3 (m_2 + m_3)$$

$$+ K_1 K_3 (m_2 + m_3) + K_2 K_3 (m_1 + m_2 + m_3)]\omega_n^2 - K_1 K_2 K_3 = 0$$

Susbtituting the values of K_1, K_2, K_3, m_1, m_2 and m_3 from step (iii), values of coefficient of ω_n^6, ω_n^4, ω_n^2 and constant term worked out as follows:

$$a = 4.338 \times 10^{15}$$

$$b = -2.0084 \times 10^{20}$$

$$c = 2.3078 \times 10^{24}$$

$$d = -3.55 \times 10^{27}$$

Solution of the equation of natural frequencies then gives:

$$\omega_{n1} = 169.79 \text{ rad/sec}$$

$$\omega_{n2} = 125.12 \text{ rad/sec}$$

$$\omega_{n3} = 42.58 \text{ rad/sec}$$

Note: ω_{n3} is the smaller natural frequency.

(v) Amplitudes

$$\omega_{na}^2 = \frac{K_2}{m_2}$$

$$\omega_{na} = 110.74 \text{ rad/sec}$$

$$V_{Ti} = \eta\sqrt{2gh} = 0.65\sqrt{2 \times 9.81 \times 0.8} = 2.575 \text{ m/s}$$

$$V_a = \frac{1+e}{1+\dfrac{W_a}{W_T}} V_{Ti} = (1.5/21) \times 2.575 = 0.184 \text{ m/s}$$

Substituting the values of ω_{na}, ω_{n1}, ω_{n2}, ω_{n3}, K_1, K_2 and K_3 in Eqns. (12.46) to (12.51), we get,

$a_1 = -0.74$ $\quad$ $b_1 = -0.152$

$a_2 = -3.615$ $\quad$ $b_2 = 1.99$

$a_3 = 1.173$ $\quad$ $b_3 = 4.55$

$$Z_{1m} = \frac{V_a (b_1 - b_2)}{\omega_{n3}[a_3 b_3 (b_1 - b_2) - a_1 b_1 (b_3 - b_2) - a_2 b_2 (b_1 - b_3)]}$$

$$= \frac{0.184(-0.152-1.99)}{42.58[1.173\times4.55(-0.152-1.99)-(0.74\times-0.152)(4.55-1.99)-(-3.615\times1.99)(-0.152-4.55)]}$$

$Z_{1m} = 2.0316 \times 10^{-4}$ m

$$Z_{2m} = \frac{V_a(b_1-b_2)b_3}{\omega_{n3}[a_3b_3(b_1-b_2)-a_1b_1(b_3-b_2)-a_2b_2(b_1-b_3)]}$$

$$= \frac{0.184(-0.152-1.99)\times4.55}{42.58[1.173\times4.55(-0.152-1.99)-(0.74\times-0.152)(4.55-1.99)-(-3.615\times1.99)(-0.152-4.55)]}$$

$Z_{2m} = 9.244 \times 10^{-4}$ m

$$Z_{3m} = \frac{V_a(b_1-b_2)a_3b_3}{\omega_{n3}[a_3b_3(b_1-b_2)-a_1b_1(b_3-b_2)-a_2b_2(b_1-b_3)]}$$

$$= \frac{0.184(-0.152-1.99)\times4.55\times1.173}{42.58[1.173\times4.55(-0.152-1.99)-(0.74\times-0.152)(4.55-1.99)-(-3.615\times1.99)(-0.152-4.55)]}$$

$Z_{3m} = 1.084 \times 10^{-3}$ m

Comparison:

Model	*Maximum amplitude of foundation block (mm)*	*Lowest natural frequency (rad/sec)*
Single degree freedom system	0.457	83.5
Two degree freedom system	0.599	76.88
Three degree freedom system	0.9244	42.58

Example 12.3

Solve the Example 12.2 using the pine wood of thickness 0.5 m below the anvil and below the foundation block. The modulus of elasticity of pad material 5×10^5 kN/m², and allowable compressive stress in pad is 3500 kN/m².

Solution

Single degree freedom system:

$W_a = 20 \times W_T = 800$ kN

$W_f = 80 \times W_T = 3200$ kN

$W_f = 3200 \times 10^3 = 25 \times 10^3 \times$ Volume of foundation block

Volume of foundation block = 128 m³

Assuming the thickness of foundation block as 2.25 m

Base area of foundation block $A_1 = 128/2.25 = 56.88$ m²

$C_u' = 2 \times 2.5 \times 10^7 = 5 \times 10^7$ N/m³

$$\omega_n = \sqrt{\frac{C_u' A_1 g}{(W_a + W_f)}} = \sqrt{\frac{5\times10^7\times56.88\times9.81}{(800+3200)\times10^3}}$$

$= 83.515$ rad/sec

$$V_{Ti} = \eta\sqrt{2gh} = 0.65\sqrt{2\times9.81\times0.8} = 2.575 \text{ m/s}$$

$$V_a = \frac{1+e}{1+\frac{W_a+W_f}{W_T}} V_{Ti}$$

$$= \frac{1+0.5}{1+\frac{800+3200}{40}} \times 2.575$$

$$= 0.0382 \text{ m/sec}$$

$$Z_{1m} = \frac{V_a}{\omega_n}$$

$$= 0.0382/83.515$$

$$Z_{1m} = 0.000457 \text{ m} = 0.457 \text{ mm}$$

Two degree freedom system:

$$W_a = 20 \times W_T = 800 \text{ kN}$$

$$W_f = 80 \times W_T = 3200 \text{ kN}$$

$$W_f = 3200 \times 10^3 = 25 \times 10^3 \times \text{Volume of foundation block}$$

Volume of foundation block = 128 m^3

Assuming the thickness of foundation block as 2.25 m

Base area of foundation block, A_1 = 128/2.25 = 56.88 m^2

(i)

$$C_u' = 2 \times 2.5 \times 10^7 = 5 \times 10^7 \text{ N/m}^3$$

$$K_1 = C_u' A_1$$

$$= 5 \times 10^7 \times 56.88$$

$$K_1 = 2.84 \times 10^9 \text{ N/m}$$

$$K_2 = ((5 \times 10^5)/0.5) \times 10.20$$

$$= 10.2 \times 10^9 \text{ N/m}$$

$$m_1 = (3200 \times 10^3)/9.81 = 326.19 \times 10^3 \text{ kg}$$

$$m_2 = (800 \times 10^3)/9.81 = 81.54 \times 10^3 \text{ kg}$$

$$\mu_m = m_2/m_1 = 0.25$$

$$\omega_{na}^2 = \frac{K_2}{m_2} = (10.2 \times 10^9 / 81.54 \times 10^3)$$

$$\omega_{na} = 353.68 \text{ rad/sec}$$

$$\omega_{nl}^2 = \frac{K_2}{m_1+m_2} = 2.84 \times 10^9/((326.19+81.54) \times 10^3)$$

$$\omega_{nl} = 83.45 \text{ rad/sec}$$

(ii) Natural frequencies

$$\omega_{n1,2}^2 = \frac{1}{2}\left[(1+\mu_m)(\omega_{na}^2+\omega_{nl}^2) \pm \sqrt{\{(1+\mu_m)(\omega_{na}^2+\omega_{nl}^2)\}^2 - 4(1+\mu_m)(\omega_{na}^2\ \omega_{nl}^2)}\right]$$

Substituting the vaues of ω_{na}, ω_{nl} and μ_m in the above expression, we get

$$\omega_{n1} = 397.72 \text{ rad/sec}$$

$$\omega_{n2} = 82.97 \text{ rad/sec}$$

Note: ω_{n2} is the smaller natural frequency.

(iii) Amplitudes:

$$V_{Ti} = \eta\sqrt{2gh} = 0.65\sqrt{2\times 9.81\times 0.8} = 2.575 \text{ m/s}$$

$$V_a = \frac{1+e}{1+\dfrac{W_a+W_f}{W_T}} V_{Ti} = (1.5/21)\times 2.575 = 0.184 \text{ m/s}$$

$$Z_{1m} = \frac{(\omega_{na}^2-\omega_{n1}^2)(\omega_{na}^2-\omega_{n2}^2)}{(\omega_{n1}^2-\omega_{n2}^2)\,\omega_{na}^2}\left[-\frac{1}{\omega_{n2}}\right]V_a$$

$$Z_{2m} = \frac{1}{(\omega_{n1}^2-\omega_{n2}^2)}\left[-\frac{(\omega_{na}^2-\omega_{n1}^2)}{\omega_{n2}}\right]V_a$$

Substituting the values of ω_{na}, ω_{n1}, ω_{n2} and V_a in the expressions of Z_{1m}, Z_{2m}, we get,

$$Z_{1m} = \frac{(353.68^2-397.72^2)(353.68^2-82.97^2)}{(397.72^2-82.97^2)353.68^2}\left[-\frac{1}{82.97}\right]\times 0.184 = 4.58\times 10^{-4} \text{ m}$$

$$Z_{2m} = \frac{(353.68^2-397.72^2)}{(397.72^2-82.97^2)}\left[-\frac{1}{82.97}\right]\times 0.184 = 4.58\times 10^{-4} \text{ m}$$

Three degree freedom system:

(i) $W_a = W_3 = 20 \times W_T = 800$ kN

$W_f = W_2 = 80 \times W_T = 3200$ kN

$W_{tr} = W_1 = 40 \times W_T = 1600$ kN (assumed)

(ii) $W_a = 800 \times 10^3 = 78.5 \times 10^3 \times$ Volume of anvil block

Volume of anvil block = 10.20 m^3

Assuming the thickness of anvil block as 1.0 m

Base area of anvil = 10.20 m^2

$W_f = 3200 \times 10^3 = 25 \times 10^3 \times$ Volume of foundation block

Volume of foundation block = 128 m^3

Assuming the thickness of foundation block as 2.25 m

Base area of foundation block, $A_2 = 128/2.25 = 56.88\ m^2$

Let the base area of through, A_1 = 1.25 times the base area of foundation block

$$= 1.25 \times 56.88$$

$$= 71.11\ m^2$$

$$K_1 = C_u' A_1$$

$$= 2 \times 2.5 \times 10^7 \times 71.11$$

$$K_1 = 3.55 \times 10^9\ N/m$$

$$K_2 = \frac{E}{b} \times A_2 = ((5\times10^5)/0.5)\times 56.88 = 56.88\times10^9\ N/m$$

$$K_3 = \frac{E}{b} \times A_3 = ((5\times10^5)/0.5)\times 10.20 = 10.2\times10^9\ N/m$$

$$m_1 = (1600 \times 10^3)/9.81 = 163.09 \times 10^3\ kg$$

$$m_2 = (3200 \times 10^3)/9.81 = 326.20 \times 10^3\ kg$$

$$m_3 = (800 \times 10^3)/9.81 = 81.55 \times 10^3\ kg$$

(ii) Natural frequencies

$$m_1m_2m_3\omega_n^6 - [m_2m_3(K_1+K_2)+m_1m_3(K_2+K_3)+m_1m_2K_3]\omega_n^4 + [K_1K_2m_3+K_1K_3(m_2+m_3)$$
$$+ K_2K_3(m_1+m_2+m_3)]\omega_n^2 - K_1K_2K_3 = 0$$

Substituting the values of K_1, K_2, K_3, m_1, m_2 and m_3 from step (iii), values of coefficient of ω_n^6, ω_n^4, ω_n^2 and constant term worked out as follows:

$$a = 4.338 \times 10^{15}$$

$$b = -2.042 \times 10^{21}$$

$$c = 3.624 \times 10^{26}$$

$$d = -2.059 \times 10^{30}$$

Solution of the equation of natural frequencies then gives:

$$\omega_{n1} = 742.47\ rad/sec$$

$$\omega_{n2} = 379.48\ rad/sec$$

$$\omega_{n3} = 77.32\ rad/sec$$

Note: ω_{n3} is the smaller natural frequency.

(iv) Amplitudes $\qquad \omega_{na}^2 = \dfrac{K_2}{m_2}$

$$\omega_{na} = 353.66\ rad/sec$$

$$V_{Ti} = \eta\sqrt{2gh} = 0.65\sqrt{2\times9.81\times0.8} = 2.575\ m/s$$

$$V_a = \frac{1+e}{1+\frac{W_a}{W_T}} V_{Ti} = (1.5/21) \times 2.575 = 0.184 \text{ m/s}$$

Substituting the values of ω_{na}, ω_{n1}, ω_{n2}, ω_{n3}, K_1, K_2 and K_3 in Eqns. (12.46) to (12.51), we get,

$$a_1 = -0.293$$

$$a_2 = -6.607$$

$$a_3 = 1.05$$

$$b_1 = -0.518$$

$$b_2 = 0.649$$

$$b_3 = 1.045$$

Substituting the values of a_1, a_2, a_3, b_1, b_2, b_3, V_a and ω_{n3} in equations (12.52), (12.53) and (12.54) we get

$$Z_{1m} = \frac{V_a(b_1 - b_2)}{\omega_{n3}[a_3 b_3 (b_1 - b_2) - a_1 b_1 (b_3 - b_2) - a_2 b_2 (b_1 - b_3)]}$$

$$= \frac{0.184(-0.518 - 0.649)}{77.32[1.05 \times 1.045(-0.518 - 0.649) - (0.293 \times (-0.518))(1.045 - 0.649) - (-6.607 \times 0.649)(-0.518 - 1.045)]}$$

$= 3.63 \times 10^{-4}$ m

$$Z_{2m} = \frac{V_a(b_1 - b_2)b_3}{\omega_{n3}[a_3 b_3 (b_1 - b_2) - a_1 b_1 (b_3 - b_2) - a_2 b_2 (b_1 - b_3)]}$$

$$= \frac{0.184(-0.518 - 0.649) \times 1.045}{77.32[1.05 \times 1.045(-0.518 - 0.649) - (0.293 \times -0.518)(1.045 - 0.649) - (-6.607 \times 0.649)(-0.518 - 1.045)]}$$

$= 3.79 \times 10^{-4}$ m

$$Z_{3m} = \frac{V_a(b_1 - b_2)a_3 b_3}{\omega_{n3}[a_3 b_3 (b_1 - b_2) - a_1 b_1 (b_3 - b_2) - a_2 b_2 (b_1 - b_3)]}$$

$$= \frac{0.184(-0.518 - 0.649) \times 1.045 \times 1.05}{77.32[1.05 \times 1.045(-0.518 - 0.649) - (0.293 \times -0.518)(1.045 - 0.649) - (-6.607 \times 0.649)(-0.518 - 1.045)]}$$

$= 1.084 \times 10^{-3}$ m

Comparison:

Model	*Maximum amplitude of foundation block (mm)*	*Lowest natural frequency (rad/sec)*
Single degree freedom system	0.457	83.5
Two degree freedom system	0.458	82.97
Three degree freedom system	0.379	77.32

References

Barkan, D.D. (1962), "Dynamics of bases and foundations", McGraw-Hill, New York.
Novak, M. (1983), "Foundations for shock producing machines", Can. Geotech. J. 20 (1), pp. 141-158.

Practice Problems

12.1 Discuss with neat sketches the various possible arrangements of a hammer foundation to minimise the vibrations.

12.2 Considering a two-degree-freedom model, derive the expression of amplitudes of anvil and foundation of a hammer.

12.3 A 20 kN forging hammer is proposed to install in an industrial Complex. The hammer has the following specifications:

Weight of tup without die	=	12 kN
Maximum tup stroke	=	900 mm
Weight of the upper half of the die	=	50 kN
Area of piston	=	0.15 m^2
Supply steam pressure	=	700 kN/m
Weight of anvil block	=	400 kN
Total weight of anvil and frame	=	500 kN
Bearing area of anvil	=	2.1 m × 2.1 m
Permissible vibration amplitude for anvil	=	1.0 mm
Permissible amplitude for foundation	=	0.8 mm

It is proposed to use a pine wood pad of thickness 0.5 m below the anvil. The modulus of elasticity of pad material is 6 × 10 kN/m^2, and allowable compressive stress in pad is 4000 kN/m^2.

A vertical resonance test was conducted on a 1.5 m × 0.75 m × 0.70 m high concrete block at the proposed depth of foundation. The data obtained are given below:

S. No.	*θ (Deg.)*	*F_{nz} (Hz)*	*Amplitude at resonance (microns)*
1.	36	40	14
2.	72	38	26
3.	108	35	33
4.	144	29	41

The soil at the site is sandy in nature and water table lies at a depth of 2.0 m below ground surface. Allowable soil pressure = 200 kN/m^2. Design a suitable foundation.

CHAPTER

13

Foundations of Rotary Machines

13.1 General

The unprecedented burden cast by the importance of oil can be relieved only by exploiting indigenous energy resources efficiently. Major power energy resources, in long term power plan, incorporate an optimal mix of thermal, hydel and nuclear generation. Power intensity is relatively high in our country due to various reasons including the substantial substitution among the forms of energy in the various important sectors along with the accelerated programme in rural electrification and assured power supply for agriculture sector etc. The aim of planning to generate power higher than the demand calls for a coordinated development of the power supply industry.

The turbogenerator unit is most expensive, vital and important part in a thermal power plant. The operating speeds of turbogenerators may range from 3000 rpm to 10000 rpm. Auxiliary equipments such as condensers, heat exchanger, pipe lines, air vents and ducts for electric wiring are essential features of a turbogenerator installation. Frame foundations are commonly used for turbogenerator because of four reasons:

(i) auxiliary equipment can be arranged more conveniently,
(ii) the inspection of and access to all parts of the machine become more convenient,
(iii) less liable to cracking due to settlement and temperature changes, and
(iv) more economical due to the saving in material and freedom to add more members to stiffen if needed.

The frame foundation is the assemblage of columns, longitudinal and transverse beams. The transverse beams may be often eccentric with respect to the column centre lines and generally have varying cross-section due to several opening in the top deck and haunches at the junction with columns. The isometric view of a typical frame foundation is shown in Fig. 13.1.

In a power-plant, the long term satisfactory performance of the turbogenerator is affected by their foundations, hence there is vital need to adequately design these foundations for all possible combinations of static and dynamic loads. Interaction with the mechanical engineer is also required for any adjustment in the layout of machinery and auxiliary fittings IS 2974-2015.

13.2 Special Considerations

For better performance of a T.G. foundation, following points may be kept in view:

(i) The entire foundation should be separated from the main building in order to isolate the transfer of vibrations from the top deck of the foundation to the building floor of the machine room. A clear gap should be provided all around.
(ii) Other footings placed near to the machine foundation should be checked for non-uniform stresses imposed by adjacent footings. The pressure-bulbs under the adjacent footings should not interfere significantly with each other.
(iii) All the junctions of beams and columns of the foundation should be provided with adequate haunches in order to increase the general rigidity of the frame foundation.

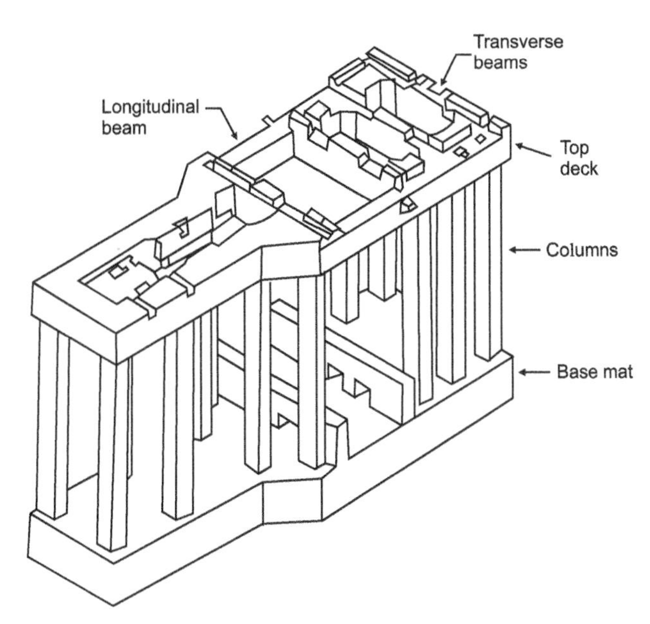

Fig. 13.1. Typical frame foundation for a turbo-generator

(iv) The cross-sectional height of the cantilever elements at the embedment point should not be less than 60 to 75 percent of its span, being susceptible to excessive local vibrations.
(v) The transverse beams should have their axes vertical below the bearings to avoid torsion. For the same reason the axes of columns and transverse beams should lie in the vertical plane.
(vi) The upper platform should be as rigid as possible in the plane.
(vii) Permissible pressure on soil may be reduced by 20 percent to account for the vibration of the foundation slab. This slab has much smaller amplitudes of vibration than the upper platform.
(viii) The lower foundation slab should be sufficiently rigid to resist non-uniform settlement and heavy enough to lower the common centre of gravity of the machine and foundation. It is therefore made thicker than required by static computations. For 25 MW machine its thickness is 2m and increases with the power of the machine to a maximum of 4m. Its weight should not be less than the weight of the machine plus the weight of the foundation excluding the base slab and condensers.
(ix) Special reinforcement detailing as laid down in the code IS-2974 Pt III should be followed.
(x) Special care in construction is called for, to avoid cracking of concrete. The foundation slab-should be completed in one continuous pouring. In this case the joint between the two concretes, preferably at one-third column height, is specially treated to ensure 100 percent bond.
(xi) Piles may be provided to meet the bearing capacity requirement but then consideration of subgrade effect is essential.
(xii) As far as possible the foundation should be dimensioned such that the centre of gravity of the foundation, with the machine, should be in vertical alignment with that of the base area in contact with the soil.
(xiii) The ground-water table should be as low as possible and deeper by at least one-fourth of the width of foundation below the base plane. This limits the vibrations propagation, as ground-water is a good conductor to wave transmission.
(xiv) Soil-profile and characteristics of soil upto at least thrice the width of the turbine foundation or till hard stratum is reached or upto pile depth, if piles are provided, should be investigated.

13.3 Design Criteria

The design of a T. G. foundation is based on the following design criteria:

(i) From the point of view of vibration, the natural frequencies of foundation system should preferably be at a variance of at least 30 percent from the operating speed of the machine as well as critical speeds of the rotor. Thus resonance is avoided. An uncertainity of 10 to 20 percent may be assumed in the computed natural frequencies. However, it may not be necessary to avoid resonance in higher modes, if the resulting resonant amplitude is relatively insignificant.
It is preferable to maintain a frequency separation of 50 percent.

(ii) The amplitude of vibration should be within permissible limits. Values of permissible amplitudes are given in Table 10.2.

13.4 Loads on a T.G. Foundation

The loads acting on a turbogenerator are as given below:

13.4.1 Dead Loads (DL)

These include the self weight of the foundation and dead weight of the machine.

13.4.2 Operation Loads (OL)

These loads are supplied by the manufacturer of the machine and include frictional forces, power torque, thermal elongation forces, vaccum in the condenser, piping forces etc.

The load due to vaccum in condenser, if not supplied by the manufacturer, can be obtained from the following equation:

$$P_C = A(p_a - p_c) \tag{13.1}$$

where, P_C = Condenser vaccum load
A = Cross-sectional area of the connecting tie between the condenser and turbine
p_a = Atmospheric pressure
p_c = Vaccum pressure

The value of $(p_a - p_c)$ may be taken as 100 kN/m^2.

The magnitude of the torque depends upon the operational speed and power output capacity of the turbines. For a T.G. unit having multistage turbine (Fig. 13.2), the torque may be calculated as below:

$$T_A = \frac{105 P_A}{N} \text{ kNm} \tag{13.2a}$$

$$T_B = \frac{105 (P_B - P_A)}{N} \text{ kNm} \tag{13.2b}$$

$$T_C = \frac{105 (P_C - P_B)}{N} \text{ kNm} \tag{13.2c}$$

$$T_g = \frac{105 P_C}{N} \text{ kNm} \tag{13.2d}$$

where, T_A = Torque due to high-pressure (H.P.) turbine in kNm
T_B = Torque due to intermediate-pressure (I.P.) turbine in kNm
T_C = Torque due to low-pressure (L.P.) turbine in kNm

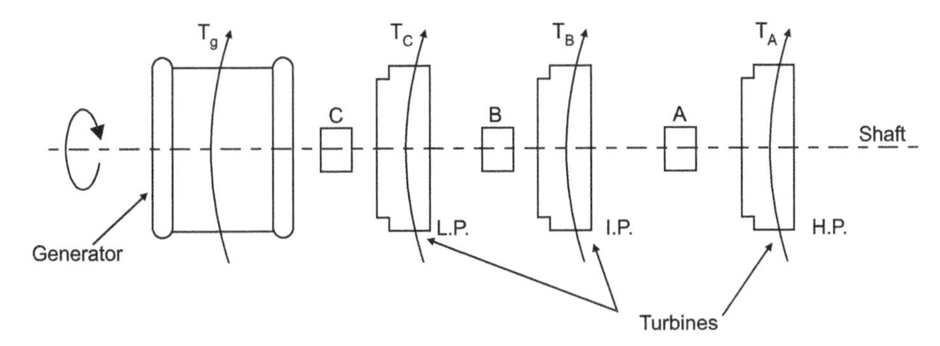

Fig. 13.2. Torque due to normal operation of a multistage turbine-generator unit

T_g = Torque due to generator in kNm
P_A, P_B and P_C = Power transferred by couplings A, B and C respectively in kW
N = Operating speed in rpm

13.4.3 Normal Machine Unbalanced Load (NUL)

As mentioned in Sec. 10.2, rotary machine are balanced before erection. However, in actual operation some unbalanced always exists. The unbalance is specified as the distance between the axis of the shaft and mass centre of gravity of rotor, and is known as effective eccentricity. The magnitude of unbalanced forces can be obtained using Eqs. (10.8) and (10.9).

For the case of rotors shown in Fig. 13.3a, the resultant unbalanced forces due to the two masses at any time cancel out, but there is a resulting moment M given by

$$M = m_e \, e \, \omega^2 \cdot l \tag{13.3}$$

where l = Distance between the mass centre of gravities of rotors.

The components of the moment M in vertical and horizontal directions are given by

$$M_V = m_e \, e \, \omega^2 \, l \sin \omega t \tag{13.4a}$$

$$M_H = m_e \, e \, \omega^2 \, l \cos \omega t \tag{13.4b}$$

When masses have an orientation as shown in Fig. 13.4b, the machine operation will give rise to both an unbalanced force and a moment. The unbalanced force is given by

$$F = 2 \, m_e \, e \, \omega^2 \tag{13.5}$$

The unbalanced moment can be computed using Eq. (13.3).

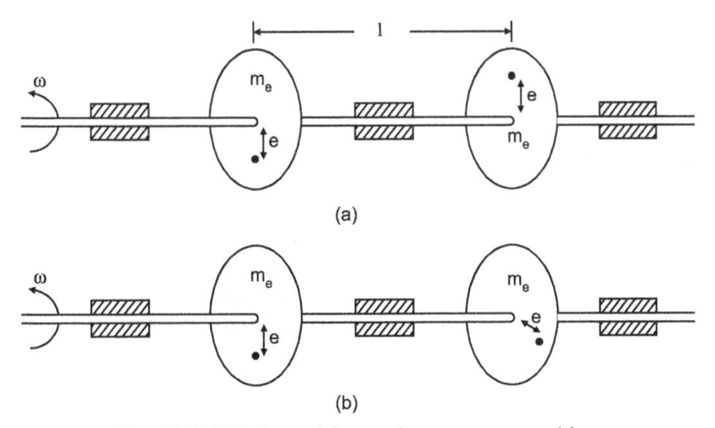

Fig. 13.3. Unbalanced forces due to rotary machines

For more than two rotors on a common shaft, combined unbalanced forces and moments can be computed in similar manner.

13.4.4 Temperature Loads in the Foundation (TLF)

The effect of differential thermal expansion and shrinkage should be considered in the design of frame foundations. In the absence of the exact data, a differential temperature of 20° may be assumed between the upper and lower slabs. Besides, a differential temperature of 20° may be assumed between the inner and outer faces of the upper slab. The upper slab should be treated as a horizontal closed frame and analysed for the induced moments due to differential temperature.

To account for the shrinkage of the upper slab relative to the base slab, a temperature fall of 10° C to 15°C may be assumed.

13.4.5 Short Circuit Forces (SCF)

Short circuit condition imposes moment on the turbogenerator foundation. A fault of this type occurs when any two of the three generator phase terminals are shorted. The shock, which is in the form of couple known as "shortcircuit moment", tends to break the stator off the foundation, and this imposes vertical loads on the longitudinal beam supporting the generator stator.

If accurate information is not available from the manufacturer, the short circuit moment (M_{SC}) may be taken emipirically as four times the rated capacity (in MW) of turbogenerator unit.

Major (1980) has suggested the following formula for estimating the short circuti moment:

$$M_{sc} = 10\, r\, W_r \text{ kNm} \tag{13.6}$$

where W_r = Capacity of T. G. unit in MW
r = Radius of the rotor in m.

13.4.6 Loss of Blade Unbalance (LBL) or Bearing Failure Load (BFL)

One of the buckets or blades of the turbine rotor may break during the operation of turbo-generator unit. It will increase the unbalanced force. This additional unbalanced force will depend on the weight of the bucket, the distance of its centre of gravity from the axis of rotation and operational speed.

13.4.7 Seismic Load (EQL)

The horizontal seismic force is considered both in longitudinal and transverse directions separately. It may be computed from the following equation (IS 1893-1984):

$$F_S = \alpha_h\, I\, \beta\, C\, S\, W \tag{13.7}$$

where F_S = Horizontal seismic force
α_h = Seismic zone coefficient
I = Importance factor
β = Soil-foundation factor
C = Numerical base shear coefficient
S = Numerical site structure response coefficient
W = Vertical load due to weight of all permanent components.

When earthquake forces are considered in design, the permissible stresses in materials and the allowable soil pressure may be increased as per IS 1893-1984.

13.4.8 Construction Loads (CL)

Construction loads occur only when the machine is being erected. As such they are not to be considered as acting simultaneously with dynamic loads which occur only during the operation of the machine. The construction loads are generally taken as uniformly distributed load varying from 10 kN/m^2 to 30 kN/m^2 depending on the size of T.G. unit.

The design of a T.G. unit should be checked for the following load combinations:

(a) Operation condition
DL + OL + NUL + TLF

(b) Short circuit condition
DL + OL + NUL + TLF + SCF

(c) Loss of blade condition/bearing failure condition
DL + OL + TLF + LBL/BFL

(d) Seismic condition
DL + OL + NUL + TLF + EQL

13.5 Methods of Analysis and Design

In the case of a frame foundation, it is necessary to check the frequencies and amplitudes of vibration and also to design the members of frame from structural considerations. The methods for carrying out dynamic analysis may be divided into two categories:

(a) Two-dimensional analysis
(b) Three-dimensional analysis

The two-dimensional analysis is based on the following assumptions:

(i) The difference between the deformations of individual frame columns is insignificant.
(ii) The deformation of the longitudinal and transverse beams is almost identical.
(iii) The torsional resistance of the longitudinal beams is insignificant in relation to the deformation of the transverse beams.
(iv) The vertical vibrations of the frames can be determined for each frame individually.
(v) The weight transmitted from the longitudinal beam can be considered as a load supported by the column head, even in case where the transverse beam is eccentrically placed with respect to the centre line of the column.
(vi) Both the columns and beams can be replaced by weightless elements with the masses lumped at a few points by equating the kinetic energies of the actual and the idealised systems.
(vii) The effect of elasticity of subsoil is neglected, it being relatively much flexible.
(viii) When considering horizontal displacement the upper slab is regarded as a rigid plate in its own plane.

The two dimensional analysis may be carried out by the following methods:

1. Resonance method (Rausch, 1959)
2. Amplitude method (Barkan, 1962)
3. Combined method (Major, 1980)

In subsequent sections, salient features of the above methods are given.

13.6 Resonance Method

In this method, the frame foundation is idealized as a single-degree freedom system, and consideration is given only to natural frequencies of the system in relation to the operating speed of the machine. The amplitudes of vibration are not computed in this method.

13.6.1 Vertical Frequency

For obtaining vertical frequency, each transverse frame that consists of two columns and a beam perpendicular to main shaft of the machine, is considered separately (Fig. 13.4a).

The loads acting on this frame are:

(i) Dead load of the machine and bearing, W_1
(ii) Load transferred to the columns by longitudinal beams, W_2
(iii) Uniformly distributed load due to self weight of cross beam, q per unit length
(iv) Unbalanced vertical force due to machine operation, $F_Z \sin w t$

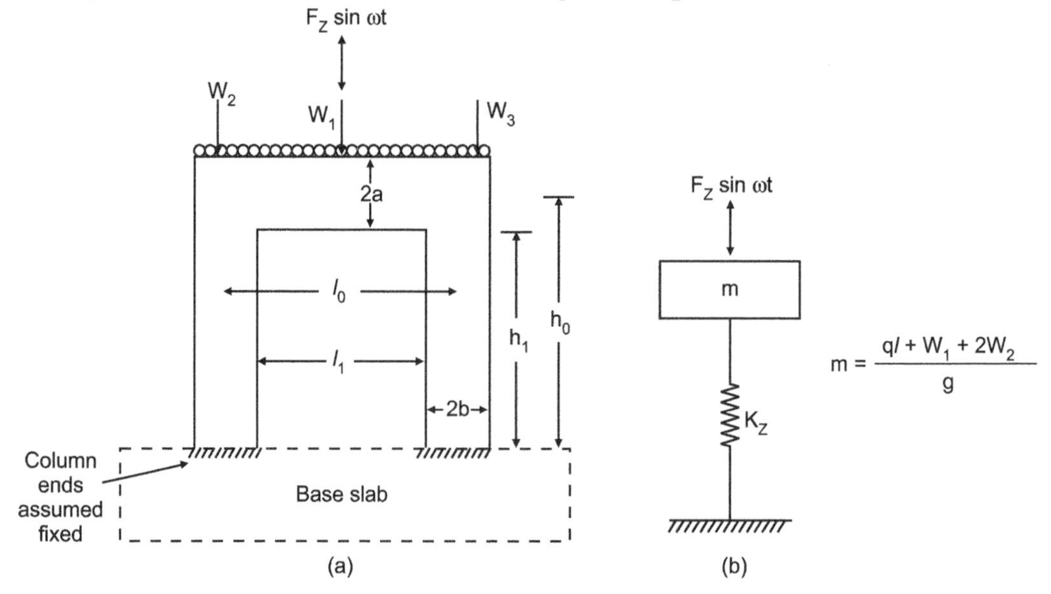

Fig. 13.4. (a) Typical transverse frame; (b) Idealised model

The frame is modelled as mass-spring system as shown in Fig. 13.4b. The stiffness of equivalent spring (K_Z) is computed as the combined stiffness of the beam and columns acting together. It is given by

$$K_z = \frac{W}{\delta_{st}} \tag{13.8}$$

where W = Total load on the frame

or
$$W = W_1 + W_2 + q \cdot l \tag{13.9}$$

l = Effective span

δ_{st} = Total vertical deflection at the centre of the beam due to bending action of beam and axial compression in columns.

$$= \delta_1 + \delta_2 + \delta_3 + \delta_4 \tag{13.10}$$

where δ_1 = Vertical deflection of beam due to load W_1

δ_2 = Vertical deflection of beam due to the distributed load q

δ_3 = Vertical deflection of the beam due to shear

δ_4 = Axial compression in column

The magnitudes of δ_1, δ_2, δ_3 and δ_4 can be obtained using following expressions:

$$\delta_1 = \frac{W_1 l^3}{96 EI_b} \cdot \frac{2K+1}{K+2} \tag{13.11}$$

$$\delta_2 = \frac{q l^4}{384 EI_b} \cdot \frac{5K+2}{K+2} \tag{13.12}$$

$$\delta_3 = \frac{3}{5}\frac{l}{EA_b}\left(W_1 + \frac{ql}{2}\right) \tag{13.13}$$

$$\delta_4 = \frac{h}{EA_c}\left(W_2 + \frac{W_1 + ql}{2}\right) \tag{13.14}$$

$$K = \frac{I_b}{I_c} \cdot \frac{h}{l} \tag{13.15}$$

where A_b = Cross-sectional area of beam
A_c = Cross-sectional area of column
I_b = Moment of inertia of beam about the axis of bending
I_c = Moment of inertia of column
E = Young's modulus of concrete
K = Relative stiffness factor
l = Effective span of frame
h = Effective height of frame

Values of l and h are obtained as below :

$$l = l_o - 2\,\alpha\, b \tag{13.16}$$

$$h = h_o - 2\,\alpha\, a \tag{13.17}$$

where l_o = Centre to centre distance between columns (Fig. 13.4a)
h_o = Height of the column from the top of the base slab to the centre of the frame beam (Fig. 13.4a)
a = One-half of the depth of the beam for a frame without haunches (Fig. 13.4a) or the distance as shown in Fig. 13.5 for a frame with haunches.
b = One-half of the column width for a frame without haunches (Fig. 13.4a) or the distance as shown in Fig. 13.5 for a frame with haunches.

Knowing the values of h_o, l_o and b, α can be obtained from Fig. 13.6.

The natural frequency of a transverse frame in vertical vibrations is given by

$$\omega_{nz} = \sqrt{\frac{K_z \cdot g}{W}} \tag{13.18}$$

Average vertical natural frequency of the T.G. foundation is taken as:

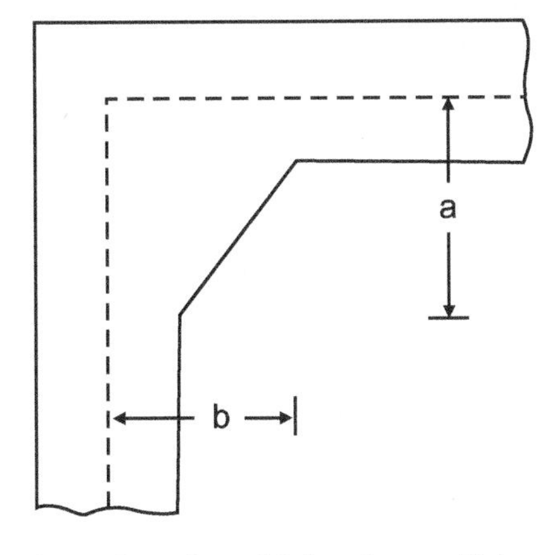

Fig. 13.5. Values of *a* and *b* for a frame with haunches

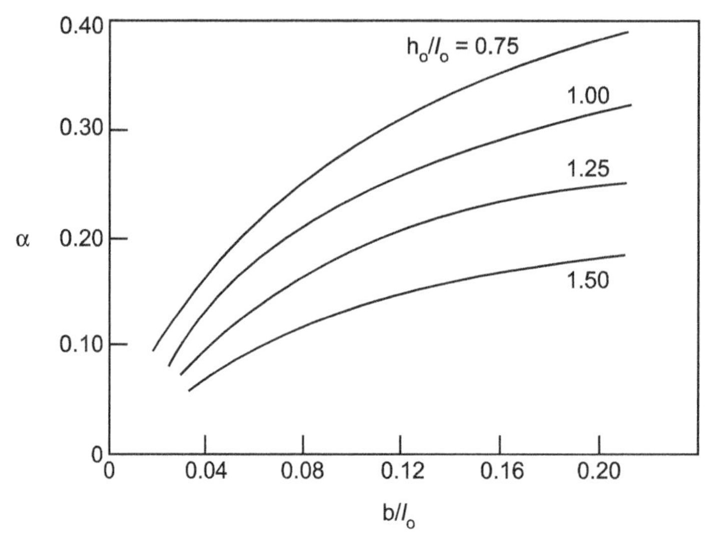

Fig. 13.6. α versus b/l_o

$$\omega_{nza} = \frac{\omega_{nz1} + \omega_{nz2} + \ldots + \omega_{nzn}}{n} \tag{13.19}$$

where ω_{nz1}, ω_{nz2},... = Vertical frequencies of individual transverse frames.

The average value of vertical amplitude of T.G. foundation may be computed as

$$A_{za} = \frac{\Sigma F_z}{(\Sigma K_z)\sqrt{\left[1-\left(\frac{\omega}{\omega_{nza}}\right)^2\right]^2 + \left(\frac{2\xi\omega}{\omega_{nza}}\right)}} \tag{13.20}$$

where A_{za} = Average vertical amplitude of T.G. foundation

$\sum F_z$ = Total vertical imbalance force

$\sum K_z$ = Sum of the stiffness of the individual frames

ξ = Damping ratio

For under-tuned foundation, i.e. $\omega < \omega_{nza}$, $\omega = \omega_{nza}$ should be used in Eq. (13.20). Then

$$A_{za} = \frac{\Sigma F_z}{(\Sigma K_z)(2\xi)} \tag{13.20a}$$

13.6.2 Horizontal Vibrations

In a T.G. frame formulation, the deck slab undergoes horizontal vibration in the direction perpendicular to the main shaft of the machine. The spring stiffness is provided by the columns due to their bending action, and for any transverse frame it is given by

$$K_x = \frac{12EI_c}{h^3}\left(\frac{6K+1}{3K+2}\right) \tag{13.21}$$

where K_x = Lateral stiffness of an individual transverse frame

If $\sum K_x$ = Sum of the lateral stiffness of all the transverse frames

W_T = Total weight of deck slab and machine

Then the natural frequency of the T.G. frame foundation is given by

$$\omega_{nxa} = \sqrt{\frac{(\Sigma K_x)g}{W_T}} \tag{13.22}$$

The average horizontal amplitude of the foundation may be computed as follows:

$$A_{xa} = \frac{\Sigma F_x}{(\Sigma K_x)\sqrt{\left[1-\left(\frac{\omega}{\omega_{nxa}}\right)^2\right]^2 + \left(\frac{2\xi\omega}{\omega_{nxa}}\right)^2}} \tag{13.23}$$

For under-tuned foundation, *i.e.* $\omega < \omega_{nxa}$
$\omega = \omega_{nxa}$ should be used in Eq. (13.23). Thus

$$A_{xa} = \frac{\Sigma F_x}{(\Sigma K_x)\,(2\xi)} \tag{13.23a}$$

As mentioned earlier, in this method only the possibility of resonance is checked i.e. the natural frequencies computed from Eqs. (13.19) and (13.22) should differ by atleast 30 percent from the operating speed of the machine. Equations (13.20), (13.20a) and (13.23) for determining amplitudes are given to be used further in combined method.

Resonance method based on idealising each transverse frame to single mass-spring system is an oversimplification of a complex problem. Therefore the values of natural frequencies computed by this method are very approximate.

13.7 Amplitude Method

In this method also, the vibration analysis is carried out for each transverse frame independently. However, the frame has been idealised as a two-degree-freedom system (Fig. 13.7). The main criterion for design is that the amplitudes due to forced vibrations are within permissible limits (Barkan, 1962).

13.7.1 Vertical Vibration

For the vertical frequency a two-degree-spring-mass system shown in Fig. 13.7b is adopted. Mass m_1 lumped over the columns is given by

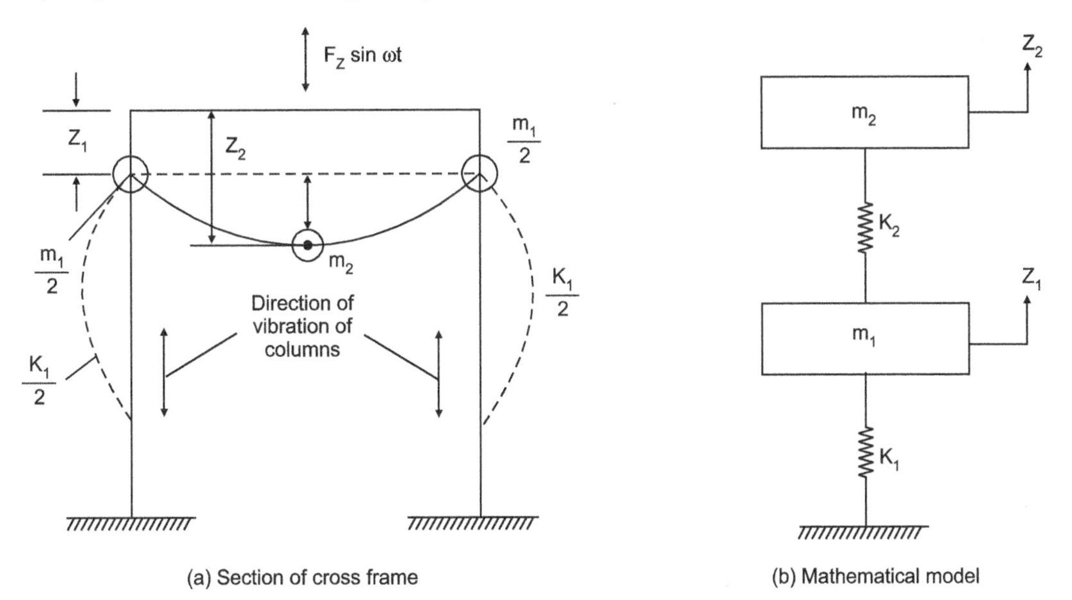

Fig. 13.7. (a) Vertical vibration of a cross frame as a two-degree-of-freedom system; (b) Mass-spring model

$$m_1 = \frac{W_1 + W_2 + 0.33W_3 + 0.25W_4}{g} \quad (13.24)$$

Mass m_2 acting at the centre of the cross beam is given by

$$m_2 = \frac{W_1 + 0.45W_4}{g} \quad (13.25)$$

where W_1 = Dead load of the machine and bearing
W_2 = Load transferred to the columns by longitudinal beams
W_3 = Weight of two columns constituting the transverse frame
W_4 = Weight of the transverse beam

The stiffness K_1 of both the columns of a transverse frame is given by

$$K_1 = \frac{2EA_c}{h} \quad (13.26)$$

The stiffness K_2 of the frame beam is given by

$$K_2 = \frac{1}{\delta_{st}} \quad (13.27)$$

$$\delta_{st} = \frac{l(1+2K)}{96EI_b(2+K)} + \frac{3l}{8GA_b} \quad (13.28)$$

where G = Shear modulus of beam material
E = Young's modulus of the material of columns
A_c = Cross-sectional area of a column
h = Effective height of the column
l = Effective span of the beam
A_b = Cross-sectional area of the beam
I_b = Moment of inertia of the beam

K is defined by Eq. (13.15).

The system shown in Fig. 13.7b is identical to the system shown in Fig. 2.28, and therefore can be analysed by the procedure explained in Sec. 2.8. The equations of motion in free vibration will be:

$$m_1\ddot{Z}_1 + K_1Z_1 - K_2(Z_2 - Z_1) = 0 \quad (13.29)$$

$$m_2\ddot{Z}_2 + K_2(Z_2 - Z_1) = 0 \quad (13.30)$$

The solution of above equations are:

$$Z_1 = A_1 \sin \omega_{nt} \quad (13.31)$$

$$Z_2 = A_2 \sin \omega_{nt} \quad (13.32)$$

Substituting Eqs. (13.31) and (13.32) into Eqs. (13.29) and (13.30), on simplification, we get

$$\omega_n^4 - (1+\mu_m)(\omega_{n11}^2 - \omega_{12}^2) + (1+\mu_m)\,\omega_{n11}^2\,\omega_{n12}^2 = 0 \quad (13.33)$$

where

$$\omega_{n11} = \sqrt{\frac{K_1}{m_l + m_2}} \quad (13.34)$$

$$\omega_{n12} = \sqrt{\frac{K_2}{m_2}} \quad (13.35)$$

$$\mu_m = \frac{m_2}{m_1} \tag{13.36}$$

The two natural frequencies of the system can be obtained by solving Eq. (13.33). In forced vibration, the equations of motion will be:

$$m_1 \ddot{Z}_1 + K_1 Z_1 + K_2 (Z_2 - Z_1) = 0 \tag{13.37}$$

$$m_2 \ddot{Z}_2 + K_2 (Z_2 - Z_1) = F_z \sin \omega t \tag{13.38}$$

The solution of the above equations can be presented as

$$Z_1 = A_{Z1} \sin \omega t \tag{13.39}$$

$$Z_2 = A_{Z2} \sin \omega t \tag{13.40}$$

Substituting Eqs. (13.39) and (13.40) into Eqs. (13.37) and (13.38), and then solving them we get

$$A_{z1} = \frac{\omega_{n12}^2 \cdot F_z}{m_1 \left[\omega^4 - (1+\mu_m)(\omega_{n11}^2 + \omega_{n12}^2)\,\omega^2 + (1+\mu_m)\,\omega_{n11}^2 + \omega_{n12}^2 \right]} \tag{13.41}$$

and

$$A_{z2} = \frac{\left[(1+\mu_m)\,\omega_{n11}^2 + \mu_m \omega_{n12}^2 - \omega^2 \right] \cdot F_z}{m_2 \left[\omega^4 - (1+\mu_m)(\omega_{n11}^2 + \omega_{n12}^2)\,\omega^2 + (1+\mu_m)\,\omega_{n11}^2 + \omega_{n12}^2 \right]} \tag{13.42}$$

13.7.2 Horizontal Vibration

For analysing the frame foundation under horizontal vibration as a two degree of freedom problem, the upper and lower foundation slabs are assumed to be infinitely rigid. The columns are taken to act as leaf springs. The stiffness of a leaf spring is considered equal to the lateral stiffness of the individual transverse frame.

Figure 13.8 shows a typical mathematical model for a two bays frame foundation. The equivalent mass m_i lumped over the spring i (representing frame i) is given by:

$$m_i = m_{mi} + m_{bi} + 0.33\, m_{ci} + m_{gi} \tag{13.43}$$

where m_{mi} = Mass of machine resting on cross beam of ith frame
m_{bi} = Mass of cross beam of ith frame
m_{ci} = Mass of columns of ith frame
m_{gi} = Mass transferred from longitudinal girders on either side

In Fig. 13.8, points G_1 and G_2 represent the centre of masses (*i.e.* m_1, m_2, and m_3) and centre of stiffnesses (*i.e.* K_{x1}, K_{x2} and K_{x3}) respectively. Line A_1B_1 shows the initial position of deck slab. The final displaced position of the deck slab is represented by the line A_3B_3. The deck slab rotates about the mass centre G_1. d_{k1}, d_{k2} and d_{k3} are the distances of different masses from point G_2. The distances of different masses from point G_1 are shown as d_{m1}, d_{m2} and d_{m3}. e represents the distance between G_1 and G_2.

The equations of motion for the system shown in Fig. 13.8 will be:

$$(\Sigma m_1)\ddot{x} + \Sigma[K_{xi}(x + d_{mi}\psi)] = F_x \sin \omega t \tag{13.44}$$

$$(\Sigma m_1 d_{mi}^2)\ddot{\psi} + \Sigma[K_{xi}(x + d_{mi}\psi)d_{mi}] = M_z \sin \omega t \tag{13.45}$$

where F_x = Horizontal unbalanced force
M_z = Unbalanced moment

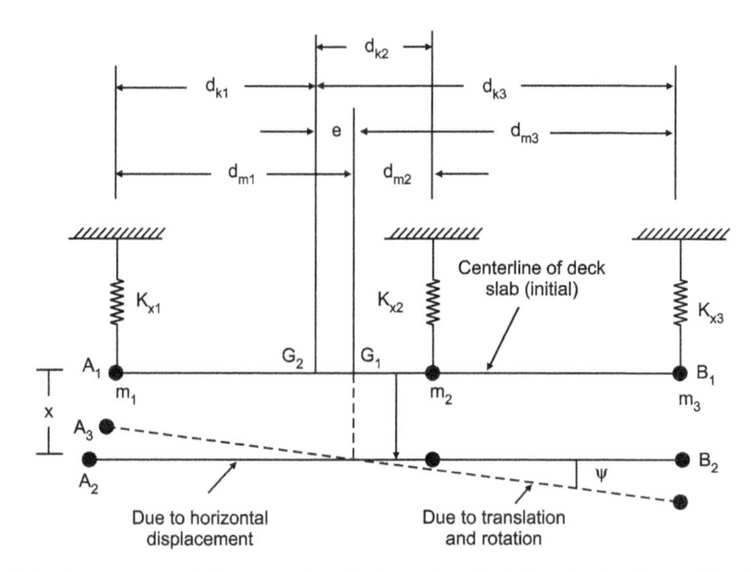

Fig. 13.8. Spring-mass model for combined horizontal and rotational vibrations of the deck slab

Denoting

$$m = \sum m_i = \text{Total mass} \tag{13.46}$$

$$M_{mz} = \Sigma m_i \, d_{mi}^2 = \text{Polar mass moment of inertia of all the masses about the vertical axis through } G_1 \tag{13.47}$$

$$\Sigma K_{xi} \, (x + d_{mi}\psi) = x \Sigma K_{xi} + (\Sigma K_{xi} \, d_{mi})\psi$$
$$= x K_x + K_x \cdot e\psi = K_x \, (x + e\psi) \tag{13.48}$$

K_x represents the total lateral stiffness.

$$K_{xi} = \frac{1}{\delta_{hi}} \tag{13.49}$$

$$\delta_{hi} = \frac{h^3}{12EI_c} \frac{3k+2}{6k+2} + \frac{6h}{5EA_c}\left[1 + \frac{A_c h}{A_b I} \frac{18k^2}{(6k+1)^2}\right] + \frac{h^3}{EI_c l^2} \frac{18k^2}{(6k+1)^2} \tag{13.50}$$

$$\Sigma K_{xi}(x + d_{mi}\psi)d_{mi} = (\Sigma K_{xi} d_{mi})x + (\Sigma K_{xi} d_{mi}^2)\psi$$
$$= K_x \cdot e \cdot x + \left[\Sigma K_{xi}(e^2 + d_{ki}^2)\right]\psi$$
$$= K_x \cdot ex + K_x e^2 \psi + (\Sigma K_{xi} \, d_{ki}^2)\psi$$
$$= K_x (x + e\psi)e + K_\psi \cdot \psi \tag{13.51}$$

K represents the equivalent torsional spring stiffness for the frame columns and is given by

$$K_\psi = \Sigma K_{xi} d_{ki}^2 \tag{13.52}$$

It may be noted that $\sum K_{xi} \, d_{ki} = 0$

Making the substitutions from Eqs. (13.46) to (13.50) in Eqs. (13.44) and (13.45), we get

$$m\ddot{x} + K_x x + K_x \cdot e\psi = F_x \sin \omega t \tag{13.53}$$

$$M_{mz}\,\ddot{\psi} + K_x ex + (K_x e^2 + K_\psi)\psi = M_z \sin \omega t \tag{13.54}$$

The solution is being given by the equation:

$$\omega_n^4 - (\alpha\omega_{nx}^2 + \omega_{n\psi}^2)\,\omega_n^2 + \omega_{nx}^2 \cdot \omega_{n\psi}^2 = 0 \tag{13.55}$$

where

$$\omega_{nx} = \sqrt{\frac{K_x}{m}} \tag{13.56}$$

$$\omega_{n\psi} = \sqrt{\frac{K_\psi}{M_{mz}}} \tag{13.57}$$

$$\alpha = 1 + \frac{e^2}{r^2} \tag{13.58}$$

$$r = \sqrt{\frac{M_{mz}}{m}} \tag{13.59}$$

The amplitude of vibration in translation and rotation are given by

$$A_x = \frac{\left[\dfrac{e^2}{r^2}\omega_{nx}^2 + \omega_{n\psi}^2 - \omega^2\right]\dfrac{F_x}{m} - \omega_{nx}^2 \cdot \dfrac{M_z}{M_{mZ}}}{\Delta(\omega^2)} \tag{13.60}$$

$$A_\psi = \frac{\dfrac{e^2}{r^2}\omega_{nx}^2 \dfrac{F_x}{m} - (\omega_{n\psi}^2 - \omega^2)\dfrac{M_z}{M_{mz}}}{\Delta(\omega^2)} \tag{13.61}$$

where

$$\Delta(\omega^2) = \omega^4 - (\alpha\omega_{nx}^2 + \omega_{n\psi}^2)\omega^2 + \omega_{nx}^2\omega_{n\psi}^2 \tag{13.62}$$

The net amplitude A_h is given by

$$A_h = A_x + y\,A\psi \tag{13.63}$$

where y = Distance of the point at which the amplitude is being calculated from the centre of gravity of the system.

13.8 Combined Method

The resonance method and the amplitude method are not complete by themselves and are complimentary. Major (1962) developed a combined method which is also known as "extended resonance" method. In the combined method the possibility of resonance is investigated, attempts are made at the same time to determine the amplitude. For the under-tuned case, the maximum dynamic forces occurring both during acceleration and deceleration have been considered and the maximum resonance peak is regarded as the maximum operational amplitude.

A typical cross frame subjected to loads is shown in Fig. 13.9.

q: distributed load acting on the beam (including weight of the beam, floor supported by the frame and other structures)

P: weight of the machine with all the masses and components

N: weight of the (i) longitudinal beams acting on the columns, (ii) part of the mechanical equipment carried by the longitudinal beam resting on the column (omitting the dynamic factor), (iii) the upper third of the column, and (iv) cantilevers and the load carried on the upper third of the column.

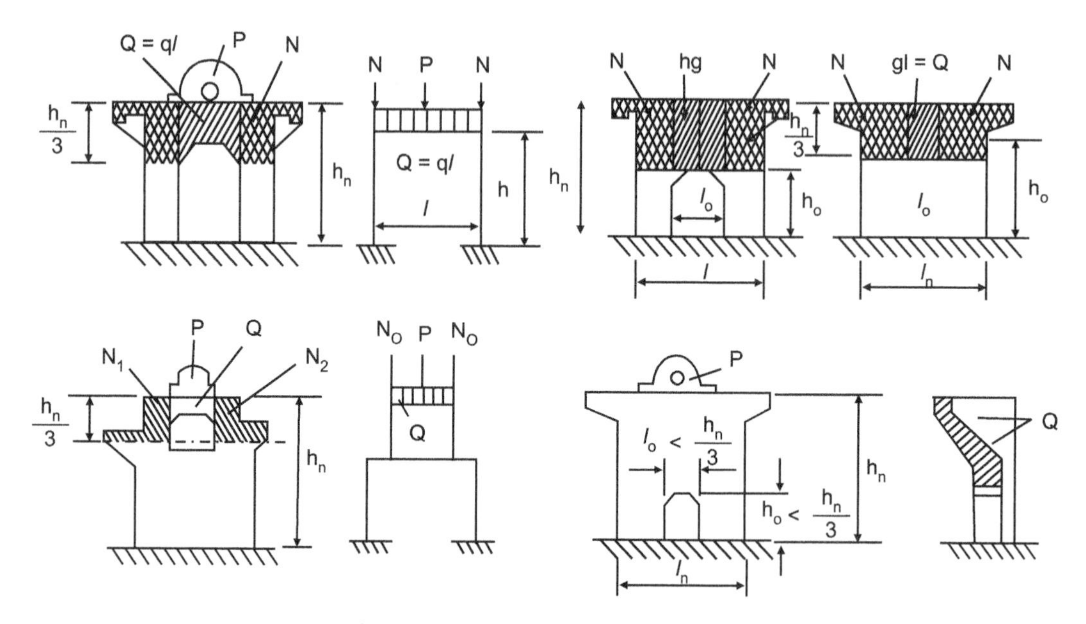

Fig. 13.9. Loads acting on frame columns

If the frame beam is larger than one third of the column height, column is considered with a height corresponding to that of the frame if the clear height and width of the opening are smaller than one third of the total height and width, respectively. If the loads N acting on the two frame columns differ, then their mean value is taken for the analysis.

13.8.1 Vertical Natural Frequencies

For obtaining vertical natural frequency, each transverse frame that consists of two columns and a beam perpendicular to main shaft of the machine, is considered separately. The complete procedure is already described in Sec. 13.6.1 is applicable in combined method also.

The natural frequency of the framed foundation can be expressed as

$$f_{nz} = \frac{1}{2\pi}\sqrt{\frac{K_z g}{W}} = \frac{1}{2\pi}\sqrt{\frac{W \cdot g}{\delta_{st} \cdot W}} = \frac{30}{\sqrt{\delta}} \tag{13.64}$$

The dynamic magnification factor, M is computed as

$$M = \frac{1}{\sqrt{\left\{1-\left(\frac{f_m}{f_n}\right)^2\right\}^2 + \left(\frac{\Delta}{\pi}\right)^2\left(\frac{f_m}{f_n}\right)^2}} \tag{13.65}$$

where Δ : logarithmic decrement due to damping, which may be taken as 0.4 for concrete foundations

f_m : operating frequency of the machine, and

f_n : natural frequency of the machine foundation soil system

For undertuned foundations, i.e., $f_m > f_n$, above equation can be simplified taking $f_m/f_n \approx 1$ as

$$M = \frac{\pi}{\Delta}$$

$$= 7.85 \text{ taking } \Delta = 0.4$$

The centrifugal force C $(= \sum C_i)$ due to rotation of unbalanced force can be expressed as

$$C = \sum C_i \tag{13.66}$$

where C_i is the centrifugal force in the ith frame.

For undertuned case C_i is given by

$$C_i = \alpha R \left(\frac{f_n}{f_m} \right)^2 \tag{13.67}$$

where $\alpha = 0.20$ for normally balanced machine with rpm ≥ 3000
$= 0.16$ for normally balanced machine with rpm ≥ 1500
$= 0.10$ for normally balanced machine with rpm ≤ 750

$$R = \sum (R_b + R_c)$$

where R_c = Rotating weights on column frames
R_b = Rotating weights on cross beams

For overtuned case, i.e., $f_n > f_m$

$$C_i = \alpha R \tag{13.68}$$

The vertical amplitude may now be obtained as

$$a_v = M.\delta_v \tag{13.69}$$

where

$$\delta_v = \frac{C}{E}\left[\frac{l^2}{96 I_b} \cdot \frac{2k+1}{k+2} + \frac{3}{5} \cdot \frac{l}{A_b} + \frac{1}{2} \cdot \frac{h}{A_c} \right] \tag{13.70}$$

13.8.2 Horizontal Natural Frequencies

Horizontal natural frequencies for the entire structure have to be obtained, as cross frames are connected by rigid elements as shown in Fig. 13.10.

Horizontal natural frequency is given by,

$$f_{nh} = 30 \left[\alpha_o \pm \left(\alpha_o^2 - \frac{\sum k_{hi}}{\sum W_i} \cdot \frac{I_H}{I_G} \right)^{1/2} \right]^{1/2} \tag{13.71}$$

where f_{nh} = horizontal natural frequency
k_{hi} = lateral stiffness of cross frame 'i'
W_i = total weight on frame 'i' including machine weight, weight of transverse beam, and weight transmitted by longitudinal girders.

$$I_G = \sum W_i X_{Gi}^2 = W_A X_{GA}^2 + W_B X_{GB}^2 + \ldots \tag{13.72}$$

X_{Gi} = distance of weight W_i from the vertical axis through the centre of gravity,

$$I_H = \sum k_{hi} X_{Hi}^2 = k_{hA} X_{HA}^2 + k_{hB} X_{HB}^2 + \ldots \tag{13.73}$$

X_{Hi} = distance of cross frame 'i' from the centre of rigidity H,

$$\alpha_o = \frac{1}{2}\left[e^2 \frac{\sum k_{hi}}{I_G} + \frac{\sum k_{hi}}{\sum W_i} + \frac{I_H}{I_G} \right] \tag{13.74}$$

e = distance between centre of gravity G and centre of rigidity H.

The dynamic magnification factor, M is computed as

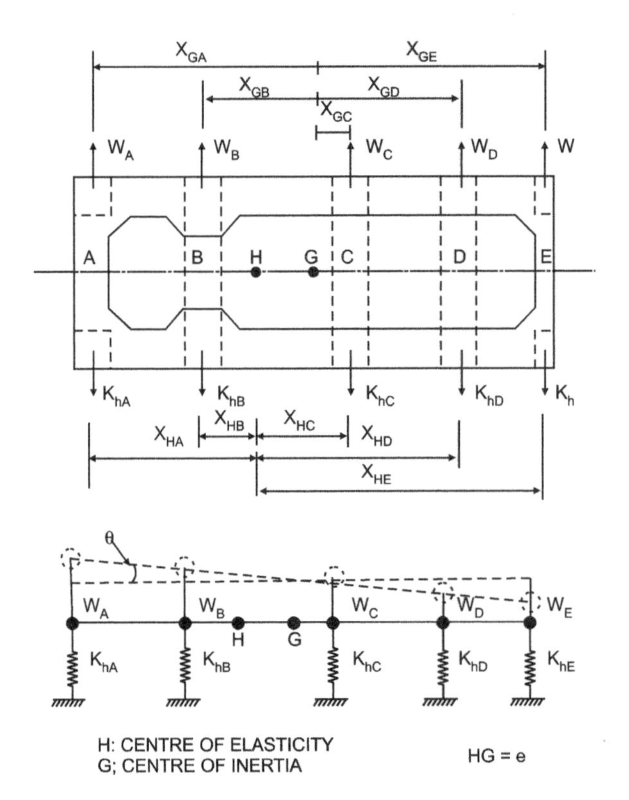

Fig. 13.10. Model system for horizontal vibrations (amplitudes and combined method)

$$M = \frac{1}{\sqrt{\left\{1 - \left(\frac{f_m}{f_n}\right)^2\right\}^2 + \left(\frac{\Delta}{\pi}\right)^2 \left(\frac{f_m}{f_n}\right)^2}} \tag{13.75}$$

where Δ = logarithmic decrement due to damping, which may be taken as 0.4 for concrete foundations

f_m = operating frequency of the machine, and

f_n = natural frequency of the machine foundation soil system.

For undertuned foundations, i.e., $f_m > f_n$, above equation can be simplified taking $f_m/f_n \approx 1$ as

$$M = \frac{\pi}{\Delta}$$

$$= 7.85 \text{ taking } \Delta = 0.4$$

The centrifugal force C $(= \sum C_i)$ due to rotation of unbalanced force can be expressed as

$$C = \sum C_i \tag{13.76}$$

where C_i is the centrifugal force in the ith frame.

For undertuned case C_i is given by

$$C_i = \alpha R \left(\frac{f_n}{f_m}\right)^2 \tag{13.77}$$

where α = 0.20 for normally balanced machine with rpm $\geq$ 3000

= 0.16 for normally balanced machine with rpm = 1500

= 0.10 for normally balanced machine with rpm $\leq$ 750

For overtuned case, i.e., $f_n > f_m$

$$C_i = \alpha R \tag{13.78}$$

For computation of α_h, the total horizontal centrifugal force, C (= $\sum C_i$) is distributed to various frames in proportion to their lateral rigidities as

$$C_i = C\frac{k_{hi}}{\sum k_{hi}} + e_1\frac{Ck_{hi}\,X_{Hi}}{I_h} \tag{13.79}$$

where

$$I_h = \sum k_{hi}\,X_{Hi}^2 \tag{13.80}$$

C = total centrifugal force = $\sum C_i$,
e_1 = eccentricity of the resultant centrifugal forces from the centre of elasticity

$$= X_H - X_C \tag{13.81}$$

X_H = centre of elasticity (centre of gravity of $\sum k_{hi}$),
X_C = centre of gravity of centrifugal forces of the rotating weights,
$\sum k_{hi}$ = sum of the lateral rigidities of cross frames, and
X_{Hi} = distance of each frame '*i*' from centre of elasticity.

Then the lateral deflection δ_{hi} of the frame '*i*' due to static influence of C_i is given by

$$\delta_{hi} = \frac{C_i}{k_{hi}} \tag{13.82}$$

The amplitude of horizontal motion of frame '*i*', i.e., a_{hi} can now be obtained as

$$a_{hi} = M\delta_{hi} \tag{13.83}$$

Since the horizontal natural frequency (f_n) is very low as compared to operating frequency (f_m), the maximum value of M = 7.85 may be taken for the computation of horizontal amplitudes.

13.9 Three Dimensional Analysis

For turbogenerator foundations of more than 100 MW capacity, a three-dimensional space frame model is preferred for analysis. The modelling should take into account the basic characteristics of the system, that is, mass, stiffness and damping. Special attention is required while idealising the points of excitation.

Nodes are specified to all bearing points, beam-column junctions, mid-points and quarter points of beams and columns and where the member cross-sections change significantly. Generally the number of nodes specified on any member should be sufficient to calculate all the modes having frequencies less than or equal to the operating speed.

Lumped-mass approach is used having lumped masses at the node points. The machine shall be modelled to lump its mass together with the mass of the foundation. Equivalent sectional properties of beams and columns are used. The computation of equivalent mass moment of inertia of the frame members pose some difficulty since these depends upon the deflection shape in each mode. These may be discretised in the first step and considered data in an iterative manner if desired. The columns may be assumed to be fixed at the base, disregarding the base mat.

A typical space frame model is shown in Fig. 13.11.

The dynamic analysis of the frame foundation requires the calculation of Eigen values of the system. The problem can be handled in a systematic manner in the matrix notation. The structure is idealised into a skeleton system which retains the properties of the original structure. The stiffness matrix of the structure as a whole is assembled from the stiffness matrices of individual members. The resulting equations are then solved for the time periods and amplitudes.

Illustrative Example

Example 13.1

Design a turbo-generator frame foundation shown in Figs. 13.12 and 13.13 with the following data:

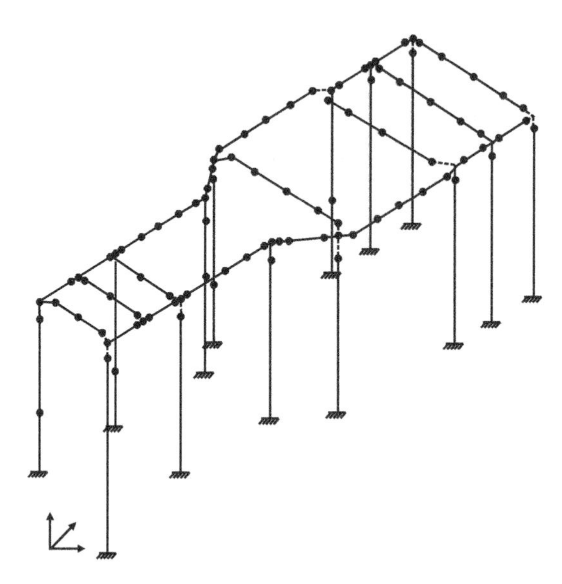

Fig. 13.11. Space frame model of the foundation shown in Fig. 13.1

Cross sectional dimensions of each column = 0.8 m × 0.8 m
Cross sectional dimensions of each cross beam = 0.7 m × 0.7 m
Cross sectional dimensions of each longitudinal beams = 0.6 m × 0.6 m
Weight on cross beam of each frame:

$$W_{11} = 40 \text{ kN};\ W_{12} = 80 \text{ kN};\ W_{13} = 80 \text{ kN};\ W_{14} = 50 \text{ kN}$$

Weight of rotating parts acting on each cross frame:

$$W_{r1} = 10 \text{ kN};\ W_{r2} = 15 \text{ kN};\ W_{r3} = 15 \text{ kN};\ W_{r4} = 20 \text{ kN}$$

Eccentricity = 0.05×10^{-3} m
Young's Modulus for concrete = 3×10^{7} kN/m^2
Damping ratio = 0.02; $\mu = 0.15$
Operating speed = 4500 rpm
Permissible vertical amplitude = 0.3 mm
Permissible horizontal amplitude = 0.5 mm

Solution

$$h_o = 4.0 \text{ m};\ l_o = 3.5 \text{ m}$$

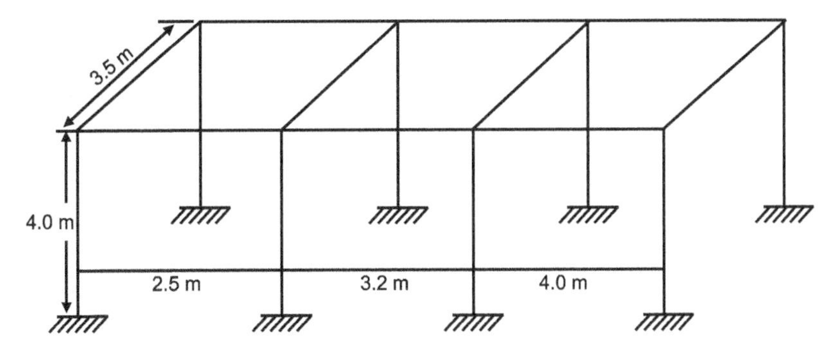

Fig. 13.12. Dimensions of turbo-generator frame

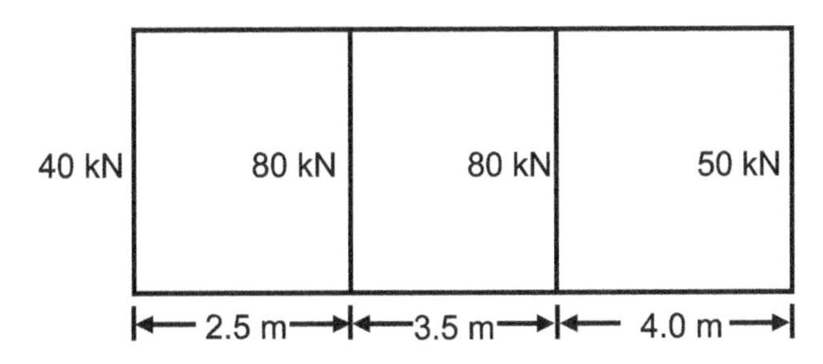

Fig. 13.13. Figure showing the weight on cross beam of each frame

Cross section of typical cross beam

$2a_b$ = 0.7 m

2a = 0.7 m

Cross section of typical column

2b = 0.8 m

$2b_c$ = 0.8 m

$A_b = 0.7 \times 0.7 = 0.49 \text{ m}^2$

$$I_b = \frac{(2a).(2a_b)^3}{12}$$

$= 0.020 \text{ m}^4$

$A_c = 0.8 \times 0.8 = 0.64 \text{ m}^2$

$$I_c = \frac{(2b).(2b_c)^3}{12}$$

$= 0.03413 \text{ m}^4$

$b/l_o = 0.4/3.5 = 0.114$

$h_o/l_o = 4/3.5 = 1.143$

$\alpha = 0.19$ (From charts Fig. 13.6)

$l = l_o - 2.\alpha.b$

$= 3.5 - 0.19 \times 0.8 = 3.348 \text{ m}$

$h = h_o - 2.\alpha.a$

$= 4.0 - 0.19 \times 0.7 = 3.867 \text{ m}$

$$k = \frac{I_b}{I_c} \cdot \frac{h}{l} = \frac{0.020}{0.03413} \cdot \frac{3.867}{3.348}$$
$$= 0.677$$

Resonance Method: Vertical vibrations:
For Ist frame:

$$W_1 = 40 \text{ kN}$$
$$W_2 = (24 \times 0.6 \times 0.6 \times 2.5)/2 = 10.8 \text{ kN}$$
$$q = 0.7 \times 0.7 \times 24 = 11.76 \text{ kN/m}$$

$$\delta_1 = \frac{W_1 l^3}{96 EI_b} \frac{2K + 1}{K + 2}$$

$$= \frac{40 \times (3.348)^3}{96 \times 3 \times 10^7 \times 0.02} \times \frac{2 \times 0.667 + 1}{0.677 + 2}$$

$$= 229.167 \times 10^{-7} \text{ m}$$

$$\delta_2 = \frac{ql^4}{384 EI_b} \frac{5K + 2}{K + 2}$$

$$= \frac{11.76 \times (3.348)^3}{384 \times 3 \times 10^7 \times 0.02} \times \frac{5 \times 0.677 + 1}{0.677 + 2}$$

$$= 129.004 \times 10^{-7} \text{ m}$$

$$\delta_3 = \frac{3l}{5EA_b}\left(W_1 + \frac{ql}{2}\right)$$

$$= \frac{3 \times 3.348}{5 \times 3 \times 10^7 \times 0.49}\left(40 + \frac{11.76 \times 3.348}{2}\right)$$

$$= 81.563 \times 10^{-7} \text{ m}$$

$$\delta_4 = \frac{h}{EA_c}\left(W_2 + \frac{W_1 + ql}{2}\right)$$

$$= \frac{3.867}{3 \times 10^7 \times 0.64}\left(10.8 + \frac{40 + 11.76 \times 3.348}{2}\right)$$

$$= 101.682 \times 10^{-7} \text{ m}$$

$$\delta_{st} = \delta_1 + \delta_2 + \delta_3 + \delta_1$$
$$= (229.167 + 129.004 + 81.563 + 101.682) \times 10^{-7}$$
$$= 541.416 \times 10^{-7} \text{ m}$$

$$W = W_1 + 2.W_2 + q.l$$
$$= 40 + 2 \times 10.8 + 11.76 \times 3.348$$
$$= 100.972 \text{ kN}$$

$$K_z = \frac{W}{\delta_{st}}$$

$$= \frac{100.972}{541.416 \times 10^{-7}}$$

$$= 0.1865 \times 10^7 \text{ kN/m}$$

$$\omega_{nz} = \sqrt{\frac{K_z.g}{W}}$$

$$= \sqrt{\frac{0.1865 \times 10^7 \times 10^3 \times 9.81}{100.972 \times 10^3}}$$

$$= 425.665 \text{ rad/sec}$$

Proceeding as illustrated above, natural frequencies of other three frames were obtained. Details of computations have been summarised below in Table 13.1.

Table 13.1. Details of Computations of Natural Frequencies in Vertical Vibration

Frame number	W_1 *(kN)*	W_2 *(kN)*	*q (kN/m)*	$\delta_1(m) \times 10^{-7}$	$\delta_2(m) \times 10^{-7}$	$\delta_3(m) \times 10^{-7}$	$\delta_4(m) \times 10^{-7}$	$\delta_{st}(m) \times 10^{-7}$	K_z *(kN/m)* $\times 10^7$	ω_{nz} *(rad/sec)*
I	40	10.8	11.76	229.167	129.004	81.563	101.682	541.416	0.1865	425.665
II	80	24.62	11.76	458.334	129.004	136.224	169.806	893.364	0.1887	331.33
III	80	31.10	11.76	458.334	129.004	136.224	182.857	906.419	0.2003	328.98
IV	50	17.28	11.76	286.459	129.004	95.228	124.804	635.495	0.195	392.897

$$\omega_{nz} = \frac{\sum_{i=1}^{n} \omega_{nzi}}{n}$$

where n is the number of frames.

$$\omega_{nza} = \frac{\omega_{nz1} + \omega_{nz2} + \omega_{nz3} + \omega_{nz4}}{4}$$

$$= \frac{425.665 + 331.33 + 328.98 + 392.897}{4}$$

$$= 369.718 \text{ rad/sec.}$$

Operating frequency of the machine = 4500 rpm = 471.24 rad/sec

$$\frac{\omega}{\omega_{nza}} = \frac{471.24}{369.718} = 1.2745 \qquad [> 1.2, \text{ hence ok}]$$

Rotating parts falling on each cross frame,

$$W_{r1} = 10 \text{ kN};\ W_{r2} = 15 \text{ kN};\ W_{r3} = 15 \text{ kN};\ W_{r4} = 20 \text{ kN}$$

$$e = 0.05 \times 10^{-3} \text{ m}$$

$$F_{z1} = me\omega^2 = 10 \times 0.05 \times 10^{-3} \times (471.238)^2/9.81 = 11.3069 \text{ kN}$$

$$F_{z2} = me\omega^2 = 15 \times 0.05 \times 10^{-3} \times (471.238)^2/9.81 = 16.9603 \text{ kN}$$

$$F_{z3} = me\omega^2 = 15 \times 0.05 \times 10^{-3} \times (471.238)^2/9.81 = 16.9603 \text{ kN}$$

$$F_{z4} = me\omega^2 = 20 \times 0.05 \times 10^{-3} \times (471.238)^2/9.81 = 22.6138 \text{ kN}$$

Table 13.2. Values of W, K_z, δ_{st} and F_z for Different Cross Frames

Frame no.	*W (kN)*	K_z *(kN/m)* × 10^7	δ_{st} *(m)* × 10^{-7}	F_z *(kN)*
I	100.972	0.1865	541.416	11.3069
II	168.62048	0.1887	893.364	16.9603
III	181.58048	0.20033	906.419	16.9603
IV	123.932	0.195	635.495	22.6138

$$\sum K_z = 0.77053 \times 10^7 \text{ kN/m}$$
$$\sum F_z = 67.8413 \text{ kN}$$

Since, $\omega > \omega_{nza}$, hence

$$\text{Maximum amplitude} = \frac{\sum F_z}{(\sum K_z)(2\xi)}$$
$$= \frac{67.8413}{(0.77053 \times 10^7)(2 \times 0.02)}$$
$$= 22.0112 \times 10^{-5} \text{ m} = 0.22 \text{ mm} \qquad [< 0.25 \text{ mm, safe}]$$

Horizontal vibrations:

Table 13.3. Values of I_b, I_c, K and K_X for different cross frames

Frame no.	I_b (m^4)	I_c(m^4)	*k*	K_x(kN/m)
I	0.020	0.03413	0.677	266856.42
II	0.020	0.03413	0.677	266856.42
III	0.020	0.03413	0.677	266856.42
IV	0.020	0.03413	0.677	266856.42

$$\sum K_x = 1067425.70 \text{ kN/m}$$

Assuming the thickness of deck slab as 0.7 m

$$W_T = 0.7 \times (2.5 + 3.2 + 4 + 0.7)(3.5 + 0.6) \times 24 + 40 + 80 + 80 + 50$$
$$= 966.352 \text{ kN}$$

$$\text{Natural frequency in horizontal direction} = \omega_{nxa} = \sqrt{\frac{(\sum K_x)g}{W_T}} = \sqrt{\frac{1067425.70 \times 10^3 \times 9.81}{966.352 \times 10^3}}$$
$$= 104.0964 \text{ rad/sec}$$

$$\text{Maximum value of force coming on foundation} = (F_x)_{\max} = 15\frac{\sum W_x \omega}{314}$$

$$\sum W_r = 10 + 15 + 15 + 20 + = 60 \text{ kN}$$

$$\omega = \frac{2\pi f}{60} = \frac{2.\pi.4500}{60} = 471.238 \text{ rad/sec}$$

$$(F_x)_{\max} = \frac{15 \times 60 \times 471.238}{314} = 1350.682 \text{ kN}$$

$$\text{Maximum amplitude} = A_{xa} = \frac{\sum F_x}{(\sum K_x)\sqrt{\left[1-\left(\frac{\omega}{\omega_{nxa}}\right)^2\right]^2+\left(\frac{2\xi\omega}{\omega_{nxa}}\right)}}$$

$$= \frac{1350.682}{(1067425.7006)\sqrt{\left[1-\left(\frac{471.238}{104.0964}\right)^2\right]^2+\left(\frac{2\times0.02\times471.238}{104.0964}\right)^2}}$$

$= 0.24633 \times 10^{-5}$ m = 0.00246 mm [<< 0.5 mm, safe]

Amplitude Method: Vertical vibration:
For 1st frame:

$W_1 = 40$ kN
$W_2 = (24 \times 0.6 \times 0.6 \times 2.5 \times 2)/2 = 21.6$ kN
$W_3 = 0.8 \times 0.8 \times 4 \times 24 \times 2 = 122.88$ kN
$W_4 = 0.7 \times 0.7 \times (3.5 + 0.7) \times 24 = 49.392$ kN

$$m_1 = \frac{W_1 + W_2 + 0.33W_3 + 0.25W_4}{g}$$

$$= \frac{40 + 21.6 + 0.33\times122.88 + 0.25\times49.392}{9.81} = 11.67 \text{ kNs}^2\text{/m}$$

$$m_2 = \frac{W_1 + 0.45W_4}{g}$$

$$= \frac{40 + 0.45\times49.392}{9.81} = 6.343 \text{ kNs}^2\text{/m}$$

$$K_1 = \frac{2EA_C}{h}$$

$$= \frac{2\times3\times10^7\times0.64}{3.867} = 0.993 \times 10^7 \text{ kN/m}$$

$$G = \frac{E}{2(1+\mu)} = \frac{3\times10^7}{2(1+0.15)} = 1.3\times10^7 \text{ kN/m}^2 \qquad \text{(Assuming } \mu = 0.15)$$

$$\delta_{st} = \frac{l^3(1+2K)}{96EI_b(2+K)} + \frac{3l}{8GA_b}$$

$$= \frac{(3.348)^3(1+2\times0.67)}{96\times3\times10^7\times0.02(2+0.677)} + \frac{3\times3.348}{8\times1.3\times10^7\times0.49}$$

$= 7.7 \times 10^{-7}$ m/kN

$$K_2 = \frac{1}{\delta_{st}} = \frac{1}{7.7\times10^{-7}} = 0.13 \times 10^7 \text{ kN/m}$$

$$\omega_{n11} = \sqrt{\frac{K_1}{m_1 + m_2}}$$

$$= \sqrt{\frac{0.993\times10^7\times10^3}{(11.67+6.343)\times10^3}} = 742.47 \text{ rad/sec}$$

$$\omega_{n12} = \sqrt{\frac{K_2}{m_2}}$$

$$= \sqrt{\frac{0.13 \times 10^7 \times 10^3}{6.343 \times 10^3}} = 452.714 \text{ rad/sec}$$

$$\mu_m = \frac{m_2}{m_1} = \frac{6.343}{11.67} = 0.544$$

$$\omega_n^4 - (1+\mu_m)(\omega_{n11}^2 + \omega_{n12}^2)\omega_n^2 + (1+\mu_m)\omega_{n11}^2 \cdot \omega_{n12}^2 = 0$$

$$\omega_n^4 - (1+0.544)(452.714)^2 + (742.47)^2)\omega_n^2 + (1+0.544)(742.47)^2 \cdot (452.714)^2 = 0$$

$$\omega_{n1} = 419.348 \text{ rad/sec}$$

$$\omega_{n2} = 995.86 \text{ rad/sec}$$

$$A_{Z1} = \frac{\omega_{n12}^2 \cdot F_z}{m_1\left[\omega^4 - (1+\mu_m)(\omega_{n11}^2 + \omega_{n12}^2)\omega^2 + (1+\mu_m)\omega_{n11}^2\omega_{n12}^2\right]}$$

$$= \frac{(452.714)^2(11.3069)}{11.67\left[(471.238)^4 - (1+0.544)((742.47)^2 + (452.714)^2)(471.238)^2 + (1+0.544)(742.47)^2(452.714)^2\right]}$$

$$= 5.5895 \times 10^{-6} M$$

$$A_{Z2} = \frac{\left[(1+\mu_m)\omega_{n11}^2 + \mu_m\omega_{n12}^2 - \omega^2\right].F_z}{m_2\left[\omega^4 - (1+\mu_m)(\omega_{n11}^2 + \omega_{n12}^2)\omega^2 + (1+\mu_m)\omega_{n11}^2\omega_{n12}^2\right]}$$

$$= \frac{\left[(1+0.544)(742.47)^2 + (0.544)(452.714)^2 - (471.238)^2\right].(11.3069)}{6.343\left[(471.238)^4 - (1+0.544)((742.47)^2 + (471.238)^2)(471.298)^2 + (1+0.544)(742.47)^2(452.714)^2\right]}$$

$$= 3.7197 \times 10^{-6} \text{ m}$$

Similarly, for other frames:

Table 13.4. Details of Computations of Natural Frequencies and Amplitudes

Frame Number	W_1 *(kN)*	W_2 *(kN)*	W_3 *(kN)*	W_4 *(kN)*	m_1 *(kNs²/m)*	m_2 *(kNs²/m)*	$\delta_{st}(m) \times 10^{-7}$	K_1 *(kN/m)* $\times 10^7$	K_2 *(kN/m)* $\times 10^7$
I	40	21.6	122.88	49.392	11.67	6.343	7.7	0.993	0.13
II	80	49.248	122.88	49.392	18.57	10.421	7.7	0.993	0.13
III	80	62.208	122.88	49.392	19.8885	10.4206	7.7	0.993	0.13
IV	50	34.56	122.88	49.392	14.0121	7.3625	7.7	0.993	0.13

Frame no.	ω_{nz1} *(rad/sec)*	ω_{nz2} *(rad/sec)*	$A_{z1}(m) \times 10^{-6}$	$A_{z2}(m) \times 10^{-6}$
I	419.348	995.86	5.5895	3.7197
II	327.41	788.8	2.481	1.3561
III	326.99	763.197	2.566	1.3451
IV	388.844	909.242	6.682	4.1722

It may be noted that average natural frequency ($\omega_{nz1} \approx 365$ rad/sec) is quite away from the operating frequency. Amplitude of each frame is much less than the permissible amplitude.

Horizontal vibration:

$m_{mi} = W_1/g;\ m_{bi} = W_4/g;\ m_{ci} = W_3/g;\ m_{gi} = W_2/g$

$m_i = m_{mi} + m_{bi} + 0.33m_{ci} + m_{gi}$

$$K_{xi} = \frac{1}{\delta_{hi}}$$

$$\delta_{hi} = \frac{h^3}{12EI_c}\frac{3k+2}{6k+2} + \frac{6h}{5EA_c}\left[1 + \frac{A_c h}{A_b l}\frac{18k^2}{(6k+1)^2}\right] + \frac{h^3}{EI_c l^2}\frac{18k^2}{(6k+1)^2}$$

$$= \frac{(3.867)^3}{12\times 3\times 10^7\times 0.03413}\cdot\frac{3\times 0.677+2}{6\times 0.677+1} + \frac{6\times 3.867}{5\times 3\times 10^7\times 0.64}$$

$$\left[1 + \frac{0.64\times 3.867}{0.49\times 3.348}\frac{18\times(0.677)^2}{(6\times 0.677+1)^2}\right] + \frac{3.867^3}{3\times 10^7\times 0.64\times 3.348^2}\times\frac{18\times(0.677)^2}{(6\times 0.677+1)^2}$$

$= 37.4778 \times 10^{-7} + 3.59 \times 10^{-7} + 0.865 \times 10^{-7}$

$= 41.9328 \times 10^{-7}$ m

$$K_{xi} = \frac{1}{\delta_{hi}} = \frac{1}{41.9328\times 10^{-7}} = 238476.801 \text{ kN/m}$$

Table 13.5. The Values of Masses and Horizontal Stiffness

Frame no.	m_{mi} *(kN-s²/m)*	m_{hi} *(kN-s²/m)*	m_{ci} *(kN-s²/m)*	m_{gi} *(kN-s²/m)*	m_i *(kN-s²/m)*	K_{xi} *(kN/m)*
I	4.0775	5.0349	12.526	2.2018	15.4477	238476.801
II	8.1549	5.0349	12.526	5.0202	22.3436	238476.801
III	8.1549	5.0349	12.526	6.3413	23.6647	238476.801
IV	5.0968	5.0349	12.526	3.5229	17.7882	238476.801

$$d_{m1} = \frac{15.4477\times 0 + 22.3436\times 2.5 + 23.6647\times(2.5+3.2) + 17.7882\times(2.5+3.2+4)}{15.4477 + 22.3436 + 23.6647 + 17.7882}$$

$= 4.5848$ m

$d_{k1} = 4.475$ m

$e = d_{m1} - d_{k1} = 0.1098$ m

$d_{m2} = 2.5 - 4.5848 = -2.0848$ m

$d_{m3} = 2.5 + 3.2 - 4.5848 = 1.1152$ m

$d_{m4} = 2.5 + 3.2 + 4 - 4.5848 = 5.1152$ m

$d_{k2} = 2.5 - 4.475 = -1.975$ m

$d_{k3} = 2.5 + 3.2 - 4.475 = 1.225$ m

$d_{k4} = 2.5 + 3.2 + 4 - 4.475 = 5.225$ m

$m = \sum m_i = 15.4477 + 22.3436 + 23.6647 + 17.7882$

$= 79.2442$ kNs²/m

$M_{mz} = \sum m_i d_{mi}^2 = (15.4477 \times 4.5848^2) + (22.3436 \times 2.0848^2) + (23.6647 \times 1.1152^2) + (17.7882 \times 5.1152^2)$

$= 916.69$ kNs²/m

$K_x = \sum K_{xi} = 4 \times 238476.801 = 953907.204$ kN/m

$K_\psi = \sum K_{xi} d_{ki}^2$

$= (4.475^2 + 1.975^2 + 1.225^2 + 5.225^2) \times 238476.801$
$= 12574285.52$ kN/m

The value of natural frequency in horizontal direction $(\omega_{nx}) = \sqrt{\dfrac{K_x}{m}} = \sqrt{\dfrac{953907.204}{79.2442}}$
$= 109.7158$ rad/sec

The value of natural frequency in rocking direction $(\omega_n) = \sqrt{\dfrac{K_\psi}{M_{mz}}} = \sqrt{\dfrac{12574285.52}{916.69}}$
$= 117.1198$ rad/sec

$$r = \sqrt{\frac{M_{mz}}{m}} = \sqrt{\frac{916.69}{79.2442}} = 3.401$$

$$\alpha = 1 + \frac{e^2}{r^2} = 1 + \frac{(0.1098)^2}{(3.401)^2} = 1.001$$

Natural frequencies

$$\omega_n^4 - \left[1.001 \times 109.7158^2 + 117.1198^2\right]\omega_n^2 + 109.7158^2 \times 117.1198^2 = 0$$

Value of first natural frequency $(\omega_{n1}) = 109.343$ rad/sec
Value of second natural frequency $(\omega_{n2}) = 117.519$ rad/sec

$$\Delta(\omega^2) = \omega^4 - (\alpha\omega_{nx}^2 + \omega_{n\psi}^2)\omega^2 + \omega_{nx}^2\omega_{n\psi}^2$$

$= (471.238)^4 - (1.001 \times (109.7158)^2 + (117.1198)^2)(471.238)^2 + (109.7158)^2(117.1198)^2$
$= 4.375 \times 10^{10}$

$M_z = 11.3069 \times 4.5848 - 16.9603 \times 2.0848 + 16.9603 \times 1.1152 + 22.6138 \times 5.1152$
$= 151.0693$ kN.m

$F_x = 1350.682 \times 0.5 = 675.341$ kN

$$A_x = \frac{\left[\dfrac{e^2}{r^2}\omega_{nx}^2 + \omega_{n\psi}^2 - \omega^2\right]\dfrac{F_x}{m} - \omega_{nx}^2 \cdot \dfrac{M_z}{M_{mz}}}{\Delta(\omega^2)}$$

$$= \frac{\left[\left(\dfrac{0.1098}{3.401}\right)^2 \times (109.7158)^2 + (117.1198)^2 - (471.238)^2\right]\dfrac{675.341}{79.2442} - (109.7158)^2\dfrac{151.0693}{916.69}}{4.375 \times 10^{10}}$$

$= -4.0628 \times 10^{-5}$ m

$$A_\psi = \frac{\left[\dfrac{e^2}{r^2}\omega_{nx}^2\dfrac{F_x}{m} - (\omega_{n\psi}^2 - \omega^2)\dfrac{M_z}{M_{mz}}\right]}{\Delta(\omega^2)}$$

$$= \frac{\left[\left(\dfrac{0.1098}{3.401}\right)^2 \times (109.7158)^2 \times \dfrac{675.341}{79.2442} - \{(109.7158)^2 - (471.238)^2\}\dfrac{151.0693}{916.69}\right]}{4.375 \times 10^{10}}$$

$= 7.936 \times 10^{-6}$ rad

$A_h = A_x + yA_\psi$
$= 4.0628 \times 10^{-5} + 2.5 \times 7.936 \times 10^{-6}$
$= 6.0428 \times 10^{-5}$ m

Operating frequency of the machine is much higher than the compound natural frequencies. Amplitude, A_h is very small in comparison to permissible amplitudes.

Combined Method: Vertical vibration:

For 1st column:

$W_1 = 40$ kN

$W_2 = (24 \times 0.6 \times 0.6 \times 2.5)/2 = 10.8$ kN

$q = 0.7 \times 0.7 \times 24 = 11.76$ kN/m

$$\delta_1 = \frac{W_1 l^3}{96WI_b}\frac{2K+1}{K+2}$$

$$= \frac{40 \times (3.348)^3}{96 \times 3 \times 10^7 \times 0.02} \times \frac{2 \times 0.667 + 1}{0.677 + 2} = 229.167 \times 10^{-7} \text{ m}$$

$$\delta_2 = \frac{ql^4}{384EI_b}\frac{5K+2}{K+2}$$

$$= \frac{11.76 \times (3.348)^3}{384 \times 3 \times 10^7 \times 0.02} \times \frac{5 \times 0.677 + 1}{0.677 + 2} = 129.004 \times 10^{-7} \text{ m}$$

$$\delta_3 = \frac{3l}{5EA_b}\left(W_1 + \frac{ql}{2}\right)$$

$$= \frac{3 \times 3.348}{5 \times 3 \times 10^7 \times 0.49}\left(40 + \frac{11.76 \times 3.348}{2}\right) = 81.563 \times 10^{-7} \text{ m}$$

$$\delta_4 = \frac{h}{EA_c}\left(W_2 + \frac{W_1 + ql}{2}\right)$$

$$= \frac{3.867}{3 \times 10^7 \times 0.64}\left(10.8 + \frac{40 + 11.76 \times 3.348}{2}\right) = 101.682 \times 10^{-7} \text{ m}$$

$\delta_{st} = \delta_1 + \delta_2 + \delta_3 + \delta_4$

$= (229.167 + 129.004 + 81.563 + 101.682) \times 10^{-7} = 541.416 \times 10^{-7}$ m

$W = W_1 + 2W_2 + q \cdot l$

$= 40 + 2 \times 10.8 + 11.76 \times 3.348 = 100.972$ kN

$$f_{nz} = \frac{30}{\sqrt{\delta}}$$

$$= \frac{30}{\sqrt{541.416 \times 10^{-7}}} = 4077.6658 \text{ rpm} = 425.566 \text{ rad/s}$$

Amplitude calculations:

$\Delta = 0.4$ for concrete foundations (Assumed)

$f_m = 4500$ rpm

$M = 7.85$ (as $f_m > f_n$) for all frames

Computation of static displacements δ_v

Rotating weights on cross beams (R_b) are given by:

$(R_b)_{A1} = 10$ kN; $(R_b)_{A2} = 15$ kN; $(R_b)_{A3} = 15$ kN; $(R_b)_{A4} = 20$ kN

Rotating weights on column frames (R_c, coming from long frames and cross frames A_1, A_2, A_3, A_4 taken together) are given by

$$(R_c)_{A1} = (R_c)_{A2} = (R_c)_{A3} = (R_c)_{A4} = 0$$

Centrifugal forces on cross beams (C_b) and frames column (C_c) are given by

$$C_b = 0.2R_b\left(\frac{f_n}{f_m}\right)^2$$

$$C_c = 0.2R_c\left(\frac{f_n}{f_m}\right)^2$$

$$\delta_v = \frac{C}{E}\left[\frac{l^2}{96I_b}\cdot\frac{2k+1}{k+2}+\frac{3}{5}\cdot\frac{l}{A_b}+\frac{1}{2}\cdot\frac{h}{A_c}\right]$$

$$a_v = M.\delta_v$$

Table 13.6. Values of Natural Frequency and Amplitudes of Different Cross Frames

Frame Number	*W_l (kN)*	*W_2 (kN)*	*q (kN)*	*δ_1 (m) × 10^{-7}*	*δ_2 (m) × 10^{-7}*	*δ_3 (m) × 10^{-7}*	*δ_4 (m) × 10^{-7}*	*δ_{st} (m) × 10^{-7}*
I	40	10.8	11.76	229.167	128.95	81.563	101.682	541.416
II	80	24.62	11.76	458.15	128.95	136.224	169.81	891.34
III	80	31.10	11.76	458.15	128.95	136.224	182.86	906.19
IV	50	17.28	11.76	286.35	128.95	95.228	124.804	635.33

Frame Number	*Nat. Freq (rpm)*	*W(kN)*	*C_b*	*C_c*	*δ_v (m) × 10^{-7}*	*Vert. Ampl, a_v (m) × 10^{-7}*
I	4077.6658	100.9725	1.6416	0	13.289	104.318
II	3174.4016	168.6205	1.49	0	12.073	94.77305
III	3151.4595	181.5805	1.471	0	11.9195	93.568
IV	3763.7943	123.9325	2.797	0	22.664	177.9124

Horizontal vibrations:

Table 13.7. Values of W_i and K_{xi}

Frame no.	*W (kN), Table 13.2*	*K_{xi} (kN/m)*
I	100.9725	237.586×10^3
II	168.6205	237.586×10^3
III	181.5805	237.586×10^3
IV	123.9325	237.586×10^3

The distance of the resultant of the weight from the axis of first frame

$$X_g = \frac{168.6205 \times 2.5 + 181.5805 \times (2.5 + 3.2) + 123.9325 \times (2.5 + 3.2 + 4)}{575.10496}$$

$$= 4.623 \text{ m}$$

$$\delta_{hi} = \frac{h^2}{12EI_c}\frac{3k+2}{6k+2} + \frac{6h}{5EA_c}\left[1 + \frac{A_c h}{A_b l}\frac{18k^2}{(6k+1)^2}\right] + \frac{h^3}{EI_c l^2}\frac{18k^2}{(6k+1)^2}$$

$$= \frac{(3.867)^3}{12\times 3\times 10^7\times 0.03413}.\frac{3\times 0.677+2}{6\times 0.677+2} + \frac{6\times 3.867}{5\times 3\times 10^7\times 0.64}$$

$$\left[1 + \frac{0.64\times 3.867}{0.49\times 3.348}\frac{18\times(0.677)^2}{(6\times 0.677+1)^2}\right] + \frac{3.867^3}{3\times 10^7\times 0.64\times 3.348^2}\times\frac{18\times(0.677)^2}{(6\times 0.677+1)^2}$$

$= 37.4778 \times 10^{-7} + 3.59 \times 10^{-7} + 0.865 \times 10^{-7}$

$= 41.9328 \times 10^{-7}$ kN/m

$\delta_{h1} = \delta_{h2} = \delta_{h3} = \delta_{h4}$

Centre of elasticity of the frames from first frame:

Rigidity factors:

$k_{hA1} = 1/\delta_{h1} = 237.586 \times 10^3$ kN/m $= k_{hA2} = k_{hA3} = k_{hA4}$

$\sum k_h = 950.34 \times 10^3$ kN/m

$$X_h = \frac{237.586\times 10^3(0+2.5+5.7+9.7)}{950.34\times 10^3}$$

$= 4.475$ m

Eccentricity $(e) = X_g - X_h = -0.1480$ m

$I_G = 100.972 \times (4.623)^2 + 168.62048 \times (4.623 - 2.5)^2 + 181.58048 \times (4.623 - 5.7)^2 + 123.932 \times (4.623 - 9.7)^2 = 6321.44$ kN/m^2

$I_H = 238.46 \times 10^3 \times (4.491)^2 + 238.46 \times 10^3 \times (4.491 - 2.5)^2 + 238.46 \times 10^3 \times (4.491 - 5.7)^2 + 238.46 \times 10^3 \times (4.491 - 9.7)^2 = 12575349.5983$ kN/m^2

Now, $\alpha_a = \frac{1}{2}\left[e^2\frac{\sum k_{hi}}{I_G} + \frac{\sum k_{hi}}{\sum W_i} + \frac{I_H}{I_G}\right] = 1825.4536$

The horizontal natural frequency is given by

$$f_{nh} = 30\left[\alpha_o \pm \left(\alpha_o^2 - \frac{\sum k_{hi}}{\sum W_i}.\frac{I_H}{I_G}\right)^{1/2}\right]^{1/2} = 1344.2568 \text{ rpm}$$

Amplitude:

$$R = \sum (R_b + R_c)$$

$$C_i = \alpha R\left(\frac{f_n}{f_m}\right)^2$$

$$C = \sum C_i$$

$\sum C_h = 0.2 \times 60 \times (1222.724/4500)^2$

$= 0.866$ kN

$$\text{Lateral deflection} = \frac{\sum C_h}{\sum K_h} = \frac{0.866}{950.34\times 10^3} = 9.29\times 10^{-7} \text{ m}$$

Horizontal amplitude $(a_h) = a_{hi} = M\,\delta_{hi} = 7.85 \times 9.29 \times 10^{-7}$

$= 7.292 \times 10^{-6}$ m [<< 0.5 mm, safe]

References

Barkan, D. D. (1962), "Dynamics of bases and foundations," McGraw-Hill Book Co. Inc., New York.
IS 2974 (Pt. III-2015), "Foundations for rotary-type machines (Medium and high frequency)"
Major., A. (1962), Vibration analysis and design of foundation for machines and turbines, Akademial Kiado, Budapest, Collet's Holdings Limited, London.
Rausch., E. (1959), "Machinen fundamente und andere dynamisch beanspruchte Baukonstructionen," VDI Verlag, Dusseldorf.

Practice Problems

13.1 What are the special considerations for the better performance of a T.G. foundation? List the loads acting on a T.G. foundation with their brief description.

13.2 Give the salient features of designing a T.G. foundation by 'Resonance Method'.

13.3 Give the salient features of designing a T.G. foundation by 'Amplitude Method'.

13.4 Give the salient features of designing a T.G. foundation by 'Combined Method'.

13.5 Design a turbo-generator frame foundation shown in Fig. 13.12 with the following data:
Dimensions of columns: $C_1 = 0.9$ m × 0.8 m; $C_2 = 0.75$ m × 0.8 m; $C_3 = 0.8$ m × 0.8 m; $C_4 = 1.0$ m × 0.8 m
All cross beams have dimensions = 0.7 m × 0.7 m
All longitudinal beams have dimensions = 0.6 m × 0.6 m
Weight on cross beam of each frame:

$$W_{11} = 40 \text{ kN};\ W_{12} = 100 \text{ kN};\ W_{13} = 100 \text{ kN};\ W_{14} = 50 \text{ kN}$$

Weight on longitudinal beam of each frame:

$$W_{21} = 60 \text{ kN};\ W_{22} = 100 \text{ kN};\ W_{23} = 100 \text{ kN};\ W_{24} = 80 \text{ kN}$$

Weight of rotating parts acting on each frame:

$$W_{r1} = 10 \text{ kN};\ W_{r2} = 15 \text{ kN};\ W_{r3} = 15 \text{ kN};\ W_{r4} = 25 \text{ kN}$$

Eccentricity = 0.07×10^{-3} m
Young's Modulus for concrete = 3×10^7 kN/m^2
Damping ratio = 0.02
Operating speed = 3000 rpm

CHAPTER

14

Vibration Isolation and Screening

14.1 General

In machine foundations, following two types of the problem may arise:

(i) Machines directly mounted on foundation block (Fig. 14.1a) may cause objectionable vibrations.

(ii) Machine foundation suffers excessive amplitudes due to the vibrations transmitted from the neighbouring machines (Fig. 14.1b).

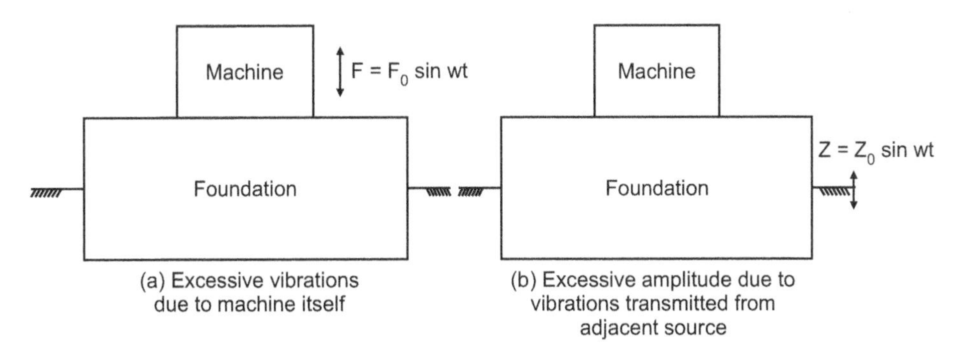

Fig. 14.1. Machine directly mounted on foundation

The first problem may be tackled by isolating the machine from the foundation through a suitably designed mounting system (Fig. 14.2) such that the transmitted force is reduced which in turn will reduce the amplitude. This type of isolation is termed as **force isolation**. This type of arrangement will also help in absorbing the vibrations transmitted from adjacent machines. The system used for this purpose is termed as **motion isolation**. For heavier machines, the isolating system may be placed between the foundation block and concrete slab as shown in Fig. 14.3. Here

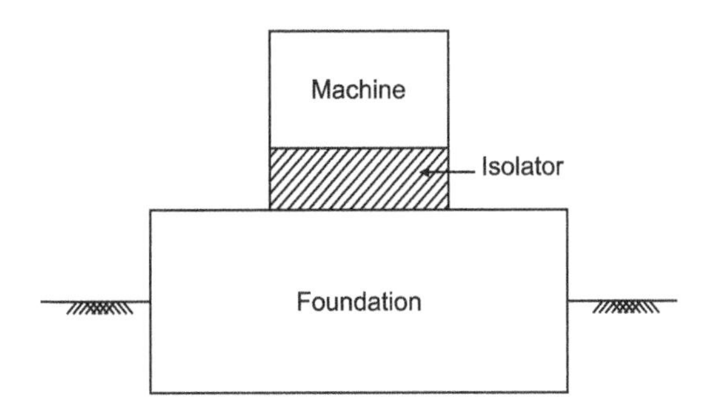

Fig. 14.2. An isolator placed between machine and foundation

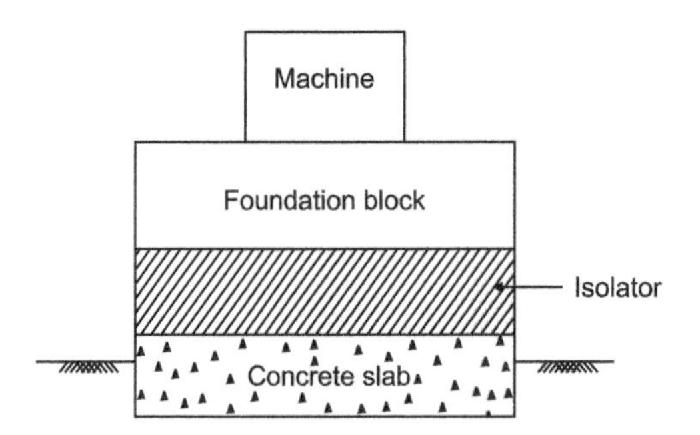

Fig. 14.3. An isolator placed between foundation block and concrete slab

the machine is rigidly bolted to the foundation block which is isolated from the concrete slab through the mounting system. The mounting system is an elastic layer which may be in form of rubber pad, timber pad, cork pad or metal springs. These have been already discussed in Sec. 2.5.

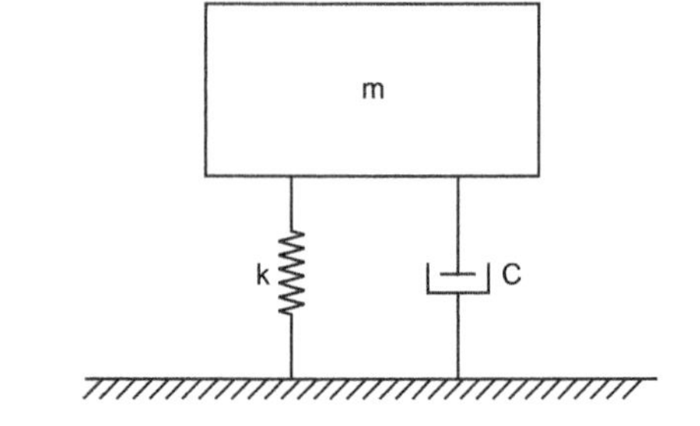

Fig. 14.4. Mathematical model

The systems shown in Figs. 14.2 and 14.3 can be represented by a simple mathematical model shown in Fig. 14.4. In this *m* represents the mass of machine (Fig. 14.2) or mass of machine plus foundation block (Fig. 14.3). The mounting system (*i.e.* the plastic layer) is characterised by a linear spring with a spring constant K and dashpot with damping constant C. This mathematical representation involves one basic assumption that the underlying soil or rock possesses infinite rigidity. This system is identical to the one shown in Fig. 2.17 (or Fig. 2.19), and the detailed analysis has already been presented in Sec. 2.5 considering both **force isolation** and **motion isolation** separately.

A more realistic model will be that in which the soil or rock is considered as an elastic medium. This will make the system as a two-degree-freedom problem, the solutions of which are presented in the next section.

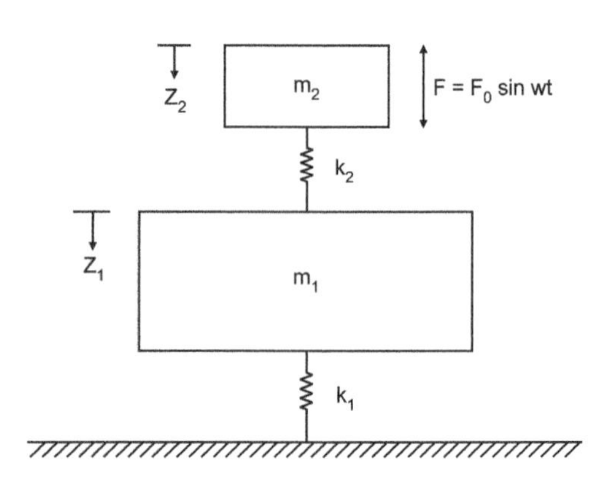

Fig. 14.5. Two degrees of freedom model

The second problem in which the vibrations are transmitted from the neighbouring machines can be solved by controlling the vibrating energy reaching the desired location. This is referred to as **vibration screening.** Effective screening of vibration may be achieved by proper interception, scattering and diffraction of surface waves by using barriers such as trenches, sheet-pile walls, and piles. If the screening devices are provided near the sources of vibration, then it is termed as **active screening** or **active isolation.** In case screening devices are used by providing barriers at a point remote from the source of disturbance but near a site where vibration has to be reduced, it is termed as **passive screening** or **passive isolation.** Both the methods of screening the vibrations have been discussed subsequently.

14.2 Force Isolation

Since the underlying soil or rock supporting the foundation block (or base slab) does not possess infinite rigidity, the foundation soil should be represented by a spring and not solely by a rigid support as was done in Sec. 10.2. Then the mathematical model becomes as shown in Fig. 14.5. The various terms used are explained below:

m_1 = Mass of foundation block or mass of base slab

m_2 = Mass of machine if isolator is introduced between machine and foundation block or mass of machine plus mass of foundation block if isolator is placed between foundation block and base slab.

K_1 = Stiffness of the soil

K_2 = Stiffness of the isolator

If the machine is subjected to a harmonic force (F_O sin ωt), the equation of motion will be:

$$m_1 \ddot{Z}_1 + K_1 Z_1 - K_2(Z_2 - Z_1) = 0 \tag{14.1}$$

$$m_2 \ddot{Z}_2 + K_2(Z_2 - Z_1) = F_O \sin \omega t \tag{14.2}$$

The solution of Eqs. (14.1) and (14.2) gives

$$Z_1 = \frac{\left(\dfrac{K_2}{m_1 m_2}\right) \sin \omega t}{\omega^4 - \left[\dfrac{K_2}{m_2} + \dfrac{(K_1 + K_2)}{m_1}\right]\omega^2 + \dfrac{K_1 K_2}{m_1 m_2}} \cdot F_o \tag{14.3}$$

$$Z_2 = \frac{\left[\dfrac{(K_1 + K_2)}{m_1 m_2} - \omega^2 m_2\right] \sin \omega t}{\omega^4 - \left[\dfrac{K_2}{m_2} + \dfrac{(K_1 + K_2)}{m_1}\right]\omega^2 + \dfrac{K_1 K_2}{m_1 m_2}} \cdot F_o \tag{14.4}$$

The principal natural frequencies of the system shown in Fig. 14.5 can be obtained by solving the following frequency equation:

$$\omega_n^4 - \left[\frac{K_2}{m_2} + \frac{(K_1 + K_2)}{m_1}\right]\omega_n^2 + \frac{K_1 K_2}{m_1 m_2} = 0 \tag{14.5}$$

If ω_{n1} and ω_{n2} represent the roots of the above equation, Eq. (14.3) can be written as

$$Z_1 = \frac{\left(\dfrac{K_2}{m_1 m_2}\right) \sin \omega t}{\Delta(\omega^2)} \cdot F_o \tag{14.6}$$

where
$$\Delta(\omega^2) = \omega^4 - \left[\frac{K_2}{m_2} + \frac{(K_1 + K_2)}{m_1}\right]\omega^2 + \frac{K_1 K_2}{m_1 m_2} \tag{14.7a}$$

or
$$\Delta(\omega^2) = (\omega^2 - \omega_{n1}^2)(\omega^2 - \omega_{n2}^2) \tag{14.7b}$$

Force transferred to the foundation block or base slab

$$F_t = K_1 Z_1$$

$$= \frac{\left(\dfrac{K_1 K_2}{m_1 m_2}\right) \sin \omega t}{\Delta(\omega^2)} \cdot F_o \tag{14.8}$$

The transmissibility of the system will be

$$T_F = \frac{F_t}{F_o \sin \omega t} \tag{14.9}$$

or,
$$T_F = \frac{\dfrac{K_1 K_2}{m_1 m_2}}{\Delta(\omega^2)} \tag{14.9a}$$

The transmissibility T_F depends on the system parameters given by Eq. (14.9). For illustration, a special case will be examined in which

$$\frac{K_1}{m_1} = \frac{K_2}{m_2} = p^2 \tag{14.10}$$

Denotng,
$$\frac{m_2}{m_1} = \text{mass ratio} = \mu_m \tag{14.11}$$

The natural frequency of machine foundation system (ω_{nz}) in which the isolating spring (k_2) is ignored, is given by

$$\omega_{nz}^2 = \frac{K_1}{m_1 + m_2} = \frac{\dfrac{K_1}{m_1}}{1 + \dfrac{m_2}{m_1}} = \frac{p^2}{1 + \mu_m} \tag{14.12}$$

Using Eqs. (14.10), (14.11) and (14.12), Eq. (14.9) can be written as

$$T_F = \frac{1}{\left[\dfrac{\omega^2}{p^2} - \left\{\left(1 + \dfrac{\mu_m}{2}\right) + \left(\mu_m + \dfrac{\mu_m^2}{4}\right)^{1/2}\right\}\right]\left[\dfrac{\omega^2}{p^2} - \left\{\left(1 + \dfrac{\mu_m}{2}\right) + \left(\mu_m + \dfrac{\mu_m^2}{4}\right)^{1/2}\right\}\right]} \tag{14.13}$$

or,
$$T_F = \frac{1}{\left[\dfrac{\omega^2}{(1 + \mu_m)\omega_{nz}^2} - \left\{\left(1 + \dfrac{\mu_m}{2}\right) + \left(\mu_m + \dfrac{\mu_m^2}{4}\right)^{1/2}\right\}\right]\left[\dfrac{\omega^2}{(1 + \mu_m)\omega_{nz}^2} - \left\{\left(1 + \dfrac{\mu_m}{2}\right) + \left(\mu_m + \dfrac{\mu_m^2}{4}\right)^{1/2}\right\}\right]}$$

Figure 14.6 shows the plots of T_F with $\left(\dfrac{\omega}{\omega_{nz}}\right)$ ratio for two values of mass ratio μ_m. It is evident from this figure that the frequency ratio (ω/ω_{nz}) must exceed a particular value (depending on the magnitude of μ_m) before the transmissibility falls below unity. The particular values of (ω/ω_{nz}) for which the transmissibility equals unity are given in Fig. 14.7.

A designer is more concerned in examining whether the isolating system reduces the amplitudes of the machine and foundation.

In case no isolating system is used, the maximum amplitude of machine foundation is given by

$$A_z = \frac{F_o}{(m_1 + m_2)(\omega_{nz}^2 - \omega^2)} \tag{14.14}$$

In most of the machines, the force F_o is frequency dependent ($2\, m_e\, e\, \omega^2$). Equation (14.14) can be written as

$$A_z = \frac{2 m_e e \omega^2}{(m_1 + m_2)(\omega_{nz}^2 - \omega^2)} \tag{14.15a}$$

or
$$A_z = \frac{2 m_e e}{m_1(1 + \mu_m)(a_1^2 - 1)} \tag{14.15b}$$

where
$$a_1 = \frac{\omega_{nz}}{\omega} \tag{14.16}$$

If ω_{nz} represents the natural frequency of mass m_2 resting on isolating spring, then

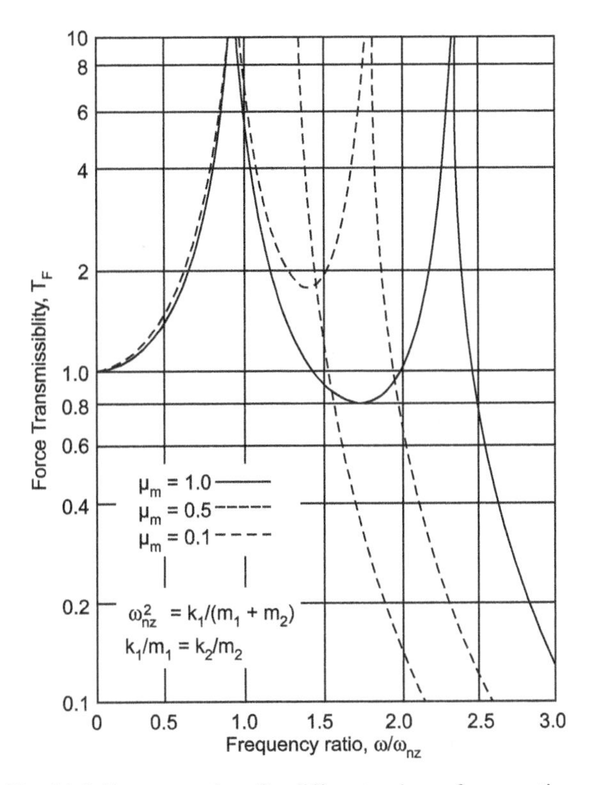

Fig. 14.6. T_F versus ω/ω_{nz} for different values of mass ratio μ_m

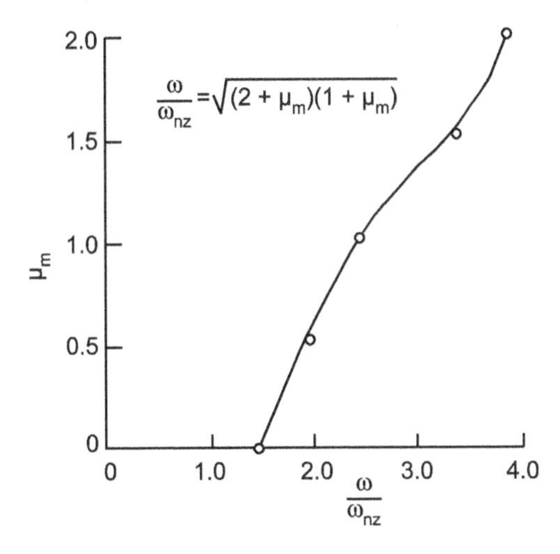

Fig. 14.7. Mass ratio versus ω/ω_{nz}

$$\omega_{na} = \sqrt{\frac{K_2}{m_2}} \tag{14.17}$$

The values of maximum amplitudes of two-degree-freedom system obtained from Eqs. (14.3) and (14.4) for frequency dependent force will be

$$A_{z1} = \frac{\omega_{na}^2\,(2m_e e\,\omega^2)}{m_1\left[\omega^4 - (1+\mu_m)(\omega_{na}^2 - \omega_{nz}^2)\omega^2 + (1+\mu_m)\omega_{na}^2\,\omega_{nz}^2\right]} \tag{14.18a}$$

or
$$A_{z1} = \frac{(2m_e e)\cdot a_2^2}{m_1\left[1-(1+\mu_m)(a_1^2+a_2^2-a_1^2 a_2^2)\right]} \tag{14.18b}$$

where
$$a_2 = \frac{\omega_{na}}{\omega} \tag{14.19}$$

Similarly,
$$A_{z2} = \frac{\left[(1+\mu_m)a_1^2+\mu_m a_2^2-1\right](2m_e e)}{m_2\left[1-(1+\mu_m)(a_1^2+a_2^2-a_1^2 a_2^2)\right]} \tag{14.20}$$

Equation (14.18b) indicates that the amplitude of vibration of foundation will be small if a_2 is small. For this ω_{na} should be small. This can be achieved by appropriate selection of the mass above the isolator spring (m_2) and stiffness of isolator spring (K_2). The efficiency of isolation system is defined as

$$\eta = \frac{A_{z1}}{A_z} \tag{14.21}$$

Values of A_z can be computed from Eq. (14.15a). The value of A_{z1} may be taken equal to the permissible amplitude. The value of η can be also expressed as given below by dividing Eq. (14.18b) by Eq. (14.15b). Thus

$$\eta = \frac{a_2^2(1+\mu_m)(a_1^2-1)}{\left[1-(1+\mu_m)(a_1^2+a_2^2-a_1^2 a_2^2)\right]} \tag{14.22}$$

Solving Eqs. (14.21) and (14.22) one can obtain the value of a_2. For this a_2, appropriate values of m_2 and K_2 are selected. It will ensure the amplitude of vibration to be within permissible limit.

Total force on the isolator, $F_a = K_2 \cdot A_{z2}$ (14.23)

F_a should be less than the allowable capacity of the isolator in compression.

14.3 Motion Isolation

Let us examine a case when a sensitive equipment of mass m_1 is placed on a foundation block of mass m_2. The spring K_1 represents the foundation soil and spring K_2 is an isolating spring which is placed between the masses m_1 and m_2, in order to minimise the transmission of vibrations from the ground to the equipment. If the ground is subjected to a periodic displacement given by $Z_o \sin \omega t$, the equations of motion will be:

$$m_1 \ddot{Z}_1 + K_1 Z_1 - K_2(Z_2 - Z_1) = K_1 Z_O \sin \omega t \tag{14.24}$$

$$m_2 \ddot{Z}_2 + K_2(Z_2 - Z_1) = 0 \tag{14.25}$$

The values of maximum amplitudes of motion are given by

$$A_{z1} = K_1 Z_o \frac{\dfrac{K_2}{m_1 m_2} - \dfrac{\omega^2}{m_1}}{\omega^4 - \left[\dfrac{(K_1+K_2)}{m_1} + \dfrac{K_2}{m_2}\right]\omega^2 + \dfrac{K_1 K_2}{m_1 m_2}} \tag{14.26}$$

$$A_{z2} = K_1 Z_o \frac{\dfrac{K_2}{m_1 m_2}}{\omega^4 - \left[\dfrac{(K_1+K_2)}{m_1} + \dfrac{K_2}{m_2}\right]\omega^2 + \dfrac{K_1 K_2}{m_1 m_2}} \tag{14.27}$$

The displacement transmissibility of the machine (T_D) is defined as the ratio of displacement amplitude of mass m_2 to the displcement amplitude of the rigid support. Then

$$T_D = \frac{A_{z2}}{Z_o} = \frac{\dfrac{K_1 K_2}{m_1 m_2}}{\omega^4 - \left[\dfrac{(K_1 + K_2)}{m_1} + \dfrac{K_2}{m_2}\right]\omega^2 + \dfrac{K_1 K_2}{m_1 m_2}} \tag{14.28}$$

Equation (14.28) is identical to Eq. (14.9), and therefore the results shown in Figs. 14.6 and 14.7 hold good in this case also.

Equation (14.28) can also be written as:

$$T_D = \frac{A_{z2}}{Z_o} = \frac{a_1^2 a_2^2 (1+\mu)}{1-(1+\mu_m)(a_1^2 + a_2^2 - a_1^2 a_2^2)} \tag{14.29}$$

If A_{z2} is taken as permissible amplitude, and Z_o is applied dynamic displacement, then the ratio A_{z2}/Z_o is known. For this value, a_2 can be determined which in turn will give the stiffness of the isolator spring i.e. K_2.

14.4 Screening of Vibrations by Use of Open Trenches

14.4.1 Active Screening

In this case the screening of vibrations is done near the source of vibration. Figure 14.8 shows a circular trench of radius R and depth H which surrounds the machine foundation that is the source of disturbance. The design of trench barriers is based on some field observations. Barkan (1962) mentioned that the reduction in vibration amplitudes occurs only when the trench dimensions are sufficiently large compared with the wave length of the surface waves generated by the source of disturbance. Dolling (1966) studied the effect of size and shape of the trench on its ability to screen the vibrations.

The first comprehensive study of screening vibrations by use of open trenches was made by Woods and Richart (1967) and Woods (1968). They conducted field tests by creating vertical vibrations with a small vibrator resting on a small pad at a prepared site. The vibrator could create a maximum force of 80 N. The soil condition at the site were as shown in Fig. 14.9. The water table was below 14.3 m depth. The depth H of trenches was varied from 150 mm to 600 mm, the radius of annular trench varied from 150 mm to 300 mm, and the angular dimension θ was varied from 90° to 360° around the source of vibration. Frequencies of 200 to 350 Hz were used in the tests. Using

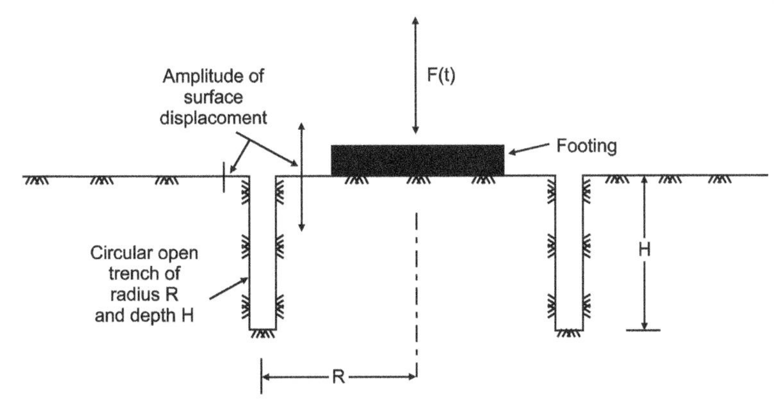

Fig. 14.8. Vibration screening using a circular trench surrounding the source of vibration – Active screning (Woods, 1968)

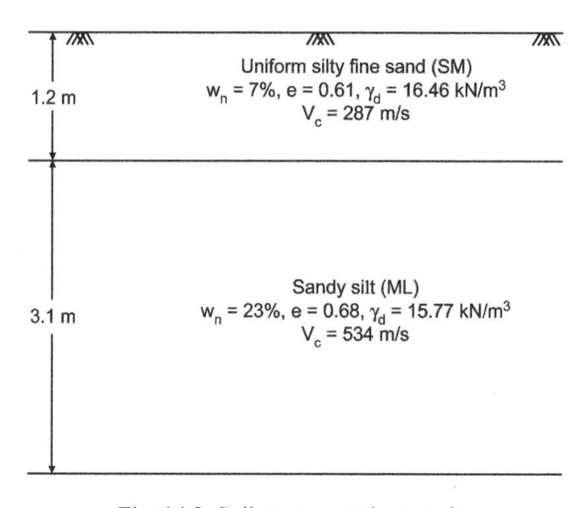

Fig. 14.9. Soil stratum at the test site

velocity transducers, the amplitudes of vertical ground motion were measured at selected points throughout the test site before installation of the trench and after installation of the trench. Woods (1968) has introduced a term **amplitude reduction factor** which is defined as

$$\text{ARF} = \text{Amplitude reduction factor}$$

$$= \frac{\text{Amplitude of vertical vibration with trench}}{\text{Amplitude of vertical vibration without trench}}$$

Some of the results of field tests conducted by Woods (1968) are shown in Fig. 14.10 in the form of ARF contour diagrams. The dimensions of the trench are expressed in non-dimensional forms by dividing H and R by the wave length λ_R of Rayleigh waves. λ_R is obtained by determining the number of waves (n) occuring at distance x from the source ($\lambda_R = x/n$). Wave lengths λ_R for different frequencies are given in Table 14.1.

Table 14.1. Wavelength and Wave Velocity for the Rayleigh Wave at the Site (Woods. 1968)

Frequency *Hz*	λ_R *mm*	V_R *m/s*
200	687	137
250	513	128
300	421	126
350	336	117

The field tests of Woods (1968) thus correspond to

$$\frac{\lambda}{\lambda_R} = 0.222 - 0.910 \text{ and } \frac{H}{\lambda_R} = 0.222 - 1.82$$

For satisfactory screening of vibrations, Woods (1968) recommended that ARF should be less than or equal to 0.25. The conclusions made on the basis of this study to keep ARF ≤ 0.25 are:

(i) For full circle trenchs ($\theta = 360°$), a minimum value of $H/\lambda_R = 0.6$ is required. The zone

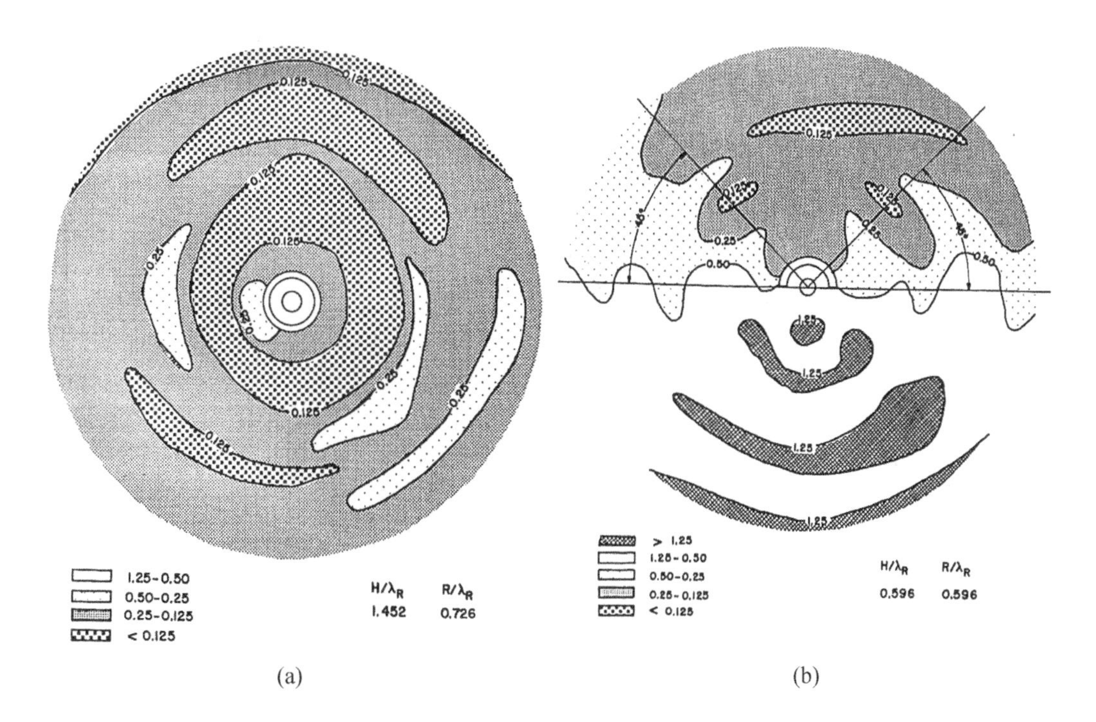

Fig. 14.10. Amplitude reduction factor contour diagrams for active screening (Woods, 1968)

screened in this case extends to a distance of atleast 10 wavelengths (10 λ_R) from the source of disturbance.

(ii) For partial circle trenches (90° < θ < 360°), the screened zone was defined as an area outside the trench extending to at least 10 wavelengths (10 λ_R) from the source and bounded on the sides by radial lines from the centre of source through points 45° from ends of trench. In this case also, a minimum value of $H/\lambda_R = 0.6$ is required.

(iii) Partial circle trenches with θ < 90°, effective screening of vibration is not achieved.

(iv) Trench width is not an important parameter.

14.4.2 Passive Screening

Woods (1968) has also performed field tests to study the effectiveness of open trenches in passive screening (Fig. 14.11). A typical layout of these tests consisting of two vibration exciters (used one at a time for the tests), 75 transducer locations, and a trench is shown in Fig. 14.12. The sizes of trenches ranged from 100 mm × 300 mm × 300 mm deep to 2440 mm × 3050 mm × 1220 mm deep. Frequencies of excitation varied from 200 to 350 Hz.

The values of H/λ_R varied from 0.444 to 3.64 and R/λ_R from 2.22 to 9.10. It was assumed in these tests that the zones screened by the trench would be symmetrical about the 0° line. Figure 14.13 shows the ARF contour diagram for one of these tests.

For satisfactory screening, Woods (1968) recommended that the ARF should be less than or equal to 0.25 in a semi-circular zone of radius (1/2) L behind the trench. The conclusion made on the basis of this field study to keep ARF ≤ 0.25 are:

(i) H/λ_R should be atleast 1.33.

(ii) To maintain the same degree of screening, the least area of the trench in the vertical direction (*i.e.* $LH = A_T$), should be as follows:

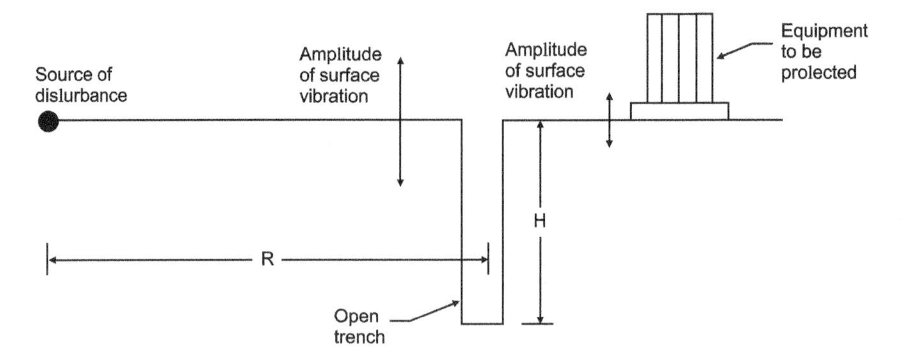

Fig. 14.11. Vibration screening using a straight trench – Passive screening (Woods, 1968)

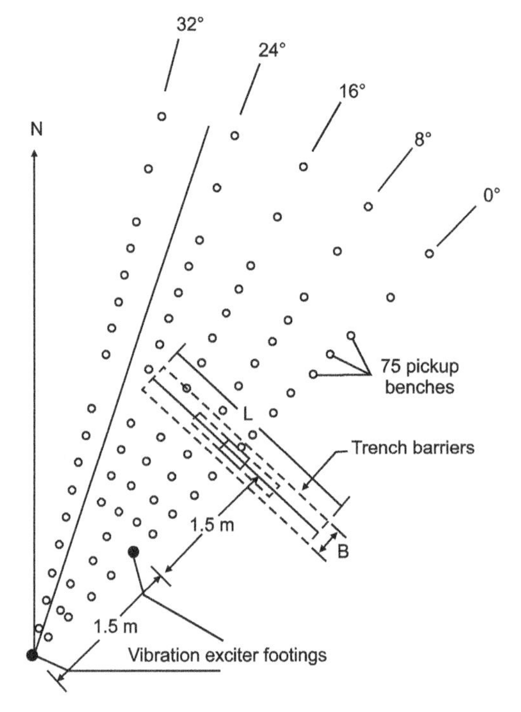

Fig. 14.12. Plan view of the field site layout for passive screening (Woods, 1968)

$$\frac{A_T}{\lambda_R^2} = 2.5 \text{ at } \frac{R}{\lambda_R} = 2.0 \tag{14.30a}$$

$$\frac{A_T}{\lambda_R^2} = 6.0 \text{ at } \frac{R}{\lambda_R} = 7.0 \tag{14.30b}$$

(iii) Trench width had practically no influence on the effectiveness of screening.

Experimental investigations of Sridharan et al. (1981) indicated that the open unfilled trenches are the most effective. However the open (unfilled) trenches may present instability problems necessitating trenches backfilled with sawdust, sand or bentonite slurry. The performance of open trench with sawdust was found better as compared with sand or bentonite slurry.

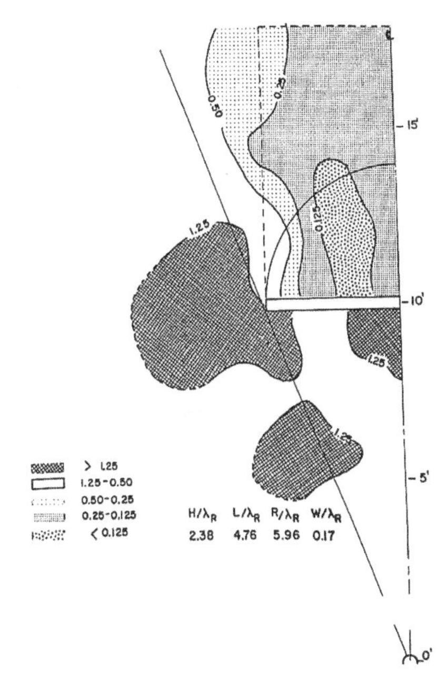

Fig. 14.13. Amplitude reduction factor contour diagrams for passive screening

14.5 Passive Screening by Use of Pile Barriers

There may be situations in which Rayleigh wavelengths may be in the range of 40 to 50 m. For such a case, the open trench will be effective if its depth range from 53 m to 66 m (i.e. 1.33 λ_R). Open trenches (filled or unfilled) with such deep depths are not practical. For this reason, possible use of rows of piles as an energy barriers has been studied by Woods et al. (1974) and Liao and Sangery (1978).

Woods et al. (1974) used the principle of holography and observed vibrations in a model half-space to evaluate the effect of void cylindrical obstacles on reduction of vibration amplitudes. A box of size 1000 mm × 1000 mm × 300 mm deep filled with fine sand constituted the model half-space (Fig. 14.14). In this figure, D is the diameter of the void cylindrical obstacle and S_n is the net space between two consecutive void holes through which energy can pass through the barrier. The numerical evaluation of barrier effectiveness was made by obtaining the average of the values of

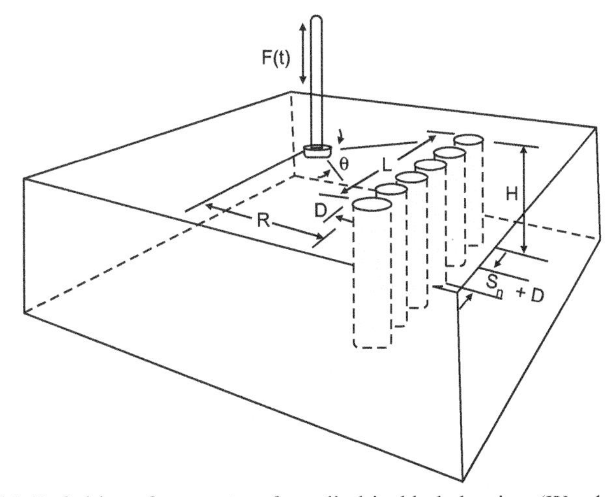

Fig. 14.14. Definition of parameters for cylindrical hole barriers (Woods et al., 1974)

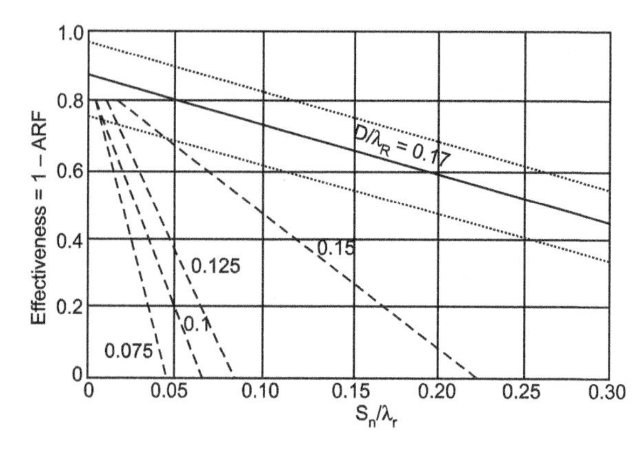

Fig. 14.15. Isolation effectiveness as a function of hole diameter and spacing (Woods et al., 1974)

ARF obtained on several lines beyond the barrier in a section ± 5° of both sides of an axis through the source of disturbance and perpendicular to the barrier. In all tests, H/λ_R and L/λ_R were kept as 1.4 and 2.5 respectively. The isolation effectiveness is defined as

$$\text{Effectiveness} = 1 - \text{ARF} \tag{14.31}$$

Using the data of different tests, a non-dimensional plot of the isolation effectiveness versus S/λ_R ratio for different values of D/λ_R was plotted as shown in Fig. 14.15. Woods et al. (1974) recommended that a row of void cylindrical holes may act as an isolation barrier if

$$\frac{D}{\lambda_R} \geq \frac{1}{6} \tag{14.32}$$

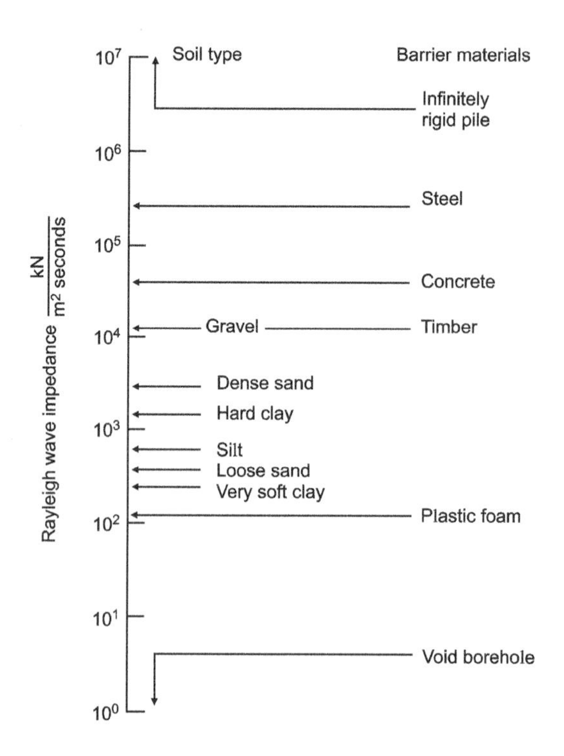

Fig. 14.16. Estimated values of Rayleigh wave impedance for various soils and pile materials (Liao and Sangrey, 1978)

and
$$\frac{S_n}{\lambda_R} < \frac{1}{4} \tag{14.33}$$

Liao and Sangrey (1978) used an acoustic model employing sound waves in a fluid medium to evaluate the possibility of the use of row of piles as passive isolation barriers. They have studied the effect of diameter, spacing and material properties of the soil pile system on the isolation effectiveness. They concluded that:

(i) Equations (14.32) and (14.33) proposed by Woods et al. (1974) are generally valid.

(ii) $S_n = 0.4\ \lambda_R$ may be the upper limit for a barrier to have some effectiveness.

(iii) The effectiveness of the barrier is significantly affected by the material of the pile and void holes. Acoustically soft piles ($IR < 1$) are more efficient than acoustically hard piles ($IR > 1$). IR is impedance ratio which is defined as

$$IR = \frac{\rho_P V_{RP}}{\rho_S V_{RP}} \tag{14.34}$$

where ρ_p = Density of pile material

ρ_s = Density of soil medium

V_{RP} = Rayleigh wave velocity in pile material

V_{RS} = Rayleigh wave velocity in soil medium

Figure 14.16 gives a general range of the Rayleigh wave impedance (ρV_R) for various soils and pile materials.

(iv) Two rows of barriers are more effective than single row barriers.

Illustrative Examples

Example 14.1

Determine the stiffness of the isolator to be kept between a reciprocating machine and the foundation shown in Fig. 14.17 to bring the vibration amplitude to less than 0.02 mm. The weight of the machine

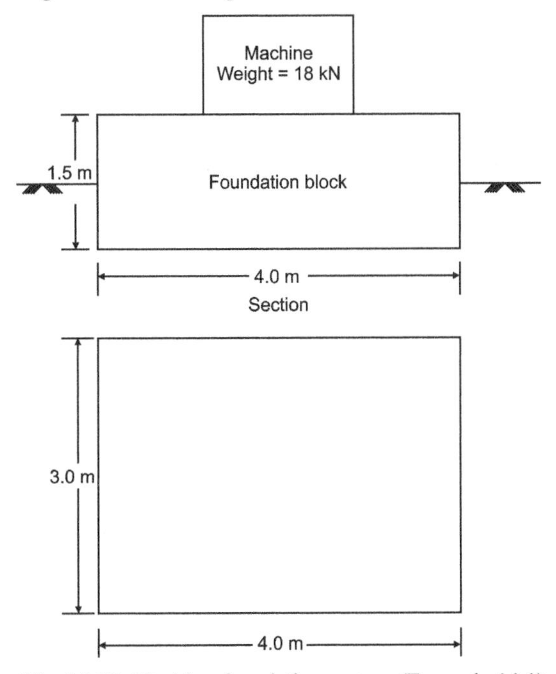

Fig. 14.17. Machine-foundation system (Example 14.1)

is 18 kN and it produces an unbalanced force of 4.0 kN when operated at a speed of 550 rpm. The dynamic shear modulus and Poisson's ratio of the soil are 2×10^4 kN/m^2 and 0.35 respectively.

Solution

1. Mass of the foundation block

$$= \frac{24 \times 4.0 \times 3.0 \times 1.5}{9.81} = 44 \text{ kNs}^2/\text{m}$$

$$\text{Mass of machine} = \frac{18}{9.81} = 1.83 \text{ kNs}^2/\text{m}$$

Total mass of machine and foundation

$= 44 + 1.83 = 45.83$ kNs^2/m

Stiffness of the soil,

$$K_z = \frac{4\,G r_o}{1-\mu}$$

$$r_o = \sqrt{\frac{A}{\pi}} = \sqrt{\frac{4 \times 3}{\pi}} = 1.95\,\text{m}$$

$$K_z = \frac{4 \times 2 \times 10^4 \times 1.95}{1-0.35} = 24 \times 10^4 \text{ kN/m}$$

Natural frequency of the whole system without any isolation will be:

$$\omega_{nz} = \sqrt{\frac{K_z}{m}} = \sqrt{\frac{24 \times 10^4}{45.83}} = 72.36\,\text{rad/s}$$

$$\omega = \frac{2\pi \times 550}{60} = 57.6\,\text{rad/s}$$

$$(A_z)_{\max} = \frac{F_0}{m(\omega_{nz}^2 - \omega^2)} = \frac{4.0}{45.83(72.36^2 - 57.6^2)}$$

$$= 4.55 \times 10^{-5} \text{ m} = 0.0455 \text{ mm}$$

The amplitude is greater than the permissible amplitude *i.e.* 0.02 mm. Hence isolator is required between machine and foundation.

2. Let the isolator be having stiffness K^2. Adopting the two-degrees-freedom system as shown in Fig. 14.5.

$m_2 = 1.83$ kNs^2/m

$m_1 = 44$ kNs^2/m

$K_1 = K_z = 24 \times 10^4$ kN/m

$\omega_{nz} = 65.0$ rad/s

$$\eta = \frac{A_{Z1}}{A_Z} = \frac{0.02}{0.0455} = 0.4395$$

$$a_1 = \frac{\omega_{nz}}{\omega} = \frac{72.36}{57.6} = 1.256$$

$$\mu_m = \frac{m_2}{m_1} = \frac{1.83}{44} = 0.0416$$

From Eq. (14.22)

$$\eta = \frac{a_2^2(1+\mu_m)(a_1^2-1)}{\left[1-(1+\mu_m)(a_1^2+a_2^2-a_1^2a_2^2\right]}$$

or
$$0.4395 = \frac{a_2^2(1+0.0416)(1.256^2-1)}{\left[1-(1+0.0416)(1.256^2+a_2^2-1.256^2\,a_2^2)\right]}$$

or
$$a_2^2 = 0.2310 \quad i.e. a_2 = 0.481$$

$$\frac{\omega_{na}}{\omega} = 0.481. \text{ It gives } \omega_{na} = 0.481 \times 57.6 = 27.7 \text{ rad/s}$$

$$K_a = m_2\,\omega_{nz}^2 = 1.83 \times 27.7^2 = 1.40 \times 10^3 \text{ kN/m}$$

3.
$$A_{z2} = \frac{\left[(1+\mu_m)a_1^2+\mu_m\,a_2^2-1\right]\left[2m_e e\omega^2\right]}{m_2\,\omega^2\left[1-(1+\mu_m)(a_1^2+a_2^2-a_1^2a_2^2\right]}$$

$$= \frac{\left[(1+0.0416)\times 1.13^2+0.0416\times 0.2310-1\right]\times(4.0)}{1.83\times 57.6^2\left[1-(1+0.0416)\,(1.13^2+0.2310-1.13^2\times 0.2310)\right]}$$

$$= -8.5 \times 10^{-4}\text{m}$$

Force in the isolator $= K_a \cdot A_{z2} = 1.40 \times 10^3 \times 8.5 \times 10^{-4}$ kN

$= 1.18$ kN

A suitable isolation system may be selected which has total stiffness of 1.40×10^3 kN/m and allowable compressive load more than 1.18 kN.

Example 14.2

Determine the stiffness of the isolation system if it is placed between the foundation block and base slab as shown in Fig. 14.3. Use the data given in example 14.1.

Solution

(i) Let the foundation block of size 4.0 m × 3.0 m × 1.0 m high be rigidly connected with machine. Isolators are placed between this block and base slab as shown in Fig. 14.18. Then

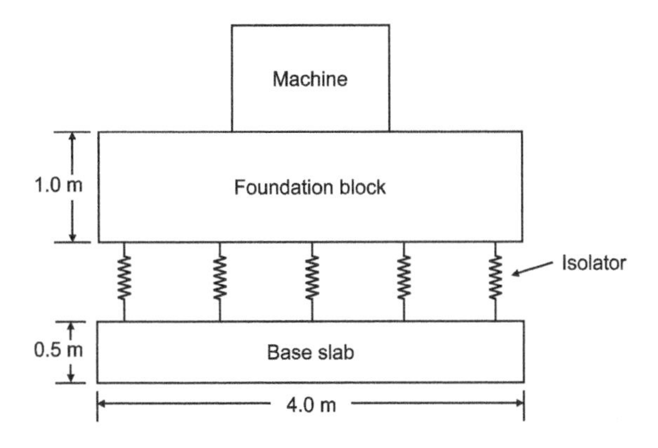

Fig. 14.18. Machine-foundation isolator system (Example 14.2)

$$m_2 = 1.83 + \frac{2}{3} \times 44 = 31.2 \text{ kNs}^2/\text{m}$$

$$m_1 = \frac{1}{3} \times 44 = 14.7 \text{ kN/s}$$

$$K_1 = 24 \times 10^4 \text{ kNs}^2/\text{m}$$
$$\omega_{nz} = 65.0 \text{ rad/s}$$
$$\eta = 0.253$$
$$a_1 = 1.13$$
$$\mu_m = \frac{31.2}{14.7} = 2.12$$

From Eq. (14.22),
$$0.253 = \frac{a_2^2(1+2.12)(1.13^2-1)}{\left[1-(1+2.12)(1.13^2+a_2^2-1.13^2\,a_2^2)\right]}$$

or
$$a_2^2 = 0.697 \; i.e. \, a_2 = 0.835$$

Hence,
$$\frac{\omega_{na}}{\omega} = 0.835. \text{ It gives } \omega_{na} = 0.835 \times 57.6 = 48 \text{ rad/s}$$
$$K_a = m_2\omega_{na}^2 = 31.2 \times 48^2 = 7.2 \times 10^4 \text{ kN/m}$$

(ii)
$$A_{Z2} = \frac{\left[(1+2.12)\times 1.13^2 + 2.12 \times 0.697 - 1\right]4.0}{31.2 \times 57.6^2\left[1-(1+2.12)\,(1.13^2+0.697-1.13^2\times 0.697)\right]}$$
$$= \frac{17.85}{103514[-2.38]}$$
$$= -7.24 \times 10^{-5} \text{ m}$$

Force in the isolator = $K_a A_{Z2} = 7.2 \times 10^4 \times 7.24 \times 10^{-5} = 5.2$ kN

On comparing with the results of example 14.1, it can be concluded that the stiffness of the isolator and forces on it depends significantly on the location of the isolator.

Example 14.3

It is planned to install a compressor having operating speed of 1000 rpm at a distance of 50 m from a precision machine. Suggest a suitable open trench barrier to provide effective vibration isolation. The velocity of shear waves at the site was found as 140 m/s.

Solution

Active screening

Operating frequency,
$$f = \frac{1000}{60} = 16.7 \text{ Hz}$$

Rayleigh wave velocity V_R may be taken approximately equal to shear wave velocity *i.e.* 140 m/s.

Therefore, Wave length
$$\lambda_R = \frac{V_R}{f} = \frac{140}{16.7} = 8.4 \text{ m}$$

Depth of the trench for active screening is given by
$$H = 0.6\,\lambda_R = 0.6 \times 8.4 = 5.04 \text{ m}$$

A partial circle trench with $\theta = 120°$ may be located at 4.0 m distance from the source (Fig. 14.19a)

Passive screening

Depth of the trench for passive isolation is given by
$$H = 1.33\,\lambda_R = 1.33 \times 8.4 = 11.2 \text{ m}$$

Let the trench be provided at a distance of 12 m from the precision machine.

$$\frac{R}{\lambda_R} = \frac{(50-12)}{8.4} = 4.52 \text{ (O.K. as lies between 2 and 7).}$$

For
$$\frac{R}{\lambda_R} = 4.52, \quad \frac{A_T}{\lambda_R^2} = 2.5 + \frac{4.52 - 2.0}{7.0 - 2.0}(6.0 - 2.5) = 4.26$$

$$\text{Length of trench} = \frac{4.26 \times 8.4^2}{11.2} = 26.8 \text{ m, say } 27 \text{ m}$$

The layout of trench with respect to compressor and precision machine is shown in Fig. 14.19b.

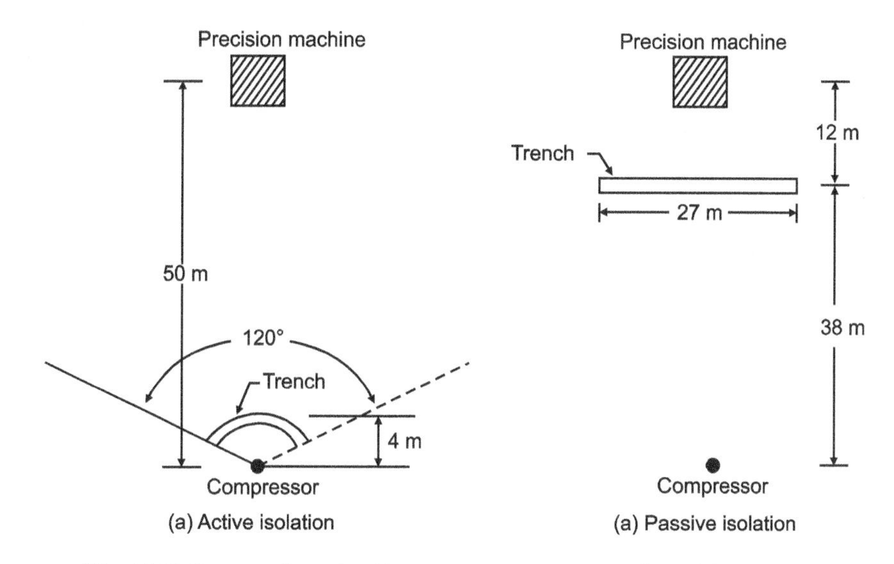

Fig. 14.19. Layout of trench with respect to compressor and precision machine

References

Barkan, D.D. (1962), "Dynamics of bases and foundations", McGraw Hill, New York.

Dolling, H.J. (1966), "Efficiency of trenches in isolating structures against vibration", Proc. Symp. Vib. Civ. Eng. Butterworth, London.

Liao, S. and Sangrey, D.A. (1978), "Use of piles as isolation barriers", J. Geotech. Engg. Div., Am. Soc. Civ. Eng., 104, (GT9), 1139-1152.

Sridharam, A., Nagendra, M.V. and Parthasarathy, T. (1981), "Isolation of machine foundations by barriers". Int. Conf. Recent Adv. Geotech. Earthquake Eng., St. Louis, Vol. 1. 279-282.

Woods, R.D. (1968), "Screening of surface waves in soils", J. Soil Mech. Foun. Div. , Proc. Am. Soc. Civ. Engg., 94 (SM-4), 951-979.

Woods, R.D. and Richart, F.E. Jr. (1967), "Screening of elastic waves by trenches", Proc. Int. Symp. Wave Propag. Dyn. Prop. Earth Mater, Albuquerque., NM, 275-284.

Woods, R.D., Barnett, N.E. and Sagessor, R. (1974), "Holography - A new tool for soil dynamics", J. Geotech. Eng. Div., Am. Soc. Civ., 100 (GT-11), 1231-1247.

Practice Problems

14.1 Explain the difference between 'force isolation' and 'motion isolation'. Sketch a suitable system for 'force isolation'. Represent it by a mathematical model and then give the procedure of getting the stiffness of the isolator.

14.2 Starting from fundamentals, derive the expression for the efficiency of isolation system.

14.3 Explain the difference between "Active screening" and "Passive screening". Give the procedure of designing the open trench barrier in both the cases.

14.4 Give the salient features of passive screening by use of pile barriers.

14.5 Design a suitable isolation system for keeping the amplitude of the foundation of a reciprocating machine less than 0.025 mm. The weight of the machine is 25 kN and it produces sinusoidally varying unbalanced force of 4 kN in the vertical direction. The operating speed of the machine is 800 rpm. The dynamic shear modulus of the soil is 2.5×10^4 kN/m^2. Assume suitably any data not given.

14.6 A compressor having an operating speed of 1300 rpm was installed in an industrial unit. Later on it was planned to place a precision machine at a distance of 60 m from it. It was felt necessary to protect this precision machine from any damaging vibration caused by the compressor. Design open trench barrier to provide effective vibration screening for the cases of (a) active and (b) passive screening. The velocity of shear waves was found as 160 m/s.

CHAPTER

15

Reinforced Soil Wall

15.1 General

Reinforced soil possesses many novel characteristics which makes it eminently suitable for construction of geotechnical structures, particularly retaining walls. A reinforced soil retaining wall has three main components, namely, (i) soil fill, (ii) reinforcement and (iii) facing unit fixed to one end of the reinforcement. The facing unit is not a vital component; however, it is necessary to prevent surface erosion and to give aesthetically acceptable external appearance. Ordinary frictional soil constitutes most of its bulk which is handled and compacted in place, as for earthen embankments. As the soil forms over 99.9 percent of the retaining wall, it is appreciably flexible and can withstand large differential settlement without stress. Also the stress concentration near the toe is small because of the relatively high base to height ratio, usually 0.8 to 1.0. As the soil is main part of a reinforced soil retaining wall, it is more economical, and suitable in poor ground conditions also. Due to above mentioned novel characteristics, a reinforced soil retaining wall is popular in use in almost all retaining structures (Vidal, 1978).

In this chapter, pseudo-static analysis of a reinforced soil wall located in a seismic region has been presented.

15.2 Analysis and Design

For the pseudo-static analysis and design of a reinforced soil retaining wall located in a seismic region, one has to consider:

(i) External stability, and
(ii) Internal stability

For external stability analysis, reinforced soil wall is considered as integral unit behaving like a rigid gravity structure and conforms the simple laws of statics.

The internal stability deals with the design of reinforcement with regards to its length and cross-section against tension failure and ensuring that it has sufficient length into the stable soil stability.

15.2.1 External Soil Stability

The possible external failure mechanisms are: (i) Sliding, (ii) Overturning, (iii) Tilting/bearing failure and (iv) Slip failure. Pictorially these are illustrated in Fig 15.1.

In Fig 15.2, a reinforced soil wall of height H and width B is shown, which will be considered as a rigid retaining wall, for checking its external stability. Usually the soil used in reinforced earth wall is better and well compacted in comparison to backfill soil. In Table 15.1, their properties used in further analysis are given.

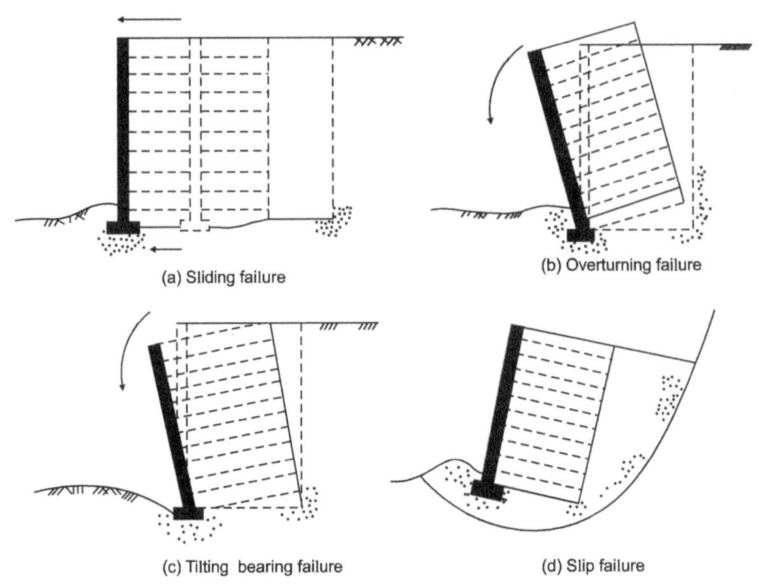

Fig. 15.1. Failure mechanism considered in external stability of reinforced soil wall

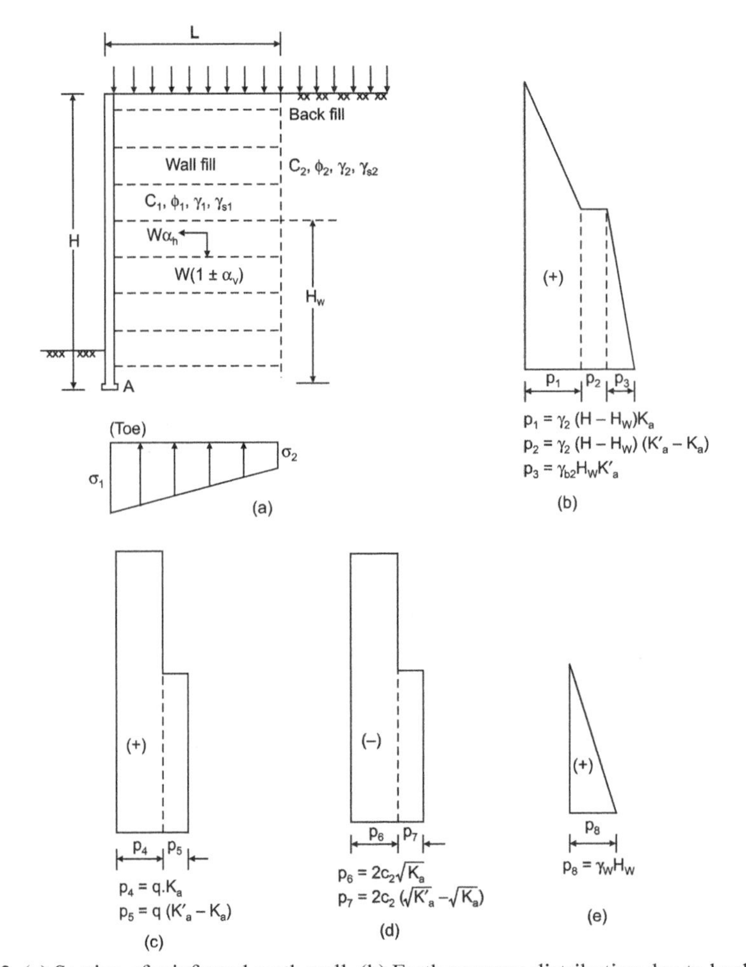

Fig. 15.2. (a) Section of reinforced earth wall, (b) Earth pressure distribution due to backfill soil, (c) Earth with pressure due to surcharge on the backfill, (d) Earth pressure due to cohesion of the backfill soil and (e) water pressure

Table 15.1. Properties of Wall and Backfill Soils

Item	*Cohesion (kN/m²)*	*Angle of internal function*	*Bulk density (kN/m³)*	*Statured unit weight (kN/m³)*	*Submerged unit weight (kN/m³)*
Wall soil	c_1	ϕ_1	γ_1	γ_{s1}	γ_{b1}
Backfill soil	c_2	ϕ_2	γ_2	γ_{s2}	γ_{b2}

Considering more general case which may occur in practice often, position of water table is considered at a height of h_w from the base of the wall, and wall located in seismic region. γ_w represents the unit weight of water (Fig. 15.2). Coefficient of interfacial friction between the wall material and backfill soil is considered δ. It may be taken as $\left(\frac{\phi_1+\phi_2}{2}\right)$. K_a and K_a' are respectively the coefficients of active earth pressure for the backfill soil above the water table and below the water table. These may be easily computed using conventional Coulomb's theory. Readers may note carefully that the value of δ should be taken as δ/2 for the fill below the water table for computing K_a'. Seismic earth pressures are computed by adding the dynamic increment to the static earth pressure. Dynamic increment means the difference of total seismic earth pressure minus the static earth pressure.

Static Earth Pressure

The static earth pressure diagrams are shown in Figs. 15.2b, c and d. In these figures value of K_a may be obtained using Eqn. (15.1) given below:

$$K_a = \frac{\cos^2\phi_2}{\cos\delta}\cdot\frac{1}{\left[1+\left\{\frac{\sin(\phi_2+\delta)\sin(\phi_2)}{\cos\delta}\right\}^{1/2}\right]^2} \tag{15.1}$$

Value of K'_a may also be obtained using Eqn. (15.1) by replacing δ by δ/2.

$$P_{a\gamma} = \frac{1}{2}\gamma_2 K_a (H-H_w)^2 + \gamma_2 K_a' H_w (H-H_w) + \frac{1}{2}\gamma_{b2} K_a' H_w^2 \tag{15.2}$$

$$P_{aq} = qK_a(H-H_w) + qK_a'H_w \tag{15.3}$$

$$P_{ac} = 2c\sqrt{K_a}(H-H_w) + 2c\sqrt{K_a'}\,.\,H_w \tag{15.4}$$

$$P_{aw} = \frac{1}{2}\gamma_w H_w^2 \tag{15.5}$$

Total static earth pressure

$$P_{Tst} = P_{a\gamma} + P_{aq} - P_{ac} + P_{aw} \tag{15.6}$$

Moment of Static Earth Pressures about Base

$$M_{a\gamma} = \frac{1}{2}\gamma_2 K_a (H-H_w)^2.\left(H_w + \frac{H-H_w}{3}\right) + \gamma_2 K_a' H_w (H-H_w).\frac{H_w}{2} + \frac{1}{2}\gamma_{b2}\,K_a'.H_w^2.\frac{H_w}{3}$$

or $$M_{a\gamma} = \frac{1}{6}\gamma_2 K_a (H + 2H_w).(H - H_w)^2 + \frac{1}{2}\gamma_{b2} K'_a (H - H_w).H_w^2 + \frac{1}{6}\gamma_{b2}\ K'_a H_w^3 \tag{15.7}$$

$$M_{aq} = qK_a (H - H_w)\left(H_w + \frac{H - H_w}{2}\right) + q.K'_a H_w.\frac{H_w}{2}$$

or $$M_{aq} = \frac{1}{2} qK_a (H^2 - H_w^2) + \frac{1}{2} q.K'_a H_w^2 \tag{15.8}$$

$$M_{ac} = c_2\sqrt{K_a}(H + 2H_w)(H^2 - H_w^2) + c\sqrt{K'_a}.H_w^2 \tag{15.9}$$

$$M_{aw} = \frac{1}{6}\gamma_w.H_w^3 \tag{15.10}$$

Total moment about base

$$M_{Tst} = M_{a\gamma} + M_{aq} - M_{ac} + M_{aw} \tag{15.11}$$

Total Seismic Earth Pressure

Value of total seismic earth pressure in vertical retaining wall for dry and moist backfill may be obtained using the following equations:

$$(P_{A\gamma})_{dyn} = \frac{1}{2}\gamma C_a.H^2 \tag{15.12}$$

where C_a is seismic active earth pressure coefficient given by Eqn (15.13).

$$C_a = \frac{(1 \pm \alpha_v)\cos^2(\phi - \lambda)}{\cos\lambda\ \cos(\delta + \lambda)} \cdot \frac{1}{\left[1 + \left\{\dfrac{\sin(\phi_2 + \delta)\sin(\phi_2 - \lambda)}{\sin(\delta + \lambda)}\right\}^{1/2}\right]^2} \tag{15.13}$$

where

$$\lambda = \tan^{-1}\frac{\alpha_h}{1 \pm \alpha_v} \tag{15.14}$$

The expression of C_a gives two values depending on the sign of α_v. For design purpose the higher value of C_a shall be taken.

For submerged backfill, the value of seismic earth pressure is obtained by Eqn (15.14) replacing γ by γ_{sub} and C_a by C'_a. C'_a may be obtained by using Eqn. (15.15) replacing λ by λ' (IS: 1893-2016):

where

$$\lambda' = \tan^{-1}\left[\left(\frac{\gamma_s}{\gamma_s - 10}\right) \cdot \frac{\alpha_h}{1 \pm \alpha_v}\right]. \tag{15.15}$$

γ_s = Saturated unit weight of submerged backfill in kN/m^3

C'_a = Seismic active earth pressure coefficient for submerged backfill.

Dynamic Increments

The ratio of the lateral dynamic increment in active pressure due to the pressure at various depth along the height of wall may be taken as shown in Fig. 15.3. Pressure distribution of dynamic increment in active pressures due to backfill may be obtained by multiplying the vertical effective pressures by the coefficient given in Fig. 15.3 at corresponding depth. C'_a and K'_a are respectively the seismic and static earth pressure coefficients in fully submerged conditions.

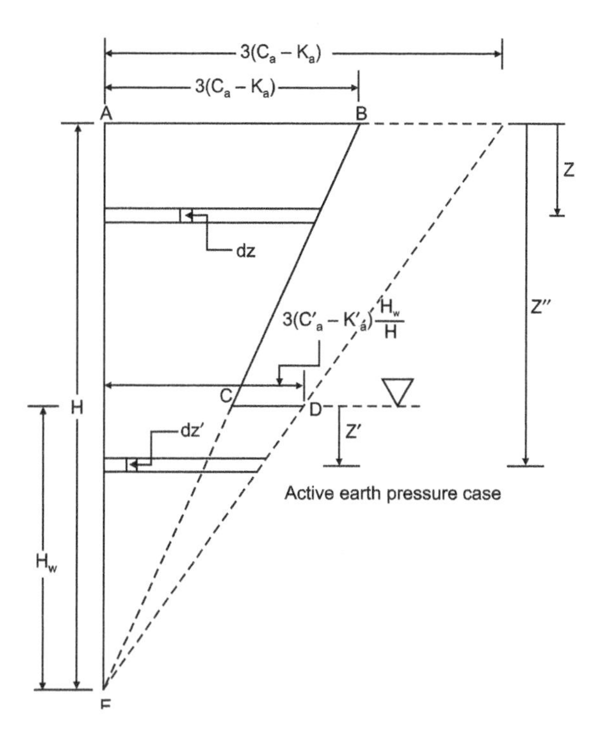

Fig. 15.3. (a) Distribution of the ratio $\dfrac{\text{Lateral dynamic increment due to backfill}}{\text{Vertical effective pressure}}$ with height of wall

The value of lateral dynamic increment due to backfill is obtained by integrating it in the portion above water level and below water level separately. By doing this, we get (Ref. Fig. 15.3a):

$$(P_{A\gamma})_{di} = \int_0^{H-H_w} 3(C_a - K_a)\frac{H-z}{H}\gamma z\,dz + \int_0^{H_w} \frac{3(C_a' - K_a').H_w}{H}\frac{H_w - z'}{H_w}[\gamma_2(H-H_w)+\gamma_{b2}z']dz'$$

or

$$(P_{A\gamma})_{di} = \frac{\gamma_2}{2H}[(C_a - K_a)(H-H_w)^2(H+2H_w)] + \frac{H_w^2}{2H}(C_a' - K_a')[3\gamma_2(H-H_w)+\gamma_{b2}H_w] \quad (15.16)$$

For getting the point of application of dynamic increment, firstly the moments of portions above water table and below water table are taken about AB (Ref. Fig. 15.3).

$$(M_{A\gamma})_{di} = \int_0^{H-H_w} 3(C_a - K_a)\cdot\frac{H-z}{H}\gamma_2 z.z dz + \int_{H-H_w}^{H} \frac{3(C_a' - K_a').H_w}{H}\cdot\frac{H-z''}{H_w}$$

$$[\gamma_2(H-H_w)+\gamma_{b2}(z''-H+H_w)]z''dz''$$

or $$(M_{A\gamma})_{di} = \frac{\gamma_2}{4H}[(C_a - K_a)(H-H_w)^3(H+3H_w)] \quad (15.17)$$

$$+\frac{\gamma_2(C_a' - K_a')}{4H}(H-H_w)[H^3 - (H-H_w)^2(H+2H_w)] + \frac{\gamma_{b2}(C_a' - K_a')}{4H}[(2H-H_w).H_w^3]$$

Height of point of application of the dynamic increment $(P_{A\gamma})_{di}$ is then given by Eqn. (15.18)

$$H_{d\gamma i} = H - \frac{(M_{A\gamma})_{di}}{(P_{A\gamma})_{di}} \tag{15.18}$$

The additional dynamic increament due to uniform surcharge of intensity q per unit area on the backfill shall be taken as Fig. 15.3b. Proceeding as above, we get

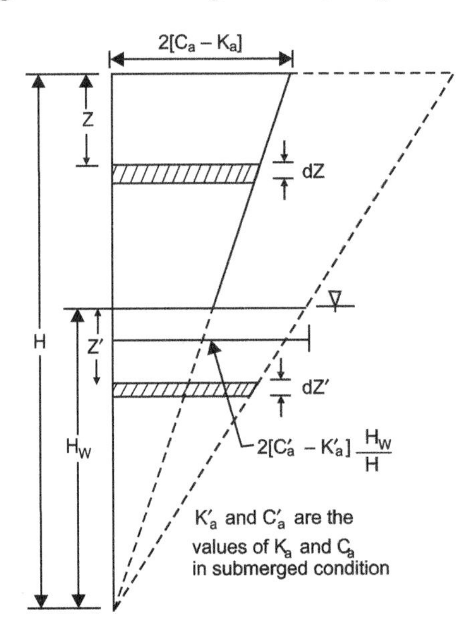

Fig. 15.3. (b) Distribution of the ratio $\frac{\text{Lateral dynamic increment due to surcharge only}}{\text{Vertical effective pressure due to surcharge only}}$ with height of wall

Similarly, $$(P_{aq})_{di} = (C_a - K_a).q.\frac{H^2 - H_w^2}{H} + (C_a' - K_a')q.\frac{H_w^2}{H} \tag{15.19}$$

$$(M_{aq})_{di} = \frac{2}{3}\left[\frac{q(C_a - K_a)(H^3 - H_w^3)}{H} + \frac{q(C_a' - K_a')H_w^3}{H}\right] \tag{15.20}$$

$$H_{dqi} = H - \frac{(M_{Aq})_{di}}{(P_{Aq})_{di}} \tag{15.21}$$

Stability Analysis

Weight of wall, $$W_w = B\{\gamma_1(H - H_w) + \gamma_{s1}H_w\} \tag{15.22}$$

Effective weight of wall,

$$W_w' = L\{\gamma_1(H - H_w) + \gamma_{s1}H_w - \gamma_w H_w\} \tag{15.23}$$

Moment of weight of wall about toe,

$$M_w = W_w \cdot \left(\frac{L}{2}\right) \tag{15.24}$$

Surcharge force on the wall, $$Q = q\,.\,L \tag{15.25}$$

Moment of surcharge force about toe,

$$M_q = \frac{(qL^2)}{2} \tag{15.26}$$

Seismic forces on the weight of wall

$$\text{Horizontal seismic force} = W \,.\, A_h \tag{15.27}$$

$$\text{Vertical seismic force} = \pm\, W \,.\, A_v \tag{15.28}$$

Moment of seismic forces acting on the weight of wall

$$M_{sw} = \alpha_h \left\{ \gamma_1 B(H - H_w)\left[H_w + \frac{H - H_w}{2} \right] + \gamma_{s1} H_w . B . \frac{H_w}{2} \right\} \pm \alpha_v W_w . \frac{B}{2}$$

or

$$M_{sw} = \frac{B}{2}[\gamma_1(H^2 - H_w^2) - \gamma_{s1} H_w^2]\alpha_h \pm W_w . \alpha_v . \frac{B}{2} \tag{15.29}$$

Seismic forces on the surcharge acting on the wall

$$\text{Horizontal seismic force} = Q \,.\, \alpha_h \tag{15.30}$$

$$\text{Vertical seismic force} = \pm\, Q \,.\, \alpha_v \tag{15.31}$$

Moment of the seismic forces on the surcharge acting on the wall about toe

$$M_{Sq} = Q.\alpha_h.H \pm Q.\alpha_v.\frac{B}{2} \tag{15.32}$$

(a) *Sliding*

Factor of safety against sliding

$$F_s = \frac{\text{Resisting force}}{\text{Sliding force}} \tag{15.33}$$

Static case:

$$F_s = \frac{\mu(W_w' + Q)}{P_{Tst}} \not< 2.0 \tag{15.34}$$

Seismic case:

$$F_s = \frac{\mu(W_w' + Q)(1 \pm \alpha_v)}{P_{Tst} + (P_{a\gamma})_{di} + (P_{aq})_{di} + (W_w + Q)\alpha_h} \not< 1.5 \tag{15.35}$$

(b) *Overturning*

Factor of safety against overturning

$$F_o = \frac{\text{Resisting moment}}{\text{Overturning moment}} \tag{15.36}$$

Static case:

$$F_o = \frac{(W_w' + Q)\frac{B}{2}}{M_{Tst}} \not< 2.0 \tag{15.37}$$

Seismic case:

$$F_o = \frac{(W_w' + Q)(1 \pm \alpha_v).\frac{B}{2}}{M_{Tst} + (M_{a\gamma})_{di} + (M_{aq})_{di} + M_{sw} + \alpha_h Q H} \not< 1.5 \tag{15.38}$$

(c) *Tilting/Bearing failure*
Static case:

$$\sigma_{max} = \frac{(W'_w + Q)}{B} + M_{Tst}.\frac{6}{B^2}. \ngtr q_a \tag{15.39}$$

$$\sigma_{min} = \frac{(W'_w + Q)}{B} - M_{Tst}.\frac{6}{B^2}. \nless 0.0 \tag{15.40}$$

Seismic case:
Net moment about the centre of base of wall

$$M_{Tn} = M_{Tst} + (M_{a\gamma})_{di} + (M_{aq})_{di} + M_{sw} + Q.\,\alpha_h.H \tag{15.41}$$

$$\sigma_{max} = \frac{(W'_w + Q)(1 \pm \alpha_v)}{B} + M_{Tn}.\frac{6}{B^2} > 1.25\, q_a \tag{15.42}$$

$$\sigma_{min} = \frac{(W'_w + Q)(1 \pm \alpha_v)}{B} - M_{Tn}.\frac{6}{B^2} < 0.0 \tag{15.43}$$

q_a = Allowable bearing pressure of base soil

15.2.2 Internal Stability

Tension Failure

The tensile force (T_i) per metre width in the grid at depth h_i is given

$$T_i = K_{aw}\, \sigma_{vi}\, S_z \tag{15.44}$$

where σ_{vi} = Maximum vertical pressure intensity

Static case

For $h_i \le (H - H_w)$

Moment of earth pressure about centre of horizontal section passing at depth h_i

$$M_1 = \frac{1}{6}\gamma_2 K_a h_i^3 + \frac{q.h_i^2 K_a}{2} - \frac{6\sqrt{K_a h_i^2}}{2} \tag{15.45}$$

$$\sigma_{vi} = (\gamma_1 h_i + q) + M_1.\frac{6}{B^2} \tag{15.46}$$

For $h_i > (H - H_w)$

Obtain $M_{a\gamma}$, M_{aq}, M_{ac} and M_{aw} by replacing H by h_i, and H_w by h_{wi} [$= h_i - (H - H_w)$]. After this compute M_2 as below:

$$M_2 = M_{a\gamma} + M_{aq} - M_{ac} + M_{aw} \tag{15.47}$$

$$\sigma_{vi} = (\gamma_1 h_i + q) + M_2\frac{6}{B^2} \tag{15.48}$$

For satisfactory action

$$T_i \ngtr R_T$$

where R_T = Permissible design strength of reinforcement

Seismic case

For $h_i \leq (H - H_w)$

Obtain M_3 and M_4 by replacing H and H_w by h_i and 0 in the expressions $M_{a\gamma i}$ and M_{aqi}.

Then $$M_{Tni} = M_1 + M_3 + M_4 + \alpha_h h_i \left(\frac{\gamma_1 B h_i}{2} + Q \right) \tag{15.49}$$

$$\sigma_{vi} = (\gamma_1 h_i + q)(1 \pm \alpha_v) + M_{Tni} \cdot \frac{6}{B^2} \tag{15.50}$$

For $h_i > (H - H_w)$

Obtain M_5 and M_6 by replacing H by h_y and H_w by h_{wi} [$= h_i - (H - H_w)$] in the expressions of $M_{a\gamma i}$ and M_{aqi}.

$$M_{Tni} = M_2 + M_5 + M_6 + \alpha_h B. \left[\gamma_1 (H - H_w) \left\{ h_i - \frac{H - H_w}{2} \right\} + \gamma_{s1} \frac{(h_i - H + H_w)^2}{2} \right] + Q.\alpha_h H \tag{15.51}$$

$$\sigma_{vi} = [\gamma_1 (H - H_w) + \gamma_{s1} (h_i - H + H_w)](1 \pm \alpha_v) + M_{Tni} \cdot \frac{6}{B^2} \tag{15.52}$$

For satisfactory action

$$T_i \ngtr 1.25 R_T \tag{15.53}$$

Wedge/Pull out Failure

In addition to investigating the internal mechanism of tension, it is necessary to consider the possibility of inclined failure planes passing through the wall forming unstable wedge of soil bounded by the front face of the wall, the top ground surface and the potential failure plane. However, in most of the cases, failure plane passing through toe gives critical case.

Static case

Figure 15.4 indicates the loads and forces to be considered in the analysis, namely:

(i) Self weight of the fill in the wedge, W_w

(ii) Uniformly distributed surcharge, q

(iii) Resultant force (R) on the potential failure plane

(iv) Total tension (T) in the reinforcements intercepted by the potential failure plane.

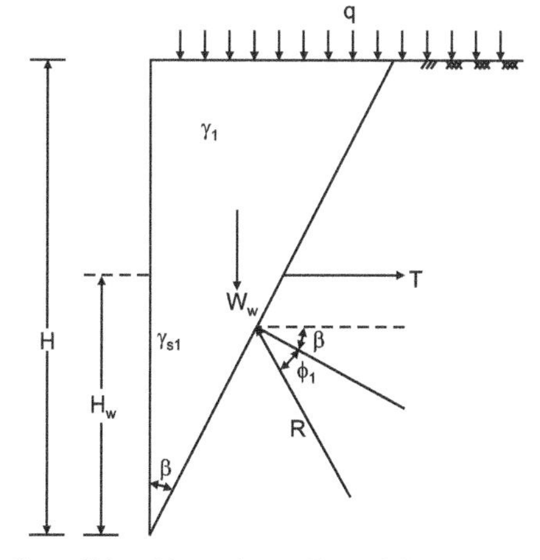

Fig. 15.4. Consolidated in wedge pull out failure in static condition

$$W_w = \frac{1}{2}\gamma_s H_w^2 \tan\beta + \gamma_1 (H - H_w) H_w \tan\beta + \frac{1}{2}\gamma_1 (H - H_w)^2 \tan\beta \tag{15.54}$$

or $$W_w = a_1 \tan\beta \tag{15.55}$$

where $$a_1 = \frac{1}{2}\gamma_s H_w^2 + \gamma_1 (H - H_w)\left[H_w + \frac{H - H_w}{2}\right]$$

$$= \frac{1}{2}[\gamma_s H_w^2 + \gamma_1 (H^2 - H_w^2)] \tag{15.56}$$

Considering the equilibrium of the wedge,

$$R\cos(\phi_1 + \beta) = T \tag{15.57}$$

$$R\sin(\phi_1 + \beta) = a_1 \tan\beta + q.H\tan\beta = a_2 \tan\beta \tag{15.58}$$

where $$a_2 = a_1 + qH \tag{15.59}$$

Solving Eqn. (15.57) and (15.59), we get

$$\frac{T}{a_2 \tan\beta} = \frac{\cos(\phi_1 + \beta)}{\sin(\phi_1 + \beta)}$$

or $$T = a_2 \cdot \frac{\tan\beta}{\tan(\phi_1 + \beta)} \tag{15.60}$$

For having maximum value of T,

$$\frac{dT}{d\beta} = a_2\left[\frac{\sec^2\beta}{\tan(\phi_1 + \beta)} - \frac{\tan\beta.\sec^2(\phi_1 + \beta)}{\tan^2(\phi_1 + \beta)}\right] = 0 \tag{15.61}$$

or $$\sec^2\beta\tan(\phi_1 + \beta) = \tan\beta\,.\,\sec^2(\phi_1 + \beta)$$

or $$\frac{\sin(\phi_1 + \beta)}{\cos(\phi_1 + \beta)}\cdot\frac{1}{\cos^2\beta} = \frac{\sin\beta}{\cos\beta}\cdot\frac{1}{\cos^2(\phi_1 + \beta)}$$

or $$\sin(\phi_1 + \beta)\cos(\phi_1 + \beta) - \sin\beta\cos\beta = 0$$

or $$\sin 2(\phi_1 + \beta) = \sin 2\beta$$

or $$2\beta = 180 - 2(\phi_1 + \beta)$$

or $$4\beta = 180 - 2\phi_1$$

or $$\beta = 45 - \left(\frac{\phi_1}{2}\right) \tag{15.62}$$

Therefore, $$T_{max} = a_2 \cdot \frac{\tan\left(45 - \frac{\phi_1}{2}\right)}{\tan\left(45 + \frac{\phi_1}{2}\right)} \tag{15.63}$$

Maximum tension per reinforcement $$= \frac{T_{max}}{n} \tag{15.64}$$

Seismic case

Figure 15.5 indicates the loads and forces to be considered on the wedge for analyzing it in seismic case. Additional forces which act in this case are (i) $W_w\ \alpha_h$, (ii) $\pm\ W_w.\alpha_v$, (iii) $q.H\ \tan\beta\ \alpha_h$ and (iv) $\pm\ qH\tan\beta\ .\ \alpha_v$

Considering the equilibrium of the wedge,

$$R\cos(\phi_1+\beta)+q.H\ \tan\beta.\alpha_h+a_1\tan\beta\ \alpha_h=T \tag{15.65a}$$

$$R\sin(\phi_1+\beta)\ =\ a_1\tan\beta(1\pm\alpha_v)+q.H\tan\beta(1\pm\alpha_v) \tag{15.65b}$$

Substituting
$$a_3=\alpha_h(a_1+qH)=\alpha_h.a_2 \tag{15.66}$$

$$a_4=a_2.(1\pm\alpha_v) \tag{15.67}$$

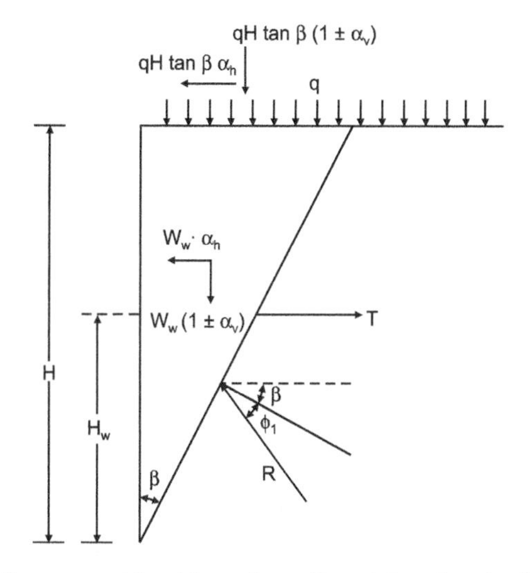

Fig. 15.5. Forces considered in wedge pullout failure in seismic condition

Solving Eqn. (15.64) and (15.65) for T, we get,

$$\frac{T-a_3\tan\beta}{a_4\tan\beta}=\frac{\cos(\phi_1+\beta)}{\sin(\phi_1+\beta)} \tag{15.68}$$

or
$$T=\frac{a_4\tan\beta}{\tan(\phi_1+\beta)}+a_3\tan\beta \tag{15.69}$$

For having maximum value of T,

$$\frac{dT}{d\beta}=a_4\left[\frac{\sec^2\beta}{\tan(\phi_1+\beta)}-\frac{\tan\beta.\sec^2(\phi_1+\beta)}{\tan^2(\phi_1+\beta)}\right]+a_3\sec^2\beta=0 \tag{15.70}$$

or
$$a_4\sec^2\beta\tan(\phi_1+\beta)-a_4\tan\beta.\sec^2(\phi_1+\beta)+a_3\sec^2\beta.\tan^2(\phi_1+\beta)=0 \tag{15.71}$$

Multiplying Eqn. (15.71) by $\cos^2(\phi_1+\beta).\cos^2\beta$, we get

or
$$a_4\sin(\phi_1+\beta)\cos(\phi_1+\beta)-a_4\sin\beta\cos\beta+a_3\sin^2(\phi_1+\beta)=0 \tag{15.72}$$

or
$$a_4\sin2(\phi_1+\beta)-a_4\sin2\beta+a_3[1-\cos2(\phi_1+\beta)]=0$$

or
$$a_4[\sin2\phi_1\cos2\beta+\cos2\phi_1\sin2\beta]-a_4\sin2\beta+a_3-a_3\cos2\phi_1.\cos2\beta$$
$$+\ a_3\sin2\phi_1\sin2\beta=0$$

or $$(a_4 \cos 2\phi_1 - a_4 + a_3 \sin 2\phi_1) \sin 2\beta + (a_4 \sin 2\phi_1 - a_3 \cos 2\phi_1) \cos 2\beta + a_3 = 0 \quad (15.73)$$

Substituting $$a_5 = a_4 \cos 2\phi_1 - a_4 + a_3 \sin 2\phi_1 \quad (15.74)$$

and $$a_6 = a_4 \sin 2\phi_1 - a_3 \cos 2\phi_1 \quad (15.75)$$

Equation (15.73) can be written as

$$a_5 \sin 2\beta + a_6 \cos 2\beta = 0 \quad (15.76)$$

or $$2a_5 \sin\beta \cos\beta + a_6(1 - 2\sin^2\beta) + a_3 = 0$$

or $$-2a_6 \sin^2\beta + 2a_5 \sin\beta \cos\beta + (a_3 + a_6) = 0$$

$$-2a_6 \tan^2\beta + 2a_5 \tan\beta + (a_3 + a_6)(1 + \tan^2\beta) = 0$$

or $$(a_3 - a_6)\tan^2\beta + 2a_5 \tan\beta + (a_3 + a_6) = 0$$

It gives $$\tan\beta = \frac{-2a_5 \pm \sqrt{4a_5^2 - 4(a_3^2 - a_6^2)}}{2(a_3 - a_6)} \quad (15.77)$$

Equation (15.77) gives the value of β at which T is maximum, denoting it by β_c. Therefore,

$$T_{max} = \frac{a_4 \tan\beta_c}{\tan(\phi_1 + \beta_c)} + a_3 \tan\beta_c \quad (15.78)$$

$$\text{Maximum tension per reinforcement} = \frac{T_{max}}{n} \quad (15.79)$$

Static case

$$L_{ip} = \frac{T_i F_{ss}}{2\alpha \tan\phi_1(\gamma_1 h_1 + q)} \quad (15.80)$$

where h_i = Depth of first layer of reinforcement from top of wall

F_{ss} = Factor of safety, 2.0

$T_i = \dfrac{T_{max}}{n}$ from Eqn. (15.64)

α = Coefficient of interaction

L_{ip} = Anchorage length of reinforcement at depth h_1

Seismic case

$$L_{ip} = \frac{T_i F_{sd}}{2\alpha \tan\phi_1(\gamma_1 h_1 + q)}$$

where F_{sd} = Factor of safety, 1.5

$T_i = \dfrac{T_{max}}{n}$ from Eqn. (15.79)

Required length of reinforcement = $L_{ip} + (H - h_1)\tan\beta$

15.3 Construction of a Reinforced Soil Wall

The construction of a reinforced soil wall is carried out as discussed in the following steps:

1. A reinforced soil structure is built in series of successive stages, each stage consisting of the assembly of new layer of facing elements, and the placing of corresponding earthfill, followed by a new layer of reinforcing strips/grids/sheets.
2. To ensure stability when first layer of metal facing elements or first layer of concrete panels is positioned temporary supports are installed on the facing side of the wall. They are removed as the backfill is placed. The purpose of these supports for initial lifts is to keep the facing being displaced in order to obtain a perfectly vertical wall facing.
3. The reinforcing strips/grids/sheets are placed perpendicular to the wall face, with the broad side of the strip resting on the previously placed layer of compacted earthfill. The metal strips are attached to the facing elements by means of high strength bolts which must be of the same metal as the reinforcing strips. The attachment of the geogrids to the facing panels is mechanical by having hooks cast into the panels during their construction. The geogrids can be connected directly or by means of a steel dowel running lengthwise behind hooks and attaching to the ends of the geogrids.
4. In order to avoid stresses that could be caused by the earth moving equipment; the fill should be spread by the moving machinery parallel to the wall facing. It is advisable that trucks and other earth moving machineries do not go closer than 2 m to the face of wall. The area closer to the facing should be compacted with light compacting equipment.

15.4 Drainage

Backfill in a reinforced earth structure generally has good frictional and drainage characteristics. The reinforced earth mass is therefore essentially self-draining, and water does not accumulate within the mass. Special precautions should be taken in the cases of backfill of poor drainage characteristics. In the case of walls supporting cut slopes, a chimney drain of about 1 m thick is frequently placed between the reinforced earth structure and embankment as shown in Fig. 15.6.

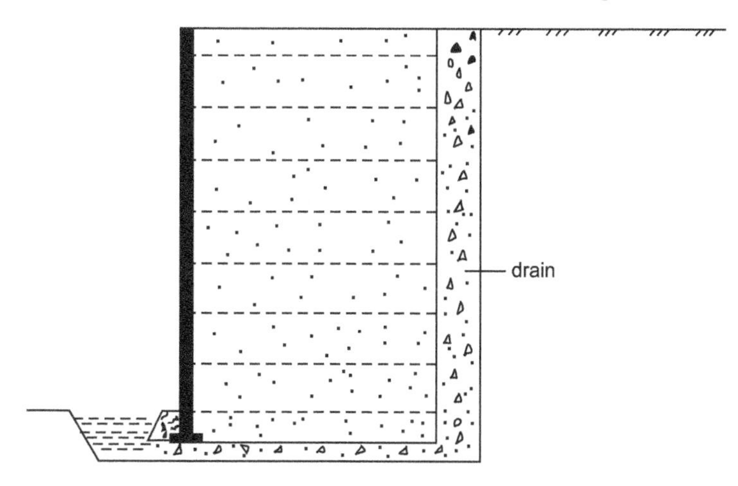

Fig. 15.6. Chimney drain

If water can infiltrate into the reinforced earth mass, which is reasonably permeable due to materials used a drain must be provided at the base of the wall. In this case the skin units at the bottom are perforated to let the water out.

Illustrative Example

Example 15.1

Design a suitable layout of Tensar SR2 grid reinforcement for an 8 m high, vertical soil wall. The properties of wallfill and backfill are given below:

Wallfill	*Backfill*
$c_1 = 1.5$ kN/m^2	$c_2 = 2$ kN/m^2
$\phi_1 = 40°$	$\phi_2 = 38°$
$\gamma_1 = 19$ kN/m^3	$\gamma_2 = 18$ kN/m^3
$\gamma_{s1} = 21$ kN/m^3	$\gamma_{s2} = 20$ kN/m^3

The water table is 2.0 m above the base of wall. The wall is situated in seismic region ($\alpha_h = 0.1$, $\alpha_v = 0.05$). The backfill and wall has a surcharge of 8.0 kN/m^2. Permissible tensile strength of grid reinforcement is 16.5 kN/m.

Solution

Earth pressure coefficients due to backfill

Assume $\delta = \frac{2}{3}\phi = \left(\frac{2}{3}\right) \times 38 = 25.34°$

Below water level

$$K_a = \frac{\cos^2 38}{\cos 25.34} \frac{1}{\left[1 + \left\{\frac{\sin(38+25.34)\sin 38}{\cos 25.34}\right\}^{1/2}\right]^2}$$

$$= 0.217$$

$$\lambda = \tan^{-1}\frac{\alpha_h}{1 \pm \alpha_v} = \tan^{-1}\left(\frac{0.1}{1 \pm 0.05}\right) = 5.44° \text{ (+sign)}$$

$$= 6°(- \text{ sign})$$

(+) α_v

$$C_a = \frac{(1+0.05)\cos^2(38-5.44)}{\cos 5.44 \cos(25.34+5.44)} \cdot \frac{1}{\left[1 + \left\{\frac{\sin(38+25.34)\sin(38-5.44)}{\cos(25.34+5.44)}\right\}^{1/2}\right]^2}$$

$$= 0.285$$

(–) α_v

$$C_a = \frac{(1-0.05)\cos^2(38-6)}{\cos 6 \cos(25.34+6)} \cdot \frac{1}{\left[1 + \left\{\frac{\sin(38+25.34)\sin(38-6)}{\cos(25.34+6)}\right\}^{1/2}\right]^2}$$

$$= 0.264$$

Adopt $C_a = 0.285$

Below water level

As mentioned in the text, δ for submerged backfill will be half of that for moist/dry backfill, *i.e.* $\frac{25.34}{2} = 12.67°$

From Eqn. (15.13)

$$\lambda = \tan^{-1}\frac{\gamma_s}{\gamma_s - 10}\cdot\frac{\alpha_h}{1 \pm \alpha_v}$$

$$\lambda = \tan^{-1}\frac{20}{20-10}\cdot\frac{1}{1 \pm 0.05} = 10.8°(+\text{sign})$$

$$= 11.9°(-\text{ sign})$$

$$C'_a = \frac{(1+0.05)\cos^2(38-10.8)}{\cos(12.67+10.8)} \frac{1}{\left[1+\left\{\frac{\sin(38+12.67)\sin(38-10.8)}{\cos(12.67+10.8)}\right\}^{½}\right]^2}$$

$$= 0.351$$

(–) Sign

$$C'_a = \frac{(1-0.05)\cos^2(38-11.9)}{\cos(11.9)\cos(12.67+11.9)}\cdot\frac{1}{\left[1+\left\{\frac{\sin(38+12.67)\sin(38-11.9)}{\cos(12.67+11.9)}\right\}^{½}\right]^2}$$

$$= 0.331$$

Adopt $C'_a = 0.351$

Total earth pressure and moments

$$P_{a\gamma} = \frac{1}{2}\gamma_2 K_a (H - H_w)^2 + \gamma_2 K'_a (H - H_w) + \frac{1}{2}\gamma_{b2} K'_a H_w^2$$

$$= \frac{1}{2} \times 18 \times 0.217(6-2)^2 + 18 \times 0.221(2)(6-2) + \frac{1}{2} \times 10 \times 0.221 \times 2^2$$

$$= 31.25 + 31.82 + 4.42 = 67.49 \text{ kN/m}$$

$$P_{aq} = qK_a(H - H_w) + qK'_a H_w$$

$$= 8 \times 0.217(6-2) + 8 \times 0.221 \times 2$$

$$= 6.94 + 3.54 = 10.48 \text{ kN/m}$$

$$P_{ac} = 2c\sqrt{K_a}\,(H - H_w) + 2c\sqrt{K'_a}\cdot H_w$$

$$= 2 \times 2 \times \sqrt{0.217}\,(6-2) + 2 \times 2\sqrt{0.221} \times 2$$

$$= 7.45 + 3.76 = 11.21 \text{ kN/m}$$

$$P_{aw} = \frac{1}{2}\gamma_{w}.H_w^2 = \frac{1}{2} \times 10 \times 2^2 = 20 \text{ kN/m}$$

$$P_{Tst} = 67.49 + 10.48 - 11.21 + 20.0 = 86.76 \text{ kN/m}$$

$$M_{a\gamma} = \frac{1}{2}\gamma_2 K_a (H + 2H_w)(H - H_w)^2 + \frac{1}{2}\gamma_2 K'_a (H - H_w)\cdot H_w^2 + \frac{1}{6}\gamma_{b2} K'_a H_w^3$$

$$= \frac{1}{6} \times 18 \times 0.217 \times (6 + 2 \times 2)(6 - 2)^2 + \frac{1}{2} 18 \times 0.221 \times (6-2) \times 2^2$$

$$+ \frac{1}{6} \times 10 \times 0.221 \times 2^3$$

$$= 104.16 + 31.82 + 2.95 = 138.93 \text{ kN-m/m}$$

$$M_{aq} = \frac{1}{2} qK_a (H^2 - H_w^2) + \frac{1}{2} qK'_a H_w^2$$

$$= \frac{1}{2} \times 8 \times 0.217(6^2 - 2^2) + \frac{1}{2} \times 8 \times 0.22 \times 2^2$$

$$= 27.77 + 3.54 = 31.32 \text{ kN-m/m}$$

$$M_{ac} = c\sqrt{K_a}(H^2 - H_w^2) + c\sqrt{K'_a}\cdot H_w^2$$

$$= 2 \times \sqrt{0.217}(6^2 - 2^2) + 2\sqrt{0.221} \times 2^2$$

$$= 29.81 + 3.76 = 33.57 \text{ kN-m/m}$$

$$M_{aw} = \frac{1}{6}\gamma_{w}.H_w^2 = \frac{1}{6} \times 10 \times 2^2 = 13.33 \text{ kN-m/m}$$

$$M_{Tst} = 138.93 + 31.32 - 33.57 + 13.33$$

$$= 150.01 \text{ kN-m/m}$$

Dynamic increments

$$(P_{a\gamma})_{di} = \frac{1}{2}\left[\gamma_2 (C_a - K_a)\frac{(H - H_w)^2}{H}(H + 2H_w) + (C'_a - K'_a)\frac{H_w^2}{H}\{3\gamma_2 (H - H_w) + \gamma_{b2} H_w\}\right]$$

$$= \frac{1}{2}\left[18(0.285 - 0.217)\frac{(6 - 2)^2}{6}(6 + 2 \times 2) + (0.351 - 0.221)\frac{2^2}{6} \times \{3 \times 18(6 - 2) + 10 \times 2\}\right]$$

$$= \frac{1}{2}[32.64 + 20.45] = 26.55 \text{ kN/m}$$

$$(P_{aq})_{di} = \left[(C_a - K_a)\frac{H^2 - H_w^2}{2H} + (C'_a - K'_a)\frac{H_w^2}{H}\right]$$

$$= 8\left[(0.264 - 0.217)\frac{6^2 - 2^2}{2 \times 6} + (0.351 - 0.221)\frac{2^2}{2 \times 6}\right]$$

$$= 28[0.125 + 0.0433] = 1.35 \text{ kN/m}$$

$$(M_{a\gamma})_{di} = \frac{\gamma_2(C_a - K_a)(H - H_w)^2}{4H}(H^2 + 2HH_w + 3H_w^2)$$

$$+\frac{(C_a' - K_a')}{4H} \cdot H_w^3[4\gamma_2(H - H_w) + \gamma_{b2}H_w)$$

$$= \frac{18(0.264 - 0.217)^2 (6 - 2)^2}{4 \times 6}[6^2 + 2 \times 6 \times 2 + 3 \times 2^2]$$

$$+ \frac{(0.351 - 0.221)2^3}{4 \times 6}[4 \times 18(6 - 2) + 10 \times 2]$$

$$= 40.60 + 13.35 = 53.95 \text{ kN-m/m}$$

$$(M_{aq})_{di} = \frac{2}{3}\left[\frac{q(C_a - K_a)(H^3 - H_w^3)}{H} + \frac{q(C_a' - K_a')H_w^3}{H}\right]$$

$$= \frac{2}{3}\left[\frac{8(0.264 - 0.217)(6^3 - 2^3)}{6} + \frac{8(0.351 - 0.221)2^3}{6}\right]$$

$$= 10.54 \text{ kN-m/m}$$

Forces and moments related to wall and surcharge
Let the length of reinforcement be 5.0 m

$$W_w = B[\gamma_1(H - H_w) + \gamma_{s1} H_w]$$

$$= 5.0[18(6 - 2) + 20 \times 2] = 560 \text{ kN/m}$$

$$W'_w = 5.0[18(6 - 2) + 10 \times 2] = 460 \text{ kN/m}$$

$$M_w = W_w.\left(\frac{B}{2}\right) = 560 \times \left(\frac{5}{2}\right) = 1400 \text{ kN-m/m}$$

$$Q = q.B = 8 \times 5 = 40 \text{ kN/m}$$

$$M_q = \frac{qB^2}{2} = \frac{(8 \times 5^2)}{2} = 100 \text{ kN-m/m}$$

Horizontal seismic force on the wall

$$= W_w.A_h = 560 \times 0.1 = 56 \text{ kN/m}$$

Vertical seismic force on the wall

$$= \pm W_w.A_h = \pm 560 \times 0.05 = \pm 28 \text{ kN/m}$$

Moment of seismic forces acting on the weight of wall

$$M_{sw} = \frac{B}{2}[\gamma_1(H^2 - H_w^2) + \gamma_{s1}H_w^2]A_h \pm W_w.A_v.\frac{B}{2}$$

$$= \frac{5}{2}[18(6^2 - 2^2) + 20 \times 2^2] \times 0.1 + 560 \times 0.05 \times \frac{5}{2}$$

$$= 164 + 70 = 234 \text{ kN-m/m}$$

Horizontal seismic force on the surcharge acting on the wall

$$= 8 \times 5 \times 0.1 = 4 \text{ kN/m}$$

Vertical seismic force on the surcharge acting on the wall

$$= \pm 8 \times 5 \times 0.05 = \pm 2 \text{ kN/m}$$

$$M_{sq} = 4 \times 6 \pm 2\left(\frac{5}{2}\right) = 29 \text{ kN-m/m}$$

(Assuming that the force is acting upwards)

External Stability

(a) Sliding

Static case:

$$F_s = \frac{\mu(W'_w + Q)}{P_{Tst}}$$

$$= \frac{0.5(460 + 40)}{86.76} = 3.46 \quad [> 2.0, \text{ safe}]$$

Seismic case:

$$F_s = \frac{\mu(W'_w + Q)(1 \pm \alpha_v)}{P_{Tst} + (P_{a\gamma})_{di} + (P_{aq})_{di} + (W_w + Q)\alpha_h}$$

$$= \frac{0.5(460 + 40)(1 - 0.05)}{86.75 + 26.55 + 1.35 + (560 + 40) \times 0.1}$$

$$= 1.36 \; [< 1.5, \text{ unsafe}]$$

(b) Overturning

Static case:

$$F_v = \frac{(W'_w + Q)\frac{B}{2}}{M_{Tst}}$$

$$= \frac{(460 + 40)\frac{5}{2}}{244.97} = 5.10 \quad [> 2.0, \text{ safe}]$$

Seismic case:

$$F_v = \frac{(W'_w + Q)(1 \pm \alpha_v)\frac{B}{2}}{M_{Tst} + (M_{a\gamma})_{di} + (M_{aq})_{di} + M_{sw} + M_{sq}}$$

$$= \frac{(460 + 40)(1 - 0.05)\frac{5}{2}}{150.01 + 53.95 + 10.54 + 234 + 29}$$

$$= 2.48 \; [> 1.5, \text{ safe}]$$

(c) Tilting/Bearing Failure

Static case:

$$\sigma_{max} = \left(\frac{W'_w + Q}{B}\right) + M_{Tst}\frac{6}{B^2}$$

$$= \left(\frac{460 + 40}{5}\right) + 150.01\frac{6}{5^2}$$

$$= 100 + 36 = 136 \text{ kN/m}^2 \qquad [< 200 \text{ kN/m}^2\text{, safe}]$$

Seismic case:

$$\sigma_{max} = \frac{(W_w' + Q) + (W_w + Q)\alpha_v}{B} + M_{Tn}\frac{6}{B^2}$$

$$M_{Tn} = M_{Tst} + (M_{aq})_{di} + (M_{aq})_{di} + M_{sw} + M_{sq}$$

$$= 150.01 + 53.95 + 10.54 + 234 + 29$$

$$= 477.5 \text{ kN-m/m}$$

$$\sigma_{max} = \frac{(460 + 40) + (560 + 40) \times 0.05}{5.0} + \frac{477.5 \times 6}{5^2}$$

$$= 106 + 115 = 221 \text{ kN/m}^2 \qquad [< 1.25 \times 200 = 250 \text{ kN/m}^2\text{, safe}]$$

$$\sigma_{min} = 106 - 115 = -9 \text{ kN/m}^2 \qquad [< 0.0\text{, unsafe}]$$

Increase the length of reinforcement to 6.0 m.
The only change will be in M_{sw} and M_{sq}

$$M_{sw} = 234 \times \left(\frac{6}{5}\right) = 281 \text{ kN-m/m}$$

$$M_{sq} = 29 \times \left(\frac{6}{5}\right) + \left(\frac{6}{2}\right) = 40.8 \text{ kN-m/m}$$

$$M_{tn} = 150.01 + 53.97 + 10.54 + 281 + 90.8 = 536.32 \text{ kN-m/m}$$

$$\sigma_{max} = 106 + \left(\frac{536.32}{6^2}\right) \times 6 = 106 + 83.4$$

$$= 189.4 \text{ kN/m}^2 \qquad [< 1.25 \times 200 = 250 \text{ kN/m}^2\text{, safe}]$$

$$\sigma_{min} = 106 - 83.4 = 22.6 \text{ kN/m}^2 \qquad [> 0.0\text{, safe}]$$

Internal stability

$$K_{aw} = \frac{\cos^2 \phi_1}{\cos \delta_1} \frac{1}{\left[1 + \left\{\frac{\sin(\phi_1 + \delta_1)\sin \phi_1}{\cos \delta_1}\right\}^{1/2}\right]^2}$$

$$\phi_1 = 40^\circ;\ \delta_1 = \frac{2}{3}\phi_1 = \frac{2}{3} \times 40 = 26.67^\circ$$

$$K_{aw} = \frac{\cos^2 40}{\cos 26.67} \frac{1}{\left[1 + \left\{\frac{\sin(40 + 26.67)\sin(40)}{\cos(26.67)}\right\}^{1/2}\right]^2} = 0.20$$

$$\psi = \tan^{-1}\frac{\alpha_h}{1 \pm \alpha_v}$$

$$= \tan^{-1}\frac{0.1}{1 \pm 0.05} = 5.44^\circ(+\text{sign})$$

$$K_{adw} = \frac{(1 \pm \alpha_h)\cos^2(\phi_1 + \lambda)}{\cos\lambda\ \cos(\delta_1 + \lambda)}\frac{1}{\left[1 + \left\{\frac{\sin(\phi_1 + \delta_1)\sin(\phi_1 - \lambda)}{\cos(\delta_1 + \lambda)}\right\}^{1/2}\right]^2} = 0.265$$

Below Water Level

$$\phi = 40^\circ;\ \delta = \frac{26.67}{2}\ 13.35^\circ$$

$$K'_{aw} = \frac{\cos^2 40}{\cos 13.55}\frac{1}{\left[1 + \left\{\frac{\sin(40 + 13.55)\sin 40}{\cos 13.55}\right\}^{1/2}\right]^2} = 0.202$$

$$\lambda = \tan^{-1}\frac{\gamma_{s1}}{\gamma_{s1} - 10}\cdot\frac{\alpha_h}{1 \pm \alpha_v}$$

$$= \tan^{-1}\frac{21}{21 - 10}\cdot\frac{0.1}{1 \pm 0.05} = 10.30^\circ$$

$$\delta = 13.35$$

$$K_{adw} = \frac{(1 \pm 0.05)\cos^2(40 - 10.30)}{\cos 10.30\ \cos(13.35 + 10.30)}\frac{1}{\left[1 + \left\{\frac{\sin(40 + 13.35)\ \sin(40 - 10.30)}{\cos(13.35 + 10.30)}\right\}^{1/2}\right]^2}$$

$$= 0.309$$

Static case:
For $h_i \leq H - H_w$

$$M_1 = \frac{1}{6}\gamma_2 K_a h_i^3 + \frac{qh_i^2}{2}K_a - \frac{c\sqrt{K_a}\cdot h_i^2}{2}$$

$$= \frac{1}{6}\times 18 \times 0.217 h_i^3 + \frac{8h_i^2}{2}\times 0.217 - \frac{c\sqrt{0.217}\cdot h_i^2}{2}$$

$$= 0.651 h_i^3 + 0.4022 h_i^2$$

$$\sigma_{vi} = (\gamma_i h_i + q) + M_1 \times \frac{6}{B^2}$$

$$= (18h_i + 8) + (0.651 h_i^3 + 0.4022 h_i^2)\times\frac{6}{6^2}$$

$$= 0.1085 h_i^3 + 0.0670.h_i^2 + 18h_i + 8$$

Safe design strength = 16.5 kN/m

In limiting case : $K_{aw}\sigma_{vi}S_z = R_T$

$$(0.20)(0.1085h_i^3 + 0.0670h_i^2 + 18h_i + 8)S_z = 16.5$$

h_i (m)	S_z (m)
0	10.3
1.0	3.15
2.0	1.83
3.0	1.26
4.0	0.94

For $h_i > (H - H_w)$

$$h_{wi} = h_i - (H - H_w) = h_i - 4$$

$$M_{a\gamma} = \frac{1}{6}\gamma_2 K_a (h_i + 2h_i - 8)\,(4)^2 + \frac{1}{2}\gamma_2 K'_a (4)(h_i - 4)^2 + \frac{1}{6}\gamma_{b2} K'_a (h_i - 4)^3$$

$$= \frac{1}{6} \times 18 \times 0.217(3h_i - 8) \times 16 + \frac{1}{2} \times 18 \times 0.221(4)(h_i - 4)^2 + \frac{1}{6} \times 10 \times 0.221(h_i - 4)^3$$

$$= 10.42(3h_i - 8) + 7.96(h_i - 4)^2 + 0.368(h_i - 4)^3$$

$$= 31.26h_i - 83.36 + 7.96h_i^2 - 63.68h_i^2 + 127.36h_i^3 + 0.368h_i^3 - 4.42h_i^2 + 17.66h_i - 23.55$$

$$= 0.368h_i^3 + 3.54h_i^2 - 14.76h_i + 20.45$$

$$M_{aq} = \frac{1}{2}qK_a(4)(3h_i - 8) + \frac{1}{2}qK'_a(h_i - 4)^2$$

$$= \frac{1}{2} \times 8 \times 0.217 \times 4(3h_i - 8) + \frac{1}{2} \times 8 \times 0.221(h_i - 4)^2$$

$$= 3.472(3h_i - 8) + 0.884(h_i - 4)^2$$

$$= 10.416h_i - 27.776 + 0.884h_i^2 - 7.072h_i + 14.144$$

$$= 0.884h_i^2 + 3.344h_i - 13.632$$

$$M_{ac} = c\sqrt{K_a}(3h_i - 8)\,(4) + c\sqrt{K'_a}\,(h_i - 4)^2$$

$$= 2 \times \sqrt{0.217}\,(3h_i - 8)\,(4) + 2\sqrt{0.221}\,(h_i - 4)^2$$

$$= 11.18h_i - 14.91 + 0.94h_i^2 - 7.52h_i + 15.04$$

$$= 0.94h_i^2 + 3.66h_i + 0.13$$

$$M_{aw} = \frac{1}{6}\gamma_w (h_i - 4)^3$$

$$= \frac{1}{6} \times 10(h_i^3 - h_i^2 + 48h_i - 64)$$

$$= 1.67h_i^3 - 20h_i^2 + 80h_i - 106.67$$

$$M_2 = M_{a\gamma} + M_{aq} + M_{ac} + M_{aw}$$

$$= 2.04h_i^3 - 16.52h_i^2 + 64.92h_i - 99.98$$

$$\sigma_{\gamma i} = \gamma_1(H - H_w) + \gamma_{s1}(h_i - H + H_w) + M_2 \times \frac{6}{B^2}$$

$$= 19 \times 4 + 21(h_i - 4) + \frac{2.04h_i^3 - 16.52h_i^2 + 64.42h_i - 99.98}{6^2} \times 6$$

$$= 21h_i - 8 + 0.3h_i^3 - 2.75h_i^2 + 18.82h_i - 16.66$$

$$= 0.34h_i^3 - 2.75h_i^2 + 39.82h_i - 24.66$$

$$K'_a \sigma_{vi} S_z = 16.5$$

$$0.221(0.34h_i^3 - 2.75h_i^2 + 39.82h_i - 24.66) \cdot S_z = 16.5$$

h_i (m)	S_z (m)
5.0	0.50
6.0	0.39

Usually the spacing between reinforcing layers is not kept more than 0.5 m. It is therefore recommended that the reinforcement may be laid throughout the height of wall uniformly at a vertical spacing (S_z) of 0.4 m.

Seismic case:

For $h_i \leq (H - H_w)$

$$M_3 = \frac{\gamma_2(K_{ad} - K_a)h_i^2}{4h_i}(h_i^3)$$

$$= \frac{18(0.264 - 0.217)}{4} h_i^3 = 0.2115h_i^3$$

$$M_4 = \frac{q(K_{ad} - K_a)h_i}{h_i}(h_i^2)$$

$$= 8(0.264 - 0.217)h_i^2 = 0.376h_i^2$$

$$M_{Tni} = M_1 + M_3 + M_4 + \alpha_h h_i \left(\frac{\gamma_1 L h_i}{2} + Q\right)$$

$$= 0.651h_i^3 + 0.4022h_i^2 + 0.2115h_i^3 + 0.376h_i^2 + 0.1h_i \times \left(\frac{19 \times 6 \times h_i}{2} + 8 \times 6\right)$$

$$= 0.8625h_i^3 + 6.4872h_i^2 + 4.8h_i$$

$$\sigma_{vi} = (\gamma_1 h_i + q)(1 \pm \alpha_v) + M_{Tni} \cdot \frac{6}{B^2}$$

$$= (19h_i + 8)(1 + 0.05) + (0.8625h_i^3 + 6.4872h_i^2 + 4.8h_i) \times \frac{6}{6^2}$$

$$= 0.14375\,h_i^3 + 1.0812h_i^2 + 20.75h_i + 8.4$$

In limiting case:

$$K_{ad}\sigma_{vi}S_z = 1.25R_T$$

$$0.264(0.14275h_i^3 + 1.0812\,h_i^2 + 20.75\,h_i + 8.4)\cdot S_z = 1.25 \times 16.5$$

or

$$(h_i^3 + 7.521h_i^2 + 144.35h_i + 58.43)\cdot S_z = 543.48$$

h_i (m)	S_z (m)
0	9.0
1.0	2.57
2.0	1.43
3.0	0.96
4.0	0.66

For $h_i > (H - H_w)$

$$M_5 = \frac{\gamma_2(K_{ad} - K_a)[h_i(h_i - 4)]^2}{4h_i}[h_i^2 + 2h_i(h_i - 4) + 3(h_i - 4)^2]$$

$$+\left(\frac{K'_{ad} - K'_a}{4h_i}\right)(h_i - 4)^3\,[4\gamma_2\{h_i - 4\} + \gamma_{b2}(h_i - 4)]$$

$$M_5 = \frac{18(0.264 - 0.217)\times 16}{4h_i}[6h_i^2 + 32h_i + 48] + \left(\frac{0.351 - 0.221}{4h_i}\right)$$

$$\times [h_i^3 - 12h_i^2 + 48h_i - 64][4 \times 18 \times 4 + 10h_i - 40]$$

$$= 20.30h_i - 108.29 + \frac{162.53}{h_i} + \frac{0.26}{h_i} \times [1.25h_i^4 + 16h_i^3 - 312h_i^2 + 1408h_i - 1984]$$

$$= 0.325h_i^3 + 4.16h_i^2 - 60.816h_i + 257.79 - \frac{353.41}{h_i}$$

$$M_{Tni} = M_2 + M_5 + M_6 + A_h.B \times \left[\gamma_1(H - H_w)\left\{h_i - \frac{H - H_w}{2}\right\} + \frac{\gamma_{s1}(h_i - H + H_w)^2}{2}\right] + Q\cdot A_h H$$

$$= 2.04h_i^3 - 16.52h_i^2 + 64.92h_i - 99.98 + 0.325h_i^3 + 4.16h_i^2 - 60.816h_i + 257.79$$

$$-\frac{353.41}{h_i} + 1.04h_i^2 - 7.968h_i + 31.872 - \frac{42.50}{h_i} + 0.1 \times 6\left[19 \times 4(h_i - 2) + \frac{21(h_i - 4)^2}{2}\right]$$

$$+ 8 \times 6 \times 0.1 \times 6$$

$$= 2.04h_i^3 - 5.02h_i^2 - 8.664h_i + 228.08 - \frac{310.91}{h_i}$$

$$\sigma_{vi} = \left[19(6-2) + 21(h_i - 6 + 2)(1 + 0.05) + M_{Tni}\frac{6}{6^2}\right]$$

$$= 22.05h_i - 8.40 + 0.34h_i^3 - 0.837h_i^2 - 1.444h_i + 38.01 - \frac{51.81}{h_i}$$

$$= 0.34h_i^3 - 0.837h_i^2 + 20.606h_i + 29.61 - \frac{51.81}{h_i}$$

In limiting case:

$$K'_{ad}\sigma_{vi}S_z = 1.25R_T$$

$$0.351\left[0.35h_i^3 - 0.837h_i^2 + 20.606h_i + 29.61 - \frac{51.81}{h_i}\right]S_z = 1.25 \times 16.5$$

or

$$\left[h_i^3 - 2.39h_i^2 + 58.87h_i + 84.67 + \frac{148.03}{h_i}\right]S_z = 167.88$$

h_i (m)	S_z (m)
5.0	0.405
6.0	0.336

In selecting practical spacing of grids, the designer should ensure that these are sufficient to provide equilibrium for all potential wedge failures. Let us assume that the arrangement as shown in Fig.15.7 below is selected on the basis of above calculation.

Author's experience indicated that normally tension in each of the participating grid is maximum when the full height of wall is considered.

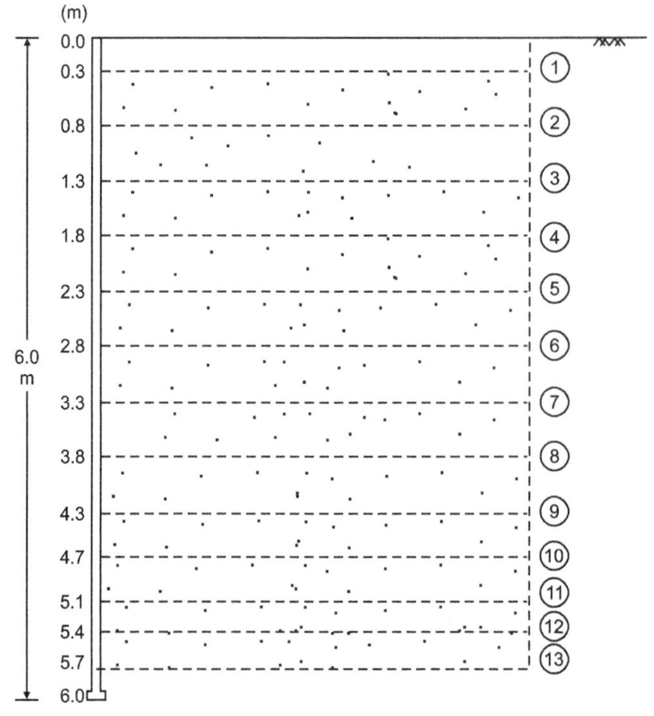

Fig. 15.7. Arrangement of reinforcement (Example 15.1)

Static case:

$$a_1 = \frac{1}{2}[\gamma_s H_w^2 + \gamma_1(H^2 - H_w^2)]$$

$$= \frac{1}{2}[21 \times 2^2 + 19(6^2 - 2^2)] = 346 \text{ kN}$$

$$a_2 = a_1 + qH = 346 + 8 \times 6 = 394 \text{ kN}$$

$$T = a_2 \frac{\tan\left(45 - \frac{\phi_1}{2}\right)}{\tan\left(45 + \frac{\phi_1}{2}\right)}$$

$$= 394 \frac{\tan\left(45 - \frac{40}{2}\right)}{\tan\left(45 + \frac{40}{2}\right)} = 85.67 \text{ kN}$$

$$\text{Tension per grid} = \frac{85.67}{13} = 6.6\,\text{kN} \quad [<14.5\,\text{kN, safe}]$$

Seismic case:

$$a_3 = A_h\, a_2 = 0.1 \times 394 = 39.4 \text{ kN}$$

$$a_4 = a_2(1 \pm A_v) = 394(1 + 0.05) \text{ [Adopting (+) sign]}$$

$$= 413 \text{ kN}$$

$$a_5 = a_4 \cos 2\phi_1 - a_4 + a_3 \sin 2\phi_1$$

$$= 413 \cos 80^\circ - 413 + 39.4 \sin 80^\circ$$

$$= 71.71 - 413 + 38.8 = -302.48$$

$$a_6 = a_4 \sin 2\phi_1 - a_3 \cos 2\phi_1$$

$$= 413 \sin 80^\circ - 39.4 \cos 80^\circ$$

$$= 406.72 - 6.84 = 399.88$$

$$\tan\beta = \frac{-2a_5 \pm \sqrt{4a_5^2 - a(a_3^2 - a_6^2)}}{2(a_2 - a_6)}$$

$$\tan\beta = \left(\frac{-2(-302.48) \pm \sqrt{4(302.48)^2 - 4(39.4^2 - 399.88^2)}}{2(39.4 - 399.88)}\right)$$

$$= \left(\frac{604.96 \pm \sqrt{365976 + 683406.6}}{-720.96}\right)$$

$$= \frac{604.96 \pm 999.69}{-720.96} = 0.5475$$

$$\beta = 28.7^\circ$$

$$T = \frac{a_4 \tan\beta}{\tan(\phi_a + \beta)} + a_3 \tan\beta$$

$$= \frac{413 \tan 28.7^\circ}{\tan(40^\circ + 28.7^\circ)} + 39.4 \tan 28.7^\circ$$

$$= 88.15 + 21.57 = 109.72 \text{ kN}$$

$$\text{Tension per grid} = \frac{109.72}{13} = 8.44 \text{ kN} \quad [<1.25 \times 14.5 = 18 \text{kN, safe}]$$

Pullout failure

Static case:

$$L_{ip} = \frac{T_i F_{ss}}{2\alpha \tan\phi_1 (\gamma_1 h_1 + q)}$$

$$= \frac{6.6 \times 2.0}{2 \times 0.9 \tan 40^\circ (19 \times 0.5 + 8)} = 0.5 \text{ m}$$

Seismic case:

$$L_{ip} = \frac{8.44 \times 1.5}{2 \times 0.9 \tan 40^\circ (19 \times 0.5 + 8)} = 0.478 \text{ m}$$

Required length of reinforcement $= L_{ip} + (H - h_i)\tan\beta$

$$= 0.5 + (6.0 - 0.5) \tan 25^\circ$$

$$= 0.5 + 2.56 = 3.06 \text{ m} \quad [< 6.0 \text{ m, Hence safe}]$$

References

IS: 1893: 2016, "Earthquake Resistant Design of Structures", ISI, New Delhi

Vidal, H. (1978), "The Development and Future of Reinforced Earth", Key Note Address, Symposium on Earth Reinforcement, ASCE, Pittsburgh, pp. 1-66.

Practice Problems

15.1 Describe stepwise the procedure of designing a reinforced earth wall.

15.2 Design a reinforced soil wall for retaining a 7.0 m high cohesionless backfill. The soil in the wall and backfill has a density of 18 kN/m^3 with angle of internal friction of 35°. The allowable soil pressure is 150 kN/m^2. The wall is situated in a seismic region having horizontal seismic coefficient as 0.15. Use grid reinforcement having permissible tensile strength as 15 kN/m.

CHAPTER

16

Walls Having Reinforced Sand Backfills

16.1 General

The technique of reinforcing the soil has been widely accepted as an economical alternative construction technique for earth retaining structures and improving the poor ground. The usefulness of the patented reinforced earth retaining wall of Vidal (1966) has been proved to be economical by thousands of such structures constructed over the world. But situations can be met where reinforced earth walls may not provide ideal solution. This can be true for locations with limited spaces behind the wall or for narrow hill roads on unstable slopes which may not permit use of designed length of reinforcement. In such circumstances rigid walls with reinforced backfill may appear more appropriate. In these walls, the backfill is reinforced with unattached horizontal strips/mats/nets laid normal to the wall.

In this chapter, an analysis has been presented for getting active earth pressures for retaining walls having reinforced backfill and subjected to earthquake forces.

16.2 Brief Review

Till mid seventies, all the investigations (both analytical and experimental) were directed towards the conventional reinforced earth retaining walls. Very few investigators have paid attention to the problem of retaining walls having reinforced backfill.

Broms (1977, 1987) analysed the internal and external stability of a retaining wall with attached and unattached continuous (or sheet) fabric reinforcement. The wall facing was made up of precast L-shaped units and the reinforcement was laid horizontally in the backfill. The reinforcement was assumed to increase the lateral confining pressure which results in small lateral earth pressure on L-shaped panels. Part of the weight of soil was assumed to be carried by the soil friction along the fabric and the surrounding soil. The lateral earth pressure which the backfill can withstand at any depth at distance x from the wall face is given as

$$\sigma'_h = \sigma'_{ho} e^{(2x \tan \psi / S_z \cdot K_b)} \tag{16.1}$$

where σ'_{h_o} and σ'_h are the effective lateral earth pressures at the wall face and at distance x respectively, S_Z, the vertical spacing of reinforcement, ψ, the angle of soil-fabric friction and K_b, the coefficient of lateral earth pressure which is greater than Rankine active earth pressure coefficient, K_a. The stress, σ'_{h_o} was assumed constant along the wall height, as for anchored sheet pile walls (Tezaghi and Peck, 1967), for the design purposes. For transferring the tension in the fabric from the backfill soil, anchorage length of reinforcement for both tied and untied reinforcement has been computed.

Tied reinforcement needs an anchorage length towards the rear. The untied reinforcements need anchorage length of fabric in contact with the horizontal base of the L-shaped facing elements.

Talwar (1981) developed theoretical analysis for computing earth pressure distribution, total pressure and its point of application behind a retaining wall with a vertical back and retaining cohesionless backfill reinforced with unattached strips (Fig. 16.1) . The analysis was for two different conditions of apparent coefficient of soil-reinforcement friction (f^*) namely, (a) f^* proportional to the effective length of reinforcing strip and (b) a constant value of f^*. Reinforcement properties were considered in analysis in terms of a non-dimensional parameter, D_p and $\frac{L}{H}$ ratio. L is the length of reinforcing strip and D_p is defined by an expression:

$$D_p = \frac{w \cdot f^* \cdot H}{S_x \cdot S_z} \tag{16.2a}$$

where w = Width of strip

H = Height of wall

S_x and S_z = horizontal and vertical spacing of reinforcing strips respectively

In case of mat type of reinforcement,

$$D_p = \frac{f^*.H}{S_x} \tag{16.2b}$$

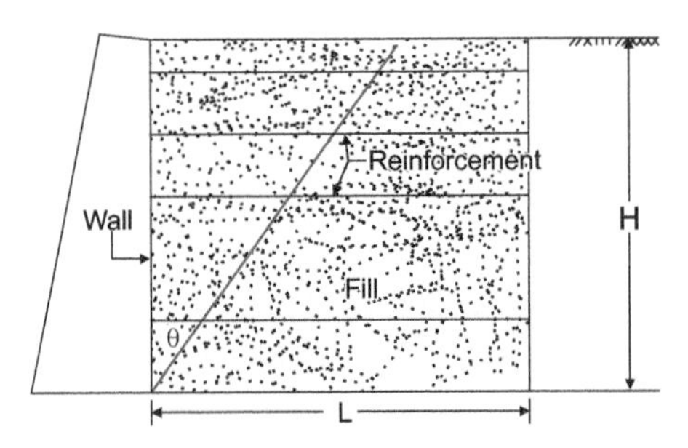

Fig. 16.1. Wall with reinforced backfill (Talwar, 1981)

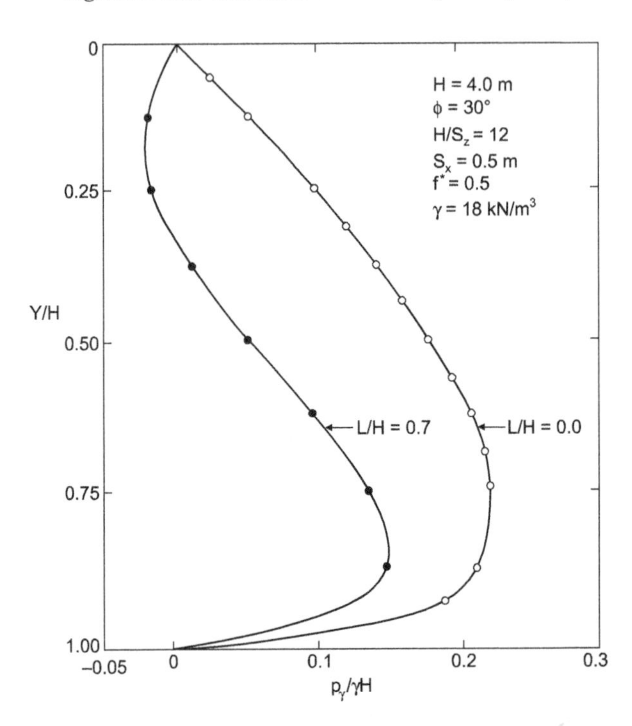

Fig. 16.2. Pressure intensity on the wall due to backfill (Talwar, 1981)

Typical pressure distribution along the wall height is shown in Fig. 16.2. It may be noted that lateral pressure intensity becomes negative when D_p is equal to 2 and $\frac{L}{H} = 0.7$. Similar trends are observed for other values of $\frac{L}{H}$ and D_p. Since earthfill cannot sustain any tension, the value of resultant pressure and its location of point of application were obtained by numerically integrating the positive portion of the pressure intensity diagrams. The results have been plotted in the form of non-dimensional parameters $\frac{P_\gamma}{\gamma H^2}$ and $\bar{H}/H$, where P_γ is the total active earth pressure and $\bar{H}$ is the height of point of application from the base of wall. γ represents the dry unit weigh of backfill soil. Two typical non-dimensional are given in Figs. 16.3 and 16.4. It is evident from these figures that resultant earth pressure reduces significantly with the increase in the L/H ratio and D_p factor. The decrease in earth pressure was found more significant upto $\frac{L}{H}$ ratio equal to 0.6 and D_p factor 1.0. The height of point of application of resultant pressure was found to increase slightly for short reinforcing strips of upto 0.4 H and reduced sharply for lengths in excess of 0.5 H for the case of

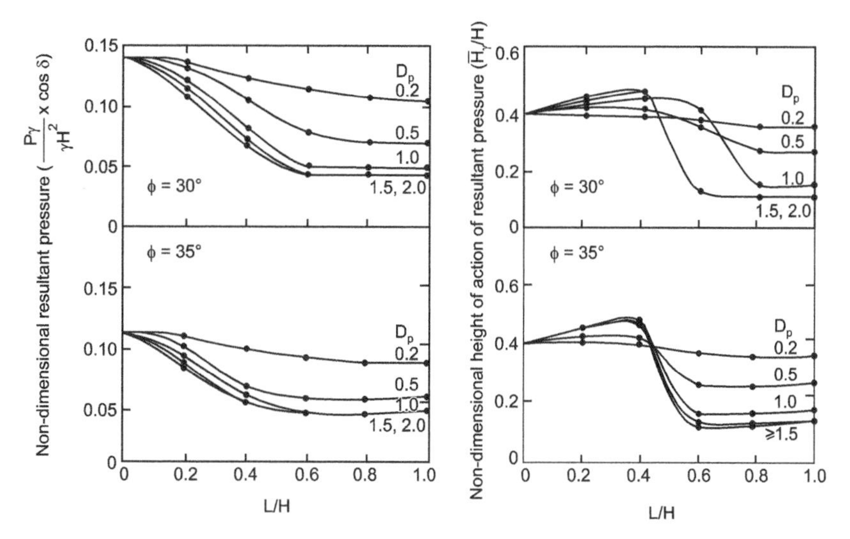

Fig. 16.3. Normalized resultant pressures and their heights of action for f^* = Constant

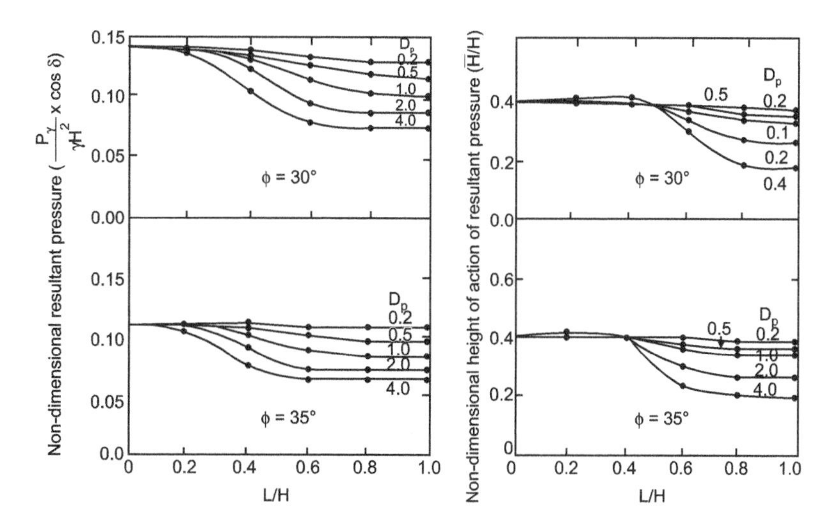

Fig. 16.4. Normalized resultant pressures and their heights of action for f^* variable (Talwar, 1981)

constant value of f^* whereas there was no increase in it for varying values of f^* for shorter strips and decreased appreciably for longer lengths of strips.

Garg (1988) extended the work of Talwar (1981) for uniformly distributed surcharge and presented the results in the form of similar non-dimensional charts considering the earth pressure due to backfill soil and surcharge separately. A typical pressure distribution along the height of wall for additional earth pressure due to surcharge is shown in Fig. 16.5. It was concluded that unattached reinforcements are quite effective in reducing the lateral earth pressure on the wall both due to backfill and surcharge loading.

Garg (1988) also developed a concept of effective placement of reinforcement (EPR) (Fig. 16.6). In this case the reinforcements are placed in such a way that these are bisected by the rupture failure line. A typical earth pressure distribution in EPR case is shown in Fig. 16.7. In the case of an effective placement of reinforcement (EPR), the total earth pressure was found much lesser than with the normal placement of reinforcement (NPR) (Table 16.1).

Table 16.1. Comparison of Total Earth Pressure (After Garg, 1988)

	NPR, $D_p = 2$				*EPR*, $D_p = 2$			
	$K_\gamma = P_\gamma/1/2\,\gamma H^2$		$K_q = P_q/qH$		$K_\gamma = P_\gamma/1/2\,\gamma H^2$		$K_q = P_q/qH$	
L/H	$\phi = 30°$	$\phi = 40°$	$\phi = 30°$	$\phi = 40°$	$\phi = 30°$	$\phi = 40°$	$\phi = 30°$	$\phi = 40°$
0.2	0.225	0.1225	0.275	0.175	0.065	0.050	0.0075	0.040
0.4	0.140	0.0625	0.215	0.0975	0.065	0.050	0.2055	0.035
0.6	0.095	0.0575	0.110	0.090	0.065	0.050	0.2055	0.035
0.8	0.095	0.0575	0.110	0.090	0.065	0.050	0.2055	0.035

NPR – Normal placement of reinforcement
EPR – Effective placement of reinforcement

Both Talwar (1981) and Garg (1988) have developed analysis for retaining wall with vertical face having reinforced backfill. Khan (1991) for the first time extended their work for inclined retaining wall having reinforced backfill with uniformly distributed surcharge on it (Fig. 16.8). Typical results obtained by Khan (1991) are reproduced in Fig. 16.9. In Fig. 16.9 P_T represents

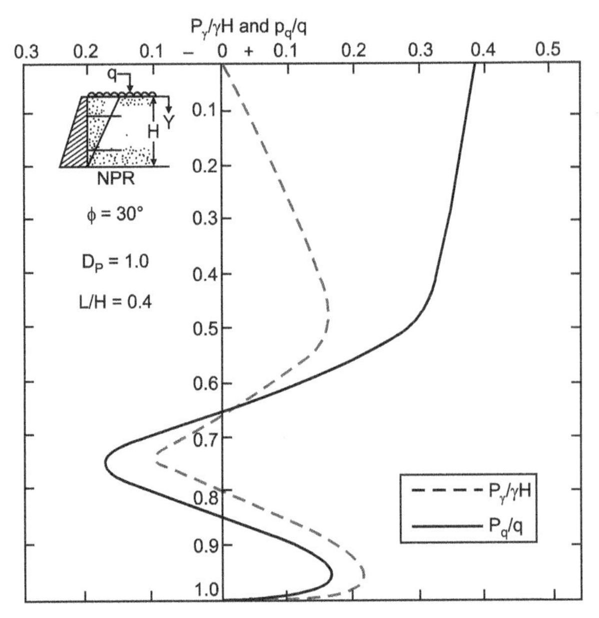

Fig. 16.5. Non-dimensional pressure intensity coefficients versus depth (NPR) (Garg, 1988)

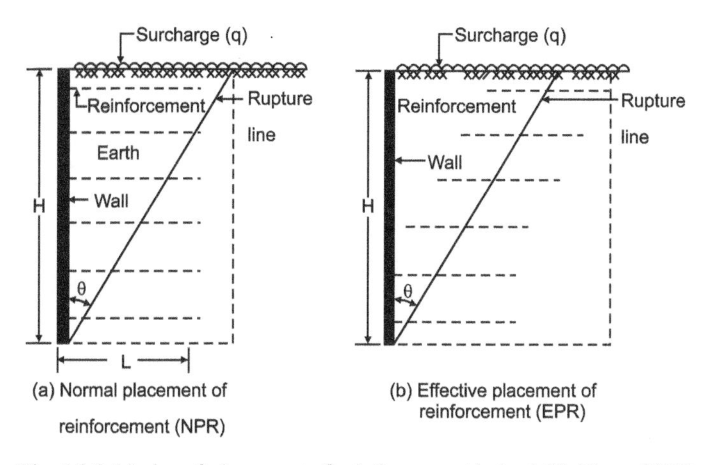

Fig. 16.6. Modes of placement of reinforcement in backfill (Garg, 1988)

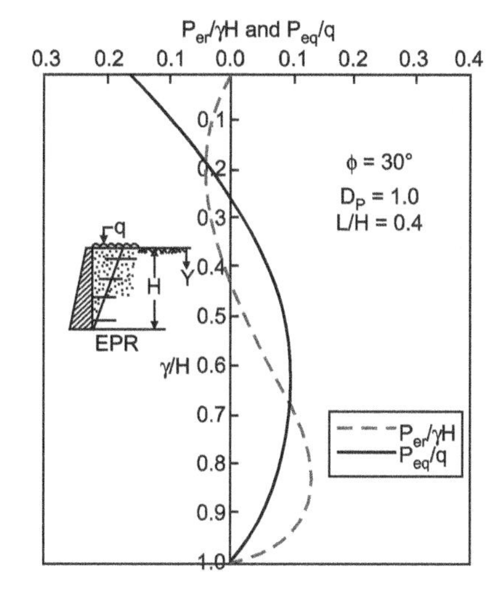

Fig. 16.7. Non-dimensional pressure intensity coefficients versus depth (EPR) (Garg, 1988)

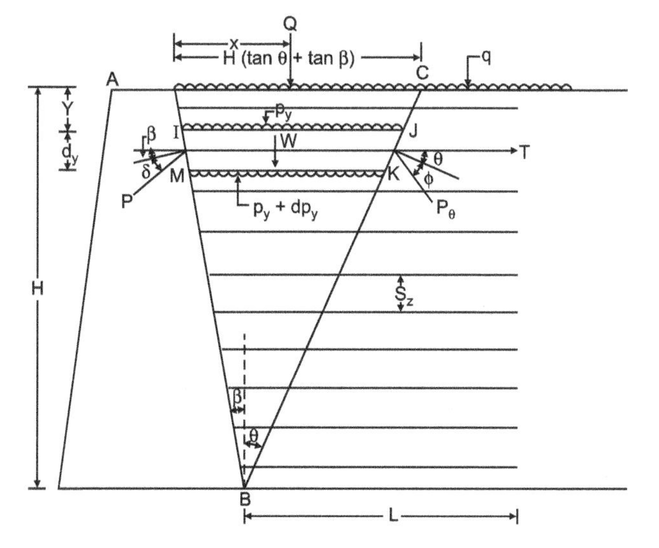

Fig. 16.8. Wall details with reinforcement (Khan, 1991)

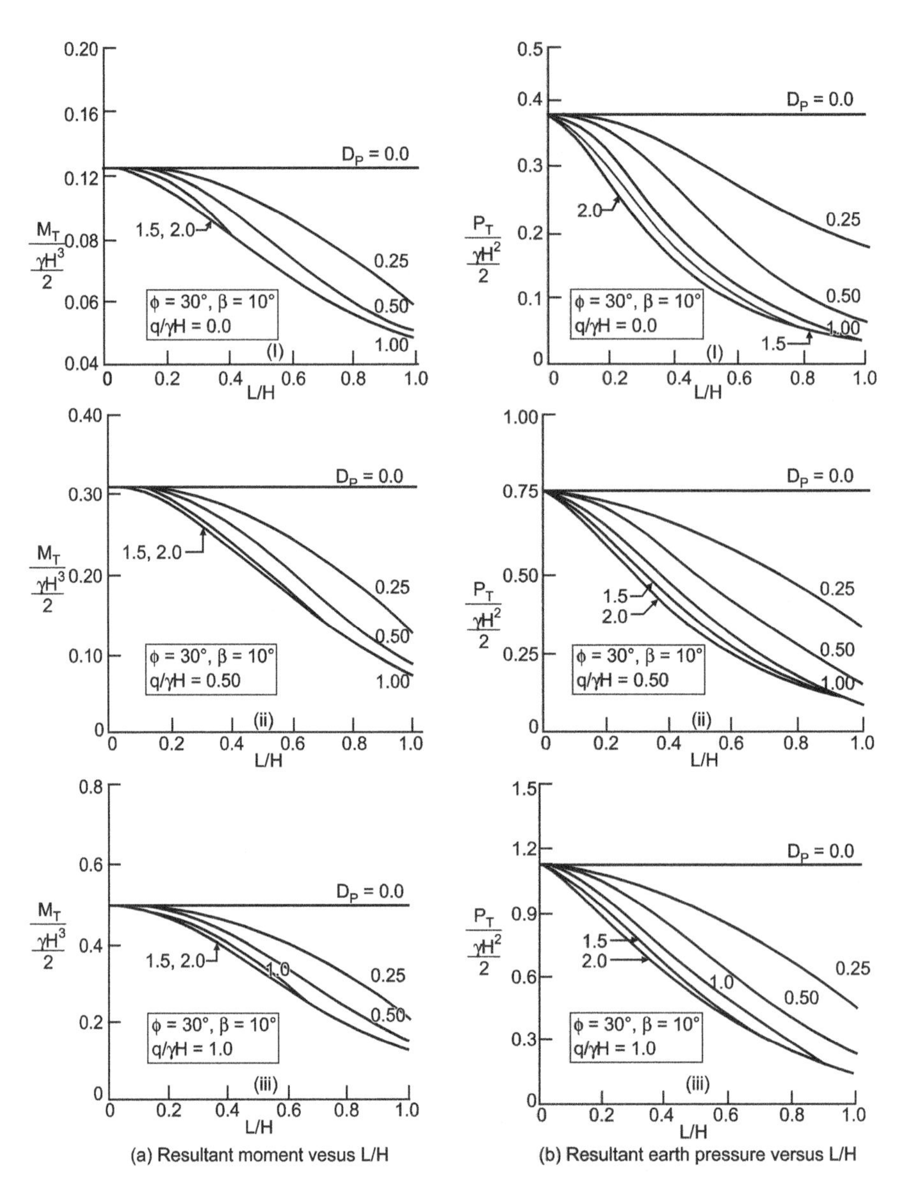

Fig. 16.9. Total earth pressure and corresponding moment about wall base (Khan, 1991)

total earth pressure and M_T, the moment of total earth pressure about the base. The main conclusion derived from this study was that the provision of reinforcement in the backfill of inclined retaining wall is equally effective as in vertical retaining walls.

In this chapter, a pseudo-static analysis of a rigid vertical retaining wall having reinforced backfill has been presented considering both vertical and horizontal acceleration in terms of seismic coefficients (α_v and α_h) (Saran, 1998).

16.3 Analysis

Assumptions

The following assumptions were made in developing a pseudo-static of a rigid retaining wall having reinforced backfill subjected to earthquake forces:

(i) The backfill is homogenous, isotropic and non-cohesive.
(ii) The failure surface is a plane passing through the heel of retaining wall.
(iii) The coefficient of friction between soil and reinforcement is independent of the overburden pressure and the length of reinforcement.
(iv) The failure plane divides the length of reinforcing strips into two zones, one that lies within failure wedge and another which lies outside. Only that part of strip which experiences movement of soil relative to itself is assumed to be contributing to frictional resistance.
(v) The retaining wall undergoes an outward movement or rotation about the base which is sufficient to cause mobilization of full frictional resistance in soil as well as the reinforcing strips.
(vi) Dynamic force caused by earthquake is replaced by equivalent static forces using seismic coefficients α_v and α_h.

Consider a retaining wall of height H with vertical back, retaining a horizontal cohesionless backfill of dry density γ, and angle of internal friction, ϕ (Fig. 16.10). It is reinforced with unattached horizontally laid strips of length, L and width, w at a vertical spacing S_z and horizontal spacing, S_x. A failure plane AC making an angle, θ with the vertical passes through the heel of the retaining wall. The frictional resistance to the lateral movement of the wedge ABC, contributed by reinforcing strips, is computed from its effective length. Effective length L_e, is the portion of the strip which experiences movement of soil relative to itself. Reinforcing strip located completely within the moving wedge will not contribute any frictional resistance to the movement of wedge.

Considering the equilibrium of an element *IJKM* of thickness, dy of failure wedge ABC, located at depth, y from the top of the wedge (Fig. 16.10). Following forces per unit length of the wall act on the element of the wedge ABC.

p_y : Pressure intensity acting on IJ in the vertical direction

$(p_y + dp_y)$: Pressure intensity acting on KM in the vertical direction

p_θ : Reaction intensity on JK acting at an angle to the normal to JK

p : Pressure intensity on IM acting at an angle with the normal to IM

W : Weight of slice $IJKM$ acting downwards

$$= \frac{\gamma dy}{2}[(H - y)\tan\theta + (H - y - dy)\tan\theta] \tag{16.3}$$

T : Tensile force in the strip assumed to be transmitted uniformly to soil layer of thickness, S_Z encompassing the strip,

$$t = \frac{T}{S_z} = \frac{2 \cdot w \cdot f^* L_e \sigma_v}{S_x S_z} \tag{16.4}$$

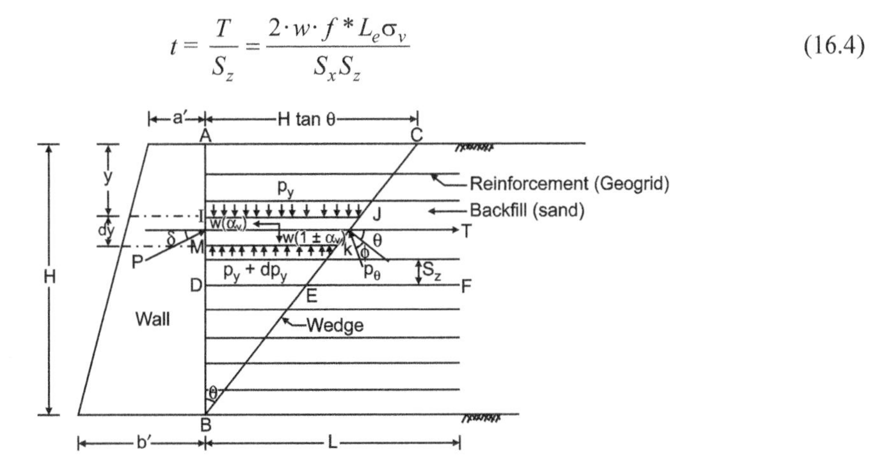

Fig. 16.10. Details of wall with reinforced earth backfill

$W \cdot \alpha_h$: Horizontal seismic force acting at the centre of gravity of slice

$\pm W \cdot \alpha_v$: Vertical seismic force acting at the centre of gravity of slice

The expressions for intensity of earth pressure, total pressure and height of point of application above the base are obtained by writing the equations of equilibrium as below.

(i) Balancing all the forces acting on the slice in horizontal direction:

$$p \cdot \cos\delta \cdot dy - p_\theta \cdot \cos(\theta+\phi) \cdot \sec\theta \cdot dy + t \cdot dy - \frac{\gamma}{2} \cdot \{(H-y)\tan\theta + (H-y-dy)\tan\theta\}\,\alpha_h \cdot dy = 0$$

which simplifies to

$$p_\theta = \frac{t + p \cdot \cos\delta - \gamma \cdot \alpha_h \left(H - y - \dfrac{dy}{z}\right) \cdot \tan\theta}{\cos(\theta+\phi) \cdot \sec\theta} \tag{16.5}$$

(ii) Balancing all the factors acting on the slice in vertical direction:

$$p_y\,(H-y)\tan\theta + W \cdot (1 \pm A_v)$$

$$= p\sin\delta \cdot dy + p_\theta \cdot \sin(\theta+\phi) \cdot \sec\theta \cdot dy + (p_y + dp_y)\,(H - y - dy) \cdot \tan\theta$$

or

$$\gamma \cdot (1 \pm \alpha_v) \cdot (H-y) \cdot dy - \frac{\gamma}{2} \cdot (1 \pm \alpha_v) \cdot (dy)^2$$

$$= \frac{p \cdot \sin\delta \cdot dy}{\tan\theta} + \frac{p_\theta \cdot \sin(\theta+\phi)}{\sin\theta} \cdot dy - p_y \cdot dy + (H-y) \cdot dp_y - dp_y \cdot dy$$

Neglecting small quantities of second order, the expression reduces to

$$\frac{dp_y}{dy} = \gamma \cdot (1 \pm \alpha_v) - \frac{p \cdot \sin\delta}{(H-y) \cdot \tan\theta} \cdot \frac{p_\theta \cdot \sin(\theta+\phi)}{(H-y) \cdot \sin\theta} + \frac{p_y}{(H-y)} \tag{16.6}$$

(iii) Taking moments of all the forces about the mid point of slice between J and K

$$p_y \cdot (H-y)\tan\theta \cdot \left[\frac{(H-y) \cdot \tan\theta}{2} - \frac{dy \cdot \tan\theta}{2}\right] - (p_y + dp_y) \cdot [(H-y-dy) \cdot \tan\theta]$$

$$\left[\frac{(H-y-dy) \cdot \tan\theta}{2} + \frac{dy \cdot \tan\theta}{2}\right] - p \cdot \sin\delta \cdot dy \left[H - y - \frac{dy}{2}\right] \cdot \tan\theta$$

$$+ \frac{\gamma dy}{2}\{(H-y) \cdot \tan\theta + (H-y-dy) \cdot \tan\theta\} \cdot \left[H - y - \frac{dy}{2}\right] \frac{\tan\theta}{2} \cdot (1 \pm \alpha_v) = 0$$

Neglecting small quantities of higher order yields:

$$\frac{dp_y}{dy} = \gamma \cdot (1 + \alpha_v) \frac{2p \cdot \sin\delta}{(H-y) \cdot \tan\theta} \tag{16.7}$$

Substituting the value of p_θ from Eq. (16.5), Eq. (16.6) reduces to

$$\frac{dp_y}{dy} = \frac{p_y + \gamma(H-y) \cdot (1 \pm \alpha_v)}{(H-y)} - \frac{p \cdot \sin\delta}{(H-y) \cdot \tan\theta} - \frac{(t + p \cdot \cos\delta)}{(H-y)} \cdot \frac{\tan(\theta+\phi)}{\tan\theta}$$

$$+ \frac{\gamma \cdot \alpha_h \cdot \left(H - y - \dfrac{dy}{2}\right) \cdot \tan(\theta+\phi)}{(H-y)} \tag{16.8}$$

Equating right hand sides of Eqs. (16.7) and (16.8), one gets

$$p = \frac{p_y \cdot \tan\theta}{\cos\delta \cdot \tan(\theta+\phi) - \sin\delta} - \frac{t \cdot \tan(\theta+\phi)}{\cos\delta \cdot \tan(\theta+\phi) - \sin\delta}$$

$$+ \frac{\gamma \cdot \alpha_h \cdot \left(H - y - \frac{dy}{2}\right) \cdot \tan\theta \cdot \tan(\theta+\phi)}{\cos\delta \cdot \tan(\theta+\phi) - \sin\delta} \tag{16.9}$$

On differentiating and adopting proper sign for dynamic increment

$$\frac{dp_y}{dy} = \frac{dp_y}{dy} \cdot \frac{\tan\theta\cos(\theta+\phi)}{\sin(\theta+\phi-\delta)} - \frac{dt}{dy} \cdot \frac{\sin(\theta+\phi)}{\sin(\theta+\phi-\delta)}$$

$$+ \frac{\gamma \,.\, \alpha_h \tan\theta.\, \sin(\theta+\phi-\delta)}{\sin(\theta+\phi-\delta)}$$

Substituting for $\frac{dp_y}{dy}$ from Eq. (16.7) and simplifying, one gets

$$\frac{dp}{dy} = \frac{\gamma \cdot (1+\alpha_v) \cdot \tan\theta \cdot \cos(\theta+\phi)}{\sin(\theta+\phi-\delta)} - \frac{2p \cdot \sin\delta \cdot \cos(\theta+\phi)}{(H-y) \cdot \sin(\theta+\phi-\delta)}$$

$$- \frac{dt}{dy} \cdot \frac{\sin(\theta+\phi)}{\sin(\theta+\phi-\delta)} + \frac{\gamma \cdot \alpha_v \cdot \tan\theta \cdot \sin(\theta+\phi)}{\sin(\theta+\phi-\delta)} \tag{16.10}$$

Equation (16.10) can be reduced to

$$\frac{dp}{dy} = -C_1 \frac{p}{H-y} + C_2 \cdot \gamma \cdot - C_3 \frac{dt}{dy} \tag{16.11}$$

where

$$C_1 = \frac{2\sin\delta \cdot \cos(\theta+\phi)}{\sin(\theta+\phi-\delta)} \tag{16.12a}$$

$$C_2 = \frac{\tan\theta}{\sin(\theta+\phi-\delta)}[(1 \pm \alpha_v)\cos(\theta+\phi) + \alpha_h \cdot \sin(\theta+\phi)] \tag{16.12b}$$

$$C_3 = \frac{\sin(\theta+\phi)}{\sin(\theta+\phi-\delta)} \tag{16.12c}$$

At limiting equilibrium, the tension T in the strip can be assumed as:

$$T = \frac{2.w.f^* \sigma_v L_e}{S_x} \tag{16.12d}$$

where L_e = Effective length of strip

σ_v = Vertical stress on the strip

f^* = Apparent coefficient of soil-reinforcement friction

w = Width of the strip

S_x = Horizontal spacing of reinforcement strips

Assuming $$\sigma_v = \left(y + \frac{dy}{2}\right).\gamma, \tag{16.13}$$

$$t = \frac{T}{S_z} = \frac{2 \cdot w \cdot f^* \sigma_v L_e}{S_x \cdot S_z} = \frac{2 \cdot w \cdot f^* \gamma \cdot \left(y + \frac{dy}{2}\right)}{S_x \cdot S_z} \cdot L_e \tag{16.14}$$

where S_z = Vertical spacing of reinforcing strips.

The values of L_e will vary from strip to strip and will depend on the angle, θ and length, L of the strip. Three cases may arise (Fig. 16.11), viz.

Case 1: $H \tan \theta \leq L/2$

$$L_e = (H - y) \cdot \tan \theta, \text{ for all strips} \tag{16.15}$$

Case 2: $H \tan \theta \leq L/2 < L$

$$L_e = L - (H - y) \cdot \tan \theta, \text{ for } y \leq Z_1 \text{ and} \tag{16.16a}$$

$$L_e = (H - y) \cdot \tan \theta, \text{ for } y > Z_1 \tag{16.16b}$$

Case 3: $H \tan \theta > L$

$$L_e = 0 \quad \text{for } y \leq Z_2 \tag{16.17a}$$

$$L_e = L - (H - y) \cdot \tan \theta, \quad \text{for } Z_2 \leq y \leq Z_3 \tag{16.17b}$$

$$L_e = (H - y) \cdot \tan \theta, \quad \text{for } y > Z_3 \tag{16.17c}$$

For Z_1, Z_2 and Z_3 refer to Fig. 16.11.

The expression for pressure intensity p for the three cases are determined below.

Pressure Intensity on the Wall

Case 1: $H \tan \theta \leq L/2$

$$t = \frac{2 \,.\, w \,.\, f^* \gamma \left(y + \frac{dy}{z}\right).\tan \theta}{S_x \,.\, S_z} \tag{16.18}$$

[because, $L_e = (H - y).\tan \theta$ for all strips]

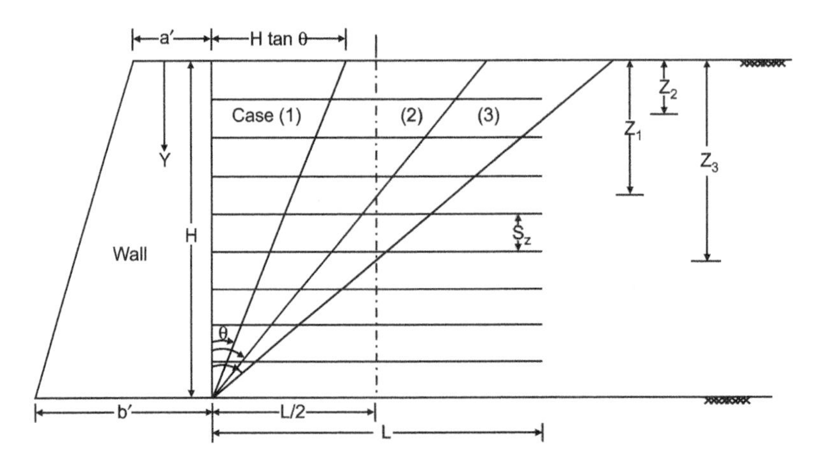

Fig. 16.11. Schematic representation of three cases of analysis

After neglecting small quantities of second order,

$$\frac{dt}{dy} = \frac{2 \cdot w \cdot f^{*}\gamma \cdot (H-2y) \cdot \tan\theta}{S_x \cdot S_z} = k\,(H-2y) \tag{16.19}$$

where
$$k = \frac{2 \cdot w \cdot f^{*}\gamma \cdot \tan\theta}{S_x \cdot S_z} \tag{16.19a}$$

Equation (16.11) becomes:

$$\frac{dp}{dy} = -C_1 \frac{p}{(H-y)} + C_2\gamma - C_4(H-2y) \tag{16.20}$$

where
$$C_4 = C_3\,k \tag{16.20a}$$

The solution of the above differential equation for the boundary condition $p = 0$ at $y = 0$ would be

$$p_1 = -\frac{C_z\gamma H}{1-C_1}\cdot\left[\left(1-\frac{y}{H}\right)-\left(1-\frac{y}{H}\right)^{C_1}\right]+\frac{C_4H^2}{1-C_1}\cdot\left[\left(1-\frac{y}{H}\right)^{C_1}-\left(1-\frac{y}{H}\right)\right]$$

$$-\frac{2C_4H^2}{2-C_1}\cdot\left[\left(1-\frac{y}{H}\right)^{C_1}-\left(1-\frac{y}{H}\right)^{2}\right]$$

or
$$p_1 = -\left[\frac{C_2\gamma H}{1-C_1}-\frac{C_4H^2}{2-C_1}\right]\cdot\left[\left(1-\frac{y}{H}\right)-\left(1-\frac{y}{H}\right)^{C_1}\right]$$

$$+\frac{2C_4H^2}{2-C_1}\left[\left(1-\frac{y}{H}\right)^{2}-\left(1-\frac{y}{H}\right)^{C_1}\right] \tag{16.21}$$

where p_1 is the pressure intensity corresponding to case 1.

Case 2: $L/2 \le H \tan\theta < L$

For $y \le Z_1$ the effective length of strip to be considered is $L - (H - y)\cdot\tan\theta$. The corresponding expressions for t and dt/dy are:

$$t = k\left(y+\frac{dy}{2}\right)\left[\frac{L}{\tan\theta}-(H-y)\right] \tag{16.22}$$

On differentiating above expression and neglecting derivatives of second order

$$\frac{dt}{dy} = k\left[\frac{L}{\tan\theta}-(H-2y)\right] \tag{16.23}$$

Equation (16.11) therefore becomes:

$$\frac{dp_2}{dy} = -C_1\frac{p_2}{(H-y)} + C_2\cdot\gamma - C_4\cdot\left[\frac{L}{\tan\theta}-(H+2y)\right] \tag{16.24}$$

The solution of Eq. (16.24) is:

$$p_2 = \frac{H}{1-C_1}\left[\left(-C_2\gamma+\frac{C_4L}{\tan\theta}+C_4\cdot H\right)\right]\cdot\left[\left(1-\frac{y}{H}\right)-\left(1-\frac{y}{H}\right)^{C_1}\right]$$

$$-\frac{2C_4H^2}{2-C_1}\left[\left(1-\frac{y}{H}\right)^2-\left(1-\frac{y}{H}\right)^{C_1}\right] \tag{16.25}$$

For $y > Z_1$, the effective length L_e will be $(H-y)\tan\theta$ and pressure intensity p_2' can be determined by integrating Eq. (16.20) over the domain $y = Z_1$ to $y = H$ for the boundary condition that at $y = Z_1$, $p_2' = (p_2)$ where p_2' represents the pressure intensity at any depth $Z_1 < y < H$.

The expression for p_2' has been obtained as

$$p_2' = \frac{H}{1-C_1}(-C_2\gamma - C_4\cdot H)\cdot\left(1-\frac{y}{H}\right)+\frac{2C_4H^2}{2-C_1}\cdot\left(1-\frac{y}{H}\right)^2+D_1\cdot H^{C_1}\cdot\left(1-\frac{y}{H}\right)^{C_1} \tag{16.26}$$

where

$$D_1 = \left\{\left[\frac{H}{1-C_1}\left(-C_2\gamma+\frac{C_4L}{\tan\theta}+C_4\cdot H\right)\right]\left[\left(1-\frac{Z_1}{H}\right)^2-\left(1-\frac{Z_1}{H}\right)^{C_1}\right]\right.$$

$$-\frac{2C_4H^2}{2-C_1}\left[\left(1-\frac{Z_1}{H}\right)^2-\left(1-\frac{Z_1}{H}\right)^{C_1}\right]\cdot\frac{1}{H^{C_1}\cdot\left(1-\frac{Z_1}{H}\right)^{C_1}}$$

$$\left.-\frac{1}{H^{C_1}}\cdot\left\{\left[-\frac{C_2H\gamma}{1-C_1}-\frac{C_4H^2}{1-C_1}\right]\left(1-\frac{Z_1}{H}\right)^{1-C_1}+\frac{2C_4H^2}{2-C_1}\left(1-\frac{Z_1}{H}\right)^{1-C_1}\right\}\right. \tag{16.26a}$$

Case 3: $H\tan\theta > L$

The pressure intensity, p_3 for this case will have three different expression for the three domains, namely:

$y = 0$ to $y = Z_2$, $y = Z_2$ to $y = Z_3$ and $y = Z_3$ to $y = H$.

For the domain, $y = 0$ to $y = Z_2$, the failure surface will not cut any reinforcing strip and pass only through soil. In other words, C_4 will be equal to zero and p_3 will be given as:

$$p_3 = -\frac{C_2\,.\,\gamma\,.\,H}{1-C_1}\left[\left(1-\frac{y}{H}\right)-\left(1-\frac{y}{H}\right)^{C_1}\right] \tag{16.27}$$

For the region $y = Z_2$ to $y = Z_3$, the equation for pressure intensity, p_3' can be evaluated by integrating Eq. (16.24) with the boundary condition that at $y = Z_2, p_3' = (p_3)_y = Z_2$. Ultimately

$$p_3' = \left(-C_2\gamma+\frac{C_4L}{\tan\theta}+C_4\cdot H\right)\cdot\frac{H}{1-C_1}\cdot\left(1-\frac{y}{H}\right)-\frac{2C_4H^2}{2-C_1}\left(1-\frac{y}{H}\right)$$

$$+D_2\cdot H^{C_1}\left(1-\frac{y}{H}\right) \tag{16.28}$$

where

$$D_2 = -\frac{C_2\cdot\gamma\cdot H}{1-C_1}\left[\left(1-\frac{Z_2}{H}\right)-\left(1-\frac{Z_2}{H}\right)^{C_1}\right]\cdot\frac{1}{H^{C_1}\cdot\left(1-\frac{Z_2}{H}\right)^{1-C_1}}$$

$$-\frac{1}{H^{C_1}}\cdot\left[-\frac{C_2\cdot\gamma\cdot H}{1-C_1}+\mathrm{L}\cdot\mathrm{H}\cdot\frac{C_4/\tan\theta}{1-C_1}+\frac{C_4H^2}{1-C_1}\right]\cdot\left(1-\frac{Z_2}{H}\right)^{1-C_1}$$

$$+\frac{2C_4H^2}{2-C_1}\cdot\frac{1}{H^{C_1}}\cdot\left(1-\frac{Z_2}{H}\right)^{2-C_1} \tag{16.28a}$$

For the domain $y = Z_3$ to $y = H$, the pressure intensity p_3'' can be computed by solving the Eq. (16.20) for the boundary condition that at $y = Z_3$, $p_3' = (p_3')_y = Z_3$. The solution is

$$p_3'' = \left[-\frac{C_2\cdot\gamma\cdot H}{1-C_1}-\frac{C_4H^2}{2-C_1}\right]\left(1-\frac{y}{H}\right)+\frac{2C_4H^2}{2-C_1}\left(1-\frac{y}{H}\right)^2+D_3\cdot H\left(1-\frac{y}{H}\right) \tag{16.29}$$

where

$$D_3 = \left\{\left(-C_2\gamma+\frac{C_4L}{\tan\theta}+C_4\cdot H\right)\cdot\frac{H}{1-C_1}\left(1-\frac{Z_3}{H}\right)-\frac{2C_4H^2}{2-C_1}\left(1-\frac{Z_3}{H}\right)+D_2\cdot H^{C_1}\left(1-\frac{Z_3}{H}\right)^{C_1}\right\}$$

$$\cdot\frac{1}{H^{c_1}\cdot\left(1-\frac{Z_3}{H}\right)^{C_1}}-\left[-\frac{C_2\cdot\gamma\cdot H}{1-C_1}\cdot\left(1-\frac{Z_3}{H}\right)^{1-C_1}-\frac{C_4H}{1-C_1}\left(1-\frac{Z_3}{H}\right)^{1-C_1}+\frac{2C_4H^2}{2-C_1}\left(1-\frac{Z_3}{h}\right)^{2-C_1}\right]$$

$$\cdot\frac{1}{H^{C_1}}$$

Resultant Pressure Acting on the Wall

The total pressure acting on the wall corresponding to the position of failure surface in cases 1, 2 and 3 can be obtained by integrating the equations for pressure intensity over their respective domains.

Case 1: The total pressure, P_1 is given by:

$$p_1 = \int_0^H p_1 dy = \frac{1}{1+C_1}\left[\frac{C_2\cdot\gamma\cdot H^2}{1-C_1}+\frac{C_4\cdot H^3}{1-C_1}-\frac{2C_4\cdot H^3}{2-C_1}\right]+\frac{2C_4\cdot H^3}{3(2-C_1)}$$

$$-\frac{C_2\cdot\gamma\cdot H^2}{2(1-C_1)}-\frac{C_4\cdot H^3}{2(1-C_1)} \tag{16.30}$$

Case 2: The total pressure, P_2 is given by:

$$P_2 = \int_0^{Z_1} p_2 dy + \int_{z_1}^{H} p_2' dy$$

The expressions for p_2 and p_2' are given by Eqs. (16.25) and (16.26)

Thus,

$$p_2 = \frac{Q^{1+C_1}}{1+C_1}(K_1-K_2+D_1\cdot H^{1+C1})+\frac{Q^2}{2}(K_4-K_1+K_5)+\frac{Q^3}{3}\cdot 2K_2$$

$$+\frac{1}{1+C_1}(K_2-K_1)+\frac{K_1}{2}-\frac{K_2}{3} \tag{16.31}$$

where

$$Q = \frac{L}{2H}.\cot\theta \tag{16.31a}$$

$$K_1 = \left[-C_2 \cdot \gamma + \frac{L}{C_4 \tan\theta} + C_4 H\right]\frac{H^2}{1-C_1} \tag{16.31b}$$

$$K_2 = \frac{2C_4 H^3}{2-C_1} \tag{16.31c}$$

$$K_4 = \frac{-C_2 \cdot \gamma H^3}{1-C_1} \tag{16.31d}$$

$$K_5 = \frac{C_4 \cdot H^3}{1-C_1} \tag{16.31e}$$

$$K_6 = K_2 \tag{16.31f}$$

Case 3: The total pressure, P_3 has been evaluated as

$$P_3 = \int_0^{Z_2} p_3 dy + \int_{Z_2}^{Z_3} p_3' dy + \int_{Z_3}^{H} p_3'' dy \tag{16.32}$$

The final expression obtained is

$$P_3 = \frac{(2Q)^{1-C_1}}{1+C_1}\left[K_4 + D_2 \cdot H^{1+C_1}\right] + \frac{Q^{1+c_1}}{1+C_1} \cdot (D_3 - D_2) \cdot H^{1+C1}$$

$$+\frac{Q^2}{2} \cdot (3K_1 - 3K_4 - K_5) - \frac{Q^3}{3} 6K_2 + K_4 \cdot \left(\frac{1}{2} - \frac{1}{1+C_1}\right) \tag{16.32a}$$

Point of Action of Resultant Pressure

The moment acting on the wall can be determined if the point of action of resultant earth pressure, P is known. The height of point of action can be determined in each of the three cases by the relation:

$$\bar{H} = H - \frac{\int_0^H p \cdot y \cdot dy}{\int_0^H p \cdot dy} \tag{16.33}$$

where H = height of point of action above base.

Case 1: *H*. tan θ ≤ *L*/2

$$\overline{\mathbf{H}_1} = \mathbf{H} - \frac{\int_0^H p_1 \cdot y \cdot dy}{\int_0^H p_1 \cdot dy} = H - \frac{NR_1}{P_1} \tag{16.34}$$

NR_1 has been computed as:

$$NR_1 = \frac{1}{2+C_1} \cdot (M_4 - M_5 + M_6) - \frac{1}{1+C_1}(M_4 - M_5 + M_6) + \left(\frac{M_4}{6} - \frac{M_5}{6} + \frac{M_6}{12}\right) \tag{16.34a}$$

where

$$M_4 = K_4 \cdot H \tag{16.34b}$$

$$M_5 = K_5 \cdot H \quad \text{and} \tag{16.34c}$$

$$M_6 = K_6 . H \tag{16.34d}$$

Case 2: *L*/2 ≤ *H*. tan θ < *L*

$$\overline{\mathbf{H_2}} = \mathbf{H} - \frac{\int_0^{Z_1} p_2 \cdot y \cdot dy + \int_{Z_1}^{H} p_2' \cdot y \cdot dy}{\int_0^{Z_1} p_2 \cdot dy + \int_{Z_1}^{H} p_2' \cdot dy} = H - \frac{NR_2}{P_2} \tag{16.35}$$

NR_2 is obtained as:

$$NR_2 = \frac{Q^{2+C_1}}{2+C_1} \cdot [M_2 - D_1 \cdot H^{2+C_1} - M_1] + \frac{Q^{1+C_1}}{1+C_1} \cdot [M_1 - M_2 + D_1 \cdot H^{2+C_1}]$$

$$-\frac{Q^2}{2}(M_1 - M_4 + M_5) + \frac{Q^3}{3}(M_1 + 2M_2 - M_4 + M_5) - \frac{Q^4}{4} \cdot 2M_6$$

$$+\frac{1}{2+C_1}(M_1 - M_6) - \frac{1}{1+C_1}(M_1 - M_6) + \frac{M_1}{6} + \frac{M_2}{12} \tag{16.35a}$$

where

$$M_1 = K_1 . H \tag{16.35b}$$

$$M_2 = K_2 . H \tag{16.35c}$$

and other terms are as already defined.

Case 3: H.tan θ > *L*

$$\overline{\mathbf{H_3}} = \mathbf{H} - \frac{\int_0^{Z_2} p_3 \cdot y \cdot dy + \int_{Z_2}^{Z_3} p_3' \cdot dy + \int_{Z_3}^{H} p_3'' \cdot y \cdot dy}{\int_0^{Z_2} p_3 \cdot dy + \int_{Z_2}^{Z_3} p_3' \cdot dy + \int_{Z_3}^{H} p_3'' \cdot dy} = H - \frac{NR_3}{P_3} \tag{16.36}$$

where

$$NR_3 = \frac{(2Q)^{2+C_1}}{2+C_1}(-M_4 - D_2 H^{2+C_1}) + \frac{(2Q)^{2+C_1}}{1+C_1}(M_4 + D_2 H^{2+C_1})$$

$$+\frac{Q^{2+C_1}}{2+C_1} \cdot (D_2 - D_3) \cdot H^{2+C_1} - \frac{Q^{1+C_1}}{1+C_1} \cdot (D_2 - D_3) + \frac{Q^2}{2} \cdot (-3M_4 + 3M_1 - M_5)$$

$$+\frac{Q^3}{3} \cdot [7(M_4 - M_1) - 6M_2 + M_5] + 14M_2 \cdot \frac{Q^4}{4} + M_4\left(\frac{1}{6} - \frac{1}{1+C_1} + \frac{1}{2+C_1}\right) \tag{16.36a}$$

where M_1, M_2, M_4 and M_5 have been defined earlier in Eqs. (16.35b), (16.35c), (16.34b) and (16.34c) respectively.

Non-dimensional Pressure and its Point of Action

The expressions for pressure intensities given in Eqs. (16.21), (16.25), (16.26), (16.27), (16.28) and (16.29) can also be written in non-dimensional form by dividing both sides of equation by γH and expressing the properties of reinforcement by a non-dimensional design parameter, D_p. Equation (16.21) for case 1 can be re-written as:

Case 1: H.tan θ ≤ *L*/2

$$\frac{p_1}{\gamma H} = D_p \left\{ \begin{array}{c} \left[\left(1-\frac{y}{H}\right)-\left(1-\frac{y}{H}\right)^{C_1}\right]\left[-\frac{2\cdot C_3\cdot\tan\theta}{1-C_1}-\frac{C_2}{1-C_1}\cdot\frac{1}{D_p}\right] \\ +\frac{4C_3\tan\theta}{2-C_1}\left[\left(1-\frac{y}{H}\right)^2-\left(1-\frac{y}{H}\right)^{C_1}\right] \end{array} \right\} \tag{16.37}$$

where $$D_p = \frac{f^*\cdot w\cdot H}{S_x\cdot S_z} \tag{16.37a}$$

and $$C_4 = \frac{2\cdot C_3\cdot w\cdot f^*\cdot\gamma\cdot\tan\theta}{S_x\cdot S_z} \tag{16.37b}$$

The non-dimensional design parameter describes all the characteristics of reinforcement except its length.

Case 2: $L/2 \le H.\tan\theta < L$

$$\frac{p_2}{\gamma H} = D_p\cdot \left\{ \begin{array}{c} \left(-\frac{C_2}{1-C_1}\cdot\frac{1}{D_p}+\frac{2\cdot C_3}{1-C_1}\cdot\frac{L}{H}+\frac{2\cdot C_3\tan\theta}{1-C_1}\right)\left[\left(1-\frac{y}{H}\right)-\left(1-\frac{y}{H}\right)^{C_1}\right] \\ -\frac{4\cdot C_3\cdot\tan\theta}{2-C_1}\left[\left(1-\frac{y}{H}\right)^2-\left(1-\frac{y}{H}\right)^{C_1}\right] \end{array} \right\} \tag{16.38}$$

and

$$\frac{p_2'}{\gamma H} = D_p \left\{ \begin{array}{c} \left(-\frac{C_2}{1-C_1}\cdot\frac{1}{D_p}-\frac{2\cdot C_3\tan\theta}{1-C_1}\cdot\left(1-\frac{y}{H}\right)+\frac{4.C_3\cdot\tan\theta}{2-C_1}\cdot\left(1-\frac{y}{H}\right)^2\right) \\ +D_1'\left(1-\frac{y}{H}\right)^{C_1} \end{array} \right\} \tag{16.39}$$

where

$$D_1' = \frac{D_1.H^{C_1}}{D_p.\gamma.H} \tag{16.39a}$$

Case 3: $\tan\theta > L/H$

$$\frac{p_3}{\gamma\cdot H} = D_p\left\{-\frac{C_2}{1-C_1}\cdot\frac{1}{D_p}\left[\left(1-\frac{y}{H}\right)-\left(1-\frac{y}{H}\right)^{C_1}\right]\right\} \tag{16.40}$$

$$\frac{p_3'}{\gamma H} = D_p \left\{ \begin{array}{c} \left[-\frac{C_2}{1-C_1}\cdot\frac{1}{D_p}+\frac{2.C_3}{1-C_1}\cdot\frac{L}{H}+\frac{2\cdot C_3\tan\theta}{1-C_1}\right]\left(1-\frac{y}{H}\right) \\ -\frac{4\cdot C_3\cdot\tan\theta}{2-C_1}\cdot\left(1-\frac{y}{H}\right)^2+D_2'\cdot\left(1-\frac{y}{H}\right)^{C_1} \end{array} \right\} \tag{16.41}$$

where

$$D_2' = \frac{D_2 \cdot H^{C_1}}{D_p \cdot \gamma \cdot H} \tag{16.41a}$$

$$\frac{p_3''}{\gamma H} = D_p \left\{ \begin{aligned} &\left[-\frac{C_2}{1-C_1} \cdot \frac{1}{D_p} - \frac{2 \cdot C_3 \tan\theta}{1-C_1} \right] \left(1 - \frac{y}{H}\right) \\ &+ \frac{4 \cdot C_3 \cdot \tan\theta}{2-C_1} \cdot \left(1 - \frac{y}{H}\right)^2 + D_3' \cdot \left(1 - \frac{y}{H}\right)^{C_1} \end{aligned} \right\} \tag{16.42}$$

where

$$D_3' = \frac{D_3 \cdot H^{C_1}}{D_p \cdot \gamma \cdot H} \tag{16.42a}$$

The expression for normalized resultant pressure (i.e. total pressure) P can be similarly framed by dividing Eqs. (16.30), (16.31), and (16.32) by $\left(\frac{1}{2}\gamma \cdot H^2\right)$ and expressing the characteristics of reinforcement in terms of D_p. The equation for position of point of action of the resultant pressure can also be handled accordingly to express the normalized value $\bar{h}$.

16.4 Design Charts

The normalised horizontal pressure and height of its point of application in non-dimensional form considering f^* as constant for different values of α_h (0 to 0.15) and φ (30°, 35° and 40°) are given in Figs. 16.12 to 16.23. Value of seismic coefficient α_v is taken as $\frac{\alpha_h}{2}$ in preparing these charts. Further, computations have been carried out adopting (+) α_v and (–) α_v separately. It was found that (+)α_v case is critical and therefore charts for this are given here. For $\alpha_v = 2/3\alpha_h$, values of $P/\frac{1}{2}\gamma \cdot H^2$ and $\bar{H}/H$ (Figs. 16.12 to 16.23) may be increased by ten percent, and for in between value linear interpolation may be done. It can be inferred from these charts that the values of non-dimensional

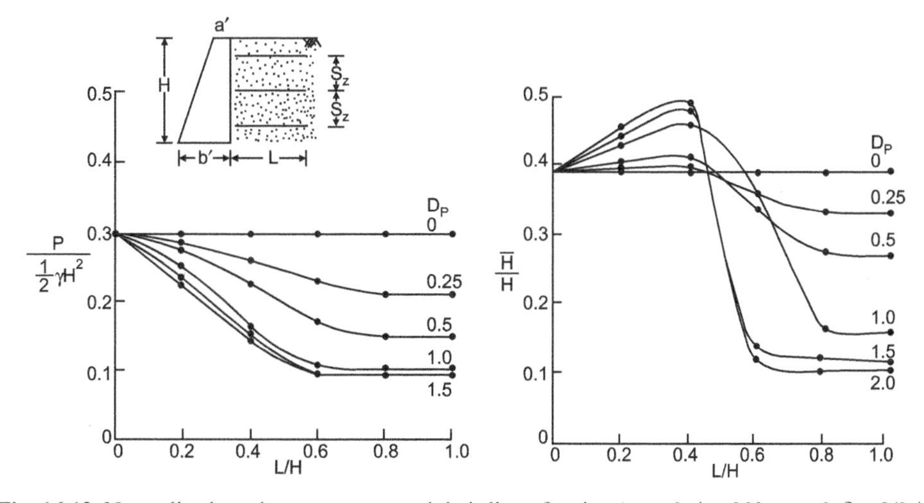

Fig. 16.12. Normalized resultant pressures and their line of action ($\alpha_h = 0$, $\phi = 30°$, $\alpha_v = 0$, $\delta = 2/3\ \phi$)

coefficient, $P/\left(\frac{1}{2}\gamma \cdot H^2\right)$ reduces sharply upto $\frac{L}{H}$ ratio of 0.8 and thereafter either it remains constant or reduces marginally depending upon the value of ϕ and D_p. It can also be seen that the value of non-dimensional coefficient $P/\left(\frac{1}{2}\gamma \cdot H^2\right)$ reduces with increase in the value of angle of shearing resistance for the same value of seismic coefficients. Based on these design charts, it seems that the length of reinforcing strips equal to 0.8 times the height of wall can be taken as an optimum length for field applications. These curves also reveal that major reduction in earth pressure is achieved for a value of D_p equal to 1.0. There is no significant reduction in the value of $P/\left(\frac{1}{2}\gamma \cdot H^2\right)$ with an increase in the value of D_p from 1.0 to 1.5 and beyond 1.5. $P/\left(\frac{1}{2}\gamma \cdot H^2\right)$ becomes almost constant for values of D_p greater than 1. This is because of ignoring negative pressures along the wall. Figures

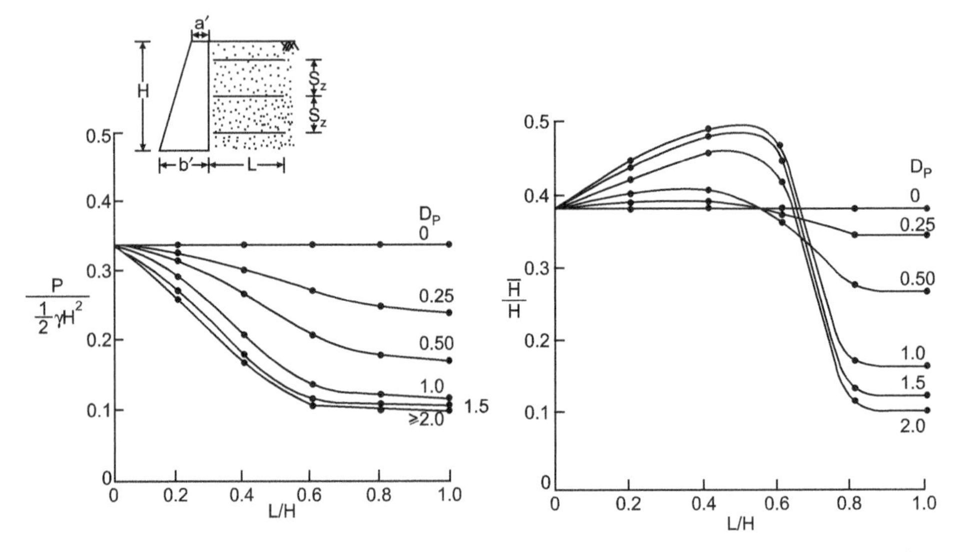

Fig. 16.13. Normalized resultant pressures and their line of action (α_h = 0.05, ϕ = 30°, α_v = + α_h /2, δ = 2/3 ϕ)

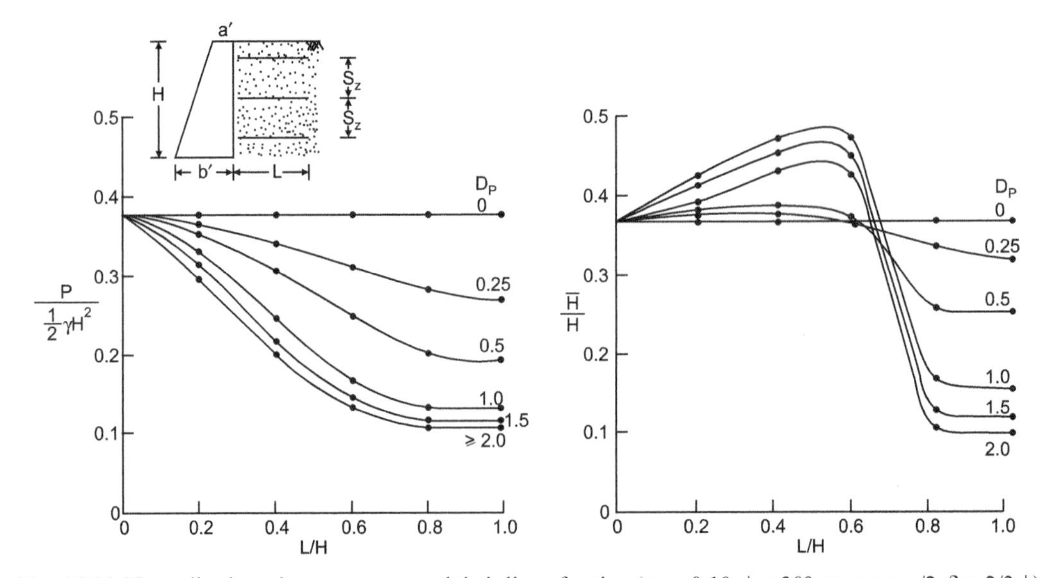

Fig. 16.14. Normalized resultant pressures and their line of action (α_h = 0.10, ϕ = 30°, α_v = + α_h /2, δ = 2/3 ϕ)

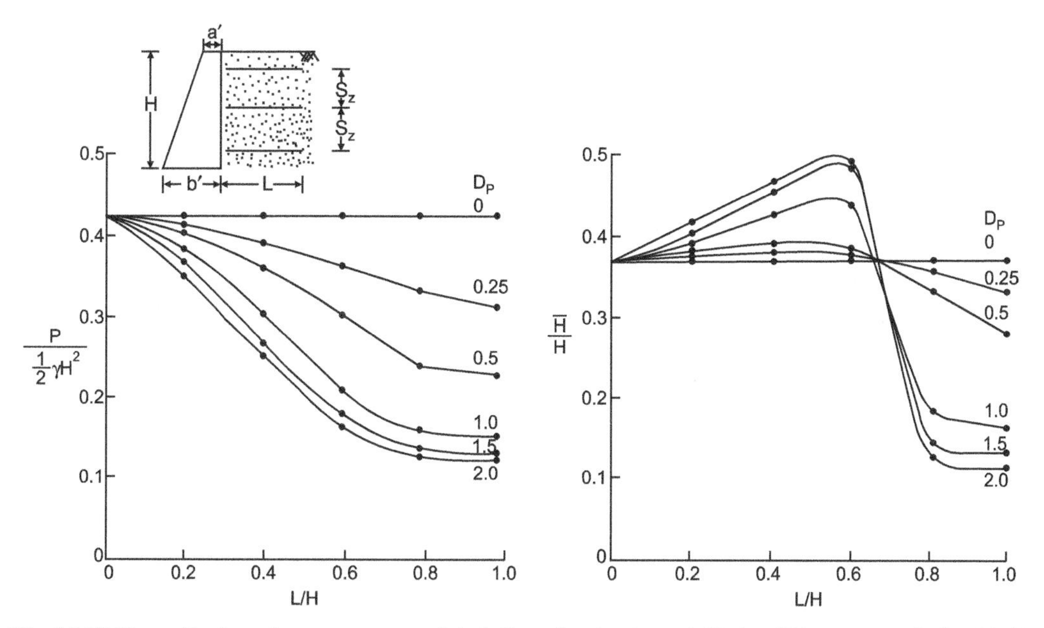

Fig. 16.15. Normalized resultant pressures and their line of action ($\alpha_h = 0.15$, $\phi = 30°$, $\alpha_v = + \alpha_h /2$, $\delta = 2/3\ \phi$)

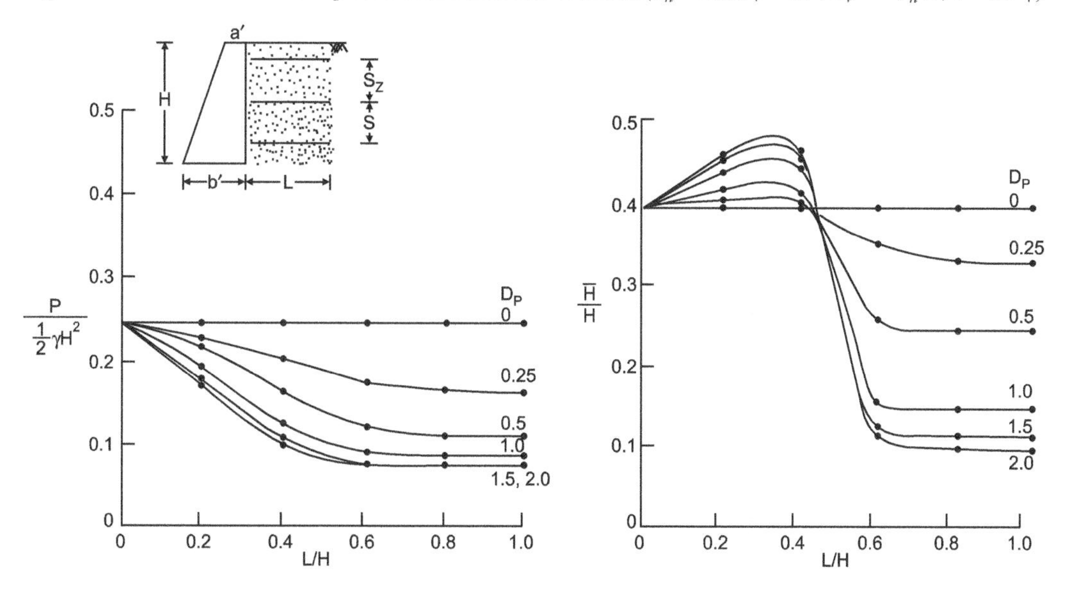

Fig. 16.16. Normalized resultant pressures and their line of action ($\alpha_h = 0$, $\phi = 35°$, $\alpha_v = 0$, $\delta = 2/3\ \phi$)

16.12 to 16.23 also show the variation of point of application of total earth pressure with $\frac{L}{H}$ ratio. It can be observed from these plots that the point of application of earth pressure rises above the base of wall with respect to unreinforced case for $\frac{L}{H}$ ratio upto 0.6; beyond that value, point of application comes down significantly. This is because the effect of increasing $\frac{L}{H}$ ratio, for a particular value of θ and D_p is to shift the zone of negative pressure distribution towards the upper portion of the wall height whereas increasing the reinforcement i.e., D_p reduces negative distribution towards lower portion of wall, ϕ and $\frac{L}{H}$ ratio remaining constant.

Illustrative Example

Example 16.1

A 5 m high wall is required to retain a horizontal cohesionless backfill with the following characteristics:

Density of backfill soil, $\gamma = 18$ kN/m^3

Angle of internal friction, $\phi = 30°$

Angle of wall friction, $\delta = 20°$

The wall is situated in a seismic area with horizontal seismic coefficient, $\alpha_h = 0.10$.

The foundation soil has an allowable soil pressure, q_a of 150 kN/m^2 and the coefficient of sliding friction at base is 0.45.

Design the wall adopting suitable reinforcement in the backfill. Compare the design with the wall having unreinforced backfill.

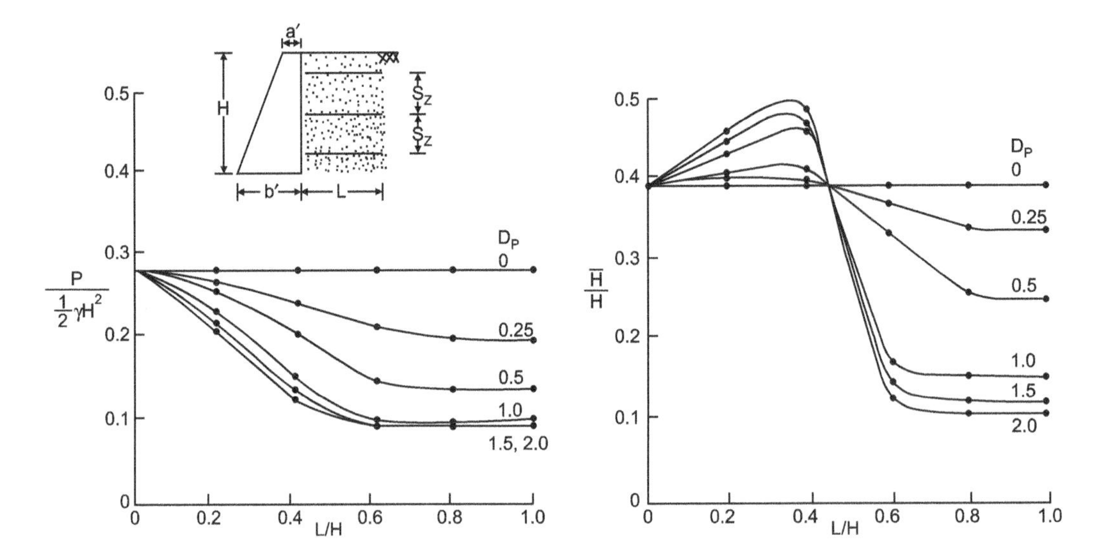

Fig. 16.17. Normalized resultant pressures and their line of action ($\alpha_h = 0.05$, $\phi = 35°$, $\alpha_v = +\alpha_h/2$, $\delta = 2/3\ \phi$)

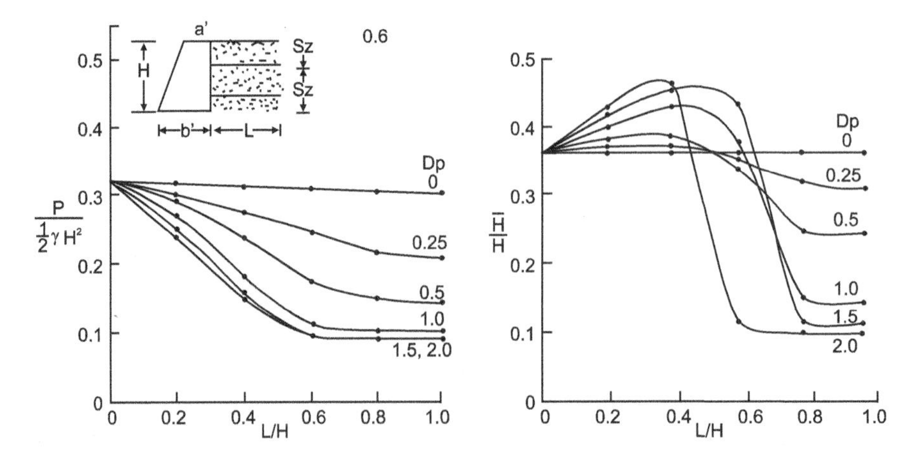

Fig. 16.18. Normalized resultant pressures and their line of action ($\alpha_h = 0.10$, $\phi = 35°$, $\alpha_v = +\alpha_h/2$, $\delta = 2/3\ \phi$)

Solution

Earth Pressure

Let the wall backfill be reinforced with geogrid Netlon SS-20 mats which are provided at 1.0 m interval as shown in Fig. 16.24.

Therefore vertical spacing of reinforcement, $S_z = 1.0$ m.

$$f^* = 0.4 \text{ (Assumed)}$$

Unit weight of masonry = 22 kN/m^3 (Assumed)

Length of reinforcement = $0.8 \times H$ (Assumed)

Vertical seismic coefficient, $\alpha_v = \dfrac{\alpha_h}{2} = \dfrac{0.10}{2} = 0.05$

Let the top thickness of gravity wall be 0.25 m (Fig. 16.24)

From the characteristics of reinforcement,

$$D_p = \frac{f^* H}{S_z} = \frac{0.4 \times 5.0}{1.0} = 2$$

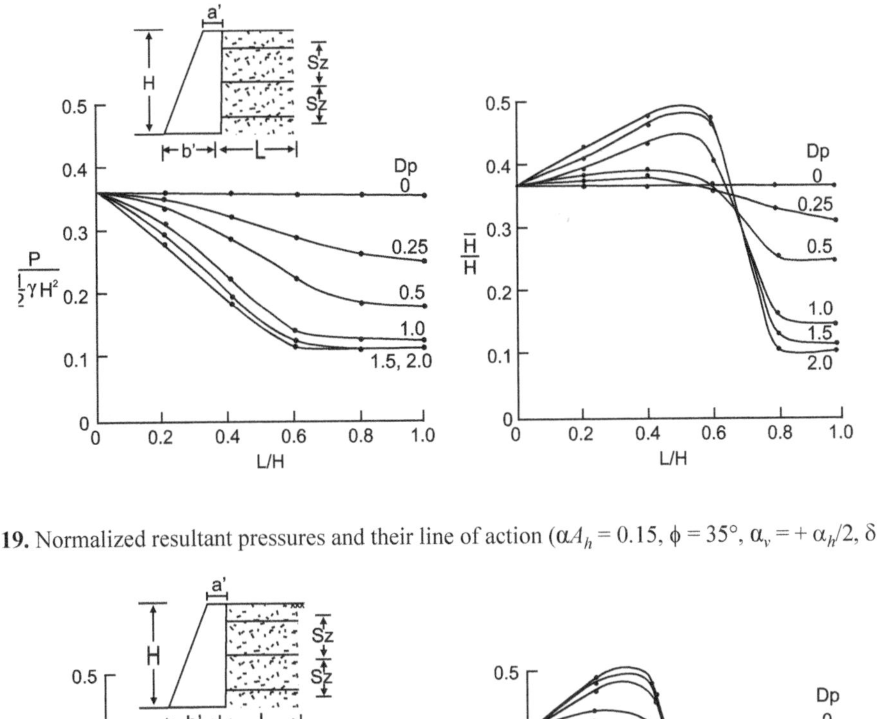

Fig. 16.19. Normalized resultant pressures and their line of action ($\alpha A_h = 0.15$, $\phi = 35°$, $\alpha_v = + \alpha_h/2$, $\delta = 2/3\ \phi$)

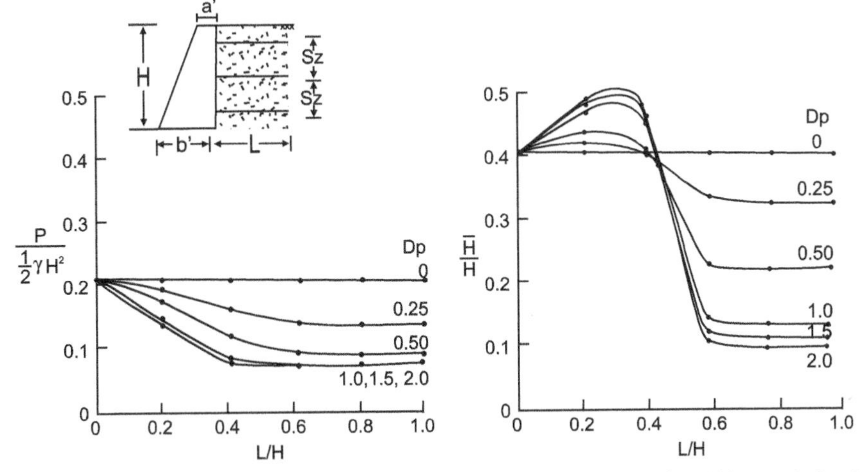

Fig. 16.20. Normalized resultant pressures and their line of action ($\alpha_h = 0$, $\phi = 40°$, $\alpha_v = 0$, $\delta = 2/3\ \phi$)

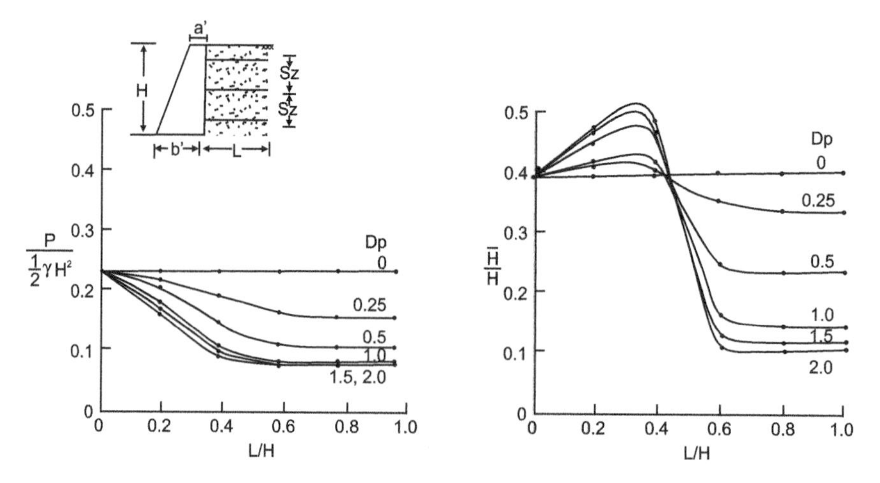

Fig. 16.21. Normalized resultant pressures and their line of action ($\alpha_h = 0.05$, $\phi = 40°$, $\alpha_v = + \alpha_h/2$, $\delta = 2/3\ \phi$)

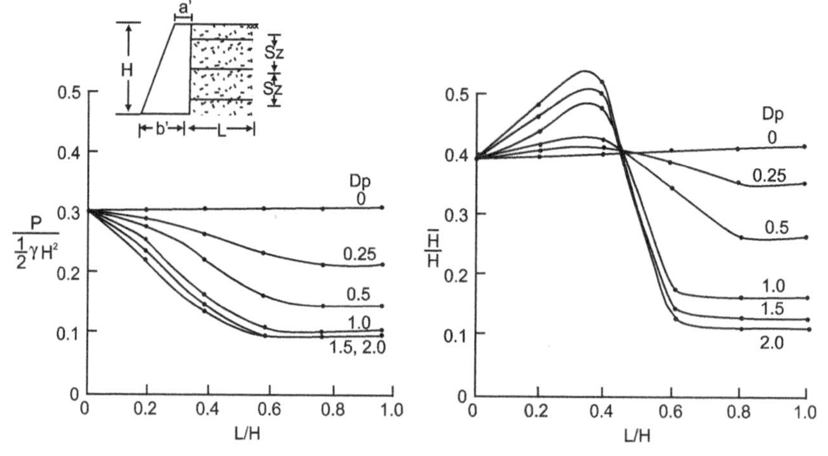

Fig. 16.22. Normalized resultant pressures and their line of action ($\alpha_h = 0.10$, $\phi = 40°$, $\alpha_v = + \alpha_h/2$, $\delta = 2/3\ \phi$)

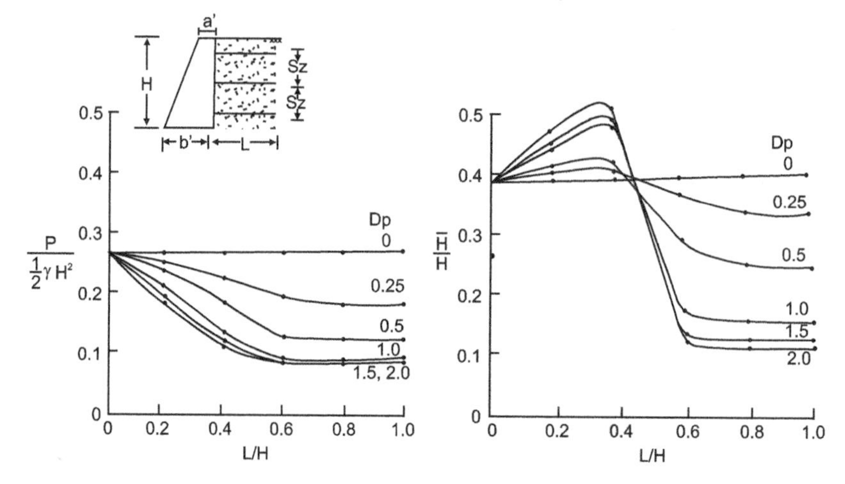

Fig. 16.23. Normalized resultant pressures and their line of action ($\alpha_h = 0.15$, $\phi = 40°$, $\alpha_v = + \alpha_h/2$, $\delta = 2/3\ \phi$)

Consider one metre length of wall.
From design chart (Fig. 16.14)

$$\frac{P_A}{\frac{1}{2}\gamma H^2} = 0.1185 \text{ and } \frac{\bar{H}}{H} = 0.1191$$

Therefore, total (resultant) earth pressure, $P_A = \frac{1}{2} \times 18 \times 0.1185 \times (5)^2$

or $$P_A = 26.66 \text{ kN/m}$$

and $$\bar{H} = 0.1191 \times 5 = 0.59 \cong 0.6 \text{ m from the base of wall}$$

In Fig. 16.24b

$$P_{AH} = P_A \cos \delta = 25.05 \text{ kN/m}$$

$$P_{AV} = P_A \sin \delta = 9.12 \text{ kN/m}$$

Refer Fig. 16.24b, W_1 and W_2 represent the weight of portions *AECD* and *BCE* of the wall.
Weight W_1 = Weight of wall portion *AECD* = 22 × 0.25 × 5 = 27.50 kN
If a factor of safety of 1.5 against sliding is desired, then

$$[W_1 (1 + \alpha_v) + W_2 (1 + \alpha_v) + P_A \sin \delta] \times \mu = [P_A \cos \delta + (W_1 + W_2) \alpha_h] \times 1.5$$

$$\therefore [(27.5 + W_2)(1 + 0.05) + 9.12] \times 0.45 = [25.05 + (27.5 + W_2) \times 0.10] \times 1.5$$

It gives:

$$W_2 = 76.30 \text{ kN}$$

$\therefore$ $$b = BE = \frac{76.30}{\frac{1}{2} \times 22 \times 5} = 1.3875 \text{ m, say } 1.40 \text{ m}$$

Therefore, $$AB = 0.25 + 1.4 = 1.65 \text{ m}$$

and weight $$W_2 = \frac{1}{2} \times 22 \times 5 \times 1.4 = 77.00 \text{ kN}$$

Factor of safety against overturning about toe:

$$F_o = \frac{27.5(1+0.05)\left(1.65 - \frac{0.25}{2}\right) + 77(1+0.05) \times \frac{2}{3}(1.65 - 0.25) + 9.12 \times 1.65}{(27.5 \times 0.1 \times 2.5) + (77.0 \times 0.1 \times (5/3)) + (25.05 \times 0.12 \times 5.0)}$$

$$= \frac{134.54}{34.62} = 3.88 \; [> 1.5, \text{ safe}]$$

Base Pressures

$$R_V = \Sigma \text{ Vertical forces} = (W_1 + W_2)(1 + \alpha_v) + P_A \sin \delta$$

$$= (27.5 + 77)(1 + 0.05) + 9.12 = 118.85 \text{ kN}$$

Net moment about point B = Overturning moment – Stabilising moment

$$= 134.54 - 34.62 = 99.22 \text{ kN-m}$$

$$\text{Eccentricity, } e = \frac{1.65}{2} - \frac{99.92}{118.85} = -0.015 \text{ m}$$

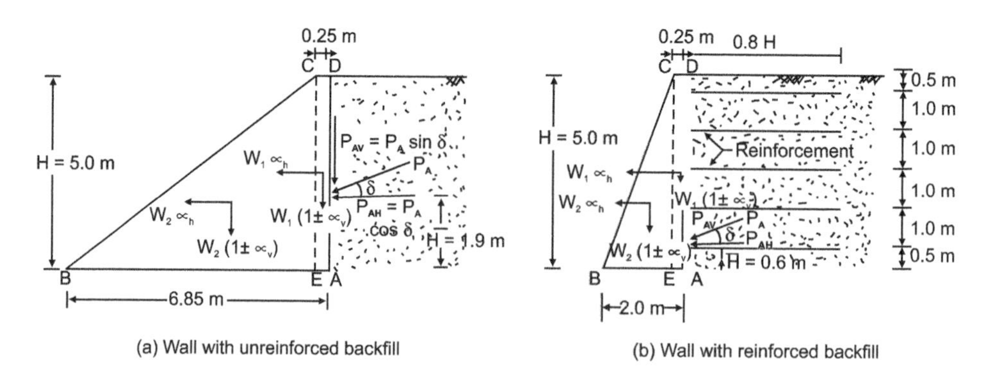

Fig. 16.24. Comparison of section of retaining wall

Maximum base pressure

$$p_1 = \frac{R_v}{b}\left(1-\frac{6.e}{b}\right) = \frac{118.85}{1.65}\left(1-\frac{\{6\times(-0.015)\}}{1.65}\right) = 76 \text{ kN/m}^2 < 150 \text{ kN/m}^2 \text{ (Hence, safe)}$$

Minimum base pressure

$$p_2 = \frac{R_v}{b}\left(1-\frac{6.e}{b}\right) = \frac{118.85}{1.65}\left(1+\frac{\{6\times(-0.015)\}}{1.65}\right) = 68.09 \text{ kN/m}^2 > 0.0 \text{ kN/m}^2$$

Therefore, no tension (Hence, safe)

The above problem was also solved considering (–) value of A_v. The factor of safety against sliding was kept same i.e. as 1.5. Following results were obtained:

Top width = 0.25 m

Base width = 2.0 m

Factor of safety against overturning, $F_O = 4.43$

Maximum base pressure = 75.8 kN/m^2

Minimum base pressure = 51.5 kN/m^2

Hence the case with (+) α_v is more critical

The problem was also solved considering unreinforced backfill. The purpose is to compare the two designs. In this case the values of $\dfrac{P_A}{\left(\dfrac{1}{2}\gamma H^2\right)}$ and $\dfrac{\bar{H}}{H}$ were taken from Fig. 16.14 corresponding to $\dfrac{L}{H} = 0$. The factor of safety against sliding was kept same (i.e. 1.5), and the base width, F_O; maximum and minimum base pressures were computed in similar manner as illustrated above.

The results are summarised below:

Top width = 0.25 m

Base width = 5.85 m

Factor of safety against overturning, $F_O = 7.3$

Maximum base pressure = 103.58 kN/m^2

Minimum base pressure = 26.56 kN/m^2

Cost economy of rigid retaining wall having two types of backfill has also been worked out and is given in Table 16.2.

Table 16.2. Comparison of Results and Cost of Rigid Retaining Wall Having Two Types of Backfill

	Unreinforced backfill	*Reinforced backfill*
$\frac{P}{\frac{1}{2}.\gamma H^2}$	0.3819	0.1185
$\frac{\bar{H}}{H}$	0.3794	0.1191
Top width (m)	0.25	0.25
Base width (m)		
F_S	1.50	1.50
F_O	7.3	3.88
$q_{(max)}$ (kN/m^2)	103.58	75.8
$q_{(min)}$ (kN/m^2)	26.56	51.5
Quantity of concrete (m^3)	17.75	5.62

Hence quantity of concrete used in wall with unreinforced backfill is about three times more than wall with reinforced backfill.

References

Broms, B.B. (1977), "Polyester Fabric as Reinforcement in Soils", Proc. International Conference on the use of Fabrics in Geotechnics, Paris, France, Vol. I, pp. 129-135.

Broms, B.B. (1987), "Fabric Reinforced Soils", Proc. International Symposium on Geosynthesis – Geotextiles and Geomembranes, Koyoto, Japan, pp. 13-54.

Garg, K.G. (1988), "Earth Pressure Behind Retaining Walls with Reinforced Backfill", Ph.D. Thesis, Department of Civil Engineering, University of Roorkee, Roorkee, India.

Khan, I.N. (1991), "A Study of Reinforcement Earth Wall and Retaining Wall with Reinforced Backfill", Ph.D. Thesis, Department of Civil Engineering, University of Roorkee, Roorkee, India.

Saran, S.K. (1998), "Seismic Earth Pressures and Displacement Analysis of Rigid Retaining Walls having Reinforced Sand Backfill, Ph.D. Thesis, Department of Civil Engineering, University of Roorkee, Roorkee, India.

Terzaghi, K. and Peck, R.B. (1967), "Soil Mechanics in Engineering Practice", Second Edition, John Wiley & Sons, New York, pp. 729.

Talwar, D.V. (1981), "Behaviour of Reinforced Earth in Retaining Structures and Shallow Foundations", Ph.D. Thesis, Department of Civil Engineering, University of Roorkee, Roorkee, India.

Vidal, H. (1906), "La Terre Armee", Annales de I institut Technique du Batiments, et des Travaux Publics, Paris, France.

Practice Problems

16.1 How is a rigid wall with reinforced backfill different from a reinforced earth wall? What are the appropriate situations for providing rigid wall with reinforced backfill?

16.2 Give the salient features of a method of analyzing and designing a vertical retaining wall with reinforced backfill.

16.3 Design a 6.0 m high retaining wall with vertical back retaining horizontal backfill. Other data are given as below:

Unit weight of backfill soil = 17 kN/m^3

Angle of internal friction of backfill soil = 30°

Angle of wall friction = 20°

Coefficient of sliding for foundation soil interface = 0.45

Allowable soil pressure = 275 kN/m^2

Horizontal Seismic coefficient = 0.1

Use geogrids as reinforcement which have tensile strength of 175 kN/m.

Index

W

Y

Z